P9-APN-910

ADVANCED PROJECT MANAGEMENT

ADVANCED PROJECT MANAGEMENT

Best Practices on Implementation

SECOND EDITION

HAROLD KERZNER, Ph.D.

Division of Business Administration
Baldwin-Wallace College
Berea, Ohio

John Wiley & Sons, Inc.

This book is printed on acid-free paper. ♾

Copyright © 2004 by John Wiley & Sons, Inc. All rights reserved

Published by John Wiley & Sons, Inc., Hoboken, New Jersey
Published simultaneously in Canada

No part of this publication may be reproduced, stored in a retrieval system or transmitted in any form or by any means, electronic, mechanical, photocopying, recording, scanning or otherwise, except as permitted under Sections 107 or 108 of the 1976 United States Copyright Act, without either the prior written permission of the Publisher, or authorization through payment of the appropriate per-copy fee to the Copyright Clearance Center, Inc., 222 Rosewood Drive, Danvers, MA 01923, (978) 750-8400, fax (978) 750-4744, or on the web at www.copyright.com. Requests to the Publisher for permission should be addressed to the Permissions Department, John Wiley & Sons, Inc., 111 River Street, Hoboken, NJ 07030, (201) 748-6011, fax (201) 748-6008, e-mail: permcoordinator@wiley.com.

Limit of Liability/Disclaimer of Warranty: While the publisher and author have used their best efforts in preparing this book, they make no representations or warranties with respect to the accuracy or completeness of the contents of this book and specifically disclaim any implied warranties of merchantability or fitness for a particular purpose. No warranty may be created or extended by sales representatives or written sales materials. The advice and strategies contained herein may not be suitable for your situation. You should consult with a professional where appropriate. Neither the publisher nor author shall be liable for any loss of profit or any other commercial damages, including but not limited to special, incidental, consequential, or other damages.

For general information on our other products and services or for technical support, please contact our Customer Care Department within the United States at (800) 762-2974, outside the United States at (317) 572-3993 or fax (317) 572-4002.

Wiley also publishes its books in a variety of electronic formats. Some content that appears in print may not be available in electronic books. For more information about Wiley products, visit our web site at www.wiley.com.

Library of Congress Cataloging-in-Publication Data:

Kerzner, Harold.
Advanced project management : best practices on implementation / Harold Kerzner.—2nd ed.
p. cm.
Rev. ed. of: Applied project management. c2000.
Includes bibliographical references and index.
ISBN 0-471-47284-0 (cloth)
1. Project management. I. Kerzner, Harold. Applied project management. II. Title.

HD69.P75K467 2004
658.4′04--dc21

2003053805

Printed in the United States of America.

10 9 8 7 6 5 4 3

To
my wife, JO ELLYN,
who showed me that excellence
can be achieved in
marriage, family, and life,
as well as at work

Contents

Preface

For almost 35 years, project management was viewed as a process that might be nice to have, but not one that was necessary for the survival of the firm. Companies reluctantly invested in some training courses simply to provide their personnel with basic knowledge on planning and scheduling. Project management was viewed as a threat to established lines of authority and, in most cases, only partial project management was used. This half-hearted implementation occurred simply to placate lower- and middle-level personnel.

During this 35 year period, we did everything possible to prevent excellence in project management from occurring. We provided only lip service to empowerment, teamwork, and trust. We hoarded information because the control of information was viewed as power. We placed personal and functional interests ahead of the best interest of the company in the hierarchy of priorities. And we maintained the faulty belief that time was luxury rather than a constraint.

By the mid-1990s, this mentality began to subside, largely due to two recessions. Companies were now under severe competitive pressure to create quality products in a shorter period of time. The importance of developing a long-term trusting relationship with the customers had come to the forefront. Businesses were now being forced by the stakeholders to change for the better. The survival of the firm was now at stake.

Today, businesses have changed for the better. Trust between the customers and contractors is at an all-time high. All of these factors have allowed a multitude of companies to achieve some degree of excellence in project management. Business decisions are now being emphasized ahead of personal decisions.

Words that were commonplace six years ago have taken on new meanings today. Change is no longer being viewed as being entirely bad. Today, change implies continuous improvement. Conflicts are no longer seen as detrimental. Conflicts managed well can be beneficial. Project management is no longer viewed as a system entirely internal to the organization. It is now a competitive weapon that brings higher levels of quality and increased value added to the customer.

Companies that were considered excellent in the past may no longer be regarded as excellent today, especially with regard to project management. Consider the book entitled *In Search of Excellence,* written by Tom Peters and Robert Waterman in 1982. How many of those companies identified in their book are still considered as excellent today? How many of those companies have won the prestigious Malcolm Baldrige Award? How many of those companies are excellent in project management?

The differentiation between the first 30 years of project management and the last six years is in the implementation of project management. For more than 30 years, we emphasized the quantitative and behavioral tools of project management. Basic knowledge and primary skills were emphasized. However, within the past six years, emphasis has been on implementation. What was now strategically important was how to put 30 years of basic project management into practice. Today it is the implementation of project management that constitutes advanced project management. Subjects such as earned value analysis, situational leadership, and cost and change control are part of basic project management courses today, whereas 15 years ago they were considered as advanced topics in project management. So, what constitutes applied project management today? Topics related to project management implementation are advanced project management concepts.

This book covers the advanced project management topics necessary for implementation of and excellence in project management. The book contains numerous quotes from people in the field who have benchmarked best practices in project management and are currently implementing these processes within their own firms. The quotes are invaluable because they show the thought process of these leaders and the direction in which their firms are heading. These companies have obtained some degree of excellence in project management, and what is truly remarkable is the fact that this happened in less than five or six years. Best practices in implementation will be the future of project management well into the twenty-first century.

An instructor's manual is available to college/university faculty members only by contacting John Wiley & Sons.

Seminars and correspondence courses on project management principles and best practices in project management are available using this text and my text, *Project Management: A Systems Approach to Planning, Scheduling, and Controlling*. Seminars on advanced project management are also available using this text. Information on these courses, e-learning courses, and on in-house and public seminars can be obtained by contacting the author at:

Phone: 216-765-8090
Fax: 216-765-8090 (same as the phone number)
E-mail: hkerzner@hotmail.com

Harold Kerzner

The Growth of Project Management

1.0 INTRODUCTION

You have attended seminars on project management. You have taken university courses on the fundamentals of project management. You have read the project management body of knowledge (PMBOK® Guide) prepared by the Project Management Institute (PMI), and you have even passed its national certification exam for project management. Now that you have learned the basics, how do you put theory into practice? That's advanced project management.

Basic project management teaches you the theory and principles of project management. Advanced project management discusses how to turn theory into practice. Simply stated, advanced project management deals with the implementation of project management. It is through implementation that excellence in project management is achieved.

1.1 UNDERSTANDING PROJECT MANAGEMENT

To understand project management, you must first recognize what a project is. A project is an endeavor that has a definable objective, consumes resources, and operates under time, cost, and quality constraints. In addition, projects are generally regarded as activities that may be unique to the company. Any company could manage repetitive activities based on historical standards. The challenge is managing activities that have never been attempted in the past and may never be repeated in the future. In today's world, projects seem to be getting larger and more complex. Some people contend that a project should also be defined as a multifunctional activity since the role of the project manager has become more of an integrator than a technical expert. *Project management* can be defined as the planning, scheduling, and controlling of a series of integrated tasks such that the objectives of the project are achieved successfully and in the best interest of the project's stakeholders. The business world has come to recognize the importance of project management for the future as well as the present. According to Thomas A. Stewart[1]:

> Projects package and sell knowledge. It doesn't matter what the formal blueprint of an organization is—functional hierarchy, matrix, or the emerging process-centered [or horizontal] organization, whose lines of communication and power are drawn along end-to-end business processes. . . . Routine work doesn't need managers; if it cannot be automated, it can be self-managed by workers. It's the never-ending book of projects—for internal improvement or to serve customers—that creates new value. It draws information together and does something with it—that is, formalizes, captures, and leverages it to produce a higher-valued asset.
>
> Consequently, if the old middle managers are dinosaurs, a new class of managerial mammals—project managers—is evolving to fill the niche they once ruled. Like his biological counterpart, the project manager is more agile and adaptable than the beast he is replacing, more likely to live by his wits than by throwing his weight around.
>
> People who lead or work on winning projects will get the first crack at the next hot gig. The best project managers will seek out the best talent, and the best talent—offered a choice as it often will be—will sign on with the best managers. Seniority matters less than what-have-you-done-for-me-lately. . . .
>
> Not every one can or should be a project manager, but those who can will be winners. When an organization ceases to be defined by its functional departments, and becomes a portfolio of projects and processes, it's much easier to claim credit for success—the results are obvious. Conversely, it's harder to blame "them" for failure because "they" are on your cross-functional project team.

Effective project management requires extensive planning and coordination. As a result, work flow and project coordination must be managed horizontally, not vertically as in traditional management. In vertical management, workers are organized along top-down chains of command. As a result, they have little opportunity to work with other functional areas. In horizontal management, work is organized across the various functional groups that work with each other. This results in improved coordination and communication among employees and managers.

Horizontal work flow generates productivity, efficiency, and effectiveness. Corporations that have mastered horizontal work flow are generally more profitable than those corporations that continue to use vertical work flow exclusively.

When project managers are required to organize their work horizontally as well as vertically, they learn to understand the operations of other functional units and how functional units interface. This knowledge results in the development of future general managers who understand more of the total operations of their company than do their counterparts who came up through a single vertical chain of command. Project management has become a training ground for future general managers who will be capable of making total business decisions.

1. From *Intellectual Capital* by Thomas A. Stewart. Copyright © 1997 by Thomas A. Stewart. Used by permission of Doubleday, a division of Random House, Inc.

No two companies manage projects in the same way. Project management implementation must be based on the culture of the organization. Some organizations have tried to accelerate learning to achieve excellence in project management by creating a center for excellence (COE) that constantly benchmarks against the best practices of those companies recognized worldwide.

The world is finally recognizing the importance of project management and its impact on the profitability of the corporation. The changes necessary for successful project management implementation are now well documented in the literature. Linda D. Anthony, manager, project management, at General Motors, provides us with her personal views of how organizations are now perceiving project management:

> Companies that have embraced a mature project management operating philosophy and practice are more likely to succeed in the competitive race of first to market than those that have not. The discipline of project management forces the attention to detail that is required for successful execution of projects. No longer can we simply manage the business by arts and charts and intuitive experience. It is imperative that we clearly understand the mission, scope, objective, and deliverable of every project at the onset.
>
> Organizations must realize that "management" experience alone does not fully equip a leader to be proficient in project management. Project managers must be trained and experienced in the fundamental principles of project management. Having completed one or two academic classes in college 15 years ago does not qualify a manager as trained and experienced.
>
> There are a wealth of project management education and training certification programs that are available to meet the growing demand for this knowledge and expertise. Companies that are serious about winning the race will incorporate this training as part of the development plans for current and future project managers.
>
> It is equally important that organizations establish an office of project management. The function of the "office" is to focus on the development of the company's project management current and future vision, deployment of project management principles, and to ensure common and consistent project management execution throughout the enterprise. This activity cannot be effectively administered by generalists or on a catch-as-catch-can basis. Success in this area of the business requires commitment and dedication to remaining current and up-to-date on methodology and technology.
>
> In today's complex and competitive business world, we must recondition our perceptions about the value of project management. Excellence in project management can only be achieved when organizations place a high priority on project management and make more prudent project management investment decisions.

According to Linda Kretz, senior consultant with the International Institute for Learning, project management as a professional discipline is undergoing significant change. Many companies use the term *project management* to include a number of different functions, some of which might be better described as expediting techniques or command/control management. Real project management differs from these techniques in timing of assignments and authority of the project manager.

Today, most of the managers described as *project managers* (sometimes called PMs for short) are assigned after project planning is complete. Charged with overseeing the implementation or execution of the project, they have no input into the budgeting process and are merely informed about the contractual constraints on the project. They are assigned to the project but are not informed of market analyses or revenue projections for the project. They have no idea how—or even whether—the planned project fits in with the corporation's overall strategic goals. Such concerns, if they are addressed at all, are usually handled by executive managers and held confidential. Ironically, many project managers working today are held accountable for the results of their projects without being privy to critical information. Two questions seem obvious: How accountable can project managers be for fulfilling someone else's plan? And with virtually no control over the project budget, how can they be held responsible for keeping the project on budget and on schedule?

In the future, project managers will be recognized for the value they contribute to the corporation's bottom line. Shooting the messenger will no longer be necessary because the message will fulfill everyone's expectations. No longer will project managers be informed of the company's financial margin at the end of the fiscal year along with all the other nonexecutive employees of the company. They will be given the authority to address potential problems by proactively managing their projects rather than reacting to ongoing risk factors.

Future project managers will be empowered to act as catalysts for corporate change and quality improvement. They will play central roles in meeting the company's financial goals. Their ability to evaluate the financial justification for projects will be recognized throughout their organizations, and they will be empowered to contribute to feasibility studies and project budgets.

1.2 RESISTANCE TO CHANGE

Why was project management so difficult for companies to accept and implement? The answer lies in Figure 1–1. Historically, project management resided only in the project-driven sectors of the marketplace. In these sectors, the project managers were given the responsibility for profit and loss. This profit and loss (P&L) responsibility virtually forced companies to treat project management as a profession.

In the non–project-driven sectors of the marketplace, corporate survival was based upon products and services, rather than upon a continuous stream of projects. Profitability was identified through marketing and sales, with very few projects having an identifiable P&L. As a result, project management in these firms was never viewed as a profession.

In reality, most firms that believed that they were non–project-driven were actually hybrids. Hybrid organizations are typically non–project-driven firms with one or two divisions that are project-driven. Historically, hybrids have functioned as though they were non–project-driven, as shown in Figure 1–1, but today they are functioning like project-driven firms. Why the change? Management has come to the realization that they could effectively run their organization on a "management by project" basis and achieve the benefits of both a project management organization and a traditional organization. The rapid growth and acceptance of project management during the past ten years has taken place in the non–project-driven/hybrid sectors. Now, project management is being promoted by marketing, engineering, and production, rather than just by the project-driven departments (see Figure 1–2).

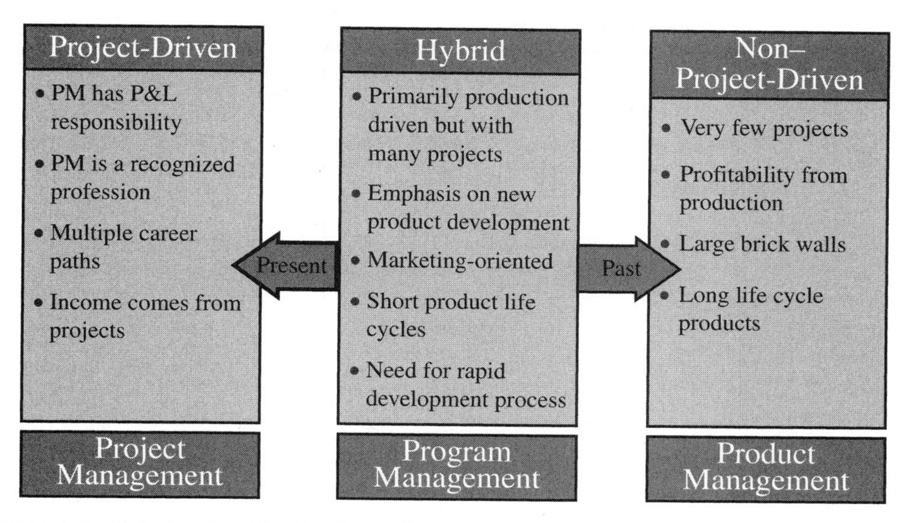

FIGURE 1–1. Industry classification (by project management utilization).

A second factor contributing to the acceptance of project management was the economy, specifically the recessions of 1979–1983 and 1989–1993. This can be seen from Table 1–1. By the end of the recession of 1979–1983, companies recognized the benefits of using project management but were reluctant to see it implemented. Companies returned to the "status quo" of traditional management. There were no allies or alternative management techniques that were promoting the use of project management.

The recession of 1989–1993 finally saw the growth of project management in the non–project-driven sector. This recession was characterized by layoffs in the white

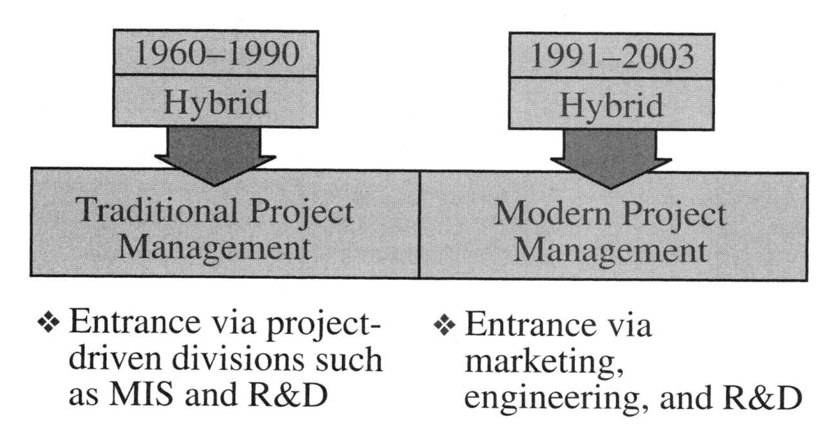

FIGURE 1–2. From hybrid to project-driven.

TABLE 1–1. RECESSIONARY EFFECTS

	Characteristics				
Recession	**Layoffs**	**R&D**	**Training**	**Solutions Sought**	**Results of the Recessions**
1979–1983	Blue collar	Eliminated	Eliminated	Short-term	• Return to status quo • No project management support • No allies for project management
1989–1993	White collar	Focused	Focused	Long-term	• Changed way of doing business • Risk management • Examine lessons learned

collar/management ranks. Allies for project management were appearing and emphasis was being placed upon long-term solutions to problems. Project management was now here to stay.

The allies for project management began surfacing in 1985 and continued throughout the recession of 1989–1993. This is shown in Figure 1–3.

- *1985:* Companies recognize that competition must be on quality as well as cost. There exists a new appreciation for total quality management (TQM). Companies begin using the principles of project management for the implementation of TQM. The first ally for project management surfaces with the "marriage" of project management and TQM.
- *1990:* During the recession of 1989–1993, companies recognize the importance of schedule compression and being the first to market. Advocates of concurrent engineering begin promoting the use of project management to obtain better scheduling techniques. Another ally for project management is born.
- *1991–1992:* Executives realize that project management works best if decision-making and authority are decentralized. Executives recognize that control can still be achieved at the top by functioning as project sponsors.
- *1993:* As the recession of 1989–1993 comes to an end, companies begin "reengineering" the organization, which really amounts to elimination of organizational "fat." The organization is now a "lean and mean" machine. People are asked to do more work in less time and with fewer people; executives recognize that being able to do this is a benefit of project management.
- *1994:* Companies recognize that a good project cost control system (i.e., horizontal accounting) allows for improved estimating and a firmer grasp of the real cost of doing work and developing products.
- *1995:* Companies recognize that very few projects are completed within the framework of the original objectives without scope changes. Methodologies are created for effective change management.

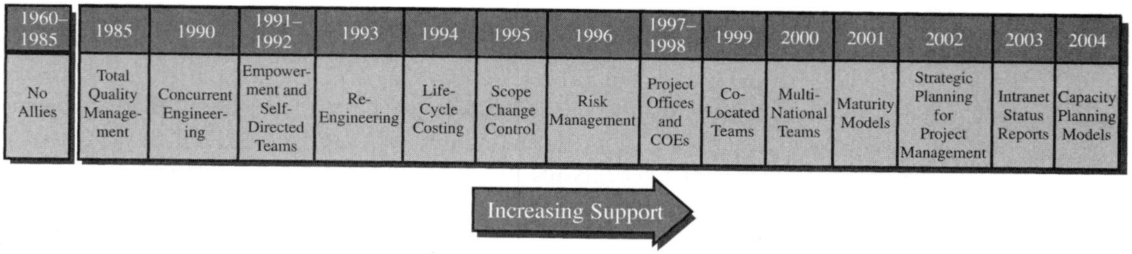

1960–1985	1985	1990	1991–1992	1993	1994	1995	1996	1997–1998	1999	2000	2001	2002	2003	2004
No Allies	Total Quality Management	Concurrent Engineering	Empowerment and Self-Directed Teams	Re-Engineering	Life-Cycle Costing	Scope Change Control	Risk Management	Project Offices and COEs	Co-Located Teams	Multi-National Teams	Maturity Models	Strategic Planning for Project Management	Intranet Status Reports	Capacity Planning Models

Increasing Support →

FIGURE 1–3. New processes supporting project management.

- *1996:* Companies recognize that risk management is more than padding an estimate or a schedule. Risk management plans are now included in the project plans.
- *1997–1998:* The recognition of project management as a professional career path mandates the consolidation of project management knowledge and a centrally located project management group. Benchmarking for best practices forces the creation of COEs in project management.
- *1999:* Companies that recognize the importance of concurrent engineering and rapid product development find that it is best to have dedicated resources for the duration of the project. The cost of overmanagement may be negligible compared to risks of undermanagement. More and more organizations can be expected to use co-located teams all housed together.
- *2000:* Mergers and acquisitions are creating more multinational companies. Multinational project management will become the major challenge for the next decade.
- *2001:* Corporations are under pressure to achieve maturity as quickly as possible. Project management maturity models help companies reach this goal.
- *2002:* The maturity models for project management provide corporations with a basis to perform strategic planning for project management. Project management is now viewed as a strategic competency for the corporation.
- *2003:* Intranet status reporting comes of age. This is particularly important for multinational corporations that must exchange information quickly.
- *2004:* Intranet reporting provides corporations with information on how resources are being committed and utilized. Corporations develop capacity planning models to learn how much additional work the organization can take on.

As project management continues to grow and mature, more allies will appear. In the twenty-first century, second and third world nations will come to recognize the benefits and importance of project management. Worldwide standards for project management will occur.

The reason for the early resistance to project management was that the necessity for project management was customer-driven rather than internally driven, despite the existence of allies. Project management was being implemented, at least partially, simply to placate customer demands. But by 1995, project management had become internally driven and a necessity for survival. Project management benchmarking was commonplace, and companies recognized the importance of achieving excellence in project management.

If a company wishes to achieve excellence in project management, then the company must go through a successful implementation process. The speed by which implementation occurs will dictate how quickly the full benefits of project management will be realized. This can be illustrated with Situation 1–1.

Situation 1–1: The aerospace division of a Fortune 500 company had been using project management for over 30 years. Everyone in the organization had attended courses in the principles of project management. From 1985 to 1994, the division went through a yearly ritual of benchmarking themselves against other aerospace and defense organizations. At the end of the benchmarking period, they would hug and kiss one another believing that they were performing project management as well as could be expected.

In 1995, the picture changed. The company decided to benchmark themselves against organizations that were not in the aerospace or defense sector. They soon learned that there were companies that had been using project management for less than five or six years but whose skills at implementation had surpassed the aerospace/defense firms who had been using project management for more than 30 years. It was a rude awakening to see how quickly several non–profit-driven firms had advanced in project management.

Another factor that contributed to a resistance to change was senior management's preference for the status quo. More often than not, this preference was based upon what is in the executives' best interest rather than the best interest of the organization as a whole. This led to frustration for those in the lower and middle levels of management who supported the implementation of project management for the betterment of the firm.

It was also not uncommon for someone to attend basic project management programs and then discover that his or her organization would not allow full project management to be implemented. To illustrate this problem, consider Situation 1–2.

Situation 1–2: The largest division of a Fortune 500 company recognized the need for project management. Over a three-year period, 200 people were trained in the basics of project management, and 18 people passed the national certification exam for project management. The company created a project management division and developed a methodology for project management. As project management began to evolve in this division, the project managers quickly realized that the organization would not allow their "illusions of grandeur" to materialize. The executive vice president made it clear that the functional areas rather than the project management division, would have budgetary control. Project managers would *not* be empowered with authority or critical decision-making opportunities. Simply stated, the project managers were being treated as expediters and coordinators, rather than real project managers. There were roadblocks that had to be overcome before theory could be turned into practice. How to overcome these obstacles had not been discussed in the basic courses on project management.

Even though project management has been in existence for more than 40 years, there are still different views and misconceptions about what project management really is. Textbooks on operations research or management science still have chapters entitled "Project Management" that discuss only PERT scheduling techniques. A textbook on organizational design recognized project management as simply another organizational form. Even among educators, differing views are still prevalent.

All companies sooner or later understand the basics of project management. But those companies that have achieved excellence in project management have done so through successful implementation and execution of processes and methodologies. Because of this relationship, throughout this text, reference is made to those companies who have achieved some degree of excellence in project management through their understanding of the advanced concepts of project management.

When a company's stock is depressed, or even if a company is in bankruptcy, it by no means indicates poor project management. In fact, project management may actually improve. During favorable economic conditions, a company might be able to tolerate a $50 million loss on a project. But during unfavorable economic times, companies are more selective of the projects they work on, risk management techniques are greatly improved, cost monitoring and earned value measurement techniques are implemented, estimating techniques are improved, and productivity will probably increase. Although this may not occur in all companies, there are indications that many of these changes for the better do occur.

1.3 STRATEGIC IMPERATIVES FOR PROJECT MANAGEMENT

Try to name one company, just one, that has given up on project management after implementing it. Probably you couldn't. Every company that has adopted project management is still using it. Why? Because it works. Once a company has gone over to project management, the only question becomes: When will we achieve the full benefits of the system?

The strategic imperatives behind achieving excellence in project management come from two sources: internal and external. Internally, senior managers may discover the benefits of project management as they monitor general business trends in their industry or when they compare their company's results with those of its competitors. Internal champions of project management recognize potential overall improvements in both efficiency and effectiveness. They may also understand that project management can create future general managers versed in the operations of virtually every functional unit.

External pressures may force a company to accept the need for change in the way it does business. For example,

- *Competition:* Customers expect lower cost and the use of project management on their projects.

- *Quality Standards:* Customers expect high quality, fewer failures, and fewer service calls.
- *Financial Outcomes:* Customers expect contractors to accept lower profit margins.
- *Legal Concerns:* Customers expect uniform project management systems that adhere to legal and regulatory boundaries (from the Environmental Protection Agency, for example).
- *Technological Factors:* Customers expect state-of-the-art technology at reasonable prices.
- *Social Concerns:* Employees want a system that allows them to do more work in less time in order to reduce the amount of overtime required.
- *Political Factors:* Companies compete in a global economy that requires uniform project management processes.
- *Economic Pressures:* Companies need to perform more work in less time and at a lower cost to reduce the impact of monetary exchange rates and the cost of borrowing money.
- *Stockholders' Concerns:* Stockholders want internal growth and external expansion through mergers and acquisitions, which must be executed quickly and cost effectively.

The benefits of project management have been demonstrated by numerous corporations. For example:

- Hewlett-Packard had shown increased sales, customer satisfaction, and with the use of integrated project/product teams, were able to lower project cost, shorten development time, and improve system integration and product design.
- During the 1990s, Radian International had garnered more repeat business and happier clients; it also reduced cost overruns and write-offs.
- Battelle (PNNL) had achieved better on-time and on-budget product delivery.
- 3M had reduced its product development time and, with the use of integrated project/product teams, outperformed their cost goals and improved performance.
- DaimlerChrysler lowered product cost, decreased development time, and improved vehicle designs.
- The Department of Defense used integrated project/product teams to achieve a lower than expected product cost, have ahead-of-schedule deliveries, and demonstrate an increase in performance.

Another strategic benefit of project management is that it can be integrated successfully with other management systems. The four most relevant management systems today are concurrent engineering (see Figure 1–4), total quality management, risk management, and change management. The combinations have produced synergistic results, as illustrated in Figure 1–5.

Combining project management with concurrent engineering may produce the following benefits:

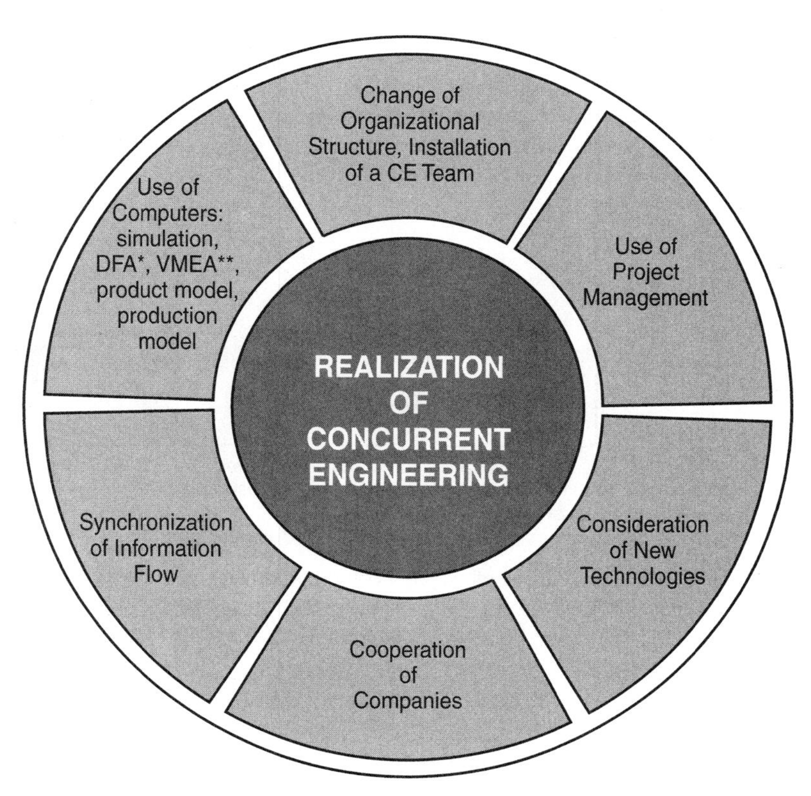

FIGURE 1–4. Combination of project management and concurrent engineering. *Source:* Reprinted from H. Kerzner, *In Search of Excellence in Project Management.* New York: Wiley, 1998, p. 8. Adapted from I. W. Eversheim, "Trends and Experience in Using Simultaneous Engineering." *Proceedings of the 1st International Conference on Simultaneous Engineering,* London, December 1990, p. 18.

- New product development time is reduced.
- Average life of the product is increased.
- Sales are increased.
- Revenues are increased.
- The number of customers is increased.

The Department of Defense estimates that concurrent engineering, combined with project management, produces these additional benefits:

- Design changes are reduced by at least 50 percent.
- Product lead times are reduced by more than 50 percent.
- Scrap and rework are reduced by 50 to 75 percent.

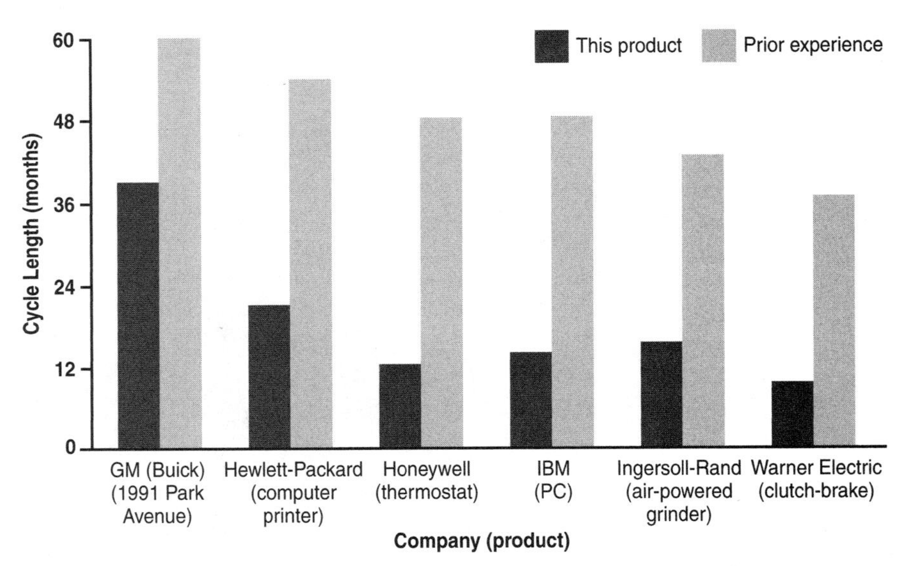

FIGURE 1–5. Results of six large corporations using product management in conjunction with other management systems to achieve improvements in product development time. *Source:* P. G. Smith and D. G. Reinertsen, *Developing Products in Half the Time,* Second Edition. New York: Wiley, 1997.

During the 1980s, Digital Equipment Corporation was a staunch supporter of both project management and concurrent engineering. According to John Hartley, Digital Equipment had identified the following improvements over a three-year period[2]:

- Time to market for new products was reduced from 30 months to 15 months.
- Product costs were reduced by 50 percent.
- Sales increased by 100 percent.
- The break-even point was reached earlier, about six months earlier on average.
- Profitability increased nine to ten times over what it had been five years before.

Combining project management with total quality management may produce the following benefits:

- Higher product quality
- Happier clients
- Fewer internal and external failures

2. From *Concurrent Engineering: Shortening Lead Times, Raising Quality, and Lowering Costs* by John Hartley. Copyright © 1992 by Productivity Press, P.O. Box 13390, Portland, OR 97213–0390, (800) 394-6868. Reprinted by permission.

- Reduced amounts of scrap
- Fewer product recalls and warranty problems

The benefits of combined risk management and project management may include:

- Better risk identification procedures
- Better risk quantification procedures
- Improved processes for responding to risk
- Improved decision-making processes
- Increased tolerance for accepting risk
- Clarified contractual identification of which parties are to bear which risks

Project management combined with change management may yield these benefits:

- The ability to respond to customers' change requests rapidly
- Decreased impact of changes on budget and schedule
- Increased value-added efforts on behalf of the customers
- Good customer relations
- Happier clients

1.4 PROJECT MANAGEMENT LIFE CYCLE

Achieving excellence in project management can be accomplished in a few years or a few decades. Excellence can't be achieved without change, and the speed of change is critical. Project management is like total quality management: both are management systems that require extensive education and training. And the educational process must start with senior managers. Why would any employee support change that is not supported from the top down?

Factors such as economic recession, dwindling market share, competition, low profitability, and poor employee morale may drive senior management's commitment to change. Executives must be committed to the change to project management and recognize the value it adds to the corporation before project management can succeed. Ultimately, they must understand that the change to project management will benefit every stakeholder in the company.

Since the early 1990s, the search for excellence in project management has taken on more and more importance. The benefits of project management today are obvious to both customers and contractors. In fact, excellence in project management has become a competitive weapon that attracts new business and retains existing customers.

During the past three decades, the Project Management Institute (PMI) has grown from 3,000 members to more than 100,000. The greatest growth has occurred since the late 1980s, probably owing to the certification process established for project personnel. Project management now has a career path. Customers are even requesting that contractors assign certified project managers to their projects. Even corporate executives are taking the exam with the goal of functioning better as executive project sponsors and exhibiting to customers that their company's senior management supports project management totally.

The life cycle that virtually every company goes through in establishing the foundations for excellence is discussed in detail in Chapters 2 and 3. (The phases of the life cycle are shown in Table 2–1.)

The fastest way to establish a foundation for excellence is to implement training and education programs. Table 1–2 displays selected industry types identified by the number of years each company has been using project management and the level of training courses in project management that each offers. Project-driven industries generate most of their income from individual projects. Non–project-driven industries generate most of their income from products and services. Put another way, non–project-driven industries take on projects that support the organization's products and services; in project-driven industries, the organization exists to support its projects. Some industries are made up of organizations that are not predominantly project-driven but include several divisions that are project-driven. These are called hybrid organizations.

The characteristics of the three types of industries are summarized here:

- The aerospace, defense, and large construction industries are the project-driven stars of yesterday and today. Hundreds of millions of dollars have already been

TABLE 1–2. AMOUNT OF PROJECT MANAGEMENT TRAINING OFFERED, BY INDUSTRY TYPES

Level of Project Management Training	Hybrid	Project-Driven	Project-Driven
High	Automobiles	Automotive subcontractors	Aerospace
	Health care	Computers	Defense
	Machinery	Electronics	Large construction
	Mining		
	Hybrid	**Hybrid**	**Project-Driven**
Medium	Beverages	Banking	Leisure
	Chemicals	Pharmaceuticals	Amusement
	Paper	Oil and gas	Nuclear utilities
		Telecommunications	
	Hybrid	**Hybrid**	**Non–Project-Driven**
Low	Insurance	Food	Commodity manufacturing
	Publishing	Railroads	Metals
	Retail	Tobacco	
	Transportation		
	1–5	**5–10**	**15 or more**
	Years of Project Management Experience		

Source: Reprinted from H. Kerzner, *In Search of Excellence in Project Management.* New York: Wiley, 1998, p. 11.

spent in the development of quantitative tools that support project management in these industries. The organizations in these industries prefer to develop their own tools rather than use canned software packages. Several of these companies now enter signed licensing agreements with other companies for the use of their once-proprietary project management software. The project management systems used in these industries are excellent, but their effectiveness is hampered by the formality and number of policies and procedures still in place. The formality of their systems has been driven by their customers, many of them federal agencies.

- The automotive subcontractor, computer, and electronics industries will be the project-driven stars of tomorrow. Several companies in these industries have already achieved excellence in project management. Driven by consumers' demands for greater quality and shorter product development time in the near future, these industries should easily surpass aerospace, defense, and large construction in their project management capabilities. Automotive subcontractors, computer developers, and electronics companies tend to employ young project managers and executives who are willing to accept risk and to reduce the amount of bureaucracy involved in project management.
- Some project-driven companies in the leisure, amusement, and nuclear utility industries have slowly and methodically achieved some degree of excellence in project management. More consistent excellence will be achieved once these industries recognize that they will not be able to survive without fully implementing project management.
- The automotive, health care, machinery, and mining industries are the hybrid stars of today and tomorrow. Although these industries were slow to adopt project management during the 1980s, they are rapidly embracing it now, and many companies have already achieved some level of excellence. Excellence will be attainable for all the companies in these industries once they recognize the effects of changing legislation and new consumer demands. Today, policies and procedures are being streamlined. In addition, subcontractor management practices are being improved significantly, thus fostering trust among contractors and subcontractors. Project management certification is being encouraged.
- Perhaps the most rapid changes in project management are occurring in the banking, pharmaceutical, oil and gas, and telecommunications industries to make them the hybrid stars of tomorrow as well as today. These industries have accomplished more in the past few years than other industries have achieved in 10 years. The need for project management excellence in these rapidly growing industries has been driven by mergers, acquisitions, and legislation. In all likelihood, these industries will surpass others in their ability to use project management as a vehicle for risk assessment.
- Other industries providing a combination of project-driven and non–project-driven products and services are slowly taking up project management, and the need for it has not yet been recognized. These industries include beverages, chemicals, paper, insurance, publishing, retail, transportation, food, railroads, and tobacco. They are often dominated by politics, and projects tend to be driven by schedule and quality concerns. At the end of a project, the project managers in these industries usually have no idea how much money was actually spent.

- The commodity manufacturing and metals industries include predominantly non–project-driven companies that have very few projects separate from production-driven manufacturing services. The full adoption of project management in these industries may not occur until well into the next century.

1.5 EXCELLENCE IN PROJECT MANAGEMENT

The difference between the average company and the company that has achieved excellence in project management is the way that the growth and maturity phases of the project management life cycle are implemented. This is where advanced project management has a major impact on project management excellence. Figure 1–6 shows the six areas in which successful companies excel in project management. These six areas are discussed in Chapters 9 through 15.

Companies that decided to embark on a project management methodology soon found that there were more potential benefits than originally believed possible. This is shown in Table 1–3. As more and more benefits became attainable, the quest for excellence began. Companies were realigning their thought processes.

Lack of executive buy-in is the principal reason project management so often fails to reach its full potential in some companies. Simply because executives recognize that changes are needed does not mean that change will take place. Executives must realize that success and excellence in project management require decentralization and that executives

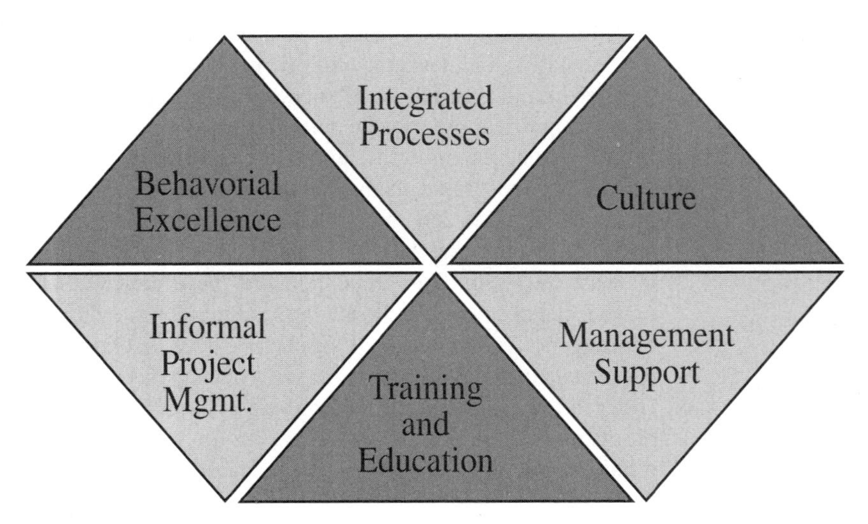

FIGURE 1–6. The six components of excellence. *Source:* Reprinted from H. Kerzner, *In Search of Excellence in Project Management.* New York: Wiley, 1998, p. 14.

TABLE 1–3. BENEFITS OF PROJECT MANAGEMENT

Past View	Present View
• Project management will require more people and add to the overhead costs	• Project management allows us to accomplish more work in less time and with less people
• Profitability may decrease	• Profitability will increase
• Project management will increase the amount of scope changes	• Project management will provide better control of scope changes
• Project management creates organizational instability and increases conflicts	• Project management makes the organization more efficient and effective through better organizational behavior principles
• Project management is really "eyewash" for the customer's benefit	• Project management will allow us to work more closely with our customers
• Project management will create problems	• Project management provides a means for solving problems
• Only large projects need project management	• All projects will benefit from project management
• Project management will increase quality problems	• Project management increases quality
• Project management will create power and authority problems	• Project management will reduce power struggles
• Project management focuses on suboptimization by looking at only the project	• Project management allows people to make good company decisions
• Project management delivers products to a customer	• Project management delivers solutions
• The cost of project management may make us noncompetitive	• Project management will increase our business

must surrender critical information and partial control of expenditures to project managers. Because control of information and funding is a source of power for executives, many are reluctant to relinquish all of their power, and they are also reluctant to commit themselves fully to project management.

There are other roadblocks to executive commitment to project management. Some companies still resist the full implementation of project management because they assume that project management is unnecessary. After all, if employees were performing their assignments correctly in the first place, why would project management be needed? Project management is mistakenly dismissed as "checking the checkers." Sometimes it is lumped together with internal auditing.

Professional project managers do not hesitate to tell the full story behind their projects. Unfortunately, this news is not always welcomed by senior managers. It's not what they want to hear. Information takes on a negative aspect given the nature of command control coordination and management. Instead, accurate project information should be accepted as the payoff to proactive professional project management. In the future, senior managers will need to recognize the contributions of project managers in the analysis of market considerations, financial planning, and technical assessments.

1.6 SELECTION OF COMPANIES AS EXAMPLES

More than 300 companies were contacted during the preparation of this book. They were identified through

- Published literature
- Survey questionnaires
- Privileged knowledge (consulting and lecturing by the author)
- External trainers and consultants

The majority of the companies were provided three sets of questionnaires. Follow-up interviews were then conducted in many instances to verify the quality of the information and to show the interviewees the exact format and context of their responses. Authorization was required for reprinting verbatim quotations from corporate executives and project managers.

The initial intent was to identify at least one or two companies from each major industry without consideration of company size. Many of the companies that have achieved excellence in project management according to my criteria refused to participate. These companies believed that their competitive edge might be compromised by the release of sensitive information.

A second group of companies climbing the ladder to excellence declined participation for fear of being benchmarked against more successful companies in their industries. Some of these organizations recognized from the survey questionnaires that they still have a long way to go to achieve excellence. For example, one executive responded, "We don't do anything you're asking about. Perhaps we should."

A third group initially responded to the surveys but could not secure authorization to release the information. A fourth group of companies that are on track for excellence had second thoughts about seeing their names in print for fear that their customers would have expectations of them higher than they would be able to achieve.

The companies considered in this book as having achieved excellence or as at least being on the right track are these:

3M
ABB
American Greetings
Antares
Armstrong World Industries
Battelle (Pacific Northwest National Laboratories)
BellSouth
Boeing
CIGNA
Computer Associates
Cooper Standard Automotive
DaimlerChrysler
Defense Acquisition University
Department of Defense
Detroit Energy
Diebold
Edelca
EDS
Eli Lilly
Ericsson
Exel
FirstEnergy
General Electric
General Motors
Hewlett-Packard

Humana
Intel
International Institute for Learning (IIL)
I-Think
Johnson & Johnson
Johnson Controls
Lear
Kinetico
Lear
Lincoln Electric
MahindraBT
Metzeler
Microsoft
Middough Consulting
Mindjet
Motorola
National City Bank
Nortel
Noveon
PDVSA
Raytheon
Roadway Express
Rockwell Automation
School of Project Management (SPM)
Sherwin-Williams
Star Alliance
StoneBridge Group
Sun Microsystems
Swiss Re
Texas Instruments
USAA
Virginia Department of Transportation
Walt Disney
Westfield Group
Xerox

Small companies have project management cultures that permeate the entire organization. Large companies have pockets of project management. Some pockets may be highly successful in project management while others still have a long way to go. This holds true even for companies that have won the prestigious Malcolm Baldrige Award. Attempts were made to get responses from those divisions that have demonstrated excellence. Similarly, the responses made by individuals do not necessarily reflect the project management practices of the entire company.

Not all companies achieve excellence in all six of the areas shown in Figure 1–6. The companies that have come closest are identified in this book. Some companies identified may excel in two or three components and are included because they are headed in the right direction. These organizations will see the light at the end of the tunnel in the near future.

Unfortunately, there are not many companies that have actually achieved excellence. Roadblocks exist and must be compensated for. Dr. Al Zeitoun, Chief Projects Officer at the International Institute for Learning, believes that:

> Project management is here to stay. The world that is going 250 miles an hour with reengineering and continuously changing processes and approaches is making the need for project management and project managers most evident. This group of key players in our organizations will continue to be the only group that can make sense of all these changes, the group that will maintain the ability to see the right amount of details without losing sight of the big picture.
>
> There are, in my opinion, several indicators that project management excellence will have a long way to go:
>
> 1. The number of organizations that truly excel in understanding and implementing project management is only a limited few.
> 2. There are a great number of organizations that talk about project management and have project managers and yet haven't provided the sponsorship required for successful

implementation. We are seeing a struggle in defining authority, deliverables, and accountability, just to name a few. There are still myths, such as the myth that there would be no need for project charters if we had good job descriptions. Organizations are still managing around who they have instead of what needs to be done, and we are still seeing the negotiations for specific people rather than for specific deliverables.

3. There is hardly any reasonable level of education and research in project management. Universities, colleges, and institutions are finally coming up with some reasonable programs. Research, however, is barely starting.
4. The standards efforts in project management have a long path to follow. The PMBOK® Guide by the Project Management Institute (PMI), among other international bodies of knowledge, provides a solid base for standards. As strong as the document is, there are still several standards issues on the table. One of the most key open issues is the global standards issue. What is the global standard going to be like? Is the recent effort of ISO 10006 a step in that direction? There is still the crucial need to develop a comprehensive international standard that encompasses all the key disciplines of project management and cuts across industries and global cultures and boundaries.
5. The certification issue has a similar path to follow. On the global side, we have to decide on the need and value of global certification, the proper certification procedure, the certifying body(ies), the recertification requirements, and other multinational complex issues. On the national side, open issues pertain to the PMP [Project Management Professional] exam format, whether it's going to be an overall exam that covers all project life cycle phases or stay in the same eight modules format. The ongoing improvement of the points system for certification/recertification will have to continue with the increased concern and demand. Since certification doesn't have a legal implication, are we going to proceed with licensing similar to other professions? Are we going to consider certification and/or licensing based on specific industries? Several open questions still exist.
6. Global cooperation between key project management organizations has only started to be evident. Previous concerns about which organization should take the global leadership, like the challenge between PMI and the International Project Management Association (IPMA), is beginning to subside. The key focus is becoming: How can we work together globally to benefit from mutual experiences? For a while the issue of forming a global federation was surfacing.
7. As more and more organizations are beginning to manage their business by projects (MOBP) and rely on the strong cooperation between project managers and resource providers, there is a need for a different skill set for project managers. The mentoring side of a project manager's nature is shining. The need for training that addresses this different set of skills is clear. There is a much greater need for the soft skills than there is for the hard skills. A continuous improvement effort for training programs will be a requirement for training providers to stay in business.

Based on the above seven indicators, among several others, I believe that in project management we have only taken the first key steps. We are starting to crawl and will shortly begin to walk, talk, and walk the talk. A strong revolution to truly understand project management and the potential return on investment (ROI) it can bring is on the way. Organizations will continue to pursue better lessons learned so as to excel in the way projects are managed successfully over and over again. The continued increase in membership of project management organizations and the number of certified professionals will have no limits.

> The global scene is going to be where project management excellence will be in greatest demand. Career paths for program managers as key individuals for achieving this global excellence will continue to be a very crucial issue. Those individuals are going to be strong candidates for the strong senior management seats of the future. Senior managers might actually take back the role of providing the vision that ties their organizations together.
>
> Virtual teams will continue to develop, enabling the fast exchange of ideas, minimization of paperwork, and more efficient and effective project work. There is still going to be the need for one-on-one meetings and face-to-face team meetings. This will enable the team to handle key conceptual issues, address handoff challenges, and solve critical issues.
>
> The true integration of systems and processes will continue to be a strong direction in the global marketplace. Minimizing and/or eliminating the redundancy between those systems and processes is the goal of this integration effort. A continuous stream of new project management ideas will appear, and the organizations today, as they become more projectized, will wonder about how strange yesterday was.

The twenty-first century will bring new meaning to the word *globalization.* If history has taught us anything, it is the fact global success may very well be based more upon managerial skills than on the products offered and markets served. A good project management methodology can provide a consistent framework for global projects. Combining a standard methodology with good managerial talent dramatically improves a firm's chances for global success.

Suzanne Zale, global program manager at EDS, pointed out:

> Driven by the world economy, there is a tendency toward an increasing number of large-scale global or international projects. Project managers who do not have global experience tend to treat these global projects as large national projects. However, they are completely different. A more robust project management framework will become more important for such projects. Planning up front with a global perspective becomes extremely important. As an example, establishing a team that has knowledge about the geographic regions relevant to the project will be critical to the success of the projects. Project managers must also know how to operate in those geographical areas. It is also essential that all project team members are trained and understand the same overall project management methodology.
>
> Globalization and technology will make sound project management practice even more important.

Some people contend that predicting the future for project management may be difficult because of the rate at which changes are being made. Adrian Lammi, president of the Northeast Ohio Chapter of the Project Management Institute, gives us his views of the twenty-first century:

> In the twenty-first century, project management will retain its identity as a unique discipline because of the skills required for project management. Project management will continue to attract those rare individuals who have a high tolerance for details and who can successfully integrate them into a viable plan. Still more will be required of them. The balance of emphasis will not shift away from scope, time, cost, and quality. Rather, human

resources management and communications management will achieve parity as success factors. Project managers will be prized for their ability to persuade, to influence, to inspire, and to negotiate. They will be required to do so across geographic and cultural boundaries. Increasingly the arena in which project managers operate will shrink. The general practitioner will become as extinct as his medical counterpart. Project sponsors will demand more than certification. They will seek individuals who possess endorsements reflecting geopolitical and industry experience. The abilities to manage detail and to be technically competent, however, will hold project managers captive within their own profession. Thus the percentage of project managers promoted to top management positions will remain relatively constant. Project management will support management rather than evolving into a core component of general management.

MULTIPLE CHOICE QUESTIONS

1. Basic project management deals with:
- **A.** The principles of project management
- **B.** Scheduling techniques
- **C.** Cost control techniques
- **D.** Working with executive sponsors
- **E.** All of the above

2. Advanced project management deals with:
- **A.** Turning theory into practice
- **B.** Implementation techniques
- **C.** How to achieve excellence in project management
- **D.** All of the above
- **E.** A and C only

3. Excellence in project management is most frequently achieved through:
- **A.** A comprehensive knowledge of the principles
- **B.** Implementation of the principles
- **C.** A compression of the schedules
- **D.** Higher product quality
- **E.** A lowering of costs

4. The classical definition of a project is an activity that:
- **A.** Is unique or one-of-a-kind
- **B.** Has a definable objective
- **C.** Operates under constraints of time, cost, and quality
- **D.** Consumes human and nonhuman resources
- **E.** All of the above

5. Effective project management requires extensive planning and coordination because:
- **A.** Coordination is now horizontal rather than only vertical
- **B.** The project's constraints are loose
- **C.** Employees cannot work efficiently without a plan
- **D.** Written communication is mandated in a project management environment
- **E.** None of the above

6. Project management is often said to generate future general managers primarily through:
- **A.** Horizontal work flow
- **B.** Control of project costs

C. The management of tight constraints
D. All of the above
E. B and C only

7. According to Linda Kretz, many companies make the often fatal mistake of:
A. Providing too large a budget for the project
B. Providing too small a budget for the project
C. Providing too few resources
D. Not assigning an executive sponsor
E. Not assigning the project manager until after the project has been planned

8. In the future, we can expect companies to recognize the contributions project managers make to the:
A. Bottom line of the company
B. Planning and control function
C. Quality improvement efforts
D. Software enhancement efforts
E. None of the above

9. During the early years of project management growth, the driving force that caused executives to recognize project management as a profession was:
A. Increased quality requirements
B. Schedule compression
C. Profit and loss responsibility
D. All of the above
E. A and B only

10. The growth of project management occurred initially in companies that were classified as:
A. Non–project-driven
B. Project-driven
C. Quasi-project-driven
D. Purely functional
E. Hybrids

11. Since the last recession of 1989–1993, the greatest growth for project management occurred in the __________ sector.
A. Project-driven
B. Non–project-driven
C. Hybrid
D. Quasi-project-driven
E. Purely functional

12. The two primary factors contributing to the recent rapid acceptance of project management were:
A. P&L responsibility and recessions
B. Executive sponsorship and total quality management
C. Executive sponsorship and university course work
D. Graduate degrees in project management and PMI certification
E. P&L responsibility and PMI certification

13. During the recession of 1979–1983, allies for project management began to appear.
A. True
B. False

14. Which of the following is not normally recognized as an ally of project management?
A. Total quality management

B. Increased span of control
C. Concurrent engineering
D. Executive empowerment/self-directed teams
E. Corporate reengineering

15. Today, project management implementation is viewed as being __________, whereas historically it was __________.
A. Customer-driven; internally driven
B. Internally driven; customer-driven
C. Part-time; a full-time effort
D. Part-time; a profession
E. None of the above

16. Once an organization recognizes the need for project management, the speed by which implementation occurs is dictated by the:
A. Number of project management training courses
B. Educational level of the project manager
C. Senior management's desire to decentralize authority and decision-making
D. Line manager's ability to manage projects better
E. All of the above

17. Which of the following is a "strategic imperative" that may force the organization to accept project management?
A. Competition
B. Quality standards
C. Legal concerns
D. Stockholder concerns
E. All of the above

18. A strategic benefit of project management is that it can be integrated successfully with other management techniques.
A. True
B. False

19. Which of the following is a benefit of combining project management with concurrent engineering?
A. New product development time is reduced
B. Sales are increased
C. The number of customers is increased
D. Average product life is increased
E. All of the above

20. Combining project management with total quality management may produce which benefit?
A. Higher product quality
B. Happier clients
C. Fewer internal and external failures
D. Reduced amounts of scrap
E. All of the above

21. The fastest way to establish a foundation for excellence in project management is through:
A. Education programs
B. Assigning project managers with advanced degrees

C. Allowing line managers to serve as a project manager
D. Allowing executives to manage projects
E. An apprenticeship period of ten years on-the-job training

22. A project-driven industry that will most likely be one of the stars of the future will be:
A. Automotive subcontractors
B. Health care
C. Mining
D. State governments
E. All of the above

23. Which of the following is a "present view" of the benefits of project management?
A. Power and authority problems will increase
B. Customers will receive solutions rather than products
C. The administrative cost of project management will increase
D. Project management will increase the number of scope changes
E. Profitability may decrease

24. Which of the following is not one of the hexagon of excellence components?
A. Integrated processes
B. Training and education
C. Recessionary survival
D. Informal project management
E. Culture

25. A common roadblock in getting executives to accept project management is that:
A. True project status is not always welcomed
B. Increased profitability can be detrimental
C. Customer expectations may not be met
D. Schedule slippages may occur more frequently
E. All of the above are possible

DISCUSSION QUESTIONS

1. Why is it that during favorable economic times, companies are less likely to implement project management?
2. In the early days of project management during the 1950s and 1960s, the aerospace and defense industries had very little difficulty in implementing project management. How do we explain this?
3. Why were allies for project management needed before non–project-driven organizations were willing to accept project management?
4. Explain the difference between a hybrid organization and a non–project-driven organization.
5. You work for a non–project-driven company. What project management support might exist, if any, from each of the following: senior management, functional management, and employees?
6. Is excellence in project management achieved more through knowledge and education or through implementation? Explain your answer.

7. One of the "strategic" imperatives promoting the acceptance of project management is social concerns. What topics other than what was identified in this chapter could have been included in this category?
8. Was the acceptance of project management in project-driven organizations more customer-driven or internally driven?
9. During the 1970s and early 1980s, public seminars on project management began to surface. Was the information taught in these seminars more behaviorally focused or technically focused?
10. If the aerospace and defense industries have been using project management for more than 40 years, then why do we still have cost overruns, perhaps as much as 200 to 300 percent? Do these cost overruns indicate poor project management?

Across

3. Part of P&L
4. Project constraint
6. Economic forces
7. Hexagon part
9. Project management ally
10. Total quality ___________
13. Project constraint

Down

1. Hexagon part
2. Project constraint
3. ________ -driven
5. Ally: ___________ engineering
7. ___________ -driven
8. Opposite of basic
11. Benefit: future ________ managers
12. Type of company

Success, Maturity, and Excellence: Definitions

2.0 INTRODUCTION

To embark upon a journey, one needs a destination and a plan to get there. The destination must be clearly identified or else we have no idea when the journey is over. To some degree, the same holds true for management practices. We must be able to define, with some reasonable degree of accuracy, what constitutes project success, project management maturity, and project management excellence.

Some people contend that to achieve maturity and excellence in project management is a never-ending journey. It is true that when the destination is defined in terms of a management process, there will always exist the opportunity for continuous improvement. However, we still have to provide a framework to know when we are approaching the destination point and when we have arrived.

2.1 EARLY DEFINITIONS OF SUCCESS

When was the last time that you were assigned as a project manager and were informed as to what constituted *success* on your project? Unfortunately, very few companies define project success for their project managers, and those that do usually provide poor definitions at best. However, companies are now recognizing how important a good definition of success is. Should success be measured as customer satisfaction, profitability, new business won, or market share achieved? Should success be defined as meeting or exceeding customers' specifications/expectations? Exceeding expectations can be an expensive process.

Over the years, our definition of project success has changed. In the early days of project management, success was measured in technical terms only. Either the product worked or it didn't work. Many years ago I was managing a project where one of the work packages was estimated at 800 hours by the engineer who would be performing the work. He completed the activity using 1,300 hours. When I asked him why he required an additional 500 hours, he commented, "What do you care how much I spent or how long it took? I got the job done, didn't ?" This type of mentality existed because neither the customers nor the contractors were under any pressure for cost containment. Cost control was lip service rather than reality. As a result, success was measured in technical terms only.

The real problem here rested with senior management, who persisted in defining success in technical terms only rather than in both technical terms and business terms. Perhaps the most important function of the project sponsor during the initiation phase of a project is the clear definition of both business and technical success.

As companies began to understand more and more about project management, and costs were monitored more closely, our definition of success changed. Success was now defined as accomplishing the effort on time, within cost, and at the desired level quality, with quality being defined by the customer rather than the contractor. Unfortunately, this definition was still incomplete.

The literature on project management started to accumulate during the early 1980s. Articles described successful projects and lessons learned. Contractors began modeling their project management processes after published success stories under the mistaken belief that what worked for one company would work equally well for theirs. This misconception was further reinforced when companies achieved success on one or two projects and concluded incorrectly that they had found a management panacea.

A 1985 paper I coauthored with David Cleland, Ph.D., identified some of the factors common to successful projects. Hindsight has proven, however, that the results we reported were more indicative of experience than of success or excellence. The problems with our research were twofold. First, the research was conducted during the early 1980s, before publications describing numerous successful projects appeared. Second, and perhaps more important, several executives we surveyed said that they considered less than 10 percent of the completed projects to be successfully managed even though the projects' ultimate objectives had been achieved.

> Any project can be driven to success using executive meddling, brute force, and a big stick. This by no means implies that a successfully completed project is the result of excellence in project management.

2.2 MODERN DEFINITIONS OF SUCCESS

The problem with defining success as having a project completed on time, within cost, and at the desired quality is that this is an internal definition of success. The ultimate customer should have some say in the definition of success. Today, our definition of success is measured in terms of primary and secondary factors such as:

- *Primary Factors:* On time; within cost; at the desired quality
- *Secondary Factors:* Accepted by the customer; customer allows you to use the customer's name as a reference

A typical list of secondary success factors might include:

- Customer reference
- Follow-on work
- Financial success
- Technical superiority
- Strategic alignment
- Regulatory agency relationships
- Health and safety
- Environmental protection
- Corporate reputation
- Employee alignment
- Ethical conduct

Today, we recognize that quality is defined by the customer, not by the contractor. The same holds true for project success. There must be customer acceptance. You can complete a project internally within your company within time, within cost, and within quality limits, and yet find that the project is not accepted by the customer. The ultimate definition of success might very well be when the customer is so pleased with the project that the customer is willing to be named as a reference.

I-Think

It is not uncommon for companies to change their definition of project success as time progresses. Sometimes these changes are necessitated by a changing business environment or by changing customer demands. Kentaro Ito, President and CEO of I-Think Corporation, comments on some of the changes taking place in Japan:

> In Japan, in the past, it was a common practice to believe that the "Customer is God." High quality meant high customer satisfaction, and this was thought to be the highest priority. Therefore, in the definition of project success, quality was believed to lead to customer satisfaction and was thus the first priority. Customer satisfaction would lead to repeat business because the company's brand image was enhanced. The more repeat business you received, the greater your brand image.
>
> Quality was perceived to be more important than costs or profits. Japanese companies would seldom cancel a project, even though they might have to incur a significant loss, because of the desire to maintain customer satisfaction. There was a mistaken belief that this deficit could be recovered from other projects within the organization.
>
> During the long recession in Japan in the 1990s, the room to carry out continuous deficit projects began to disappear. Customers were no longer considered as Gods. Customers and customer satisfaction were no longer the primary components for project success, but were regarded as one of the components. Quality and customer satisfaction were now just as important as cost, schedule and risk. Japanese companies had to re-think their definition of

> success and be more selective of the projects undertaken and the customers with which to partner.
>
> With the survival of Japanese companies at stake, the definition of quality has changed along with the development and acceptance of project management. Although improving quality is still important, the highest priority today seems to be the integration of quality and customer satisfaction with cost, schedule and risk. For many Japanese companies, this concept of project management that focuses on this integration is relatively new. Hopefully, through project management training and education in Japan, more companies will understand the importance of integration for the survival of their business.

The definition of success can also change based upon whether you are project or non–project-driven. In a project-driven firm, the entire business of the company is projects. But in a non–project-driven firm, projects exist to support the ongoing business of production or services. In a non–project-driven firm, the definition of success also includes completion of the project *without* disturbing the ongoing business of this firm. It is possible to complete a project within time, within cost, and within quality and, at the same time, cause irrevocable damage to the organization. This occurs when the project manager doesn't realize that *this* project is secondary in importance to the firm's ongoing business.

Not all companies have the same primary factors for defining success. A brewery in Venezuela defines success as within time, cost, quality, and scope. Walt Disney defines success as within time, cost, quality, and meeting safety requirements, with the latter being most important. Brian Vannoni, formerly of General Electric Plastics, defines success as follows:

> The technical aspects, timing, and costs were [in the past] the three critical areas of performance measurement for our project managers. In today's world, that is not sufficient. We have to also be concerned with environmental and safety regulations, the quality, customer satisfaction, and delivering the productivity for the manufacturing operations. So a project now has at least eight measurables and critical parameters around which we gauge success.

Some companies define success in terms of not only critical success factors (CSFs), but also key performance indicators (KPIs). Critical success factors identify those factors necessary to meet the desired deliverables of the customer. Typical CSFs include

- Adherence to schedules
- Adherence to budgets
- Adherence to quality
- Appropriateness and timing of signoffs
- Adherence to the change control process
- Add-ons to the contract

CSFs measure the end result, usually as seen through the eyes of the customer. KPIs measure the quality of the process used to achieve the end results. KPIs are internal measures and can be reviewed on a periodic basis throughout the life cycle of a project. Typical KPIs include

- Use of the project management methodology
- Establishment of control processes
- Use of interim metrics
- Quality of resources assigned versus planned for
- Client involvement

Key performance indicators answer such questions as: Did we use the methodology correctly? Did we keep management informed, and how frequently? Were the proper resources assigned and were they used effectively? Were there lessons learned that could necessitate updating the methodology or its use? Excellent companies measure success both internally and externally using KPIs and CSFs.

ABB

Key performance indicators and critical success factors are now used as metrics to measure the successful growth and maturity of project management within a company. As an example, consider the following remarks from James M. Triompo, Group Senior Vice President, Process Area Project Management at ABB:

> ABB was driven to strive for excellence in project management because of an increasingly tough market situation that demanded elimination of all margin erosion. ABB is implementing a common ABB PM process, with emphasis on appropriate PM skills in project sales to ensure value added solutions, a solid project definition, costing and risk management. The implementation started a few years ago in ABB organizations where the process is now being used. Implementation is now underway in many other parts of the ABB organization. ABB's definition of excellence is Predictable Project Management executed as booked, or better, with complete customer satisfaction. Our critical success factors for project success and excellence in project management include project margin development, cash flow, on-time delivery, and customer satisfaction metrics.

Diebold

When a company wants to grow in project management, there is always the fundamental question of how the company knows that they are doing things the right way. Sometimes, the critical success factors can be identified upfront, especially if the organization performed benchmarking studies; at other times the critical success factors become readily apparent as the organization begins the implementation of project management. It is probably easier to identify the critical success factors as you proceed because the organization can see what seems to be working well. According to Mark Smith, Director, Global Design Excellence at Diebold Corporation:

> Formalized Project Management is relatively new within our development organization. Yet we have quickly made some major strides in the adoption of project management processes and practices. I feel we have been able to accomplish this quickly primarily for the following reasons:
>
> - *Strong Senior Management Support:* We have strong support for project management from our senior management. Our senior management knows the benefits that project management can achieve and are supporting us throughout the organization to achieve

these benefits in a reasonable amount of time. We believe that best practices in project management begin with support at the senior levels of management and we have that. Senior management within the engineering organization mandated that we would use good project management practices.

- *Training:* We are beginning to see a return on our investment for the training dollars spent on project management education. This training permeates all levels of project and line management and will certainly help us maintain the straight and narrow path toward best practices in project management. The people we have trained thus far are helping us develop a project management methodology. Through training and education in project management, we expect to continuously learn and implement best practices in project management. We plan on accelerating our learning process by not only learning from our mistakes, but also from the mistakes made by other companies.
- *We Saw the Need:* We are constantly looking for ways to improve our development processes and reduce our development cycle times. Based on our experience, and after project management training, we realized that project management would provide us with the next level of continuous improvement. Project management has helped to link the "vertical silos" within our development organization.

2.3 PROJECT MANAGEMENT MATURITY

It is important to understand that all organizations go through a maturity process, and that this maturity process must precede excellence. The learning curve process for maturity is measured in years. Companies that are committed to project management may be lucky to achieve maturity in two years or so, whereas the average firm may take up to five years. The critical questions, of course, are: What is maturity? and How do I know when I have reached it?

In certain cultures, maturity is defined as gray hair, baldness, and age. In project management, this definition is furthest from the truth. Maturity in project management is the development of systems and processes that are repetitive in nature and provide a high probability that each project will be a success. Repetitive processes and systems do not guarantee success. They simply increase the probability of success.

A company can be mature in project management but not be excellent in project management. The definition of excellence goes well beyond maturity. When companies develop mature systems and processes, two additional benefits emerge. First, the work is accomplished with minimum scope changes, and second, the processes are designed such that there will be minimal disturbance to the ongoing business.

Table 2–1 shows the life cycle phases for project management maturity. Virtually every company that achieves some level of maturity goes through these phases. The culture of the organization and the nature of the business will dictate the amount of time spent in each of the phases.

Embryonic Phase

In the embryonic phase, both middle management and senior management must recognize the need for, benefits of, and applications of project management. This recognition is more than simply providing "lip service" by telling people that project management should be used to achieve project objectives. Senior management *must* understand that excellence in project management will affect the corporate bottom line. Virtually no company anywhere in the world has gone to project management and then has given it up. The reason for this is that it works!

TABLE 2–1. THE FIVE PHASES OF THE PROJECT MANAGEMENT LIFE CYCLE

Embryonic	Executive Management Acceptance	Line Management Acceptance	Growth	Maturity
Recognize need	Get visible executive support	Get line management support	Recognize use of life cycle phases	Develop a management cost/schedule control system
Recognize benefits	Achieve executive understanding of project management	Achieve line management commitment	Develop a project management methodology	Integrate cost and schedule control
Recognize applications	Establish project sponsorship at executive levels	Provide line management education	Make the commitment to planning	Develop an educational program to enhance project management skills
Recognize what must be done	Become willing to change way of doing business	Become willing to release employees for project management training	Minimize creeping scope Select a project tracking system	

Once executives and managers realize that project management not only affects the corporate bottom line but also is a necessity for survival, the maturity process is accelerated. Unfortunately, it may take months or even years to recognize the true impact on the bottom line. Even after an Ohio company had spent three years training employees one of their directors asked why they should continue funding project management training courses because this director did not see any return on investment on training dollars. During a review meeting, however, the director began to realize that his company was now having more successes than in the past, realizing or getting closer to achieving project objectives, receiving more concise reports with more accurate information, and finding that fewer conflicts came up to the director level for resolution. The director was so involved in projects that he had not seen the changes that had occurred over the past three years. When he was asked to take a step backwards and look at the big picture, however, the changes became apparent, and the support for project management training continued.

Executive Management Acceptance Phase

The second phase of the project management life cycle shown in Table 2–1 is the executive management acceptance phase. In this phase, it is critical that executives visibly identify their support if the organization is to become mature in project management. The key word here is "visibly." In a large appliance manufacturer, the executives met away from the company for a week in order to identify the strategic initiatives necessary to remain competitive in the twenty-first century.

At the end of the week-long meeting, 20 strategic initiatives were identified. Number 3 on the list was project management. The executives returned to their offices the following week and sent out memos to all managers beneath them identifying the importance of project management. The managers filed the memo in the trash can, recognizing correctly that this was simply lip service and no changes would be made, nor would executive support for project management be forthcoming. Attempts were made over a three-year period by middle management to develop a one-hour executive briefing on what executives could do in order to accelerate the maturity process of project management. The executive briefing never took place as the executives continuously found excuse after excuse as to why they could not attend a briefing. This reinforced line management's belief that project management support was simply lip service rather than reality.

How do executives convince lower-level personnel that they, the executives, actually understand project management? Perhaps the best way is for the executives to function as a project sponsor. This not only demonstrates support for the process and a desire for project management maturity, but it also shows an understanding of project management.

Lack of visible executive support is the biggest detriment to achieving maturity and excellence in project management. During the 1980s and early 1990s, the telecommunications industry struggled with the problem of how to achieve maturity and excellence in project management without having visible executive support. In this industry, the top levels of management were politically astute, and executive-level appointments were based upon politics. As a result, senior managers were very reluctant to act as a project sponsor for fear that if they were the sponsor on a project that failed, it may be the end of their political career. Simply stated their attitude was: never identify yourself with an activity the failure of which can damage your political career. Fortunately, this mentality is now changing, but at a slow pace.

Line Management Support

The third phase in the maturity model is line management support. The biggest obstacle to obtaining line management support is the previous phase, executive management acceptance. What line manager would eagerly accept and support project management while realizing that his or her superiors would not support the process?

Line managers do not necessarily need a strong understanding of the project management tools, but they must understand the principles of project management since it is the line managers who are responsible for the staffing of the project. Understanding of the principles is a necessity for line managers to provide visible support and commitment for the process.

Growth Phase

The fourth phase is the growth phase. This phase can actually begin as early as the embryonic phase and run in parallel with the first three phases. However, the three preceding phases *must* be completed before this phase can be completed.

Senior management's knowledge of project management and support can accelerate the growth phase. Perhaps one of the best managed health management organizations is

ChoiceCare/Humana of Cincinnati, Ohio. During 1995 a live video conference on "How to Achieve Maturity in Project Management," ChoiceCare/ Humana's vice president for information systems, Byron Smith, responded on what senior management had done at ChoiceCare/Humana to bring in a mature project management system.

> I think if you're talking to most executives today, they will tell you that they want to do more, better and faster. And I think that very well categorizes our goals for project management at ChoiceCare/Humana. You talked earlier in your presentation about companies that had been pressured to adopt project management capabilities as a result of recessions. We have been fortunate to be at the other end of the economic environment. Our business has been very positive over the past few years. In fact, we've been able to grow by a 60 percent rate over that period. We believe that our success story has been a result of aggressively pursuing opportunities to lower medical cost while improving the quality of care that our enrollees receive. Senior management at ChoiceCare/Humana feels that it is critical for us to take advantage of as many opportunities as we possibly can to have, in essence, that stream of continuously successful projects that you discussed. As a relatively small organization, we have approximately 500 associates. We feel that means we need to leverage project management capabilities throughout our entire organization, and that we, in fact, need to take advantage of the skills and talents of all our associates. We have done three things to accomplish that. First, we have created a standard methodology for our teams to follow. This lets our teams know what to do, what deliverables to produce, and who to involve in the process. Secondly, we developed a series of cross-functional steering teams around our strategic areas. The role of these teams is to provide guidance to our various project initiatives. I think it's interesting to point out that this allows us to use the experience of our senior managers to guide a large number of projects rather than be hands-on with only a few. Finally, we have clearly defined our corporate values and desired behaviors. This helps the individual associates understand how to make decisions and how to interact when they participate in our projects.

Christine Dombrowski, a director at ChoiceCare/Humana, commented on what happened at the middle-management levels during the early 1990s:

> I think three things are going on at ChoiceCare/Humana that support our going toward maturity in project management. First is the model that Bryon mentioned. It was developed by a cross-functional group. ChoiceCare/Humana invested the time in that group in developing what we call our team management model. It gives us a common language and set of expectations for projects. It clarifies roles and responsibilities so that team members and stakeholders know what and when to expect things from one another. And it includes the perspectives of the individuals who hold our coverage, our employer group customers, as well as our participating physicians. So, all this wraps together to give us a comfort level for doing projects and that allows project teams the freedom to do their work. In addition, we have a lot of required training. Twenty-five percent of ChoiceCare/Humana associates have been trained in project management, as well as our team management model. And when I say 25 percent, I include our vice president and our executive management.
>
> And interpersonal skills training is another element that supports our quality project management. Finally, associates on projects at ChoiceCare/Humana know that they have senior management support. I've even heard that our executive vice president from

Marketing required that outside consultants who work with us learn our model and use it to work effectively with us.[1]

During the growth phase, project management systems are developed and refined for control and standardization. Such systems reflect a company's commitment to quality and planning, as well as the need to minimize scope changes (also called scope creep). Scope creep sometimes occurs during the planning or execution of a project, most often during execution. Scope creep results when features or functions are added to the project. Such changes drive up costs and lengthen the schedule. Although most scope creep changes are small, added together they can endanger the project. You might say that scope creep can happen when you are building a new home. Each design change made after construction has started adds to the cost of the house, as well as to how long it will take to build.

There is a mistaken belief, especially among young project managers, that perfect planning can be achieved, thus eliminating scope creep. In excellent companies, scope creep is expected and planned for. According to Frank T. Anbari, Professor of Project Management at George Washington University:

> Scope definition is extremely important to the success of any project. Scope "creep" and scope "leap" are often the root cause of project failure. That much we know. However, in some projects, it is extremely difficult, if not impossible, to have complete definition of project scope before execution is started. This is particularly true in the case of high, new, and emerging technologies projects, such as certain projects in manufacturing and construction. The customer may not be able to visualize the organization's needs and to what extent these needs can be satisfied. Technological advances may surpass implementation pace. Devices originally envisioned and selected in project design may become obsolete, or even unavailable, due to the introduction of more advanced models.
>
> In such cases, it may become imperative to plan the project "from the middle out." This requires heavy involvement of the customer/user in initial project definition, general planning, design, and prototype development. Flexibility and adaptability become essential in project execution to accommodate changes in technology, refinements of requirements, and project replanning. To succeed in these projects, it is important to pull the "middle" of the project to the earliest possible time. This can be accomplished by profound understanding of the business needs of the customer and by helping the customer understand current and expected technological capabilities. Yet, even with this approach, scope creep is still possible.

The final element in the growth phase is the selection of a software package for project planning and control. In the 1970s and 1980s, companies spent millions of dollars developing their own packages. With the quality of the off-the-shelf packages available today, however, there is no need to reinvent the wheel. During the early 1990s, Ford Motor Company had contracted with Microsoft to custom-design Microsoft Project for specific use at Ford.

Ford Motor Company's decision to use Microsoft Project as the standard was not a spur-of-the-moment decision nor was it based upon intuition. Instead, it was the result of

1. From this point forward, ChoiceCare/Humana will be referred to as Humana Health Plans of Ohio.

a methodical process to select the best project management software available that was *compatible* with Ford's project management methodology.

Ford's approach was correct. They designed a methodology and then selected the software package that best supported the methodology, not vice versa. Some firms foolishly do it backwards. They purchase software and then design a methodology to fit the software. An executive steering committee once asked me what I thought about a PC-based software package that, with licensing agreements, had cost about $600,000. I informed the company that since no methodology existed as yet, they may very well have thrown away $600,000, and that the purchasing decision was premature. The company realized the mistake too late and after finally designing their own methodology, selected a different software package.

Maturity Phase

Most firms make it through the first four phases sooner or later. Twelve to 24 months would be a reasonable time frame for aggressive companies to get into the growth phase. The maturity phase, however, is another story. The maturity phase mandates that the company understand the importance of integrating time and cost. You cannot determine the status of a project just by looking at the schedule. Likewise, you cannot determine project status from cost alone. Cost and schedule must be integrated. In a large division of one of the big three in Detroit, the cost control people sat on one side of the building and the scheduling people on the other side. The cost people had no idea what the schedule was, the scheduling people had no idea what the costs were, and nobody knew the real status of the project. Spending 30 percent of the budget does *not* mean that you are 30 percent complete (in performance). Likewise, being 30 percent complete in performance does not mean that you have spent 30 percent of the budget.

The difficulty with integrating time and cost usually requires a revamping of the cost accounting system to include earned value measurement. Also required is an *accurate* determination of how much money is being spent on each activity. Employees are reluctant to support systems requiring that they fill out time cards on a daily basis. Employees feel that "big brother" is looking over their shoulder.

All companies have "vertical" accounting systems. However, what we are discussing here is a "horizontal" accounting system to find out accurately how, when, and where the money is being utilized. There exists tremendous resistance to horizontal accounting systems, especially at the lower levels of management and at the executive levels. Lower level managers resist horizontal accounting because it quickly and accurately identifies which managers have been providing accurate estimates and which have not. Executives resist horizontal accounting because it may show that their target end date was too optimistic and their target budget is too low. Executives like to establish a launch date and budget even before a project plan has been laid out. Horizontal accounting will show that this method fails more often than it succeeds.

In the early 1980s, I was invited to consult for RJR Tobacco Company (now RJR-Nabisco). The company had laid out plans to spend more than $2 billion to double the production capacity for cigarettes. When I asked one of the vice presidents why they needed me, he said, "The last capital project we managed came in 20 percent *below* budget. What happens if we come in 20 percent over budget instead? Over budget by 20 percent on

$2 billion is a lot of money." I had never seen a capital project come in 20 percent below budget before. I asked the vice president to prove to me that he came in 20 percent below budget. He picked up the phone and asked the procurement people to gather together all of the material purchase requisitions for this project. I then informed the vice president that I was not interested in the material costs, but was interested in seeing the direct labor hours. The vice president informed me that in the Winston-Salem plant, the employees do not fill out time cards, especially the white collar workers. I then asked the vice president if it were possible that RJR Tobacco Company actually came in 20 percent over budget rather than under budget, and simply did not know it? Today, the organization fills out time cards. After all, how does a firm improve its estimating on future projects if it does not know how much it spent on previous activities?

To integrate cost and schedule requires the use of earned value measurement. In most firms, this requires a major change to the archaic cost accounting systems that have existed for centuries. Cost accounting personnel are notoriously resistant to change. It means learning new systems and new practices, and even involves a change in culture. Cost accounting personnel will come up with 1,000 excuses as to why the organization should not go to earned value measurement.

Since this phase is where the supporting tools are eventually put in place, there is also the risk that either excessive tools will be installed or overly heavy reliance will be placed upon the tools such that the tools are running the project. According to Lieutenant General Hans H. Driessnack, USAF (retired): "Never let the tool control the handle that uses it!" This implies that the zealots experienced with the tools used in scheduling, cost estimating, risk management, and earned value measurement systems should not end up controlling the program managers. The tail should not end up wagging the dog.

The final item in the maturity phase is the development of a long-term educational program so that the organization can maintain its maturity position. Without a sustained, long-term educational program, the organization can revert from maturity to immaturity very quickly. Long-term educational programs to support project management demonstrate to employees that the organization is committed to project management.

The best educational programs are those based upon documented "lessons learned" files. In mature organizations, the project team is required to prepare a lessons learned file, which is then integrated into the appropriate training programs.

> Without "documented" lessons learned, a company can quickly revert from maturity to immaturity in project management. Knowledge is lost and past mistakes are repeated.

ABB

As an example of the benefits of lessons-learned reviews, consider the following remarks from James M. Triompo, Group Senior Vice President, Process Area Project Management at ABB:

> The ABB PM process calls for a Project Closeout Review at the end of each project. The objective is to identify what went well and areas for improvement on future contracts. Feedback is also provided on the process effectiveness with findings provided to the process owner to facilitate further improvements. The meeting is conducted by the PM and involves sales and the project team, plus management of PM, Engineering, Manufacturing, and Supply Management as applicable. Experience/feedback is solicited from all involved parties, how internal/external suppliers have performed, project financials, engineering reuse, what we've done well and what can be improved. Most typically, the recommendations are used to enhance future bidding of similar projects. Suggested process improvements are considered by the ABB PM process owner (in the project office) for possible process updates. One of the lessons that have been learned includes the importance of involving project execution people in project sales. Another lesson learned is that change/scope management starts with a solid project baseline accompanied by well defined scope. The importance of project meetings, internal and external, is another important lesson learned in achieving customer satisfaction.

Is it possible to complete the maturity phase of the project management life cycle without having gone through the other phases? Some say that it is possible, but it's definitely not usual. If a company has completed the maturity phase effectively, it will be able to answer yes to the following questions:

- Has your company adopted a project management system and used it consistently?
- Has it implemented a philosophy that drives it toward project management success?
- Has your company made a serious commitment to project planning at the onset of each new project?
- Has it minimized the number of scope changes by committing itself to realistic objectives?
- Does it recognize that cost and schedule control are inseparable?
- Has your company selected the right people as project managers?
- Do executive managers receive project sponsor–level information rather than too much project manager–level information?
- Have executives strengthened the involvement of line managers and supported their efforts?
- Does your company focus on deliverables rather than resources?
- Does it cultivate and reward effective communication, cooperation, teamwork, and trust?
- Do senior managers share recognition for successful projects with the entire project team and line managers?
- Does your company focus on identifying and solving problems early, quickly, and cost-effectively?
- Do project staff use project management software as a tool rather than as a substitute for effective planning and interpersonal communications?
- Has your company instituted an all-employee training program based on documented lessons learned?

According to Byron Smith of ChoiceCare/Humana:

> We have a great success story to tell in that we have very rapidly been able to move up the maturity curve that you've talked about. In just two years we've gone from a very immature environment to what I would characterize as a growth stage in your model, and our hope is to, in the not too distant future, get to that mature stage. I simply think it is possible for companies to go into maturity without going through immaturity first, but you're talking now more of an exception than the rule.

2.4 CRITICAL SUCCESS FACTORS IN PROJECT MANAGEMENT

Table 2–1 listed the critical success factors for achieving a fully developed project management system. Critical failure factors that create obstacles to effective project management are just as informative. Typical critical failure factors are listed in Table 2–2.

TABLE 2–2. CRITICAL FACTORS IN PROJECT MANAGEMENT LIFE CYCLE

Critical Success Factors	Critical Failure Factors
Executive Management Acceptance Phase	
Consider employee recommendations	Refuse to consider ideas of associates
Recognize that change is necessary	Unwilling to admit that change may be necessary
Understand the executive role in project management	Believe that project management control belongs at executive levels
Line Management Acceptance Phase	
Willing to place company interest before personal interest	Reluctant to share information
Willing to accept accountability	Refuse to accept accountability
Willing to see associates advance	Not willing to see associates advance
Growth Phase	
Recognize the need for a corporate-wide methodology	View a standard methodology as a threat rather than as a benefit
Support uniform status monitoring/reporting	Fail to understand the benefits of project management
Recognize the importance of effective planning	Provide only lip service to planning
Maturity Phase	
Recognize that cost and schedule are inseparable	Believe that project status can be determined from schedule alone
Track actual costs	See no need to track actual costs
Develop project management training	Believe that growth and success in project management are the same

2.5 DEFINITION OF EXCELLENCE

An experienced company can use project management routinely and for years and still not achieve excellence in project management. An organization that practices project management is not guaranteed that excellence will follow automatically. The last phase in the life cycle may be based on repetitive practices or rigid policies and procedures. It may be achieved when all employees simply understand their job descriptions. The definition of project management *excellence* must extend well beyond experience and success. Organizations excellent in project management create an environment in which there exists a *continuous* stream of successfully managed projects, where success is measured by having achieved performance that is in the best interest of the whole company, as well as having completed a specific project.

First Part of "Excellence" Defined

This definition can be broken down into two parts. First, excellence in project management requires a continuous stream of successfully managed projects. This in no way implies that the projects are successful, just that the projects were successfully managed. Remember: successful implementation of project management does not guarantee that individual projects will be successful. As Figure 2–1 indicates, companies excellent in project management still have their share of project failures. Should a company find that 100 percent of its projects are successful, then that company is not taking enough business risks. That's to say, excellent companies take risks; they simply know which risks are worth taking and which are not. The same argument can be made of management. Any executive who always makes the right decision probably is not making enough decisions.

> Early termination of a project, under the right circumstances, can be viewed as successful when the resources initially dedicated to that project are reassigned to more profitable activities or the technology needed for that project does not exist and cannot be invented cost-effectively within a reasonable time period.

Second Part of "Excellence" Defined

The second part of the definition requires that decisions made on individual projects must take into account the best interest of both the project and the company as a whole. Take the case of a project manager who fights for the best resources available, knowing full well that his or her project has been assigned an extremely low priority. Companies excellent in project management develop cultures in which project managers are taught and encouraged to make decisions based on sound business judgment and not internal parochialism.

Companies that have achieved excellence in project management also realize that excellence is ongoing. Complacency opens doors for the competition.

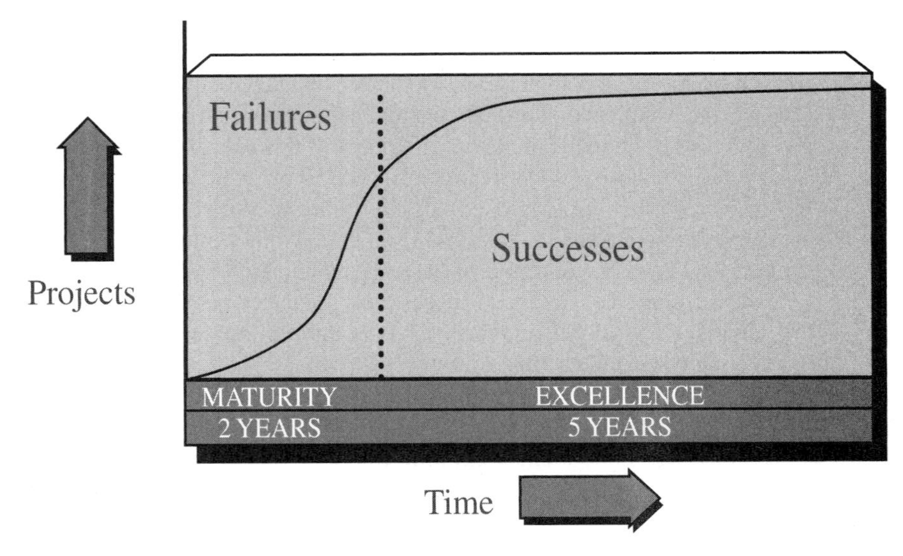

FIGURE 2–1

Chris Hansen, vice president for engineering for Kinetico, has this comment on the need for continuous improvement in project management:

> Project management never stops. It can always [get] better. We've put procedures into place that we've never used before. To be successful and competitive and to meet market needs on a timely basis, this is the only way our company will survive. People in our company are being educated about the benefits [of project management excellence] and are becoming more positive. They are trying harder to work together as a team. They know it is the best way to accomplish our goal.

ABB

Effective continuous improvement in project management can lead to consistency, which, in turn, can lead to excellence in project management. This is especially true in companies that have come to this realization. Such was the case at ABB, which embarked upon the quest for consistency and excellence in project execution. The following remarks were part of a published presentation made by James M. Triompo, Group Senior Vice President, Process Area Project Management at ABB:

> The Utility division has been working with Group Processes to implement an action plan for consistent and successful project execution, which promises to save well over US$ 20 million per year. Jim Triompo Head of Process Area Project Management explains the background.
>
> Delivery, delivery, delivery—on time and within budget—every time! That's what our customers demand and that's what we have to achieve to ensure our future success. But often as we work to meet the needs of our customers, our success is undermined as gross margins are eroded by inefficiencies, penalties and poor quality. Did you know, for exam-

ple, that the cost of poor quality sometimes accounts for 15 percent of project revenue? This is something we cannot allow at a time when ABB is being forced to bid tighter and tighter to win contracts and maintain our competitiveness. So we are aiming to address the problem of margin erosion by implementing an action plan for consistent and successful project execution through disciplined and rigorous project management.

The action plan for the Utility division is derived from Group Processes' philosophy, which aims to develop a common and consistent approach to ABB's global business. The emphasis is on consistency throughout the company, but this initiative really needs to be nurtured and implemented in each division—not imposed from outside. For this reason, our division head, Richard Siudek, asked me, "a Utilities insider," to map out a plan, giving me the discretion to apply the Group Processes approach to the areas where we believe it will truly offer the most benefit for the Utility Division. We surveyed several critical countries and we prioritized on the following initiatives: Project Management, Change Management, Risk Management, Document Management and Installation and Commissioning (including civil works). These are the areas where we feel the biggest savings can be achieved by improving the way we work.

In the first phase of this project, which began in March 2001, we have concentrated on mapping the key countries where our efforts should be focused to achieve the maximum initial impact. These countries were defined as Priority 1, 2 and 3 according to both order intake and number of employees. Currently, we are focusing on the 21 countries identified as Priority 1, which represent around 90 percent of our business. A group process manager has been appointed for each of these 21 countries and each of them is now engaged in deploying, implementing and training the local project managers and installation and commissioning managers who will ultimately be responsible for the success of this initiative.

Our aim for 2002 is to "define and ingrain" the fundamentals for successful project execution into the Utility division organization. We will measure our success in two ways.

First there are implementation metrics. These will establish that the project execution process, tools, templates and procedures we are providing are in place and that people are using them. This requires that the local organization adapt to a common and proven way of working.

Secondly, there are business metrics, which will focus on the impact on our business. We are aiming for a 2 percent improvement in gross margin and intend to reduce the cost of poor quality (due to claims, penalties, re-work, supplier non-compliance, liquidated damages, etc.) by 50 percent, while improving cash flow. This will enable us to save over US$ 20 million in the first year and we expect to be able to more than double this in the future.

The project execution plan will also have a major impact in boosting customer satisfaction. Our increased efficiency will help us to meet and even beat delivery schedules and match the ever-increasing expectations of our customers. We will be in a better position to ensure they get their assets on-line fast, generating revenue as quickly as possible. In fact, customers have been involved in helping ABB to develop our Group Processes and we have been able to share experiences with other companies.

We realize that we are seeking a major cultural shift in making some major changes in the way the Utility division conducts business. And we expect some resistance from those who believe that they are already operating at maximum efficiency and do not need to change. Unfortunately our past performance proves that we must change. Our targets are ambitious—we must drive continuous improvement if we are to maintain our competitive advantage and deliver outstanding service to our customers and increasing value to our stakeholders.

2.6 BEST PRACTICES IN PROJECT MANAGEMENT

Companies often consider some of the outstanding critical success factors and key performance indicators as best practices. *Best practices* are reusable activities or processes that continuously add value to the deliverables of the projects. Best practices can also increase the likelihood of success of each and every project. But while all of this sounds like motherhood and apple pie, there still exists the fundamental question of who defines what is or is not a best practice.

Best practices are defined internally within the company by looking at what worked well for the company and what is most likely to work well in the future if this practice is repeated on every project and for multiple customers. One company identified a critical issue in the relationship between quality and customer satisfaction. The company focused heavily on project quality and discovered that they were sacrificing customer satisfaction to improve quality. When the managers refocused their attention on customer satisfaction, they discovered that quality also improved. Therefore, the company concluded that customer satisfaction was now a company best practice. Everyone in the company was then instructed to focus heavily on customer satisfaction.

What works well as a best practice for one company may not work equally well for another. For example, if you were benchmarking against the above-mentioned company, you may erroneously conclude that customer focus rather than quality focus should be a best practice in your company. However, your company may see degradation in quality rather than improvement in quality.

Best practices can appear in working relationships, design of templates, and the way that project management methodologies are used and implemented. Best practices can occur anywhere. The best practices identified in this book are not necessarily universal best practices but more like acceptable best practices within a specific company. Some companies have Web sites where all best practices are identified. The sites are updated continuously as knowledge is gained from both ongoing and completed projects.

2.7 A STRUCTURED APPROACH

To achieve maturity and excellence in project management should not be left to chance or to trial and error. Instead, it should be a structured process whereby workers can see the light at the end of the tunnel. If the critical success factors and key performance indicators can be identified upfront, there is a good chance that a process for maturity and excellence can be defined.

Sometimes, a simple model or process such as that shown in Table 2–1 can be used. Some companies use a model such as this as a framework for designing their own approach to achieving maturity, and end up with a superb approach. Such was the case for the three companies identified below.

Texas Instruments Companies often make the fatal mistake of believing that the development of a project management methodology is the solution to their ailments. Although this may be true in some circumstances, the excellent companies realize that people execute methodologies and that the best practices in project management might be achieved more quickly if the focus is initially on the people rather than the tools. This requires the identification of people-oriented critical success factors and people-oriented key performance indicators. In the author's opinion, this is the most important component of initially implementing project management and often the most neglected.

Texas Instruments recognized the importance of focusing on people as a way to accelerate project success. The Success Pyramid developed by Texas Instruments for managing global projects is shown in Figure 2–2. A spokesperson at Texas Instruments describes the development and use of the Success Pyramid for managing global projects at Texas Instruments:

> By the late '90's, the business organization for Sensors & Controls had migrated from localized teams to global teams. I was responsible for managing 5–6 project managers who were in turn managing global teams for NPD (New Product Development). These teams typically consisted of 6–12 members from North America, Europe and Asia. Although we were operating in a global business environment, there were many new and unique difficulties that the teams faced. We developed the Success Pyramid to help these project managers in this task.

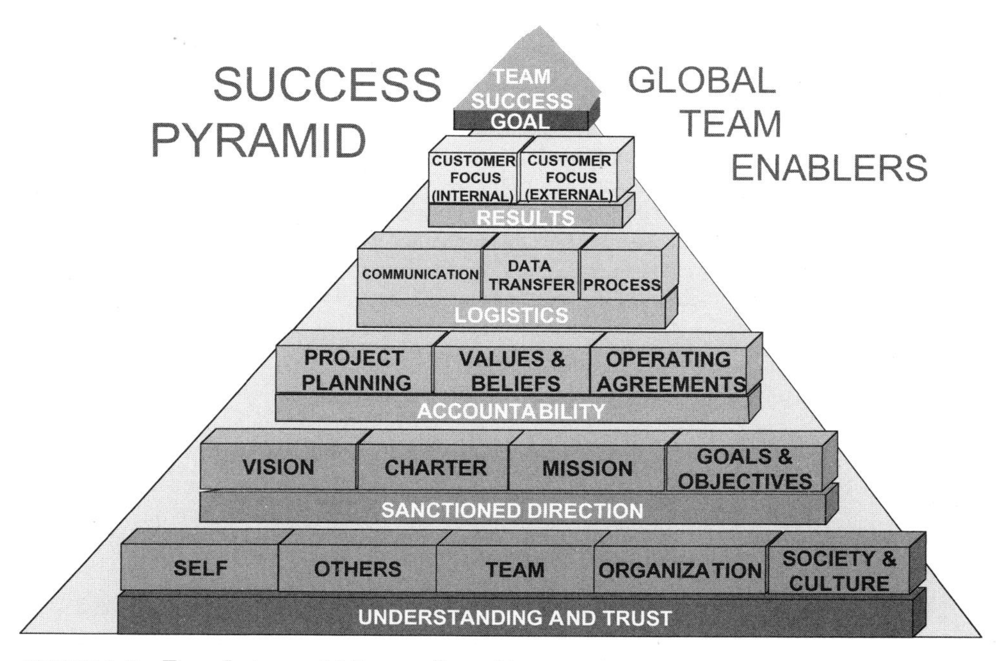

FIGURE 2–2. Texas Instruments' Success Pyramid.

Although the message in the pyramid is quite simple, the use of this tool can be very powerful. It is based on the principle of building a pyramid from the bottom to the top. The bottom layer of building blocks is the *foundation* and called "Understanding and Trust." The message here is that for a global team to function well, there must be a common bond. The team members must have trust in one another, and it is up to the project manager to make sure that this bond is established. Within the building blocks at this level, we provided additional details and examples to help the project managers. It is common that some team members may not have ever met prior to the beginning of a project, so this task of building trust is definitely a challenge.

The second level is called "Sanctioned Direction." This level includes the team charter and mission, as well as the formal goals and objectives. Since these are virtual teams that often have little direct face time, the message at this level is for the project manager to secure approval and support from all the regional managers involved in the project. This step is crucial in avoiding conflicts of priorities from team members at distant locations.

The third level of the pyramid is called "Accountability." This level emphasizes the importance of including the values and beliefs from all team members. On global teams, there can be quite a lot of variation in this area. By allowing a voice from all team members, the project planning cannot only be more complete, but everyone can directly buy into the plan. Project managers using a method of distributed leadership in this phase usually do very well. The secret is to get people to transition from attitude of obligation to a willingness of accepting responsibility.

The next level, called "Logistics," is where the team lives for the majority of the project and conducts the day-to-day work. This level includes all of the daily, weekly and monthly communications and is based on an agreement of the type of development process that will be followed. At TI, we have a formal process for NPD projects, and this is usually used for this type of project. The power of the pyramid is that this level of detailed work can go very smoothly, provided there is a solid foundation below it.

Following the execution of the lower levels in the pyramid, we can expect to get good "Results," as shown in the fifth level. This is driven in the two areas of internal and external customers. Internal customers may include management, or it may include business center sites that have financial ownership of the overall project.

Finally, the top level of the pyramid, which shows the overall goal, is labeled as "Team Success." Our experience has shown that a global team that is successful on a 1–2 year project is often elevated to a higher level of confidence and capability. This success breeds added enthusiasm and positions the team members for bigger and more challenging assignments. The ability of managers to tap into this higher level of capability provides competitive advantage and leverages our ability to achieve success.

DTE Energy

There are several maturity models available in the marketplace to assist companies in their quest for growth, excellence, and success in project management. Table 2–1, discussed previously, is one such model. The purpose of the model is simply to provide some sort of structure to the maturity process. Tim Menke, PMP, is assigned to the ITS Project Management Office at DTE Energy. Tim Menke describes the growth process at DTE Energy:

DTE Energy's Information Technology Services (ITS) organization embarked on a formal drive to improve our project management maturity in 2000. Based on the survey in the book *In Search of Excellence in Project Management* by Dr. Harold Kerzner, our organization has achieved the Embryonic Stage, has made progress in the Executive Management Stage, but has work to do in the Line Management Stage.

Primary drivers for increasing our project management maturity were to:

- Improve cost and schedule predictability on our IT Projects
- Improve relations with our business partners (internal customers)
- Adherence to Capability Maturity Model (CMM) Level 2 practices
- Improve our competitiveness (industry de-regulation)
- Repeat successes realized in our Y2K endeavors

Accomplishments to date include:

- Established and funded a Project Management Office (PMO)
- Developed a Project Portfolio for ITS
- Recruited skilled/certified PMs
- Established a project management training curriculum to develop PM skills and enable attainment of PMP certification
- Established organizational goal that all IT projects be managed within +/− 10% on CPI, Cost, and Schedule
- Creation of a Planning and Management Table along with weekly meetings to review/approve new projects, review progress reporting, serve as a change control board, and review/approve project closeouts complete with lessons learned
- Established a project prioritization committee that meets quarterly to review, authorize, and prioritize new projects (based on NPV, IRR and other economic considerations set forth via the business case)
- Established organizational definition of project success
- Established critical success factors for projects be identified at project initiation
- Developed/communicated project management requirements, checklists, and work aids for both PM practices and enterprise tool
- Focusing on developing a method to assess PM process adherence for use in rewards/compensation
- Capability Maturity Model Level 2 Key Practices combined with PMI best practices form the basis for organizationally required PM practices
- Established a forum for stakeholder involvement
- Developed job roles for Program Manager, Project Manager, PMO Coach, and Planner/Controller
- Attained 66% compliance with CMM Level 2 practices for Software Project Planning and 70% compliance with CMM Level 2 practices for Software Project Tracking and Oversight
- Included Earned Value Analysis performance targets in PM goals and job performance evaluations

Future focus areas include:

- Establish formal job codes/salary ranges for Project Managers and Senior Project Managers
- Achieve 90% compliance on CMM Level 2 Software Project Planning and Tracking and Oversight topics by mid '03

- Develop and implement a project health check to assess project management process adherence
- Establish requirement that large projects or programs will require a PMP certification in the future

Exel

Previously, we stated that one of the characteristics of a well-managed company is not only early recognition that excellence in project management is needed, but also development of a structured approach to achieve some degree of maturity and excellence in project management in a reasonable time frame. Any structured approach to implementing project management must be done first and foremost for the benefit of improving the business. Therefore, there exists a relationship between strategic planning and project management. When an organization performs some degree of strategic planning for excellence in project management, and also has visible executive support, the road to maturity and excellence can be shortened. This occurred at Exel. According to Phil Trabulsi, Sr. Director, Solutions Design, Americas, for Exel:

> Beginning in 1997, Exel set out on a journey to understand the business needs of the organization with regard to projects and how projects were managed. This journey has spanned nearly 7 years and has evolved into what Exel, today, calls Enterprise Project Management, thus developing project management a global core competency.
>
> The following list is a summary of major project management changes that have occurred in recent years that indicate a direction leading toward maturity and excellence in project management:
>
> - Operations to project-centric mindset
> - Defined project management processes and tools
> - Defined project management methodology
> - Senior management directive for project management and support
> - Project management value and maturity awareness and tracking
> - Global PM rollout
> - Global training curriculum
> - Centralized PM center of excellence (EPM group)
> - Strong PM marketing and communication efforts
> - Establishment of a Project Manager career path
> - Multi-lingual tools and training
>
> The four phases listed below outline this journey toward project management growth and maturity. This is shown in Figure 2–3.
>
> **Phase I: Business Needs Analysis (1997)**
>
> *Situation*
>
> - Exel is operationally focused
> - Poor project execution
> - Inconsistent project management approach
> - Rely heavily on individual heroics

FIGURE 2–3. Exel Project Management journey.

Barriers

- Perception of the organization by management as an operating company vs. a project-centric organization
- No project management expertise demand from customers
- Undefined PM processes, roles and responsibilities
- Organizational perception of project management as non-value added

Critical Success Factors and Accomplishments

- Senior management directive for development of single, global, business delivery process
- Investment in outside consultancy to facilitate change program over 9 month period
- Project management begins to be perceived as enabler to change
- Informal Project Management Office (PMO) established as internal thought leadership team
- Assessment and tracking of PM value and maturity

Phase II: Process and Methodology (1997–1998)

Situation

- Exel establishes formal Business Delivery Process (BDP)
- 3 Phases (Strategy, Business Delivery, Supporting Infrastructure)
- 9 steps from Pursuit through Implementation to Closing

- Identified three new focus areas (strategic alignment, business delivery, supporting infrastructure)
- No singular PM methodology established

Barriers

- Multiple organizational "agendas"
- Ad hoc PM approach by department
- Buy-in across business units
- No available proof source for PM
- No reference point for 3PL industry

Critical Success Factors and Accomplishments

- Project management process must fit overall Business Delivery Process (BDP)
- Simple is better to enable rapid engagement
- Previous project failures utilized to reinforce need for consistent processes across business units/divisions
- Methodology standardized
- Executive sponsorship obtained

Phase III: Tools, Training and Organizational Structure (1998–2001)

Situation

- Defined PM process with basic tools
- Small PMO—thought leadership focus
- Divisional-based PM teams forming
- Limited engagement in organization

Barriers

- Identification of internal PM proponents
- Shift to Matrix team structure
- Reluctance to invest in PM development
- Quantification of PM value

Critical Success Factors and Accomplishments

- Project Management defined as a high priority within the organization.
- Acceptance of project-centric organization
- Appeal to broad audience
- Initial participants highly supportive
- PM tools and training must connect with all stakeholders and Business Delivery Process
- Global training curriculum
- Develop KPI's to illustrate progress/value
- Divisional teams responsible for tactical application and implementation
- Centralized PM center of excellence (EPM group)
- Avoidance of mandated tools that may not be globally acceptable/appropriate
- Ensure training of tools prior to release and utilization
- Determine appropriate level of technology that support processes
- Emphasis on professional certification in project management

Phase IV: Visibility, Collaboration and Globalization (2001–2003)

Situation

- Well established business delivery processes
- Global merger/organization
- Geographically and culturally diverse organization
- Varying degrees of PM maturity
- Multi-industry service offerings
- More complex customer expectations
- Track record of reliability and predictability

Barriers

- Geography and time zones
- Culture and languages
- Multi-cultural project teams
- Global PM visibility
- Geographically dispersed project teams
- Project resource constraints (internal and customer)
- Variations in cultural issues

Critical Success Factors and Accomplishments

- Enterprise Project Management group (EPM) established and linked to divisional and regional PMO's
- EPM focused solely on strategic development, training, tools, processes, and knowledge management
- Strong, global communication and broad marketing efforts
- Centralized knowledge base (e.g., Lotus Notes)
- Collaboration among all divisions/departments via enterprise management software (PlanView), and collaboration software (Lotus SameTime)
- Global collaboration on PM tool development and deployment
- Consistent, multi-lingual PM tools with standard Exel "look-n-feel"
- Globally consistent PM training curriculum

For growth and maturity in project management to occur in a reasonable time frame, there must exist not only a structured approach, such as the Exel journey shown above, but also strong, visible executive support. This requires that senior management visibly spearhead the journey and define their expectations at journey's end. This was done at Exel by Bruce Edwards, CEO. Mr. Edwards provided Exel with a clear definition of excellence in project management. According to Mr. Edwards, this included, in no specific order:

- Project-centric organization
- Project-based culture
- Strong organizational and leadership support for project management
- Matrix team structure
- Focus on project management skill development and education
- Emphasis on project management skill track
- Globally consistent PM training curriculum

- Globally consistent project management processes and tools
- Template-based tools versus procedures
- Multi-lingual tools and training
- Acknowledgment and support of project management professionals advanced certification in project management (PMP, CAPM)
- Internal PMP and CAPM support programs for associates
- Strong risk management
- Project management knowledge sharing
- Organizational visibility to portfolio of projects and status via enterprise software (PlanView)

One of the reasons for Exel's success was that Exel viewed its journey to growth and maturity in project management as a strategic planning effort. The strategic planning process for project management excellence at Exel is explained by Todd Daily, PMP, Project Manager, Enterprise Project Management, Americas:

> Recently, Exel's EPM group, with guidance from Dr. Harold Kerzner's paper on *Strategic Planning for Project Management and the Project Office,* has put together a strategic planning matrix for each of the critical activities identified in the paper. Using these identified activities and a few Exel-specific categories, the EPM group conducted a gap analysis of project management strengths and weaknesses that can be used for planning purposes.
>
> The critical activities are:

- Project Management Information System (PMIS)
- Project Failure Information System (PFIS)
- Dissemination of information
- Mentorship
- Development of standards and templates
- Project management benchmarks
- Business case development
- Customized training in project management
- Stakeholder and relationship management
- Continuous improvement
- Capacity planning/resource management
- Reporting and organizational structure
- Internal/functional projects
- Project management maturity

An extract from the strategy-planning matrix is shown in Table 2–3.

According to Todd Daily, there were significant driving forces that provided an indication that excellence in project management was needed:

- *Increasing Project Size:* As projects become more complex and involved, the level of control and management is essential to minimize risk and improve quality.
- *Customer Demand for Faster Implementations:* Exel provides full service supply chain solutions for our customers that often lead to competing with smaller, niche players.

TABLE 2–3. EXEL PROJECT MANAGEMENT—STRATEGIC PLANNING MATRIX

Critical Activities	Description	Kerzner's Guidelines	Exel's Current Status	Rank	Gaps / Suggestions	Risks, Challenges, Next Steps
Project Management Information System (PMIS)	Process and tools for capturing intellectual information relating to value measurement and risk management	• Company PM Intranet • Project Web sites and DB's • PMIS/enterprise systems • PM databases	• Campus-interactive • EPM database • PlanView	◑	• More detailed Information gathering tools required	• New version of PlanView • Not standard across the organization • Needs vary by sector
Performance Failure Information System (PFIS)	Method for gathering project performance information to enhance future project successes	• Lesson's learned activities • Postmortem analysis	• Lessons learned activities • Failure modes and effects analysis (auto sector) • Operational readiness reviews	◔	• EPM group could conduct Lessons Learned activities for projects	• Reluctance to relinquish power by sector • Some information considered confidential and proprietary
Dissemination of information	Communication of critical project management information such as KPI's, CSF's, data, news, and training	KPI's • Line mgmt support • Senior management support • Methodology CSF • Time • Cost • Quality • Scope	• Newsletter • EPM brochure • Press releases • Training • PM Process Guides • Campus-Interactive	◕	• Need more line mgmt support • Qualitative information available • Lack of quantitative information	• Continue to market and promote PM aggressively • Work directly with operations for support and to improve line management support (i.e., project reviews)
Membership	Project office assumes mentorship role for inexperienced PM's seeking advice and guidance in the practice of project management	• PM guidance and mentorship of functional PM's • Dotted-line reporting structure	• Provide guidance to developing markets (Mexico, South America, U.K., Asia-Pacific) • Support sector PM's	◑	• More involvement with project teams • Periodic project reviews	• Viewed by some as overstepping bounds of authority • Perceived as watchdog activity • Some sectors don't see EPM as knowledge center
Development of standards and templates	Design and Implementation of project management standards and tools that foster teamwork by creating a common "pm" language	• Customized templates, forms, and checklists • Limited formalized policies/procedures • Simple and dynamic • Managed and maintained by project office	• Developed and own PM tools and templates for the organization • Manage updates ensuring consistency across the organization	◕	• Formalized processes, standards, and templates in United States • Introduction of tools to Mexico, Latin America, U.K. and Asia	• Acceptance • Adoption • Requires continued "face-time" with all project teams • Continue to ensure tools are optimal for our business

Demands a rapid implementation approach. Project management enables better control, more efficient implementations, and higher quality deliverables.

- *Customer Demand for Project Management Expertise and Proficiency:* Today, more and more customers, like Exel, are realizing the importance of project management in their business. It is strategically advantageous for Exel to possess project management excellence when being considered as a supply chain solution provider for customers.
- *Globalization of the Organization:* Over the past two years, Exel has completed a significant global merger and a number of strategic acquisitions that have integrated a number of different approaches with regard to project management. As Exel incorporates these new organizations, and broadens its service offerings, a consistent, go-to-market project management approach is essential. Geographical and cultural diversity is a key characteristic of Exel's business. With excellence in project management Exel can establish globally diverse project teams, working in any region, with one common project management methodology and a common understanding of all tools. It also enables the delivery of a consistent message to our global customers.
- *Varying Degrees of PM Maturity:* Exel is structured in a sector-based (or divisional) format by industry type: for example, automotive, technology, consumer products, healthcare, retail, chemical, etc. Based on assessments of each sector, it was apparent there were varying levels of project management maturity by sector. Assessments of each sector enabled Exel to develop an approach to balance project management awareness, knowledge, and expertise.
- *Senior Management Mandate for Consistent PM Processes.*
- *Shared Success Stories among and across Sectors (lessons learned).*
- *MBO's Tied to PM Training and Advancement.*
- *Business Development and Account Teams Using PM as a Selling Tool to Clients.*

2.8 MYTHS

There are several myths about excellence in project management. Some organizations have been making mistakes so long that it has become standard practice where employees believe they are doing some good. Unfortunately, they appear to be supporting myths rather than sound project management practices. These typical myths include:

- Profitability does not guarantee excellence.
- Continuously slashing project budgets by 10 percent does not lead to excellence in project management.
- The amount of time spent looking at project estimates does not improve the quality of the estimate, nor does it lead to excellence in project management.
- Walk-the-halls management by executives does not lead to excellence in project management, nor does executive invisibility.
- Project managers who always make the right decisions are not making enough decisions.
- The span of time, in years, needed to achieve excellence in project management is ten times the number of project management policies and procedures in place. Additional executive involvement will increase the span of time by one century.

- Excellence in project management is rarely achieved the first time around.
- Customer satisfaction is not a guarantee that the project was managed with excellence.

MULTIPLE CHOICE QUESTIONS

1. When a company has gone over to project management, the critical question becomes:
   A. How long before the full benefits will be achieved?
   B. How long before functional sponsorship can be used?
   C. How long before cost control can be implemented?
   D. How long before profitability will exceed 25 percent?
   E. All of the above.
2. Project management implementation is most commonly based upon:
   A. Vertical work flow
   B. Horizontal work flow
   C. The culture
   D. Executive sponsorship
   E. Creation of a center of excellence in project management
3. In the early days of project management, the most common definition of success was:
   A. Did it work?
   B. Within time
   C. Within cost
   D. Customer satisfaction
   E. Nonexistent
4. The most common definition of success within excellent companies is:
   A. Within time
   B. Within cost
   C. At the desired technical/quality level
   D. Accepted by the customer
   E. All of the above
5. External measures of success are called:
   A. Critical success factors
   B. Critical performance indicators
   C. Key success factors
   D. Key performance indicators
   E. All of the above
6. Internal measures of success are called:
   A. Critical success factors
   B. Critical performance indicators
   C. Key success factors
   D. Key performance indicators
   E. All of the above
7. At General Electric Plastics, there are __________ factors analyzed to determine project success.
   A. Three
   B. Four

- **C.** Five
- **D.** Seven
- **E.** Eight

8. According to the definition of key performance indicators (KPIs), a typical KPI would be:
- **A.** Adherence to budgets
- **B.** Consistent use of project management systems
- **C.** Use of interim metrics
- **D.** All of the above
- **E.** B and C only

9. Which of the following is not a life cycle maturity phase?
- **A.** Embryonic
- **B.** Executive management acceptance
- **C.** Growth
- **D.** Maturity
- **E.** Implementation

10. Developing a methodology for project management appears in the __________ phase of the life cycle maturity model.
- **A.** Initiation
- **B.** Executive management acceptance
- **C.** Growth
- **D.** Maturity
- **E.** Implementation

11. During favorable economic times, changes in management styles occur __________ during unfavorable conditions.
- **A.** Just as fast as
- **B.** Faster than
- **C.** Slower than
- **D.** All of the above are possible
- **E.** Cannot be determined

12. Project objectives should be defined in:
- **A.** Technical terms only
- **B.** Business terms only
- **C.** Both technical and business terms
- **D.** Whatever terms are acceptable to the sponsor
- **E.** Whatever terms are acceptable to the line managers

13. The need to shorten product development time is referred to as:
- **A.** Synchronous management
- **B.** Acceleration management
- **C.** Concurrent engineering
- **D.** Time-dependent management
- **E.** Rapid-development management

14. A common method for controlling scope creep is through:
- **A.** Termination of configuration control boards
- **B.** Continuously allowing changes to occur
- **C.** Rigid executive sponsorship
- **D.** Frozen specifications
- **E.** Enhancement projects

15. Some people contend that maturity and excellence in project management:
- **A.** Can be obtained in two years or less
- **B.** Can be obtained in three years or less
- **C.** Can be obtained in four years or less
- **D.** Can be obtained in five years or less
- **E.** Is a never-ending journey

16. Historically, senior management defined the project's objectives but did not define:
- **A.** What constitutes success
- **B.** The size of the project office
- **C.** The role of the executive sponsor
- **D.** Customer communication channels
- **E.** All of the above

17. If quality and project management are, in fact, similar:
- **A.** The customer develops the work breakdown structure (WBS)
- **B.** The customer defines both the quality and the success of the project
- **C.** Both require a project manager
- **D.** Both are independent functions reporting to executive management
- **E.** None of the above

18. In the early days of project management, success was defined in __________ terms only.
- **A.** Business
- **B.** Technical
- **C.** Market share
- **D.** All of the above
- **E.** A and B only

19. Client involvement, use of interim metrics, and adherence to the project management methodology are examples of:
- **A.** Critical success factors (CSFs)
- **B.** Key performance indicators (KPIs)
- **C.** Customer critical success factors (CSFs)
- **D.** All of the above
- **E.** A and B only

20. Once a company makes the decision to accept project management, the impact on the bottom line may take months or years to be recognized.
- **A.** True
- **B.** False

21. Which phase of the project management life cycle maturity model can occur in parallel with other phases?
- **A.** Embryonic
- **B.** Executive Management Acceptance
- **C.** Line Management Acceptance
- **D.** Growth
- **E.** Maturity

22. The *biggest* detriment to achieving maturity and excellence in project management is:
- **A.** Lack of functional support
- **B.** Lack of visible executive support
- **C.** Lack of training for employees

D. Lack of a horizontal accounting system
E. Lack of sponsorship by the board of directors

23. The purpose of the growth phase of the project management maturity model is to lay the foundation for standardization and control.
A. True
B. False

24. Which of the following is the last to be performed?
A. Establishment of a project management methodology
B. Definition of the life cycle phases
C. Policies/procedures for effective planning
D. Policies/procedures for handling scope creep
E. Selection of a project management software package

25. Good project managers make project decisions based upon what is in the best interest of the:
A. Project
B. Project manager
C. Project sponsor
D. Company
E. None of the above

DISCUSSION QUESTIONS

1. Why do some people believe that excellence in project management is never reachable?
2. Why is it important for the project manager to have a clear definition of what the executives or customers view as success?
3. In the early years of project management, why was success defined in technical terms only (i.e., type of industries, education, background, etc.)?
4. As companies begin to mature in project management, would you expect the key performance indicators (KPIs) to change? Explain your answer.
5. Explain the most likely impact on the project management life cycle maturity model if executive management support is not visible.
6. Why is it advisable to select a project management software package *after* the methodology is developed?
7. In the life cycle maturity model for project management, why is it so difficult to complete the last phase (i.e., maturity phase)?
8. Would you expect a project-driven or non–project-driven company to complete the phases of the maturity model more quickly?
9. All companies have primary success factors of time, cost, and quality. However, some companies enhance their definition of success with other factors such as safety. Prepare a list of secondary factors and, if possible, identify the industry that identifies these factors.
10. A company goes through the life cycle phase model and achieves excellence. Ten years later, an entirely new management team takes over, and the new team does not support project management. Explain what may occur.

Across

1. Project management ___________
3. Life cycle phase
4. Part of CSF
6. Life cycle phase
8. Scope _____
10. Part of KPI
11. Part of KPI
12. Life cycle phase: __________ management acceptance
13. Part of CSF
14. ___________ accounting

Down

1. Life cycle phase
2. Life cycle phase: ____ management acceptance
5. Life _____ phase
7. CSF: customer ___________
9. Project management goal
12. Life cycle phase

The Driving Forces for Maturity

3.0 INTRODUCTION

The first phase in the maturity model for project management described in Chapter 2 was the embryonic phase. In this phase, senior and middle management must not simply recognize the need for project management, but must share a burning passion for wanting the project management system to become mature. Simply using project management systems and tools, even frequently, does not guarantee that effective project management will become a permanent way of doing business.

Fortunately, there are driving forces that point management in the right direction and push the organization toward maturity in project management. The driving forces originate with real business problems and opportunities that need to be addressed through solid business practices. The driving forces are normally based upon some business need where solid project management practices must be used rather than just paid lip service.

The seven most common forces for achieving project management maturity are:

- Capital projects
- Customer expectations
- Competitiveness
- Executive understanding and buy-in
- New product development
- Efficiency and effectiveness
- Survival

3.1 CAPITAL PROJECTS

Even the smallest manufacturing organization can conceivably spend millions of dollars each year on capital projects. Without good estimating, good cost control, and good schedule control, *capital projects* can strap the organization's cash flow, force the organization to lay off workers because the capital equipment either was not available or was not installed properly, and irritate customers with late shipment of goods. In non–project-driven organizations and manufacturing firms, capital projects are driving forces for maturity.

Lincoln Electric Sometimes it takes capital projects for a company that is manufacturing oriented to recognize the need for project management. General Electric's Lamp Division and Tungsten Division realized this in the early 1980s as the number of capital projects increased. Another example is a company that embarks on a massive capital project that is larger than any it has attempted in the past, as in the case of Lincoln Electric when it upgraded its former motor business.

Lincoln Electric, the leading producer of arc welding products, is located in Cleveland, Ohio. All of its projects are performed for internal customers, and the projects range in value from $10,000 to $30 million. Because the size of its projects was growing, the company needed a better project management system. According to Jim Nelson, senior plant engineer at Lincoln Electric:

> The $30 million motor department expansion was so large and complex that better control was needed. After a slow, more "seat-of-the-pants" start, a professional scheduler was hired and a more project management atmosphere was created.

The subfactors that Nelson identified as drivers for maturity included:

- Deadlines or milestones were set and management wanted reliable information on the progress toward those objectives.
- The number of simultaneous activities required the organization to prevent or coordinate overlapping responsibilities.
- Costs can quickly run out of control if not monitored and managed.

When asked what lessons were learned from this project, Nelson stated:

> Most of the projects that I worked on throughout my career here at Lincoln Electric were my own projects. I believe that I've learned from critical self-evaluation and have formed strong opinions throughout the years about how projects should be run. The EP3 project, which I managed, gave me the opportunity to expand those opinions on a grand scale. I believe that this project was one of the best organized that I have worked on at Lincoln Electric, and the best organized that Lincoln Electric has ever had. I believe that the main reason for success was that I spent a lot of time gathering and disseminating information to the right people, who then had the authority to make their own intelligent decisions, as long as the general guidelines were being followed.

Nelson's comments deserve further discussion. First, Lincoln Electric did not burden the project managers with rigid policies and procedures but simply used general guidelines. This is a characteristic of excellent companies. Second, the majority of his time was being spent as a communicator and integrator. Third, Lincoln Electric had empowered team members with the authority to make their own intelligent decisions. Lincoln Electric is certainly on the right track to achieving excellence in project management.

3M

Because of the amount of funding needed for capital projects, companies usually realize that a standard methodology for project management may be necessary. At the Cynthiana plant at 3M, capital projects range from $5,000 to $15 million and with time frames of two months to two year. In 1997, Randy Canham comments on the evolution of project management for capital projects at the Cynthiana plant at 3M:

> I joined my current organization about seven years ago (1990). At that time, they [3M] were doing projects without any methodology. Procedures existed for the approval of money and authorization of projects, but no standard way to do a project.
>
> About 1991, a list of commonly desired results was developed for equipment design and acquisition projects. The list included recommended spare parts lists, operator and maintenance manuals, and training and other information necessary to insure proper operation and maintenance of the equipment.
>
> Shortly thereafter (~1993), a series of checklists were developed. These checklists were the beginning of a procedure and the initial involvement of nonengineering groups in the development of process machinery. This set of checklists primarily addressed the functional side of project management. Very little was done about the schedule and cost aspects of project management.
>
> Currently, we are using a second set of checklists that includes cost and schedule considerations. This set was developed in 1996. It is very similar to the first checklist set. Almost all functionality, cost, and schedule control is done on an informal basis at this time. However, I see our company continuing to evolve a system that continues to get closer and closer to true project management.

3.2 CUSTOMERS' EXPECTATIONS

Customers' expectations can be another driving force. Today, customers expect contractors not only to deliver a quality product or quality services, but also to manage this activity using sound project management practices. This includes effective periodic reporting of status, timely reporting of status, and overall effective customer communications. It should be no surprise that low bidders may not be rewarded contracts because of poor project management practices on previous projects undertaken for the client.

Hewlett-Packard Company

Hewlett-Packard Company (HP) is a company that has mastered customer expectations. Between 1988 and 1994, revenue at Hewlett Packard doubled. During that period of performance, especially in 1992, Hewlett Packard's

Worldwide Customer Support Organization recognized that customer needs were becoming increasingly customized and complex. Customers needed smooth transitions as they were implementing new environments, and they looked to their vendors to provide solutions. Support services were becoming more critical and were viewed as a key decision factor in winning both the product and support orders.

Hewlett-Packard's management made the decision to expand its custom Support Sales Organization and focus support resources on developing maturity and excellence in project management. A new group of dedicated project resources was formed within the support organization and given the charter to become professional project management "experts." The group was and still is composed of individuals who have extensive backgrounds in field service, including support and technical problem management. Hewlett-Packard established an aggressive project management training program, as well as an informal "mentor" program where senior project managers would provide guidance and direction for the newly assigned people. In addition to the existing internal training courses, new project management courses were developed. When necessary, these courses were supplemented with external programs that provided comprehensive education on all aspects of project management. Efforts to achieve industry recognized certification in project management became a critical initiative for the group.

Hewlett-Packard recognized that their business could be expanded by demonstrating superior project management skills. In large, complex solution implementations, project management was viewed as a differentiator in the sales process. Satisfied customers were becoming loyal customers. The net result was additional support and product business for Hewlett-Packard. Hewlett-Packard recognized also that their customer's either did not have or did not want to tie up their own resources, and Hewlett-Packard was able to educate customers in the value of professional project management. Simply stated, if Hewlett-Packard has the skills, then why not let Hewlett-Packard manage the project?

According to Jim Hansler (PMP), project manager at Hewlett-Packard, the following benefits were obtained:

> First, we are meeting the implementation needs of our customers at a lower cost than they can achieve. Second, we are able to provide our customers a consistent means of implementing and delivering a project through the use of a common set of tools, processes, and project methodologies. Third, we are leveraging additional sales using project management. Our customers now say, "Let HP do it!"

Hewlett-Packard recognized early on that they were no longer in the business of selling only products, but more in the business of providing "solutions" to their customers. In the past, companies like HP provided their customers with a cardboard box (i.e., a product) and the customer had the responsibility to unpack it, inspect it, install it, test it, validate it, and get it up and running. Today, HP sells solutions to their customers whereby HP takes on all of these responsibilities and many more. In the end, the customer is provided with a complete, up-and-running solution without the customer having to commit significant company resources. To do this successfully and on a repetitive basis, HP must also sell their outstanding project management capabilities. In other words, customers expect HP to have superior project management capability to deliver solutions. This is one of the requirements when customers' expectations are the driving force.

Mike Rigodanzo, Senior Vice President, HP Services Operations and Information Technology, believes that:

> In the services industry, *how* we deliver is as important as *what* we deliver. Customers expect to maximize their return on IT investments from our collective knowledge and experience when we deliver best-in-class solutions.
>
> The collective knowledge and experience of HP Services is easily accessible in HP Global Method. This integrated set of methodologies is a first step in enabling HPS to optimize our efficiency in delivering value to our customers. The next step is to know what is available and learn how and when to apply it when delivering to your customers.
>
> HP Global Method is the first step toward a set of best-in-class methodologies to increase the credibility as a trusted partner, reflecting the collective knowledge and expertise of HP Services. This also improves our cost structures by customizing pre-defined proven approaches, using existing checklists to ensure all the bases are covered and share experiences and learning to improve Global Method.

Hewlett-Packard clearly identifies their project management capabilities in their proposals. The following material is typically included in HP proposals. The material was provided by Ron Kempf, PMP, Director of PM Competency & Certification at HP Services Engagement:

HP Services' Commitment to Project Management

Why HP Services Project Management

HP Services considers strong project management a key ingredient to providing successful solutions to our customers. Our project managers are seasoned professionals with broad and deep experience in solutions, as well as managing projects. Our rigorous business processes make sure you are satisfied. A program roadmap provides an overall architecture of the project lifecycle while senior HP Services management conduct regular progress reviews to ensure quality. Our world-class project management methodology combines industry best practices with HP's experience to help keep everything on track. Our knowledge management program enables project managers and technology consultants to put our experience around the globe to work for you.

Project Manager Capability

HP Services has 2,500 experienced and qualified project managers throughout the world in 160 countries. Our project managers generally have technical degrees and 12+ years of significant industry experience. We have over 2,000 projects active worldwide and manage over 3,000 each year. Project sizes range up to $300 million.

As part of the project business infrastructure we have Project Management Offices (PMO) at worldwide, regional, and country levels to provide support and ensure consistency. HP's goal is to ensure that the highest standards are maintained and to improve both the effective management of our projects and customer satisfaction.

PM Processes and Methodology

HP Services uses rigorous processes to manage our programs. The Program Opportunity Roadmap provides an overall architecture for the project lifecycle. It includes the Solution and Opportunity Approval and Review (SOAR) process that approves new business as well as conducts implementation progress reviews to ensure quality and resolve problems quickly.

HP Services' project management methodology uses industry best practices with the added value of our experience implemented through web-based technology to allow quick updates and access throughout the world. It has over 20,000 web pages of information available to support our project teams. The methodology includes extensive knowledge management databases such as lessons learned and project experience from prior engagements that our project managers can use to help in managing their projects.

Computer Associates (CA) For consulting companies and many other service-related industries, a driving force of customer expectations means that the customer comes first. The development of processes, methodologies, tools, and techniques together with the implementation of best practices focus on the importance of the customer and the interfacing with the customer. History has shown that those companies that develop a single method of dealing with a client and try to ram this single method down the client's throat fare poorly. Companies seem to be more successful when they develop a multitude of methods, each flexible enough such that it can be adapted to a specific customer's needs. This concept of multiple flexible methods leads to repetitive business and serves as a good measurement of success. Computer Associates has been very successful with this approach. According to a spokesperson from Computer Associates:

In the spirit of CA's "Customer First" core value, CA felt the need to align services with the way customers view their IT management challenges and the specific implementation programs they use to address them. To this end, CA has aligned services with a customer-centric view of the world.

The focus of CA's Services strategy is on helping customers manage the "business of IT"—the processes and tools required to:

- build and deploy applications;
- exploit knowledge from operational processes and systems; and
- manage systems and networks.

CA offers a range of services consistent with the Services Strategy Framework shown in Figure 3–1.

By offering services at the IT Strategy and Planning layers of this model (in addition to Design and Implementation services), CA seeks to engage customers at all levels of an IT organization. This will help ensure that our "top-to-bottom" solutions address customer recognized business needs.

CA Services builds upon prior successes to make the CA Services organization a major contributor to the success of our customers. CA captures the knowledge gained from our experience with customers around the world and help customers establish effective IT infrastructure management environments that leverage CA brand technology.

In support of this strategy, the Best Practices Library [discussed in more detail in Chapter 8] currently contains 177 methods supporting the framework pyramid layers as follows: Strategy (3); Planning (32); Design (6); and Implementation (136) representing multiple types of product implementations across all six (6) CA Brand areas.

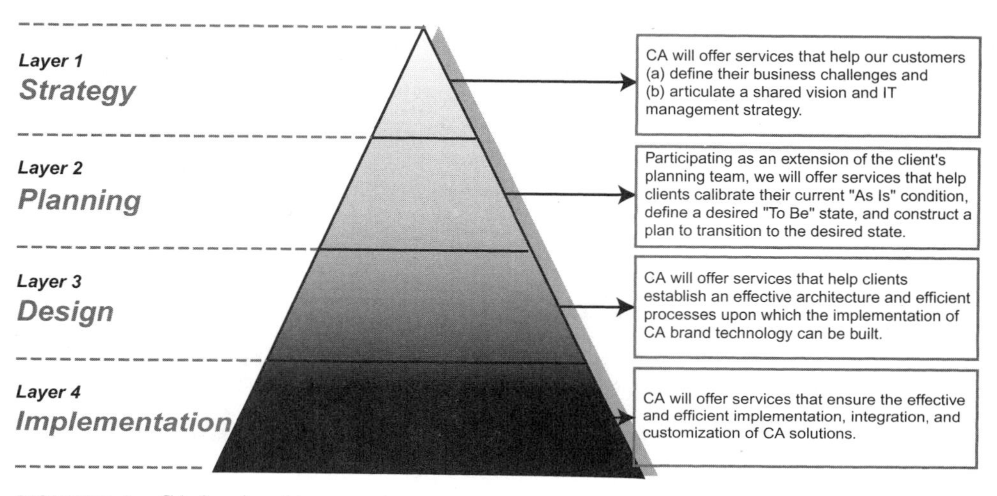

FIGURE 3–1. CA Services Framework.

When the driving force for a company is customer expectations, the best practices that are captured are used not only for providing customers with high-quality deliverables, but also in developing the best possible approach in working with the customers. At Computer Associates, this is referred to as the Engagement Management Model (EMM). According to a spokesperson at CA:

> EMM (shown in Figure 3–2) defines the way in which CA Services carries out an engagement. The Best Practices Library maintains the methods required to provide a repeatable, profitable, well managed engagement. The Project Management Method (PMM) provides project management activities and the Project Management Audit Method defines how to conduct a project audit against CA's Engagement Management Model and ISO standards.
>
> **Overview of the Engagement Management Model (EMM)**
>
> The engagement management model defines the way in which CA Services carries out an engagement from end to end—from Opportunity qualification, to Setup, then Execution and through to Closure (Figure 3–3). This high-level, web-styled, globally accessible application satisfies the requirements of the ISO9001 quality system—to have a single, coherent, all-encompassing, process-model approach to documenting our preferred method of doing business. Examples are shown in Figures 3–4 and 3–5. Each element of the EMM identifies the owner who is responsible for making sure that those elements of the engagement are carried out correctly, so that the customer's needs are satisfied and CA's interests are safeguarded.
>
> The EMM clearly documents for each of the four phases of an engagement the entry and exit criteria, and the inputs and outputs that control the progress of the engagement.

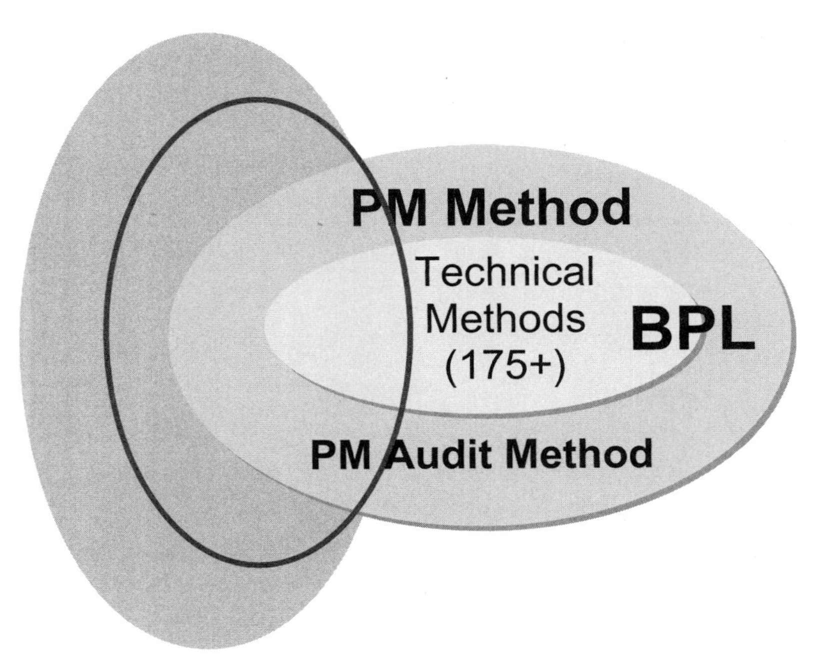

FIGURE 3–2. EMM.

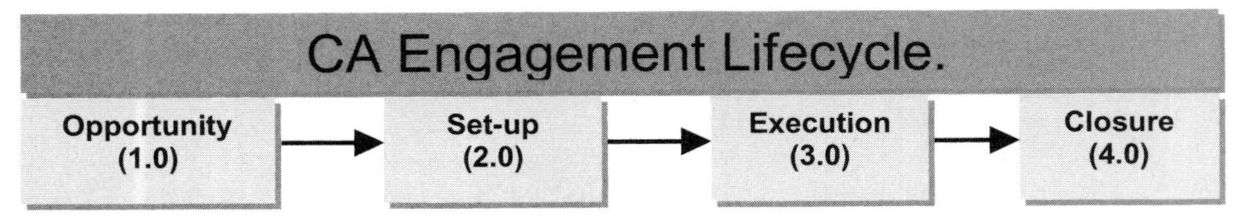

FIGURE 3–3. CA Engagement Lifecycle.

"Swim lane" diagrams of each phase of the engagement lifecycle enable drill-down from each pictured task to a listing of that task's criteria. This includes dynamic links to both global and local procedures and tools associated with that task. By default only the information appropriate to the region in which the user is based will be displayed. Information for other regions is available by "de-personalizing" the menu.

The EMM embodies elements of commercial management, internal quality control and technical methodologies. The over-arching methodology is CA's Project Management Methodology (PMM), based on the PMBOK model and augmented to ISO9001 quality certification. The other methodologies in CA's Best Practices Library are the Technical methods which detail the work breakdown structure, task descriptions, effort estimates, and deliverables that technical consultants must perform to implement and configure CA's software product set.

In essence the EMM is one of the primary elements that directly supports and embraces CA's commitment to its Core Value of *passion for quality and innovation*—a one-stop process guide and pointer to all the policies, procedures and forms that govern CA Services' business operations anywhere and everywhere in the world.

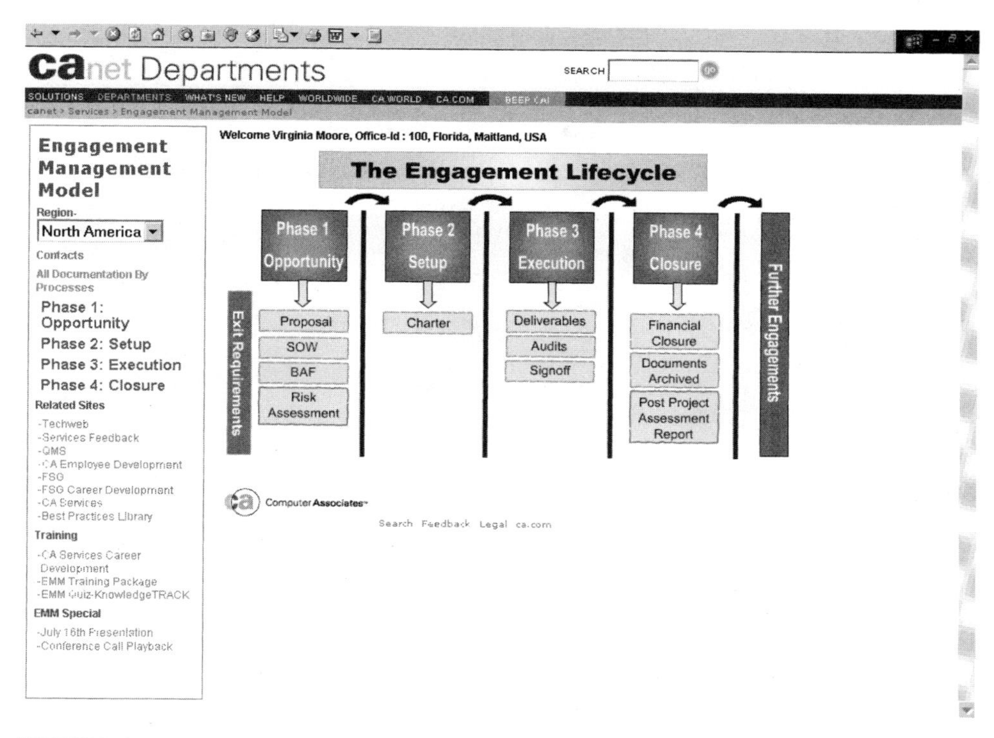

FIGURE 3–4

Nortel

An extension of customers' expectations is stakeholders' expectations. Companies must maintain a delicate balance in trying to satisfy both internal and external customers simultaneously. Jean Claude LeBigre, vice president of GSM France, Nortel Networks, illustrates this conflict:

> Nortel Networks faces stiff competition in the market, and as a by-product of this may face delivery penalties for system-level solutions. The penalties can quickly eat away the profits of a large job. The key to avoiding these penalties is to plan the projects properly and to ensure that realistic scope planning, time, and cost objectives are set. Implementing a project on time is extremely important to our customers as they have business partners to satisfy, regulatory obligations to meet, and shareholders who demand a fair return on their investments. The expectations placed on a project manager are significant, and the key to meeting our contractual obligations is the up-front planning of all projects. A project plan available upon contract award de-risks the implementation and ensures on time, on budget implementations to ensure customer satisfaction.

One of the most significant changes in project management in the twenty-first century will be the working relationship between project management and the sales department. Customers today are expecting delivery of the projects, products, and services to be provided according to the promises made by the sales group. This requires that the project

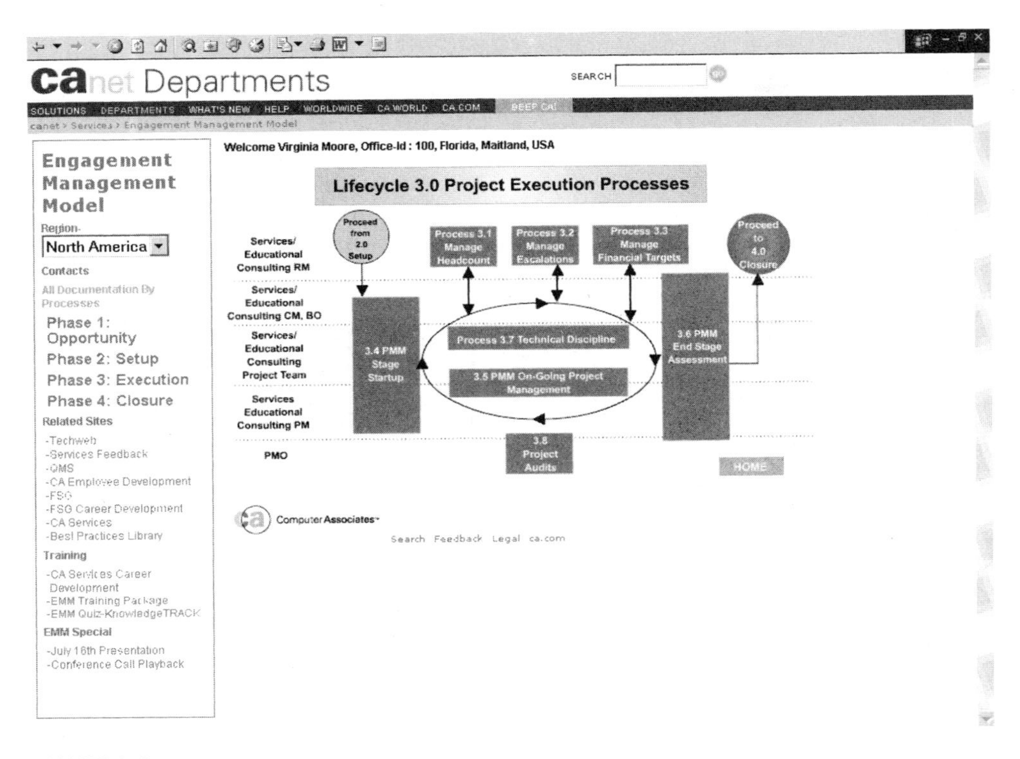

FIGURE 3–5

management group and sales group work closely together, especially if customer's expectations are the company's driving force. In Chapter 2, Jim Triompo of ABB identified this sales-project management working relationship as a best practice. In another example, Guy Grindborg, Vice President for Sales for MahindraBT, believes that:

> Project management is being introduced into areas that traditionally haven't used project managers. Project management in a sales organization has proven to be critical and should be made a central part of the sales process. It is evident that successful companies link sales people with the delivery organization, through project management, and make the two functions share the responsibility for the success and profitability of a contract.
>
> The organization I belong to now defines three distinct roles in a sales process: Sales with commercial responsibilities; Project Management with the overall delivery responsibility; and finally Sales Engineering with a technical solutions responsibility. These three functions make up the foundation for a sales team that stays together throughout the sales project. The project manager's role in the early sales process is to define the three areas of the triple constraint, Cost, Time and Scope. These things were traditionally left up to the sales person to define and, without any training in project management; the result was sometimes a "bit" unreliable.
>
> The role of the project manager changes after a contract has been awarded in the model described above. The project manager assumes a team lead role for the sales team, in par-

allel with his delivery project management function. The team with the three responsibilities stays intact throughout delivery so that new opportunities can be identified and that customer satisfaction can be managed.

3.3 COMPETITIVENESS

The third common driving force behind project management is *competitiveness.* Companies such as Nortel and Hewlett-Packard view project management as a competitive weapon. Project-driven companies that survive on contracts (i.e., income) from external companies market their project management skills through virtually every proposal sent out-of-house. The difference between winning or losing a contract could very well be based upon a firm's previous project management history of project management successes and failures.

The most common form of competitiveness is when two or more companies are competing for the same work. Contracts have been awarded based upon previous project management performance, assuming that all other factors are equal. A subset of this type of competitiveness is when a firm discovers that outsourcing is cheaper than insourcing. This can easily result in layoffs, disgruntled employees, and poor morale. This creates an environment of internal competition and can prevent an organization from successfully implementing project management. An example of this is Avalon Power and Light Company (a disguised case).

Avalon Power and Light Avalon Power and Light is a mountain states utility company that, for decades, had functioned as a regional monopoly. All of this changed in 1995 with the beginning of deregulation in public utilities. Emphasis was now being placed on cost-cutting and competitiveness.

The Information Systems Division of Avalon was always regarded as a "thorn in the side" of the company. The employees acted as prima donnas and refused to accept any of the principles of project management. Cost-cutting efforts at Avalon brought to the surface the problem that the majority of the work in the Information Systems Division could be outsourced at a significantly cheaper price than performing the work internally. Management believed the project management could make the division more competitive but would the employees now be willing to accept the project management approach?

According to a spokesperson for Avalon Power and Light:

> Two prior attempts to implement a standard application-development methodology had failed. Although our new director of information systems aggressively supported this third effort by mandating the use of a standard methodology and standard tools, significant obstacles were still present.
>
> The learning curve for the project management methodology was high, resulting in a tendency of the leads to impose their own interpretations on methodology tasks rather than in learning the documented explanations. This resulted in an inconsistent interpretation of the methodology, which in turn produced inconsistencies when we tried to use previous estimates in estimating new projects.

> The necessity to update project plans in a timely manner was still not universally accepted. Inconsistency in reporting actual hours and finish dates resulted in inaccurate availabilities. Resources were not actually available when indicated on the departmental plan.
>
> Many team leads had not embraced the philosophy behind project management and did not really subscribe to its benefits. They were going through the motions, producing the correct deliverables, but managing their projects intuitively in parallel to the project plan rather than using the project plan to run their projects.
>
> Information Systems management did not ask questions that require use of project management in reporting project status. Standard project management metrics were ignored in project status reports in favor of subjective assessments.

The Information Systems Division realized that their existence could very well be based upon how well and how fast they would be able to develop a mature project management system. By 1997, the sense of urgency for maturity in project management had permeated the entire Information Systems Division. When asked what benefits were achieved, the spokesperson remarked:

> The perception of structure and the ability to document proposals using techniques recognized outside of our organization has allowed Information Systems to successfully compete against external organizations for application development projects.
>
> Better resource management through elimination of the practice of "hoarding" preferred resources until another project needs staffing has allowed Information Systems to actually do more work with less people.
>
> We are currently defining requirements for a follow-on project to the original project management implementation project. This project will address the lessons learned from our first two years. Training in project management concepts (as opposed to tools training) will be added to the existing curriculum. Increased emphasis will be placed on why it is necessary to accurately record time and task status. An attempt will be made to extend the use of project management to non–application-development areas, such as network communications, and technical support. The applicability of our existing methodology to client-server development and Internet application development will be tested. We will also explore additional efficiencies such as direct input of task status by individual team members.
>
> We now offer project management services as an option in our service level agreements with our corporate "customers." One success story involved a project to implement a new corporate identity in which various components across the corporation were brought together. The project was able to cross department boundaries and maintain an aggressive schedule. The process of defining tasks and estimating their durations resulted in a better understanding of the requirements of the project. This in turn provided accurate estimates that drove significant decisions regarding the scope of the project in light of severe budget pressures. Project decisions tended to be based on sound business alternatives rather than raw intuition.

Deregulation of an industry can cause the driving forces to change. As an example, deregulation of the utilities industry in the United States has resulted in a drastic improvement in the way that projects are managed because the survival of the company could easily depend upon the present and future competitiveness of the organization. This same situation is now occurring across the world. According to Winston J. Sosa, Executive Sponsor for Project Management Initiatives at Edelca:

> Edelca has been in existence for 40 years and during this time the company has executed projects of great magnitude related mainly to the generation and transmission of hydroelectric energy. Edelca has accumulated valuable experience in project management for hydroelectric projects.
>
> During the last three years, the company has put forth serious effort directed at the improvement of the quality of the management of its projects. In this regard, a project called "Improvement of the Function of Project Management (MGP)" was conceived in order to improve the quality of the management of Edelca's projects by establishing and using a standard methodology, based on the best practices in project management.[1] This project, "MGP," strategically plans and develops project management improvements for Edelca in the following four areas: Methodology, Organization, Human Resources and Technology.
>
> At present, the hydroelectric industry in which Edelca functions is characterized by a sustained increase in the demand for electric power, restructuring of Edelca into different business units due to legal regulations, competition in electric sectors which are not regulated, and environmental requirements. These functions are placing a greater demand on Edelca's project management systems. Under these circumstances, the maturity and the excellence of project management at Edelca are factors of great importance in order to achieve a sustained competitive advantage with regard to other competitors in the market.
>
> The value proposition [i.e., portfolio management of projects] at Edelca is related to the following aspects: to optimize the capital investments, by selecting and prioritizing the projects that add major value to the company; to increase the performance by improving the payback period and taking advantage of the opportunities of the market; to reduce the total capital costs; to reduce the total time of project development; to obtain less variability in costs, schedules and characteristics of operation; and to increase the probability of satisfying the goals of business, environmental regulations and social responsibilities of the projects.

3.4 EXECUTIVE MANAGERS' BUY-IN

A fourth driving force toward excellence is *executive buy-in.* Visible executive support can reduce the impact of many obstacles. Typical obstacles that can be overcome through executive support include:

- Line managers who do not support the project
- Employees who do not support the project
- Employees who believe that project management is just a fad
- Employees who do not understand how the business will benefit
- Employees who do not understand customers' expectations
- Employees who do not understand the executives' decision

1. Edelca is a hydroelectric utility in Venezuela. MGP is the Spanish acronym for this project.

Roadway Express

In the spring of 1992, Roadway Express realized that its support systems (specifically information systems) had to be upgraded in order for Roadway Express to be well-positioned for the twenty-first century. Mike Wickham, the president of Roadway Express, was a strong believer in continuous change. This was a necessity for his firm, because the rapid changes in technology mandated that reengineering efforts be an ongoing process. Several of the projects to be undertaken required a significantly larger number of resources than past projects had needed. Stronger interfacing between functional departments would also be required.

At the working levels of Roadway Express, the knowledge of the principles and tools of project management was minimal at best in 1992. However, at the executive levels, the knowledge of project management was excellent. This would prove to be highly beneficial. Roadway Express recognized the need to use project management on a two-year project that had executive visibility and support and that was deemed strategically critical to the company. Although the project required a full-time project manager, the company chose to appoint a line manager who was instructed to manage his line and the project at the same time for two years. The company did not use project management continuously, and their understanding of it was extremely weak.

After three months, the line manager resigned his appointment as a project manager, citing too much stress and being unable to manage his line effectively while performing project duties. A second line manager was appointed on a part-time basis and, as with his predecessor, he found it necessary to resign as project manager.

The company then assigned a third line manager, but this time released her from *all* line responsibility while managing the project. The project team and selected company personnel were provided with project management training. The president of the company realized the dangers of quick implementation, especially on a project of this magnitude, but was willing to accept the risk.

After three months, the project manager complained that some of her team members were very unhappy with the pressures of project management and were threatening to resign from the company if necessary simply to get away from project management. But when asked about the project status, the project manager stated that the project had met every deliverable and milestone thus far. It was quickly apparent to the president, Mike Wickham, and officers of the company that project management was functioning as expected. The emphasis now was how to "stroke" the disgruntled employees and convince them of the importance of their work and how much the company appreciated their efforts.

To quell the fears of the employees, the president himself assumed the role of the project sponsor and made it quite apparent that project management was here to stay at Roadway Express. The president brought in training programs on project management and appeared at each training program.

The reinforcement by the president and his *visible* support permeated all levels of the company. By June of 1993, less than eight months after the first official use of project management, Roadway Express had climbed further along the ladder to maturity in project management than most other companies accomplish in two to three years, thanks to the visible support of senior management.

Senior management quickly realized that project management and information systems management could be effectively integrated into a single methodology. To do this, clear definitions had to be established. The following definitions were prepared and approved by senior management.

They defined a project at that time as:

- Has a sponsor
- Has a project manager
- Has a plan with well-defined scope
- Has steering committee approval
- Has dedicated people
- Is moderately to highly visible
- Is a specific request within a specific time frame with an assigned priority within the corporate strategic plan
- Will result in a change to personal or business status quo
- Has defined costs and benefits associated with it
- If cross-functional in nature, the project manager is assigned out of the Project Management and Integration Department
- Involves multiple tasks/multiple people
- Usually a large effort (cost >$25,000)
- Has a set of management approved deliverables
- May impact fixed resources

Project Manager's Responsibilities:

- Define the scope of the project
- Identify enablers—lines areas that need to do something for the project to succeed
- Identify costs and benefits
- Get approvals
- Negotiate commitments
- Monitor and report on progress (submit period reports and/or report as requested by sponsor)
- Manage identification and resolution of problems
- Negotiate scope deliverables
- Control changes (avoid scope creep)
- Plan and schedule user training—may actually participate in training
- Ensure that user manuals and training materials are developed
- Plan and carry out testing and data verification—arrange for field testing if needed
- Plan final implementation—includes notification of system users
- Plan for post-implementation user support (i.e., who will answer questions about how the system is supposed to work?)
- Provide input to the line manager and information systems (IS) manager on the performance of the project workers
- Declare it done
- Do risk assessment evaluation
- Control the work breakdown structure (WBS) (for both IT and business)
- Report period and quarterly status to project's steering committee
- Do benefit realization report within six months of completion of project

Line Manager's Responsibilities:

- Manage resources
- Estimate costs
- Commit to deliverables
- Manage ongoing responsibilities (i.e., perform maintenance, train staff, develop plans, perform administrative duties, provide for continuous improvement)
- Report status on a regular basis
- Report variances immediately
- Consult to identify cross-functional requirements and provide functional and technical expertise
- Identify and resolve problems
- Assist in development of test, training, implementation, and support plans

Although these definitions may seem fundamental to most of us, it must be remembered that eight months earlier Roadway Express had had no project management whatsoever. Roadway Express is certainly doing things right to reach maturity and excellence in project management. In 1996, a survey of the state of the art of project management at Roadway Express was conducted. The following comments were obtained during the survey:

Executive Comments:

- "We have developed a standard methodology that is easy to understand. Most of our managers have undergone some degree of training and understand the project manager/line manager relationship."
- "Our people are committed through costs and time."
- "We understand our role as executives and that we are a servant to business requirements. Our project managers understand the business we are in, where project management fits in, and when to use project management."

Bottom-up Comments:

- "Top management believes in project management and provides strong support."
- "There has been formalized training."
- "Top management has recognized that, in some instances, there is a need for full-time project manager."

Mike Wickham correctly recognized that the quicker he could convince his line managers to support the project management methodology, the quicker they would achieve maturity. According to Mike Wickham, president of Roadway Express at that time (now chairman of the Board):

> Project management, no matter how sophisticated or how well trained, cannot function effectively unless all management is committed to a successful project outcome. Before we put our current process in place, we actively involved all those line managers who thought it was their job to figure out all of the reasons a system *would never work!* Now, the steer-

ing committee says, "This is the project. Get behind it and see that it works." It is a much more efficient use of resources when everyone is focused on the same goal.

3.5 NEW PRODUCT DEVELOPMENT

Another driving force behind project management is *new product development.* The development of a new product can take months or years and may well be the main source of the company's income for years to come. The new product development process encompasses the time it takes to develop, commercialize, and introduce new products to the market. By applying the principles of project management to new product development, a company can produce more products in a shorter period of time, at lower cost than usual, with a potentially high level of quality, and still satisfy the needs of the customer.

In certain industries, new product development is a necessity for survival because it can generate a large income stream for years to come. Virtually all companies are involved in one way or another in new product development, but the greatest impact may very well be with the aerospace and defense contractors. For them, new product development and customer satisfaction can lead to multiyear contracts, perhaps for as long as 20 or more years. With product enhancement, the duration can extend even further.

Customers will pay only reasonable prices for new products. Therefore, any methodology for new product development must be integrated with an effective cost management and control system. Aerospace and defense contractors have become experts in earned value measurement systems. The cost overruns we often hear about on new government product development projects are not necessarily attributed to ineffective project management or improper cost control, but more to scope changes and enhancements.

Raytheon

Raytheon is one of most highly respected aerospace and defense contractors in the world. New produce development and cost controls are extremely important to Raytheon and have led to many best practices. According to Bob Smith, PMP, some of Raytheon's best practices include[2]:

> Program management is a core competency at Raytheon, fundamental to everything we do, domestically and internationally. We are a global technology, defense, and aerospace leader, with approximately $17 billion in annual sales. We manage about 8,000 programs, and submit over 10,000 proposals per year. PMs represent our primary day-to-day contact

2. Bob Smith, PMP, is currently managing Department of Defense programs supporting the Joint Warfare Analysis Center and the Joint Program Office—Special Technology Countermeasures at the Naval Surface Warfare Center, Dahlgren, VA. He is from the Washington Operations Office, Intelligence and Information Systems, Raytheon. These excerpts are from Mr. William Swanson's (Raytheon President) remarks made to the 14th Annual Integrated Program Management Conference (Tyson's Corner, VA) on November 18, 2002 and "Conversations on Project Management—One on One" with the Editor, Aaron Smith, in the November/December 2002 issue of *Projects @ Work* published by IMARK Publications. This material has been provided by permission of William Swanson, President, Raytheon.

with our customers—they are the face of Raytheon. The companies that are successful at PM are going to get return and referral business—past performance is taken seriously.

We measure ourselves against the strictest of measurement standards—Customer Satisfaction. We have invited customers to provide their unvarnished assessments of Raytheon's performance, good and bad, and to share their best practices. We want to DELIGHT CUSTOMERS.

There are several arrows in our program management quiver—Integrated Product Development System (IPDS), risk management, Six Sigma, Earned Value, and diversity.

IPDS captures all of the processes, methods and tools that we use on our programs.

Risk management is *not* an optional thing to be tailored out of the process. Risk areas must be identified, mitigation plans developed, and progress tracked against these plans.

Six Sigma addresses a problem by defining where we are today, using data for an objective assessment, and then determining what we want to be. Until recently, the Six Sigma efforts were primarily focused on manufacturing operations, but we are turning our efforts on "design for Six Sigma"—building Six Sigma into our products from the initial design. In the last three years, Raytheon employees have generated initiatives with gross benefits of $1 billion. After adjusting for several factors, to include the nature of our government contracts, Raytheon netted $275 million in benefits.

Diversity is not just gender and color—it includes background and thought. Good program managers realize they need the diversity of input to be successful. You can get so focused on details; you don't realize other people have good ideas.

Last but not least is the Earned Value Management System, or EVMS. EVMS is perhaps our most critical management tool. EVMS will be implemented on all programs to which it can add value, whether the customer requires it or not. These programs are those of at least six months' duration and a dollar value of over $1 million for development or $5 million for production.

EVMS forces us to look at our programs through the eyes of the customers. Although not a silver bullet, EVMS provides an easy and accurate way to evaluate how we are doing on a program. It helps to assure our CEO and CFO that we have a handle on our programs. It gives us a common language for communication—and allows for open communication with our customers. Our customers in the defense world see us through the metrics of EVM, through CPIs, SPIs and through TCPIs.[3] It is not a good thing for a program manager to come to a review without knowing what the CPI or SPI or TCPI is. EVMS is used to actually manage the program versus using it to report status. This is the simplest sanity check in the world. Customers want more for less. Everyone is trying to become faster, better, cheaper. EVM WILL HELP YOU DELIGHT CUSTOMERS.

A specific program that puts all of this into context is the state-of-the-art radar system for the F/A-18, designated the APG–79 AESA system. The U.S. Navy and Boeing picked Raytheon three years ago to develop this six-year, $105 million program. The program has hit every significant milestone, using the "three Ps" of program management: process, practices and people. The processes include integrated product teams and Six Sigma. The practices include the application of EVM and periodic reviews. The people element revolves around communication, communication, and more communication. You can never communicate enough.

3. These acronyms are part of the EVMS. CPI is the cost performance index, SPI is the schedule performance index, and TCPI is the time to complete performance index.

Raytheon has achieved 100% on-time hardware delivery, maintained the staffing budget within 5% of plan, and avoided costs that add up to $8 million. To foster buy-in, recognition, and unity throughout the program, the Program Manager, Tom Kennedy, has an awards ceremony to celebrate major accomplishments. This is a way to take the time to stop, thank people for their hard work, and give them recognition. It makes them feel good about what they have developed.

Everyone, from leadership to the rank and file, play an important role in DELIGHTING THE CUSTOMER.

3.6 EFFICIENCY AND EFFECTIVENESS

Improvement in overall *efficiency and effectiveness* of the company is difficult. It often requires change in the corporate culture, and culture changes are always painful. The speed at which such changes accelerate the implementation of project management often depends on the size of the organization. The larger the organization, the slower the change.

Kinetico Inc.

One example of a company that has successfully been climbing the ladder to maturity is Kinetico. Kinetico Inc. operates within the highly competitive water treatment industry. The company consists primarily of three core divisions: the Consumer Products Division (CPD), the Engineered Systems Division (ESD), and the Community Water Systems Division (CWSD). For the purpose of this discussion, we will focus on the Consumer Products Division. Kinetico Inc. is a relatively small, privately held company. The number of employees within the CPD is approximately 200. The CPD is a manufacturer of residential products that are designed, manufactured, and distributed from one location. The company was started in 1970 by two engineers. Their influence is still seen in the heavy engineering/technical approach taken in their high quality and innovative products. Kinetico's products are among the most expensive in the industry and widely regarded to be the highest quality on the market.

One of the greatest strengths of Kinetico is its strong culture instilled since the start. The owners to this day insist on a family type atmosphere. This is shown in the encouragement of employees to continue to better themselves either through education or promotion from within. Historically, project management at Kinetico has been on an informal basis. A very relaxed environment also exists at Kinetico. This can be seen in the pastoral setting at Kinetico and in the unique layout of the building to take advantage of the setting. Windows are plentiful throughout the building and management encourages all employees to be located so that they may have the ability to enjoy the outdoors! The culture seems to allow for a better relationship between fellow employees and helps to promote teamwork.

Since starting small, the owners actually designed the revolutionary products and hired the first employees. This helped to create a very close relationship between the owners and the employees. Projects were naturally managed with everyone being involved in the process. This helped foster communication that led to the company's early success.

As Kinetico continued to grow, their close knit culture was kept intact. However, as more departments began to form, they became more structured. This eventually led to a breakdown in communication between departments. During the mid 1980s, Kinetico continued to grow, and the management of projects began to decline. Projects were still well managed within the departments. However, there was difficulty in communicating across the departmental lines. The result was the formation of dominant departments within Kinetico. This was most evident in the research and development department. Because of the owners'—and hence the company's—heavy emphasis on strong engineering, the R&D Department housed some of the company's most influential employees. R&D became the dominant department and took the majority of control of which new products would be developed and released. R&D also had control of how these new products would be manufactured and influenced the price. Most of these new products were developed with minimal input from the sales department. A formal marketing department did not exist, and the function was shared by many, including the president of the company.

In the early years, Kinetico adopted a "get it out" philosophy when it came to introducing new products. Kinetico would pick a release date and, in most cases, would meet that release date despite the circumstances. Often this meant working overtime and cutting testing programs short. Also in these early years, project management was very basic. It existed only at the upper management level and was not very detailed.

During the 1980s, Kinetico began to grow and departmentalized project management began to change. The company went into a phase known as "pass it on" style of project management, which was linear engineering as opposed to the concurrent engineering used today. The projects would be broken down into elements. When a particular department would finish its task, it would pass the project on to the next department. This caused problems for many obvious reasons, such as a lack of efficiency resulting in redundancy of work. Projects were continually finishing over budget and behind schedule.

In the 1990s, formal project management began to emerge. The inefficiency of projects in the 1980s led to the hiring of managers with experience in project management. Project teams were now headed by experienced project leaders. The teams usually consisted of one or more members from each of the functional departments. The departments usually include: marketing, manufacturing, engineering, legal, quality, R&D, compliance, accounting, and purchasing.

Lessons Learned During the last 25 years, Kinetico has gone through some very distinct stages of project management. These various stages were promoted by problems that typically occurred and forced Kinetico to change. These problems that occurred allowed Kinetico's management to learn from their mistakes.

During the early years at Kinetico, much of its success can be attributed to informal project management. Kinetico was small and efficient and communicated very effectively. It is ironic that this early success led to a deterioration of the way in which projects were handled. As the company became larger, there was less communication, which resulted in an increasing number of project difficulties and inefficiencies.

In the late 1980s through the early 1990s, the company was finding new challenges in the area of project management. Each department seemed to have a different agenda and they did not work well together. There were various examples of products that did not meet sales expectations in the marketplace. There were several reasons, including no real market, too high a cost of development, and quality problems.

Chris Hansen, vice president for engineering at Kinetico, made the following comments:

> The marketing and research and development departments in our company were very free to put forth ideas. Research was working on ideas and so was marketing. The two departments never seemed to connect, and when they did connect, the price had never been put forth. Requirements were never laid out. We'd come together with a complete product and marketing would say it was too expensive, so the project would never get off the ground.
>
> The mass of confusion created wasted time, wasted energy, and wasted resources. People would get excited about new ideas, but once their ideas came to fruition, the project wouldn't go anywhere. Involvement from the marketing, manufacturing, quality, finance, accounting, and purchasing departments was not adequate or constant. Ideas, products, prototypes, and actual production tooling would be ordered, and none of these groups were involved. This created dissension, problems, people questioning moves, backseat drivers, and rightfully so. When people aren't involved, that is not the right way to conduct business. I wholeheartedly agree with a saying a friend of mine has that pertains to project engineering. He says, "People equally informed seldom disagree."
>
> So we created a new program and organized a project flow that not only fit the management style, but also fit our manufacturing capabilities at Kinetico Incorporated. We developed, introduced, and implemented the program and communicated to people that this program is very flexible. The program is not carved in stone, and we have made a lot of changes and adjustments to it from people's input. We are always after input and continually state that if something becomes too cumbersome, difficult, or if something doesn't make sense, we are flexible to change. That is a point that has to be executed; you have to be able to meet change. Things change every day. You have to be able to adapt. We're going to get faster and better.
>
> While it's yet a little immature, within the last year I believe project management has become much more focused. We can develop things in a shorter time. This will give us a competitive edge regarding being able to utilize the report tracking and market studies for product development. It should result in lower product costs simply from the standpoint that we won't be developing products three or four times. We can utilize the team engineering concept, go into this with all the different functional areas of the company, and do it right the first time. There will always be design, prototyping, and redesign, but at least we'll be on track with requirements. We'll be developing products with higher market potential.

3.7 COMPANY SURVIVAL

Obviously, the most powerful force behind project management excellence is *survival.* It could be argued that all of the other forces are tangential to survival (see Figure 3–6). In some industries, such as aerospace and defense, poor project management can quickly lead to going out of business. Smaller companies, however, certainly are not immune.

Defcon Corporation

A defense contractor, which wishes to remain nameless (we'll call it "Defcon Corporation"), had survived for almost 20 years on fixed-price, lump sum government contracts. A characteristic of a fixed-price contract is that the

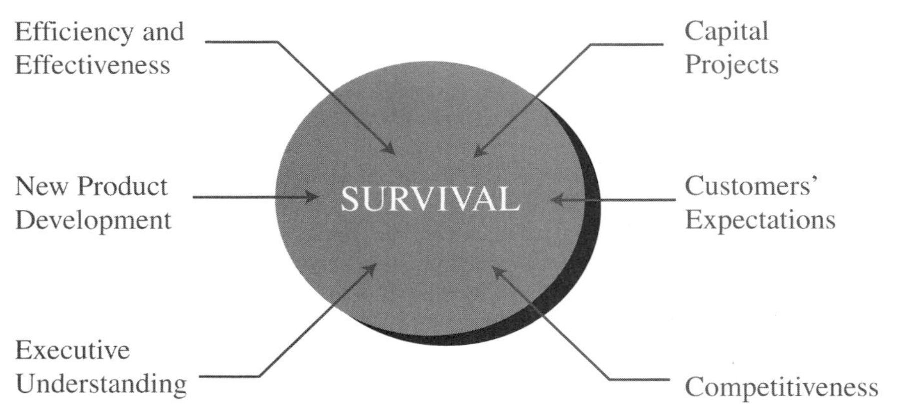

FIGURE 3–6. The components of survival. *Source:* Reprinted from H. Kerzner, *In Search of Excellence in Project Management.* New York: Wiley, 1998, p. 51.

customer does not audit your books, costs, or perhaps even your project management system. As a result, the company managed its projects rather loosely between 1967 and 1987. As long as deliveries were on time, the capabilities of the project management system were never questioned.

By 1987, the government subcontracting environment had changed. There were several reasons for this:

- The Department of Defense (DoD) was undergoing restructuring.
- There were cutbacks in DoD spending, and the cutbacks were predicted to get worse.
- DoD was giving out more and more cost-reimbursable contracts.
- DoD was pressuring contractors to restructure from a traditional to product-oriented organizational form.
- DoD was pressuring contractors to reduce costs, especially the overhead rates.
- DoD was demanding higher quality products.
- DoD was now requiring in their proposals that companies demonstrate higher quality project management practices.

Simply to survive, Defcon had to bid on cost-reimbursable contracts. Internally, this mandated two critical changes. First, the organization had to go to more formal rather than informal project management. Second, the organization had to learn how to use and report earned value measurement. In order to be looked upon favorably by the government for the award of a cost-reimbursable contract, a company *must* have its earned value cost control/reporting system validated by the government.

A manager within one such company that was struggling made the following comments on how "survival" had forced the organization to climb the ladder to maturity over the past ten years:

Formal project management began with the award of the first major government program. There was a requirement to report costs by contract line item and to report variances at specific contract levels. A validated system was obtained to give us the flexibility to submit proposals on government programs where cost schedule reporting was a requirement of the RFP [request for proposal].

We have previous experience in PERT [program evaluation and review technique] Networking, work breakdown structures, and program office organizations. Management was also used to working in a structured format because of our customer's requirements for program reviews. After system validation in 1987, it took six months to a year to properly train and develop the skills needed in cost account managers and work package supervisors. As you move along in a program, there is the need to retrain and review project management requirements with the entire organization.

We visited other companies and divisions of our own company who had prior experience in project management. We sent people to seminars and classes held by experts in the field. We conducted internal training classes and wrote policies and procedures to assist employees with the process of project management. Later we purchased canned reporting packages to reduce the cost of internal programming of systems.

We established dedicated teams to a contract/program. We have program office organizations for large programs to follow through and coordinate information to internal management and our customer. We adjusted our systems and reports to meet both our internal and external customers' needs.

Implementation of integrated systems will provide data on a more timely basis. This data will allow management to react quickly to solve the problem and minimize the cost impact.

Project management has allowed us to better understand costs and variances by contract/program. It provides us with timely data, makes tracking of schedule issues, budget issues, and earned values more manageable. Project management has given us visibility into the programs that is useful in implementing cost reductions and process improvements. Having a validated system allows us to remain competitive for bidding on those programs that require formal costs schedule control systems.

Kombs Engineering

The company described above was very fortunate to have identified the crises and taken the time to react properly. Some companies are not so fortunate. Consider the Michigan-based Kombs Engineering (name of the company is disguised at company's request).

In June 1993, Kombs Engineering had grown to a company with $25 million in sales. The business base consisted of two contracts with the Department of Energy (DoE), one for $15 million and one for $8 million. The remaining $2 million consisted of a variety of smaller jobs for $15,000 to $50,000 each.

The larger contract with DoE was a five-year contract for $15 million per year. The contract was awarded in 1988 and was up for renewal in 1993. DoE had made it clear that, although they were very pleased with the technical performance of Kombs, the follow-on contract must go through competitive bidding by law. Marketing intelligence indicted that DoE intended to spend $10 million per year for five years on the follow-on contract with a tentative award date of October 1993.

On June 21, 1993, the solicitation for proposal was received at Kombs. The technical requirements of the proposal request were not considered to be a problem for Kombs.

There was no question in anyone's mind that on technical merit alone, Kombs would win the contract. The more serious problem was that DoE required a separate section in the proposal on how Kombs would manage the $10 million/year project, as well as a complete description of how the project management system at Kombs functioned.

When Kombs won the original bid in 1988, there had been no project management requirement. All projects at Kombs were accomplished through the traditional organizational structure. Only line managers acted as project leaders.

In July 1993, Kombs hired a consultant to train the entire organization in project management. The consultant also worked closely with the proposal team in responding to the DoE project management requirements. The proposal was submitted to DoE during the second week of August. In September 1993, DoE provided Kombs with a list of questions concerning their proposal. More than 95 percent of the questions involved project management. Kombs responded to all questions.

In October 1993, Kombs received notification that it would not be granted the contract. During a post-award conference, DoE stated that they had no "faith" in the Kombs project management system. Kombs Engineering is no longer in business.

Kombs Engineering is an excellent case study to give students in project management classes. It shows what happens when a subcontractor does not recognize how smart the customer has become in project management. Had Kombs been in close contact with its customers, the company would have had five years rather than one month to develop a mature project management system.

Mason & Hanger Corporation Some companies have learned well the importance of staying close to the customer and the danger of allowing the customer to become more knowledgeable than they are in project management. Mason & Hanger Corporation has a mature project management system. When asked what caused Mason & Hanger to accelerate the process to maturity, John Drummond, division manager for quality, commented (in 1997):

> Continuing demand by the Department of Energy (DoE) to "do more with less" caused us to rethink the methods we used to allocate resources to our work. We found ourselves beginning to overcommit resources in many areas, especially in our process engineering and risk management departments. This became especially noticeable in the early 1990s. We believe that our strong project management culture enables us to better "partner" with our customer by doing a better job of aligning our resources with their expectations!

Williams Machine Tool Company The strength of a culture can not only prevent a firm from recognizing that a change is necessary, but can also block the implementation of the change even after need for it is finally realized. Such was the situation at Williams Machine Tool Company (another disguised case).

For 75 years, the Williams Machine Tool Company had provided quality products to its clients, becoming the third largest U.S.-based machine tool company by 1980. The company was highly profitable and had an extremely low employee turnover rate. Pay and benefits were excellent.

Between 1970 and 1980, the company's profits soared to record levels. The company's success was due to one product line of standard manufacturing machine tools. Williams spent most of its time and effort looking for ways to improve its "bread and butter" product line rather than to develop new products. The product line was so successful that other companies were willing to modify their production lines around these machine tools, rather than asking Williams for major modifications to the machine tools.

By 1980, Williams Company was extremely complacent, expecting this phenomenal success with one product line to continue for 20 to 25 more years. The recession of 1979–1983 forced management to realign their thinking. Cutbacks in production had decreased the demand for the standard machine tools. More and more customers were asking either for major modifications to the standard machine tools or for a completely new product design.

The marketplace was changing and senior management recognized that a new strategic focus was necessary. However, attempts to convince lower level management and the workforce, especially engineering, of this need were meeting strong resistance. The company's employees, many of them with over 20 years of employment at Williams Company, refused to recognize this change, believing that the glory days of yore would return at the end of the recession.

In 1986, the company was sold to Crock Engineering. Crock had an experienced machine tool division of its own and understood the machine tool business. Williams Company was allowed to operate as a separate entity from 1985 to 1986. By 1986, red ink had appeared on the Williams Company balance sheet. Crock replaced all of the Williams senior managers with their own personnel. Crock then announced to all employees that Williams would become a specialty machine tool manufacturer and the "good old days" would never return. Customer demand for specialty products had increased threefold in just the last 12 months alone. Crock made it clear that employees who would not support this new direction would be replaced.

The new senior management at Williams Company recognized that 85 years of traditional management had come to an end for a company now committed to specialty products. The company culture was about to change, spearheaded by project management, concurrent engineering, and total quality management.

Senior management's commitment to project management was apparent by the time and money spent in educating the employees. Unfortunately, the seasoned 20+ year veterans still would not support the new culture. Recognizing the problems, management provided continuous and visible support for project management in addition to hiring a project management consultant to work with the people. The consultant worked with Williams from 1986 to 1991.

From 1986 to 1991, the Williams Division of Crock Engineering experienced losses in 24 consecutive quarters. The quarter ending March 31, 1992, was the first profitable quarter in over six years. Much of the credit was given to the performance and maturity of the project management system. In May 1992, the Williams Division was sold. More than 80 percent of the employees lost their jobs when the company was relocated over 1,500 miles away.

Williams Machine Tool Company did not realize until too late that the business base had changed from production-driven to project-driven. Living in the past is acceptable only if you want to be a historian. But for businesses to survive, especially in a highly competitive environment, they must look ahead and recognize that change is inevitable.

3.8 OTHER DRIVING FORCES

Not all companies are driven by one single force. One paint manufacturer, Sherwin-Williams Company, can point to four driving forces behind its implementation of project management during the 1990s:

- Rapid growth through acquisitions
- Emphasis on new product development
- Emphasis on being able to measure the effectiveness of R&D
- ISO 9000 certification process

MULTIPLE CHOICE QUESTIONS

1. A common mistake made by executives during the initial stages of project management implementation is the belief that:
   A. Project management is a "fad" and will disappear
   B. Executive sponsorship is never needed
   C. Project management, even on large projects, can be done with a part-time project manager
   D. Project management implementation can be accomplished without any training
2. Companies that have achieve excellence in project management soon recognize that the singular driving force is:
   A. Executive buy-in
   B. New product development
   C. Efficiency and effectiveness
   D. Customer expectations
   E. Survival
3. The driving forces for the implementation of project management are related to which life cycle phase of the project management maturity model?
   A. Embryonic
   B. Executive Management Acceptance
   C. Line Management Acceptance
   D. Growth
   E. Maturity
4. Forces that push organizations to (successful) implementation of project management are called __________ forces.
   A. Excellence
   B. Maturity
   C. Performance
   D. Driving
   E. Hybrid

5. Manufacturing companies usually respond to which driving force?
   A. Capital projects
   B. Customer expectations
   C. Executive buy-in
   D. Efficiency and effectiveness
   E. Competitiveness
6. Small companies with growing pains respond to which driving force?
   A. Capital projects
   B. Customer expectations
   C. Executive buy-in
   D. Efficiency and effectiveness
   E. Competitiveness
7. A company with a large, talented R&D group will most likely respond to which driving force?
   A. Survival
   B. New product development
   C. Efficiency and effectiveness
   D. Customer expectations
   E. Executive buy-in
8. If your customers are using project management and they expect you to use it as well, the driving force is:
   A. Executive buy-in
   B. New product development
   C. Survival
   D. Efficiency and effectiveness
   E. Capital projects
9. Lincoln Electric's driving force was:
   A. Capital projects
   B. Customer expectations
   C. Competitiveness
   D. Executive buy-in
   E. New product development
10. Avalon's driving force was:
    A. Capital projects
    B. Customer expectations
    C. Competitiveness
    D. Executive buy-in
    E. New product development
11. Hewlett-Packard's driving force was:
    A. Capital projects
    B. Customer expectations
    C. Competitiveness
    D. Executive buy-in
    E. New product development
12. Roadway Express identified __________ as the driving force for project management maturity.
    A. Capital projects
    B. Customer expectations

C. Competitiveness
D. Executive buy-in
E. New product development

13. Raytheon identified __________ as the driving force for project management maturity.
    A. Capital projects
    B. Customer expectations
    C. Competitiveness
    D. Executive buy-in
    E. New product development
14. Kinetico's driving force was:
    A. Survival
    B. Efficiency
    C. New product development
    D. Customer expectations
    E. Executive buy-in
15. It is possible for a company to have more than one driving force.
    A. True
    B. False

Questions 16–25 are difficult. There may be more than one correct answer to each question. Select all correct answers.

16. The most likely driving force(s) for the plastics division of General Electric would be:
    A. Capital projects
    B. Customer expectations
    C. Competitiveness
    D. Executive understanding
    E. New product development
    F. Efficiency and effectiveness
    G. Survival
17. The most likely driving force(s) for Microsoft would be:
    A. Capital projects
    B. Customer expectations
    C. Competitiveness
    D. Executive understanding
    E. New product development
    F. Efficiency and effectiveness
    G. Survival
18. The most likely driving force(s) for Dell Computer would be:
    A. Capital projects
    B. Customer expectations
    C. Competitiveness
    D. Executive understanding
    E. New product development
    F. Efficiency and effectiveness
    G. Survival
19. The most likely driving force(s) for Walt Disney's motion picture division would be:
    A. Capital projects

B. Customer expectations
C. Competitiveness
D. Executive understanding
E. New product development
F. Efficiency and effectiveness
G. Survival

20. The most likely driving force(s) for Walt Disney's theme parks division would be:
A. Capital projects
B. Customer expectations
C. Competitiveness
D. Executive understanding
E. New product development
F. Efficiency and effectiveness
G. Survival

21. The most likely driving force(s) for Arthur Anderson's Consulting Group would be:
A. Capital projects
B. Customer expectations
C. Competitiveness
D. Executive understanding
E. New product development
F. Efficiency and effectiveness
G. Survival

22. The most likely driving force(s) for the power plants of a nuclear utility would be:
A. Capital projects
B. Customer expectations
C. Competitiveness
D. Executive understanding
E. New product development
F. Efficiency and effectiveness
G. Survival

23. The most likely driving force(s) for the information systems division of Merrill Lynch would be:
A. Capital projects
B. Customer expectations
C. Competitiveness
D. Executive understanding
E. New product development
F. Efficiency and effectiveness
G. Survival

24. The most likely driving force(s) for Northern Telecom would be:
A. Capital projects
B. Customer expectations
C. Competitiveness
D. Executive understanding
E. New product development
F. Efficiency and effectiveness
G. Survival

25. The most likely driving force(s) for Intel would be:
    A. Capital projects
    B. Customer expectations
    C. Competitiveness
    D. Executive understanding
    E. New product development
    F. Efficiency and effectiveness
    G. Survival

DISCUSSION QUESTIONS

1. What are the differences between driving forces and critical success factors?
2. Is it possible for an organization to have more than one driving force?
3. Can driving forces be defined according to the type of industry?
4. Can the type of driving force for maturity and excellence change over time?
5. Can the type of driving force for maturity and excellence change during each phase of the life cycle maturity model?
6. Which level of management usually identifies the driving forces?
7. Can a driving force turn out to be a detriment?
8. What environmental factors pressure an organization to identify driving forces?
9. Is it necessary for an organization to recognize the existence of driving forces before committing to project management?
10. What influence can large customers exert in pressuring an organization to accept project management?

Across

2. New _ _ _ _ _ _ _ _ development
3. _ _ _ _ _ _ _ _ _ _ _ and effectiveness
5. Driving force
6. Driving forces for _ _ _ _ _ _ _ _ _
7. Driving force: _ _ _ _ _ _ _ _ projects
9. Theoretical number of driving forces
10. _ _ _ _ _ _ _ _ _ _ sponsorship

Down

1. Driving _ _ _ _ _
3. Customer _ _ _ _ _ _ _ _ _ _ _ _ _
4. _ _ _ _ _ _ _ _ _ _ sponsorship
8. Actual driving forces

Project Management Methodologies

4.0 INTRODUCTION

In Chapter 2 we described the life cycle phases for achieving maturity in project management. The fourth phase was the growth phase, which included the following:

- Establish life cycle phases
- Develop a project management methodology
- Base the methodology upon effective planning
- Minimize scope changes
- Select the appropriate software to support the methodology

The importance of a good *methodology* cannot be understated. Not only will it improve your performance during project execution, but it will also allow for better customer relations and customer confidence.

Creating a workable methodology for project management is no easy task. One of the biggest mistakes made is developing a different methodology for each type of project. Another is failing to integrate the project management methodology and project management tools into a single process. When companies develop project management methodologies and tools in tandem, two benefits emerge. First, the work is accomplished with fewer scope changes. Second, the processes are designed to create minimal disturbance to ongoing business operations.

This chapter discusses the components of a project management methodology and some of the most widely used project management tools. Three detailed examples of systems at work are also included.

4.1 EXAMPLES OF METHODOLOGY DEVELOPMENT

Simply having a project management methodology and following it does not lead to success and excellence in project management. The need for improvements in the system may be critical. External factors can have a strong influence on the success or failure of a company's project management methodology. Change is a given in the current business climate, and there's no sign that the future will be any different. The rapid changes in technology that have driven changes in project management over the past two decades are not likely to subside. Another trend, the increasing sophistication of consumers and clients, is likely to continue, not go away. Cost and quality control have become virtually the same issue in many industries. Other external factors include rapid mergers and acquisitions, and real-time communications.

Project management methodologies need to change as the organization changes in response to the ever-evolving business climate. Such changes, however, require that managers on all levels be committed to the changes and develop a vision that calls for the development of project management systems along with the rest of the organization's other business systems.

Rockwell Automation

Today, companies are managing their business by projects. This is true for both non–project-driven and project-driven organizations. Virtually all activities in an organization can be treated as some sort of project. Therefore, it is only fitting that well-managed companies regard a project management methodology as a way to manage the entire business rather than just projects. The following comments on Rockwell Automation's view of project management were made by Mike Waldron, Logix™ Program Manager, and Jeff Moranski, Engineering Program Manager:

> Rockwell Automation is a leading industrial automation company focused to be the most valued global provider of power, control and information solutions. With a focus on automation solutions that help customers meet their productivity objectives, the company brings together leading brands in industrial automation.
>
> Project management is coming under increased focus in Rockwell Automation. As Rockwell shifted its focus to enterprise-wide solutions, the complexity of project management has increased significantly. Some examples of enterprise-wide solutions include Logix™ and ViewAnyWare™ products. Customers expect to use a common configuration software package and have a consistent user experience across the wide variety of products and platforms offered. Despite the necessary diversity of products, all must fit into a common framework. This requires a high degree of coordination amongst the various product businesses; project managers are called on to make it all happen.
>
> Project management for product development at Rockwell Automation has standardized on a common tool set that provides a consistent approach from the project manager all the way through the organization to senior management. Benefits from this common tool set are common metrics, a fast stage-gated approval process, standard language, and a consistent process.
>
> One of the key tenets of collaborative project management at Rockwell Automation is that projects within a business have to be successfully managed before projects across the

> business can be successfully managed. Investments in common process and standards for projects have been made at Rockwell Automation in order to ensure that the proper building blocks are in place for cross-business collaboration.
>
> The Project Management process at Rockwell Automation has built-in flexibility, allowing us to work with partners and to align our process to their organization specific methods and structure, empowering them to run their business via their own business strategies.
>
> Rockwell Automation's project management curriculum helps to prepare Project Managers to successfully handle the needs and challenges of our changing environment. By employing a strategy of skills gap analysis, training sessions, and follow-up metrics, Rockwell Automation Project Management is meeting the challenges head-on.

Developing a standard project management methodology is not for every company. For companies with small or short-term projects, such formal systems may not be cost-effective or appropriate. However, for companies with large or ongoing projects, developing a workable project management system is mandatory.

Let's look at an example. A company that manufactures home fixtures had several project development protocols in place. When they decided to begin using project management systematically, the complexity of the company's current methods became apparent. The company had multiple systems development methodologies based on the type of project. This became awkward for employees who had to struggle with a different methodology for each project. The company then opted to create a general, all-purpose methodology for all projects. The new methodology had flexibility built into it. According to one spokesman for the company:

> Our project management approach, by design, is not linked to a specific systems development methodology. Because we believe that it is better to use a [standard] systems development methodology than to decide which one to use, we have begun development of a *guideline* systems development methodology specific for our organization.
>
> We have now developed *prerequisites* for project success. These include:
>
> - A well-patterned methodology
> - A clear set of objectives
> - Well-understood expectations
> - Thorough problem definition

> The moral here is do not reinvent the wheel; do not develop the methodology from scratch.

During the late 1980s, merger mania hit the banking community. With the lowering of costs due to economies of scale and with the resulting increased competitiveness, the banking community recognized the importance of using project management for mergers and acquisitions. The quicker the combined cultures became one, the less the impact on the corporation's bottom line.

The need for a good methodology became apparent, according to a spokesperson at one bank:

> The intent of this methodology is to make the process of managing projects more effective: from proposal, to prioritization, to approval through implementation. This methodology is not tailored to specific types or classifications of projects, such as system development efforts or hardware installations. Instead, it is a commonsense approach to assist in prioritizing and implementing successful efforts of any jurisdiction.

In 1996, the Information Services Division of one bank formed an IS Reengineering Team to focus on developing and deploying processes and tools associated with project management and system development. The mission of the IS Reengineering Team was to improve performance of IS projects, resulting in increased productivity, cycle time, quality, and satisfaction of the projects' customers.

According to a spokesperson at the bank, the process began as follows:

> Information from both current and previous methodologies used by the bank was reviewed, and the best practices of all these previous efforts were incorporated into this document. Regardless of the source, project methodology phases are somewhat standard fare. All projects follow the same steps, with the complexity, size, and type of project dictating to what extent the methodology must be followed. What this methodology emphasizes are project controls and the tie of deliverables and controls to accomplishing the goals.

To determine the weaknesses associated with past project management methodologies, the IS Reengineering Team conducted various focus groups. These focus groups concluded that there was a:

- Lack of management commitment
- Lack of a feedback mechanism for project managers to determine the updates and revisions needed to the methodology
- Lack of adaptable methodologies for the organization
- Lack of training curriculum for project managers on the methodology
- Lack of focus on consistent and periodic communication on the methodology deployment progress
- Lack of focus on the project management tools and techniques

Based on this feedback, the IS Reengineering Team successfully developed and deployed a project management and system development methodology. Beginning June 1996 through December 1996, the target audience of 300 project managers became aware and applied a project management methodology and standard tool (MS-Project).

The bank did an outstanding job of creating a methodology that reflects guidelines rather than policies and provides procedures that can easily be adapted on any project in the bank. Listed below are the selected components of the project management methodology.

Organizing With any project, you need to define what needs to be accomplished and decide how the project is going to achieve those objectives. Each project begins with an idea, vision, or business opportunity, a starting point that must be tied to the organization's business objectives. The Project Charter is the foundation of the project and forms the contract with the partied involved. It includes a statement of business needs, an agreement of what the project is committed to deliver, an identification of project dependencies, the roles and responsibilities of the team members involved, and the standards for how project budget and project management should be approached. The Project Charter defines the boundaries of the project.

Planning Once the project boundaries are defined, sufficient information must be gathered to support the goals and objectives and to limit risk and minimize issues. This component of project management should generate sufficient information to clearly establish the deliverables that need to be completed, define the specific tasks that will ensure completion of these deliverables, and outline the proper level of resources. Each deliverable affects whether or not each phase of the project will meet its goals, budget, quality, and schedule. For simplicity sake, some projects take a four-phase approach:

- *Proposal:* Project initiation and definition
- *Planning:* Project planning and requirements definition
- *Development:* Requirement development, testing, and training
- *Implementation:* Rollout of developed requirements for daily operation

Each phase contains review points to help ensure that project expectations and quality deliverables are achieved. It is important to identify the reviewers for the project as early as possible to ensure the proper balance of involvement from subject matter experts and management.

Managing Throughout the project, management and control of the process must be maintained. This is the opportunity for the project manager and team to evaluate the project, assess project performance, and control the development of the deliverables. During the project, the following areas should be managed and controlled:

- Evaluate daily progress of project tasks and deliverables by measuring budget, quality, and cycle time.
- Adjust day-to-day project assignments and deliverables in reaction to immediate variances, issues, and problems.
- Proactively resolve project issues and changes to control scope creep.
- Aim for client satisfaction.
- Set up periodic and structured reviews of the deliverables.
- Establish a centralized project control file.

Reporting Two essential mechanisms for successfully managing projects are solid status reporting procedures and issues and change management procedures. Status reporting is necessary for keeping the project on course and in good health. The status report should include the following:

- Major accomplishment to date
- Planned accomplishments for the next period

- Project progress summary:
  - Percent of effort hours consumed
  - Percent of budget costs consumed
  - Percent of project schedule consumed
- Project cost summary (budget versus actual)
- Project issues and concerns
- Impact to project quality
- Management action items

Issues and change management protects project momentum while providing flexibility. Project issues are matters that require decisions to be made by the project manager, project team, or management. Management of project issues needs to be defined and properly communicated to the project team to ensure the appropriate level of issue tracking and monitoring. This same principle relates to change management because inevitably the scope of a project will be subject to some type of change. Any change management on the project that impacts the cost, schedule, deliverables, and dependent projects is reported to management. Reporting of issue and change management should be summarized in the status report denoting the number of open and closed items of each. This assists management in evaluating the project health.

Simply having a project management methodology and using it does not lead to maturity and excellence in project management. There must exist a "need" for improving the system toward maturity. Project management systems can change as the organization changes. However, management must be committed to the change and have the vision to let project management systems evolve with the organization.

General Motors Powertrain For companies with small or short-term projects, project management methodologies may not be cost-effective or appropriate. For companies with large projects, however, a workable methodology is mandatory. General Motors Powertrain Group is another example of a large company achieving excellence in project management. The company's business is based primarily on internal projects, although some contract projects are taken on for external customers. The size of the group's projects ranges from $100 million to $1.5 billion. Based in Pontiac, Michigan, the GM Powertrain Group has developed and implemented a four-phase project management methodology that has become the core process for its business. The company decided to go to project management in order to get its products out to the market faster. According to Michael Mutchler, former vice president and group executive:

> The primary expectation I have from a product focused organization is effective execution. This comprehends disciplined and effective product program development, implementation, and day-to-day operations.
>
> Product teams were formed to create an environment in which leaders could gain a better understanding of market and customer needs, to foster systems thinking and cross-functional, interdependent behavior, and to enable all employees to understand their role in executing GM Powertrain strategies and delivering outstanding products. This organi-

zational strategy is aimed at enabling a large organization to be responsive and to deliver quality products that customers want and can afford.

The program management process at GM Powertrain is based upon common templates, checklists, and systems. The following lists several elements that were common across all GM Powertrain programs during the 1990s:

- Charter and contract
- Program team organizational structure with defined roles and responsibilities
- Program plans, timing schedules, and logic networks
- Program-level and part-level tracking systems
- Four-phase product development process
- Change management process

Two critical elements of the GM Powertrain methodology are the program charter and program contract. The program charter defines the scope of the program with measurable objectives, including:

- Business purpose
- Strategic objective
- Results sought from the program
- Engineering and capital budget
- Program timing

The program contract specifies how the program will fulfill the charter. The contract becomes a shared understanding of what the program team will deliver and what the GM Powertrain staff will provide to the team in terms of resources, support, and so on.

Today, companies seem to be promoting the use of the project charter concept, but not all companies create the project charter at the same point in the project life cycle, as shown in Figure 4–1. The three triangles in Figure 4–1 show possible locations where the charter can be prepared.

- In the first triangle, the charter is prepared immediately after the feasibility study is completed. In this point, the charter contains the results of the feasibility study as well as documentation of any assumptions and constraints that were considered. The charter is then revisited and updated once this project is selected.
- In the second triangle, which seems to be the preferred method, the charter is prepared after the project is selected and the project manager has been assigned.
- In the third method, the charter is prepared after detail planning is completed. The charter contains the detailed plan. Management will not sign the charter until after detail planning is approved by senior management. Then, and only then, is the project officially sanctioned by the company. Once management signs the charter, the charter becomes a legal agreement between the project manager and all involved line managers as to what deliverables will be met, and when.

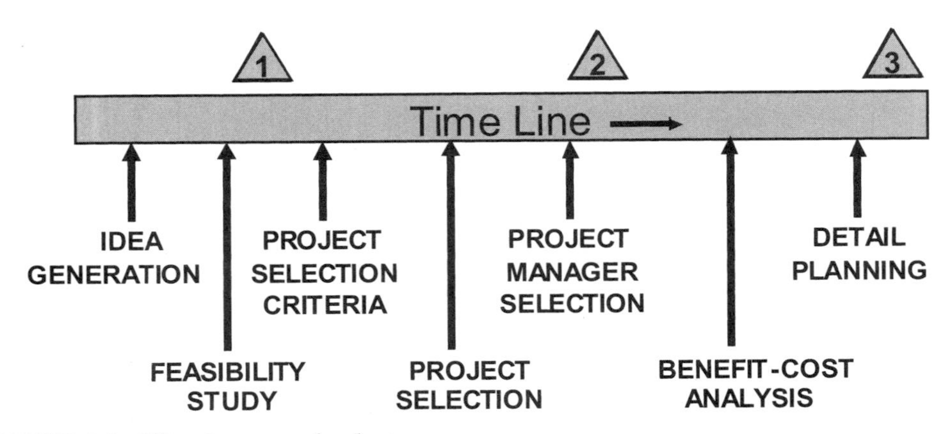

FIGURE 4–1. When to prepare the charter.

Ericsson Telecom AB

General Motors Corporation and the bank are examples of project management methodologies that are internal to the organization (i.e., internal customers). For Ericsson Telecom AB, the problem is more complicated. The majority of Ericsson's projects are for external customers, and Ericsson has divisions all over the world. Can a methodology be developed to satisfy these worldwide constraints?

In 1989, Ericsson Telecom AB developed a project management methodology called PROPS.[1] Though it was initially intended for use at Business Area Public Telecommunications for technical development projects, it has been applied and appreciated throughout Ericsson worldwide, in all kinds of projects.

New users and new fields of applications have increased the demands on PROPS. Users provide lessons-learned feedback so that their shared experiences can be used to update PROPS. In 1994, a second generation of PROPS was developed, including applications for small projects, concurrent engineering projects, and cross-functional projects, and featuring improvements intended to increase quality on projects.

PROPS is generic in nature and can be used in all types of organizations, which strengthens Ericsson's ability to run projects successfully throughout the world. PROPS can be used on all types of projects, including product development, organizational development, construction, marketing, single projects, large and small projects, and cross-functional projects.

PROPS focuses on business, which means devoting all operative activities to customer satisfaction and securing profitability through effective use of company resources. PROPS uses a tollgate concept and project sponsorship to ensure that projects are initiated and procured in a business-oriented manner and that the benefits for the customer, as well as for Ericsson, are considered.

---

1. The acronym for PROPS is in Swedish. For simplicity sake, it is referred to as PROPS throughout this book.

The PROPS model is extremely generic, which adds flexibility to its application to each project. The four cornerstones of the generic project model are

- Tollgates
- The project model
- The work models
- Milestones

Tollgates are superordinate decision points in a project at which formal decisions are made concerning the aims and execution of the project, according to a concept held in common throughout the company. In PROPS, five tollgates constitute the backbone of the model. The function and position of the tollgates are standardized for all types of projects. Thus the use of PROPS will ensure that the corporate tollgate model for Ericsson is implemented and applied.

The project sponsor makes the tollgate decision and takes the overall business responsibility for the entire project and its outcome. A tollgate decision must be well prepared. The tollgate decision procedure includes assessment and preparation of an executive summary, which provides the project sponsor with a basis for the decision. The project and its outcome must be evaluated from different aspects: the project's status, its use of resources, and the expected benefit to the customer and to Ericsson. At the five tollgates, the following decisions are made:

- Decision on start of project feasibility study
- Decision on execution of the project
- Decision on continued execution, confirmation of the project or revision of limits, implementation of design
- Decision on making use of the final project results, handover to customer, limited introduction on the market
- Decision on project conclusion

The project model describes which project management activities to perform and which project documents to prepare from the initiation of a prestudy to the project's conclusion. The project sponsor orders the project and makes the tollgate decisions while most of the other activities described in the project model are the responsibility of the project manager. The project model is divided into four phases: prestudy, feasibility study, execution, and conclusion phases.

The purpose of the prestudy phase is to assess feasibility from technical and commercial viewpoints based on the expressed and unexpressed requirements and needs of external and internal customers. During the prestudy phase a set of alternative solutions is formulated. A rough estimate is made of the time schedule and amount of work needed for the project's various implementation alternatives.

The purpose of the feasibility study phase is to form a good basis for the future project and prepare for the successful execution of the project. During the feasibility study, different realization alternatives and their potential consequences are analyzed, as well as their potential capacity to fulfill requirements. The project goals and strategies are defined, project plans are prepared, and the risks involved are assessed. Contract negotiations are initiated, and the project organization is defined at the comprehensive level.

The purpose of the execution phase is to execute the project as planned with respect to time, costs, and characteristics in order to attain the project goals and meet the customer's requirements. Technical work is executed by the line organization according to the processes and working methods that have been decided on. Project work is actively controlled; that is, the project's progress is continuously checked and the necessary action taken to keep the project on track.

The purpose of the conclusion phase is to break up the project organization, to compile a record of the experiences gained, and to see to it that all outstanding matters are taken care of. During the conclusion phase, the resources placed at the project's disposal are phased out, and measures are suggested for improving the project model, the work models, and the processes.

Besides describing the activities that will be performed to arrive at a specific result, the work model also includes definitions of the milestones. However, to get a complete description of the work in a specific project, one or more work models should be defined and linked to the general project model. A work model combined with the general project model is a PROPS application. If there are no suitable work models described for a project, it is the project manager's responsibility to define activities and milestones so that the project plan can be followed and the project actively controlled.

A milestone is an intermediate objective that defines an important, measurable event in the project and represents a result that must be achieved at that point. Milestones link the work models to the project model. Clearly defined milestones are essential for monitoring progress, especially in large and/or long-term projects. Besides providing a way of structuring the time schedule, milestones will give early warning of potential delays. Milestones also help to make the project's progress visible to the project members and the project sponsor. Before each milestone is reached, a milestone review is performed within the project in order to check the results achieved against the milestone criteria. The project manager is responsible for the milestone review.

Ericsson's worldwide success can be partially attributed to the acceptance and use of the PROPS model. Ericsson has shown that success can be achieved with even the simplest of models and *without* the development of rigid policies and procedures.

Nortel

Large companies with multiple product lines can still have a single/standard methodology for project management, but the methodology will have different "process engines." One company that possesses a world-class methodology with multiple engines is Nortel. In 1999, Dave Hudson, vice president for Time-to-Market Product Development at Nortel Networks, commented:

> Nortel Networks is currently deploying Time-to-Market in the product development area. Time-To-Market encompasses the concept of integrated project teams for every substantial product development we undertake. Our project teams use a standard methodology, albeit one with flexibility in the right places. For instance, we use three different "process

engines" for the three types of products using the Wheelwright model—derivative products, platform replacements, and breakthrough products. Our methodology uses standard tools, metrics, and executive reporting techniques across all of our R&D projects. Templates and specific best practice examples are available for most steps, along with process guidelines for all steps. We have found senior level active participation to be necessary and use the concept of a general manager-led team, called the Portfolio Management Team. Their role is to own and direct the evolution of the business unit's portfolio, manage the projects throughout their life cycle using a disciplined "Business Decision Review" process and, finally, to manage the design pipeline (the people resources), including selection of project leaders and project managers and project starts and stops.

4.2 OVERCOMING DEVELOPMENT AND IMPLEMENTATION BARRIERS

Making the decision that the company needs a project management methodology is a lot easier than actually doing it. There are several barriers and problems that surface well after the design and implementation team begins their quest. Typical problem areas include:

- Should we develop our own methodology, or benchmark best practices from other companies and try to use their methodology in our company?
- Can we get the entire organization to agree upon a singular methodology for all types of projects, or must we have multiple methodologies?
- If we develop multiple methodologies, how easy or difficult will it be for continuous improvement efforts to take place?
- How should we handle a situation where only part of the company sees a benefit in using this methodology and the rest of the company wants to do its own thing?
- How do we convince the employees that project management is a strategic competency and that the project management methodology is a process to support this strategic competency?
- For multinational companies, how do we get all worldwide organizations to use the same methodology? Must it be intranet-based?

These are typical questions that plague companies during the methodology development process. These challenges can be overcome, and with great success, as illustrated by the companies identified below.

DTE Energy

Every company usually develops its own methodology for project management. Some methodologies are created entirely within the company, while others are modified from benchmarking studies but adapted to the specific internal situations expected to be encountered. In either event, most companies prefer to end up with a custom-designed methodology specific to their own types of projects. In excellent companies, the methodology is viewed as a "living" document, subject to enhancements and changes. Beverly Jeffries, PMP, Supervisor, describes the methodology used in the ITS Project Management Office at DTE Energy:

> The project management methodology used by Information Technology Services (ITS) project managers at DTE Energy was developed by the ITS Project Management Office. It is based on best practices within DTE, other organizations, PMI's Project Management Body of Knowledge (PMBOK®) and SEI's Capability Maturity Model (CMM).
>
> The PM methodology includes processes, procedures, checklists and templates to promote standardization, consistency and to facilitate performance reporting for the ITS portfolio of projects. The methodology covers the entire project life cycle from project initiation to project closeout. The degree of formality of the project management methodology can be scaled based on project criteria, such as size, costs, and risks.
>
> The PM methodology is maintained in an asset library and is available to the entire enterprise via the intranet. Future efforts will provide the capability of delivering the methodology through the project management information tool. The project work products are also maintained by the PMO and used to create a database of lessons learned and best practices to be used by Project Managers and identify opportunities for improvement.
>
> This year, an enterprise project management tool was selected, acquired and deployed to support the methodology and to enable enterprise project reporting. The tool and supporting processes have increased the capability of the PMO to provide senior management and our business partners with project performance and status information and thereby, enables informed decision-making.
>
> Enhancements are made to the methodology based on opportunities identified through an internal process improvement effort, to support emerging software developing methodologies and lessons learned. Recent endeavors to implement an iterative software development methodology have resulted in enhancements to the project management methodology. Also, users throughout the organization are also encouraged to identify process improvement opportunities. These opportunities are submitted to the PMO for review, and often result in an enhancement to the process. This process has proven to be an effective means of getting feedback on the project management processes.

Texas Instruments Some people believe that it is possible to develop a singular methodology for project management that encompasses both new product development and IT projects. While some companies have been successful at this, others have found that having a separate methodology for IT-related projects is best. The reason for the separate methodology for IT is because its compatibility to the corporate infrastructure may require different life cycle phases, phase gate reviews, templates, and closeout procedures. In any event, compatibility with the corporate infrastructure is mandatory. Such was the case at Texas Instruments. According to a spokesperson at Texas Instruments:

> I will tell you about our Information Technology Readiness (ITR) Process we have developed and have been using very successfully here in Information Technology Services at Texas Instruments. This process was designed to parallel a development methodology but concentrate on the readiness of the IT Infrastructure and Operations.
>
> This process was developed in 1997 and is designed to provide project managers with a process to ensure all infrastructure and operations components are ready for an application

implementation. We found that most projects failed, not because the application did not work as designed (you push this key and you got the right answer), but the operations and infrastructure were not in place to support the application (i.e., production servers not being monitored or backed up, the help desks not knowing about the application deployment, the performance and application availability goals not mutually agreed upon by customer and program manager, plus numerous other infrastructure and operations care-abouts).

This process ensures all infrastructure and operations stakeholders are involved early. We have formal reviews with standard templates that guide the project manager through a discussion with ITS management to ensure all operations and infrastructure components are involved and assisting the project for a successful implementation.

CIGNA

Companies appear reluctant to implement project management across an entire company using the same methodology within each functional division. Instead, each division may be given the right to develop its own form of project management and to custom design it for their particular use. Although on the surface this practice appears to have some merit, problems can occur during continuous improvement efforts when decisions are made for what is in the best interest of a particular functional area rather than for the company as a whole. The maximization of benefits occurs when continuous improvement efforts in project management benefit the company as a whole.

Michael J. Leser, Vice President for Project Management Practice at CIGNA Corporation, comments on his experiences at CIGNA:

> Coming up through the Project Management ranks over the last 20 years, I recognized one key element for all PM's—the notion of 90% of your time spent communicating is true. That leaves precious little time to effectively administer the project. I quickly realized the value of standards, document templates and consistent, reusable processes.
>
> CIGNA had set up a Program Management Practice (PMP) in 2001, led by Claudia Piccirilli, and staffed with subject matter experts to establish consistent methodology and oversight processes along with a group to build the project management discipline and establish a community feeling. When I took over the PM Methodology program within the PMP, as the case in most large companies, our business organizations ran and managed projects differently. My challenge was to work with our various business organizations to derive and enhance our internal best practices and deliver a usable solution for all our project managers.
>
> Working with a set of knowledge areas that are closely aligned to the Project Management Institute's (PMI), we started at the highest level of project management methodology, and developed four basic project process flows. These flows, as depicted in Figure 4–2, along with standard project phases, form the core task structure for all project work plans.
>
> These processes were then decomposed into sub-processes and activities, and when combined with the necessary roles, responsibilities, inputs and outputs, formed a complete process guidebook for the project manager to reference when needed. We then set up our PMP website to deliver our four process guidebooks as interactive web pages, along with standard templates and instructions for activities requiring deliverables.
>
> Having all of this content was great, but we didn't want another "shelfware" binder or seldom accessed website. By taking the necessary level of detail needed for a project workplan (or schedule), we delivered the methodology right to where project managers spend most of their desk time—in the project workplan as seen in Figure 4–3. We used the task notes to store detailed task descriptions for each task, and added a hyperlink to our

The PM Methodology is structured to provide project managers access to detailed process information by four major **Process Flows**:

- Define & Initiate Project
- Start Project
- Monitor & Control Project
- Close Project

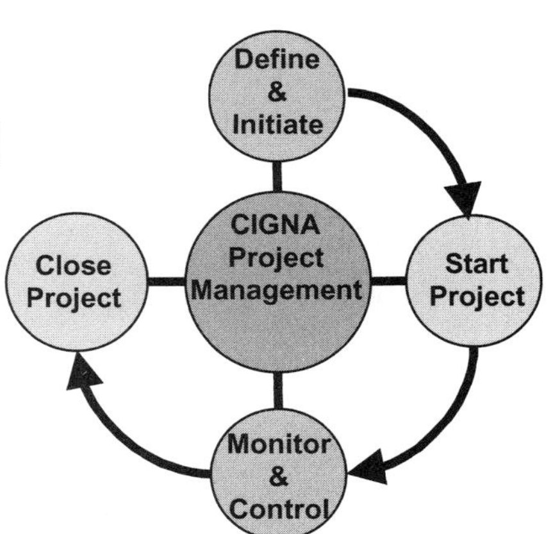

FIGURE 4–2. Project Management processes.

website to quickly access reference information and detailed process and procedure flows. Additional links were also added to access any tools or templates needed for a particular task. To further improve the usability of the planning tool, we added a custom toolbar with easy to use macros to add/change data and produce reports.

By following this standard plan template, we give our project managers a common structure for their own project that has built-in corporate required deliverables. As they define their unique deliverables and work breakdown structure, they simply add to this core set of tasks. To assist in the implementation of this standard, the PMP has built a staff of "Green Berets." These are skilled project management professionals that can come in and work as advisors or full-time project managers.

Our next step is to integrate our project execution processes, some of which were already defined as part of our CIO Andrea Anania's major restructuring effort, into this same model. Standard processes execution flows for requirements, design, construction and implementation were defined and are being incorporated with the PM methodology tasks into a base systems project workplan.

Figures 4–2 and 4–3 illustrate that a good methodology need not be overly complicated, but needs to be useful. The CIGNA project template was developed by customizing the MS project global template file with a custom toolbar and then using a standard work plan template that contains the PMP standard views, tables, filters, and hyperlinks to the process documentation and tools on their Web site.

This methodology provides the CIGNA project managers with a standard project task list format in an easy-to-use MS project work plan to help manage projects and promote the consistency needed to deliver the PMP's vision.

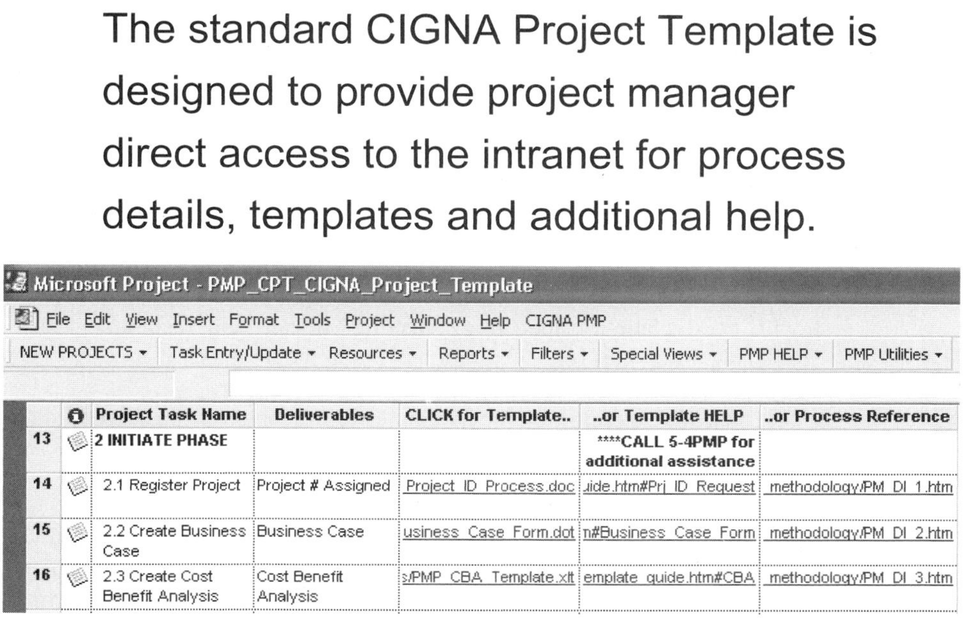

The standard CIGNA Project Template is designed to provide project manager direct access to the intranet for process details, templates and additional help.

Microsoft Project - PMP_CPT_CIGNA_Project_Template

File Edit View Insert Format Tools Project Window Help CIGNA PMP

NEW PROJECTS ▾ Task Entry/Update ▾ Resources ▾ Reports ▾ Filters ▾ Special Views ▾ PMP HELP ▾ PMP Utilities ▾

		Project Task Name	Deliverables	CLICK for Template..	..or Template HELP	..or Process Reference
13		2 INITIATE PHASE			****CALL 5-4PMP for additional assistance	
14		2.1 Register Project	Project # Assigned	Project_ID_Process.doc	uide.htm#Prj_ID_Request	methodology/PM_DI_1.htm
15		2.2 Create Business Case	Business Case	usiness_Case_Form.dot	n#Business_Case_Form	methodology/PM_DI_2.htm
16		2.3 Create Cost Benefit Analysis	Cost Benefit Analysis	s/PMP_CBA_Template.xlt	emplate_guide.htm#CBA	methodology/PM_DI_3.htm

FIGURE 4–3. Project workplan integration.

Swiss Re[2]

When companies realize the benefits that project management can provide for their organizations, they often struggle with determining the best approach by which all the benefits can be achieved in the shortest possible time. Today, well-managed companies are focusing on project management training and education and the implementation of a project management organization (PMO).[3]

Swiss Re is a company that has done both of these quite well. Swiss Re spends approximately a third of its annual budget on projects. Projects are therefore an integral part of Swiss Re's change processes. Having recognized the importance of projects and the impact their success (or failure) has on Swiss Re's overall performance, top management decided to make project management a core competency of Swiss Re's staff. In September 1999, senior management stated that Swiss Re's project execution capabilities needed to be radically improved.

To reach that aim, Swiss Re decided to introduce a singular project management methodology as a standard for all business groups. After careful evaluation of several of the well-known methodologies, the basis for what later became the Best Practice Project Management (BPPM) was chosen and adapted to Swiss Re's specific needs. BPPM

2. The Swiss Re material was provided by Marianne L. Hofmann-Luchsinger of Group PMO at Swiss Reinsurance Company.

3. Swiss Re refers to PMO as a project management organization, whereas other companies refer to PMO as a project management office. However, the PMO at Swiss Re may or may not also include the standard project office.

enhances project management maturity by setting standards, by offering diverse documentation and supporting tools, and by creating a recognized common language for project-related issues. In detail, BPPM includes:

- A project life cycle work breakdown structure
- Checklists of what is important in each life cycle phase
- Skeleton templates for the most important documents needed in the project life cycle with guidelines on how they have to be filled in and what their content should be
- Step-by-step process descriptions (e.g., Project Planning Process)
- Technique papers (e.g., Meeting Management Technique)
- Examples (e.g., Operational Guidelines)
- Project management-related glossary of terms
- Role definitions for major players involved in projects (project sponsor, project beneficiary, project manager, etc.)

BPPM recognizes five phases within the project life cycle and a sixth phase within the product life cycle. This is shown in Figure 4–4. In the first phase, the vision phase, the development of a business case is highlighted and benefits to management introduced. In the product life cycle, the last phase, the benefit realization should become reality.

Within each of the four middle phases, BPPM ensures management of the following areas:

- *Scope:* What is included in the project, what isn't included in the project
- *Time:* Not only for the aim of the project, but for all key deliverables right down to each work stream
- *Costs:* Budget and continuous monitoring of costs

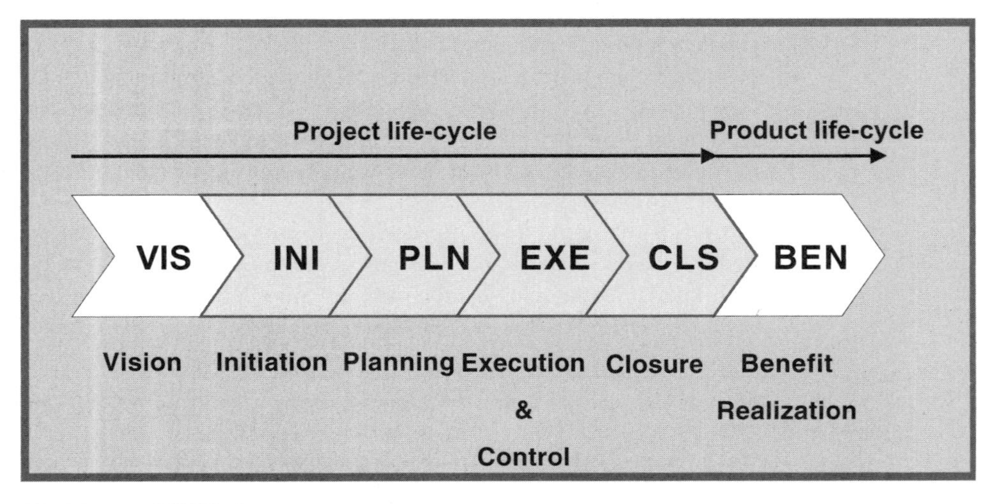

FIGURE 4–4. BPPM phases.

- *Human Resources:* Roles and responsibilities, but also resource planning
- *Progress:* Performance analysis on all important levers of a project
- *Communication:* Not only within the project, but also to all the stakeholders
- *Risks:* Prioritization and assessment
- *Quality:* Definition and control mechanisms
- *Benefits:* Calculations and monitoring

Each phase is divided into subphases with clearly defined tasks. Here, only the most important questions should be addressed, together with the most important task within that area.

The PMO offers courses on BPPM:

- BPPM training for project novices
- Advanced project manager training (APM) for experienced project managers
- Sponsor training for project sponsors

The PMO also offers tailormade workshops on project management. There are additional duties assigned to the PMO other than training. The PMO is tasked with fostering a professional project management culture across all of Swiss Re. The PMO fulfills this task by enhancing project management maturity by setting standards, by offering diverse documentation and supporting tools, through training, but also by being directly involved in various projects as project coaches, leaders, or team members, thus creating a recognized common language for project-related issues. The PMO developed the Best Practice Project Management (BPPM) methodology as the standard best suited to Swiss Re's project management needs across the organization.

In order to measure success, the PMO is tasked by top management with a governance function, overseeing Swiss Re's project portfolio on a defined number of measurements. Through all these various efforts, the PMO optimizes the value potential within the firm in terms of information and know-how exchange and develops cost synergies through common processes and platforms.

In all project management–related areas, the PMO supports Swiss Re's key priorities through its actions (see Figure 4–5):

- Operational excellence is enhanced by better project management.
- BPPM enables common processes through predefined standards and supporting tools.
- The same measurements are applied across the entire group.

To ensure that the PMO combines project management knowledge with the requirements of the business, the PMO is organized as a network. Each division and geographical area has independent, dedicated PMO teams. The core responsibilities of the PMO at Swiss Re, shown in Figure 4–6, include:

- *Support:* The PMO offers all projects the possibility of having one of the PMO's specialists on the team. The PMO members work mostly as project leaders, but sometimes also as team members.

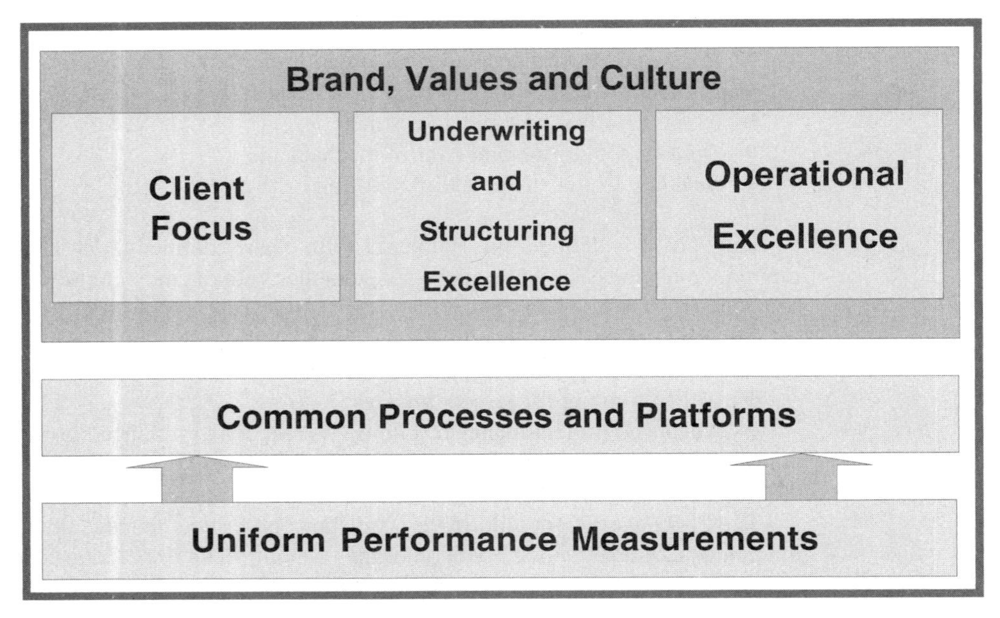

FIGURE 4–5. PMO actions.

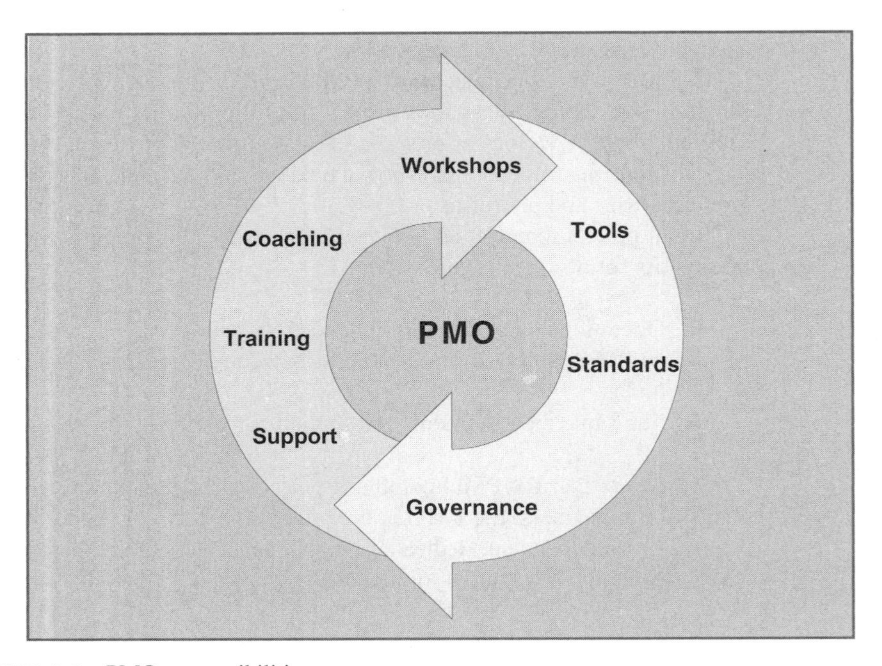

FIGURE 4–6. PMO responsibilities.

- *Training:* The PMO offers various courses on project management, covering all the managing roles within a project. The BPPM training teaches the BPPM methodology to project management's novices. The Advanced Project Manager course highlights the softer issues of people management and addresses the more experienced project managers. The Project Sponsor course addresses project sponsors and clarifies their important role within a project.
- *Coaching:* The PMO coaches projects either throughout the entire life cycle of a project or through a difficult phase. As a project leader, you can use the PMO project coach as a sparring partner.
- *Workshops:* At strategic moments in a project, you might want to make sure that the setup of your project is optimized. A tailormade workshop fitting your exact requirements can be set up by PMO.
- *Tools/Standards:* This includes a repository of intellectual property collected on projects, project reporting guidelines, and project interdependency management techniques.
- *Governance:* Through the Project Progress Report (PPR), the PMO gathers the data on Swiss Re's project portfolio. The PMO is tasked to present these results to senior management once a month.

Exel

Exel was placed in a position of having to develop a methodology that would be global in nature and company intranet-based. Exel's project management methodology is described by Phil Trabulsi, Sr. Director, Solution Design, Americas:

> Exel has an established project management methodology (DePICT™),[4] which was introduced in 1997, based on a detailed assessment of the project management processes. This methodology is based on specific needs of the organization with strong influence from the Project Management Institute's Project Management Body of Knowledge (PMBOK®).
>
> The acronym, DePICT, signifies the 5 phases of an Exel project; Define, Plan, Implement, Control and Transition.
>
> Additionally, Exel has a business delivery methodology—The Exel Way™,[5] which was introduced to formalize a structured approach to all projects. This is shown in Figure 4–7.
>
> The methodology is intranet-based to some extent. There are two main on-line sources (internally) that are accessible to the organization with information, tools, and training pertaining to project management and the DePICT™ methodology. First, Exel has an on-line, computer-based training program known as Exel Campus, which provides a number of training courses in various subjects. This program is accessible to all associates worldwide.
>
> In mid-2002, a Project Management Primer was developed and rolled out. This course provides project management associates and new associates the opportunity to participate in a 30-minute, on-line tutorial focusing on the core concepts of project management and the Exel methodology.

---

4. DePICT is a trademark of Exel plc.
5. The Exel Way is a trademark of Exel plc.

- *Business orientation*
- *Supply chain mapping*
- *Benchmarking*
- *Business value drivers*
- *Opportunity identification*

- *Current state assessment*
- *Future state evaluation*
- *Cost of change*
- *Risks, sensitivities, contingencies*
- *Business case*

- *Project management*
- *Change management*
- *Operational realization*
- *Project back-check*
- *Continuous improvement*

- *Detailed testing of solution*
- *Operational process design*
- *Implementation planning*
- *Contingency planning*
- *Budget development*

FIGURE 4–7. The Exel Way™.

Secondly, the Enterprise Project Management group manages a database, which resides in Lotus Notes, and contains the PM toolkit, procedures, training material, guides, and newsletters. It is accessible to all employees and can be linked to their Lotus Notes home page.

Finally, an EPM mailbox is set up for all inquiries from project managers, companywide and is accessible by all members of the Enterprise Project Management group.

In Exel's case, the methodology is stable and has not undergone any significant modifications in original format. However, concepts, approaches, tools, and training information are revised on a continual basis, based on demands from project managers, customers, and new industry trends. This includes the ongoing enhancement of tools to support efficient and consistent business practices.

Exel receives customer feedback with respect to its project management methodology/approach through various modes. The most common is direct feedback on project team performance from customers to Exel account owners and customer lessons learned exercises. In many cases, customers will provide feedback on Exel to various media outlets and industry publications. Today, most lessons-learned exercises are conducted by individual project teams and archived by each specific sector/department. Exel is evaluating methods of collecting, disseminating, and distributing key lessons learned across all sectors.

4.3 CRITICAL COMPONENTS

It is almost impossible to become a world-class company with regard to project management without having a world-class methodology. Companies such as General Motors Powertrain Group, Motorola, Ericsson, Key Services, Johnson Controls Automotive, Lear, ABB, Exel, Computer Associates, and Nortel all have world-class project management methodologies.

The characteristics of a world-class methodology include:

- Maximum of six life cycle phases
- Life cycle phases overlap

- End-of-phase gate reviews
- Integration with other processes
- Continuous improvement (i.e., hear the voice of the customer)
- Customer oriented
- Company-wide acceptance
- Use of templates (level 3 work breakdown structure [WBS])
- Critical path scheduling (level 3 WBS)
- Simplistic, standard bar chart reporting (standard software)
- Minimization of paperwork

Generally speaking, each life cycle phase of a project management methodology requires paperwork, control points, and perhaps special administrative requirements. Having too few life cycle phases is an invitation for disaster, and having too many life cycle phases may drive up administrative and control costs. Most companies prefer a maximum of six life cycle phases.

Historically, life cycle phases were sequential in nature. However, because of the necessity for schedule compression, life cycle phases today will overlap. The amount of overlap will be dependent upon the magnitude of the risks the project manager will take. The more the overlap, the greater the risk. Mistakes made during overlapping activities are usually more costly to correct than mistakes during sequential activities. Overlapping life cycle phases require excellent up-front planning.

End-of-phase gate reviews are critical for control purposes and verification of interim milestones. With overlapping life cycle phases, there are still gate reviews at the end of each phase, but they are supported by intermediate reviews during the life cycle phases.

Intel Corporation

Good methodologies always stress clear identification of deliverables and a clear definition of the requirements. But how many companies reevaluate the deliverables during the life cycle of the project to make sure that the original deliverables were realistic, the timing of the deliverables are realistic, the cost of the deliverables are within tolerance, and the quality of the deliverables will either meet or exceed customer expectations? This reevaluation process occurs at Intel. Esteri Hinman, PMP, Product Development Project Controls Group Manager, explains the process at Intel:

> Intel Corporation has solved the problem of bottoms-up schedule building through the use of "Map Days." In its most traditional form, Map Days bring representatives of every team involved in a project or program face to face for one, or more, days of planning. A typical Map Day agenda includes the following:
>
> - Review the scope, requirements, and expected benefits of the project or program
> - Review the high level (top down) schedule and major milestones (including definitions and measures)
> - Presentations by the teams of the Deliverables they expect to complete and the Receivables they require to do their work
> - "Mapping" exercise where all the Deliverables and Receivables are matched, and teams come to agreement on what and when Deliverables will be delivered
> - Review of existing, and identification of new risks and issues, including initial prioritization

The "Mapping" exercise can be handled in one of several ways. The most low-tech solution includes the use of many colors of sticky notes and butcher paper gridded with time along the length and teams along the width. Teams write their Deliverables and Receivables on the sticky notes. Place their deliverables on their own line and Receivables on the line of the team they expect to receive the item from. The Map Day Facilitator walks the entire group through a matching exercise, ensuring there is agreement between the deliverer and the receiver on dates and content. This method tends to be best for project types that are new, or teams that do not have a history of working together.

Some well-established teams have developed an "Electronic Map Day" which supports the Mapping exercise over a computer network as part of the Map Day pre-work. This allows the team to focus on the gaps between deliverer and receiver, and ignore the items that are not at issue during their face-to-face. Using this method, teams have reduced Map Day times from eight hours to one hour. This method works best for teams with well-defined deliverables using templates.

The value of the Map Day is directly related to the quality of the pre-work and preparation that is done prior to the event. Map Day facilitators send out templates for each participant to complete and bring with them. The template contains all the information the participant will be expected to present. Often, the most difficult part of the Map Day is getting the teams to spend significant time completing their prework! Map Day agendas and pre-work are highly customized to best meet the needs of the project.

The output from the Map Day is a bottom-up schedule that supports (or doesn't) the ability of the team to achieve the project's objectives and high level timeframes. This reconciliation between top-down guidance and bottom-up planning can be extremely useful to the project manager when discussing the project with management. In addition, the project manager receives valuable insight into project risks and issues that will need to be dealt with during the project.

World-class project management methodologies are integrated with other management processes such as change management, risk management, total quality management, and concurrent engineering. This produces a synergistic effect, which minimizes paperwork, minimizes the total number of resources committed to the project, and allows the organization to perform capacity planning to determine the maximum workload that the organization can endure.

World-class methodologies are continuously enhanced through key performance indicator (KPI) reviews, lessons learned updates, benchmarking, and customer recommendations. The methodology itself could become the channel of communication between the customer and contractor. Effective methodologies foster customer trust and minimize customer interference in the project.

Project management methodologies must be easy for workers to use as well as covering most of the situations that can arise on a project. Perhaps the best way is to have the methodology placed in a manual that is user-friendly. As an example of a user-friendly methodology, look at the table of contents shown below for the ABB project management methodology:

Contents

- Project Management
  - Objective
  - Benefits

  - Scope
  - Process Overview Diagram
  - Description
  - Key Changes
  - Process Performance Metrics
  - Dependencies and Prerequisites
  - RACI Chart
- Order Preparation
  - Contract Analysis
  - Order Entry and Acknowledgment
  - Basic Project Planning
- Expand Project Team
- Internal Project Kickoff
- Customer Project Kickoff
- Project Startup Process Review
- Customer Satisfaction Management
  - Customer Project Review
  - Customer Satisfaction Surveys
  - Customer Communication
- Project Team Management
  - Project Team Progress Review
  - Resource Conflict Resolution
  - Performance Assessment
- Financial and Cost Management
  - Monitor Project Costs
  - Cash Flow Management
  - Invoicing to Customer
  - Contract Management
- Reporting
  - Internal Project Report/Meeting
  - Customer Project Report
- Change Management
- Project Opportunity Management
- Project Procurement Management
  - Selection of Suppliers
  - Purchase Orders and Expediting
  - Inspection and Test Coordination
  - Packing, Shipping, and Storage
  - Purchase Order Close-out
- Risk Management
- Planning and Schedule Management
- Acceptance/Takeover
- Warranty/Aftercare
- Project Closeout Review

Excellent methodologies try to make it easier to plan and schedule projects. This is accomplished by using templates for the top three levels of the work breakdown structure. Simply stated, using a WBS level 3 template, standardized reporting with standardized terminology exists. The differences between projects will appear at the lower levels (i.e., levels 4 to 6) of the WBS. This also leads to a minimization of paperwork.

Edelca

Companies have discovered that project management methodologies work best with templates, forms, checklists, and guidelines rather than policies and procedures. Also, as companies begin using the templates, there is always the recognition that more templates are needed for the work flow to become more efficient and effective. At Edelca, Marcelino Diez, PMP, Project Management Consultant, believes that:

> In Edelca some templates are used, but the company is convinced that there must be an increase in their use with the intention of obtaining uniformity in the communication and systemization of some of the critic tasks. The templates are incorporate in all the phases of the project life cycle (Planning, Alternatives Analysis, Definition, Implementation and Evaluation). Some of the templates we use include the Need or Opportunity Statement, Business Case, Project Execution Plan, Project Justification, Front End Loading Package, as well as others related to specific tasks such as Lessons Learned, Contracting Strategy Plan, Risks Identification, and Checklists.

ABB

The difficulty with templates, especially for a multinational organization, is finding a set of templates that everyone will agree to and use. Having too few templates may be of limited value, and having too many templates may create limitations on the project manager. Some form of compromise is needed. Benny Nyberg, Manager, Process Unit Project Management at ABB, comments on the use of templates:

> ABB has about 10 project management templates standardized across the group. They are used throughout a project, starting with a Risk Review Report intended to be used during Project Sales, continuing with a Handover Checklist (to document the transfer from Sales to Execution). Another important document is a Process Plan used to document how the ABB PM process applies to a specific project or business. ABB also has templates for meetings.

Exel

As companies begin the development of project management methodologies, emphasis seems to be on developing templates first. Exel Corporation has achieved success through their project management templates. According to Todd Daily, PMP, Project Manager, Enterprise Project Management, Americas:

> Exel uses a project management toolkit comprised of 20 templates that support the phases of a project from Initiation through Closure. The use of the templates enables the project manager to satisfy the specific needs of his/her project without overburdening them with unnecessary paperwork.
>
> These templates have been created in MS Excel and are contained in one individual file, or briefcase, that is transferable and can be easily modified to meet the unique informa-

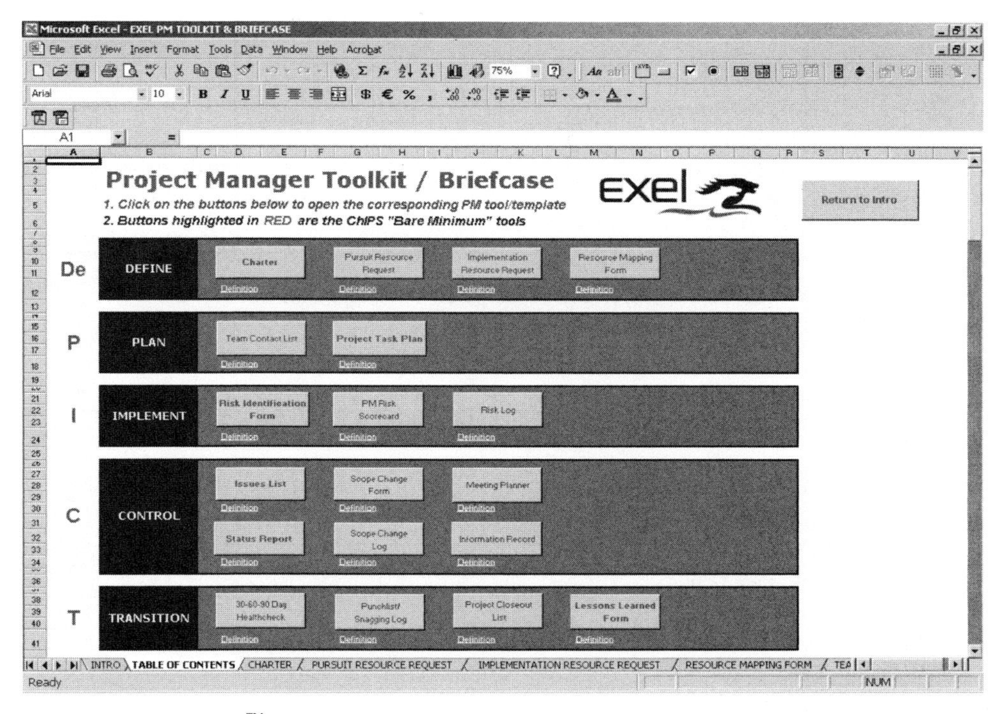

FIGURE 4–8. DePICT™'s five phases.

tional requirements of a project. It is macro-driven using navigation buttons for ease of use and contains a glossary with information specific to each tool.

The layout of the toolkit is designed to support Exel's project management methodology DePICT™.[6] Tools specific to each phase are categorized accordingly. DePICT™ stands for the 5 phases of a project—Define, Plan, Implement, Control, and Transition. This is shown in Figure 4–8.

Additionally, a subset of tools has been identified as the bare minimum requirement for each project. These are known as the ChIPS™—Charter, Issues List, Project Plan, and Status Report. The use of templates increases understanding of the tool functions, provides a sufficient level of guidance for less experienced project managers, enables flexible/dynamic adaptation to all projects—pursuit or implementation, reduces training time, and provides quicker adaptability for associates.

Exel uses templates for all phases of projects. Typical templates appear in Figures 4–9 through 4–14. The two primary types of projects at Exel are Pursuit Projects, which involves the pursuit of new business via RFP responses and solution design; and Implementation Projects, which are much larger, involving the implementation and delivery of the designed solution for our clients. Pursuit projects range in length from a couple

6. DePICT and ChIPS are trademarks of Exel plc.

PROJECT CHARTER			
Project Name:		exel	
Project Sponsor:			
Customer Name:			
Customer Sponsor:		Sector/Dept:	
Project Manager:		Date:	
Project Owner:		Project Number:	

Executive Summary: (Brief description of situation and need for the project)

Project Objectives:

Project Scope:

Mgt. Approach: (generally descriptional project resource structure-e.g., PM, 2 IT developers, DBA, etc.)

Project Risks:

Assumptions:

Date/Version:	Attachments:

FIGURE 4–9. Project charter template.

PROJECT ISSUES LIST

Project Name:	0	exel
Project Manager:	0	

#	Category/ Function	Issue Description	Issue Origin	Issue Owner	Origin Date	Date Due	Date Completed	Status	Scope Change (Y/N)	Comments

FIGURE 4–10. Project issues list template.

PROJECT TASK PLAN

Project Name:	0	exel	
Project Manager:	0		
Customer Name:	0		
Sector/Dept.:	0	Project Start Date:	
Project Owner:	0	Target Project End Date:	
Project Cost Ctr. Code:		Actual Project End Date:	

OK	#	Project Task	Resource Name	Start Date:	Target Finish	Duration	Actual Finish	Predecessor

FIGURE 4–11. Project task plan template.

PROJECT STATUS REPORT

Project Name:	0	exel
Project Owner:	0	
From: (Project Manag	0	
To:		
Reporting Period:		
SUBJECT:		

Functional Area	**Status Update**

Items in Progress:

#	Task	Responsibility

Major Issues:

#	Event & Resolution	Responsibility

Major Milestones:

Milestone	Date

Distribution:

Attachments:

FIGURE 4–12. Project status report template.

RISK IDENTIFICATION FORM			
Project Name:	0	exel	
Project Manager:	0		
Customer Name:	0		
Sector/Dept:	0	Change Originator:	
Project Owner:	0	Change Request #:	
Risk Event:		Change Request Date:	

Description of Risk Event

Probability of Occurence	High	Medium	Low

Impact	High	Medium	Low

	Impact (must be qualifiable)	Budget	Schedule	Customer

	Probable Causes

	Preventative Steps & Actions

	Contingency Plan & Priority	Trigger

	Risk Response

FIGURE 4–13. Risk identification template.

Project Risk log

Project Name:		exel	
Project Manager:			
Project Number:			
Customer Name:		Risk Log Controller:	
Sector/Dept.:		Risk Log #:	
Project Owner:		Date:	

No.	Key Risk	Impact 1-3	Prob 1-3	Score	Risk Mitigation Plan (summary)	Risk form no.	By When	Owner(s)	Status	Comment
1										
2										
3										
4										
5										
6										
7										
8										
9										
10										
11										
12										
13										
14										
15										
16										
17										
18										

FIGURE 4–14. Risk log template.

of weeks to more than a month. Implementation projects can run from a couple of months to more than one year, depending on complexity.

IT-specific projects tend to follow a traditional System Development Life Cycle format with links to general project management tools, such as a Charter and Issues List. A large implementation project, which usually involves an IT component, may also have systems specific project management tools and templates.

4.4 BENEFITS OF A STANDARD METHODOLOGY

For companies that understand the importance of a standard methodology, the benefits are numerous. These benefits can be classified as both short- and long-term benefits. Short-term benefits were described by one company as:

- "Decreased cycle time and lower costs
- Realistic plans with greater possibilities of meeting time frames
- Better communications as to 'what' is expected from groups and 'when'
- Feedback: lessons learned"

These short-term benefits focus on KPIs or, simply stated, the execution of project management. Long-term benefits seem to focus more upon critical success factors (CSFs) and customer satisfaction. Long-term benefits of development and execution of a world-class methodology include:

- Faster "time to market" through better scope control
- Lower overall program risk
- Better risk management, which leads to better decision-making
- Greater customer satisfaction and trust, which lead to increased business and expanded responsibilities for the tier one suppliers
- Emphasis on customer satisfaction and value added rather than internal competition between functional groups
- Customer treating the contractor as a "partner" rather than as a commodity
- Contractor assisting the customer during strategic planning activities
- Benchmarking/continuous improvement made easier and quicker

Perhaps the largest benefit of a world-class methodology is the acceptance and recognition by your customers. If one of your critically important customers develops its own methodology, that customer could "force" you to accept it and use it in order to remain a supplier. But if you can show that your methodology is superior or equal to the customer's, your methodology will be accepted, and an atmosphere of trust will prevail.

One contractor recently found that its customer had so much faith in and respect for its methodology that the contractor was invited to participate in the customer's strategic planning activities. The contractor found itself treated as a partner rather than as a commodity or just another supplier. This resulted in sole-source procurement contracts for the contractor.

Developing a standard methodology that encompasses the majority of a company's projects and is accepted by the entire organization is a difficult undertaking. The hardest part might very well be making sure that the methodology supports both the corporate culture and the goals and objectives set forth by management. Methodologies that require changes to a corporate culture may not be well accepted by the organization. Nonsupportive cultures can destroy even seemingly good project management methodologies.

During the 1980s and 1990s, several consulting companies developed their own project management methodologies, most frequently for information systems projects, and then pressured their clients into purchasing the methodology rather than helping their clients develop a methodology more suited to the client's needs. Although there may have been some successes, there appeared to be significantly more failures than successes. Previously, we discussed a hospital that purchased a $130,000 project management methodology with the belief that this would be the solution to their information system needs. Unfortunately, senior management made the purchasing decision without consulting the workers who would be using the system. In the end, the package was never used.

Another company purchased a similar package, discovering too late that the package was inflexible and the organization, specifically the corporate culture, would need to change to use the project management methodology effectively. The vendor later admitted that the best results would occur if no changes were made to the methodology.

These types of methodologies are extremely rigid and based on policies and procedures. The ability to custom-design the methodology to specific projects and cultures was nonexistent, and eventually these methodologies fell by the wayside—but after the vendors made significant profits. Good methodologies must be flexible.

Microsoft Corporation

There are training programs that discuss how to develop a good methodology. These programs focus on "best practices" in methodology development rather than on the use of a single methodology. Microsoft has developed a family of processes that embody the core principles of, and best practices in, project management. These processes are called the Microsoft Solutions Framework (MSF). The foundational principles for all MSF courses include:

- Clear accountability, shared responsibility
- Empowering team members
- Focusing on business value
- Sharing project vision
- Staying agile, expecting change
- Fostering open communications
- Learning from all experiences
- Investing in quality

One such course that is included under MSF is the Principles of Application Development.[7] The course focuses on best practices using three important project planning tools:

- *Team Model for Application Development:* The MSF team model for application development advocates a small team of peers working in interdependent multidisciplinary roles. It is designed so that its six team roles map directly to six quality goals that MSF considers necessary for project success. The team model is specifically designed to overcome the communication issues typically experienced by the more traditional hierarchical teams.
- *Process Model for Application Development:* The MSF process model for application development is a flexible approach to project management that improves project control, minimizes risk, and improves product quality, while increasing development speed. It is based upon four major milestones, which are review and synchronization points that allow the project team to adjust the scope of the project to reflect changing customer requirements or to react to risks that may materialize during the course of the project.
- *Proactive Risk Management Process:* The MSF approach to proactive risk management is a five-step process that enables the project teams to assess risks continuously and use them for decision-making in all phases of the project. Using the process, project teams carry the risks forward and deal with them until they are resolved or until they turn into problems and are handled as such.

Microsoft continuously updates its MSF courseware to include current best practices in project management. This course, in addition to other courses that are part of the MSF family, uses two important project management best practices. First, the course is represented as a framework rather than a rigid methodology. Frameworks are based upon tem-

7. The remainder of this section has been adapted from the Principles of Application Development Course; reproduced by permission of Microsoft Corporation.

plates, checklists, forms, and guidelines instead of the more rigid policies and procedures. Inflexible processes are one of the root causes of project failure.

The second-best practice's characteristic is its focus on people rather than technology. The core for successful project management is people. Effective implementation of project management is a series of good processes with emphasis on people and their working relationships: namely, communication, cooperation, teamwork, and trust. Failure to communicate and work together is another root cause of project failure.

MSF courseware emphasizes the importance of people and teamwork. This includes:

- A team is developed whose members relate to each other as equals.
- Each team member is provided with specific roles and responsibilities.
- The individual members are empowered in their roles.
- All members are held accountable for the success of their roles.
- The project manager strives for consensus-based decision-making.
- The project manager gives all team members a stake in the success of the project.

The team model for application development is shown in Figure 4–15. The model defines the functional job categories or skill sets required to complete the project work as well as the roles and responsibilities of each team member. The team model represented in Figure 4–15 focuses on teamwork rather than a strong reliance on the organizational structure.

The overall process model for application development is shown in Figure 4–16. The model combines the benefits of both a milestone-based process and a flexible and iterative

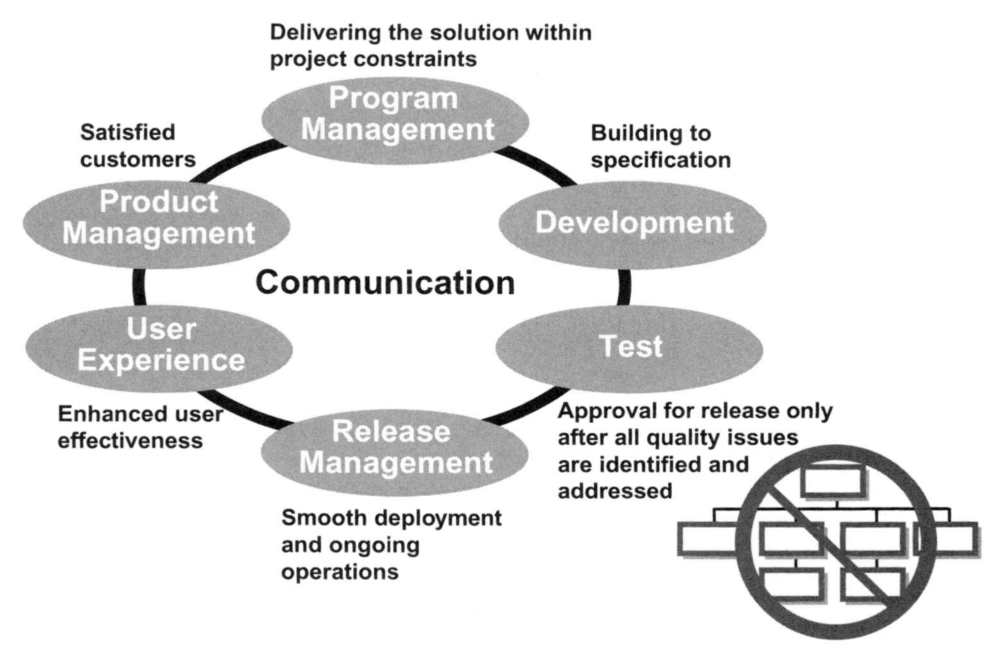

FIGURE 4–15. MSF team model.

FIGURE 4–16. MSF process model.

process. For each of the five milestones in the process model, MSF team roles can be defined as illustrated in Table 4–1.

Realistic milestones serve as review and synchronization points. Milestones allow the team to evaluate progress and make midcourse corrections where the costs of the corrections are small. Milestones are used to plan and monitor project progress as well as to schedule major deliverables. Using milestones benefits projects by:

- Helping to synchronize work elements
- Providing external visibility of progress and quality
- Enabling midcourse corrections
- Focusing reviews on goals and deliverables
- Providing approval points of work being moved forward

There are two types of milestones: major and interim. Major milestones represent team and customer agreement to proceed from one phase to another. An example of the major milestone "Vision Approved" is shown in Figure 4–17. Interim milestones indicate progress within a phase and divide large efforts into workable segments. Figure 4–18 shows the interim milestones for the Vision Approved Major Milestone as well as for the other four major milestones.

For each of the major milestones and phases, Microsoft defines a specific goal and team focus. For example, the goal for the Envisioning Phase is to create a high-level view of the project's goals, constraints, and solution. The team focus for the Envisioning Phase is to:

- Identify the business problem or opportunity
- Identify the team skills required

TABLE 4–1. MSF TEAM ROLES THROUGH THE PHASES

	Envisioning Phase	Planning Phase	Developing Phase	Stabilizing Phase	Deploying Phase
Product Management	• Overall goals • Identify customer requirements • Vision / scope document	• Conceptual design • Business requirements analysis • Communications plan	• Customer expectations	• Communications plan execution • Launch planning	• Customer feedback, assessment signoff
Program Management	• Design goals • Solution concept • Project structure	• Conceptual and logical design • Functional specification • Master project plan • Master project schedule • Budget	• Functional specification management • Project tracking • Plan updating	• Project tracking • Bug triage	• Solution / scope comparison • Stabilization management
Development	• Prototypes • Development and technology options • Feasibility analysis	• Technology evaluation • Logical and physical design • Development plan / schedule • Development estimates	• Code development • Infrastructure development • Configuration documentation	• Bug resolution • Code optimization	• Problem resolution • Escalation support
User Experience	• User Performance needs and implications	• Usage scenarios / use cases • User requirements • Localization / accessibility requirements • User documentations, training plans and schedules	• Training • Training plan updates • Usability testing • Graphic design	• User documentation stabilization • Training materials	• Training • Training schedule management
Test	• Testing approach • Test acceptance criteria	• Design evaluation • Testing requirement • Test plan and schedule	• Functional testing • Issues identification • Documentation testing • Updated test plan	• Testing • Bug reporting and status • Configuration testing	• Performance testing • Problem resolution
Release Management	• Deployment implications • Operations management and supportability • Operations acceptance criteria	• Design evaluation • Operations requirements • Pilot and deployment plan and schedule	• Rollout checklists • Rollout and pilot plan updates • Site preparation checklists	• Pilot setup and support • Deployment planning • Operations and support training	• Site development management • Change approval

❖ **Deliverables**
- **— Vision/scope document**
- **— Project structure document**
- **— Initial risk assessment document**

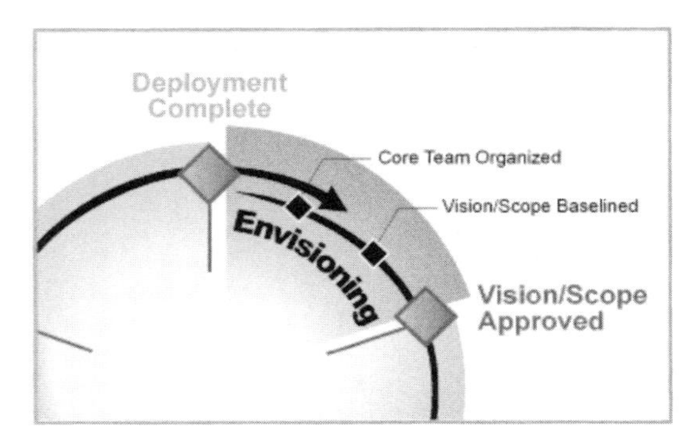

FIGURE 4–17. MSF envisioning phase milestones and deliverables.

- Gather the initial requirements
- Create the approach to solve the problem
- Define goals, assumptions, and constraints
- Establish a basis for review and change

Previously, we stated that one of the characteristics of successful organizations is that they provide the project team with a clear definition of success. MSF provides success criteria for each process phase. For example, the success criteria for the Envisioning Phase are:

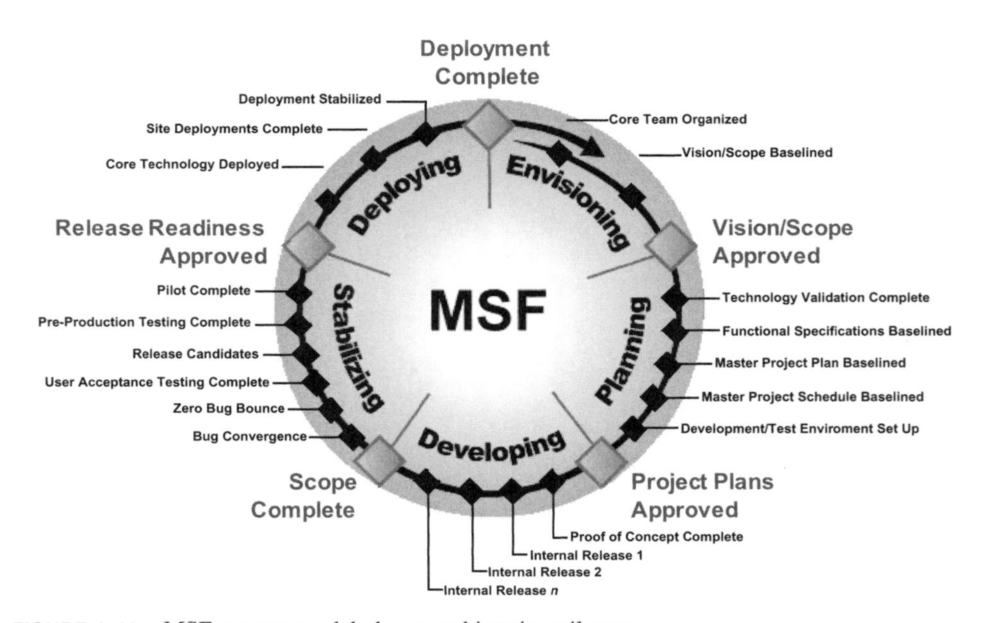

FIGURE 4–18. MSF process model phases and interim milestones.

- Agreement by the stakeholders and team has been obtained on:
  - Motivation for the project
  - Vision of the solution
  - Scope of the solution
  - Solution concept
  - Project team and structure
- Constraints and goals have been identified.
- Initial risk assessment has been done.
- Change control and configuration management processes have been established.
- Formal approval has been given by the sponsors/and or key stakeholders.

It is often said that development projects can go on forever. The application development course encourages baselining documents as early as possible, but freezing the documents as late as possible. This also requires a structured change control process combined with the use of versioned releases, as shown in Figure 4–19. The benefits of versioned releases include:

- Forcing closure on project issues
- Setting clear and motivational goals for all team members
- Effective management of uncertainty and change in project scope
- Encouraging continuous and incremental improvement
- Enabling shorter delivery time

One of the strengths of the MSF courseware is the existence of templates to help create project deliverables in a timely manner. The templates provided by MSF can be

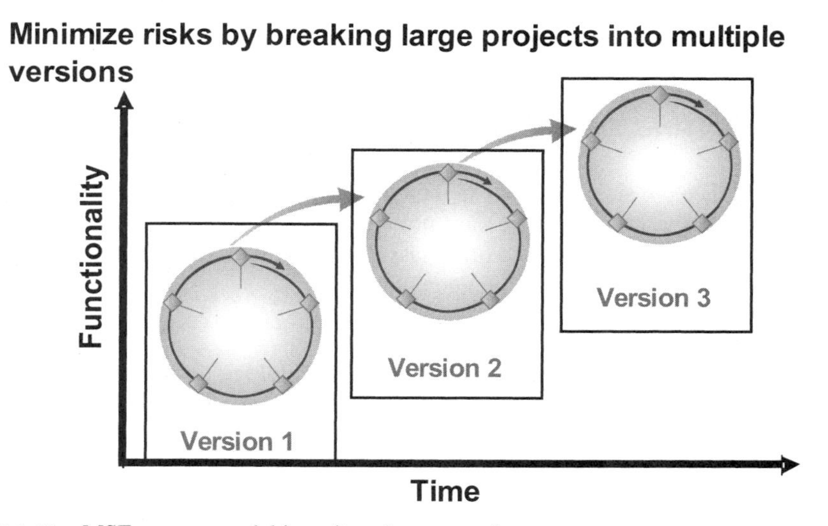

FIGURE 4–19. MSF process model is an iterative approach.

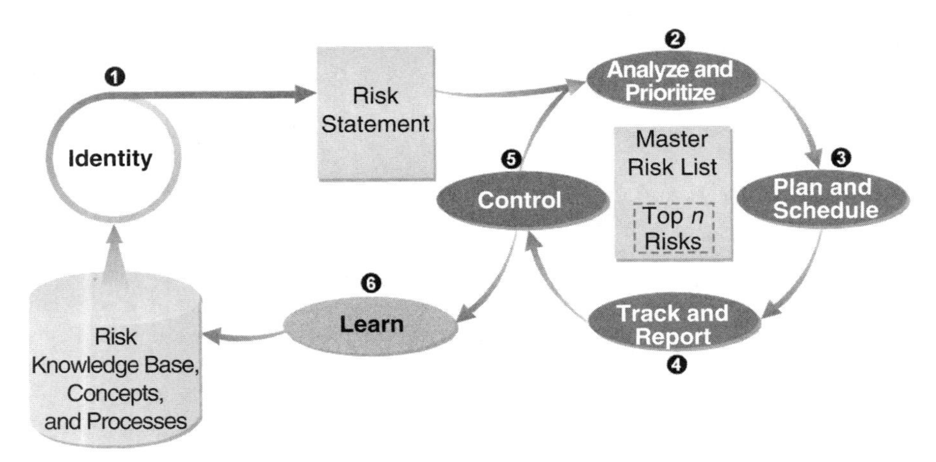

FIGURE 4–20. MSF risk management process.

custom-designed to fit the needs of a particular project or organization. Typical templates for application development include:

- Project schedule template
- Risk factor chart template
- Risk assessment matrix template
- Postmortem template

Today, companies have come to realize the importance of risk management as a core process for successful project management. The MSF process for risk management is shown in Figure 4–20. Because of the importance of risk management today, it has become an important component of all project management training programs.

4.5 IMPLEMENTING THE METHODOLOGY

The physical existence of a methodology does not convert itself into a world-class methodology. Methodologies are nothing more than pieces of paper. What converts a standard methodology into a world-class methodology is the culture of the organization and the way the methodology is implemented.

> The existence of a world-class methodology does not by itself constitute excellence in project management. The *corporate-wide acceptance* and *use* of it does lead to excellence. It is through excellence in execution that an average methodology becomes a world-class methodology.

One company developed an outstanding methodology for project management. About one-third of the company used the methodology and recognized its true long-term benefits. The other two-thirds of the company would not support the methodology. The president eventually restructured the organization and mandated the use of the methodology.

The importance of execution cannot be underestimated. One characteristic of companies with world-class project management methodologies is that they have world-class managers throughout their organization.

Rapid development of a world-class methodology mandates an executive champion, not merely an executive sponsor. Executive sponsors are predominantly on an as-needed basis. Executive champions, on the other hand, are hands-on executives who drive the development and implementation of the methodology from the top down. Most companies recognize the need for the executive champion. However, many companies fail to recognize that the executive champion position is a life-long experience. One Detroit company reassigned their executive champion after a few successes were realized using the methodology. As a result, no one was promoting continuous improvement to the methodology.

Good project management methodologies allow you to manage your customers and their expectations. If customers believe in your methodology, then they usually understand it when you tell them that no further scope changes are possible once you enter a specific life cycle phase. One automotive subcontractor carried the concept of trust to its extreme. The contractor invited the customers to attend the contractor's end-of-phase review meetings. This fostered extreme trust between the customer and the contractor. However, the customer was asked to leave during the last 15 minutes of the end-of-phase review meetings when project finances were being discussed.

Project management methodologies are an "organic" process, which implies that they are subject to changes and improvements. Typical areas of methodology improvement might include:

- Improved interfacing with suppliers
- Improved interfacing with customers
- Better explanation of subprocesses
- Clearer definition of milestones
- Clearer role delineation of senior management
- Recognition of need for additional templates
- Recognition of need for additional metrics
- Template development for steering committee involvement
- Enhancement of the project management guidebook
- Ways to educate customers on how the methodology works
- Ways of shortening baseline review meetings

Sherwin-Williams

There are several ways that a company can develop a methodology for project management. Outsourcing the development process to another company can be beneficial. Some companies have template methodologies that can be used as a basis for developing your own methodology. This can be beneficial if the template methodology has enough flexibility to be adaptable to your organization. The downside is that this approach may have the disadvantage that the end result may not fit the needs of your organization or your company's culture. Hiring outside consultants may improve the situation a little, but the end result may still be the same unfavorable result as well as being more costly. This approach may require keeping contractors on your payroll for a long time such that they can fully understand your culture and the way you do business.

Benchmarking may be effective, but by the time the benchmarking is completed, the company could have begun the development of its own methodology. Another downside risk of benchmarking is that you may not be able to get all of the information you need or the supporting information to make the methodology work for you.

Companies that develop their own methodology internally seem to have greater success, especially if they incorporate their own best practices and lessons learned from other activities. This occurred in companies such as General Motors, Lear, Johnson Controls, Texas Instruments, Exel, Sherwin-Williams, and many other organizations. Rinette Scarso, PMP, Project Manager, discusses the Sherwin-Williams situation with a case study.

Company Background

The Sherwin-Williams Company is engaged in the manufacture, distribution and sale of coatings and related products to professional, industrial, commercial and retail customers primarily in North and South America. The company's Paint Stores Segment consists of 2,488 company-operated specialty paint stores. The Consumer Segment develops, manufactures and distributes a variety of paint, coatings and related products to third-party customers and the Paint Stores segment. The Automotive Finishes Segment develops, manufactures and distributes a variety of motor vehicle finish, refinish and touch-up products. The International Coatings Segment develops, licenses, manufactures and distributes a variety of paint, coatings and related products worldwide. The Administrative Segment (Corporate Division) includes the administrative expenses of the company's and certain consolidated subsidiaries' headquarters sites.

The Corporate Information Technology (IT) Department for The Sherwin-Williams Company provides shared services support for the five operating divisions, described above, that make up the organization.

Case Study Background

During the summer of 2002, the Corporate IT Department engaged in activities surrounding the conversion of international, interstate, intrastate and local telecommunications services from the company's present voice telecommunications carrier to a new carrier. Project management disciplines and best practices, using a structured project management methodology, were utilized on this project, ultimately leading to a successful project outcome.

The project was implemented using a phased approach consisting of the major phases as described below. The phases were established to include many of the principles stated in the PMBOK®, and also included many of the best practices that had been developed previously at The Sherwin-Williams Company. The phases could overlap, if necessary, allowing for a gradual evolvement from one phase to the next. The overlapping also allowed us to accelerate schedules, if need be, but possibly at an additional risk. Project reviews were held at the end of each phase to determine the feasibility of moving forward into the next phase, to make "go/no-go" decisions, to evaluate existing and future risks, and to determine if course corrections are needed.

- *Initiate:* The first phase is the Initiate phase where the project team is formed, a project kickoff meeting is held, needs and requirements are identified, and roles and responsibilities are defined.
- *Planning:* The Planning phase is the next phase, and regarded by most project managers as the most important phase. Most of the project's effort is expended in the Planning phase, and it is believed that the appropriate time and effort invested in this phase ensures the development of a solid foundation for the project. Management wholeheartedly supports the efforts we put forth in this phase because this is where many of our best practices have occurred. Also, a solid foundation in this phase allows for remaining phases of the project to be accomplished more efficiently giving senior management a higher degree of confidence in the ability of our project managers to produce the desired deliverables and meet customer expectations.

  A series of meetings are typically held throughout this phase to identify at the lowest level the project needs, requirements, expectations, processes and the activities/steps for the processes. The result of these meetings are several deliverables, including a needs and requirements document, a project plan, a risk management plan, an issue log and action item list. Additional documents maintained include quality management and change management plans. Together these documents provide management with an overview of the entire project and the effort involved to accomplish the goal of transitioning services by the target date established by our management.
- *Execution:* The third phase in implementation is Execution. This phase is evolved into gradually once the majority of planning has been completed. All activities outlined in the processes during the planning phase come to fruition at this time as actual communication line orders began to take place, as well as the installation of equipment where necessary. Services began to be transitioned by Division/Segment and implementation moved forward aggressively for this project due to a stringent timeframe. It is of vital importance that activities in this phase be monitored closely in order to facilitate the proactive identification of issues that may negatively impact the timeline, cost, quality or resources of the project.

  To facilitate monitoring and control of the project, weekly status meetings were held with the vendor and the project team, as well as short internal daily meetings to review activities planned for each day. Ad hoc meetings also occurred as necessary.
- *Closure:* The final phase of the project is Closure. In this phase, there is typically a closure meeting to identify any remaining open issues and to determine the level of client satisfaction. This phase also included any "clean-up" from the project, administrative close-out, the communication of post-implementation support procedures and a review of lessons learned.

Best practices that worked notably well for The Sherwin-Williams Company included the establishment of success criteria, consisting of project objectives and a needs/requirements analysis, regular communications both within the project team and with stakeholders, dedicated resources, defined roles and responsibilities, knowledge transfer between cross-functional teams, teamwork, the development of a fun, synergistic working environment and reviewing lessons learned.

One of the best practices in project management is that maturity and excellence in project management can occur quickly when senior management not only actively supports project management but also articulates to the organization their vision of where they expect project management to be in the future. This vision can motivate the organization to excel, and best practices improvements to a project management methodology seem to occur at a rapid rate. Such was the case at The Sherwin-Williams Company. Tom Lucas, Vice President of IT at The Sherwin-Williams Company, comments on his vision for The Sherwin-Williams Company:

> The future of project management at The Sherwin-Williams Company includes the integration of project management disciplines and best practices through the establishment of a Virtual Project Management Office (PMO), combined with portfolio management techniques to deliver high value project results on a consistent basis. The Sherwin-Williams Company anticipates that the use of a Virtual PMO will not only instill the best practices of project management as core competencies, but also aid in the growth of the organization's project management maturity.
>
> We define a Virtual PMO as a function that has permanent full-time members at its core. This core group would be responsible for PMO office operation and administration crossing all IT operating groups. This core group would be the standards setting body, provide for best practices identification and sharing, and coordinate the work of the "practicing" project managers, which reside in individual operating groups. These practicing/virtual project managers would be assigned to the PMO office as the project workload requires, and when not attached to the PMO they would reside in the various functional IT workgroups to assist with tactical project management within these functions. The intent is to expand and contract PMO resources as you would expand and contract project teams as the workload dictates.
>
> One goal is to unify the goals and objectives of individual departments by applying a universal yet flexible project management framework in pursuit of better across-the-board results. The Sherwin-Williams Company desires to learn from past successes as well as mistakes, make processes more efficient, and develop people's skills and talents to work more effectively through the establishment of standardized procedures within the company.
>
> In order to complete projects successfully on a consistent basis, The Sherwin-Williams Company realizes the value of establishing a Virtual PMO. A Virtual PMO will facilitate the alignment of projects with strategic organizational goals and objectives through project portfolio management. PMO best practices will encompass the establishment of standard project management guidelines and templates, centralized communications, training and mentoring of project managers and a centralized repository for lessons learned.
>
> The Sherwin-Williams Company understands that this undertaking will take time to accomplish based on the current level of project management maturity in the organization and on the fact that the PMO structure will continually evolve over time as the level of project management maturity increases. The Sherwin-Williams Company plans to persevere in their efforts in anticipation of the many benefits that will come as the organization moves up the maturity ladder, including recognition by customers and industry peers.

Antares Management Solutions

Some companies have found that some of the readily available methodologies that can be purchased or leased have enough flexibility to satisfy their needs. This is particularly true in the IT area. The following information was provided by John Frohlich, Director, Information Systems Development at Antares, and Dan Halicki, Planning Coordinator at Antares:

Introduction

Industry analysts believe that only 19% of IT projects are delivered on time, and fewer are within budget. In response to this problem, various project management approaches have emerged with the primary objective of ensuring project success. The business challenge is to find the proper balance between rigorous methodology requirements and the realities of containing administrative project overhead. More is not always better in the world of project management.

Since 1989, the systems development team of Antares Management Solutions has managed all IT projects by utilizing selected aspects of the Navigator Systems Series methodology originally developed by Ernst & Young (now Cap Gemini Ernst & Young). Navigator was chosen, in large part, because its flexibility and adaptability provided the most customization at a good price. Over the past fourteen years, this approach has a solid track record of project successes that have helped Antares to grow and prosper in the competitive world of IT outsourcing.

What, then, are the best practices that Antares has followed to bring the right amount of discipline to its project management approach? The following discusses the techniques and deliverables that constitute the standard project management elements of structure, plan and control. Additionally, it is important to look beyond the elements of the methodology itself to include the support system that makes it all work, such as, executive management buy-in, training and follow-up activity.

- *Make use of project management concepts and terminology throughout the enterprise, not just in Information Systems.* Antares uses a strategic planning process that requires collaboration between client management and IT planning to effectively align business vision with technical solutions. To achieve the desired goals, related project proposals are drafted using the standardized project charter format. Consequently, the client management team has a clear idea of how a project is to be defined and structured from the very beginning.
- *Provide ongoing project management training throughout the enterprise.* The term *project management* can mean different things to different people. An instructor-led class explains how Antares applies the CGE&Y "Navigator Project Management Methodology" to ensure that all project participants, including the customer community, have a common understanding of the process. The class provides instruction and a hands-on workshop on the major components of the project management process, including terms, diagrams, deliverables, charter writing, plans, milestones and changes.
- *Structure every project in a consistent manner, including scope, responsibilities, risks and high-level milestones.* Central to the project management process is the creation of the project charter, which documents and formalizes the agreement of all concerned parties, including business sponsors, project manager and project team members. The charter spells out what will be accomplished, who is responsible for getting it done, known problems to be addressed, and major milestones to be met. The project plan details out the tasks, timeframes, dependencies, and resources needed to deliver the project within agreed-upon timeframes and budget.
- *Communicate and update project plans on an ongoing basis, using on-line tools when appropriate.* Once an approved project charter is in place, the project plan becomes the next focus of attention. The plan and the regular status reports that document progress of the plan are the primary communication vehicles for keeping management and staff informed. Since 2000, Antares has instituted an on-line project status center to keep project planning and status information both current and readily available.

- *Maintain an official project issues list, including who is responsible, what are the potential impacts, and how the issue is being resolved.* As problems arise during the execution of a project, it is vital to maintain an accurate "issues log" to properly document the problems and what is being done to solve them. In project meetings, the log serves as a meaningful tool that not only assures concerned areas that their problems have not been forgotten, but also to prevent similar problems from recurring. Along with the charter and the regular progress reports, the "issues log" can be accessed via the online project status center.
- *Use a formal change control process, including an executive steering committee to resolve major changes.* As issues arise that could require additional resources, the nemesis known as "scope creep" can begin to surface and impede a smooth-running project. A formal change control process is essential to help the project manager with the predicament of pleasing the customer or going over budget or missing milestones. For situations that become threatening to the project's ultimate success, the assistance of an executive steering committee can determine the best course of action.
- *Periodically audit the project management process and selected projects to determine how well the methodology is working and to identify opportunities for improvement.* With the establishment of project management standards and the sizable investment in training, it makes sense to step back occasionally and assess if the overall objectives are being met. Where projects ran into difficulty, would a different approach have helped? Where projects went well, what aspects of the project execution ought to be shared by other projects? These are the kinds of questions the audit exercise tries to address.
- *Conclude every major project with an open presentation of the results to share knowledge gained, demonstrate new technology, and gain official closure.* Many organizations rush through the project conclusion phase in order to get on with the next assignment. However, taking the time to fully document accomplishments, outstanding issues, and deferred deliverables is a worthwhile exercise that gives proper credit for successful completions, reduces misunderstandings over omissions and facilitates transitions to follow-up projects, and ensures knowledge transfer useful for future project management success.

Conclusion

One of the major reasons that projects lose control is because project managers operate reactively rather than anticipating possible stumbling blocks. When the stumbling block finally appears, they are hesitant to make effective decisions, which often results in situations that can kill a project. Antares has found that utilizing the eight best practices above, the company has been able to foster a more proactive environment in partnership with the customer, resulting in successful project delivery. The outlined approach strikes a balance that requires minimum project overhead while obtaining maximum benefit.

Westfield Group

Developing a project management methodology may not be as complicated as one believes. There are four essential building blocks that, if present, allows for the development of an effective project management methodology in a reasonably short period of time. The four building blocks are:

- Recognition at the executive levels of the need for the methodology
- Executive-level support combined with a clear vision

- Establishing an organization, reporting to senior management, committed to this development process
- Using the PMBOK® as a starting point for methodology development

Kathy Rhoads, Director, Project Management COE for the Westfield Group, provides us with some insight on the growth of their project management organization and methodology:

> Westfield Insurance recognized that the needs of our customers had changed with regard to information technology and information systems projects. Also, the relationship with our customers had changed. Our customers were now expecting complete solutions to their business needs rather than just products. We needed an organization that was designed to "partner" with our business customers and specialize in providing high quality business solutions. We replaced our information systems group with a Business Solutions Center (BSC). Senior management's vision for the BSC was to provide solutions to Westfield Insurance that would exceed customer expectation. Our focus was on adding value.
>
> There are three units within the BSC, as shown in Figure 4–21:
>
> - The Shared Service Center activities include:
>   - Dedicated support to existing systems
>   - Answering customers' questions
>   - Providing technical support
> - The Delivery Engagement Center activities include:
>   - Use of "Virtual Teams"
>   - A focus on project delivery
>   - Use of business Subject Matter Experts (SMEs)
>   - The Program Management Office activities include:
>   - Defining the scope for programs under consideration
>   - Ensuring consistent application or use the program framework
>   - Providing enterprise technology direction

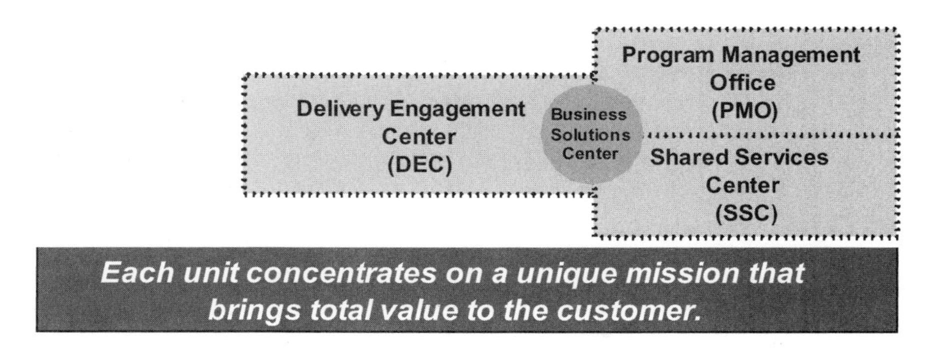

FIGURE 4–21. The units within the BSC: SSC, day to day; DEC, projects; PMO, programs.

Each of the units within BSC can have one or more Centers of Excellence for activities within that unit. A typical Center of Excellence is:

- Composed of a team of people
- Coached by a COE Director
- Specialized in a particular discipline
- Focused on deliverables
- Empowered

Some high-level accomplishments in building the BSC Project Management COE:

- *2000:*
  - Built the technical skills of internal project managers
  - Established the beginnings of the Westfield Insurance Project Management Methodology (PMM)
  - Introduced the Corporate Prioritization Process
- *2001:*
  - Recruited senior project managers
  - Refined the PMM
  - Refined the Corporate Prioritization Process, emphasizing executive level partnerships
- *2002:*
  - Established Enterprise Capacity Planning and Prioritization
  - Recognition of technically sound and experienced project management team
  - Increased successful delivery of projects
  - PMM applied more consistently by the project managers and project team members

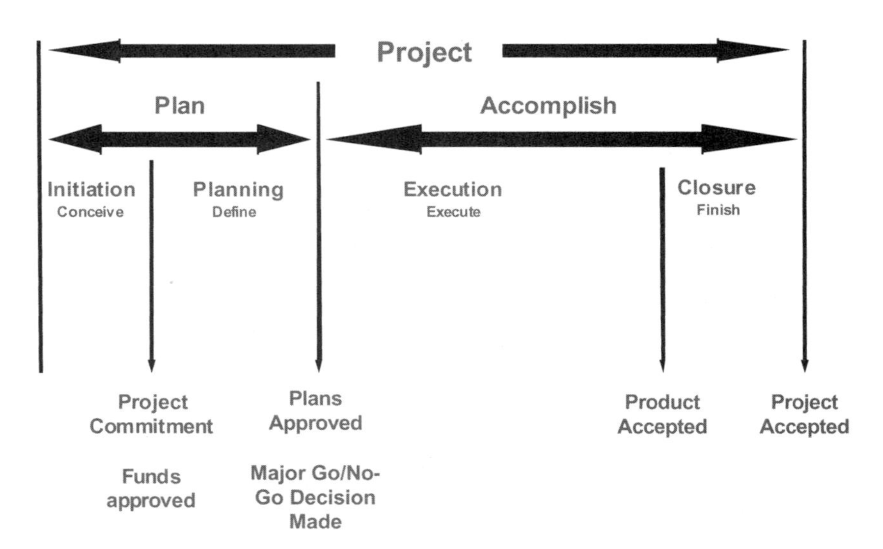

FIGURE 4–22. Life cycle phases.

- *2003:*
  - Refining Enterprise Capacity Planning and Prioritization
  - Establishing strategies for maintaining a proficient project management team
  - Establishing measurable standards for applying the PMM
  - Establishing continuous improvement strategies for PMM

When it came time to develop our Project Management Methodology, we decided to use the PMBOK® as a guideline to provide some degree of structure to the development process. Figure 4–22 shows the life cycle phases of our methodology. Within each of these phases, we then identified the activities to be performed and cross-referenced the activities according to the nine process areas of the PMBOK®. This is shown in Figure 4–23.

Each of the activities in Figure 4–23 can then be exploded into a flowchart showing the detailed activities needed for accomplishment of the deliverables for this activity. For example, the "Project Startup Process" activity in Figure 4–23 is cross–listed under Human Resource Management. The detailed activities are shown by the flowchart in Figure 4–24.

Kathy Rhoads, Director, Project Management COE, summarizes Westfield Insurance's experience with project management:

Putting the right building blocks in place created the foundation necessary for moving ahead. Currently, we are experiencing significant increased success in delivery and

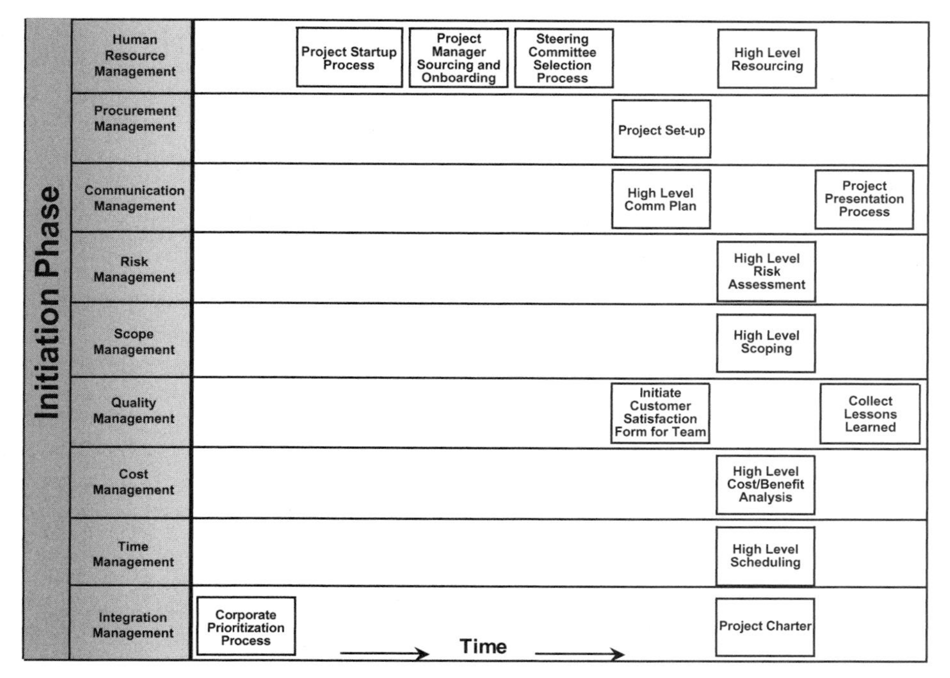

FIGURE 4–23. Activities according to the PMBOK® process areas.

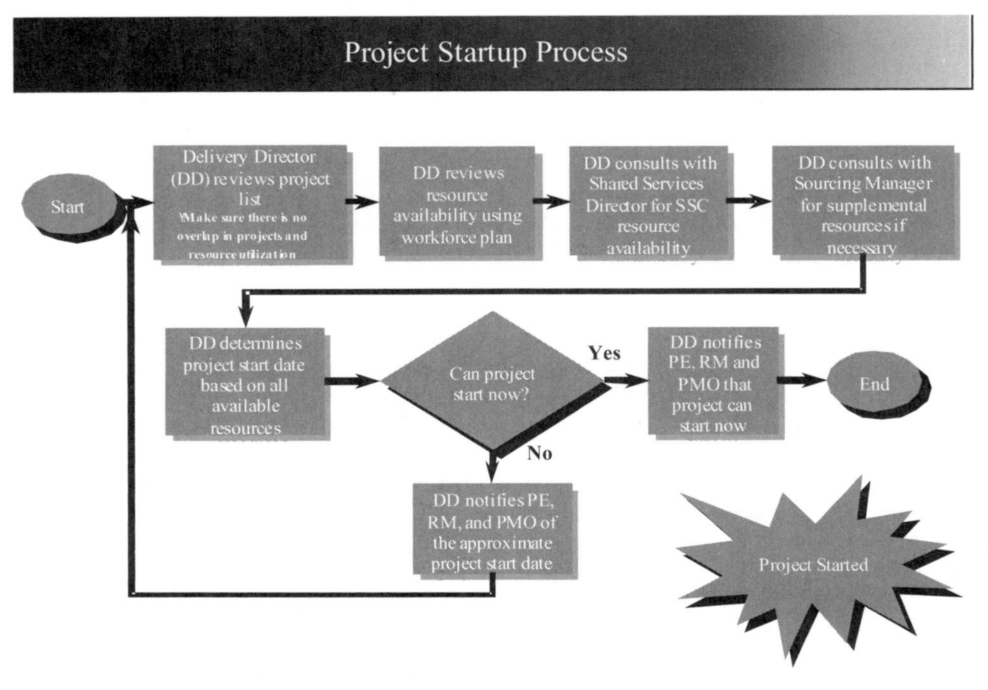

FIGURE 4–24. Project Startup Process.

improved overall results with key business initiatives. Our results validate the need for project management. Increased support from all levels of the organization is evident as more business customers request partnerships with teams led by our staff project managers. We are focusing on continuous improvement and realize a great deal of work remains. Building towards excellence in project management continues at Westfield Insurance.

4.6 PROJECT MANAGEMENT TOOLS

Project management methodologies require software support systems. As little as five years ago, many of the companies described in this book had virtually no project management capabilities. How did these companies implement project management so fast? The answer came with the explosion of personal computer-based software for project planning, estimating, scheduling, and control. These were critical for methodology development.

Until the late 1980s, the project management tools in use were software packages designed for project scheduling. The most prominent were:

- Program evaluation and review technique (PERT)
- Arrow diagramming method (ADM)
- Precedence diagramming method (PDM)

These three networking and scheduling techniques provided project managers with computer capabilities that far surpassed the bar charts and milestone charts that had been in use. The three software programs proved invaluable at the time.

- They formed the basis for all planning and predicting and provided management with the ability to plan for the best possible use of resources to achieve a given goal within schedule and budget constraints.
- They provided visibility and enabled management to control one-of-a-kind programs.
- They helped management handle the uncertainties involved in programs by answering such questions as how time delays influence project completion, where slack exists among elements, and which elements are crucial to meeting the completion date. This feature gave managers a means for evaluating alternatives.
- They provided a basis for obtaining the necessary facts for decision-making.
- They utilized a so-called time network analysis as the basic method of determining manpower, material, and capital requirements, as well as providing a means for checking progress.
- They provided the basic structure for reporting information

Unfortunately, scheduling techniques can't replace planning. And scheduling techniques are only as good as the quality of the information that goes into the plan. Criticisms of the three scheduling techniques in the 1980s included the following:

- Time, labor, and intensive effort were required to use them.
- The ability of upper-level management to contribute to decision-making may have been reduced.
- Functional ownership of the estimates was reduced.
- Historical data for estimating time and cost were lost.
- The assumption of unlimited resources was inappropriate.
- The amount of detail required made full use of the scheduling tools inappropriate.

Advancements in the memory capabilities of mainframe computer systems during the 1990s eventually made it possible to overcome many of the deficiencies in the three scheduling techniques being used in project management in the 1970s and 1980s. There emerged an abundance of mainframe software that combined scheduling techniques with both planning and estimating capabilities. Estimating then could include historical databases, which were stored in the mainframe memory files. Computer programs also proved useful in resource allocation. The lessons learned from previous projects could also be stored in historical files. This improved future planning, as well as estimating processes.

The drawback was that mainframe project management software packages were very expensive and user-unfriendly. The mainframe packages were deemed more appropriate for large projects in aerospace, defense, and large construction. For small and medium-sized companies, the benefits did not warrant the investment.

The effective use of project management software of any kind requires that project teams and managers first understand the principles of project management. All too often, an organization purchases a mainframe package without training its employees how to use it in the context of project management.

For example, in 1986, a large, nationally recognized hospital purchased a $130,000 mainframe software package. The employees in the hospital's information systems department were told to use the package for planning and reporting the status of all projects. Less than 10 percent of the organization's employees were given any training in project management. Training people in the use of software without first training them in project management principles proved disastrous. The morale of the organization hit an all-time low point, and eventually no one even used the expensive software.

Generally speaking, mainframe software packages are more difficult to implement and use than smaller personal computer–based packages. The reason? Mainframe packages require that everyone use the same package, often in the same way. A postmortem study conducted at the hospital identified the following common difficulties during the implementation of its mainframe package:

- Upper-level managers sometimes didn't like the reality of the results.
- Upper-level managers did not use the packages for planning, budgeting, and decision-making.
- Day-to-day project planners sometimes didn't use the packages for their own projects.
- Some upper-level managers sometimes didn't demonstrate support and commitment to training.
- Clear, concise reports were lacking.
- Mainframe packages didn't always provide for immediate turnaround of information.
- The hospital had no project management standards in place prior to the implementation of the new software.
- Implementation highlighted the inexperience of some middle managers in project planning and in applying organizational skills.
- Neither the business environment nor the organization's structure supported the hospital project management/planning needs.
- Sufficient/extensive resources (staff, equipment, etc.) were required.
- The business entity didn't determine the extent of, and appropriate use of, the systems within the organization.
- The system was viewed by some employees as a substitute for the extensive interpersonal skills required of the project manager.
- Software implementation didn't succeed because the hospital's employees didn't have sufficient training in project management principles.

Today, project managers have a large array of personal computer–based software available for planning, scheduling, and controlling projects. Packages such as Microsoft Project have almost the same capabilities as mainframe packages. Microsoft Project can import data from other programs for planning and estimating and then facilitate the difficult tasks of tracking and controlling multiple projects.

The simplicity of personal computer–based packages and their user-friendliness has been especially valuable in small and medium-sized companies. The packages are so affordable that even the smallest of companies can master project management and adopt a goal of reaching project management excellence.

Clearly, even the most sophisticated software package can never be a substitute for competent project leadership. By itself, such packages can't identify or correct task-related problems. But they can be terrific tools for the project manager to use in tracking the many interrelated variables and tasks that come into play in contemporary project management. Specific examples of such capabilities include the following:

- Project data summary; expenditure, timing, and activity data
- Project management and business graphics capabilities
- Data management and reporting capabilities
- Critical path analyses
- Customized as well as standardized reporting formats
- Multiproject tracking
- Subnetworking
- Impact analysis
- Early-warning systems
- Online analyses of recovering alternatives
- Graphical presentations of cost, time, and activity data
- Resource planning and analyses
- Cost and variance analyses
- Multiple calendars
- Resource leveling

Mindjet

During the 1980s and most of the 1990s, software for project management focused heavily on information analysis for project monitoring, schedule tracking, and performance reporting. The software was used more often for project execution and reporting than for project planning. As companies began seeing the potential benefits of an intranet tracking and reporting system, companies then recognized the importance of having more information available on the screen at one time for effective decision-making to take place. In the marketplace, there appeared more software being developed for the front end of the project management life cycle. This included software to assist in brainstorming, conceptual planning, creativity in solutions, and quality upfront planning. The display of information was now becoming significantly important to senior managers because properly displayed information allowed senior management to review more projects in the same amount of time as well as making quicker and more effective project decisions.

One such software package to support project management is MindManager from Mindjet. The package focuses on the mapping of information that is especially useful in the early phases of a project. According to Hobart Swan, External Affairs Manager at Mindjet:

> Mindjet's MindManager visual project-planning software is not a tool for continuing to do what you have done in the past. Used well, the software actually changes the way you think about and communicate the complex relationships that exist in a project. MindManager's unique contribution to project management is in giving project teams the ability to:
>
> - Literally *see* where a task or piece of information belongs in the project. Our brains function better, are more creative, and are better able to spot relationship and associations among pieces of information when that information is presented visually.

- Understand the reasons for project details, including changes to the original plan. The ability to link any and all background or supporting information to any branch means that valuable details are easy to locate. Key information that may otherwise be lost or difficult to locate in pure scheduling applications is captured by MindManager in an intuitive, easy-to-find visual structure.
- Quickly see how a change in one area may affect other parts of a project.
- Use visualization to review a project and highlight problem areas. MindManager's Level of Detail feature enables stakeholders to view the project from 50,000 feet, or drill down to all details.
- Zoom in to work in a selected area. Level of Detail and Power Select enable stakeholders to quickly view customized "slices" of the project.

Seeing Project Information in Context

In MindManager, pieces of information are never isolated. They are always put into context. MindManager's map interface creates a structure that puts each piece of information into relationship with another piece of information. Putting information in context makes it easier to see where something belongs, why it belongs there, and what else it relates to. This makes it easy to see the two things that are hardly ever written down by anyone: assumptions and implications. Assumptions and implications are frequently the sources of trouble down the line in projects, and the earlier you catch them, the better your chances of staying in control.

Create Maps with a Clear Purpose in Mind

Some examples of purpose include maps designed to:

- Brainstorm ideas and do initial thinking
- Define scope and goals as part of early project planning
- Communicate and explain project elements to stakeholders
- Explain to participants what their part in the project will be and how their work will tie into the greater whole
- Prepare the basic project schedule and/or task lists for export to another tool (e.g., Microsoft Project)
- Serve as a concise progress report, to be updated as the project proceeds

Mindjet's MindManager may be representative of the software to be developed in the first decade of the twenty-first century. Companies are undertaking more projects than ever before and the complexity of the projects is increasing. More data are available and required for effective decision-making to occur. Time has evolved from a simple constraint to a severe constraint. Software to support these situations should certainly be well accepted in the project management marketplace.

According to Bettina Jetter, CEO Mindjet LLC:

> Mindjet's focus is to develop innovative software that can significantly increase the productivity of the business professionals. Mindjet for the Tablet PC and Mindjet's visual tool for planning and brainstorming allows people the mobility and versatility to meet with others, capture thoughts, share ideas naturally and produce structured results. Used in the planning phases of common business processes, such as creating new project plans and

defining or streamlining business processes, the product enables business teams to deliver results through faster consensus, and quicker and better decisions in less time than is normally spent in meetings.

MULTIPLE CHOICE QUESTIONS

1. Good project methodologies improve project execution as well as allowing for:
 A. Better customer relations
 B. Improved customer confidence
 C. Improved supplier delivery
 D. All of the above
 E. A and B only

2. A common mistake made by companies is the failure to:
 A. Integrate the methodology and tools into a single process
 B. Develop a different methodology for each type of project
 C. Reduce paperwork for project management
 D. All of the above
 E. A and B only

3. A benefit of a good project management methodology is fewer scope changes.
 A. True
 B. False

4. A good project management methodology will force a company to become excellent in project management.
 A. True
 B. False

5. External factors that can influence the success or failure of a project management methodology include the:
 A. Current business climate
 B. Rapid changes in technology
 C. Sophistication level of clients
 D. Sophistication level of consumers
 E. All of the above

6. Every company can benefit from a standard project management methodology.
 A. True
 B. False

7. Successful project management methodologies are based upon:
 A. Policies
 B. Procedures
 C. Guidelines
 D. All of the above
 E. A and B only

8. Which of the following is not one of the four major components of the project management methodology of the bank discussed in this chapter?
 A. Planning
 B. Organizing
 C. Staffing

D. Managing
E. Reporting

9. Project management methodologies should not change even when the organization changes.
A. True
B. False

10. Long-term use of a good project management methodology will:
A. Create or enhance the corporate culture
B. Create more paperwork
C. Generate more scope changes on projects
D. Eliminate the need for a project sponsor
E. Eliminate the need for a project champion

11. The program management process at General Motors Powertrain Group is based upon:
A. No executive sponsor
B. No definable customer
C. Common templates, checklists, and systems
D. Common suppliers
E. Colocated teams

12. The primary reason General Motors Powertrain Group decided to go to project management was to:
A. Benchmark against other competitors
B. Speed up continuous improvement efforts
C. Increase product quality
D. Get their products to their customers faster
E. All of the above

13. Which of the following companies have project management methodologies designed primarily for internal customers?
A. General Motors Powertrain Group
B. Key Services Corporation
C. Ericsson Telecom A-B
D. All of the above
E. A and B only

14. Which of the following is *not* one of the four cornerstones of PROPS?
A. Tollgates
B. The Project Model
C. The Work Models
D. Milestones
E. The Scope Management plan

15. World-class methodologies for project management have a maximum of __________ life cycle phases.
A. Four
B. Five
C. Six
D. Seven
E. Eight

16. Which of the following is *not* a characteristic of a world-class methodology?
 A. Overlap of no more than one life cycle phase
 B. End-of-phase gate reviews
 C. Integrated with other processes
 D. Customer oriented
 E. Minimization of paperwork

17. Which characteristic of a world-class methodology is most often impacted by concurrent engineering activities?
 A. The maximum number of life cycle phases
 B. End-of-phase gate reviews
 C. Minimization of paperwork
 D. Templates
 E. Integration with other processes

18. Typical templates in a world-class methodology usually go down to level {4pirule} of the work breakdown structure.
 A. 1
 B. 2
 C. 3
 D. 4
 E. 5

19. The amount of overlap of the life cycle phases is most often dependent upon:
 A. The number of life cycle phases
 B. The title of the executive sponsor
 C. The profitability of the project
 D. The strategic importance of the project
 E. The amount of risk the project manager is willing to accept

20. World-class methodologies are continuously enhanced through:
 A. Key performance indicators reviews
 B. Lessons learned updates
 C. Benchmarking efforts
 D. Customer recommendations
 E. All of the above

21. The two most commonly used ways to measure the benefits of a standard methodology are:
 A. Short-term and long-term
 B. Strategic and operational
 C. Internal and external
 D. Hybrid and project-driven
 E. Priority and maintenance

22. World-class methodologies allow for faster time to market through:
 A. Better scope control
 B. Effective executive sponsorship
 C. Colocated teams
 D. Continuous benchmarking activities
 E. All of the above

23. The ultimate benefit of a world-class methodology may very well exist when the customer:
  A. Maximizes the contractor's profit
  B. Stops interfering with the contractor's work
  C. Provides non–sole-source contracts
  D. Treats the contractor as a partner
  E. Requests that an executive sponsor not be assigned

24. Converting a standard methodology into a world-class methodology requires:
  A. An executive-level sponsor
  B. The assignment of senior project managers
  C. Company-wide acceptance
  D. Colocated teams
  E. A minimum of five years or longer using the methodology

25. Which individual(s) is(are) most critical for updating the methodology and driving its acceptance?
  A. Executive sponsor
  B. Executive champion
  C. Project managers
  D. Functional managers
  E. Customers

DISCUSSION QUESTIONS

1. What is a standard methodology?
2. What are the differences between a standard methodology and a world-class methodology?
3. Why are executive champions needed for implementation of a methodology? Does it depend upon the size of the company?
4. What differences might exist between the project management methodologies of project-driven, non–project-driven, and hybrid organizations?
5. How much influence or say should a customer have in how the contractors implement their methodologies?
6. Under what conditions would templates for a methodology go down to level 5 of the WBS?
7. Is it possible for companies with world-class methodologies *not* to be competitive in a given market segment?
8. What types of projects would *not* require the use of a project management methodology?
9. Can a methodology that is used successfully in one company be implanted into another company successfully?
10. Who should attend training classes on how to use a project management methodology?

Across

1. Type of acceptance
5. PROPS cornerstone
7. Methodology benefit: customer _ _ _ _ _ _ _ _ _ _ _ _
8. Informal policies/procedures
10. System
12. Treatment by customer
13. Continuous improvement

Down

1. Executive _ _ _ _ _ _ _ _
2. Methodology benefit customer _ _ _ _ _ _ _ _ _ _
3. PROPS cornerstone
4. Type of methodology
6. Maximum number of phases
9. Commonly used process
11. Concurrent _ _ _ _ _ _ _ _ _ _ _

Strategic Planning for Excellence in Project Management

5.0 INTRODUCTION

Strategic planning for excellence in project management needs to consider all aspects of the company: from the working relationships among employees and managers and between staff and management, to the roles of the various players (especially the role of executive project sponsors), to the company's corporate structure and culture. Other aspects of project management must also be planned. Strategic planning is vital for every company's health. Effective strategic planning can mean the difference between long-term success and failure. Even career planning for individual project managers ultimately plays a part in a company's excellence in project management or its mediocrity. All of these subjects are discussed in this chapter.

5.1 INFLUENCE OF ECONOMIC CONDITIONS

During favorable economic times, changes in management style and corporate culture move very slowly, but favorable economic conditions don't last forever. The period between recognizing the need for change and garnering the ability to manage change is usually measured in years. As economic conditions deteriorate, change occurs more and more quickly in business organizations, but not fast enough to keep up with the economy.

Before the recession of 1989–1993, U.S. companies were willing to accept the implementation of project management at a tedious pace. Corporate managers in general believed that their guidance was sufficient to keep their companies healthy, and outside consultants were brought in primarily to train production workers in the principles of project management. Executive training sessions, even very short ones, were rarely offered.

During the recession, senior managers came to realize that their knowledge of project management was not as comprehensive as they had once believed (see Table 5–1). The table shows how the recession affected the development of project management systems.

By the end of the recession in 1993, companies finally recognized the importance of both strategic planning and project management, as well as the relationship between them. The relationship between project management and strategic planning can best be seen from Figure 5–1. Historically, a great deal of emphasis was placed on strategic formulation with little emphasis on strategic implementation. Now companies were recognizing that the principles of project management could be used for the implementation of strategic plans, as well as operational plans.

Another factor promoting project management was the acceptances of strategic business units (SBUs). The strategic business units were usually less resistant than the parent corporation to the use of project management, and more aware of the necessity to obtain horizontal as well as vertical work flow. This is shown in Figure 5–2. Project management was now recognized as a vehicle for the implementation of just about any type of plan for any type of project.

This relationship has become evident in many companies. At Edelca, Marcelino Diez, PMP, Project Management Consultant, believes that:

> Companies must reach a consensus about the strategic importance of project management, the fact that it has become an organizational core competency, and the importance of the role that the project manager plays in achieving the success of the project. Also, there must exist a more explicit definition of the relation between the projects and the organization's success, among the projects and the strategic plan of the business, and the company as a whole.

To address the far-reaching changes in the economic environment, senior managers began to ask a fundamental question: How do we plan for excellence in project management? In answering this question, it would be futile to expect managers to implement immediately all of the changes needed to set up modern project management in their companies. (The principles of modern project management are discussed in Chapter 6.) What senior managers needed was a plan like the one shown in Table 5–2, expressed in terms of

TABLE 5–1. EFFECTS OF THE 1989–1993 RECESSION ON THE IMPLEMENTATION OF PROJECT MANAGEMENT

Factor	Prior to the Recession	After the Recession
Strategic focus	Short-term	Long-term
Organizational structuring	To secure power, authority, and control	To get closer to customers
Management focus	To manage people	To manage work and deliverables
Sponsorship	Lip service sponsorship	Active
Training emphasis	Quantitative	Qualitative/behavioral
Risk analysis	Minimal effort	Concerted effort
Authority	In writing	Implied
Team building	Functional teams	Cross-functional teams

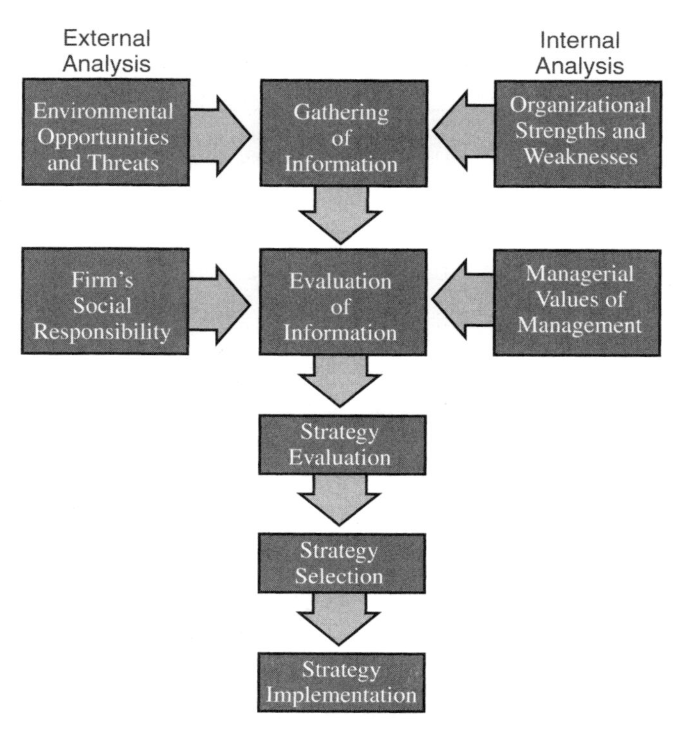

FIGURE 5–1. Traditional strategic planning model. *Source:* P. Rea and H. Kerzner, *Strategic Planning.* New York: Wiley, 1997, p. 3.

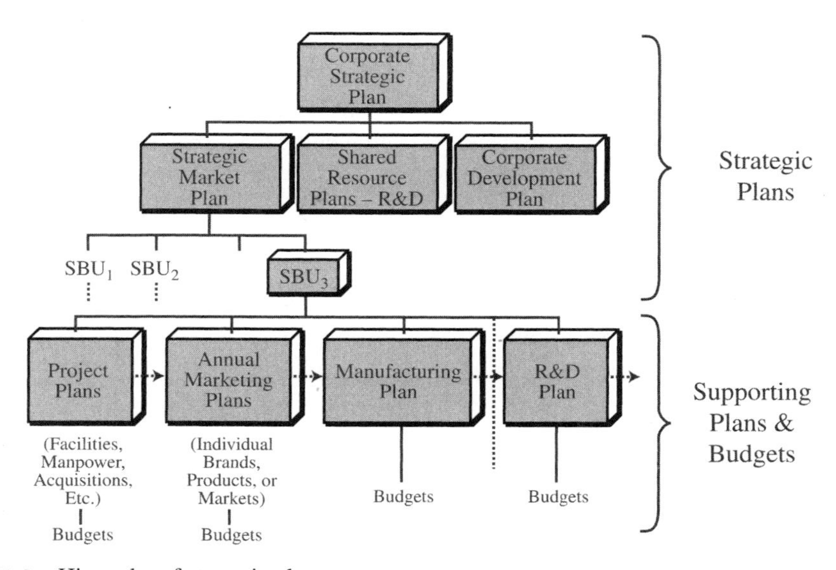

FIGURE 5–2. Hierarchy of strategic plans.

TABLE 5–2. STRATEGIC FACTORS IN ACHIEVING EXCELLENCE

Factor	Short-Term Applications	Long-Term Implications
Qualitative	Provide educational training Dispel illusion of a need for authority Share accountability Commit to estimates and deliverables Provide visible executive support and sponsorship	Emphasize cross-functional working relationships and team building
Organizational	Deemphasize policies and procedures	Create project management career path
	Emphasize guidelines	Provide project managers with reward/penalty power
	Use project charters	Use nondedicated, cross-functional teams
Quantitative	Use a single tool for planning, scheduling, and controlling	Use estimating databases

three broad, critical success factors: qualitative factors, organizational factors, and quantitative factors. These factors are discussed in later sections.

5.2 WHAT IS GENERAL STRATEGIC PLANNING?

Strategic planning for a business environment is the process of formulating and implementing decisions about an organization's future direction. This can be seen from Figure 5–1. This process is vital to every organization's survival because it is the process by which the organization adapts to its ever-changing environment and is applicable to all levels and types of organizations. The formulation process is the process of deciding where you want to go, what decisions must be made, and when they must be made in order to get there. It is the process of defining and understanding the business you are in. The outcome of this process results in the organization doing the right thing by producing goods or services for which there is a demand or need in the external environment. When this occurs, we say the organization has been effective as measured by market response, such as sales and market shares. All organizations must be effective or responsive to their environments to survive in the long run.

The formulation process is performed at the top levels of the organization. Here, top management values provide the ultimate decision template for directing the course of the firm.

Formulation

- Scans the external environment and industry environment for changing conditions.
- Interprets the changing environment in terms of opportunities or threats.
- Analyzes the firm's resource base for asset strengths and weaknesses.

- Defines the mission of the business by matching environmental opportunities and threats with resource strengths and weaknesses.
- Sets goals for pursuing the mission based on top management values and sense of responsibility

Implementation translates the formulated plan into policies and procedures for achieving the grand decision and involves all levels of management in moving the organization toward its mission. The process seeks to create a fit between the organization's formulated goal and its ongoing activities. Because it involves all levels of the organization, it results in the integration of all aspects of the firm's functioning. Middle- and lower-level managers spend most of their time on implementation activities. Effective implementation results in stated objectives, action plans, timetables, policies and procedures, and results in the organization moving efficiently toward its mission.

5.3 WHAT IS STRATEGIC PLANNING FOR PROJECT MANAGEMENT?

Strategic planning for project management is the *development of a standard methodology for project management,* which can be used over and over again, and which will produce a high likelihood of achieving the project's objectives. Although strategic planning and execution of the methodology does not guarantee profits or success, it does improve the chances of success.

One primary advantage of developing an implementation methodology is that it provides the organization with a consistency of action. As the number of interrelated units in organizations have increased, so have the benefits from the integrating direction afforded by the project management implementation process.

Methodologies need not be complex. Figure 5–3 shows the "skeleton" for the development of a simple project management methodology. The methodology begins with a project definition process that is broken down into a technical baseline, functional and management baseline, and financial baseline. The technical baseline includes, at a minimum:

- Statement of work (SOW)
- Specifications
- Work breakdown structure (WBS)
- Timing (i.e., schedules)
- Spending curve (S curve)

The functional and management baselines indicate how you will manage the technical baseline. This includes:

- Résumés of the key players
- Project policies and procedures
- The organization for the project
- Responsibility assignment matrices (RAMs)

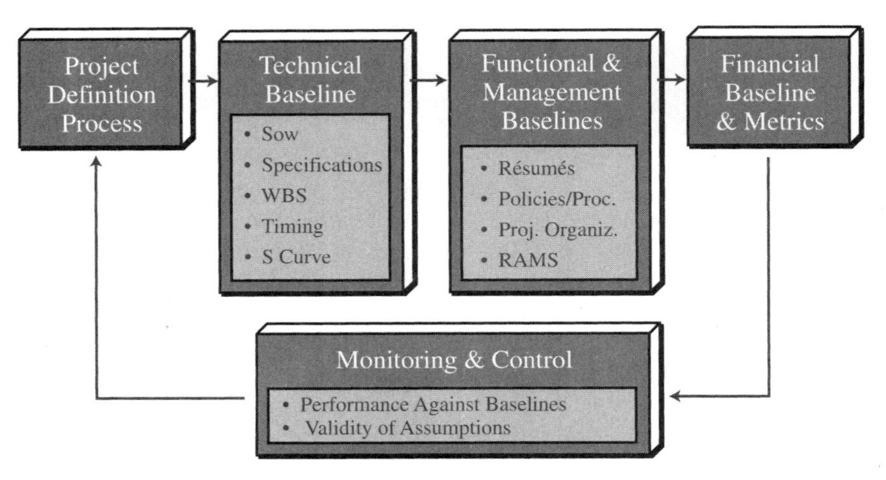

FIGURE 5–3. Methodology structuring.

The financial baseline identifies how costs will be collected, analyzed, variances explained, and reports prepared. This is covered separately in Section 5.8. Altogether, this is a simplistic process that can be applied to each and every project.

Without this repetitive process, subunits tend to drift off in their own direction without regard to their role as a subsystem in a larger system of goals and objectives. The objective-setting and the integrating of the implementation process using the methodology assure that all of the parts of an organization are moving toward the same common objective. The methodology gives direction to diverse activities.

Another advantage of strategic project planning is that it provides a vehicle for the communication of overall goals to all levels of management in the organization. It affords the potential of a vertical feedback loop from top to bottom, bottom to top, and functional unit to functional unit. The process of communication and its resultant understanding helps reduce resistance to change. It is extremely difficult to achieve commitment to change when employees do not understand its purpose. The strategic project planning process gives all levels an opportunity to participate, thus reducing the fear of the unknown and eliminating resistance.

The final and perhaps the most important advantage is the thinking process required. Planning is a rational, logically ordered function. Many managers caught up in the day-to-day action of operations will appreciate the order afforded by a logical thinking process. Methodologies can be based upon sound, logical decisions. Figure 5–4 shows the logical decision-making process that could be part of the competitive bidding process for an organization. Checklists can be developed for each section of Figure 5–4 to simplify the process.

In the absence of an explicit project management methodology, decisions are made incrementally. A response to the crisis of the moment may result in a choice that is unrelated and perhaps inconsistent with the choice made in the previous moment of crisis.

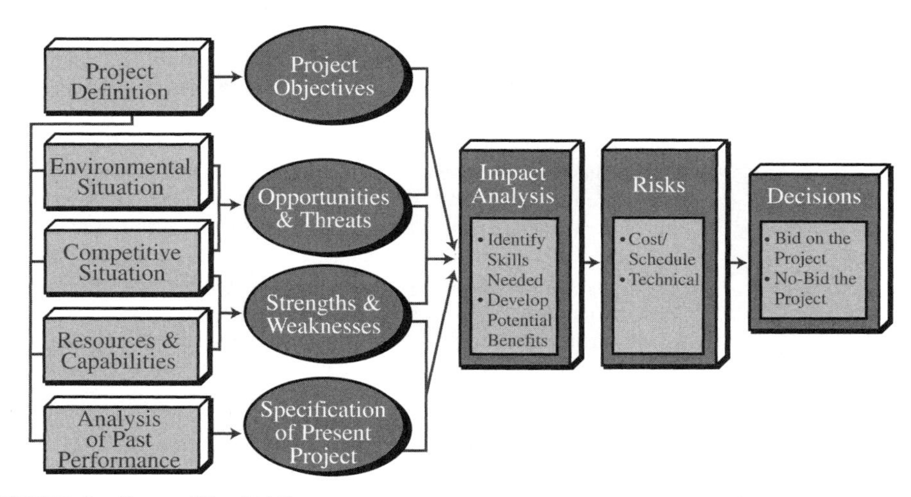

FIGURE 5–4. Competitive bidding process.

Discontinuous choices serve to keep the organization from moving forward. Contradictory choices are a disservice to the organization and cause its demise. These discontinuous and contradictory choices occur when decisions are made independently to achieve different objectives. When the implementation process is made explicit, however, objectives, missions, and policies become visible guidelines that produce logically consistent decisions.

Small companies usually have an easier time in performing strategic planning for project management excellence. Large companies with highly diversified product lines and multiple management styles find that institutionalizing changes in the way projects are managed can be very complex. Innovation and creativity in project management can be a daunting, but not impossible, task.

Bob Storeygard, Advanced Project Leadership Specialist with 3M Company, comments on the changes taking place at 3M:

> Work at 3M has begun and will continue on several related fronts that are taking our PM [project management] principles and their bottom-line effect into the future. These are:
>
> - Continued fostering of our PMSIG (PM Special Interest Group), which has well over 3,000 people in 3M worldwide participating in it. This is a conduit to share best practice across the corporation and is one of the only groups in 3M that reaches into every business unit. (This fostering consists of monthly, topical, videotaped presentations on PM topics, a Web site and repository for various 3M best practices and PM information, a Project Office Forum, as well as other special events to enable networking and PM communications throughout the company.)
> - Increased networking and partnerships with both 3M internal forums and SIGs (e.g., Marketing, ODSIG, IT, etc.), as well as external business (other companies) and professional society partners (PMI, PDMA, etc.) to capture new and refine our existing best practices in PM.

- Work on better defining the value, construction, and use of Project Offices through 3M . . . whether they be primarily focused on specific large projects, programs, organizations, or the whole company. This work resulted in the Project Office Implementation Kit.
- The establishment and rollout of a service that helps organizations begin to first recognize all the pieces involved in a total healthy Project Management Environment and then begin to create or recreate, deploy, and sustain this environment into the future. This is called our PM Deployment Process and consists of several of the following pieces:
- Defining and communicating what a healthy PM environment looks like
- Assessing the particular business unit's strengths and weaknesses according to that environment
- Customizing standard processes and procedures for both portfolio and project levels within the organization
- Piloting these new portfolio and project-level processes and procedures with "real" bottom-line projects
- Rolling these new ways out to the rest of the organization's projects and programs

Effective strategic planning for project management is a never-ending effort. The two most common, continuous supporting strategies are the integration opportunities strategy and the performance improvement strategy. Figure 5–5 is the strategy for opportunities to integrate or combine the existing methodology with other types of management opportunities currently in use by the company. Such other methodologies available for integration include concurrent engineering, total quality management, scope management, and risk management. Integrated strategies provide a synergistic effect. Integrated strategies are discussed in Chapter 7.

The generic performance improvement strategy, shown in Figure 5–6, is designed to improve the efficiency of the existing methodology and to find new applications for the

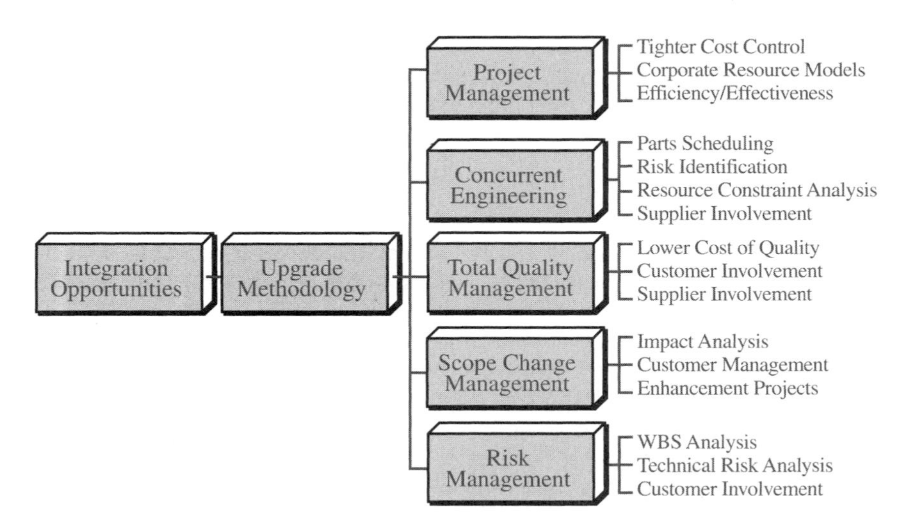

FIGURE 5–5. Generic integrated process strategies.

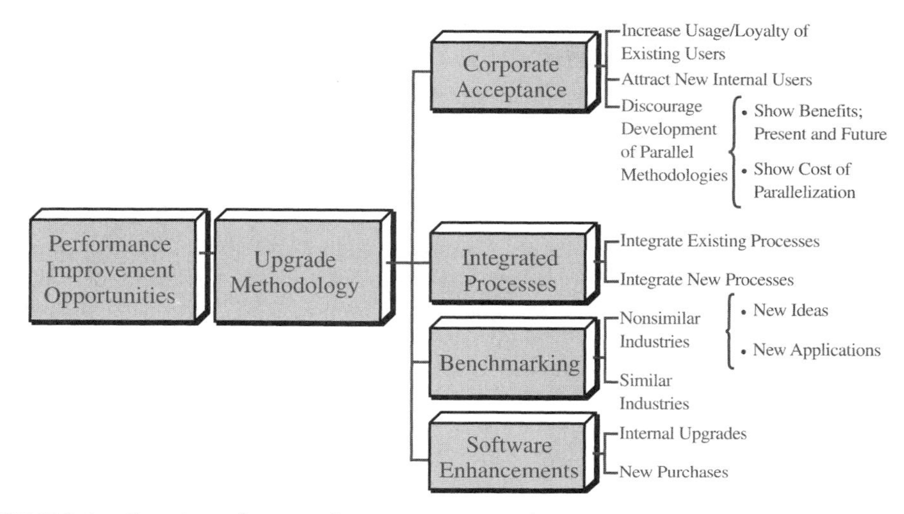

FIGURE 5–6. Generic performance improvement strategies.

methodology. The integrated process strategies of Figure 5–5 are also part of the process improvement strategies. Process improvement will be discussed in Section 5.8.

The goal of most organizations is to be more profitable than their competitors. There are both internal and external factors that are regarded as contributing sources to profitability. These are shown in Figure 5–7. Project management methodologies contribute to

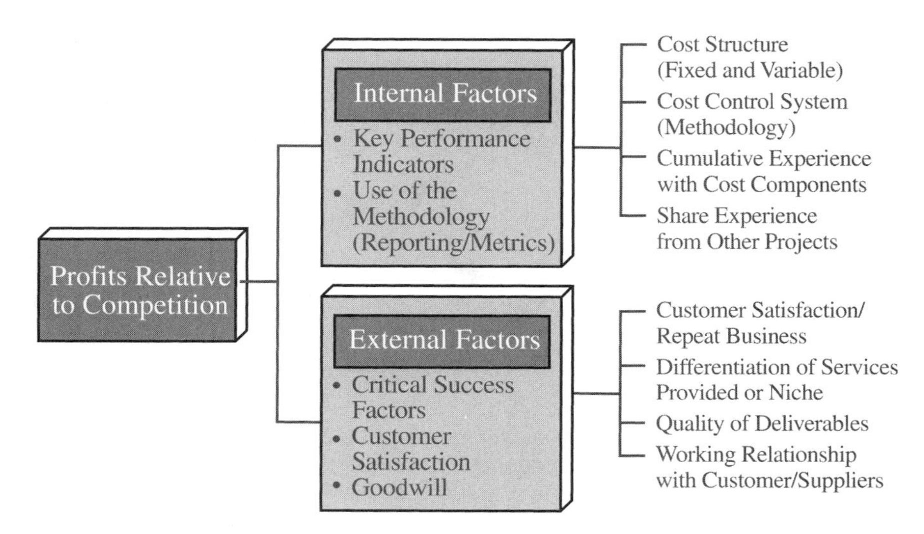

FIGURE 5–7. Sources of profitability.

profitability through more efficient execution of the project and implementation of the methodology. This is another valid reason mandating continuous strategic planning.

A good project management methodology will serve all of its stakeholders. Stakeholders are people who have a vested interest in the company's performance and who have claims on its performance. Figure 5–8 shows six commonly used categories of stakeholder: suppliers, customers, employees, creditors, shareholders, and even competitors. Organizations serve multiple stakeholders as customers, suppliers, government officials, shareholders, employees, and society at large. Methodologies for project management may indicate, when a problem exists, the order in which stakeholders will be satisfied. Good methodologies also include "standard practices" sections, which discuss morality and ethics when dealing with stakeholders.

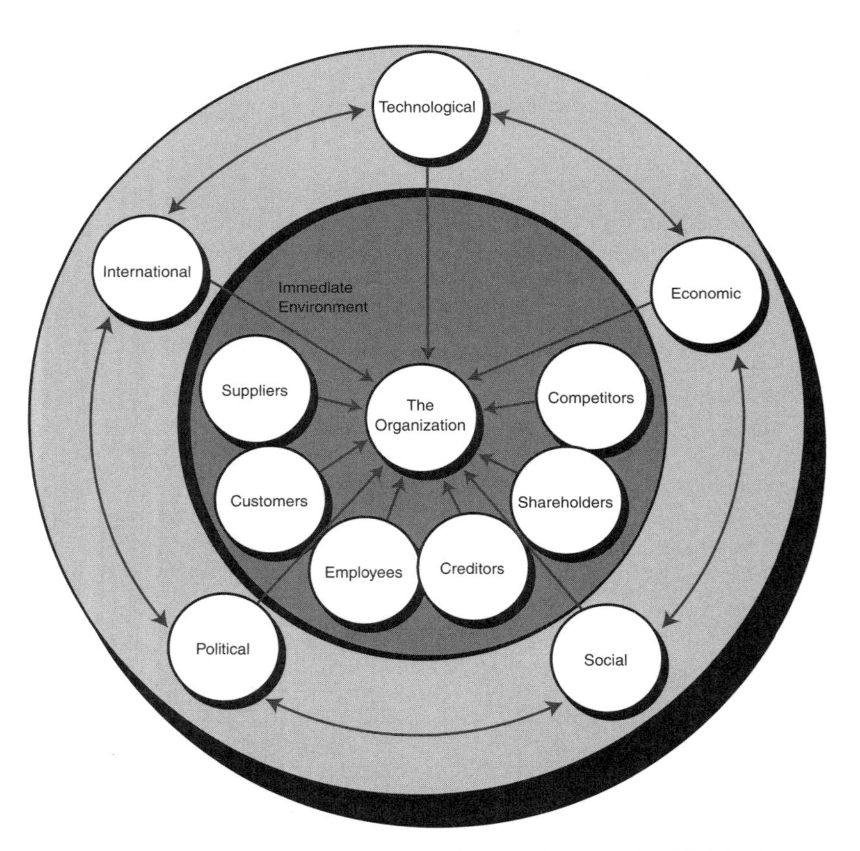

FIGURE 5–8. The macroenvironment of business. Stakeholders are identified in the "Immediate Environment" circle. (From *The Changing Environment of Business, A Managerial Approach,* 4th edition by Grover Starling. © 1995. Reprinted with permission of South-Western, a division of Thomson Learning: www.thomsonrights.com. Fax 800 730-2215.)

5.4 CRITICAL SUCCESS FACTORS FOR STRATEGIC PLANNING

Critical success factors for strategic planning for project management include those activities that must be performed if the organization is to achieve its long-term objectives. Most businesses have only a handful of critical success factors. However, if even one of them is not executed successfully, the business's competitive position will be threatened.

The critical success factors in achieving project management excellence apply equally to all types of organizations, even those that have not fully implemented their project management systems. Though most organizations are sincere in their efforts to fully implement their systems, stumbling blocks are inevitable and must be overcome. Here's a list of common complaints from project teams.

- There's scope creep in every project and no way to avoid it.
- Completion dates are set before project scope and requirements have been agreed upon.
- Detailed project plans identifying all of the project's activities, tasks, and subtasks are not available.
- Projects emphasize deadlines. We should emphasize milestones and quality and not time.
- Senior managers don't always allow us to use pure project management techniques. Too many of them are still date-driven instead of requirements-driven. Original target dates should be used only for broad planning.
- Project management techniques from the 1960s are still being used on most projects. We need to learn how to manage from a plan and how to use shared resources.
- Sometimes we are pressured to cut estimates low to win a contract, but then we have to worry about how we'll accomplish the project's objectives.
- There are times when line personnel not involved in a project change the project budget to maintain their own chargeability. Management does the same.
- Hidden agendas come into play. Instead of concentrating on the project, some people are out to set precedents or score political points.
- We can't run a laboratory without equipment, and equipment maintenance is a problem because there's no funding to pay for the materials and labor.
- Budgets and schedules are not coordinated. Sometimes we have spent money according to the schedule but are left with only a small percentage of the project activities complete.
- Juggling schedules on multiple projects is sometimes almost impossible.
- Sometimes we filter information from reports to management because we fear sending them negative messages.
- There's a lot of caving in on budgets and schedules. Trying to be a good guy all the time is a trap.

With these comments in mind, let's look at the three critical success factors in achieving project management excellence.

Qualitative Factors

If excellence in project management is a continuous stream of successfully completed projects, then our first step should be to define *success.* As discussed in Chapter 2, success in projects has traditionally been defined as achieving the project's objectives within the following constraints:

- Allocated time
- Budget cost
- Desired performance at technical or specification level
- Quality standards as defined by customers or users

In experienced organizations, the four preceding parameters have been extended to include the following:

- With minimal or mutually agreed upon scope changes
- Without disturbing the organization's corporate culture or values
- Without disturbing the organization's usual work flow

These last three parameters deserve further comment.

Organizations that eventually achieve excellence are committed to quality and upfront planning so that minimal scope changes are required as the project progresses. Those scope changes that are needed must be approved jointly by both the customer and the contractor. A well-thought-out process for handling scope changes is in place in such organizations. Even in large profit-making, project-driven industries, such as aerospace, defense, and large construction, tremendous customer pressure can be expected to curtail any "profitable" scope changes introduced by the contractor.

Most organizations have well-established corporate cultures that have taken years to build. On the other hand, project managers may need to develop their own subcultures for their projects, particularly when the projects will require years to finish. Such temporary project cultures must be developed within the limitations of the larger corporate culture. The project manager should not expect senior officers of the company to allow the project manager free rein.

The same limitations affect organizational work flow. Most project managers working in organizations that are only partially project-driven realize that line managers in their organizations are committed to providing continuous support to the company's regular functional work. Satisfying the needs of time-limited projects may only be secondary. Project managers are expected to keep the welfare of their whole companies in mind when they make project decisions.

> For companies to reach excellence in project management, executives must learn to define project success in terms of both what is good for the project and what is good for the organization.

Executives can support project managers by reminding them of this two-part responsibility by:

- Encouraging project managers to take on nonproject responsibilities such as administrative activities
- Providing project managers with information on the company's operations and not just information pertaining to their assigned projects
- Supporting meaningful dialogue among project managers
- Asking whether decisions made by project managers are in the best interest of the company as a whole

Organizational Factors

Organizational behavior in project management is a delicate balancing act, something like sitting on a bar stool. Bar stools usually come with three legs to keep them standing. So does project management: one leg is the project manager, one is the line manager, and one is the project sponsor. If one of the legs is lost or unusable, the stool will be very difficult to balance.

Although line managers are the key to successful project management, they will have a lot of trouble performing their functions without effective interplay with the project's manager and corporate sponsor. In unsuccessful projects, the project manager has often been vested with power (authority) over the line managers involved. In more successful projects, project and line managers share authority. The project manager negotiates the line managers' commitment to the project and works through line managers rather than around them. Project managers provide recommendations regarding employee performance. And leadership is centered around the whole project team, not just the project manager.

In successful project management systems, the following equation always hold true:

$$\text{Accountability} = \text{Responsibility} + \text{Authority}$$

When project and line managers view each other as equals, they share equally in the management of the project, and thus they share equally the authority, responsibility, and accountability for the project's success. Obviously the sharing of authority makes sharing decision-making easier. A few suggestions for executive project sponsors follow.

- Do not increase the authority of the project manager at the expense of the line managers.
- Allow line managers to provide technical direction to their people if at all possible.
- Encourage line managers to provide realistic time and resource estimates and then work with the line managers to make sure they keep their promises.
- Above all, keep the line managers fully informed.

In organizations that have created effective project management systems, the role of the executive manager has changed along with project management. Early in the implementation of project management, executives were actively involved in the everyday project management process. But as project management has come into its own and general economic conditions have changed, executive involvement has become more passive, and they concentrate on long-term and strategic planning. They have learned to trust project managers' decisions and view project management as a central factor in their company's success.

Project sponsors provide visible, ongoing support. Their role is to act as a bodyguard for the project and the project manager. Unlike other executives on the senior management team, individual project sponsors may play a more active role in projects, depending on how far along the project is. Early in the project's functioning, for example, the project sponsor might help the project manager define the project's requirements. Once that is done, the sponsor resumes a less active role and receives project information only as needed.

In successful project management systems that carry a high volume of ongoing project work, an executive sponsor may not be assigned to low-dollar-value or low-priority projects. Middle managers may fill the sponsorship role in some cases. But no matter the size or value of the project, project sponsors today are responsible for the welfare of all members of their project teams and not just the project manager.

The existence of a project sponsor implies visible, ongoing executive support for project management. And executive support motivates project personnel to excel. Executive project sponsorship also supports the development of an organizational culture that fosters confidence in the organization's project management systems.

Sometimes executives do not recognize the value of project sponsorship. This is particularly true of organizations primarily made up of white-collar professionals who hold advanced degrees. Let's look at a company we'll call Garcia Sciences Corporation, a small company headquartered in Tucson, Arizona.

In 1985, Roy Garcia, M.D., retired as vice president of research and development at a small pharmaceutical company. With more than 30 years of experience in research and development, Dr. Garcia was confident in his ability to start his own medical equipment company. By 1995, Garcia Sciences Corporation employed 120 people and generated $65 million per year in revenue. The research and development group had developed a family of new products that were well accepted by the medical community. State-of-the-art manufacturing techniques were producing high-quality, low-cost products.

Financially, Garcia Sciences Corporation was doing well. However, the company was troubled by detrimental cultural issues. Dr. Garcia began having second thoughts about the corporate culture he had created in the late 1980s. Because most of Dr. Garcia's employees held advanced degrees in engineering or science, he had given his professional employees a great deal of freedom to keep them motivated. Dr. Garcia believed strongly that scientists and engineers needed the freedom to be creative. Project sponsorship did not exist, and individual project teams were empowered to set their own objectives for the projects.

Almost every project was carried through to completion, and canceling projects was unacceptable. Status reporting was infrequent, so that executive managers often interfered in projects just to find out what was going on. Ultimately, Dr. Garcia and several senior managers began canceling projects either because the objectives didn't fit with the organization's overall goals or because the marketplace had changed. Team members became extremely unhappy when their projects were canceled before completion. The morale at Garcia Sciences weakened.

Dr. Garcia recognized that being able to see projects through to completion was what motivated project teams and managers. Through the grapevine, he learned that the project teams believed that their projects were being canceled to satisfy the personal whims of management rather than on the basis of sound business decisions.

It's easy for us to see that Dr. Garcia was operating under the mistaken assumption that advanced degrees prepared people to make intelligent business decisions. There are two kinds of intelligence operating in research and development companies: technical intelligence and business intelligence. Garcia's professional employees had the technical knowledge they needed, but they did not understand the business aspects of their projects.

Still, the problem in this case study did not lay with the employees. It lay with the corporation's senior managers. The key issue was the lack of executive project sponsorship. An executive project sponsor familiar with each project from its first stages would have been able to identify both the technical and the business objectives of the project from the beginning.

Not only must executive support be visible, it must also be "continuous" until such time as the organization reaches some level of maturity. Linda Kretz, now a project management trainer and consultant with the International Institute for Learning, describes her experiences as a project manager with US WEST:

> Project management as a formal process came about by default rather than design at US WEST. It took several years to build it up and a very little time to ultimately do away with it. I was probably the catalyst, not because I knew anything about it, but because after being titled a "Project Manager" I quickly learned that I wasn't managing to do anything but put out fires. . . .
>
> It started in 1984 with the divestiture of AT&T when all of our processes were confused and a merger was taking place between Pacific Northwest Bell, Mountain Bell, and Northwestern Bell. We decided to explore our processes from beginning to end, and I was lucky enough to be a part of the local service team in Portland, Oregon, to participate on this special "process" team. We started from the customer's perspective in the business office and were able to supervise for one month in the next department and learn their processes. It was through that participation that I learned the big picture of our business. I learned about departments I didn't even know existed.
>
> In 1985, I wanted to learn more about the technical side of telecommunications. Women still weren't positioned in technical roles, and I was told that I did not possess the skills necessary to be in our "Network" or operations departments. I went to a community college and took courses in telecommunications, which included analog and data concepts, signaling and supervision, central office maintenance and repair, installation, engineering and outside plant construction, as well as basic electricity and other related courses. I now had the book learning but no practical experience. But I got the job and was responsible for all of the data installations for the strategic accounts within the state of Oregon. I worked 16–20 hours a day and almost had a nervous breakdown.
>
> I kept a learning journal for one year on all of the issues which "bit me in the backside" and soon came to realize that I was reacting with brute force coordination, not proactively managing anything. I personally started to research project management and began studying available literature. I wrote a local manual based on the principles I learned and trained

it to my peers in Oregon. The project results in Oregon were far superior to any of the 14 other states in US WEST. The results captured the attention of headquarters. I was promoted to a second level manager and charged with establishing a framework for project management for the company.

I was allowed to benchmark against other companies, and with the cooperation of some of my larger clients (Boeing, Hewlett-Packard, Microsoft, 3M, and IBM), leveraged their expertise. The result was that in 1989 I became a Regional Staff Manager responsible for establishing seven Project Management Operations Centers (PMOC's) at US WEST. They were located in Seattle, Portland, Salt Lake City, Phoenix, Denver, Omaha, and Minneapolis. I had a staff of ten managers whose responsibilities were to provide project management support and technical training. We accomplished the following:

- Defined core competencies for project managers and established a training curriculum for them.
- Defined technical and human resource skills and established a methodology for interviewing potential candidates.
- Defined the hardware and software requirements and established a 14 state wide Macintosh network.
- Established the Center for Program Management at University College through the University of Denver. This program provides 20 hours to their Master of Technology Program if degreed or a Certificate in Project Management from the University of Denver.
- Established documented processes for project management for MIS [management information systems], product development, internal capital projects, and external customer implementations.
- Established a documented RFP [request for proposal] process, which increased our win rate considerably.
- Trained several hundred project specialists and project and program managers with a class designed to test evidence of learning through documentation of a real project throughout its normal cycle.
- Established job descriptions and a salary structure.
- Provided technical methods and procedures from a project management perspective.
- Actively mentored ongoing projects.

At this point, everyone wanted to be on the project management bandwagon. Results were exceptional. It was now 1991. Then downsizing began in earnest. We found ourselves with new leadership. Our original mentors were gone. Reengineering began with an eye toward centralization. The new "henchmen" decided that project management was a "nice to have" but not necessary. In 1992, my staff was disbanded—considered functionally obsolescent. Centralization turned out to be a major mistake. As a company, we were back to brute force coordination.

Project management lived on without staff support, and I became the project implementation manager of all of the project managers in Oregon, Washington, Utah, Idaho, and Montana. My job was to manage project managers. I once again convinced the senior team that a formal process was necessary. I became a certified project management professional (PMP) in 1994 and was the only one of my peers to do so. Approximately 78 percent of my team also received their PMPs. My boss and peers did not buy into the need for certification, but again my team proved to have very impressive results. My peers ultimately came around, but my boss would not support the process and, therefore, would not provide adequate resources for us to do the job well. He ultimately left the business.

By now I was promoted again, this time as director of project management for US WEST Multimedia Group and transferred to Atlanta, Georgia, to manage a multimillion dollar project there. I was always the "champion for the cause," but by now all of my mentors were gone and the team in Atlanta had no understanding of the process or the need for up-front planning. I inherited a project that was out of control. My role was perceived as being a passive notetaker or secretary. I eventually captured the attention of the senior team with data and facts defending brilliantly the need for a controlled process. It worked for a while.

By 1997, US WEST had purchased Continental Cable, and Atlanta found itself with an entirely new management team who didn't understand or see the need for project management. I could continue on but would not be provided with the resources that had originally been promised. I was tired of fighting the battle, [and so I] decided to retire. When I left, they once again disbanded the project management process. The project, which was under control when I left, is now in chaos again.

Some lessons learned:

- Corporations must have senior leadership buy-in. This can only happen if you plan brilliantly, present brilliantly, and defend brilliantly.
- No matter what they are talking about, they are talking about money. Always present to those who hold the purse strings and show how the money spent up-front is going to impact the bottom line.
- Strategic fit must be understood by all and decisions need to be made that support where the corporation wants to be when it grows up.
- The "not invented here" syndrome is alive and well and will remain—live with it and overcome by planning brilliantly, presenting brilliantly, and defending brilliantly.
- The "good old boy" network is also here to stay for a while. If you are a woman, particularly in the South, you must wait for them to die. In the meantime, persevere, smile, and let them think it was their idea.
- Vertical management will always hinder the process unless roles and responsibilities are well defined and understood so that they can be supported.
- Make the right decisions for the right reasons. Your moral compass will guide you.
- Even though you are singing the same tune, the audience is always changing. Make sure that the senior team continues to understand the difference between reactive brute force coordination and proactive project management, as well as the costs and savings of each.

It is interesting to me that the cycle is rekindling itself. I have just taught a class to some US WEST employees who would like to explore professional project management processes.

Conclusion: **Executive project sponsorship must exist and be visible so that the project–line manager interface is in balance.**

Recommendations for obtaining maturity include:

- Educate the executives as to the benefits of project management.
- Convince the executives of the necessity for ongoing, visible support in the capacity of a project sponsor.
- Convince executives that they need not know all the details. Provide them with the least information that tells the most.

Quantitative Factors

The third factor in achieving excellence in project management is the implementation and acceptance of project management tools to support the methodology. (See the discussion of project management tools in Chapter 4.) Some companies are quick to implement PERT/CPM tools, but many are reluctant to accept other mainframe or personal computer network software for project planning, project cost estimating, project cost control, resource scheduling, project tracking, project audits, or project management information systems.

Mainframe project management tools have been resurrected in the past few years. These new mainframe products are being used mainly for total company project control. However, executives have been slow to accept these sophisticated tools. The reasons for this are:

- Upper management may not like the reality of the output.
- Upper management uses their own techniques rather than the system for the planning, budgeting, and decision-making process.
- Day-to-day planners may not use the packages for their own projects.
- Upper management may not demonstrate support and commitment to training.
- Use of sophisticated mainframe packages requires strong internal communication lines for support.
- Clear, concise reports are lacking even though report generators exist.
- Mainframe packages do not always provide for immediate turnaround of information.
- The business entity may not have any project management standards in place prior to implementation.
- Implementation may highlight middle management's inexperience in project planning and organizational skills.
- Sufficient/extensive resources required (staff, equipment, etc.) may not be in place.
- Business environment and organizational structure may not be appropriate to meet project management/planning needs.
- Software utilization training without project management training is insufficient.
- Software may be used inappropriately as a substitute for the extensive interpersonal and negotiation skills required by project management.
- The business entity may not have predetermined the extent and appropriate use of the software within the organization.

Conclusions: Project management education must precede software education. Also, executives must provide the same encouragement and support for the use of the software as they do for project management.

The following recommendations are made to accelerate the maturity process:

- Educating people in the use of sophisticated software and having them accept its use is easier if the organization is already committed to project management.
- Executives must provide standards and consistency for the information they wish to see in the output.
- Executive knowledge (overview) in project management principles is necessary to provide meaningful support.
- Not everyone needs to become an expert in the use of the system. One or two individuals can act as support resources for multiple projects.

5.5 IDENTIFYING STRATEGIC RESOURCES

All businesses have corporate competencies and resources that distinguish them from their competitors. These competencies and resources are usually identified in terms of a company's strengths and weaknesses. Deciding upon what a company "should do" can only be achieved after assessing the strengths and weaknesses to determine what the company "can do." Strengths support windows of opportunities, whereas weaknesses create limitations. What a company "can do" is based upon the quality of its resources.

Strengths and weaknesses can be identified at all levels of management. Senior management may have a clearer picture of the overall company's position in relation to the external environment, whereas middle management may have a better grasp of the internal strengths and weaknesses. Unfortunately, most managers do not think in terms of strengths and weaknesses and, as a result, worry more about what they should do rather than what they can do.

Although all organizations have strengths and weaknesses, no organization is equally strong in all areas. Procter & Gamble, Budweiser, Coke, and Pepsi are all known for their advertising and marketing. Computer firms are known for technical strengths, whereas General Electric has long been regarded as the training ground for manufacturing executives. Large firms have vast resources with strong technical competency, but react slowly when change is needed. Small firms can react quickly but have limited strengths. The strengths and weaknesses can change over time and must, therefore, be closely monitored.

Strengths and weaknesses are internal measurements of what a company can do and must be based upon the quality of its resources. Consider the situation in Figure 5–9. A company with a world-class methodology in project management will not be able to close the gap in Figure 5–9 until the proper internal or subcontracted resources are available.

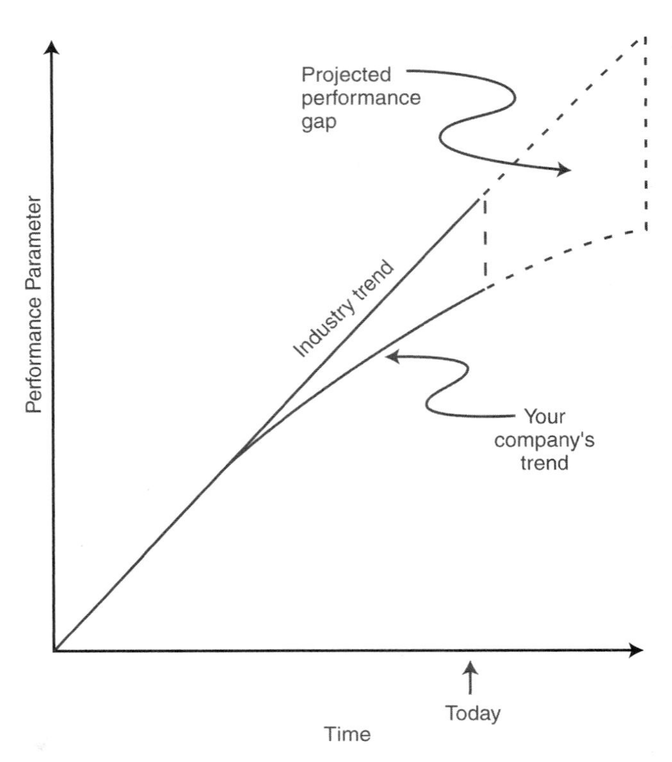

FIGURE 5–9. Projecting performance. (From *The Changing Environment of Business, A Managerial Approach,* 4th edition by Grover Starling. © 1995. Reprinted with permission of South-Western, a division of Thomson Learning: www.thomsonrights.com. Fax 800 730-2215.)

Methodologies, no matter how good, are executed by resources. Project management methodologies do not guarantee success. They simply increase the chances for success provided that (1) the objective is realistic and (2) the proper resources are available with the skills needed to achieve the objective.

Tangible Resources

In basic project management courses, the strengths and weaknesses of a firm are usually described in the terms of its tangible resources. The most common classification for tangible resources is:

- Equipment
- Facilities
- Manpower
- Materials
- Money
- Information/technology

Another representation of resources is shown in Figure 5–10. Unfortunately, these crude types of classification do not readily lend themselves to an accurate determination of internal strengths and weaknesses for project management. A more useful classification would be human resources, nonhuman resources, organizational resources, and financial resources.

Human Resources

Human resources are the knowledge, skills, capabilities, and talent of the firm's employees. This includes the board of directors, managers at all levels, and employees as a whole. The board of directors provides the company with considerable experience, political astuteness, and connections, and possibly sources of borrowing power. The board of directors is primarily responsible for selecting the CEO and representing the best interest of the diverse stakeholders as a whole.

Top management is responsible for developing the strategic mission and making sure that the strategic mission satisfies the shareholders. All too often, CEOs have singular strengths in one area of business such as marketing, finance, technology, or production.

The biggest asset of senior management is its decision-making ability, especially during project planning. Unfortunately, all too often senior management will delegate planning (and the accompanying decision-making process) to staff personnel. This may result in no effective project planning process within the organization and may lead to continuous replanning efforts.

Another important role of a senior management is to define clearly its own managerial values and the firm's social responsibility (see Figure 5–1). A change in senior management could change the organization's managerial values and social responsibility overnight. This could cause an immediate update of the project management methodology.

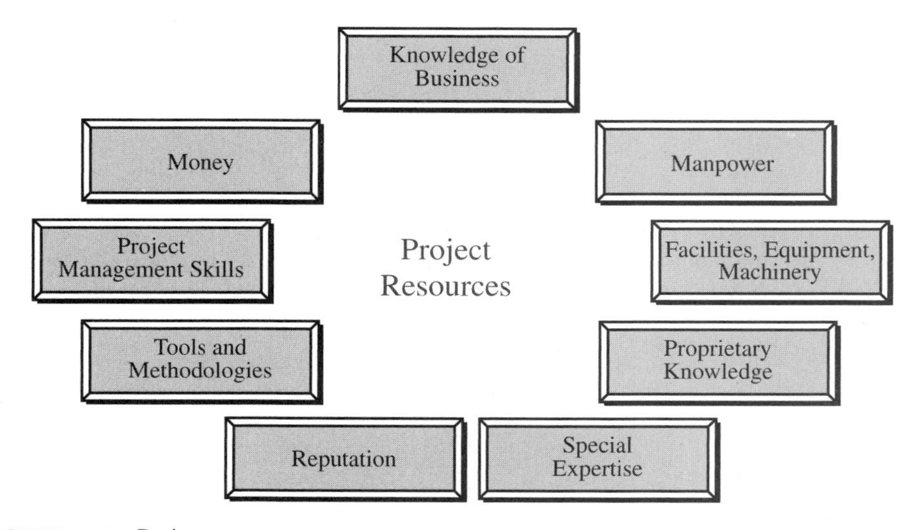

FIGURE 5–10. Project resources.

Lower and middle management are responsible for developing and maintaining the "core" technical competencies of the firm. Every organization maintains a distinct collection of human resources. Middle management must develop some type of cohesive organization such that synergistic effects will follow. It is the synergistic effect that produces the core competencies that lead to sustained competitive advantages and a high probability of successful project execution.

Nonhuman Resources

Nonhuman resources are physical resources that distinguish one organization from another. Boeing and IBM both have sustained competitive advantages but have different physical resources. Physical resources include plant and equipment, distribution networks, proximity of supplies, availability of a raw material, land, and labor.

Companies with superior nonhuman resources may not have a sustained competitive advantage without having superior human resources. Likewise, a company with strong human resources may not be able to take advantage of windows of opportunities without having strong physical resources. An Ohio-based company had a 30-year history of sustained competitive advantage on R&D projects that were won through competitive bidding. Unfortunately, the megaprofits were in production, and in order to acquire physical production resources, the organization diluted some of its technical resources. The firm learned a hard lesson in that the management of human resources is not the same as the management of nonhuman resources. The firm also had to reformulate their project management methodology to account for manufacturing operations.

Firms that endeavor to develop superior manufacturing are faced with two critical issues. First, how reliable are the suppliers? Do the suppliers maintain quality standards? Are the suppliers cost effective? The second concern, and perhaps the more serious of the two, is the ability to cut costs quickly and efficiently to remain competitive. This usually leads to some form of vertical integration.

Organizational Resources

Organizational resources are the glue that holds all of the other resources together. Organizational resources include the organizational structure, the project office, the formal (and sometimes informal) reporting structure, the planning system, the scheduling system, the control system, and the supporting policies and procedures. Decentralization can create havoc in large firms where each strategic business unit (SBU), functional unit, and operating division can have their own policies, procedures, rules, and guidelines. Multiple project management methodologies can cause serious problems if resources are shared between SBUs.

Financial Resources

Financial resources are the firm's borrowing capability, credit lines, credit rating, ability to generate cash, and relationship with investment bankers. Companies with quality credit ratings can borrow money at a lower rate than companies with nonquality ratings. Companies must maintain a proper balance between equity and credit markets when raising funds. A firm with strong, continuous cash flow may be able to fund growth projects out of cash flow rather than through borrowing. This is the usual financial-growth strategy for a small firm.

Intangible Resources

Human, physical, organizational, and financial resources are regarded as tangible resources. There are also intangible resources that include the organizational culture, reputation, brand name, patents, trademarks, know-how, and relationships with customers and suppliers. Intangible resources do not have the visibility that tangible resources possess, but they can lead to a sustained competitive advantage. When companies develop a brand name, it is nurtured through advertising and marketing and is often accompanied by a slogan. Project management methodologies can include paragraphs on how to protect the corporate image or brand name.

Social Responsibility

Social responsibility is also an intangible asset, although some consider it both intangible and tangible. Social responsibility is the expectation that the public perceives that a firm will make decisions that are in the best interest of the public as a whole. Social responsibility can include a broad range of topics from environmental protection to consumer safeguards to consumer honesty and employing the disadvantaged. An image of social responsibility can convert a potential disaster into an advantage. Johnson and Johnson earned high marks for social responsibility in the way it handled the two Tylenol tragedies in the 1980s. Nestlé, on the other hand, earned low marks for the infant-formula controversy.

5.6 STRATEGIC SELECTION OF PROJECTS

What a company wants to do is not always what it can do. The critical constraint is normally the availability and quality of the critical resources. Companies normally have an abundance of projects they would like to work on but, because of resource limitations, have to develop a prioritization system for the selection of projects.

A commonly used selection process is the portfolio classification matrix shown in Figure 5–11. Each project undergoes a SWOT assessment, a situational assessment for strengths, weaknesses, opportunities, and threats. The projects are then ranked on the nine-square grid based upon the projects' benefits and the quality of resources needed to achieve those benefits. The characteristics of the benefits appear in Figure 5–12 and the characteristics of the resources needed are shown in Figure 5–13.

Use of this matrix technique allows for proper selection of projects, as well as providing the organization with the foundation for a capacity planning model to see how much work the organization can take on. Companies usually have little trouble deciding where to assign the highly talented people. The model, however, provides guidance on how to make the most effective utilization of the average and below average individuals as well.

The boxes in the nine-square grid of Figure 5–11 can then be prioritized according to strategic importance as shown in Figure 5–14. If resources are limited but funding is adequate, the boxes identified as "high priority" will be addressed first.

Strengths &
Weaknesses
Resource Quality
Situation
Assessment
Strategic
Issues
Opportunities &
Threats
Projects' Benefits
Strong
Medium
Low
High
Medium
Low

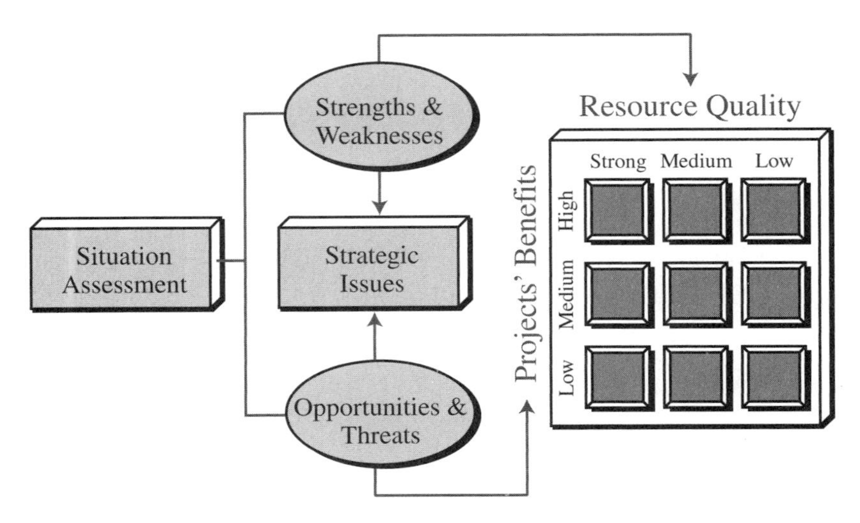

FIGURE 5–11. Portfolio classification matrix.

The nine-square grid in Figure 5–14 can also be used to identify the quality of the project management skills needed, in addition to the quality of functional employees. This is shown in Figure 5–15. As an example, the project managers with the best overall skills will be assigned to those projects that are needed to protect the firm's current position. Each of the nine cells in Figure 5–15 can be described as follows:

- *Protect Position (High Benefits and High Quality of Resources):* These projects may be regarded as essential to the survival of the firm. These projects mandate professional project management, possibly certified project managers, and the organization considers project management as a career path position. Continuous

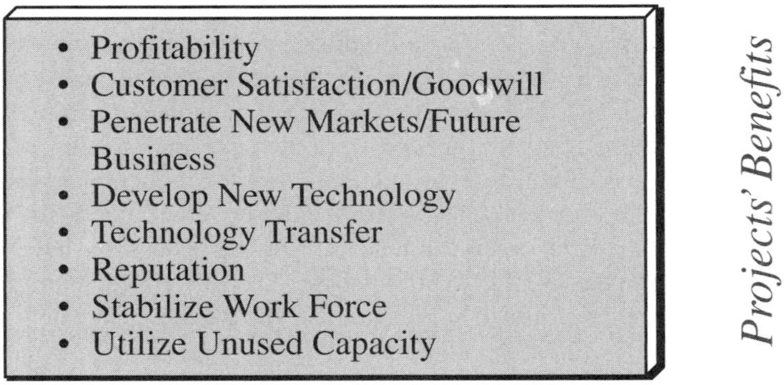

• Profitability
• Customer Satisfaction/Goodwill
• Penetrate New Markets/Future Business
• Develop New Technology
• Technology Transfer
• Reputation
• Stabilize Work Force
• Utilize Unused Capacity
Projects' Benefits

FIGURE 5–12. Characteristics of benefits.

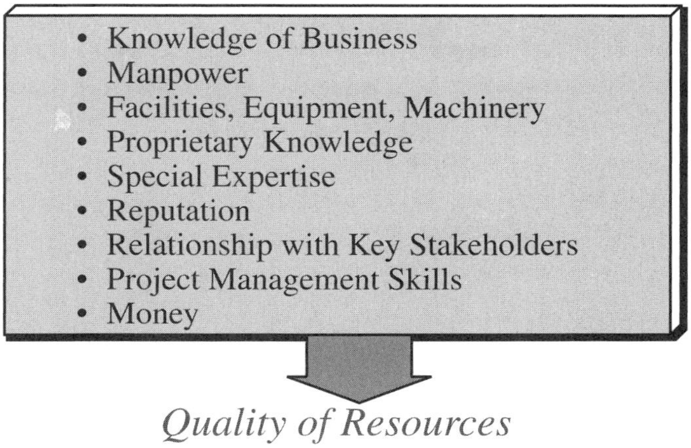

Knowledge of Business
Manpower
Facilities, Equipment, Machinery
Proprietary Knowledge
Special Expertise
Reputation
Relationship with Key Stakeholders
Project Management Skills
Money
Quality of Resources

FIGURE 5–13. Quality of resources needed.

improvement in project management is essential to make sure that the methodology is the best it can be.

- *Protect Position (High Benefits and Medium Quality of Resources):* This may require a full-time project manager but not necessarily certified. An enhanced project management methodology is needed with emphasis on reinforcing vulnerable areas of project management.

Quality of Resources Needed
Strong
Medium
Low
High
Medium
Low
Projects' Benefits
Project's Priority:
High
Medium
Low

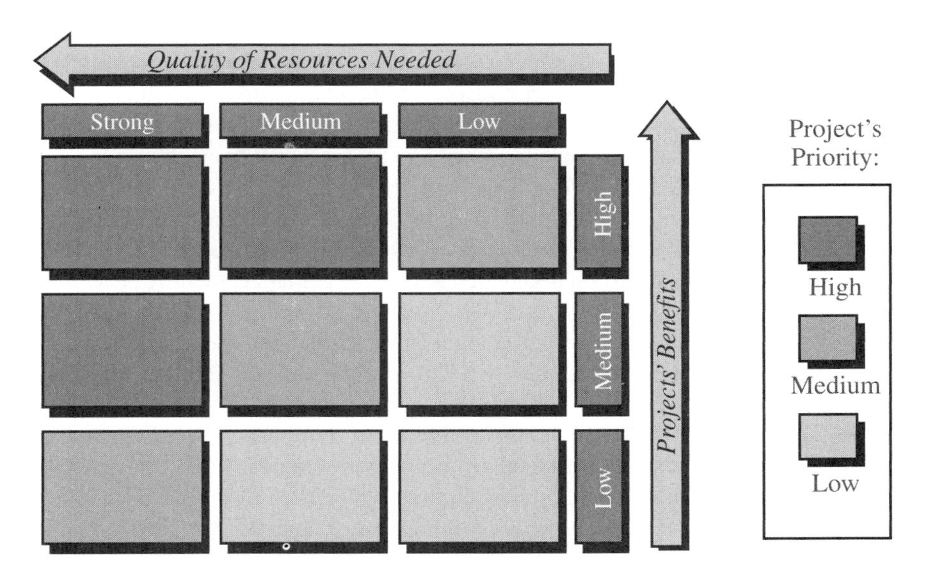

FIGURE 5–14. Strategic importance of projects.

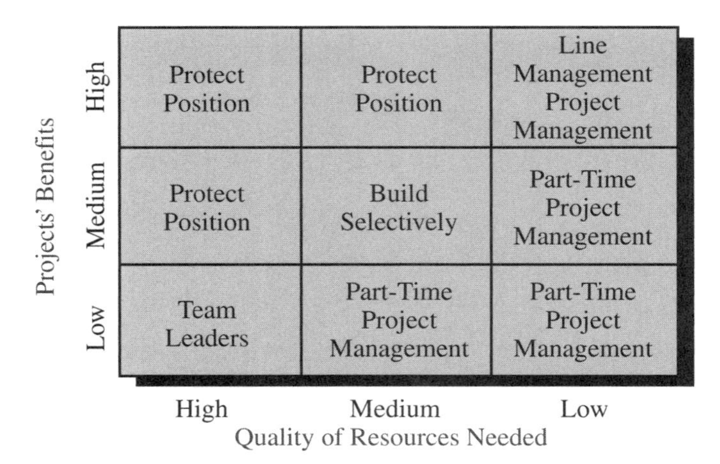

FIGURE 5–15. Strategic guide to allocating project resources.

- *Protect Position (Medium Benefits and High Quality of Resources):* Emphasis is on training project managers with emphasis on leadership skills. The types of projects here usually emphasize customer value-added efforts rather than new product development.
- *Line Management Project Management (High Benefits and Low Quality of Resources):* These projects are usually process improvement efforts to support repetitive production. Minimum integration across functional lines is necessary, which allows line managers to function as project managers. These projects are characterized by short time frames.
- *Build Selectively (Medium Benefits and Medium Quality of Resources):* These projects are specialized, perhaps repetitive, and focus on a specific area of the business. Limited project management strengths are needed. Risk management may be needed, especially technical risk management.
- *Team Leaders (Low Benefits But High Quality of Resources):* These are normally small, short-term R&D projects that require strong technical skills. Since minimal integration is required, scientists and technical experts will function as team leaders. Minimal knowledge of project management is needed.
- *Part-Time Project Management (Medium Benefits and Low Quality Resources):* These are small capital projects that require only an introductory knowledge of project management. One project manager could end up managing multiple small projects.
- *Part-Time Project Management (Low Benefits and Medium Quality of Resources):* These are internal projects or very small capital projects. These projects have small budgets and perhaps a low to moderate risk.
- *Part-Time Project Management (Low Benefits and Low Quality of Resources):* These projects are usually planned by line managers but executed by project coordinators or project expediters.

5.7 HORIZONTAL ACCOUNTING

In the early days of project management, project management was synonymous with scheduling. Project planning was simply laying out a schedule with very little regard for costs. After all, we know that costs will change (i.e., most likely increase) over the life of the project and that the final cost will never resemble the original budget. Therefore, why worry about cost control?

Recessions and poor economic times have put pressure on the average company for better cost control. Historically, costs were measured on a vertical basis only. This created a problem in that project managers had no knowledge of how many hours were actually being expended in the functional areas to perform the assigned project activities. Standards were very rarely updated and, if they were, it was usually without the project manager's knowledge.

Today, methodologies for project management mandate horizontal accounting using earned value measurement techniques. This is extremely important, especially if the project manager has the responsibility for profit and loss. Projects are now controlled through a series of charge numbers or cost account codes assigned to all of the work packages in the work breakdown structure.

Strategic planning for cost control on projects is a three-phase effort, as shown in Figures 5–16 through 5–18. The three phases are:

- *Phase I—Budget-Base Planning (Figure 5–16):* This is the development of a project's baseline budget and cash flow based upon reasonably accurate historical data. The historical databases are updated at the end of each project.
- *Phase II—Cost/Performance Determination (Figure 5–17):* This is where the costs are determined for each work package and where the actual costs are compared against the actual performance in order to determine the true project status.

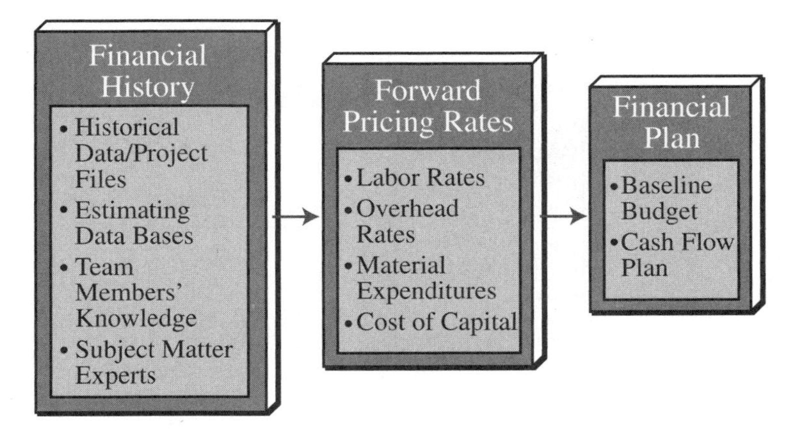

FIGURE 5–16. Evolution of integrated cost–schedule management. Phase I: budget-based planning.

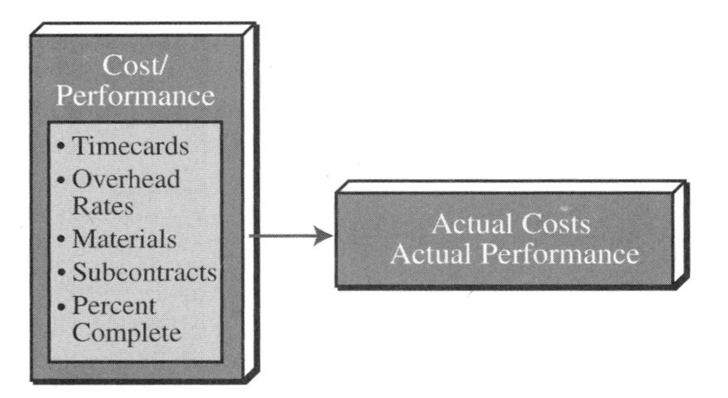

FIGURE 5–17. Evolution of integrated cost–schedule management. Phase II: cost/performance determination.

- *Phase III—Updating and Reporting (Figure 5–18):* This is the preparation of the necessary reports for the project team members, line managers, sponsors, and customer. At a minimum, these reports should address the questions of:
  - Where are we today (time and cost)?
  - Where will we end up (time and cost)?
  - What problems do we have now and will we have in the future, and what mitigation strategies have we come up with?

Good methodologies provide the framework for gathering the information to answer these questions.

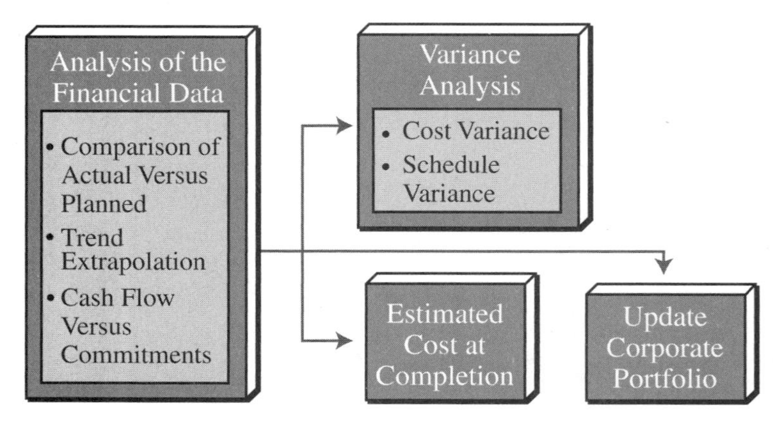

FIGURE 5–18. Evolution of integrated cost–schedule management. Phase III: updating and reporting.

5.8 CONTINUOUS IMPROVEMENT

Project management methodologies must undergo continuous improvement. This may be strategically important to stay ahead of the competition. Continuous improvements to a methodology can be internally driven by factors such as better software availability, a more cooperative corporate culture, or simply training and education in the use of the methodology. Externally driven factors include relationships with customers and suppliers, legal factors, social factors, technological factors, and even political factors.

Five areas for continuous improvement to the project management methodology are shown in Figure 5–19.

Existing Process Improvements

- *Frequency of Use:* Has prolonged use of the methodology made it apparent that changes can be made?
- *Access to Customers:* Can we improve the methodology to get closer to our customers?
- *Substitute Products:* Are there new products (i.e., software) in the marketplace that can replace and improve part of our methodology?
- *Better Working Conditions:* Can changes in the working conditions cause us to eliminate parts of the methodology (i.e., paperwork requirements?)
- *Better Use of Software:* Will new or better use of the software allow us to eliminate some of our documentation and reports?

Integrated Processes

- *Speed of Integration:* Are there ways to change the methodology to increase the speed of integrating activities?

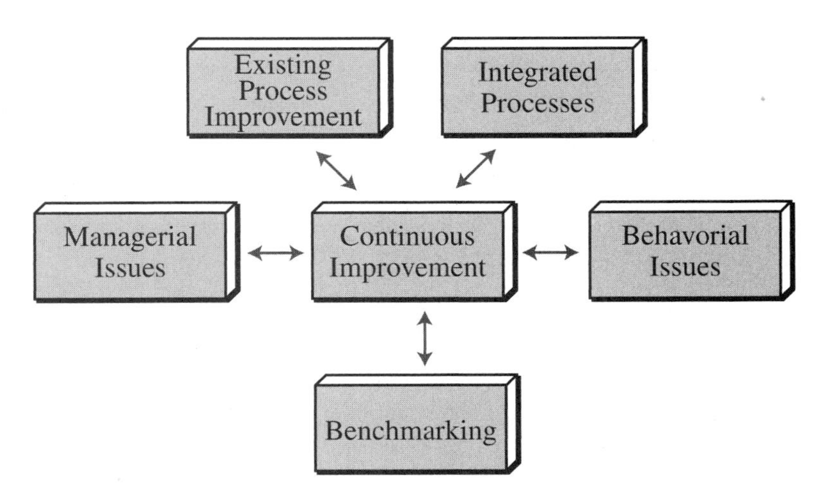

FIGURE 5–19. Factors to consider for continuous improvement.

- *Training Requirements:* Have changes in our training requirements mandated changes in our methodology?
- *Corporate-wide Acceptance:* Should the methodology change in order to obtain corporate-wide acceptance?

Behavioral Issues

- *Changes in Organizational Behavior:* Have changes in behavior mandated methodology changes?
- *Cultural Changes:* Has our culture changed (i.e., to a cooperative culture) such that the methodology can be enhanced?
- *Management Support:* Has management support improved to a point where fewer gate reviews are required?
- *Impact on Informal Project Management:* Is there enough of a cooperative culture such that informal project management can be used to execute the methodology?
- *Shifts in Power and Authority:* Do authority and power changes mandate a more rigid or more loose methodology?
- *Safety Considerations:* Have safety or environmental changes occurred that will impact the methodology?
- *Overtime Requirements:* Do new overtime requirements mandate an updating of forms, policies, or procedures?

Benchmarking

- *Creation of a Project Management Center for Excellence:* Do we now have a "core" group responsible for benchmarking?
- *Cultural Benchmarking:* Do other organizations have better cultures than we do in project management execution?
- *Process Benchmarking:* What new processes are other companies integrating into their methodology?

Managerial Issues

- *Customer Communications:* Have there been changes in the way we communicate with our customers?
- *Resource Capability Versus Needs:* If our needs have changed, what has happened to the capabilities of our resources?
- *Restructuring Requirements:* Has restructuring caused us to change our sign-off requirements?
- *Growing Pains:* Does the methodology have to be updated to include our present growth in business (i.e., tighter or looser controls)?

The five factors considered above provide a company with a good framework for continuous improvement. The benefits of continuous improvement include:

- Better competitive positioning
- Corporate unity
- Improved cost analysis

- Customer value added
- Better management of customer expectations
- Ease of implementation

Bill Marshall, formerly director of Project Management Standards for Nortel Networks, discusses how continuous improvement supports the supply chain at Nortel:

> The Nortel Networks project manager in project-driven areas of the company oversees turnkey projects as part of the supply chain. Benchmarking and ongoing monitoring are part of the success metrics required for optimization of project processes. Customer-defined measures, such as project order responsiveness and so on, are benchmarked against industry standards (best and average). The introduction of enterprise-wide project process and project tools standards in Nortel Networks and the use of success metrics to assess the capabilities to meet the customer expectations in a competitive environment provide the optimal setting for "best practices." The success metric as an ongoing benchmark monitor will assist in identifying the high performing business unit projects. Implementation of the standard project process will enable the implementation of this business units "best practice" into the standard project process for enterprise-wide improvements.

How does a company know whether or not improvements have been made or whether or not the employees believe that the system has improved? Colm Murray, vice president for Central and Latin America Project Management at Nortel Networks, describes the process he uses to measure improvement:

> In Nortel Networks, project management is our mandated method of delivering integrated solutions to our customers, managing R&D projects to the customer's satisfaction, and controlling the implementation of internal IT projects. Project management is leading the corporation in the standardization of process and tool sets to enable the corporation to step up to the increasing demands of our customers and shareholders in delivering successful and profitable projects. To ensure that Nortel Networks is on the right track, a corporate-wide survey has been undertaken. We used the survey from Dr. Harold Kerzner's book *In Search of Excellence in Project Management.* The survey was initially taken in the first quarter of 1998 and will now be an annual event. A focused set of plans to improve the metrics revealed has given objectives that will drive the standards into all areas of the corporation, encouraging benefits for all from the "lessons learned."

If a company wishes to be successful at continuous improvement for project management, it must rely upon more than one approach to achieve its objectives. According to Susan Spradley, vice president, Engineering Operations Wireless Solutions for Nortel Networks:

> Nortel Networks has used project management since early 1980 when new market entrants such as US West entered the competitive wireless telecommunications market. In the 1980s, the project manager was an expediter and a "plumber" who fixed leaks in the process. However, the mid-1990s brought a revitalization of project management within Nortel Networks. The maturing of project industry businesses and the dependence of customers for a complete solution has demanded that vendors be prepared to step up to higher

demands on the project manager. Today, Nortel Networks is focused on corporate-wide implementation of standards and sharing of "lessons learned." There are three current initiatives to improve project management: (1) creating a project culture through training, project skills improvements, standard position profiles and adherence to the PMBOK® methods of implementing projects; (2) standardizing a process definition for projects with information connectors to the rest of the standard business processes; and (3) introducing enterprise-wide standard set of tools, including project management.

Benchmarking and continuous planning are outstanding mechanisms for strategic planning for project management improvement. Motorola is one of those companies that perform benchmarking well. Martin O'Sullivan, formerly vice president and director for Business Process Management at Motorola's System Solution Group, describes what motivated Motorola to perform benchmarking and what was learned:

In 1996, we began looking at our business from the viewpoint of its core processes. In Motorola's continuing effort not only to improve our internal effectiveness and timeliness, but also to more thoroughly satisfy customers, it was clear that the way we in the System Solution Group managed product development and business delivered to customers in project form was critical. As you might expect, project management made the short list as one of the vital, core processes to which quality principles needed to be applied.

Motorola had been successful in applying quality principles to product design, manufacturing, and other processes. One of the lessons learned from these quality initiatives was the importance of benchmarking organizations who are viewed as having effective processes, both within and outside the company. While project management is as much art as process, it nevertheless seemed important to apply benchmarking to this vital area of our business.

We began the effort just prior to the publication of the PMI's Project Management Body of Knowledge. Seeing the process-oriented view presented by the PMBOK® acted as a reinforcement to the approach we had embarked upon.

Gregory M. Willits, project manager process champion of Motorola, discusses the companies being benchmarked and preliminary findings:

We benchmarked our organization against industry leaders in 1996. These included IBM, HP, TI, Disney, McDonnell Douglas, Lucent, Bechtel, and others. As a corporation, we saw a lot of "low hanging fruit." The benchmarking work was a major driver in making the case for formal project management process across the corporation. One of the primary findings was that the "best of the best" companies had a significant presence of process leadership at the top echelons of their corporation (Centers of Excellence). Some of the other best practices noted were:

- Customer satisfaction surveys by project
  - As milestones are accomplished
- Retrospective in-depth briefing in private for team
  - Lessons learned
- PMs supported in large programs by project control managers
- Value management concept for products

- Spread PM discipline to non-product disciplines
- Detailed tracking on progress of design with task level visibility
- Two week look ahead schedule
- Domain specific training for PM
- Working across organizations course
- PM's tool kit and survival kit

Martin O'Sullivan then describes his views of what was learned:

We learned that project management in the 1995–1996 time frame was experiencing what might be called a renaissance of interest, professionalism, and support within many companies. The increasing level of resources being devoted to improvement and professionalism among even commercial, non–defense-oriented companies was the most significant message. The resources being applied spoke volumes about the renewed focus of senior executives on project management.

Benchmarking resulted in our group assigning a process champion and process action team to project management. The team is devoted to examining such things as process definition, metrics, training, career paths, communication, rewards, and recognition. Specific best practices within Motorola and other companies observed during the benchmarking are one foundation for this initiative.

One early result was refreshing the educational offerings available within the company. Working with Motorola University, a more comprehensive curriculum offering certificates for accomplishment of professional education was installed.

We came away with increased resolve that giving recognition to core processes like project management is important to customer satisfaction. Every "moment of truth," to use the phrase of SAS Airlines in describing customers' experiences with a company's products and services, is the result of a process, whether consciously or unconsciously. The project management process is key to providing timely solutions to customer's problems. The project management process depends on many functions and many of our associates to be successful, not only the designated project managers. If the customer's experience is to be exceptional, the quality of project management provided by all of these participants must be both conscious and exceptional.

Steve Gregerson, vice president for product development at Metzeler Automotive Profile System, had done extensive benchmarking of best practices in project management. According to Steve Gregerson:

Program management benchmarking is a passion of mine, and as such could take up an entire book in itself. In all the companies we have reviewed, I have boiled down the best practices into what I call the "Ten Commandments of Program Management." These are ten basic elements of an excellent project management process. They are:

1. Thou shalt have senior management participation.
2. Thou shalt come together as teams.
3. Thou shalt train thy people.
4. Thou shalt know thy task.
5. Thou shalt communicate.
6. Thou shalt have value-added events.

7. Thou shalt track thy progress.
8. Thou shalt honor thy design freeze.
9. Thou shalt launch internally first.
10. Thou shalt continually improve.

Every program initially complies with these Ten Commandments, by management edict, but eventually because the team members realize the benefits of compliance. Each has a critical aspect that can be measured, therefore evaluated, as to whether a team is in compliance.

5.9 WHY DOES STRATEGIC PLANNING FOR PROJECT MANAGEMENT FAIL?

We have developed a strong case in earlier sections of this chapter for the benefits of strategic planning for project management. Knowledge about this process is growing and new information is being disseminated rapidly. Why, then, does this process often fail? Following are some of the problems that occur during the strategic planning process. Each of these pitfalls must be examined carefully if the process is to be effective.

- *CEO Endorsement:* Any type of strategic planning process must originate with senior management. They must start the process and signal their own aspirations. A failure to endorse may signal line management that the process is unreal.
- *Failure to Reexamine:* Strategic planning for project management is not a one-shot process. It is a dynamic, continuous process of reexamination, feedback, and updating
- *Being Blinded by Success:* Simply because a few projects are completed successfully does not mean that the methodology is correct, nor does it imply that improvements are not possible. Simply put, believing that "you can do no wrong" usually leads to failure.
- *Overresponsiveness to Information:* Too many changes in too short a time frame may leave employees with the impression that the methodology is flawed or that its use may not be worth the effort. The argument here is whether changes should be made continuously or at structured time frames.
- *Failure to Educate:* People cannot implement successfully and repetitively a methodology they do not understand. Training and education on the use of the methodology is essential.
- *Failure of Organizational Acceptance:* Company-wide acceptance of the methodology is essential. This may take time to achieve in large organizations. Strong, visible executive support may be essential for rapid acceptance.
- *Failure to Keep the Methodology Simple:* Simple methodologies based upon guidelines are ideal. Unfortunately, as more and more improvements are made, there is a tendency to go from informality using guidelines to formality using policies and procedures.

- *Blaming Failures on the Methodology:* Project failures are not always the result of poor implementation. Unrealistic objectives or poor executive expectations can lead to poor implementation. Good methodologies do not guarantee success, but they do imply that the project will be managed correctly.
- *Failure to Prioritize:* There can exist serious differences in the priorities assigned to strategic project objectives by different functional areas such as marketing and manufacturing. Figure 5–20 shows three projects and how they are viewed differently by marketing and manufacturing. A common, across-company prioritization system may be necessary.
- *Rapid Acquisitions:* Sometimes an organization will purchase another company as part of their long-term strategy for vertical integration. Backwards integration occurs when you purchase suppliers of components or raw materials to reduce your dependency upon them. Forward integration occurs when you purchase your

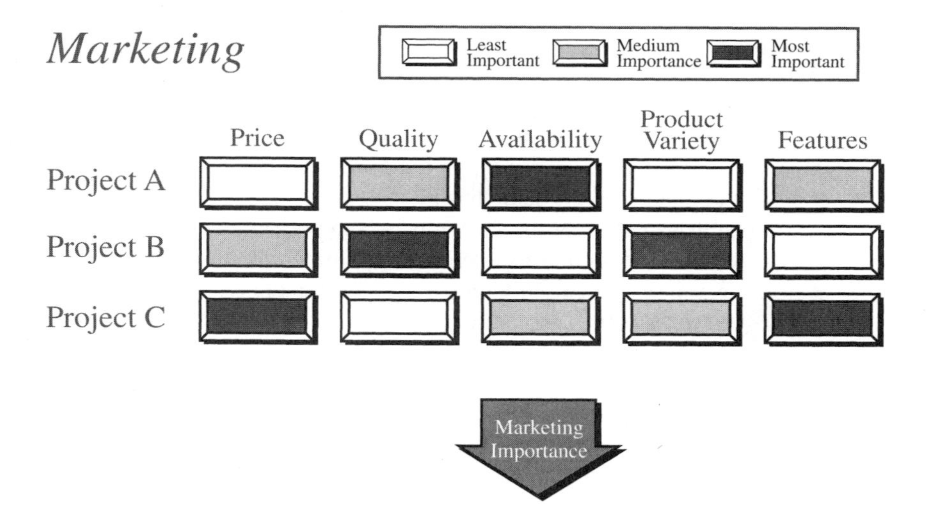

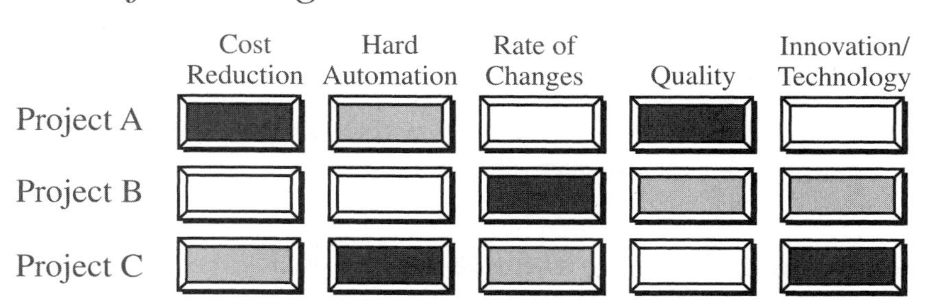

FIGURE 5–20. Differences in strategic importance as perceived by marketing.

forward channels of distribution for your products. In both cases, your projects now require more work, and this must be accounted for in the methodology. Changes may occur quickly.

5.10 STRATEGIC PLANNING IN ACTION

Recognizing that excellence in project management is an ongoing process of continuous improvement leads us to the natural conclusion that there must exist strategic planning to achieve and maintain excellence in project management. Previously, we discussed Kombs Engineering and Williams Machine Tool Company. These two companies failed because strategic planning for excellence in project management did not exist. However, there are companies that have performed strategic planning well.

Cooper-Standard Automotive

One of the best managed auto subcontractors in Detroit was Standard Products Company, formerly based in Dearborn, Michigan. Acquired by Cooper Tire & Rubber Company in 1999, the company is now known as Cooper-Standard Automotive, Inc. The company headquarters is located in Novi, Michigan.

Several years ago, a major auto producer benchmarked 19 subcontractors worldwide that competed in the same business as Standard Products. At that time, Standard Products was rated the best of the 19 companies in project management. Their project management systems were second to none. Cooper-Standard Automotive continues to be a leader in project management with their global customers.

A few years ago, Standard Products approached me to assist them in performing strategic planning to maintain their position of excellence in project management. Their need was simple. For years, the auto subcontractors have been more mature in project management than the "big three" in Detroit. Now Ford, DaimlerChrysler, and General Motors were beginning to catch up. The questions posed to me by Standard Products were: What will project management be like in the next century? What decisions must we make *today* so that we will maintain our position of excellence into the next century? How do we stay ahead of our customer in project management knowledge and applications?

Standard Products was 100 percent correct in what they wanted to do. It is always dangerous when the customer's project management knowledge exceeds that of the contractor. If this exists, three results are possible, all bad for the contractor!

- The customer will want to manage the projects themselves.
- The customer will dictate to you how to manage the projects.
- The customer will seek out other more mature contractors.

National City Bank

During in-house project management training programs, it is usually a good idea to have senior management make a brief presentation to the participants to show senior management's support for project management. Previously, we showed how effective this was at Roadway Express. During a project management training program for National City Bank in Cleveland, Ohio, the executive vice president for National City Corporation, Jon Gorney, told the class that the excellence of National City's project management systems had allowed them to acquire other banks and integrate the acquired banks into National City's culture in less time than other banks allow for mergers and acquisitions. Furthermore, Mr. Gorney said that with upcoming meetings with Wall Street analysts, he will impress upon them the importance of the project management systems and how effectively they will be used in future acquisitions and mergers. Obviously, National City views project management as an asset that has a very positive effect on the corporate bottom line. Jon Gorney's comments are absolutely correct and should serve as targets for all other banks.

General Motors Powertrain In 1995, Rose Russett and her team of program administrators at General Motors Powertrain prepared a plan for strategic improvements to be made to their project management methodology. The following reflect the status at that time:

- Timing plans
- Different procedures per program teams
- Multiple scheduling tools
- Rework due to lack of templates
- Inconsistent reporting content and format

The goal for the future was stated that all product programs would utilize the following:

- Integrated logic networks
- Standard procedures
- Standard scheduling tools
- Standard templates
- Standard reports

In 1998, significant progress was made toward achieving those goals, but since excellence in project management is an evolution, goals are always being added or refined. The current effort is to improve communications and implementation of common processes through the use of electronic program notebooks and Web-based tools. Ready access to program templates and information will allow for a more efficient and effective program management process.

Key Bank The same strategic planning process used by General Motors Powertrain Group was also used by Key Bank. In 1995, Key Bank identified the strategic benefits that they expected with the development of a standard methodology for project management. The targets were set as follows:

- Simplification of existing methodologies
- Management of Key Bank client relationships

- Improved communication with management
- Shortening of the learning curve
- Long-term expense reduction
- Overall paperwork reduction
- Compatibility with current tools

During a 1997 interview, Philip Carter, a vice president for Key Bank, made the following remarks:

> The company has already come to realize some of the benefits, such as increased quality, increased line management support, the ability to deliver the product on time, the ability to have total control of the process change, and enhanced communications of the process change.

Radian International During the 1990s, Radian International performed strategic planning for project management to get closer to their customer and to add value to their customers' projects. According to William E. Corbett, formerly senior vice president, the need for maturity in project management was apparent:

- Decision to emphasize client-focused, value-added services rather than technical skills maintenance and management.
- Decision to flatten and de-layer our line management organization to lower our costs, which necessarily shifted a lot of the burden and responsibility for managing our business to our project managers.

Radian found tangible benefits as a result of their strategic planning efforts. According to William E. Corbett:

- Increased commitment of our project managers to the success of our business
- Happier clients/more repeat business
- Increased staff involvement in and focus upon serving our clients rather than focusing on internal issues
- Reduced overruns/"write-offs"

Ericsson Telecom A-B Ericsson has been performing strategic planning for project management for years in order to maintain consistency in the way it manages its projects across 130 countries worldwide. Robert Shepherd, PMP, Change Director, Market Unit North America of Ericsson, believes that for Ericsson to be well positioned for project management throughout the twenty-first century, there will be:

> . . . a much stronger emphasis on leadership than today. The project manager will work in a cross-functional business organization. The emphasis will be to provide the customer with a *solution,* not a product. There will also be a stronger tie to sales where the project manager supports the sales phase to ensure successful proposal generation and project im-

> plementation to meet the customer's contractual requirements, which leads to customer satisfaction.

Robert Shepherd's comments show that Ericsson is emphasizing the project manager's role as a multinational integrator of work, and that the project manager will be working more closely with the salesforce to make sure that the promises sales makes to the customer can be achieved by project management. Also, Ericsson will be emphasizing providing a solution for their customer, not merely a product.

Johnson Controls Automotive

In the past 20 years, the achievements made by Johnson Controls due to their strategic planning efforts are truly remarkable. (This is discussed in Chapter 9.) According to Richard J. Crayne, Johnson Controls has realized a number of benefits.

> We have reduced our product development time and improved our ability to maintain our "best methods" for many customers' timing requirements. We have increased our customers' confidence in our ability to deliver. We are using these systems as part of our "solution provider" approach. We have used this to help win new business. By controlling our processes, we are in a much better position to do proper value analysis and value engineering (VA/VE) analysis at the right times. Future integration of our new product development process is expected to yield large project and product cost reductions.

5.11 ORGANIZATIONAL RESTRUCTURING

Effective project management cultures are based on trust, communication, cooperation, and teamwork. When the basis of project management is strong, organizational structure becomes almost irrelevant. (Culture is discussed in detail in Chapter 10.) Restructuring an organization only to add project management is unnecessary and perhaps even dangerous. Companies may need to be restructured for other reasons, such as making the customer more important. But successful project management can live within any structure, no matter how awful the structure looks on paper, just as long as the culture of the company promotes teamwork, cooperation, trust, and effective communication.

The organizations of companies excellent in project management can take almost any form. Today, small to medium-sized companies sometimes restructure to pool management resources. Large companies tend to focus on the strategic business unit as the foundation of their structures. Many companies still follow matrix management. Any can work with project management as long as they have the following traits:

- They are organized around nondedicated project teams.
- They have a flat organizational hierarchy.
- They practice informal project management.
- They do not consider the reporting level of project managers important.

The first point listed above may be somewhat controversial. Dedicated project teams have been a fact of life since the late 1980s. Although there have been many positive results from dedicated teams, there has also been a tremendous waste of manpower, coupled with duplication of equipment, facilities, and technologies. Today, most experienced organizations believe that they are scheduling resources effectively so that multiple projects can make use of scarce resources at the same time. And, they believe, nondedicated project teams can be just as creative as dedicated teams, and perhaps at a lower cost.

Although tall organizational structures with multiple layers of management were the rule when project management came on the scene in the early 1960s, today's organizations tend to be lean and mean, with fewer layers of management than ever. The span of control has been widened, and the results of that change have been mass confusion in some companies but complete success in others. The simple fact is that flat organizations work better. They are characterized by better internal communication, cooperation among employees and managers, and atmospheres of trust.

In addition, today's project management organizations, with only a few exceptions (purely project-driven companies), prefer to use informal project management. With formal project management systems, the authority and power of project managers must be documented in writing, formal project management policies and procedures are required, and documentation is required on even the simplest tasks. By contrast, in informal systems, paperwork is minimized. In the future I believe that even totally project-driven organizations will develop more informal systems.

The reporting level for project managers has fluctuated between top-level and lower-level managers. As a result, some line managers have felt alienated over authority and power disagreements with project managers. In the most successful organizations, the reporting level has stabilized, and project managers and line managers today report at about the same level. Project management simply works better when the managers involved view each other as peers. In large projects, however, project managers may report higher up, sometimes to the executive level. For such projects, a project office is usually set up for project team members at the same level as the line managers with whom they interact daily.

To sum it all up, effective cross-functional communication, cooperation, and trust are bound to generate organizational stability. Let's hope that organizational restructuring on the scale we've seen in recent years will no longer be necessary.

5.12 CAREER PLANNING

In organizations that successfully manage their projects, project managers are considered professionals and have distinct job descriptions. Employees usually are allowed to climb one of two career ladders: the management ladder or the technical ladder. (They cannot, however, jump back and forth between the two.) This presents a problem to project managers, whose responsibilities bridge the two ladders. To solve this problem, some organizations have created a third ladder, one that fills the gap between technology and management. It is a project management ladder with the same opportunities for advancement as the other two.

5.13 THE PROJECT MANAGEMENT MATURITY MODEL

All companies desire to achieve maturity and excellence in project management. Unfortunately, not all companies recognize that the time frame can be shortened by performing strategic planning for project management. The simple use of project management, even for an extended period of time, does *not* lead to excellence. Instead, it can result in repetitive mistakes and, what's worse, learning from your own mistakes rather than from the mistakes of others.

Companies such as Motorola, Nortel, Ericsson, and Compaq perform strategic planning for project management, and the results are self-explanatory. What Nortel and Ericsson have accomplished from 1992 to 1998, other companies have not achieved in twenty years of using project management.

Strategic planning for project management is unlike other forms of strategic planning in that it is most often performed at the middle-management level, rather than by executive management. Executive management is still involved, mostly in a supporting role, and provides funding together with employee release time for the effort. Executive involvement will be necessary to make sure that whatever is recommended by middle management will not result in unwanted changes to the corporate culture.

Organizations tend to perform strategic planning for new products and services by laying out a well-thought-out plan and then executing the plan with the precision of a surgeon. Unfortunately, strategic planning for project management, if performed at all, is done on a trial-by-fire basis. However, there are models that can be used to assist corporations in performing strategic planning for project management and achieving maturity and excellence in a reasonable period of time.

The foundation for achieving excellence in project management can best be described as the project management maturity model (PMMM), which is comprised of five levels, as shown in Figure 5–21. Each of the five levels represents a different degree of maturity in project management.

- *Level 1—Common Language:* In this level, the organization recognizes the importance of project management and the need for a good understanding of the basic knowledge on project management, along with the accompanying language/terminology.
- *Level 2—Common Processes:* In this level, the organization recognizes that common processes need to be defined and developed such that successes on one project can be repeated on other projects. Also included in this level is the recognition that project management principles can be applied to and support other methodologies employed by the company.
- *Level 3—Singular Methodology:* In this level, the organization recognizes the synergistic effect of combining all corporate methodologies into a singular methodology, the center of which is project management. The synergistic effects also make process control easier with a single methodology than with multiple methodologies.
- *Level 4—Benchmarking:* This level contains the recognition that process improvement is necessary to maintain a competitive advantage. Benchmarking must be

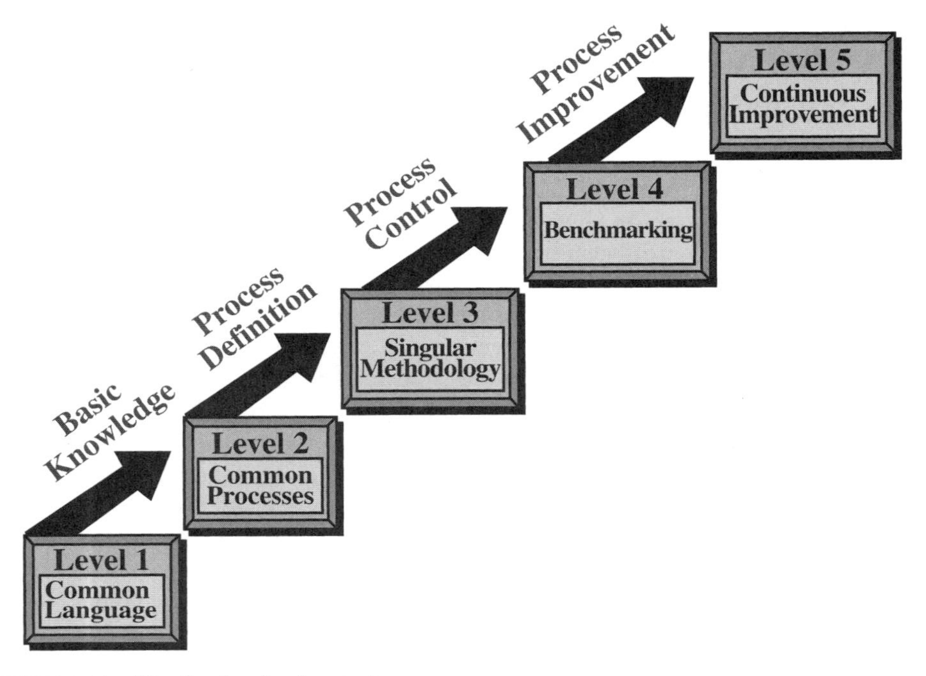

FIGURE 5–21. The five levels of maturity.

performed on a continuous basis. The company must decide whom to benchmark and what to benchmark.

- *Level 5—Continuous Improvement:* In this level, the organization evaluates the information obtained through benchmarking and must then decide whether or not this information will enhance the singular methodology.

When we talk about levels of maturity (and even life cycle phases), there exists a common misbelief that all work must be accomplished sequentially (i.e., in series). This is not necessarily true. Certain levels can and do overlap. The magnitude of the overlap is based upon the amount of risk the organization is willing to tolerate. For example, a company can begin the development of project management checklists to support the methodology while it is still providing project management training for the workforce. A company can create a center for excellence in project management before benchmarking is undertaken.

Although overlapping does occur, the order in which the phases are completed cannot change. For example, even though level 1 and level 2 can overlap, level 1 *must* still be completed before level 2 can be completed. Overlapping of several of the levels can take place, as shown in Figure 5–22.

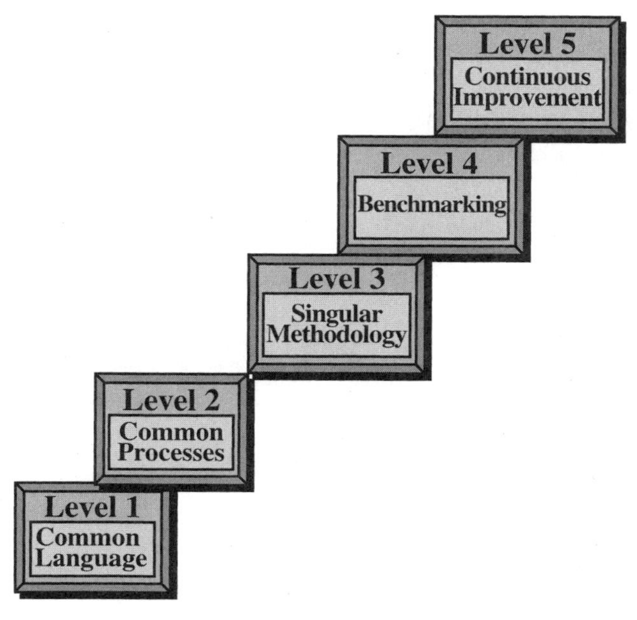

FIGURE 5–22. Overlapping levels.

- *Overlap of Level 1 and Level 2:* This overlap will occur because the organization can begin the development of project management processes either while refinements are being made to the common language or during training.
- *Overlap of Level 3 and Level 4:* This overlap occurs because, while the organization is developing a singular methodology, plans are being made as to the process for improving the methodology.
- *Overlap of Level 4 and Level 5:* As the organization becomes more and more committed to benchmarking and continuous improvement, the speed by which the organization wants changes to be made can cause these two levels to have significant overlap. The feedback from level 5 back to level 4 and level 3, as shown in Figure 5–23, implies that these three levels form a continuous improvement cycle, and it may even be possible for all three of these levels to overlap.

Level 2 and level 3 generally do not overlap. It may be possible to begin some of the level 3 work before level 2 is completed, but this is highly unlikely. Once a company is committed to a singular methodology, work on other methodologies generally terminates. Also, companies can create a center for excellence in project management early in the life cycle process, but will not receive the full benefits until later on.

Risks can be assigned to each level of the PMMM. For simplicity's sake, the risks can be labeled as low, medium, and high. The level of risk is most frequently associated with

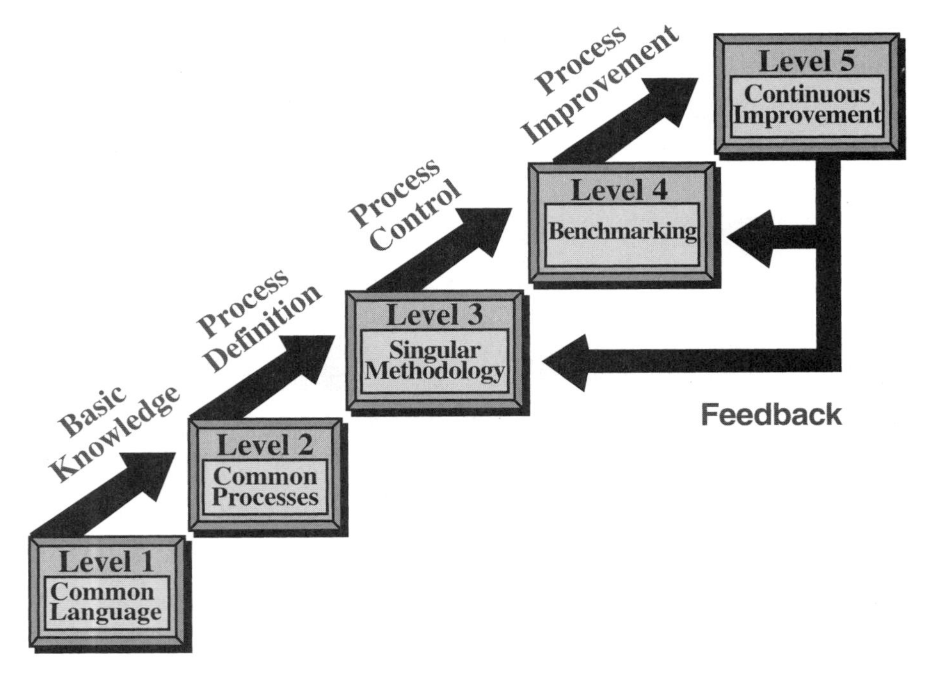

FIGURE 5–23. Feedback between the five levels of maturity.

the impact on the corporate culture. The following definitions can be assigned to these three risks:

- *Low Risk:* Virtually no impact upon the corporate culture, or the corporate culture is dynamic and readily accepts change.
- *Medium Risk:* The organization recognizes that change is necessary but may be unaware of the impact of the change. Multiple-boss reporting would be an example of a medium risk.
- *High Risk:* High risks occur when the organization recognizes that the changes resulting from the implementation of project management will cause a change in the corporate culture. Examples include the creation of project management methodologies, policies, and procedures, as well as decentralization of authority and decision-making.

Level 3 has the highest risk and degree of difficulty for the organization. This is shown in Figure 5–24. Once an organization is committed to level 3, the time and effort needed to achieve the higher levels of maturity have a low degree of difficulty. Achieving level 3, however, may require a major shift in the corporate culture.

Level	Description	Degree of Difficulty
1	Common Language	Medium
2	Common Processes	Medium
3	Singular Methodology	High
4	Benchmarking	Low
5	Continuous Improvement	Low

FIGURE 5–24. Degrees of difficulty of the five levels of maturity.

These types of maturity models will become more common in the future, with generic models being customized for individual companies. These models will assist management in performing strategic planning for excellence in project management.

5.14 HOW TO CONDUCT A PM MATURITY ASSESSMENT[1]

Once you've decided that a PM Maturity Assessment is right for your organization, you'll be faced with the somewhat daunting realization that now you've got to actually plan, organize, and implement the assessment. Where do you start? And how do you turn all that assessment data into a meaningful action plan? In this section you will be provided with some helpful guidelines and checklists.

Find Ways to Bypass the Corporate Immune System

Even though what you're doing is for the good of the organization, you may encounter cultural resistance as you prepare to implement a PM Maturity Assessment. That's because any organization is like a biological organism. It will tend to reject anything that is new and unfamiliar (like the body's

1. This section was prepared by G. Howland Blackiston, Executive Vice President, International Institute for Learning, Inc. For further information on the Project Management Maturity Assessment Instrument, contact Lori Milhaven at 212-515-5121 or through e-mail: lori.milhaven@iil.com.

immune system rejecting a lifesaving transplant). Intellectually, the organization will appreciate the value of the assessment, but the company's culture can be a troublemaker. So be sure to take specific steps to prevent "rejection" and to ensure success.

- Recognize and anticipate that there will be pockets of resistance.
- Acknowledge the fear factor: the apprehension that we may be doing things wrong.
- Identify the specific cultural issues that might cause resistance. Are there personal issues involved (concern about status or job security)?
- Are there legal restrictions regarding asking employees to take an assessment? Some countries have laws on the books that make assessments difficult.
- Squarely address each and every concern. Defuse resistance by acknowledging problems. Be honest and candid.
- Begin an assessment using volunteers who share your enthusiasm. Let their positive experiences convince the others.
- Launch an assessment effort by beginning with a part of the business that is project driven (e.g., IT; R&D, Marketing).
- Start small and scale up. Learn from your early successes and failures before you launch a company-wide assessment effort. Walk before you run.
- Clearly communicate exactly what you are doing and why (see next section).

Explain Why You're Doing This

A well-planned and executed PM Maturity Assessment is time consuming. It will involve careful planning and follow-up, and will draw on considerable resources. So be sure there's clear agreement in the organization as to why you are doing this. Here's where good communication skills come in handy. Prepare a brief and lucid document that can be shared with everyone who will participate in the PM Maturity Assessment. You've got to "sell" the importance of doing this assessment. And you've got to disarm resistance. This will help ensure buy-in and head off any problems that can derail success. As you put together this memo, make sure that you address the following issues:

- Define what is meant by "Project Management Maturity."
- Explain why it's important for our company to measure PM Maturity.
- Convey how this assessment will make the organization more competitive.
- Underscore how competitive companies create growth and job security.
- Disclose who in the company will be invited to participate in the assessment. Why are these people being chosen?
- Describe what is involved and how long it will take.
- Recognize what management will do (and not do) with the assessment information.
- Alleviate any concerns that the information will be used to judge an individual's performance (don't threaten job security).
- Explain how the organization will turn the assessment data into specific improvements.

- Communicate any plans for doing this assessment again (you will want to measure ongoing progress).

Create an Effective Welcome Message *(Here's a sample welcome message from a company using the Kerzner Project Management Maturity Online Assessment Tool. This message appeared on the home page of the Web-based assessment tool. Clear statements like this help alleviate cultural concerns about participating in what might be incorrectly regarded as a threatening "test of knowledge.")*

> Welcome to our online Project Management Maturity Assessment Tool. Project management has been recognized as playing an essential role in our organization. By participating in this assessment you will help create a strategic development plan for identifying the training curriculum that will build and improve our current capabilities. Results of this assessment will be used to set a baseline for all departments and will serve as a tool for identifying future training opportunities. Your support demonstrates your commitment to helping our project management community achieve professional recognition, continuous improvement, and productivity. In turn, this will result in higher consultant/partner/customer satisfaction. We wish to thank you for participating in this groundbreaking event and helping us realize the company's future vision. Thank you for your participation. —Corporate Project Office

Pick the Model That's Best for Your Organization

There's no lack of assessment tools on the market. There are a lot to choose from: simple or complex; generic or industry specific. Basically, they all seek similar objectives: to measure an organization's project management strengths and weaknesses, and to identify improvement opportunities. No one model will be 100 percent perfect for your organization, but some may come close. In all likelihood you'll wind up with a blend or a customization that best fits your organization. As you evaluate assessment models, consider the following:

- Is it compatible with your project management methodology?
- Does it speak your language (use similar terms and definitions)?
- Has the assessment model been validated (has it been tested and used successfully in other industries)?
- Will this tool work well for your industry? In your organization?
- How easy is the tool to use and administer?
- What delivery mechanism would be best (printed surveys, interviews, online)?
- Is the tool aligned with industry standards (e.g., the PMBOK® Guide)?
- Should global organizations determine if the tool is applicable internationally?
- Can the results of the assessment be easily mapped to your organization's business plan?
- Is the tool flexible? Does it allow for special features and customization?
- Can the tool measure professional skills of project personnel?
- What resources will be required to utilize the tool? How many employees will be involved and how long will it take?
- What will it cost to undertake the assessment?

Maturity Models: How Do They Compare?

Are you ready to embark upon an assessment? In the spirit of shameless self-promotion, I urge you to consider the Kerzner Project Management Maturity Model (the subject of Section 5.13). But in the spirit of comparative analysis, other maturity models are currently available.

The Origin of Today's Maturity Models Back in the mid to late 1980s the software industry explored formal ways to better measure and manage the quality and reliability of the processes used for software development. The industry saw value in applying the concepts of total quality management (TQM) and continuous improvement to their development processes. This prompted the Software Engineering Institute's (SEI) development (in 1990) of the Capability Maturity Model (CMM®). The tool provided the industry with a structured and objective means for measuring a software organization's development processes and for comparing these measures against optimum practices. CMM® helped software developers identify specific improvements that would allow them to become more competitive in a highly competitive industry. To utilize CMM® in other industries, the tools have been blended with project management measures and standards (e.g., the PMBOK® Guide) to serve as the foundation for many of the Project Management Maturity Models now on the market.

Create the Right Fit

Some models (like the Kerzner Project Management Maturity Model) have been designed to meet the needs of a broad array of industries and cultures. They are generic. Other models have been developed for specific industries or applications. As you select a model to use in your organization, consider to what degree (if any) the model must be tailored to fit your culture, industry, and business objectives. Some issues to consider:

- Is the model based on a project management standard that fits with what's used in your organization? Or will you have to tailor the tool to comply with your standards?
- Does the model work in your industry? Are the terms and language used familiar to your business?
- Is there a cultural fit?
- Is the model comprehensive enough to measure leadership, professional development, and management involvement?
- Will the model help you develop a corrective action plan to continuously improve project management processes and practices?
- Does the model allow you to add questions and make modifications without compromising the effectiveness of the assessment?
- Can you sort assessment results to take into account different roles and responsibilities; various departments; geographic locations; or job functions?

It's OK to Make Changes When tailoring a maturity model to better fit your organization, making changes to the model is perfectly acceptable. For an example, look at our Kerzner Model. Note that level 3 determines if the organization is using a "singular methodology" (rather than using multiple methodologies). Some organizations may intentionally use several methodologies rather than one — for example, one for IT and another for new product development. By all means tailor the model to fit the realities of your organization.

Choose an Appropriate Delivery Method

As you ponder over how you will "deliver" the assessment instrument to your audience, keep in mind that no one way is correct. There are a number of options available to you. The method you choose may depend on your audience, size of company, time available, budget, flexibility, and even technology. Regardless of the option you choose, it's helpful to clarify the time frame for completing the assessment. Let your audience know when you need it back. Don't give them too much time or they'll put it aside forever. Tell them you need the completed assessment within a couple of weeks. That way you'll have a better chance of getting what you need in a timely way.

Here are some things to consider:

- Decide if you want an informal or a formal approach. If your organization is small and straightforward, an informal assessment may be all you need.
- If you decide to use conventional questionnaires (paperwork), keep in mind the logistics of distributing, collecting, and tabulating all the data.
- Consider conducting interviews to gather the data you need. This involves some tricky scheduling issues, but if the numbers are manageable, this might be a doable option.
- Don't overlook the possibility of utilizing online technology. This is a mighty convenient way to reach a large audience in a short time. In addition, the tabulation of results is automatic and instantaneous. And the online format permits easy editing and modification of the tool itself (see below).
- Pick a model and stick with it. Using different instruments may confound your ability to take meaningful corrective action.
- How about using an outside resource (e.g., consultants)? Their objective approach can add value to the assessment results. Often, staff reports from impartial outside experts are more readily accepted than the same reports from one of your own staff.

Use Online Technology to Your Advantage Creating an interactive Web-based assessment has its advantages: easily editable; autoscoring; efficient distribution; and so on. I've included a few things to consider and some screen shots of the Kerzner Project Management Maturity Online Assessment Tool (Figures 5–25 through 5–28) to illustrate how it works. If you decide to spend the time and resources to develop your own Web-based assessment, make sure that you take the following into account:

- Create an interface that is intuitive and easy to use.
- Make the scoring function automatic.
- Provide for an autosorting of results by critical filters (key departments, divisions, job functions, hierarchy, etc.).
- Build an "Executive Dashboard" feature so that the project office and/or top management can monitor the assessment results continuously.

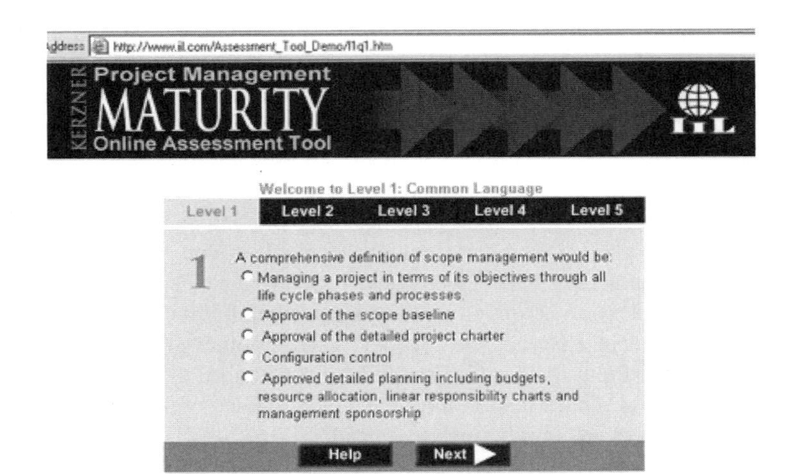

FIGURE 5–25. Participants using the Kerzner Assessment Tool answer a series of multiple-choice questions within each of the model's five levels. There are a total of 183 questions in the Kerzner assessment tool.

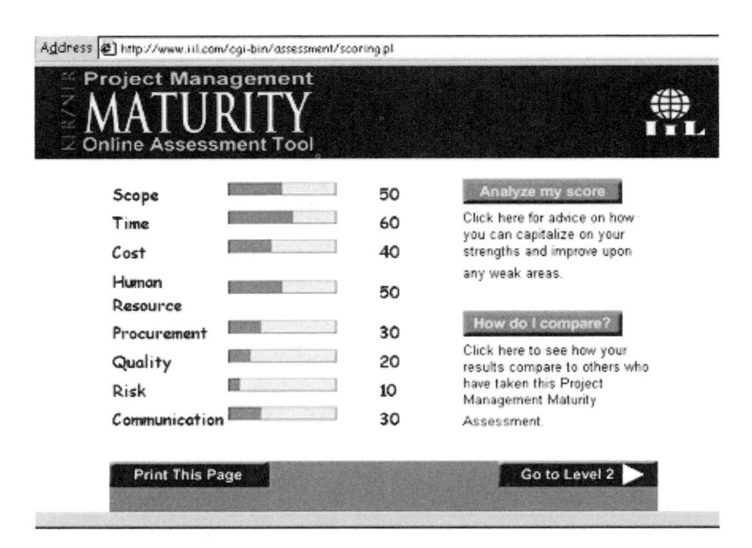

FIGURE 5–26. When completing each level of the Kerzner assessment, participants can see their own score results broken down by the subcategories within that level. This helps them recognize specific strengths and weaknesses.

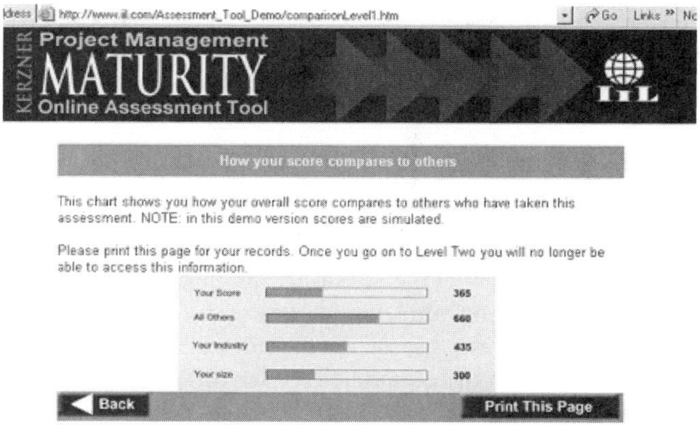

http://www.iil.com/Assessment_Tool_Demo/comparisonLevel1.htm

KERZNER Project Management MATURITY Online Assessment Tool

How your score compares to others

This chart shows you how your overall score compares to others who have taken this assessment. NOTE: in this demo version scores are simulated.

Please print this page for your records. Once you go on to Level Two you will no longer be able to access this information.

Your Score	365
All Others	660
Your Industry	435
Your size	300

Back — Print This Page

FIGURE 5–27. A click of a button compares a person's score to those of all others who have taken the Kerzner assessment—both inside and outside their organization (kind of a sanity check to see how the results compare to those of the rest of the world). Users can also see how their company's overall scores compare to other companies in similar industries. Scores are automatically broken down by whatever filtering criteria have been established by the organization. The tool also allows users to display a prescriptive narrative analysis of their results. This autogenerated report suggests what can be done to advance to higher levels of project management maturity (the suggestions are stimulated by the assessment scores).

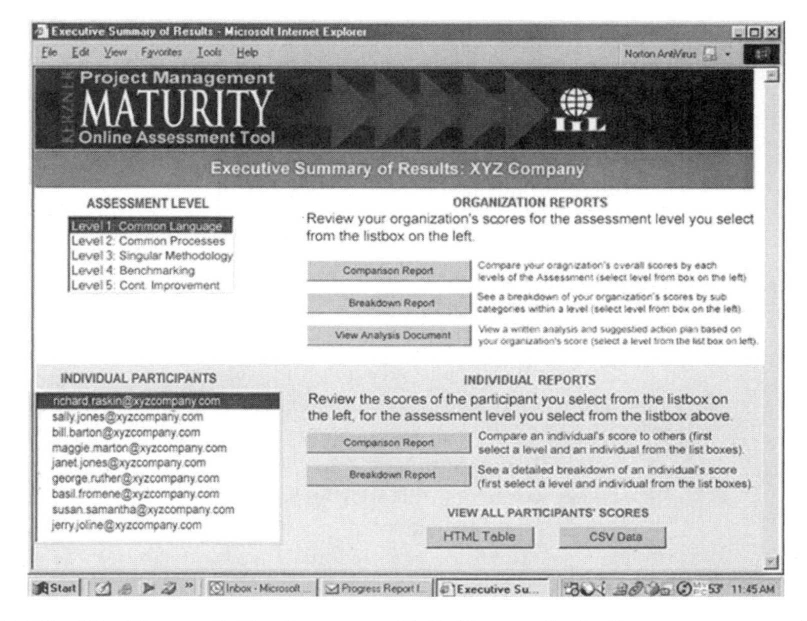

FIGURE 5–28. The Kerzner Online Assessment Tool allows authorized managers to see a live summary of their organization's results. Managers can view and compare both individual and company-wide assessment scores. They can even export the raw data into other applications (such as Excel).

Establish Responsibility In the best of all worlds, the project office (assuming that you have one) should assume overall responsibility for the project management maturity assessment. Chances are that this office reports directly to executive management. Or perhaps its membership is comprised of executive management. In either case, that link to top management will come in handy, because the assessment will identify opportunities for improvement that will rely on decisions and directives that only they can supply:

- Providing overall leadership for the assessment
- Helping with the selection/design of the assessment tool
- Identifying who should participate in the assessment
- Monitoring assessment results and aligning the resulting improvement opportunities with the organization's business plan
- Setting priorities (identifying which few actions will have the most meaningful impact)
- Developing an action plan that will allow you to achieve these improvements
- Supplying the needed resources (time, budget, personnel)
- Approving the remedial training curriculum
- Encouraging a broadening of the assessment throughout the organization
- Evaluating ongoing progress

Decide Who Should Participate Here's where you need to make some strategic decisions. Who should participate in the assessment? Ideally, you should include a broad and diverse representation of the entire company—the broader the representation, the more objective and accurate the assessment. Assuming that you are attempting to make the company more competitive in project management, you should be assessing the entire organization's project management maturity (not merely assessing the maturity of a single department).

Getting everyone involved may be your ultimate goal, but it may be easier on the company culture to begin in a receptive department before scaling up to include the entire organization. In either case, you want to be certain that you have the right people participating in the assessment. Consider the following guidelines as you prepare your invitation list:

- Use the company's business plan as a guide to identify which areas of the organization will critically rely on project management expertise. These departments, locations, or functions are obvious candidates for participating in the assessment.
- Get the right cross-representation involved (the right composition is critical).
- You will underutilize the assessment if you limit participation to experienced project managers (this narrow approach will not accurately reflect the company's actual project management maturity—it's a biased assessment of reality).
- Be sure to include key customers and stakeholders.
- Include broad representation across departments (remember that in world-class companies, all departments, functions, and levels are involved in or knowledgeable about project management efforts).

- Include representation of the entire hierarchy (executive; middle; lower; associates).
- Include a large enough sample to be statistically valid. You should invite at least 30 or more individuals from each operating unit each time you conduct the assessment.
- For large companies, your ultimate objective is to assess at least 10 percent of the population (target larger percentages in small companies).
- Create the ability to sort assessment results by key departments/divisions/job functions/hierarchy (such sorting will enhance your ability to identify project management strengths and weaknesses)
- Decide if participation should be mandatory or optional. For cultural acceptance I suggest that participation is optional at the start, and mandatory as you scale up your assessment efforts and are able to demonstrate value (see the earlier section, "Find Ways to Bypass the Corporate Immune System").

Turn the Results into an Action Plan

With the assessment results in hand, the data should be used to identify meaningful improvement activities. You've got to turn scores into corrective action plans. The prioritization and deployment of these improvements should be spearheaded by the company's project office. If no such office is in place, consider creating a PMM Assessment Action Team. Any significant effort will ultimately require the support of top management.

As you organize this effort, keep the following in mind:

- Consider using an objective outside resource to interpret and analyze the assessment results (they've got no axe to grind, and often recommendations from outsiders are more readily accepted by the culture).
- Be as specific as possible when converting assessment data into specific improvement actions. There's little value in concluding that "parts of the company need to improve their understanding of risk management." Be much more specific. <u>How</u> is the company going to achieve this goal? A better action would be "starting in December, we will schedule intensive two-day workshops in risk management for key personnel in our marketing and legal departments."
- Don't forget to identify and deal with any obstacles in the way of making improvements.
- Treat each improvement objective as an individual improvement project (keep in mind that <u>all</u> improvement takes place <u>project by project</u> and in no other way).
- When prioritizing improvements, start with some home runs (create those bellwether examples that help sell the value of what's being done).
- Focus first on those things that improve the business (prioritize actions based on the organization's business goals and objectives). Use Pareto Analysis to identify the vital few improvements that will have the greatest positive impact on the company.
- Use the assessment tool repeatedly to measure progress. Companies that are serious about improving project management should conduct Project Management Maturity Assessments at least once every quarter.

Develop a Remedial Training Curriculum

The assessment will be helpful in identifying a training curriculum that will help you "close the gaps." The analysis of scores will clearly identify where training is needed, and in what subjects. There is great economy here. It means that you only need to train those individuals, functions, departments, or locations that have been identified as needing training. It also means that you need training only in those specific subjects that have been identified as lacking. This subscribes to the concept of "just in time" training, versus "just in case" training.

The project office (or top management) should establish a task force to plan the organization's remedial training curriculum. Remedial training should be mandated (not voluntary). This group should ensure the following:

- Make certain that the assessment results have been analyzed in such a way that it's clear what must be done to make meaningful improvements. If necessary, hire outside experts to help you with the analysis.
- Establish the criteria to be followed in designing or selecting the curriculum.
- Keep the training curriculum keenly responsive to the subject needs and target audiences identified by the assessment ("just in time" vs. "just in case").
- Decide whether to buy training from outside firms or develop your own.
- Prepare local case materials as supplements to training. This keeps the tools and methodologies relevant to your business. It also helps participants understand how to apply their new knowledge to their jobs.
- Adapt interactive exercises to fit your culture and job situations.

Keep Top Management Informed

The overall purpose of doing a project management maturity assessment is to identify opportunities for making significant improvements in the way you manage projects. In turn, such improvements lead to better project outcomes, lower costs, faster results, higher quality, and greater customer satisfaction. But no significant corrective action is possible without the involvement of top management. They are the only ones who can authorize the significant time and resources needed to turn assessment results into a specific action plan. Keeping top management informed is vital if you are to win their support and leadership.

Part of your assessment plan should include a means for keeping top management informed. I know of one company that has 100 employees participate in a project management maturity assessment every month. And every month top management is given a report summarizing the assessment results. This flow of information helps management prioritize action plans, identify training needs, and measure the company's improvement progress.

Keep the reporting relevant to top management's needs. Share information that is most meaningful to them. This will vary from company to company, but you can be sure that anything expressed in the "language of money" will get their attention and stimulate action. Here are some suggestions:

- Present a detailed breakdown of scores to clearly identify the company's specific strengths and weaknesses.

- Show a comparison of scores between different departments, job functions, geographic locations, or whatever filtering criteria are most meaningful to your business objectives.
- You may wish to provide management with a breakdown of scores by individuals participating in the assessment. This information can be used to identify outstanding talent. And it can even be used to identify individuals who would benefit from remedial training. But be careful! If this information is used to reprimand, criticize, or "clean house," you will effectively crush cultural acceptance of the assessment and all will come to a grinding and hopeless halt.
- Prepare an action plan based on the assessment results. Be specific as to what corrective action is needed and why (see the next section of this chapter).

Virtual Reporting

For those utilizing an online assessment tool, a helpful option is to create an online reporting feature. I refer to this as an "Executive Dashboard." This consists of a unique URL address that allows authorized managers to see a detailed summary of their organization's assessment results. Because it's online, the information is real-time, displaying the latest, up-to-date scores each time it is accessed. Instant gratification! Managers can view and compare individual and company-wide assessment scores whenever they wish. The feature can also allow them to export the raw data into spreadsheet applications (for other reports, sorting options, etc.). Keep the interface simple and intuitive to use.

Benchmark Your Results to Others

It's helpful to compare your results with those of others who have taken the same assessment (note that the online version of the Kerzner PM Maturity Assessment has this feature). Such comparisons should be both within and outside your industry. The benchmarking of results is helpful for the following reasons:

- It provides a "sanity check" within your industry: Is your maturity level close to that of your competitor?
- It gives you a realistic target, proving that achieving higher levels of maturity is possible (after all, others have already reached higher levels).
- It avoids the deadly sin of "resting on your laurels" (if you are complacent about being best within your industry, it's a sobering jolt to see that other industries are much better than you are).
- It sells the need and urgency for improvement (management will be motivated to action if they see that other companies are outperforming your organization).

Do It Again

As helpful as a maturity assessment can be, its usefulness is minimized when it's regarded as a one-time event. That's an underutilization of a powerful tool. Sure you'll be able to identify your strengths, weaknesses, and opportunities for improvement—that will help you develop a corrective action plan. But it's only when you use the assessment on a *repetitive* basis that you can objectively measure the progress of your corrective action plan.

- Are your overall scores improving?
- Is the company achieving higher levels of project management maturity?
- How do you compare with the competition?
- Based on the latest assessment results, do you need to modify your corrective action plan?
- Have new opportunities for improvement emerged since the last assessment?
- Can you improve the assessment tool for more effective use in our organization?
- If one division is outperforming others, are there skills and methods within that exemplary division that can be applied elsewhere in the organization?

Stay nimble and in tune with the marketplace by conducting the assessment on a regular basis. Ongoing use of the tool also allows you to evaluate a larger and larger percentage of your total population. How often should you conduct the assessment? That depends on your organization. Here are some guidelines:

- If your organization is project-driven (projects are critical to your business success), you should perform an assessment every month. Vary the audience each time, striving to capture a broad and diverse representation of the organization.
- Other organizations should conduct the assessment a minimum of once a quarter to ensure that improvements are being made. Again, vary the audience each time, striving to capture a broad and diverse representation of the organization.
- Always keep top management informed.

MULTIPLE CHOICE QUESTIONS

1. Which of the following are benefits of a world-class methodology for project management?
   A. Faster new product development
   B. Lower overall project costs
   C. Improved communications
   D. Overall paperwork reduction
   E. All of the above

2. Which of the following are characteristics of a world-class methodology for project management?
   A. Maximum of eight life cycle phases
   B. Elimination of end-of-phase gate reviews
   C. Internally oriented rather than customer-oriented
   D. Overlapping life cycle phases
   E. All of the above

3. Effective long-term success in project management requires that an organization perform strategic planning for project management.
   A. True
   B. False

4. Most organizations do a much better job performing strategic planning for project management:
A. During favorable economic times
B. During unfavorable economic times
C. Prior to a recession
D. When profitability is high
E. None of the above

5. Project management is most closely aligned to which function of strategic planning?
A. Gathering information
B. Organizational strengths and weaknesses
C. Strategy selection
D. Strategy evaluation
E. Strategy implementation

6. Which organizational structure helped promote the acceptance and use of project management?
A. Classical structure
B. Strategic business units
C. Colocated teams
D. Structural teams
E. None of the above

7. The two basic processes of strategic planning are __________ and __________.
A. Formulation; conversion
B. Identification; execution
C. Formulation; implementation
D. Realization of benefits; implementation
E. Strategic evaluation; strategic execution

8. Strategic planning for project management is the:
A. Development of multiple, functional methodologies
B. Development of a standard methodology for project planning
C. Development of a standard methodology for project execution
D. Development of a standard methodology for project planning and execution
E. Selection of project management software

9. Some people consider that the major advantage of a project management methodology is that it provides a way to achieve:
A. Maximum profitability
B. Perfect cooperation between functional areas
C. Flawless execution
D. 100 percent project success
E. Consistency of action

10. A simplistic project management methodology would include:
A. A technical baseline
B. A management baseline
C. A financial baseline
D. A feedback/control process
E. All of the above

11. The Responsibility Assignment Matrix (RAM), which is also referred to as the Linear Responsibility Chart, is part of the __________ baseline.
A. Technical

B. Management
C. Responsibility
D. Financial
E. Control

12. Good methodologies act as vehicles for overall organizational communications.
A. True
B. False

13. Good methodologies provide a structure for sound, logical project decision-making.
A. True
B. False

14. Which of the following is a continuous, supporting strategy for project management?
A. Performance improvement strategy
B. Integrated process strategy
C. Education and training strategy
D. All of the above
E. A and B only

15. A typical methodology that can be integrated into the project management methodology is:
A. Total quality management
B. Risk management
C. Scope change management
D. Concurrent engineering management
E. All of the above

16. Good project management methodologies satisfy the needs of:
A. Employees
B. Suppliers
C. Customers
D. Creditors
E. All of the above

17. Methodologies satisfy the needs of stakeholders. When a problem exists, a good methodology must never indicate the order as to which stakeholder gets satisfied first.
A. True
B. False

18. In excellent companies, the definition of accountability is:
A. Authority and power
B. Authority and influence
C. Power and influence
D. Responsibility and influence
E. Authority and responsibility

19. What a company "can do" is measured by:
A. Strengths and weaknesses
B. Opportunities and threats
C. Product quality
D. Financial benefits achievable
E. Nonfinancial benefits achievable

20. The implementation of a project management methodology is most often based upon the:
A. Benefits possible

B. Experience of the project champion
C. Experience of the project sponsor
D. Quality of resources assigned
E. All of the above

21. Project management methodologies can include sections on how to "protect" the corporate image or product brand name.
A. True
B. False

22. Potential benefits and quality of resources assigned are used to:
A. Select projects to work on
B. Establish a portfolio of projects
C. Analyze whether or not to bid on a job
D. All of the above
E. A and B only

23. An organization committed exclusively to the improvement of a project management methodology is the:
A. Customer's project office
B. Corporate project office
C. Center for excellence
D. Functional benchmarking groups
E. Committee of project sponsors

24. Which of the following processes normally generates resistance to implementation of a project management methodology?
A. Continuous improvement
B. Benchmarking
C. Risk management
D. Horizontal accounting
E. Scope change control

25. Which of the following often leads to the failure of strategic planning for project management?
A. Lack of CEO endorsement
B. Lack of continuous improvement
C. Being blinded by the success of one project
D. Failure to get organizational acceptance
E. All of the above

DISCUSSION QUESTIONS

1. How does a company begin the process of creating a project management methodology?
2. How can economic conditions influence the acceptance of a project management methodology?
3. What are the differences between strategic planning in general and strategic planning for project management?
4. Why do project management methodologies work better with guidelines and checklists rather than with policies and procedures?

5. From a project management perspective, under what conditions would an organization hire only superior talented people? People with average abilities?
6. Why is continuous improvement in project management necessary?
7. How can a good project management methodology contribute to the profitability of a corporation?
8. How does one perform strategic planning for a portfolio of projects? What types of industries are you considering?
9. Can a methodology for project management become a world class methodology without a horizontal accounting system?
10. Why is total organization acceptance and use of a project management methodology so critical for long-term success?

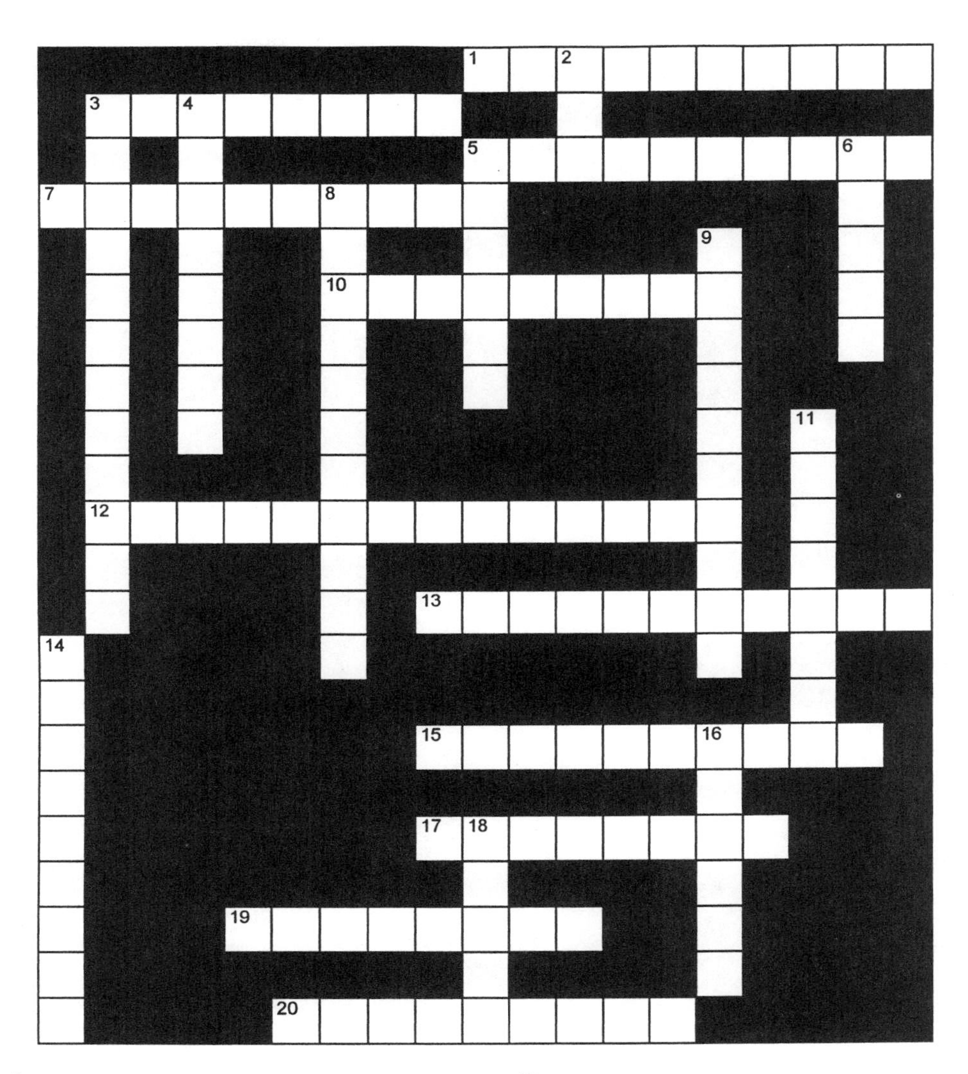

Across

1. Part of COE
3. Part of SBU
5. ___________ improvement
7. Resource type
10. Grouping
12. Part of strategic planning
13. Part of strategic planning
15. Part of SWOT
17. Type of sponsor
19. Resource type
20. Part of SWOT

Down

2. What ___ we do?
3. Continuous improvement
4. Type of methodology
5. Part of COE
6. Part of SBU
8. Continuous ___________
9. Type of accounting
11. Resource type
14. Part of SBU
16. What ______ we do?
18. Resource type

The Maturity of Modern Project Management

6.0 INTRODUCTION

Sooner or later, all organizations develop some degree of *maturity* in project management. The maturity may appear in one functional area, such as information systems, in an entire directorate, or simply in one division of a firm. The maturity may appear in the way the organization plans its efforts, human behavior, or simply project reporting practices.

Rapid maturity is sporadic at best. Some companies require months, some years, and some even decades to achieve a *first* level or maturity. How well the organization perceives the need for project management usually defines the speed by which change will occur. Companies go through changes as the project management process is accepted and begins to flourish. Unfortunately, some of us are so intimately involved in the process that we cannot see the changes that have taken place.

We are certainly managing projects today a lot differently than we did 40 years ago since the birth of project management in the fields of aerospace, defense and construction. Recently, a new phrase has appeared on the scene: modern project management (MPM). Modern project management will be used to differentiate what we do now from what we did over the past 40 years.

6.1 CLASSIFICATION OF CHANGES

For simplicity's sake, the evolution of project management can be broken down into three phases, as shown in Figure 6–1. The traditional project management period was dominated by companies in the aerospace, defense, and construction fields. Project management was used predominantly on large projects employing vast resources and with a definitive profit goal.

Traditional Project Management	Renaissance Period	Modern Project Management
1960 – 1985	1986 – 1993	1994 – 2003

FIGURE 6–1. Evolution.

Totally dedicated teams were the norm. Cost and schedule took a backseat to the development of technology. A project manager almost always was selected from the technical ranks.

Companies in other industries took a wait-and-see position, watching the aerospace and defense industries implement project management. Newspapers reported an onslaught of large projects being completed years behind schedule and 200–300 percent over budget, with the overruns and delays often the result of scope changes. Smaller projects seemed to do much better. Unfortunately, people prefer to read about bad news rather than good news, and the result was a slow acceptance of project management. This slow process would last for three decades.

During the renaissance period of 1985–1993, there appeared a great awakening among corporations, which finally understood that project management was, in fact, applicable to their industry and could improve profitability. Project management was now being applied to even small projects. All functional areas of a business began to recognize the importance of project management. Multidisciplinary teams became important and emphasis was now being placed upon company decisions rather than individual decisions. PC-based project management software gave everyone the opportunity to use the project management tools.

By the mid-1990s, the recession had taken its toll on American business. Corporate America was learning the importance of total quality management and shortening product development time. Market share was being lost. Companies finally realized that perhaps the problem all of these years had been with the way we were managing. Executives finally began taking a serious look at project management as the solution to some of their management problems. Executives historically were spending more and more time involved in the daily operations at the expense of strategic planning. Something must be done! We were sacrificing the future to manage the present. Could we use project management to decentralize corporate decision-making?

Three major factors accounted for the increasing reliance on project management during the mid-1990s. First, the tasks that organizations were facing had become more complex and demanded more sophisticated and flexible organizational approaches. Second, the size and scope of projects required the development of management systems for planning and controlling project performance, schedules, and cost/budgets. Without these systems, administrative chaos would result. The third factor responsible for wide-scale utilization of project management was that the environment within which today's organizations operated had become increasingly turbulent. The accelerated rate of external change and the uncertainty that this change brought required new management approaches that could provide rapid internal response capability. Most traditional organizations were not designed to achieve the adaptability necessary to cope with rapidly changing environments. Project management was now recognized as one approach for coping with instability in the organization's environment.

Between 1993 and 1996, companies began to recognize that both the quantitative and behavioral areas of project management were changing so significantly that it was now a necessity to differentiate between traditional (or past) project management practices and modern project management. This was prompted by organizations that were now developing some degree of maturity in project management and wanted their clients, employees, and stakeholders to recognize these improvements.

Changes were now occurring that would support projects in all sorts of industries, not merely aerospace, construction, and defense. Project management had spread to virtually all areas of business, not just for those units that were project-driven.

6.2 CLASSIFICATION OF COMPANIES

From a project management perspective, companies can be classified as either project-driven or non–project-driven. In project-driven industries such as aerospace, construction, and defense, profitability occurs predominantly from projects. In these industries, modern project management is reasonably mature. But in non–project-driven industries, where profitability is measured through functional product lines, modern project management has been slow in achieving both acceptance and maturity.

In non–project-driven companies, projects generally exist to support product lines or general business practices. Looking at a non–project-driven company under a microscope, one would see some divisions that may be project-driven, such as information technology, but the majority of the business is still non–project-driven. These types of companies are referred to as *hybrids*. In non–project-driven companies, or hybrids, modern project management normally originates in a project-driven division such as information technology, where project management and systems development methodologies can be merged. However, getting modern project management to be accepted throughout the organization is usually a difficult task even though divisions such as manufacturing and engineering may have been using some form of project management for the past five to ten years or even longer.

During the renaissance period of project management, it was actually the hybrids that began accepting the use of project management. In the past, hybrids acted more like non–project-driven companies, with very little support for project management. Once the benefits of project management became apparent, however, hybrids began acting as though they were project-driven. This is shown in Figures 6–2 and 6–3.

What makes project management mature quickly is the profit and loss (P&L) responsibility given to project managers. In project-driven organizations, P&L responsibility is "tangible" and clear-cut. In non–project-driven firms, P&L responsibility is a "fuzzy" area.

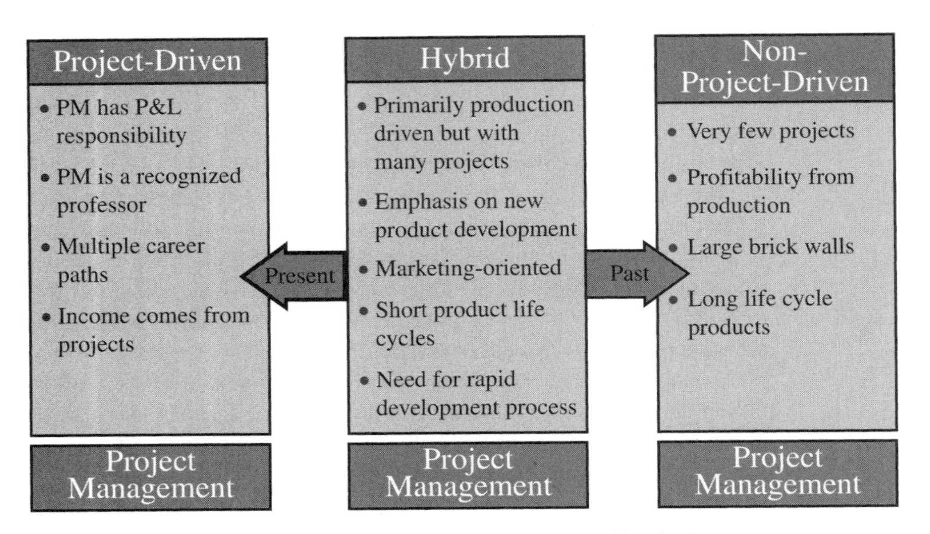

FIGURE 6–2. Industry classification (by project management utilization).

Recently, executives in non–project-driven firms have solved this problem by using the benefit–cost analysis as a replacement for P&L. The project managers are continuously asked to update the benefit–cost analysis so as to obtain continued management support. The project managers are asked to maintain a minimum benefit-to-cost ratio or run the risk of having the project terminated. Thus the benefit-to-cost ratio functions as a P&L statement that allows the hybrids to function under the characteristics of a project-driven organization.

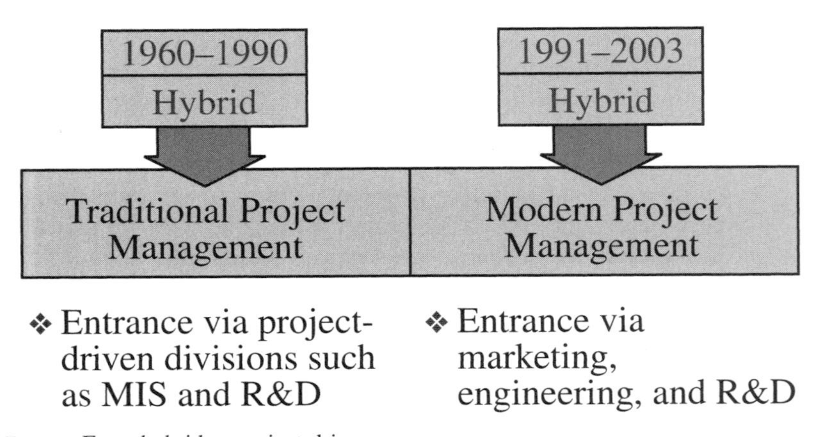

FIGURE 6–3. From hybrid to project-driven.

6.3 RECESSIONARY EFFECTS

Recessions generally create unfavorable thoughts in the minds of people. Scenarios such as downsizing, layoffs, position elimination, salary and benefit reductions, and plant closings create common fears, which eventually may become reality. There is no question that a multitude of employees will suffer during recessionary times. However, recessions can be beneficial if companies are forced to streamline, become more efficient and productive, and become more competitive in the market place. Recessions force companies to change the way they do business to assure survival.

Recessions force management to make risky decisions. Learning from the successes and failures resulting from these risky decisions can provide invaluable information in the event similar situations arise. Unfortunately, very few companies document lessons learned even for the benefit of other divisions within their own company.

The recession of 1989–1993 had different characteristics from those of the previous recession of 1979–1983. During the previous recession, emphasis was placed upon cost-cutting, short-term rather than long-term planning, layoffs predominantly in the blue collar rather than white collar ranks, and the elimination or reduction in both training and R&D dollars. Simply stated, the mind-set of management during the 1979–1983 recession focused on short-term solutions.

From a project management perspective, the benefits of modern project management had been well known for years and had been demonstrated in countless industries. These benefits included:

- Improved efficiency and increased profitability through better utilization of limited resources.
- Enhanced planning, estimating, and cost control, leading to more consistent achievement of milestones, goals, and objectives.

Despite numerous success stories, companies still refused to accept the fact that project management could help them during recessionary times. Project management implementation was falsely viewed as a slow, long-term implementation effort requiring cultural shock. There also existed the thought that the business would be disrupted rather than enhanced. Management persisted in their focus on short-term solutions and refused to recognize and admit to the usefulness of project management, especially outside of the divisions that were project-driven. When the earlier recession ended in 1983, management felt that there was no longer any reason to reconsider project management principles since the company had exited the recession without project management implementation.

Table 6–1 compares the recession of 1979–1983 with the recession of 1989–1993. During the recession of 1989–1993, deep cuts were made in the white collar ranks. Layers of management were systematically being eliminated, especially middle management. Companies were now looking for long-term solutions and a focused business strategy.

At last, organizations recognized that project management implementation could improve production and efficiency not only in the present, but also for the future. Long-term strategies for implementation were now commonplace. Project management applications now encompassed all facets of the business.

TABLE 6–1. RECESSIONARY EFFECTS

	Characteristics				
Recession	**Layoffs**	**R&D**	**Training**	**Solutions Sought**	**Results of the Recessions**
1979–1983	Blue collar	Eliminated	Eliminated	Short-term	• Return to status quo • No project management support • No allies for project management
1989–1993	White collar	Focused	Focused	Long-term	• Changed way of doing business • Risk management • Examine lessons learned

6.4 GLOBAL PRESSURES

Inaccurate views of project management persisted throughout the early 1980s. Comments such as the following were typical of that decade:

- "We've gotten this far without project management, so why do we need it at all?"
- "We are extremely profitable. Implementing project management may cause this to change."
- "Management's guidance is sufficient."
- "Project management applies to only a small portion of our business, so why disrupt the corporate culture?"
- "We are not an aerospace or construction firm. We simply don't need it."

During the late renaissance period, the recessionary indications appeared once again. However, this time the recession was accompanied by global pressures for drastically improved quality and shortened product development time. The concepts of total quality management (TQM), concurrent engineering, self-directed teams, empowerment, and life cycle costing made it quite apparent that long-term rather than short-term solutions would be necessary (see Figure 6–4). Computer companies such as IBM, Digital Equipment, and Unisys reduced manpower by over 300,000 people, the majority of whom were in the white collar ranks. The battering of computer stock prices and the competitive nature of the industry supported the premise that long-term solutions were needed. Companies were now planning for decades ahead, rather than years.

Once again, executives began reassessing the merits of modern project management, not only for the organization's financial health, but also for the future. The following comments are indicative of how companies viewed the benefits of project management during the last recession:

- "Project management is a tool for change in a rapid, complex environment."
- "It is the lifeblood for future survival and success."
- "The strategic importance of project management is now on an upward trend."

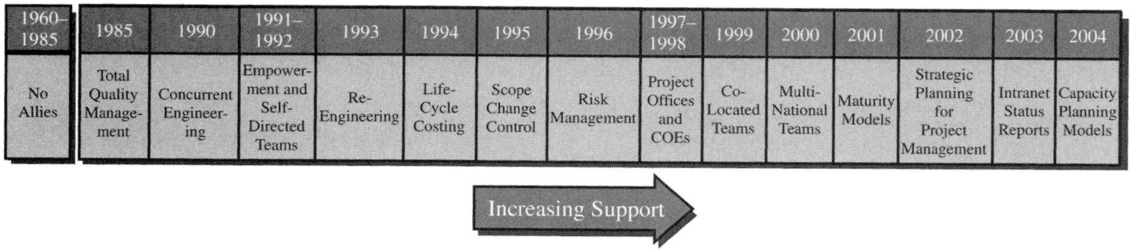

1960–1985	1985	1990	1991–1992	1993	1994	1995	1996	1997–1998	1999	2000	2001	2002	2003	2004
No Allies	Total Quality Management	Concurrent Engineering	Empowerment and Self-Directed Teams	Re-Engineering	Life-Cycle Costing	Scope Change Control	Risk Management	Project Offices and COEs	Co-Located Teams	Multi-National Teams	Maturity Models	Strategic Planning for Project Management	Intranet Status Reports	Capacity Planning Models

Increasing Support

FIGURE 6–4. New processes supporting project management.

- "It is an important management methodology and tool and is essential for future competitiveness."
- "Project management is fundamental to our business. We cannot do without it. Ultimately, the success of our business is based on our success as project managers."

It is truly a pity that recessions were needed to wake people up as to the true benefits of modern project management. The benefits are now well known and lessons-learned can be exchanged between divisions, companies, and industries.

It is questionable whether the recession alone enhanced the acceptance of project management or whether TQM and concurrent engineering had an impact. In 1986, Johnson Controls embarked upon an aggressive TQM program involving all employees. In 1987, Johnson Controls recognized that a "marriage" between project management and TQM was possible, and that project management could greatly improve the implementation of TQM. During the recession, more and more companies recognized and accepted these relationships, thus accelerating the acceptance of project management. Figure 6–4 shows the approximate time frames when the "allies" of project management appeared in the past, as well as the "allies" that we can expect in the near future.

As project management began to grow, companies began to change their opinion of project management as maturity became reachable. The differences between the views in mature and immature organizations are:

Mature

- Project management was now viewed as a tool for success, not as interference to daily objectives.
- Organizational changes in the structure promoted by project management led to a maximization of human resources utilization.
- Synergies were present in mature project management organizations.
- The entire organization supported the project manager and his/her requirements.

Immature

- Project management was viewed as intimidating and disruptive in the daily operations.

- Power struggles erupted over allocating scarce human and technical resources.
- Executives and project managers tended to micromanage rather than allow line and technical participants to offer their expertise.
- There was no planned basis for the evolution of formal versus informal project management structures. There was no thought given as to which structure would be better for a project prior to its undertaking.

6.5 CONCURRENT ENGINEERING

Shortening product development time has always been an issue for U.S. corporations. During favorable economic conditions, corporations would deploy massive resources at a problem in order to reduce development time. However, during a recession, not only are resources scarce, but time becomes a critical constraint rather than a luxury. The principles of concurrent engineering have been almost universally adopted as the ideal solution. In most organizations, concurrent engineering requires that marketing, engineering, and manufacturing all be brought on board the project early in the planning stage. In this regard, the importance of the communication skills of the project manager cannot be understated, as shown in Figures 6–5, 6–6, and 6–7. Parallel activities dramatically increase risks. However, conducting activities in parallel rather than in series mandates superior planning in order to compensate for the increased downstream risks. The old adage of "We never have enough money to plan correctly the first time but have plenty of money to replan a second or third time," was now discarded.

Since project management advocated better planning, it was no surprise that companies adopted both concurrent engineering and project management. Companies such as

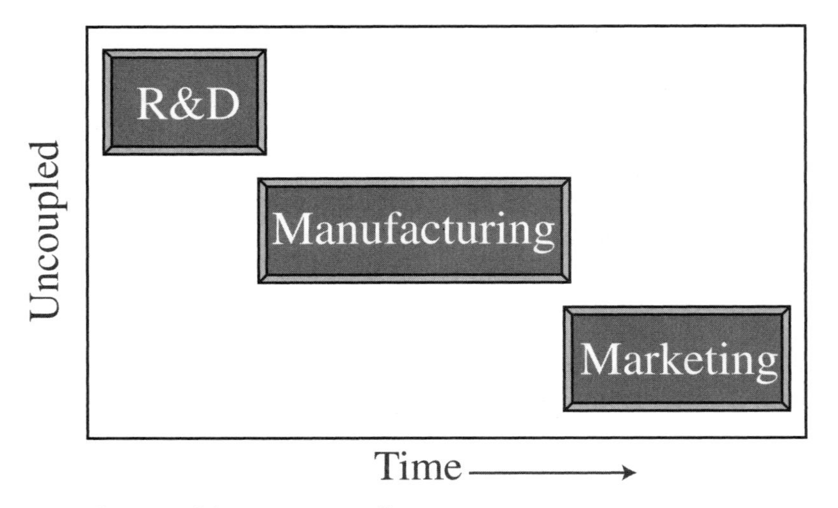

FIGURE 6–5. Degrees of downstream coupling.

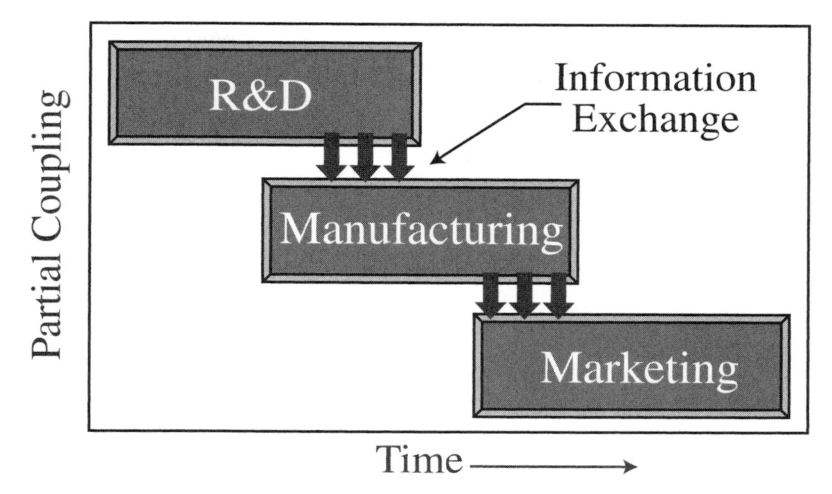

FIGURE 6–6. Degrees of downstream coupling.

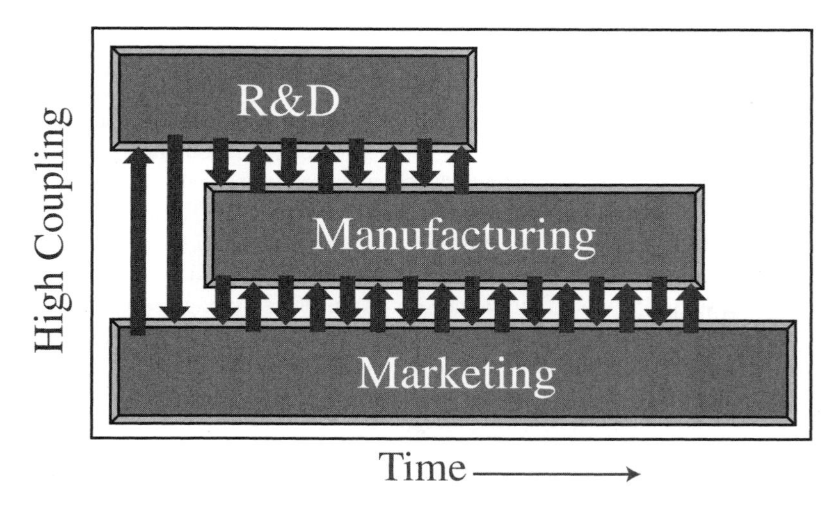

FIGURE 6–7. Degrees of downstream coupling.

Chrysler used concurrent engineering and project management to go from concept to introduction of the Viper in less than three years. Concurrent engineering is forcing organizations to consider long-term maturity implications for project management and may very well be the strongest driving force for the acceptance of modern project management.

6.6 PROJECT OBJECTIVES

In the early days of the aerospace and defense industries, the government awarded "cost-plus-percentage-of-cost" contracts. The contractors virtually had an unending supply of funds. The more money they spent, the more money they received, including a healthy profit on every dollar spent. Although this type of contract is illegal in the government today, it did accomplish its goal of encouraging the aerospace and defense industries to grow. One defense contractor went from a handful of employees to well over 12,000 in slightly more than two years.

The cost-plus-percentage-of-cost contracts were intended to beef up our technical manpower base so that we could stay ahead of the Russians in the development of sophisticated, cold-war technology. These contracts had predominantly technical objectives from both the customer's and contractor's view. Change control was not utilized well, or simply not used at all, since funding was virtually unlimited.

The thirst for technical knowledge embraced two important criteria during the traditional period of project management. First, all objectives were defined in technical terms and, second, most project managers were selected from the engineering ranks. This caused the business community to view project management as a process applicable only to engineering-related projects, and specifically large ones.

During the renaissance period of the 1980s, and because of the recession of 1979–1983, more and more organizations realized the importance of business objectives in conjunction with technical objectives. This is shown in Figure 6–8. Any engineer can design a product that cannot be produced. Any engineer can manufacture a product that has no market potential. The world was now realizing that business objectives may very well be more important than technical objectives. This led us to the realization that perhaps individuals with business backgrounds were better suited to manage projects than were technically oriented personnel.

Traditional Project Management		Renaissance Period		Modern Project Management	
• Technical	75%	• Technical	50%	• Technical	10%
• Business	25%	• Business	50%	• Business	90%

FIGURE 6–8. Project objectives. Most projects have multiple objectives. There are both technical objectives and business objectives.

As we entered the era of modern project management, most project objectives were defined by 90 percent business terms and only 10 percent technical terms. More and more project managers were coming from marketing, sales, procurement, and human resource management. The knowledge base necessary to be an effective project manager had now shifted from technical decision-making to business decision-making.

6.7 DEFINITION OF SUCCESS

Our definition of project success also appears to have changed, as shown in Figure 6–9. During the traditional period, project success was measured in technical terms only: Did it work or didn't it work? This occurred because the project's objectives were defined in technical terms only. After all, if the objectives are defined in technical terms only, then project success will be defined in technical terms only. Many years ago I had an engineer working for me who had requested 800 hours to complete a task. He was given 800 hours but completed the job in 1,300 hours. When asked why he required an additional 500 hours, he commented, "What do you care how long it took or how much I spent? I got the job done, didn't I?" This type of mentality persisted throughout the traditional period.

During the renaissance period, cost and quality became equally important as technology. Success was being defined as within time, within cost, and at the appropriate technical or quality level. As we approached the period of modern project management, we included in the definition, "accepted by the customer." This is only fitting as today we recognize that quality is defined by the customer, not the manufacturer.

Emphasis is now being placed upon the addition of two more criteria: with minimum scope changes and without disturbing the ongoing business of the company. The former criterion emphasizes quality planning, and the latter stresses making company decisions rather than project decisions. What is in the best interest of the project may not be in the best interest of the company. These two added criteria will probably become even more important as we enter the twenty-first century.

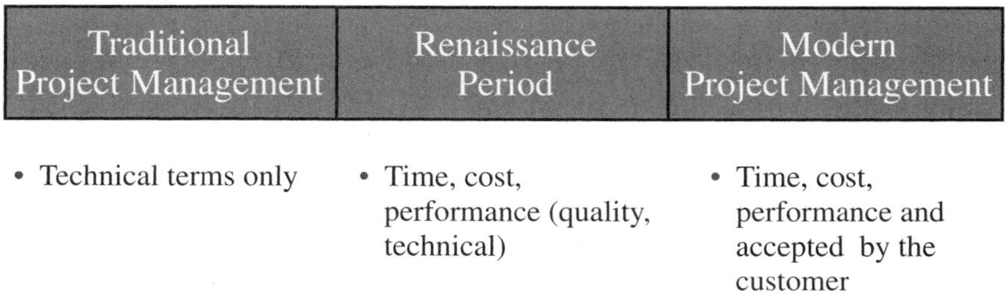

Traditional Project Management	Renaissance Period	Modern Project Management
• Technical terms only	• Time, cost, performance (quality, technical)	• Time, cost, performance and accepted by the customer

FIGURE 6–9. Definition of success.

6.8 VELOCITY OF CHANGE

If there is one thing that executives have learned from a recession, it is: "If a change is deemed necessary, then make the change as quickly as possible." Procrastination can make the situation worse. The decision to go to project management can be made quickly, but the implementation process may take years, as was evidenced during the traditional period of project management (see Figure 6–10).

Project management implementation was traditionally regarded as a slow, tedious process involving the following:

- Begin with a *small* breakthrough project where everyone can follow and easily observe project management in action.
- Keep everyone constantly informed (even overinformed, if necessary), so that project status is known by all.
- Try not to place too much pressure initially on line managers for commitments and deliverables.
- Let changes take place gradually to the point where people volunteer to be included in the project management system rather than being ordered to participate.

Unfortunately, slow change was not acceptable during the last recession of 1989–1993. To illustrate the problem and an eventual solution, a shipping company recognized the need to use project management on a two-year project that had executive visibility and support and was deemed strategically critical to the company. Although the project required a full-time project manager, the company chose to appoint a line manager who was instructed to manage his line and the project at the same time. The company did not use project management continuously, and their understanding of it was extremely weak.

Traditional Project Management	Renaissance Period	Modern Project Management
• Minimum 3–5 years • Organizational restructuring mandatory • Emphasis on power and authority • Sponsorship not critical	• Minimum 3–5 years • Organizational restructuring mandatory • Emphasis on power and authority • Sponsorship needed	• Can be done quickly (perhaps 6–24 months) • Restructuring unnecessary • Emphasis on multifunctional teamwork • Sponsorship mandatory

FIGURE 6–10. Velocity of change.

After three months, the line manager resigned his appointment as a project manager, citing too much stress and being unable to manage his line effectively while performing project duties. A second line manager was appointed on a part-time basis and, as with his predecessor, he found it necessary to resign as project manager.

The company then assigned a third line manager, but this time released her from *all* line responsibility while managing the project. The project team and selected company personnel were provided with project management training. The president of the company realized the dangers of quick implementation, especially on a project of this magnitude, and was willing to accept the risk.

After three months, the project manager complained that some of her team members were very unhappy with the pressures of project management and were threatening to resign from the company, if necessary, simply to get away from project management. When asked about the project status, the project manager stated that the project had met every deliverable and milestone to date. It was quickly apparent to the president and officers of the company that project management was functioning as expected. The emphasis shifted to focusing on how to appease the disgruntled employees and convince them of the importance of their work and how much the company appreciated their efforts.

The lesson here is that modern project management has shown that the velocity of change can be rapid with or without a recession. Perhaps because of recessionary fears or corporate pressure, change can occur more quickly than it otherwise would. As an epilogue to the shipping company situation, none of the employees actually left the project.

6.9 MANAGEMENT STYLE

The overall changes in the management style of project managers are shown in Figure 6–11. Traditionally, we rushed into projects and never spent the appropriate amount of time on upfront planning. Quality planning was performed sporadically and emphasis was, therefore,

Traditional Project Management	Renaissance Period	Modern Project Management
• Reactive management	**• Reactive management**	**• Proactive management**
(Never enough money to plan it right the first time, but plenty of money to replan it for the second and third times.)	(Project plans are organic; they feed, breathe, grow, change, etc.)	(Bring problems to the surface and we will help you. Bury the problem and your job is in jeopardy.)

FIGURE 6–11. Management style.

placed upon reactive management. Another problem we had in the early days of project management was the selection of engineers as project managers. During the traditional period, engineers were usually assigned as project managers because of their technical knowledge. These individuals were extremely optimistic, believing that the project would proceed according to whatever plan they developed. Contingency planning was deemed unnecessary. However, crisis management was employed when problems occurred.

As we entered the renaissance period, product life cycles were becoming shorter with a greater need for rapid product development. The leadership styles of project, line, and senior management had changed from reactive to proactive. There were two factors that forced proactive management. First, project management advocated proactive management through quality planning and contingency planning, and second, risk management had now become more important in corporate decision-making.

6.10 AUTHORITY AND JOB DESCRIPTIONS

The authority of the project manager has changed dramatically over the evolutionary process of modern project management, as shown in Figure 6–12. During the traditional period of project management, large projects were the norm, with the majority of the project team reporting full-time on one and only one project. With the majority of the project team full-time, the project manager acted more as a line manager than as a project manager. Formal project management job descriptions were in place, including authority levels, mainly at the request of the customer. Project managers were given complete and formal authority for all technical decisions.

As we approached the renaissance period, numerous power and authority conflicts arose. We were now placing people with a business background in charge of technical de-

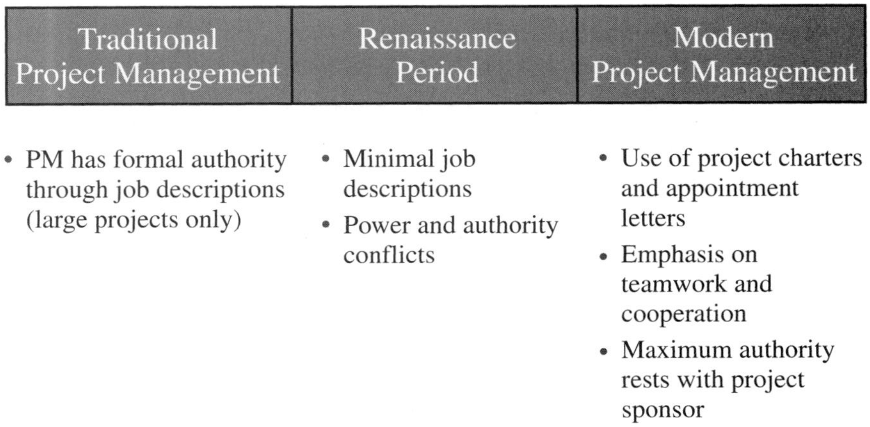

Traditional Project Management	Renaissance Period	Modern Project Management
• PM has formal authority through job descriptions (large projects only)	• Minimal job descriptions • Power and authority conflicts	• Use of project charters and appointment letters • Emphasis on teamwork and cooperation • Maximum authority rests with project sponsor

FIGURE 6–12. Authority and job descriptions.

velopment projects. The real technical experts were the line managers. Employees were sharing their time on more than one project. Customers were no longer demanding job descriptions that identified levels of authority. And most important, executives were extremely reluctant to prepare job descriptions that provided all project managers with the same level of authority. Additionally, giving authority to project managers meant taking authority away from other individuals who would certainly resist this. The perceived problems with power and authority provided additional fuel for the opponents of project management implementation. Some companies believed that the benefits/rewards of project management implementation were not worth the risks of disturbing the organizational hierarchy and the status quo.

As modern project management approached, companies found the solution to the authority problems: Maximum authority will reside with the project sponsor who, in turn, will delegate the authority he/she sees fit through a project charter or an appointment letter. This allowed organizations to assign authority on a project-by-project basis. An executive in a bank commented on the authority of a project manager:

> The project manager's authority usually exceeds that which is given to him by his normal title because he carries the weight of the sponsor with him. Therefore, his authority is very high.

The moral here is that, in project management, it is more important where the sponsor resides in the hierarchy than to whom the project manager reports. Even today, several of the excellent companies in project management still do not have project management job descriptions. Also, in many of the firms that utilize job descriptions, we find only an explanation of duties or responsibility, not the accompanying authority levels.

6.11 EVALUATION OF TEAM MEMBERS

Organizations will never implement change without considering the impact on the wage and salary administration program. This is important in the way we evaluate team members. Over the past 35 years, we have changed the way we evaluate team members. This is shown in Figure 6–13. In the traditional period of project management, the project managers had virtually no input into the evaluation of team members. Line management performed the entire evaluation process, most often without any discussions with the project managers, even when the employees were full-time on one project only!

As the renaissance period approached, we realized the importance of giving the project manager some sort of input, even if informal. It became obvious that on some projects the project manager was the person best qualified to evaluate the performance of the team members. The critical issue was that the project manager may not be on the management pay scale. As a result, employees were unhappy about having their present and future salary dependent on someone who wasn't even a corporate manager. The situation was further complicated when the project manager's team was composed of employees who were two or even three pay grades higher than the project manager.

Traditional Project Management	Renaissance Period	Modern Project Management
• PM provides virtually no input	• Informal input	• Formal input

FIGURE 6–13. Evaluation of team members.

During the 1990s, we solved the problem by mandating that all evaluations be signed jointly by the project manager and line manager. This brought forth additional problems, but not of the magnitude seen during the renaissance period.

6.12 ACCOUNTABILITY

During the traditional period of project management, we assigned our best technical experts as project managers. These individuals would then negotiate for the best technical resources. The project manager maintained single person total accountability by virtue of his technical knowledge. But during the 1980s, changes in accountability occurred, as shown in Figure 6–14. Project managers were no longer the technical experts in all phases of the project. Projects were becoming more and more complex, and accountability was now being shared with the project team. This forced project managers to be more aggressive in seeking out qualified team members. The relationship between the project and line managers had become hostile and competitive.

Traditional Project Management	Renaissance Period	Modern Project Management
• Maximum accountability with the project manager	• Accountability shared with the project team	• Accountability shared with line managers as well
• Negotiate for best resources	• Negotiate for best resources	• Negotiate for deliverables
• PM provides technical direction	• PM provides some technical direction	• Line managers provide majority of technical direction

FIGURE 6–14. Accountability.

During the period of modern project management, two important changes were made. First, real technical expertise was now recognized as being in the line, most frequently with the line manager. Second, line managers were now asked to share accountability with the project managers. These two changes altered the way project managers negotiated for resources. Since accountability is now shared between the project and line managers, project managers now negotiate for results/deliverables rather than people. The line manager must now accept accountability for promises made to the projects and must provide the daily supervision of the assigned workers.

6.13 PROJECT MANAGEMENT SKILLS

The evolution of project management has, as expected, changed the skill requirements expected of an effective project manager. In the early days of project management, emphasis was placed on technical skills, as shown in Figure 6–15. During the renaissance period, project teams involved more and more nonengineering personnel, and behavioral skills became equally important as technical skills. During this period, it became apparent that to be an effective project manager may require having an *understanding* of technology rather than being a *technical expert*. Projects were becoming so large and complex that it was simply no longer possible for the project leader to remain a technical expert in all aspects of the project. Project managers were spending more of their time scheduling, performing cost control, and monitoring progress rather than providing technical direction.

Modern project management has changed the skill base once again. Because business objectives are now perhaps more important than technical objectives, the new skill set includes knowledge of the business, risk management, and integration skills. Project managers are now expected to make business decisions rather than just technical decisions.

Perhaps the biggest skill needed for the twenty-first century will be risk management. However, in order to perform risk management adequately, a good knowledge of the business is mandatory. This implies that, in the future, project management will not be an entry level position. Most project management slots will be from within, either through lateral transfers or promotions. Individuals who are hired in from outside the company will

Traditional Project Management	Renaissance Period	Modern Project Management
• Technical skills	• Technical and behavioral skills	• Knowledge of the business, risk management, and integration skills

FIGURE 6–15. Project management skills.

be required to work for 18 to 36 months in one or more functional areas to first understand the business.

According to Rose Russett of General Motors Powertrain:

> The individuals who are chosen as program managers all have multifunctional backgrounds. Many have held assignments in engineering, manufacturing, quality systems, and so on. This [project manager] is not an entry level position. Their skills at the behavioral side of project management, such as managing and motivating teams, communication, and conflict resolution, become as important as the knowledge of the business itself when working in a project environment.

Knowledge of the business is also a prerequisite to integrate work across the organization. As projects get larger and more complex, integration skills likewise become critical. Rose Russett believes that

> They [PMs] are the integrators of all functional deliverables and must understand all of the various functions, their interrelationships, and have the ability to work within a strong matrix organization. This integration ability is a key success factor for project managers.

Integration skills require an ability to work with people, to communicate, to delegate, and to organize work, as exemplified in the following quote from a steel company executive:

> Since the project manager sits on top of the whole cake, he/she must make sure all sides are moving in harmony and in tempo. A project manager must be well organized to monitor and orchestrate this activity.

Bob Storeygard, Advanced Project Leadership Specialist with 3M Company, also believes that integration skills are critical. In a paper presented at the 1995 New Orleans Symposium, "Growing the Professional Project Leader," he commented:

> Performance is the KEY to the value of a professional project leader. It is not so much a question of WHAT they know (knowledge), although that can help; or of WHO they know (organizational connection and savvy), although that can really help; but the **INTEGRATION** of these two with the actual application (performance) that determines whether there is value.
>
> Likewise, it is also the performance of their team that will determine the success or failure of a project effort. Many a well-intentioned, well-skilled, well-resourced team has still failed due to the lack of integration of the factors stated above.

When it comes to understanding the importance of integration, most people have a parochial view and consider integration management as simply getting diverse functional groups, which are located under the same roof, to communicate with one another and focus on a common objective. For the global project manager, integration management has a more complex meaning, since there will certainly be more than one roof and each roof may be separated by thousands of miles and in different time zones. Integration management now takes on a cultural dimension.

Suzanne Zale, global program manager of EDS, commented:

For global projects, the project manager must understand how to manage in the global environment. Some of the key areas for special attention are:

- The oversight / governance of the project must be from the global perspective and have adequate global representation.
- Cultural change management needs to be an integral part of the project.
- The integration of the various components of the project must accommodate the different geographic peculiarities; e.g., differences / uniqueness in infrastructure, financial control, communications, time tracking, and so on. As an example, time tracking may not be legal or may require special government approval in some countries.
- Resource planning and communications become more complex because of time differences and geographically dispersed teams.
- Implementation of major deliverable needs to take the geography into consideration, such as how business is done locally, how to interact with local suppliers, the quality standards/practices, and so on.
- Security of information and intellectual management need to be considered early on in the project.
- Many of our clients are struggling with global growth. Project managers frequently must help clients to better manage the business on a global basis. It is important to stay flexible, as local business conditions may drive changes in requirements or increase the project risk.

All these also add time and cost to the project.

There are two subsets of management skills that need to be discussed; problem resolution involvement and decision-making criteria. These are shown in Figures 6–16 and 6–17. Historically, project managers were technical experts and were involved only with technical problems. Technical objectives dominated the decision-making process. During the renaissance period, with more and more people part-time on multiple projects, the project manager would be more involved with functional problems that impacted the project and resource allocation. But with modern project management, which is based upon shared accountability and strong business objectives, problem resolution involvement is mainly risk management and integration issues.

Traditional Project Management	Renaissance Period	Modern Project Management
• Mostly technical problems	• Mostly technical problems and some functional problems	• Mostly integration problems and risk management

FIGURE 6–16. Problem resolution involvement.

Traditional Project Management	Renaissance Period	Modern Project Management
• Technical	• Project versus professional	• Corporate credo

FIGURE 6–17. Decision-making criteria.

A corporate credo is a document that tells the stockholders and stakeholders the order of importance during decision-making. For example, at Johnson and Johnson, the credo states that the health and well-being of the consumer has first priority. Therefore, with modern project management, companies like Johnson & Johnson would stress the health, well-being, and safety of the consumer above what may be in the best interest of the project or the company.

6.14 PLANNING HOURS/DOLLARS

At last, we were recognizing the importance of up-front planning. The most critical phase of any project is the planning phase. Ideally, when the project is planned carefully, success is likely. Still, no matter how much or how well we plan, changes are needed and contingencies must be developed.

In the early days of project management, planning was performance from milestone to milestone. This approach led to suboptimization of the individual project segments and resulted in very little regard for the total effort. (Methodologies for project management were nonexistent.) Today, in successful project management organizations, all projects are broken down into similar life cycle phases. This provides consistency among projects, and it provides checkpoints at which managers can either cancel or redirect individual projects. This also has the added benefit of allowing us to establish standards for effective planning.

During the traditional period of project management, we spent only about 15 to 20 percent of the direct labor hours/dollars on planning. Replanning and massive scope changes were the norm, and most projects were severely overmanaged. As shown in Figure 6–18, the renaissance period forced us to reassess the costs of overmanagement versus the costs of undermanagement to determine proper management levels. We began evaluating the size of a project office for a given project and concluded that for non–project-driven organizations, 8–10 percent of the total labor hours would be required for project management support, versus 12–15 percent for project-driven organizations. By the 1990s, we estimated that 35–55 percent of the total project hours would be required just for effective planning, and that the project office size would be dependent upon the amount of integration required.

Traditional Project Management	Renaissance Period	Modern Project Management
• 15–20% • Overmanagement	• 25–30% • Overmanagement versus under-management (fixed for duration: 8–10% for non–project-driven to 12–15% for project-driven)	• 35–55% (planning is imperfect) • Varies over the project life cycle • Maximum time is spent in integration

FIGURE 6–18. Percent of labor hours/dollars spent on planning.

6.15 EDUCATION AND TRAINING

Project management education and training has taken on paramount importance during the last several years, as shown in Figures 6–19 and 6–20. During the traditional period of project management, there were very few training programs. Companies that did offer project management training programs emphasized the technical aspects of project management, namely planning, scheduling, and sometimes cost control. Project management was viewed quite simply as program evaluation and review technique (PERT) scheduling.

By the mid-1980s, companies began recognizing the importance of all aspects of project management. The behavioral aspects of project management were now recognized as equally important as the technical aspects. Colleges and universities, which had previously sponsored only one or two seminars on project management, were now offering sequences

Traditional Project Management	Renaissance Period	Modern Project Management
• Few courses	• Several courses and Master's/MBA programs	• Curriculums in project management (internal and external)
• No certification	• Introduction of PMP	• Refinement of PMP and company-specific PMP

FIGURE 6–19. Educational programs.

Traditional Project Management	Renaissance Period	Modern Project Management
• Technical courses	• Mostly technical but some behavioral	• Mostly behavioral but some technical

FIGURE 6–20. Project management training.

of courses, as well as Masters and MBA Programs in project management. The introduction of the project management certification program also accelerated coursework development. More and more coursework on the behavioral side of project management now appeared.

By the mid-1990s, surveys were being published that indicated that the fastest way to achieve growth and maturity in project management was through training and education. Companies like IBM and USAA were now developing internal curriculums on project management, many of which were designed to allow participants to pass the project management certification exam. Some organizations actually developed their own internal project management certification programs.

6.16 PROJECT SPONSORSHIP

During the traditional project management period, only the project-driven industries were accepting project management as a way of life. And since these projects had a definable profit goal, sponsorship existed primarily at the executive levels. However, as more and more non–project-driven companies accepted project management during the renaissance period, it became impossible for senior management to function as a sponsor for the multitude of projects that were in progress. Thus prioritization of projects was deemed necessary. Those projects that were not critical enough to require senior management sponsorship were assigned to lower- and middle-level management, as shown in Figure 6–21.

Effective project sponsorship will accelerate the decision-making process in an organization. Effective sponsorship also reduces the number of communication channels. A bank that has just begun using project management has found this to be evident. According to a spokesperson from the bank:

> The project manager has to deal with too many people within the corporation, ranging from line managers who need to be informed to committees that need to approve changes to a plan. With project sponsorship, we have found that if a critical change is necessary, the project manager and the sponsor can make the move relatively quickly, even for a large organization.

Traditional Project Management	Renaissance Period	Modern Project Management
• Executive sponsors (may change over life cycle)	• Executive and middle management sponsors	• Multiple-level sponsorship • Committee sponsorship

FIGURE 6–21. Sponsorship.

As the modern project management period time frame occurred, two situations appeared that altered our view of sponsorship. First, some of the projects were so low in the organization that even a first-line supervisor could perform the role of a project sponsor. At National City Bank in Cleveland, Ohio, there were three levels of sponsorship: working sponsors, project sponsors, and executive sponsors.

The second critical factor was concurrent engineering, which required that marketing, engineering, and production personnel all be brought on board right from day one. In the past, sponsors could change based upon where the bulk of the work was taking place. But with all disciplines on board early, the only viable solution was committee sponsorship. General Motors Powertrain Group and Rohm & Haas are examples of organizations using committee sponsorship effectively.

6.17 PROJECT FAILURES

During the evolution of project management, the causes of failure of projects seems to have changed. In the traditional period of project management, we often rushed into projects and spent an inadequate amount of time planning and estimating projects. If a project failed, the blame was always placed upon poor planning, poor estimating, improper scheduling, and inadequate control practices. This is shown in Figure 6–22. During the renaissance period, we began to realize the importance of the behavioral side of project management. We still believed that the majority of the failures were quantitative, but we did recognize that behavioral failure could exist.

By the time the 1990s approached, post-project analysis indicated that failures were more behavioral than quantitative. In fact, we were now convinced that we probably underwent behavioral failure during the traditional and renaissance periods but were unable to recognize the real culprit. Today we offer more and more behavioral project management courses because of this behavioral failure.

Traditional Project Management	Renaissance Period	Modern Project Management
Quantitative Failures • Planning • Estimating • Scheduling • Controlling	**Quantitative Failures** • Planning • Estimating • Scheduling • Controlling	**Behavioral Failures** • Poor morale • No employee commitment • No functional commitment • Poor productivity • Poor human relations

FIGURE 6–22. Project failures.

6.18 MATURITY AND IMMATURITY

As a final note, we should discuss that it is possible to go from a mature system in project management to an immature system, as seen in Figure 6–23. We have seen this happen during the modern project management period where companies stopped training younger employees in project management and even began canceling project management courses. One executive commented that they have trained enough people in project management and their training dollars can be spent more fruitfully in other areas. Although this may be true, the risk in no follow-up training is complacency. NASA was a prime example. During the 1980s, training dollars for project management were on a decline. Unfortunately, the employees that helped design the project management systems and bring the organization to maturity were now retiring. The mentors for the newly hired employees were disappearing, and many of the younger workers were making the same mistakes that their pre-

Traditional Project Management	Renaissance Period	Modern Project Management
A Quest for Knowledge	**Growth and Maturity**	**Immaturity** • Complacency • No lessons learned • Loss of knowledge • No follow-up education

FIGURE 6–23. From maturity to immaturity.

decessors had made in the previous two decades. Project management maturity can be lost without documented lessons learned and follow-up training.

MULTIPLE CHOICE QUESTIONS

1. Which of the following is not considered an evaluation phase of project management?
   A. Military project management
   B. Traditional project management
   C. Renaissance period
   D. Modern project management
   E. None of the above
2. Which of the following factors promoted the acceptance of project management?
   A. Recessions
   B. Tasks were becoming more complex
   C. The size and scope of projects were increasing
   D. The environment was becoming more turbulent
   E. All of the above
3. During the traditional period of project management, the majority of companies using project management were:
   A. Hybrids
   B. Project-driven
   C. Non–project-driven
   D. All of the above
   E. A and B only
4. During the modern project management period, the majority of companies using project management were:
   A. Hybrids
   B. Project-driven
   C. Non–project-driven
   D. All of the above
   E. A and B only
5. Project management seems to mature the quickest when project managers are given:
   A. Complete authority
   B. Complete accountability
   C. Wage and salary administration responsibility
   D. Profit and loss responsibility
   E. An astute project sponsor
6. The implementation of concurrent engineering practices requires an organization to have effective communications throughout the organization.
   A. True
   B. False
   C. Cannot be determined from the information provided
7. During the traditional period, project objectives were defined predominantly in __________ terms.
   A. Business
   B. Market share

C. Financial
D. Technical
E. Profitability

8. With modern project management, project objectives are now defined in __________ terms.
   A. Mostly technical but some business
   B. Mostly business but some technical
   C. Equal percentages of business and technical
   D. None of the above
9. Historically, project success was measured with the same terminology that was identified in the:
   A. Project's objectives
   B. Project's specifications
   C. Project's work breakdown structure
   D. All of the above
   E. A and D only
10. Project management implementation was traditionally viewed as:
   A. A slow, tedious process
   B. A very fast process if an executive sponsor existed
   C. A very fast process if only a few projects existed
   D. A very fast process if deemed necessary by the customers
   E. None of the above
11. Today, most people believe that the velocity of change can be accomplished:
   A. In 3 to 5 years
   B. In 2 to 4 years
   C. In 6 to 24 months
   D. In 3 to 6 months
   E. Immediately
12. Historically, most project managers adopted a "reactive" management leadership style because:
   A. Proactive management hadn't been developed yet
   B. The technology did not allow for contingency planning
   C. The assigned project managers were "optimistic" engineers
   D. The project's schedule was too short
   E. The project's schedule was too long
13. Contingency planning is a characteristic of a proactive management style.
   A. True
   B. False
14. With modern project management, the project manager's authority is based upon:
   A. Which line manager will relinquish authority to the project manager
   B. What is written in the job description
   C. Whether the project manager has a command of technology or an understanding of technology
   D. Whether the company is project-driven or non–project-driven
   E. Who is assigned as the executive sponsor
15. The document used today to define the project manager's authority is the:
   A. Project charter
   B. Project manager's appointment letter

   C. Corporate job description for the project manager
   D. All of the above
   E. A and B only

16. Even today, companies are resistant to giving project managers the responsibility for wage and salary administration because:
   A. The employees may be a higher pay grade than the project manager
   B. The project manager is not viewed as a professional manager (as would be the line manager)
   C. The project manager has not been trained in wage and salary administration
   D. All of the above
   E. B and C only

17. With modern project management, project managers negotiate with line managers for __________ rather than for __________.
   A. Accountability; authority
   B. Resources; results
   C. Deliverables; resources
   D. Project sponsorship; avoidance
   E. Specific people; knowledge

18. Project management today seems to work best if:
   A. The line manager has authority over the project manager
   B. The project manager has authority over the line manager
   C. The project and line managers are at significantly different pay grades
   D. The project and line managers share accountability
   E. The project and line managers are both in the same line organization as the executive sponsor

19. Most people believe that the predominant skill needed to be an effective project manager in the twenty-first century will be:
   A. A command of technology
   B. Communication skills
   C. A knowledge of the business
   D. Risk management
   E. Integration skills

20. Knowledge of the business is a major prerequisite for which of the following skills?
   A. Communication skills
   B. Decision-making skills
   C. Planning skills
   D. Risk management
   E. Scheduling skills

21. The most critical phase of any project is:
   A. Planning
   B. Execution
   C. Implementation
   D. Conversion
   E. Closure

22. Typical project management support for a labor-intensive project would be __________ of the total labor hours.
   A. 5 to 8 percent
   B. 8 to 10 percent

C. 12 to 15 percent
D. 15 to 30 percent
E. 30 to 55 percent

23. For a large project with very little risk, what percentage of the total labor hours of the project are spent in planning, estimating, and all other activities prior to manufacturing or execution?
   A. 5 to 10 percent
   B. 10 to 20 percent
   C. 20 to 35 percent
   D. 35 to 55 percent
   E. 55 to 70 percent

24. Which of the following forms of sponsorship is most frequently found in organizations using concurrent engineering?
   A. Line manager sponsorship
   B. Middle management sponsorship
   C. Executive sponsorship
   D. Committee sponsorship
   E. Sponsorship is not needed with concurrent engineering

25. Which of the following is believed today to be a major reason for project failures?
   A. Poor morale
   B. No employee commitment
   C. Poor human relations
   D. No functional commitment
   E. All of the above

DISCUSSION QUESTIONS

1. How did the last recession (1989–1993) impact the growth of modern project management?
2. In which type of organization (i.e., project-driven, non–project-driven, or hybrid) did modern project management grow the fastest? Explain your answer.
3. Explain how the educational background of the person assigned as the project manager can be impacted by the way we identify the objectives of the project.
4. Under what conditions would a project manager have complete wage and salary responsibility for the project team members?
5. What causes force the project manager to share accountability with line managers rather than have single person accountability?
6. Why do we believe that today project failures are more behavioral than quantitative?
7. Explain how a corporate credo can influence project decision-making.
8. Project management for a reasonably sized, labor-intensive project would require 12 to 15 percent of the total labor hours. Under what conditions would this percentage increase? Decrease?
9. Why is committee sponsorship perhaps better than traditional sponsorship for projects using concurrent engineering activities?
10. Under what conditions (or projects) would a project manager be expected to have a command of technology rather than an understanding of technology?

Across

3. Type of sponsorship
6. PM skill
8. ____________ letter
12. Type of support
14. CSF
15. Project _______
16. Management style
17. Shared _______________
18. Type of failure
19. 1985–1993 period

Down

1. PM skill: ____ management
2. Management style
4. ______ project management
5. 1960–1985 PM education
7. 1960–1985: ___________ period
9. CSF
10. CSF: acceptance
11. CSF
13. Type of objective

Project Portfolio Management

7.0 INTRODUCTION

Your company is currently working on several projects and has a waiting list of an additional 20 projects that they would like to complete. If available funding will support only a few more projects, how does a company decide which of the 20 projects to work on next? This is the project portfolio management process.

It is important to understand the difference between project management and project portfolio management. Debra Stouffer and Sue Rachlin have made this distinction for IT projects[1]:

> An IT portfolio is comprised of a set or collection of initiatives or projects. Project management is an ongoing process that focuses on the extent to which a specific initiative establishes, maintains, and achieves its intended objectives within cost, schedule, technical and performance baselines.
>
> Portfolio management focuses attention at a more aggregate level. Its primary objective is to identify, select, finance, monitor, and maintain the appropriate mix of projects and initiatives necessary to achieve organizational goals and objectives.
>
> Portfolio management involves the consideration of the aggregate costs, risks, and returns of all projects within the portfolio, as well as the various tradeoffs among them. Of course, the portfolio manager is also concerned about the "health" and well being of each project that is included within the IT portfolio. After all, portfolio decisions, such as whether to fund a new project or continue to finance an ongoing one, are based on information provided at the project level.

Portfolio management of projects helps determine the right mix of projects and the right investment level to make in each of them. The outcome is a better balance between ongoing and new strategic initiatives.

1. Debra Stouffer and Sue Rachlin, "A Summary of First Practices and Lessons Learned in Information Technology Portfolio Management," prepared by the Chief Information Officer (CIO) Council, Washington D.C., March 2002, p. 7.

Portfolio management is not a series of project-specific calculations such as ROI, NPV, IRR, payback period, and cash flow and then making the appropriate adjustment to account for risk. Instead, it is a decision-making process for what is in the best interest of the entire organization.

Portfolio management decisions are not made in a vacuum. The decision is usually related to other projects and several factors, such as available funding and resource allocations. In addition, the project must be a good fit with other projects within the portfolio and with the strategic plan.

The selection of projects could be based upon the completion of other projects that would release resources needed for the new projects. Also, the projects selected may be constrained by the completion date of other projects that require deliverables necessary to initiate new projects. In any event, some form of a project portfolio management process is needed.

7.1 INVOLVEMENT OF SENIOR MANAGEMENT AND STAKEHOLDERS

The successful management of a project portfolio requires strong leadership by individuals who recognize the benefits that can be accrued from portfolio management. The commitment by senior management is critical. Stouffer and Rachlin comment on the role of senior management in an IT environment in government agencies[2]:

> Portfolio management requires a business and an enterprise-wide perspective. However, IT investment decisions must be made both at the project level and the portfolio level. Senior government officials, portfolio and project managers, and other decision makers must routinely ask two sets of questions.
>
> First, at the project level, is there sufficient confidence that new or ongoing activities that seek funding will achieve their intended objectives within reasonable and acceptable cost, schedule, technical, and performance parameters?
>
> Second, at the portfolio level, given an acceptable response to the first question, is the investment in one project or a mix of projects desirable relative to another project or a mix of projects?
>
> Having received answers to these questions, the organization's senior officials, portfolio managers, and other decision makers then must use the information to determine the size, scope, and composition of the IT investment portfolio. The conditions under which the portfolio can be changed must be clearly defined and communicated. Proposed changes to the portfolio should be reviewed and approved by an appropriate decision making authority, such as an investment review board, and considered from an organization-wide perspective.

Senior management is ultimately responsible for clearly defining and communicating the goals and objectives of the project portfolio as well as the criteria and conditions considered for the portfolio selection of projects. According to Stouffer and Rachlin, this includes[3]:

- Adequately define and broadly communicate the goals and objectives of the IT portfolio.
- Clearly articulate the organization's and management's expectations about the type of benefits being sought and the rates of returns to be achieved.

---

2. See note 1, p. 8.

3. See note 1, p. 13.

- Identify and define the type of risks that can affect the performance of the IT portfolio, what the organization is doing to avoid and address risk, and its tolerance for ongoing exposure.
- Establish, achieve consensus, and consistently apply a set of criteria that will be used among competing IT projects and initiatives.

Senior management must also collect and analyze data in order to assess the performance of the portfolio and determine whether or not adjustments are necessary. This must be done periodically such that critical resources are not being wasted on projects that should be canceled. Stouffer and Rachlin provide insight on this through their interviews[4]:

According to Gopal Kapur, President of the Center for Project Management, organizations should focus on their IT portfolio assessments and control meetings on critical project vital signs. Examples of these vital signs include the sponsor's commitment and time, status of the critical path, milestone hit rate, deliverables hit rate, actual cost versus estimated cost, actual resources versus planned resources, and high probability, high impact events. Using a red, yellow, or green report card approach, as well as defined metrics, an organization can establish a consistent method for determining if projects are having an adverse impact on the IT portfolio, are failing and need to be shut down.

Specific criteria and data to be collected and analyzed may include the following:

- Standard financial measures, such as return on investment, cost benefit analysis, earned value (focusing on actuals versus plan, where available), increased profitability, cost avoidance, or payback. Every organization participating in the interviews included one or more of these financial measures.
- Strategic alignment (defined as mission support), also included by almost all organizations.
- Client (customer) impact, as defined in performance measures.
- Technology impact (as measured by contribution to, or impact on, some form of defined architecture).
- Initial project and (in some cases) operations and schedules, as noted by almost all organizations.
- Risks, risk avoidance (and sometimes risk mitigation specifics), as noted by almost all participants.
- Basic project management techniques and measures.
- And finally, data sources and data collection mechanisms also are important. Many organizations interviewed prefer to extract information from existing systems; sources include accounting, financial, and project management systems.

One of the best practices identified by Stouffer and Rachlin for IT projects was careful consideration of both internal and external stakeholders[5]:

Expanding business involvement in portfolio management often includes the following:

- Recognizing that the business programs are critical stakeholders, and improving that relationship throughout the life cycle

---

4. See note 1, p. 18.

5. See note 1, pp. 22–23.

- Establishing service level agreements that are tied to accountability (rewards and punishment)
- Shifting the responsibilities to the business programs and involving them on key decision making groups

In many organizations, mechanisms are in place to enable the creation, participation and "buy-in" of stakeholder coalitions. These mechanisms are essential to ensure the decision making process is more inclusive and representative. By getting stakeholder buy-in early in the portfolio management process, it is easier to ensure consistent practices and acceptance of decisions across an organization. Stakeholder participation and buy-in can also provide sustainability to portfolio management processes when there are changes in leadership.

Stakeholder coalitions have been built in many different ways depending on the organization, the process and the issue at hand. By including representatives from each major organizational component who are responsible for prioritizing the many competing initiatives being proposed across the organization, all perspectives are included. The approach, combined with the objectivity brought to the process by using pre-defined criteria and a decision support system, ensures that everyone has a stake in the process and the process is fair.

Similarly, the membership of the top decision making body is comprised of senior executives from across the enterprise. All major projects, or those requiring a funding source, must be voted upon and approved by this decision making body. The value of getting stakeholder participation at this senior level is that this body works toward supporting the organization's overall mission and priorities rather than parochial interests.

7.2 PROJECT SELECTION OBSTACLES[6]

Portfolio management decision-makers frequently have much less information to evaluate candidate projects than they would wish. Uncertainties often surround the likelihood of success for a project, the ultimate market value of the project, and its total cost to completion. This lack of an adequate information base often leads to another difficulty: the lack of a systematic approach to project selection and evaluation. Consensus criteria and methods for assessing each candidate project against these criteria are essential for rational decision-making. Although most companies have established organizational goals and objectives, they are usually not detailed enough to be used as criteria for project portfolio management decision-making. However, they are an essential starting point.

Portfolio management decisions are often confounded by several behavioral and organizational factors. Departmental loyalties, conflicts in desires, differences in perspectives, and an unwillingness to openly share information can stymie the project selection, approval, and evaluation processes. Much project evaluation data and information is necessarily subjective in nature. Thus, the willingness of the parties to openly share and put trust in each other's opinions becomes an important factor.

6. William Souder, *Project Selection and Economic Appraisal.* New York: Van Nostrand Reinhold, 1984, pp. 2–3.

The risk-taking climate or culture of an organization can also have a decisive bearing on the project selection process. If the climate is risk adverse, high-risk projects may never surface. Attitudes within the organization toward ideas and the volume of ideas being generated will influence the quality of the projects selected. In general, the greater the number of creative ideas generated, the greater the chances of selecting high-quality projects.

7.3 IDENTIFICATION OF PROJECTS

The overall project portfolio management process is a four-step approach, as shown in Figure 7–1. The first step is the identification of the ideas for projects and needs to help support the business. The identification can be done through brainstorming sessions, market research, customer research, supplier research, and literature searches. All ideas, regardless of merit, should be listed.

Because the number of potential ideas can be large, some sort of classification system is needed. There are three common methods of classification. The first method is to place the projects into two major categories, such as survival and growth. The sources and types of funds for these two categories can and will be different. The second method comes from typical R&D strategic planning models, as shown in Figure 7–2. Using this approach, projects

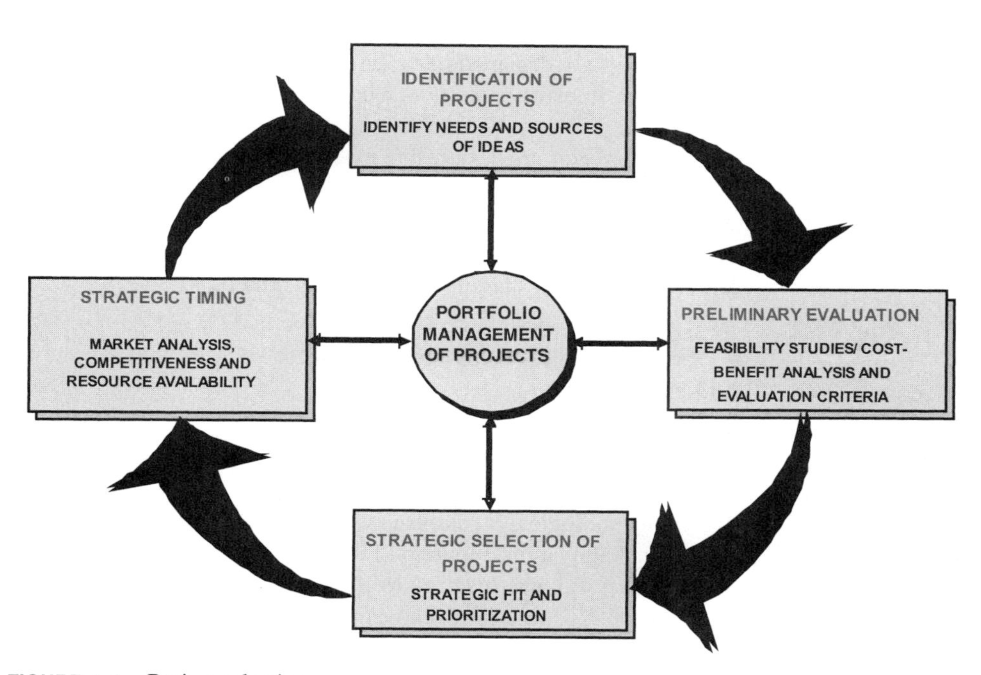

FIGURE 7–1. Project selection process.

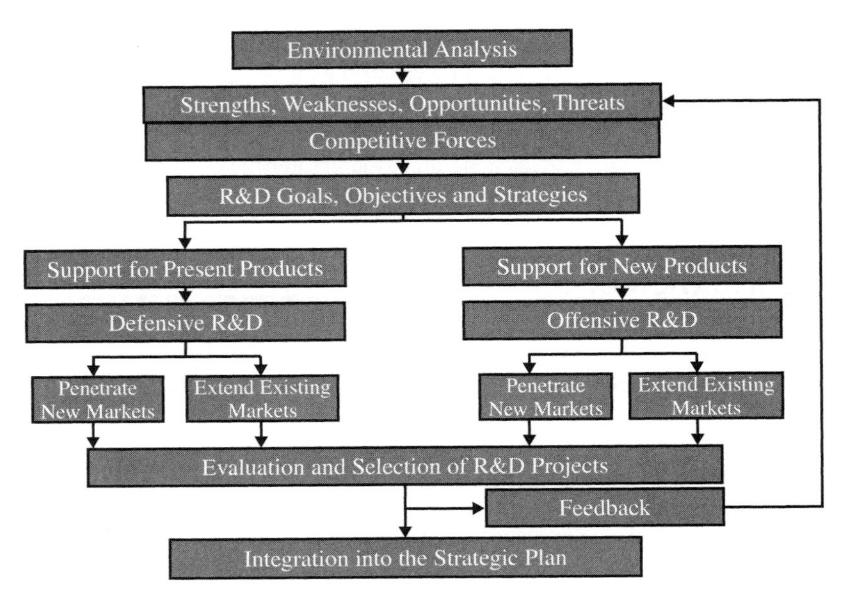

FIGURE 7–2. R&D strategic planning process.

to develop new products or services are classified as either offensive or defensive projects. Offensive projects are designed to capture new markets or expand market share within existing markets. Offensive projects mandate the continuous development of new products and services.

Defensive projects are designed to extend the life of existing products or services. This could include add-ons or enhancements geared toward keeping present customers or finding new customers for existing products or services. Defensive projects are usually easier to manage than offensive projects and have a higher probability of success.

Another method for classifying projects would be:

- Radical technical breakthrough projects
- Next-generation projects
- New family members
- Add-ons and enhancement projects

Radical technological breakthrough projects are the most difficult to manage because of the need for innovation. Figure 7–3 shows a typical model for innovation. Innovation projects, if successful, can lead to profits that are many times larger than the original development costs. Unsuccessful innovation projects can lead to equally dramatic losses, which is one of the reasons why senior management must exercise due caution in approv-

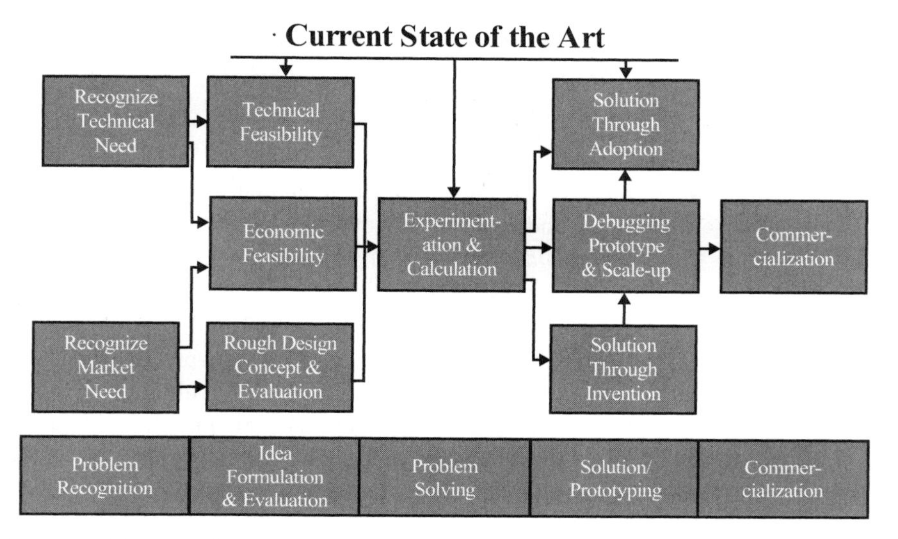

FIGURE 7–3. Modeling the innovation process.

ing innovation projects. Care must be taken to identify and screen out inferior candidate projects before committing significant resources to them.

There is no question that innovation projects are the most costly and difficult to manage. Some companies mistakenly believe that the solution is to minimize or limit the total number of ideas for new projects or to limit the number of ideas in each category. This could be a costly mistake.

In a study of the new-product activities of several hundred companies in all industries, Booz, Allen, and Hamilton[7] defined the new- product evolution process as the time it takes to bring a product to commercial existence. This process began with company objectives, which included fields of product interest, goals, and growth plans, and ended with, hopefully, a successful product. The more specifically these objectives were defined, the greater guidance would be given to the new product program. This process was broken down into six manageable, fairly clear sequential stages:

- *Exploration:* The search for product ideas to meet company objectives.
- *Screening:* A quick analysis to determine which ideas were pertinent and merit more detailed study.
- *Business analysis:* The expansion of the idea, through creative analysis, into a concrete business recommendation, including product features, financial analysis, risk analysis, market assessment, and a program for the product.

---

7. *Management of New Products:* Booz, Allen & Hamilton, Inc., 1984, pp. 180–181.

- *Development:* Turning the idea-on-paper into a product-in-hand, demonstrable and producible. This stage focuses on R&D and the inventive capability of the firm. When unanticipated problems arise, new solutions and trade-offs are sought. In many instances, the obstacles are so great that a solution cannot be found, and work is terminated or deferred.
- *Testing:* The technical and commercial experiments necessary to verify earlier technical and business judgments.
- *Commercialization:* Launching the product in full-scale production and sale; committing the company's reputation and resources.

In the Booz, Allen & Hamilton study, the new-product process was characterized by a decay curve for ideas, as shown in Figure 7–4. This showed a progressive rejection of ideas or projects by stages in the product-evolution process. Although the rate of rejection varied between industries and companies, the general shape of the decay curve is typical. It generally takes close to 60 ideas to yield just one successful new product.

The process of new-product evolution involves a series of management decisions. Each stage is progressively more expensive, as measured in expenditures of both time and money. Figure 7–5 shows the rate at which expense dollars are spent as time accumulates for the average project within a sample of leading companies. This information was based on an all-industry average and is therefore useful in understanding the typical industrial new-product process. It is significant to note that the majority of capital expenditures are concentrated in the last three stages of evolution. It is therefore very important to do a better job of screening for business and financial analysis. This will help eliminate ideas of limited potential before they reach the more expensive stages of evolution.

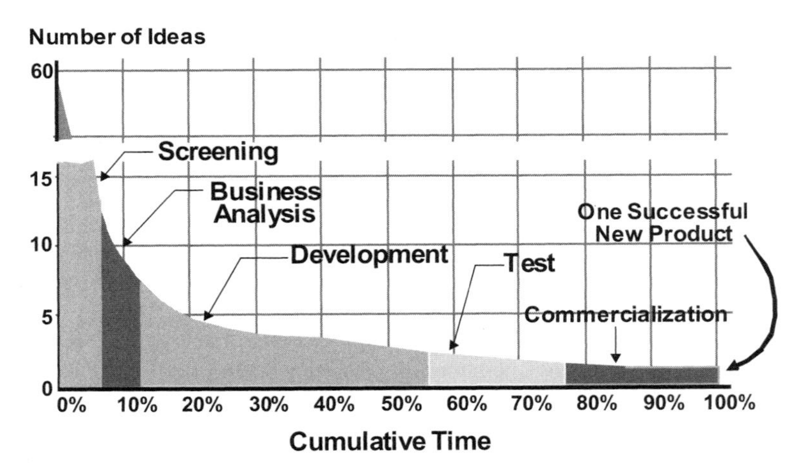

FIGURE 7–4. Mortality of new product ideas.

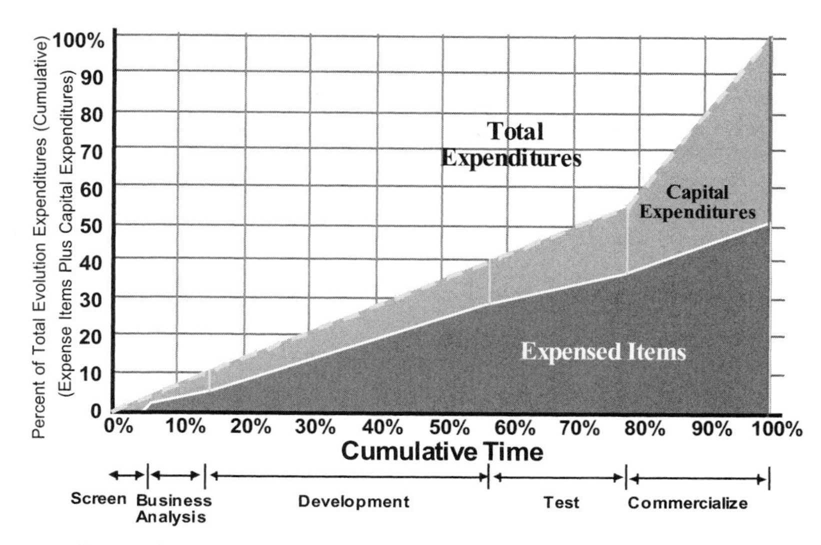

FIGURE 7–5. Cumulative expenditures and time.

7.4 PRELIMINARY EVALUATION

As shown in Figure 7–1, the second step in project selection is preliminary evaluation. From a financial perspective, preliminary evaluation is basically a two-part process. First, the organization will conduct a feasibility study to determine whether the project can be done. The second part is to perform a benefit-to-cost analysis to see whether the company should do it (see Table 7–1).

The purpose of the feasibility study is to validate that the idea or project meets feasibility of cost, technological, safety, marketability, and ease of execution requirements. It is

TABLE 7–1. FEASIBILITY STUDENTS AND COST–BENEFIT ANALYSES

	Feasibility Study	Cost–Benefit Analysis
Basic question	Can we do it?	Should we do it?
Life cycle phase	Preconceptual	Conceptual
PM selected	Not yet	Perhaps
Analysis	Qualitative • Technical • Cost • Quality • Safety • Legal • Economical	Quantitative • NPV • DCF • IRR • ROI • Assumptions • Reality
Decision criteria	Strategic fit	Benefits > cost

possible for the company to use outside consultants or subject matter experts (SMEs) to assist in both feasibility studies and benefit-to-cost analyses. A project manager may not be assigned until after the feasibility study is completed because the project manager may not have sufficient business or technical knowledge to contribute prior to this point in time.

If the project is deemed feasible and a good fit with the strategic plan, the project is prioritized for development along with other approved projects. Once feasibility is determined, a benefit-to-cost analysis is performed to validate that the project will, if executed correctly, provide the required financial and nonfinancial benefits. Benefit-to-cost analyses require significantly more information to be scrutinized than is usually available during a feasibility study. This can be an expensive proposition.

Estimating benefits and costs in a timely manner is very difficult. Benefits are often defined as:

- Tangible benefits, for which dollars may be reasonably quantified and measured
- Intangible benefits, which may be quantified in units other than dollars or may be identified and described subjectively

Costs are significantly more difficult to quantify, at least in a timely and inexpensive manner. The minimum costs that must be determined are those that are used specifically for comparison to the benefits. These include:

- The current operating costs or the cost of operating in today's circumstances.
- Future period costs that are expected and can be planned for.
- Intangible costs that may be difficult to quantify. These costs are often omitted if quantification would contribute little to the decision-making process.

There must be careful documentation of all known constraints and assumptions that were made in developing the costs and the benefits. Unrealistic or unrecognized assumptions are often the cause of unrealistic benefits. The go or no-go decision to continue with a project could very well rest upon the validity of the assumptions.

7.5 STRATEGIC SELECTION OF PROJECTS

From Figure 7–1, the third step in the project selection process is the strategic selection of projects, which includes the determination of a strategic fit and prioritization. It is at this point where senior management's involvement is critical because of the impact that the projects can have on the strategic plan.

Strategic planning and the strategic selection of projects are similar in that both deal with the future profits and growth of the organization. Without a continuous stream of new products or services, the company's strategic planning options may be limited. Today, advances in technology and growing competitive pressure are forcing companies to develop new and innovative products while the life cycle of existing products appears to be decreasing at an alarming rate. Yet, at the same time, executives may keep research groups in a vacuum and fail to take advantage of the potential profit contribution of R&D strategic planning and project selection.

There are three primary reasons by corporations work on internal projects:

- To produce new products or services for profitable growth
- To produce profitable improvements to existing products and services (i.e., cost reduction efforts)
- To produce scientific knowledge that assists in identifying new opportunities or in "fighting fires"

Successful project selection is targeted, but targeting requires a good information system, and this, unfortunately, is the weakest link in most companies. Information systems are needed for optimum targeting efforts, and this includes assessing customer and market needs, economic evaluation, and project selection.

Assessing customer and market needs involves opportunity-seeking and commercial intelligence functions. Most companies delegate these responsibilities to the marketing group, and this may result in a detrimental effort because marketing groups appear to be overwhelmed with today's products and near-term profitability. They simply do not have the time or resources to adequately analyze other activities that have long-term implications. Also, marketing groups may not have technically trained personnel who can communicate effectively with the R&D groups of the customers and suppliers.

Most organizations have established project selection criteria, which may be subjective, objective, quantitative, qualitative, or simply a seat-of-the-pants guess. The selection criteria are most often based upon suitability criteria, such as:

- Similar in technology
- Similar marketing methods used
- Similar distribution channels used
- Can be sold by current sales force
- Will be purchased by the same customers as current products
- Fits the company philosophy or image
- Uses existing know-how or expertise
- Fits current production facilities
- Both research and marketing personnel enthusiastic
- Fits the company long-range plan
- Fits current profit goals

In any event, there should be a valid reason for selecting the project. Executives responsible for selection and prioritization often seek input from other executives and managers before moving forward. One way to seek input in a quick and reasonable manner is to transform the suitability criteria shown above into rating models. Typical rating models are shown in Figures 7–6, 7–7 and 7–8.[8] These models can be used for both strategic selection and prioritization.

---

8. William Souder, *Project Selection and Economic Appraisal.* New York: Van Nostrand Reinhold, 1984, pp. 66–69.

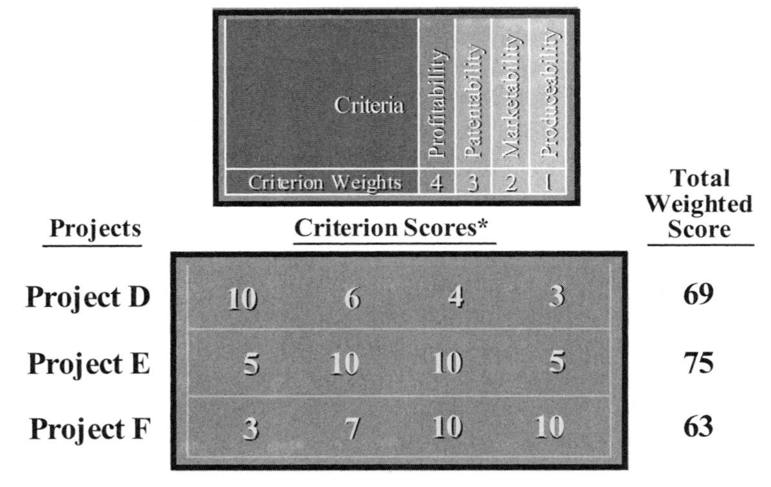

FIGURE 7–6. Scoring model.

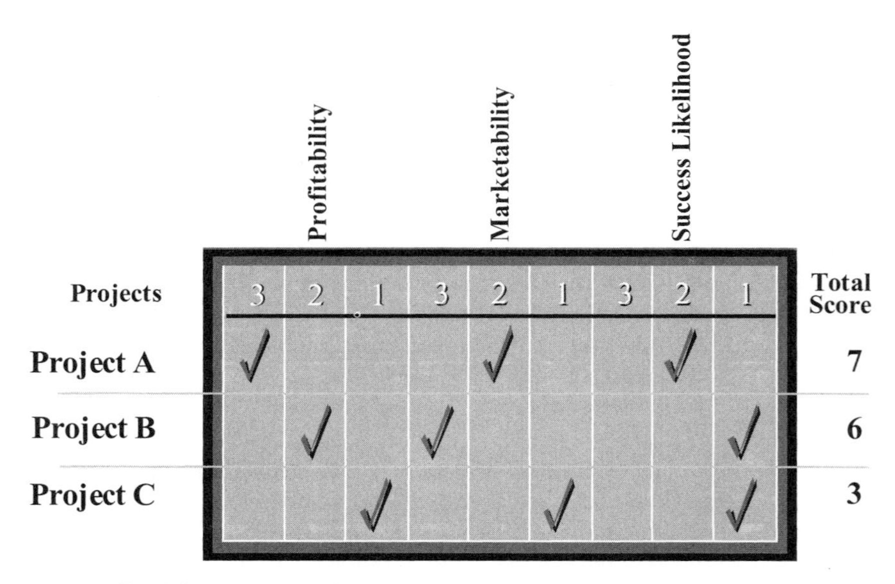

FIGURE 7–7. Checklist for three projects.

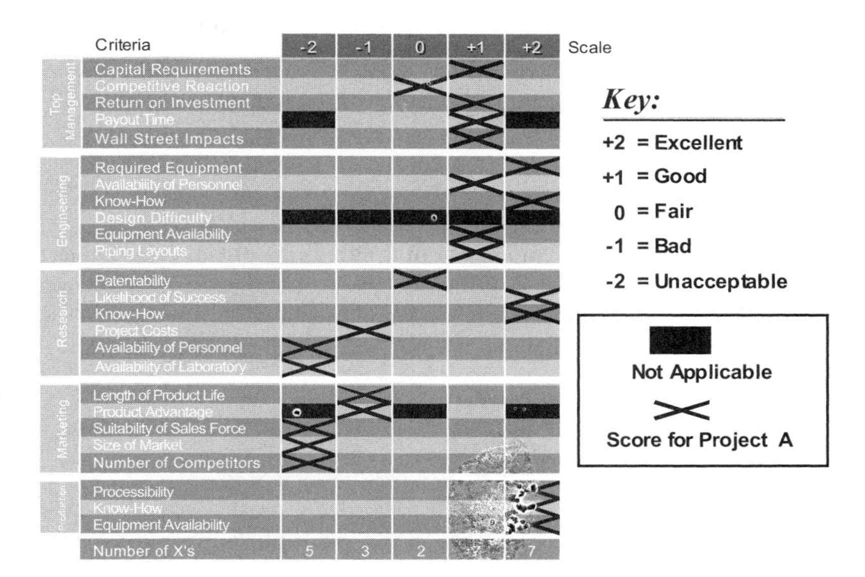

FIGURE 7–8. Scaling model for one project, Project A.

Prioritization is a difficult process. Factors such as cash flow, near-term profitability and stakeholder expectations must be considered. Also considered are a host of environmental forces, such as consumer needs, competitive behavior, existing or forecasted technology, and government policy.

Being highly conservative during project selection and prioritization could be a road map to disaster. Companies with highly sophisticated industrial products must pursue an aggressive approach to project selection or risk obsolescence. This also mandates the support of a strong technical base.

7.6 STRATEGIC TIMING

Many organizations make the fatal mistake of taking on too many projects without regard for the limited availability of resources. As a result, the highly skilled labor is assigned to more than one project, creating schedule slippages, lower productivity, less than anticipated profits, and never-ending project conflicts.

The selection and prioritization of projects must be made based upon the availability of qualified resources. Planning models are available to help with the strategic timing of resources. These models are often referred to as *aggregate planning models.*

Another issue with strategic timing is the determination of which projects require the best resources. Some companies use a risk–reward cube, where the resources are assigned based upon the relationship between risk and reward. The problem with this approach is that the time required to achieve the benefits (i.e., payback period) is not considered.

Aggregate planning models allow an organization to identify the overcommitment of resources. This could mean that high-priority projects may need to be shifted in time or possibly be eliminated from the queue because of the unavailability of qualified resources. It is a pity that companies also waste time considering projects for which they know that the organization lacks the appropriate talent.

Another key component of timing is the organization's tolerance level for risk. Here, the focus is on the risk level of the portfolio rather than the risk level of an individual project. Decision makers who understand risk management can then assign resources effectively such that the portfolio risk is mitigated or avoided.

7.7 ANALYZING THE PORTFOLIO

Companies that are project-driven organizations must be careful about the type and quantity of projects they work on because of the resources available. Because of critical timing, it is not always possible to hire new employees and have them trained in time, or to hire subcontractors that may end up possessing questionable skills.

Figure 7–9 shows a typical project portfolio.[9] Each circle represents a project. The location of each circle represents the quality of resources and the life cycle phase that the project is in. The size of the circle represents the magnitude of the benefits relative to other projects, and the pie wedge represents the percentage of the project completed thus far.

In Figure 7–9, Project A has relatively low benefits and uses medium-quality resources. Project A is in the definition phase. However, when Project A moves into the design phase, the quality of resources may change to low quality or strong quality. Therefore, this type of chart has to be updated frequently.

Figures 7–10, 7–11, and 7–12 show three types of portfolios. Figure 7–10 represents a high-risk project portfolio where strong resources are required on each project. This may be representative of project-driven organizations that have been awarded large, highly profitable projects. This could also be a company in the computer field that competes in an industry that has short product life cycles and where product obsolescence occurs six months downstream.

Figure 7–11 represents a conservative, profit portfolio where an organization works on low-risk projects that require low-quality resources. This could be representative of a project portfolio selection process in a service organization, or even a manufacturing firm that has projects designed mostly for product enhancement.

Figure 7–12 shows a balanced portfolio with projects in each life cycle phase and where all quality of resource levels are being utilized, usually quite effectively. A very delicate juggling act is required to maintain this balance.

9. This type of portfolio was adapted from the life cycle portfolio model commonly used for strategic planning activities.

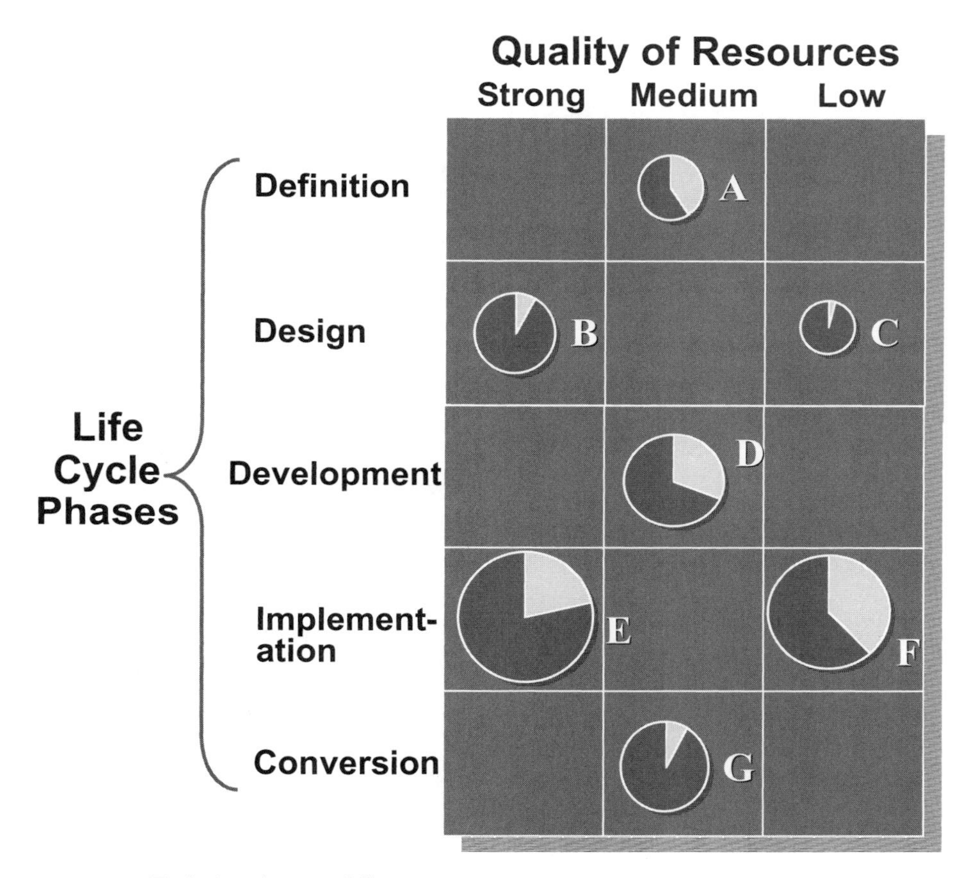

FIGURE 7–9. Typical project portfolio.

7.8 PROBLEMS WITH MEETING EXPECTATIONS

Why is it that, more often than not, the final results of either a project or an entire portfolio do not meet senior management's expectations? This problem plagues many corporations and the blame is ultimately (and often erroneously) rationalized as poor project management practices. As an example, a company approved a portfolio of 20 R&D projects for 2001. Each project was selected on its ability to be launched as a successful new product. The approvals were made following the completion of the feasibility studies. Budgets and timetables were then established such that the cash flows from the launch of the new products would support the dividends and the cash needed for ongoing operations.

Full-time project managers were assigned to each of the 20 projects and began with the development of detailed schedules and project plans. For eight of the projects, it quickly became apparent that the financial and scheduling constraints imposed by senior management were unrealistic. The project managers on these eight projects decided not to

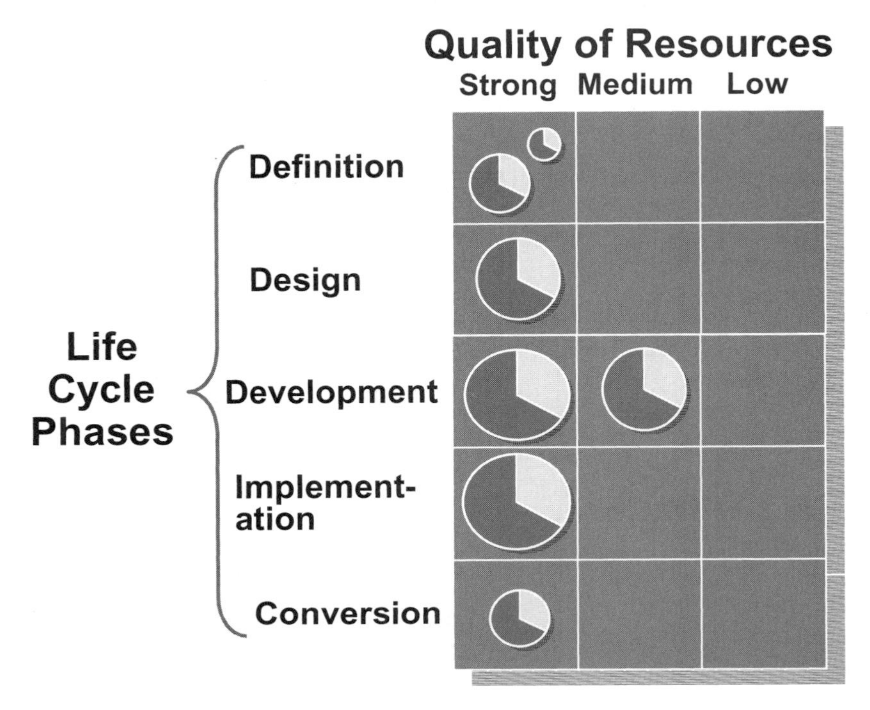

FIGURE 7–10. High-risk portfolio.

inform senior management of the potential problems but to wait awhile to see if contingency plans could be established. Hearing no bad news, senior management was left with the impression that all launch dates were realistic and would go as planned.

The eight trouble-plagued projects were having a difficult time. After exhausting all options, and failing to see a miracle occur, the project managers then reluctantly informed senior management that their expectations would not be met. This occurred so late in the project life cycle that senior management became quite irate and several employees had their employment terminated, including some of the project sponsors.

Several lessons can be learned from this situation. First, unrealistic expectations occur when financial analysis is performed from "soft" data rather than "hard" data. In Table 7–1 we showed the differences between a feasibility study and a benefit-to-cost analysis. Generally speaking, feasibility studies are based upon soft data. Therefore, critical financial decisions based upon feasibility study results may have significant errors. This can also be seen from Table 7–2, which illustrates the accuracy of typical estimates. Feasibility studies use top-down estimates that can contain significant error margins.

Benefit-to-cost analyses should be conducted from detailed project plans using more definitive estimates. Benefit-to-cost analysis results should be used to validate that the financial targets established by senior management are realistic.

Even with the best project plans and comprehensive benefit-to-cost analyses, scope changes will occur. There must be a periodic reestimation of expectations performed on a

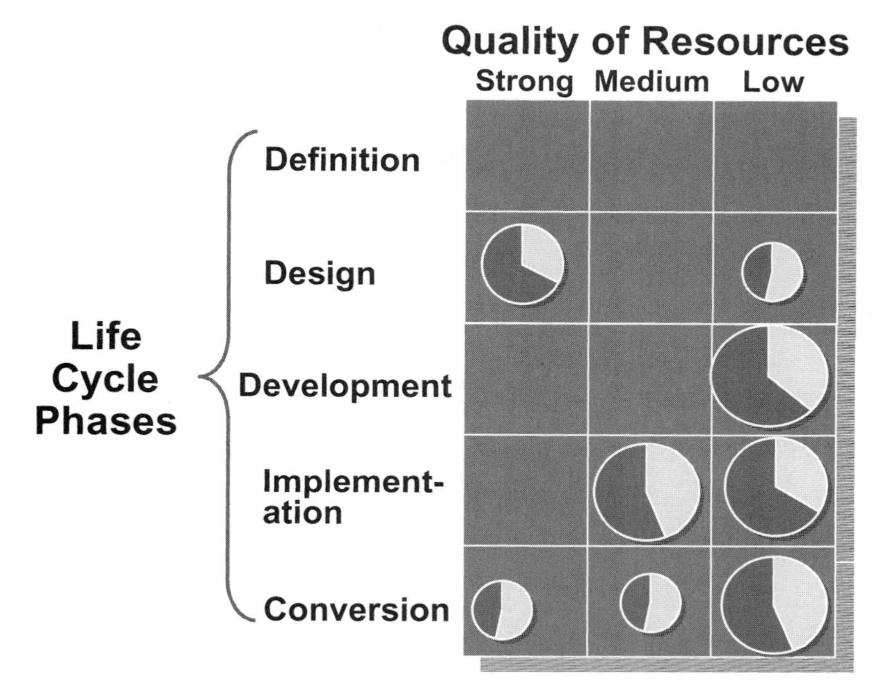

FIGURE 7–11. Profit portfolio.

timely basis. One way of doing this is by using the rolling wave concept shown in Figure 7–13. The rolling wave concept implies that as you get further along in the project, more knowledge is gained, which allows us to perform more detailed planning and estimating. This then provides additional information from which we can validate the original expectations.

Continuous reevaluation of expectations is critical. At the beginning of a project, it is impossible to ensure that the benefits expected by senior management will be realized at project completion. The length of the project is a critical factor. Based upon project length, scope changes may result in project redirection. The culprit is most often changing economic conditions, resulting in invalid original assumptions. Also, senior management must be made aware of events that can alter expectations. This information must be made known quickly. Senior management must be willing to hear bad news and have the courage to possibly cancel a project.

Since changes can alter expectations, project portfolio management must be integrated with the project's change management process. According to Mark Forman, the Associate Director for IT and e-Government in the Office of Management and Budget[10]:

> Many agencies fail to transform their process for IT management using the portfolio management process because they don't have change management in place before starting. IT

10. See note 1, p. 1.

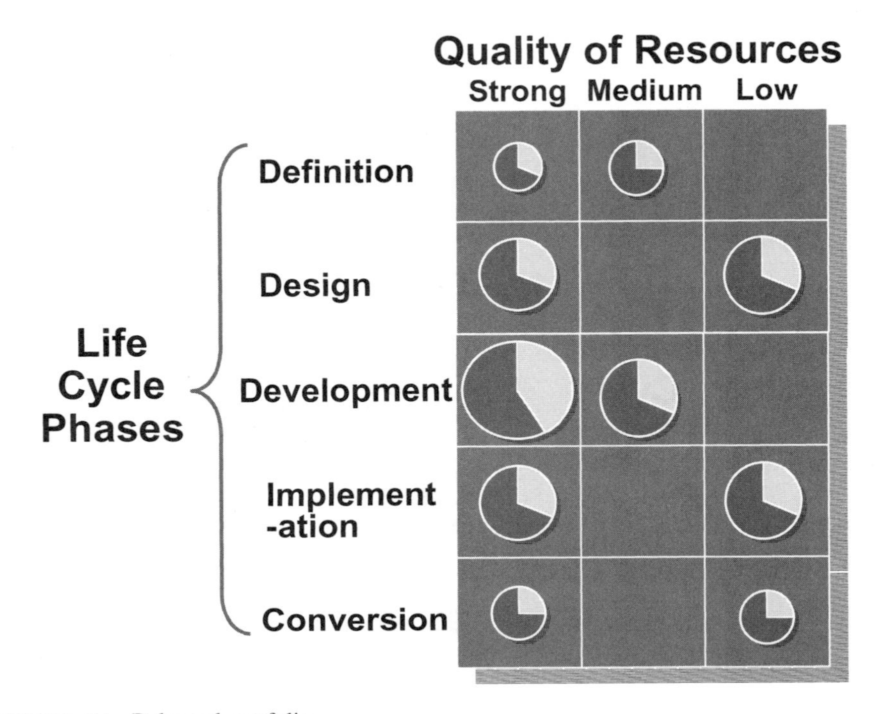

FIGURE 7–12. Balanced portfolio.

will not solve management problems—re-engineering processes will. Agencies have to train their people to address the cultural issues. They need to ask if their process is a simple process. A change management plan is needed. This is where senior management vision and direction is sorely needed in agencies.

Although the comments here are from government IT agencies, the problem is still of paramount importance in nongovernment organizations and across all industries.

TABLE 7–2. COST/HOUR ESTIMATES

Estimating Method	Generic Type	WBS Relationship	Accuracy	Time to Prepare
Parametric	ROM[a]	Top down	25% to +75%	Days
Analogy	Budget	Top down	−10% to +25%	Weeks
Engineering (grass roots)	Definitive	Bottom up	−5% to +10%	Months

[a]Rough order of magnitude.

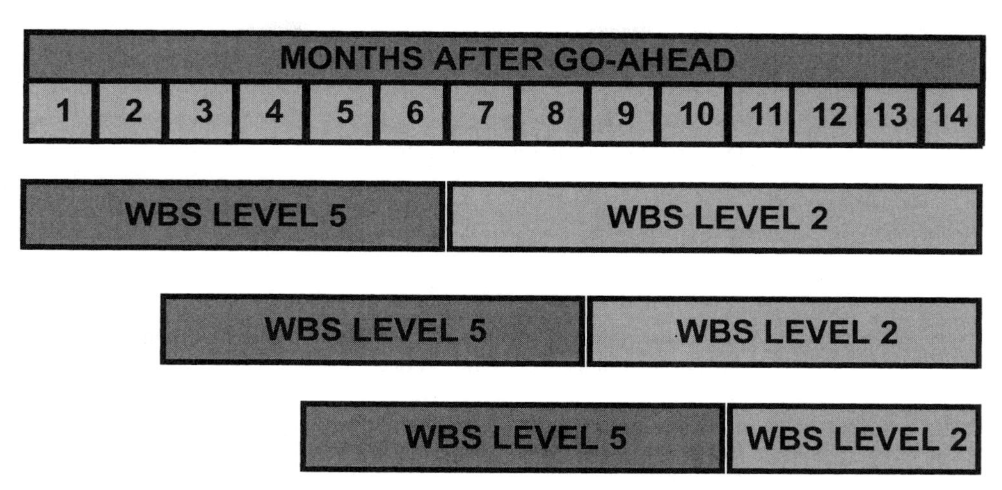

FIGURE 7–13. Rolling wave concept.

MULTIPLE CHOICE QUESTIONS

1. Portfolio management mandates that decisions be made in the best interest of the:
  A. Portfolio
  B. Individual projects
  C. Individual stakeholders
  D. Project sponsors

2. Portfolio management focuses on:
  A. The highest prioritized project
  B. The best mix of projects
  C. The priorities of the decision-makers
  D. Project-specific calculations

3. The lack of a systematic approach to project selection is most often the result of:
  A. Behavioral factors
  B. Lack of project sponsorship
  C. Lack of adequate information
  D. Lack of stakeholder involvement

4. Most project managers are brought on board the project:
  A. Too early
  B. After the feasibility study is completed
  C. After the benefit-to-cost analysis is completed
  D. At the beginning of project execution

5. Continuous improvement to a project management methodology leads to:
  A. An unstructured portfolio management process
  B. A structured portfolio management process
  C. Inconsistent sponsorship
  D. Poor decision-making

6. Which of the following is not part of the overall portfolio management process?
   A. Identification of projects
   B. Preliminary evaluation
   C. Project sponsorship
   D. Strategic selection of projects
7. Which of the following is not a common classification system for projects?
   A. Survival vs. growth
   B. Offensive vs. defensive
   C. Innovation vs. add-ons
   D. Internal vs. external
8. According to a study by Booz, Allen & Hamilton, how many ideas are needed to develop one commercially successful product?
   A. 10
   B. 30
   C. 60
   D. 100
9. According to the Booz, Allen & Hamilton study, the new product development process most frequently is represented by:
   A. An exponentially increasing curve
   B. A decay curve
   C. A straight line sloping upward
   D. A straight line sloping downward
10. Which of the following is not considered part of a feasibility study?
    A. What can be done?
    B. What should be done?
    C. Manufacturability
    D. Safety considerations
11. Which of the following is not considered part of a benefit-to-cost analysis?
    A. What can be done?
    B. What should be done?
    C. Tangible benefits
    D. Intangible benefits
12. Validation of the financial objectives usually occurs:
    A. Immediately before the feasibility study
    B. Immediately after the feasibility study
    C. Immediately before the benefit-to-cost analysis
    D. Immediately after the benefit-to-cost analysis
13. Determining which projects require the best resources is part of:
    A. Project identification
    B. Preliminary evaluation
    C. Project prioritization
    D. Timing
14. Aggregate planning models allow an organization to identify __________ of resources.
    A. The cost
    B. Overcommitment or undercommitment
    C. The desired skill level
    D. All of the above

15. Decision-makers who understand risk management can assign resources to __________ the risks.
    A. Eliminate
    B. Identify and eliminate
    C. Postpone
    D. Mitigate
16. Portfolio management has, as its primary objective, to __________ and maintain the appropriate mix of projects.
    A. Identify
    B. Select
    C. Finance
    D. All of the above
17. Portfolio management must consider the following:
    A. Aggregate costs
    B. Aggregate risks
    C. Trade-offs among projects
    D. All of the above
18. Portfolio management can determine:
    A. The best mix of projects
    B. The right level of investment for each project
    C. The appropriate relationship between maintenance expenditures and strategic initiatives
    D. All of the above
19. Consistently applying a uniform set of criteria to be used among competing projects is the responsibility of:
    A. Portfolio managers
    B. Executive management
    C. Project managers
    D. Project sponsors
20. Adjusting a portfolio's size, scope, schedule composition, and funding is the responsibility of:
    A. Stakeholders
    B. Executive management
    C. Portfolio management
    D. All of the above

DISCUSSION QUESTIONS

1. What is easier to quantify and why: the costs or the benefits?
2. Why are feasibility studies referred to as soft data?
3. Explain why strategic timing is critical during the portfolio management process.
4. What are the major differences between portfolio management and project management?
5. Why is stakeholder involvement so critical after portfolio decision-making has been completed?
6. What information should go into a rating model for project selection and prioritization?

7. What are the differences in responsibilities between executive management and portfolio management?
8. What is the relationship between strategic timing and resource availability?
9. How can strategic timing influence portfolio risk mitigation?
10. How important is it to reevaluate expected financial returns periodically during portfolio management?

Across

1. Conformance evaluation
5. Type of benefit
6. Portfolio management process
7. Type of process
8. Classification system: _ _ _ _ _ _ _ _ _ _ _ vs. add-ons
11. Classification system: _ _ _ _ _ _ _ _ _ _ vs.defensive
13. Purpose of risk management
14. Portfolio management process

Down

2. Limited _ _ _ _ _ _ _ _ _ _ _ _ _ of resources
3. New product development curve
4. Type of management
9. Type of benefit
10. _ _ _ _ _ _ _ _-to-cost analysis
12. Portfolio management process

The Project Office

8.0 INTRODUCTION

As companies begin to recognize the favorable effect that project management has on profitability, emphasis is placed upon achieving professionalism in project management using the project office concept. The concept of a project office (PO) or project management office (PMO) could very well be the most important project management activity in this decade. With this recognition of importance comes strategic planning for both project management and the project office. Maturity and excellence in project management does *not* occur simply by using project management over a prolonged period of time. Rather, it comes through strategic planning for both project management and the project office.

General strategic planning involves the determination of where you wish to be in the future and then how you plan to get there. For project office strategic planning, it is often easier to decide which activities should be under the control of the project office than determining how or when to do it. For each activity placed under the auspices of the project office, there may appear pockets of resistance that initially view removing this activity from their functional area as a threat to their power and authority.

8.1 THE PROJECT OFFICE: 1950–1990

For almost 40 years, the project office (or program office) functioned as a customer group project office and was compromised of a group of project management personnel assigned to a specific project, usually a large project. Aerospace and defense contractors created three project offices for grouping Army, Navy, and Air Force customers. Some companies created project offices exclusively to service either large projects or small projects.

The concept behind this project office approach was to get closer to the customer by setting up an organization dedicated to that customer. The project office became an organization within an organization and could function as a "real" or "virtual" organization to service a particular customer. The majority of these "projects" were actually programs that were very large in dollar value and with multiyear government funding. It was not uncommon for people to spend 10 or 15 years working on just one project.

The members of the project office had unique roles and responsibilities, but essentially worked together as a project management team. Each person in the project office may have been required to have both primary and secondary responsibilities. The secondary responsibilities included functioning as a backup for other project office personnel in case some project office personnel were reassigned to other projects, left the company, or were out sick.

Head count in the project office was not considered extremely important because the customer often paid the added costs. Technology and schedules were viewed as significantly more important than cost. Customers preferred to have more people than necessary assigned to project offices. The cost of having someone assigned full-time to the project office was viewed as an insignificant overmanagement cost compared to the risks of undermanagement, where individuals were assigned part-time but may be needed full-time. The only people who were trained in project management, and truly understood it, were the project office personnel. Project offices functioned horizontally throughout the organization and were viewed as profit centers, whereas the traditional functional hierarchies were treated as cost centers.

During the 1980s, military and government agencies became more cost-conscious. Project offices were pared down as personnel other than those assigned to the project office underwent training in project management. Line managers also underwent training and were asked to better understand project management and to share accountability with project managers for project success.

8.2 THE PROJECT OFFICE: 1990–2000

The 1990s began with a recession that took a heavy toll on white-collar ranks. Management's desire for efficiency and effectiveness led them to take a hard look at nontraditional management techniques such as project management. Project management began to expand to non–project-driven industries. The benefits of using project management, which were once seen as applicable only to the aerospace, defense, and heavy construction industries, were now recognized as being applicable for other industries.

As the benefits of project management became apparent, management understood that there might be a significant, favorable impact on the corporate bottom line. This led management to two important conclusions:

- Project management had to be integrated and compatible with the corporate reward systems for sustained project management growth.
- Corporate recognition of project management as a profession was essential in order to maximize performance.

The recognition of project management professionalism led companies to accept PMI's Certification Program as the standard and to recognize the importance of the project office concept. Consideration was being given to all critical activities related to project management to be placed under the supervision of the project office. This included such topics as:

- Standardization in estimating
- Standardization in planning
- Standardization in scheduling
- Standardization in control
- Standardization in reporting
- Clarification of project management roles and responsibilities
- Preparation of job descriptions for project managers
- Preparation of archive data on lessons learned
- Benchmarking continuously
- Developing project management templates
- Developing a project management methodology
- Recommending and implementing changes and improvements to the existing methodology
- Identifying project standards
- Identifying best practices
- Performing strategic planning for project management
- Establishing a project management problem-solving hotline
- Coordinating and/or conducting project management training programs
- Transferring knowledge through coaching and mentorship
- Developing a corporate resource capacity/utilization plan
- Assessing risks
- Planning for disaster recovery

With these changes taking place, organizations began changing the name of the project office to the center of excellence (COE) in project management. The COE was mainly responsible for providing information to stakeholders rather than actually executing projects or making midcourse corrections to a plan.

Each of these activities brought with it both advantages and disadvantages. The majority of the disadvantages were attributed to the increased levels of resistance to the new responsibilities given to the project office.

For simplicity sake, the resistance levels can be classified as low risk, moderate risk, and high risk according to the following definitions:

- *Low Risk:* Easily accepted by the organization with very little shift in the balance of power. Virtually no impact on the corporate culture.
- *Moderate Risk:* Some resistance by the corporate culture and possibly a shift in the balance of power and authority. Resistance levels can be overcome in the near term and with minimal effort.
- *High Risk:* Heavy pockets of resistance exist and a definite shift in some power and authority relationships. Strong executive leadership may be necessary to overcome the resistance.

FIGURE 8–1. Activity implementation risks.

Not every project office has the same responsibilities. Similarly, the same responsibilities implemented in two project offices can have differing degrees of risk. People tend to resist change even when they know that the change may be in the best interest of the organization.

Figure 8–1 shows typical risk levels for implementing the project office responsibilities selected. Evaluating potential implementation risks is critical. It may be easier to gain support for the establishment of a project office by implementing low-risk activities first. The low-risk activities in Figure 8–1 are operational activities to support project management efforts in the near term, whereas the high-risk activities are more in line with strategic planning responsibilities and possibly the control of sensitive information.

8.3 THE PROJECT OFFICE: 2000–PRESENT

As we entered the twenty-first century, the project office became commonplace in the corporate hierarchy. Although the majority of activities assigned to the project office had not changed, there was now a new mission for the project office:

- *The project office now has the responsibility for maintaining all intellectual property related to project management and to actively support corporate strategic planning.*

The project office was now servicing the corporation, especially the strategic planning activities, rather than focusing on a specific customer. The project office was transformed into a corporate center for control of project management intellectual property. This was a necessity as the magnitude of project management information grew almost exponentially throughout the organization.

Senior managers were now recognizing that project management and the project office had become an invaluable asset for senior management as well as for the working levels. As an example, consider the following comments from senior management at American Greetings:

- Through project management, we learned the value of defining specific projects and empowering teams to make them happen. We've embraced the program management philosophy and now we can use it again and again and again to reach our goals.
  Jim Spira, Retired President and Chief Operating Officer
- The program management office provides the structure and discipline to complete the work that needs to get done. From launch to completion, each project has a roadmap for meeting the objectives that were set.
  Jeff Weiss, President and Chief Operating Officer
- Through project management, we've learned how to make fact-based decisions. Too often in the past we based our decisions on what we thought could happen or what we hoped would happen. Now we can look at the facts, interpret the facts honestly and make sound decisions and set realistic goals based on this information.
  Zev Weiss, Chief Executive Officer

During the past 10 years, the benefits for the executive levels of management of using a project office have become apparent. They include:

- Standardization of operations
- Company rather than silo decision-making
- Better capacity planning (i.e., resource allocations)
- Quicker access to higher-quality information
- Elimination or reduction of company silos
- More efficient and effective operations
- Less need for restructuring
- Fewer meetings which rob executives of valuable time
- More realistic prioritization of work
- Development of future general managers

All of the benefits above are related to project management intellectual property either directly or indirectly. To maintain the project management intellectual property, the project office must maintain the vehicles for capturing the data and then disseminating the data to the various stakeholders. These vehicles include the company project management intranet, project Web sites, project databases, and project management information systems. Since much of this information is necessary for both project management and corporate strategic planning, there must exist strategic planning for the project office.

Computer Associates

Previously we showed that workers are sometimes fearful about the implementation of a project management office. Most of the fear is the result of a lack of understanding of the purpose of a project or project management office. This fear can often be overcome by the development of a clearly defined mission statement for a PMO and clearly stated objectives. As an example, consider the following information provided by Computer Associates:

PMO Mission Statement

The PMO will provide assistance and advice in the production and implementation of new policies and processes in conjunction with executive management. To continually improve, we will audit the compliance with policy and process. The audit information will be analyzed to provide data†for process improvement. By continually monitoring and improving processes and procedures, CA Services will improve consistency, profitability, repeatability and quality when executing engagements.

PMO Objectives

- Approving Proposals/SOWs based on Risk Analysis and application of CA's Best Practice methodologies and estimation techniques
- Mentoring project managers in the use of CA's Engagement Management Model and Project Management Methodology, forms, templates
- Auditing projects to ensure adherence to CA methodology
- Performing following paid quality assessments where stipulated in the SOW:
  - Mid-project or end-of-stage Assessments (QA Reviews), involving CA and Client staff of all seniority
  - End-of-project Readiness Assessments
  - Inspection of milestones prior to delivery to the client
- Recommending upgrades to the Best Practice Library—methodology and templates—based on feedback from post-project assessments and CAR analysis
- Providing project review and oversight
- Confirming the appropriate archival of project documentation

Virginia Department of Transportation

Sooner or later, most organizations recognize the need for a project management office or a center of excellence for project management. This recognition appears not only in the private sector but in the public sector as well.

Mal Kerley, the Virginia Department of Transportation (VDOT) Chief Engineer for Program Development, describes the VDOT Project Management Office:

Virginia Department of Transportation (VDOT) has been using some form of project management since its inception (since the early 1900s). More recently, VDOT has undertaken the development of a Project Management Office (PMO) to assist with modernizing the department's project management tools, techniques, and knowledge, as well as providing project management support to project managers. The PMO will be staffed initially with one PMO manager and three project management analysts. Currently, the PMO is developing an improved project management methodology to complement existing processes in the area of preliminary engineering. In addition, the PMO has trained several engineers in project management principles.

Roadway Express

In 1997, Kelly Baumer, Director of Project Management and Integration at Roadway Express, described the evolution and success of the center for excellence at Roadway:

Our organization has been practicing project management since 1991; however, there was no formalized structure in place. It required officer commitment in 1992, specifically because of rollover of project managers on mission critical projects. Roadway Express hired a consultant/trainer to come in and discuss the pitfalls of not profitably running projects. He trained all current business and I.S. project personnel on proper practices and guidelines and established a methodology. Additionally, senior executives were given a one-day course on how to sponsor a project, what project management can do for the company, etc. After that training, our President and CEO at the time, Mike Wickham, required all key officers to be a sponsor of a major project initiative. Finally in February of 1996, key managers/directors of the company developed a Roadway Express Project Management Guide. This guide was used as a spawning ground for establishing the Center for Excellence (C/E). The C/E is a group of peers and acts as a forum for project managers to present their projects for approval (before presenting them to the Steering Committee). In 1994, we established an I.S. Steering Committee comprised of corporate officers, the Director of Audit, the Director of Project Management & Integration, and the Director of I.S. The Steering Committee is the actual body that will approve or reject a project. The C/E, however, is a way for project managers to be critiqued by their peers before going to the Steering Committee to ensure they have covered all bases.

We feel one of the strengths was in the establishment of the Center for Excellence. The C/E is a committee established to provide consultant service and guidance to those managing projects at Roadway. The Director of Project Management facilitates the C/E. The C/E and the PMO (Project Management Office) aka the Project Management and Integration Department, ensures that the methods outlined in the Project Management Guide are followed. They review and revise methods as necessary, assist project managers in obtaining officer support, provide input to corporate strategic planning, handle project issues that do not require officer integration, guide project managers preparing presentations to the Steering Committee, and provide a line of formal communication with the Steering Committee. Projects can't officially be approved or rejected by the C/E; however, if peers do not support the project or its impacts to the organization, it is highly unlikely that the project manager will go forward to the Steering Committee. With both the Director of I.S. and the Director of Project Management sitting on the C/E as well as the Steering Committee, the officers often look to them for feedback and conformity.

Since the inception of the C/E and Steering Committee forum, the company has realized the benefit of good sound practices in project management. Therefore, in January of 1998 they formally formed the Project Management Office (PMO). The PMO is responsible

> now to run all major large cross-functional projects in the corporation, whether they be operational, financial, administration, or technological in nature. Establishment of this PMO has helped us reduce product development time, reduce turnover of project managers in the functional areas, and has allowed us as a company to realize project management as a strategic initiative in the corporation.
>
> Establishment of the PMO is an example of how Roadway Express has used project management as a strategic necessity in our industry. All projects, at least those cross functional in nature, have a senior executive as a sponsor. It is the responsibility of the line managers to fund resources necessary to accomplish the project goals and objectives. Senior executives are involved whenever conflict resolution cannot be met between the project managers (PMO) and line managers. Generally speaking, conflict management is handled at the project line-management level.
>
> The officers of the company rely on the PMO to execute many of their strategic initiatives. Because the PMO is a seasoned group of veteran project managers, we're not likely to repeat the same errors, therefore reducing some of the life cycle of the project.
>
> Additionally, we have established a "lessons learned" database for members of the C/E to use as a mentoring tool/archive. In our lessons learned file we also have letters from key senior executives on how to be an active and good sponsor of projects.

Edelca

The benefits of a project office now appear to be recognized worldwide. However, each project office may have different responsibilities, and these responsibilities can change over time. As an example, consider the following remarks from Marcelino Diez, PMP, project management consultant at Edelca:

> Edelca maintains a Corporate Project Office whose functions are being constantly redefined and updated. In the past, the project office performed some operative tasks relating to investment controls, cost variations, and payment transactions. The new basic functions proposed for the Corporate Project Office are related to the following:
>
> - To be a link between the projects and the company's strategic plans by participating in project portfolio management
> - To improve the maturity of project management in the company by the development and updating of the methodology, personnel coaching and training
> - To evaluate and incorporate an adequate information system technology
> - To provide expert assistance to the project managers and to the project teams
> - To perform assessments and audits of the projects
> - To register and centralize project information and lessons learned
> - To promote project management as a fundamental core competency in the company.

General Motors Powertrain

Some organizations require full-time assignments to the centers for excellence or project offices. However, it is possible for the assignments to be part-time and still produce effective results. General Motors Powertrain has been successful at creating a center of excellence in project management using a combination of full- and part-time support.

At General Motors Powertrain during the 1990s, Rose Russett formed a program administrator network to bring together all administrators from various programs to create a center of excellence on program management. The network's mission is to:

- Improve the management of the four-phase product development process
- Share lessons learned across all Powertrain programs
- Ensure consistent application of common processes and systems required for successful execution of programs
- Focus on process improvement while project teams focus on successful program execution

Through the program administrator network, and in combination with their highly successful committee sponsorship concept, the General Motors Powertrain is positioned for excellence in project management well into the next century. The concept of program administrators, as used by General Motors Powertrain, could very well be part of the future of project management.

The concept of the project administrator is not new. During the early years of project management, especially on large projects, the management of the project and control of information was heavily labor-intensive. Computer technology as we know it today was nonexistent. Large projects were managed in "war rooms" that contained one door and no windows. The walls were covered with schedules prepared either on large sheets of paper or on sliding magnetic boards. Everything was a manual effort.

Word processors and PowerPoint weren't invented yet. Project managers were glorified paper-pushers worrying about time and cost. The chief project engineer, who was also called the assistant project manager for engineering, made most of the technical decisions.

To get the project manager more actively involved in the project, large project teams began assigning project administrators to alleviate some of the paperwork headaches from the shoulders of the project manager. There could be a project administrator responsible for manually updating all of the schedules in the war room. Another project administrator could be responsible for handouts, reports, and recording minutes of team meetings.

As technology began to change and ease the pain of document management, the role of the project administrator changed. Rather than eliminate the job of the project administrator, companies found additional responsibilities that could be undertaken by the project administrator.

Use of a project administrator and the creation of a project administration functional area are growing. At Middough Consulting Inc., Kim Dontenville is the Project Services Manager. She comments on the role of a project administrator:

> The role of the project administrator has changed significantly over the last 20 years. It continues to evolve today, and in the future will play an even more critical role in project management.
>
> In the past, the project administrator performed tasks similar to a traditional secretary. These responsibilities included typing, filing and document reproduction on a task basis. Outside client, consultant and vendor contact was conducted only to the limit of planning meetings.

> During the 80's and 90's, the economy changed and businesses responded with "right-sizing" or "down-sizing"; however, one aspect of the organizational chart that didn't change was the existence of a project administrator role. The need to do more with less senior management fell heavily on the support staff. The role of the project administrator quickly evolved into more than an employee who would create and reproduce documents. This role was now responsible for document management during the emergence of electronic networks, e-mail, and Internet website collaboration. The project administrator was communicating with clients, consultants and vendors to manage project documentation. This produced relationship building and practical project work experience with peers, project managers and, depending upon the structure of an organization, CEO's or other high-level project sponsors.
>
> The experience gained by this expanded role and the advanced skills required are now leading the way for project administrator professionals to take on more responsibility as it relates to the office environment as well as working with project managers, clients, consultants and vendors. Today's project administrator is not only responsible for document management and project work procedures, but also for business arrangements including contract negotiations and management of office equipment, supplies, travel and accommodations.
>
> In the future, the role of the project administrator will further develop into a management role that is highly integrated with the project team and will be responsible for multiple project assistants. The responsibilities will include project planning, client protocol, connectivity and procedures that include project document management, network filing, equipment utilization and use of website collaborations. The importance of the role of project administrator will continue to increase significantly as business and industry realize the positive impact on cost, quality and relationships resulting from professional project administration in this information age.

In some companies, including General Motors' Powertrain Division and Middough Consulting, the project administrators worked together to look for ways to improve the project management methodology and to capture best practices and lessons learned. The project administrator is now an invaluable asset for both the project manager and the entire company.

8.4 TYPES OF PROJECT OFFICES

There exist three types of project offices commonly used in companies.

- *Functional Project Office:* This type of project office is utilized in one functional area or division of an organization, such as information systems. The major responsibility of this type of project office is to manage a critical resource pool (i.e., resource management).
- *Customer Group Project Office:* This type of project office is for better customer management and customer communications. Common customers or projects are clustered together for better management and customer relations. Multiple customer group project offices can exist at the same time and may end up functioning as a temporary organization. In effect, this acts like a company within a company.

- *Corporate Project Office:* This type of project office services the entire company and focuses on corporate and strategic issues rather than functional issues.

Let's look at a few examples.

StoneBridge Group

Project offices can reside anywhere in an organization. However, it is vitally important to recognize that the existence of the project office is to service the entire organization, and therefore whatever type of project office is developed, it must be networked with the business areas and share information. Brad Ruzicka, Senior Consulting Manager at the StoneBridge Group, comments on IS project offices:

> We have been contracted to establish such an entity (i.e., PMO) for our clients. In general, we recommend the project office have strategic responsibilities for project definition in partnership with business management. The project office can reside on either the IS or business side but should be jointly sponsored by IS and business management. The project office would generally have a core staff of project managers responsible for managing larger, complex, strategic projects as well as providing mentoring services to both IS and business areas.

DTE Energy

Although functional project offices can be developed anywhere in an organization, they are most common in an information systems environment. DTE Energy maintains an information systems project management office. According to Beverly Jeffries, PMP, Supervisor, ITS Project Management Office:

> The Information Technology Services (ITS) Project Management Office has been in existence at DTE Energy for approximately 4 years. It is a process-based, project management center of excellence that was developed to support the strategic IT goal of the organization to reduce fixed and variable costs for IT projects.
>
> The PMO is a function of the Process Engineering Group, which reports to the ITS Director of Process and Skills. A staff of 23 individuals provide a number of project management support services to the ITS organization. These services include:
>
> - Developing and maintaining the project management methodology
> - Maintaining the portfolio of 30–50 IT projects
> - Collection and dissemination of project data and metrics
> - Administration of project management training curriculum
> - Providing project management support to 20–30 project managers
>
> The ITS portfolio of projects consists of enterprise strategic IT projects and internal infrastructure projects. These projects have budgets that range from less than $50K to over $7M. The types of projects range from application development to preventive maintenance. Through its full implementation, the PMO has enabled the ITS organization to achieve operational efficiencies and increased effectiveness and efficiencies in all project efforts.

Currently, Project Managers who report to functional line managers manage the majority of projects. Every Project Manager is assigned a Coach from the ITS PMO. Large projects will have a Project Planner assigned also. The Coach and Project Planner are knowledgeable in PM best practices, PMO processes, project financials, and project management tools. They work with the Project Manager to coordinate project controls, performance and status reporting and to assist the Project Manager in successfully completing the project.

The success of project management in ITS has been defined as:

- Increase operational efficiencies throughout the organization.
- Achieve the goals of CMM level in the key process areas of project planning and project tracking and oversight.
- Increase the number of PMI certified project managers.
- Increase organizational capability in projecting project performance and project budget forecasting using Earned Value metrics.

ITS has been successful in achieving its strategic goals through project management. Last year, after completing a major merger, a large number of ITS employees opted for early retirement. The benefits of ITS operational effectiveness and project effectiveness and efficiencies were realized as ITS sustained past year's numbers for delivering IT solutions to our business partners. Also, the ITS PMO has served as a benchmark within DTE for functional project management offices that have been implemented to support other project initiatives. DTE and ITS continue to mature in the project management practices and the organization has recognized the benefits of a project management office.

American Greetings

Companies can champion more than one type of project office at the same time. For example, at American Greetings there exist both a functional project office and a strategic/corporate project office that work together. Previously in this chapter, we identified comments by senior management at American Greetings. Their comments indicated their understanding and support of project management. This support is illustrated in Figure 8–2, which shows the responsibilities of the strategic project office at American Greetings.

Tables 8–1 through 8–3 identify in more detail each of the three goals of the American Greetings corporate project office.

American Greetings Corporate PMO's Mission, Goals, and Guiding Principles
The PMO's mission is to drive continuous improvement through successfully managed projects delivering bottom-line results. This mission is comprised of 3 goals or levels:

- *Opportunity Identification:* Ensuring that the organization is focused on the right opportunities to improve efficiency and effectiveness. Identifying these opportunities is critical in helping the organization achieve its strategic goals.
- *Project Management:* Pursuing these improvement opportunities through standardized management processes. These processes include feasibility studies, charter documents, status and budget monitoring, issue resolution and impact analysis.
- *Change Champions:* Identifying, enabling and driving change throughout the organization. While often working "behind the scenes," the PMO fosters fact-based decision-making

FIGURE 8–2. PMO's mission, goals, and guiding principles.

and cross-organziational collaboration at all levels of the organization—within project teams, in support of project leaders, and during communications with senior management.

At the heart of the PMO are its guiding principles or values:

- Out-of-the-box thinking that challenges the status quo
- Big, hairy, audacious goals that stretch the organization
- Fact-based decision-making to get to the right recommendations
- Teamwork to develop holistic solutions
- Open and direct communication to build consensus
- Results orientation to deliver bottom-line benefits

American Greetings Issues Resolution Template

During the course of managing a project, issues may arise that jeopardize progress. Most issues can, and should, be resolved within the project team. Occasionally, however, there are issues that need to be elevated to senior management for their input and decision-making. The Issues Resolution Template, shown in Table 8–4, is designed to concisely summarize the issue, provide options and recommend a solution. This information enables senior management to make a timely and informed decision so the project can proceed.

TABLE 8–1. GOAL 1: FACILITATE IDENTIFICATION OF OPPORTUNITIES TO GROW REVENUE, REDUCE COSTS, AND OPTIMIZE ASSETS

Specific	• Identify opportunities to fuel financial performance • Conduct "deep dive" analysis as appropriate to — Validate the opportunity — Estimate ROI — Build the case for change • Nominate high potential individuals to serve as project leaders • Synthesize and present business improvement opportunities to senior management
Measurable	• Right slate of corporate projects — Bottom-line benefits — Clearly defined ROIs (either quantitatively or qualitatively) — Broad mix of organizational involvement
Attainable	• Assuming fact-based analysis to articulate opportunity and build support • Requiring balancing of resources
Relevant	• Links directly to corporate goals of growing revenue, reducing costs and optimizing assets • Enables achievement of strategic imperatives
Timing	• Preliminary slate of 200X projects by end of CY0X • Final slate of projects approved and launched by end of CY0X

TABLE 8–2. GOAL 2: DEVELOP AND INSTITUTIONALIZE PROJECT MANAGEMENT PROCESSES AND TOOLS

Specific	• Create a process by which new projects are identified, submitted, reviewed, and approved on an ongoing basis • Publish and refresh a standard set of project management tools and templates (e.g., selecting a project scheduling tool) • Monitor project quality and progress to ensure the desired impact; communicate status regularly to senior management • Establish use of ROIs • Ensure that fact-based analysis underline decision-making • Facilitate coordination of "like" projects to deliver more holistic solutions • Mentor project sponsors and leaders both in content and process • Develop and implement a comprehensive project management training curriculum targeted initially at project teams; measure feedback
Measurable	• Completed project deliverables (e.g., charter, status reports, budget, benefits) • Achieved budgets and benefits targets • Collective input and development of enhanced PMO approach • Organizational survey
Attainable	• Builds on project management foundation established during the restructuring • Recognizes that deliverables are co-owned by PMO and the project leaders
Relevant	• Enables achievement of corporate goals and strategic imperatives • Builds organizational disciplines in project management (e.g., goal setting, process management)
Timing	• Weekly, bi-monthly or monthly for status reporting • Monthly for budget and benefit updates • Fall CY0X for updates to next year's priority project management processes and tools

TABLE 8–3. GOAL 3: CHAMPION AND ENABLE CHANGE

Specific	• Improve organization awareness, understanding and support of project management through development and execution of a holistic communication strategy • Constructively encourage as well as challenge different points of view • Provide "deep dive" analysis as appropriate to surmount roadblocks • Utilize logical arguments in communications • Promote informal information sharing across organization • Create a broadening career path in/out of Delta for high performers
Measurable	• Subjective evaluation of contributions • Defined PMO career path; initiated implementation
Attainable	• Builds on the restructuring • Inherently difficult because challenges status quo
Relevant	• Drives towards more holistic, cross-functional solutions • Facilitates improved bottom-line results
Timing	• Ongoing

Team members may be hesitant to draw attention to an issue for fear it indicates a team weakness or mismanagement. However, project teams should be encouraged to use the Issues Resolution Template as soon as a project reaches an impasse so the issue can be resolved quickly. Successful project management includes keeping senior management informed of achievements as well as challenges so that expectations for the project are always realistic.

American Greetings Closing Document Template

The Closing Document officially marks the completion of a project. Guidelines for the Closing Document are shown in Table 8–5 and the template is shown in Table 8–6. The Closing Document provides objective and quantifiable data to determine how closely the achievements of the project met the original goals. If any of the goals were not met, the project team should provide an explanation and, if appropriate, recommend next steps. The document also asks the project team to provide more subjective information, such as "lessons learned," so that others can benefit from the team's experience.

TABLE 8–4. ISSUES RESOLUTION TEMPLATE

Project #(Project Number)—(Project Name)
Issue: (Issue Name)
Submitted by: (Name of Person Submitting Issue)
Date Submitted: (Date Submitting Issue Document)

- **Issue Description**
  Document a brief description of the issue.
- **Risk/Impact**
  Include key points on the risks and potential impacts associated with the issue.
- **Urgencies**
  Indicate what sense of urgency resolution of the issue should be given and why.
- **Discussion of Options**
  Briefly document the primary options considered for resolving the issue and the pros and cons of each.
- **Recommendations**
  Document your recommendation on which option should be pursued and why.
- **Plan Forward**
  Indicate what the next steps need to be, including who needs to do what.

TABLE 8–5. PROJECT CLOSING DOCUMENT GUIDELINES

Requirement Overview
All projects are required to produce a Closing Document at the conclusion of the project. You and your project sponsor will complete this document and submit it through the PMO to the Senior Team for review.
Purpose
The purpose of the Closing Document is to keep the Senior Team aware of the final status of every completed project, including final costs, benefits, deliverables achieved and objectives met. The objective of using a standard format is so that the closing documents are consistent in communicating key items.
Deliverables
One closing document should be produced for each project completed and is due within one month of close
Email your Project Closing Document to the PMO for storing out on the PMO website and for issuing to the Senior Team.
All Project Closing Documents will be consolidated with weekly status reports and forwarded to the Senior Team.

TABLE 8–6. TEMPLATE FOR REPORTING PROJECT COMPLETION

Project #(Project Number)—(Name of Project)
(Name of Project Leader)
Closing Document (Date)

- **Project Objective**
  Re-state the original objectives and state if the objectives were met.
- **Project Deliverables**
  Re-state each deliverable and state how it was met and where there were shortfalls. Identify any expected deliverables that will be accomplished at a later date or not at all.
- **Lessons Learned**
  State key lessons, both positive and negative, that you and the team have learned on the project that should be shared with others and/or applied to future projects.
  1. Positive experiences
  2. Pitfalls to avoid
  3. Processes that emerged and can be reused.
  4. Tools that emerged and can be reused.
- **Benefits**
  Attach the original benefits realization template for comparison of targeted versus achieved benefits. State the actual benefits of each listed function and/or process; list both tangible and intangible benefits. Qualify and quantify the benefits received, as this information will be used for fiscal year planning.
- **Time/Cost Estimates**
  Attach the original high levels costs template for comparison of estimated versus actual costs. State the actual cost by category listed in the original costs template. Qualify and quantify any costs that exceeded the original estimate. This information will be used for fiscal year planning.
- **Closing Document Signatures**
  Sign-off requires a review of the key stakeholder names in the Charter. Review key stakeholders listed in the Project Charter and modify the list as necessary to assure a complete, appropriate level of sign-off.

______________________________	______________________________
(Name of Sponsor/Stakeholder)	(Name of Key Customer)

Corporate or strategic project offices appear to be most common today and are the focus of the remainder of this chapter. This type of project office can also support the company on a global basis. According to Jim Triompo, Group Senior Vice President, Process Area Project Management at ABB:

> The project office does not deliver projects. The projects managed by the project management office are limited to process/tools development, implementation and training. The project management office is sometimes requested to perform reviews and to participate in division level risk reviews and operational reviews in various countries.

Exel

For multinational companies, there can exist several project offices that must function in a coordinated effort. According to Francena D. Gargaro, PMP, Director Project and Resource Management, Americas for Exel:

> Exel's Enterprise Project Management (EPM) group serves the global organization as a project management center of excellence supporting project managers from all regions and sectors.
>
> The mission of the EPM is to provide thought leadership and training of Exel's project management tools, techniques and methodology. It is also responsible for the development of a strategic business plan that will leverage strengths of both project and resource management disciplines. The EPM group provides a single source solution for Exel's internal customers (Sector PMO's and Project Managers). The Exel EPM is responsible for the following:
>
> - Center of Excellence, supporting project management tools and techniques
> - Creation and deployment of an enterprise-wide project management methodology
> - Project management training for all sectors
> - Consulting and mentoring of project managers in the Americas
> - Facilitation and support for the establishment of Project Management Offices in Latin America, Asia, Europe/UK and Canada
> - Visibility to Exel's project portfolio and resource capacity across the organization
> - Executive level strategic reporting for EPM initiatives
>
> Additionally, there are regional project management offices (PMO's), established in North America, South America, Mexico, UK/Europe, and Asia Pacific—Exel's primary theatres (Figure 8–3). Each of these regional PMO's provides dedicated project management in their respective regions and have dotted line reporting to the EPM, based in the U.S.
>
> **The Enterprise Project Management (EPM)** group serves a number of roles. Primarily, responsibility involves managing the foundation for project management in the organization. The roles of the EPM group can be categorized into three major elements: visibility, collaboration, and globalization
>
> - Visibility
>   - Access to the global project pipeline via internal opportunity/leads database
>   - Assists account teams in the establishment and support of projects
>   - Executive level reporting, and resource capacity planning via enterprise software

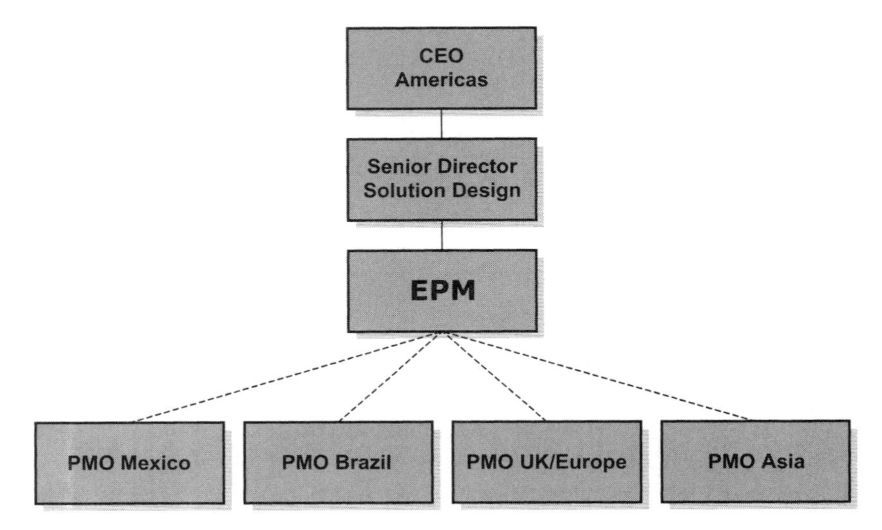

FIGURE 8–3. Regional PMO's.

- Collaboration
  - Project management strategy and customer relationship management
  - Project management support of sectors/departments
  - Project management support of internal/functional projects
  - Internal/external marketing and communications about PM practices at Exel
  - Development, maintenance, and delivery of project management training and certification assistance
  - Development, maintenance, and deployment of PM tools
  - Career path development
  - Benchmarks and metrics
- Globalization
  - Establishment and support of regional PMO's
  - Global training curriculum
  - Globally consistent tools (multi-lingual)

Regional Project Management Offices are groupings of project management associates (project managers, team members, etc.) who perform project management duties within specific regional or industry specific areas. Primary PMO responsibilities are:

- Promotion of Exel's project management methodology
- Promotion of the use of PM tools
- Project execution and delivery
- Subject matter expertise

The Enterprise Project Management group, on occasion, will manage or assist with the project management of internal/functional projects. For example, in the past 2 years, the EPM group has managed a Canadian payroll improvement project, a corporate office move project, and an enterprise project management application implementation. In these cases, a member of the EPM group will perform the duties of project manager, managing the day-to-day project activities of functional project teams.

Hewlett-Packard

Another company that has recognized the importance of a global project management office is Hewlett-Packard. According to Ron Kempf, PMP, Director PM Competency & Certification, HP Services Engagement PMO at Hewlett-Packard:

> For large, global companies the need for project management standardization and support is essential. To solve this problem, companies have developed a network of global project management offices all coordinated from a single source. At Hewlett-Packard, this network is referred to as the HP Services Project Management Office.
>
> "In the 80's our organization had spread across the world and inevitably we ran into some problems on project margins, our ability to deliver on time and to the expected budget," says Renee Speitel, Vice-president HP Services Program Management Office. "We set a goal to increase project management performance, consistency and financials". A global Program Management Office (PMO) was established to provide central management and mentorship.
>
> The characteristics of a global PMO as defined by HP Services are:
>
> - Manages across geographies and multiple projects
> - Involves organizational and business responsibility in addition to project disciplines
> - Long-term impact on organization and business
> - Responsible for the professional development of project management (PM) community of practitioners
> - Functional responsibility for PM infrastructure deployment
>
> HP Services PMO structure supports more than 2,500 project managers in 160 countries with regional offices located in Americas, Asia Pacific, Europe, Middle East, Africa, and Japan. Three focus areas are: the health of the portfolio, project management development, and processes.
>
> The health of the portfolio considers the status and profit of projects. "Portfolio tracking systems enable us to keep status on more than 2,400 active customer projects around the world" says Speitel. "A typical PMO scorecard includes customer satisfaction, portfolio financial performance (actual vs. budget), number of problem projects, number of certified project managers, and project manager utilization. The objective is to improve portfolio status year over year. PMO activities within this area include:
>
> - Managing escalations
> - Supporting project start-up activity
> - Reviewing and auditing projects regularly
> - Implementing review and approval process
> - Troubleshooting projects in difficulty
>
> Project management development involves formal training and certification as well as informal development. Project management is a core skill and competency for HP

Services. The award winning Project Management Development Program is organized by core project management courses, advanced PM topics, courses specific to HP Services practices, and professional skills training. The 35 course curriculum is taught in multiple languages. Other PMO sponsored activities that support project management development include:

- Driving PM certification programs
- Updating and managing the formal training curriculum in coordination with workforce development
- Driving and participating in major events like PMI Congresses and regional PM training/networking events
- Encouraging informal communication and mentoring
- Providing mentorship to field project managers

Project management processes include business practices, methods and tools, and rewards and recognition programs. HP Services' Opportunity Roadmap is a project life cycle architecture that defines the major business activities required to successfully pursue a customer engagement. It provides a process to determine scope and evaluate risk and price in order to win and succeed over a project lifetime.

The Opportunity Roadmap also incorporates the Solution Opportunity Approval and Review (SOAR) process, which facilitates appropriate levels of cross business unit involvement, review, and approval of global deals. This is shown in Figure 8–4. The Global Method for Program Management provides project managers with methodologies and a standardized approach using industry best practices and incorporating the added value of HP's experience. This is shown in Figure 8–5. The PMO is also responsible for defining and maintaining policies, procedures and other business practices relating to project management.

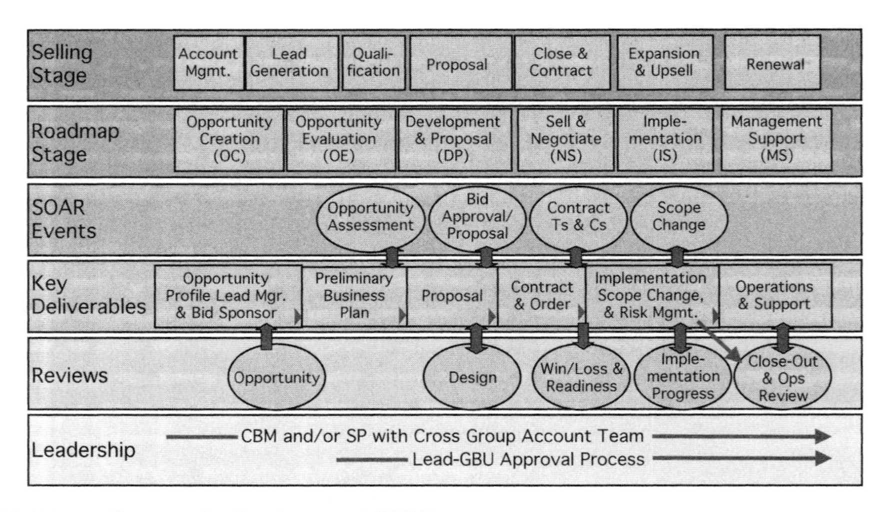

FIGURE 8–4. Opportunity Roadmap and SOAR.

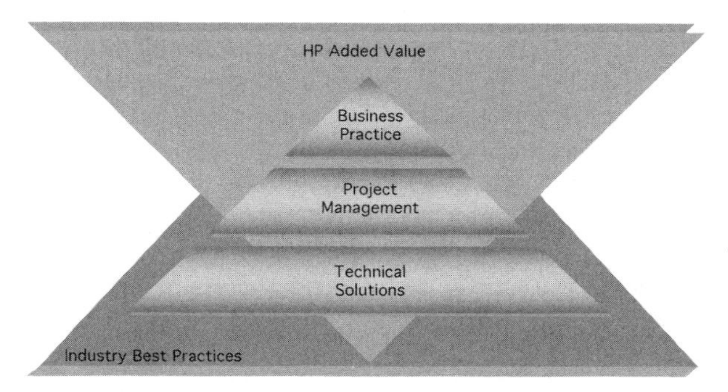

FIGURE 8–5. Global Method: Program Methodology.

Speitel summarizes: "The goals of our Program Management Offices are to deliver a quality solution, provide business value, and meet customer needs. Our project performance has improved nearly 70 percent of our projects within or under budget. This compares with an industry average of 50 percent. PMO structure and consistent approach enhances our ability to manage global projects and provides the flexibility to acquire and retain qualified project managers where we need them."

Star Alliance

The concept of a coordinating PMO will become commonplace over the next several years as the result of more mergers, acquisitions, and joint ventures. In the previous example, Exel's EPM coordinated the efforts of four regional project management offices. But what if the coordinating PMO had to coordinate the efforts of 15 or more regional PMOs? This is the situation at Star Alliance. Christian Frey, Head of the Project Management Office at Star Alliance Services GmbH, comments on some of the complexities that had to be overcome.

Star Alliance: The Airline Network for Earth

Star Alliance is a global, integrated airline network which was established on May 14, 1997 by several of the world's leading airlines. Its mission is to deliver a smooth travel experience and offer reward and recognition to passengers all over the world.

The current members of Star Alliance are Air Canada, Air New Zealand, ANA—All Nippon Airways, Austrian Airlines Group (which includes Austrian Airlines, Lauda Air and Tyrolean Airways), bmi british midland, Lufthansa German Airlines, Mexicana Airlines, SAS—Scandinavian Airlines, Singapore Airlines, Thai Airways International, United Airlines, VARIG Brazilian Airlines, Asiana Airlines, Spanair and LOT Polish Airlines.

Since its formation, Star Alliance has created a route network which carries more than 310 million passengers annually to 709 airports in 128 countries. Star Alliance customers already enjoy a number of benefits which make traveling easier, and new travel-related products are continually being developed.

Members of any Star Alliance frequent flyer program (FFP) can collect and redeem mileage points on any Star Alliance carrier. Accrued mileage also counts toward higher status in the member's FFP.

The Structure of Star Alliance Services GmbH

Star Alliance Services, located in Frankfurt/Germany, is the service provider for the Star Alliance member airlines. All projects that affect more than two airlines are facilitated and managed by Star Alliance project managers. These project managers are organized in a project manager pool and are assigned to the respective projects according to the necessary skill set and the availability from the Project Management Office (PMO).

They carry out projects for the Business Units (BU's) such as Information Technology, Marketing and Loyalty, Commercial and Sales as well as Products and Services. Within these BU's a product director, responsible for a whole range of products, acts as the owner for the Star Alliance projects.

The project managers are supported by the PMO, consisting of one PM specialist that is ultimately responsible for the management of the PM Pool, the definition and maintenance of the PM Guidelines and training and coaching of the PM's. In addition, one process manager takes care of the product development related processes and a project support analyst supports the PM's and maintains the online project database.

In Search of Excellence

Of course, acting as a service provider in the area of project management means that we must be able to define standards and fulfill the expectations that are set by the project owners and the airlines that function as our customers. The vision for Star Alliance Services GmbH is clear. We want to become the Center of Excellence for Project Management for the Star Alliance member airlines.

To achieve this goal, it is necessary not only to work out structures and processes but also to have the right people with the right skill set and to understand the challenges of managing projects with stakeholders from all over the world. In addition, it is necessary to raise awareness for Star Alliance project management, not only local but worldwide.

The Star Alliance Project Management Challenge

Managing a Star Alliance project has its own unique challenges. Consider managing 15 separate stakeholders for every project as well as handling the various vendors. These stakeholders are from different parts of the world, each with their own cultural and corporate background, as well as working behaviors. These project teams also have conflicting priorities because they must perform their own work within their respective airline and are not necessarily assigned full time to Star Alliance projects.

The project teams are not collocated. The representatives work from their respective home country, where every project team member is also responsible for servicing the airline's internal project teams. Since face-to-face meetings are very time consuming and expensive, most of the communication is done via video- or telephone conferencing.

To help deal with these issues, a web based reporting and collaboration tool was developed to ensure that all stakeholders can gather the necessary information from every part of the world and thereby contribute to the project's success. This system serves also as the main reporting tool for the stakeholders to keep them informed on the actual status of the projects.

Product Delivery Process and Star Alliance Project Management Guidelines

One of the most important prerequisites was the definition of the Product Delivery Process (PDP) as well as the Project Management Guidelines, which are a subdivision of the PDP. This PDP consists of eight product development steps, starting from Idea Generation to Business Case stage to Development, Testing, Launch and finally Life Cycle.

The Project Management Guidelines concentrate on the three stages of Development, Testing and Launch, at the time when the Star Alliance PM takes over. The Product Director is the product owner and maintains responsibility for the product throughout the entire cycle described above.

The PM Guidelines were developed in close cooperation with the PM's to ensure their buy-in to the process. The Guidelines consist of three major parts: the description of the role, a detailed description of every process step for the PM and a guide for project team members to ensure they fully understand the process and their role. The guidelines follow the PMI standards.

PM Training and Certification Programs

Setting common language and adherence to standards is critically important when dealing with a multi-cultural project management force.

The main objectives of this customized training program were the unification of the PM language and methodology as well as the certification of all PM's as Project Management Professionals (PMP) according to PMI standards.

In addition to the PM basic training, other classes were included into the training program, such as legal, negotiating and intercultural awareness. Ongoing discussion and input from the PM's ensure that the guidelines respond to their needs.

FirstEnergy[1]

When a company establishes a PMO and uses it effectively, favorable results will occur both now and in the future. FirstEnergy is on track to reap excellent benefits from their use of a PMO. The benefits can appear in several formats, ranging from better portfolio management of projects, better value added for the stakeholders, improved scheduling, better utilization of resources, and the ability to take on more projects without adding additional resources. FirstEnergy has also done an excellent job integrating portfolio management into the PMO. The need for this integration is readily apparent for the following quote from a senior officer at FirstEnergy:

> We may determine that an IT project proposed by a business unit doesn't have to be designed or purchased at all. One business unit may already be using a computer program that can be used or slightly adapted to meet another business unit's needs, which saves time and money. The PMO process is designed to find ways to meet customers' needs and use our resources most wisely as we work toward our corporate goals.

---

1. The material from FirstEnergy has been provided by Thomas J. Krise, Director, I/T Program Management Office; Jane McClellan-Renner, Operations & Integration Lead, I/T Program Management Office; and Adrian Lammi, PMP, Project Management Specialist, I/T Program Management Office.

There are many driving forces that encourage a company to establish a PMO. At FirstEnergy, portfolio management headed the driving force priority list up. The following comments illustrate this necessity:

> First you have to know the rules, and then you learn when to break them. Conventional project management suggests you build on a firm project management foundation. FirstEnergy chose to emphasize portfolio management instead. Senior Management was looking for significant savings. By ensuring that all projects supported corporate goals and by rigorously applying ROI criteria, the I/T PMO was able to achieve those savings and still deliver quality projects using the existing level of project management expertise.
>
> The Program Management Office developed procedures that enabled portfolio rationalization, quantified the ROI of projects in terms of real dollars, and provided a mechanism for ensuring projected savings were in fact realized. The Program Managers of the PMO identified potential synergies and eliminated or combined redundant projects. With Senior Management's support, the PMO was able to achieve the significant savings Senior Management sought.

Previously, we showed, with examples from Hewlett-Packard, Exel, American Greetings, and Computer Associates, that best results occur when the company has a PMO mission statement and a structured PMO process that is integrated with the project life cycle. Both of these also exist at FirstEnergy.

In order to fulfill its mission, the Program Management Office began with three fundamental questions, as shown in Table 8–7. Are we doing the right things? Are we doing enough of them? And are we doing the right things right? By analyzing the considerations

TABLE 8–7. PMO MISSION

The PMO provides an enterprise-wide approach to identify, prioritize, and successfully execute a portfolio of IT initiatives, aligned with the business units' strategic objectives.

	Consideration	PMO Tool/Process
• Are we (company, IT) doing the right things?	• Are projects aligned with strategic goals? (e.g., shareholder value, customer satisfaction, etc.) • Are resources deployed against projects with the highest strategic value?	• Objective strategic fit analysis • Effective project prioritization • Consolidated resource planning
• Are we doing enough of the right things?	• Given our strategic objective, will we meet our strategic goals? (e.g., SVA = $100M improvement) • Will we hit current fiscal year targets or should we manage expectations of stakeholders?	• Rigorous benefits tracking • Accurate portfolio reporting
• Are we doing the right things right?	• What are the necessary capabilities to effectively manage the technology portfolio? (e.g., benefit realization)	• Common tools and processes for projects (e.g., standard status reporting, issue and risk management)

involved in answering these questions, the PMO was able to identify what tools and processes were needed. Effective tools and well-defined processes enabled project managers to deliver consistent, measurable results that improved the quality and currency of data used by senior management in making decisions.

Figure 8–6 shows the integration of the PMO and project life cycle activities. FirstEnergy's I/T Program Management Office acts as a facilitator between the business unit community and the Information Technology Steering Committee (ITSC), who are responsible for approval of projects. The PMO champions the projects, helping them develop the best argument for approval and then by recommending projects for approval. The PMO establishes objective criteria that management can use in prioritizing projects. Finally, the PMO adds value to project delivery by improving resource utilization, by providing consistent, comprehensive status reporting, and by helping project managers identify and quantify realistic benefits.

Companies often make a fatal mistake of trying to add complexity to the work required by a PMO. FirstEnergy avoided this pitfall by utilizing a relatively simple approach of:

- Evaluation of projects
- Prioritization of projects
- Monitoring the delivery of the projects

Figure 8–7 shows the evaluation of projects. Standard business cases "put legs" on ideas that came in from all area of the corporation. The I/T Program Management Office designed a pro forma business case that leveled the project playing field and promoted projects that delivered the most value to the corporation. Elimination of nonvalue business

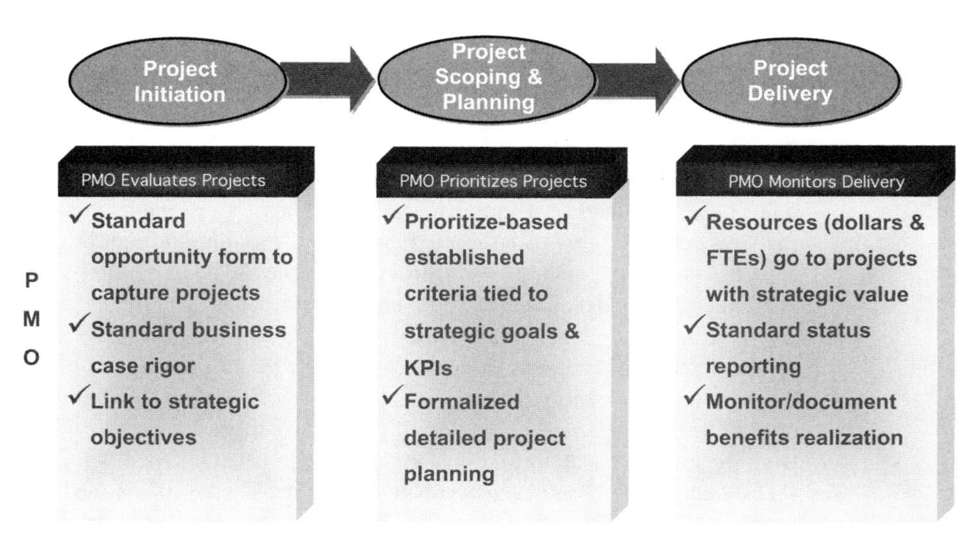

FIGURE 8–6. PMO and project life cycle.

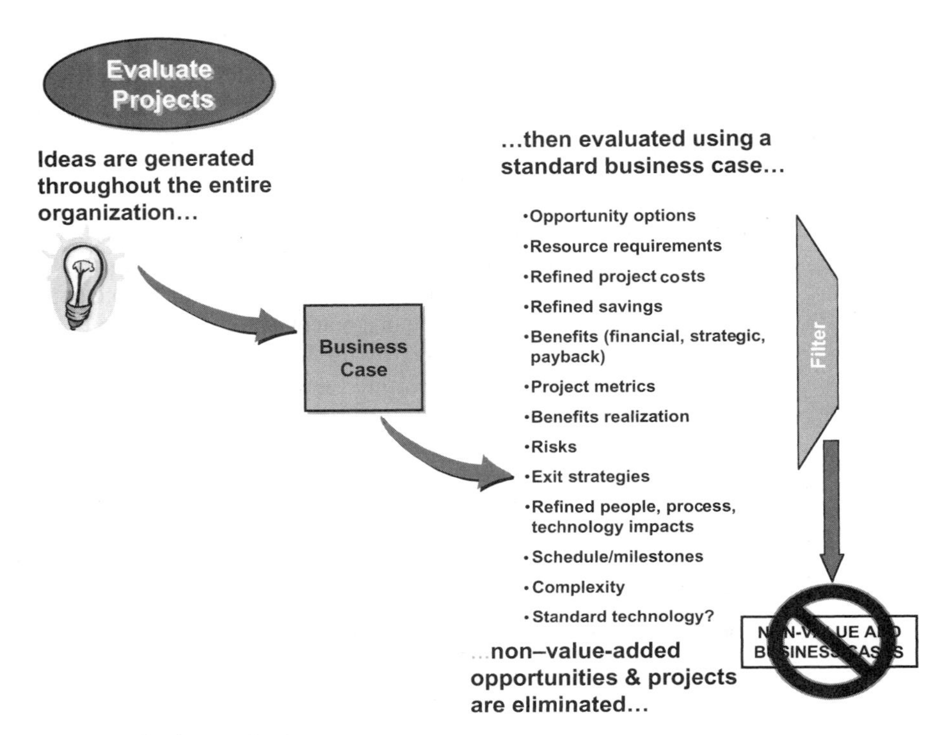

FIGURE 8–7. Project evaluation process.

cases saved real dollars and applied FirstEnergy's resources where they would produce the best payback.

The I/T Program Management Office established a prioritization process that had more than just strategic implications. This is shown in Figure 8–8. The PMO's program managers contributed their knowledge of their functional areas and the interdependencies of projects. Resource managers, known as center of excellence managers at FirstEnergy, contributed forecasts of resource availability. The process delivered a master program plan that was realistic, able to be implemented, and promoted a high level of project sponsor satisfaction. At the same time, it provided employee development and developed an appropriate, knowledgeable workforce for future projects.

During the delivery phase of the project, as shown in Figure 8–9, the I/T Program Management Office assists the project manager by monitoring the progress of the project through monthly status reports, which also tracks issues and benefits realization. Any change to the project that exceeds 10 percent of either the cost or duration of the project triggers the PMO's change control process. Senior management's I/T Steering Committee (ITSC) either approves or rejects change requests. These tools in the hands of project managers resulted in an almost perfect record of projects delivered on time and on budget to satisfied project sponsors.

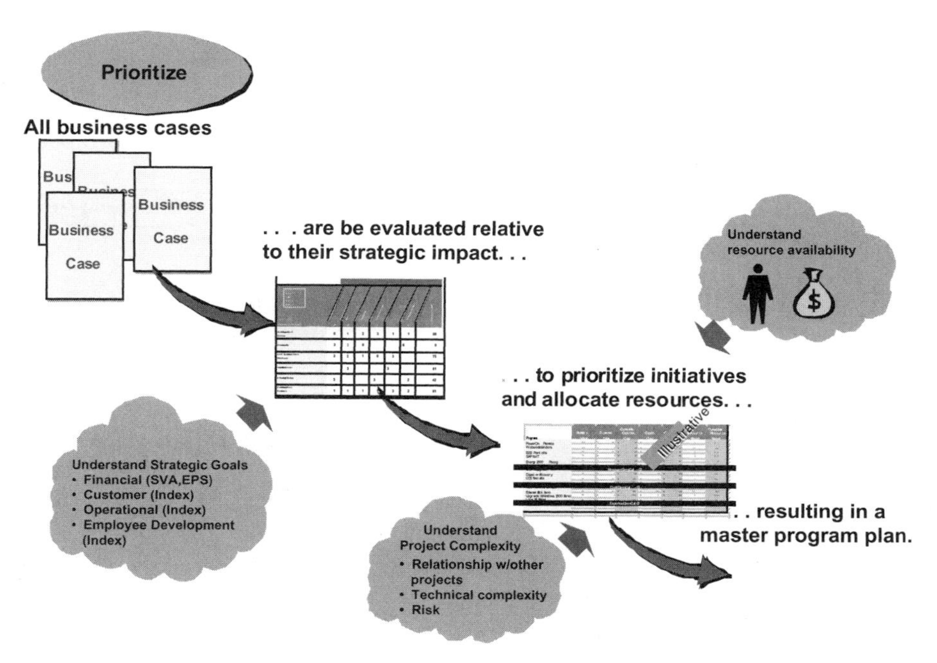

FIGURE 8–8. Prioritization process.

FirstEnergy's I/T Program Management Office developed a project methodology based on these three core processes as shown in Figure 8–10. As a potential project moves from a preliminary assessment through benefits realization, the PMO has three opportunities to recommend or not recommend that the project proceed. These recommendations provide tighter project controls and improve the probability that senior management will reap expected results. The various support processes help the project manager build quality into management of their project and improve their chances of delivering the project successfully.

One of the challenges facing companies is the identification of people with the appropriate skills necessary to function in a project management environment. FirstEnergy already had people with prior project management experience. The question was which of these people were best qualified for managing future projects. The I/T Program Management Office developed selection criteria which recognized that the skill set for successful project managers was different from a business unit subject matter expert or technical expert. This skill set includes soft skills such as team building and attitude that are as important to project success as the hard skills.

The skill set is shown in Figure 8–11. Most companies look only at soft skills and hard skills. FirstEnergy understood the importance of selecting people who understood the culture and would work well within the guidelines established by senior management and the PMO.

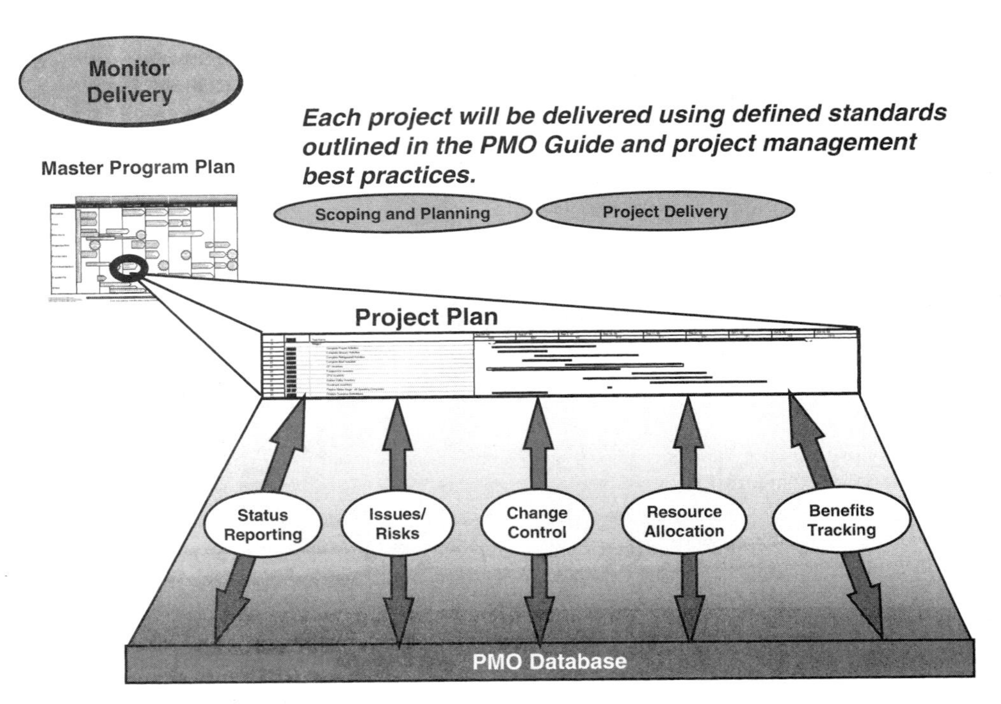

FIGURE 8–9. Monitoring.

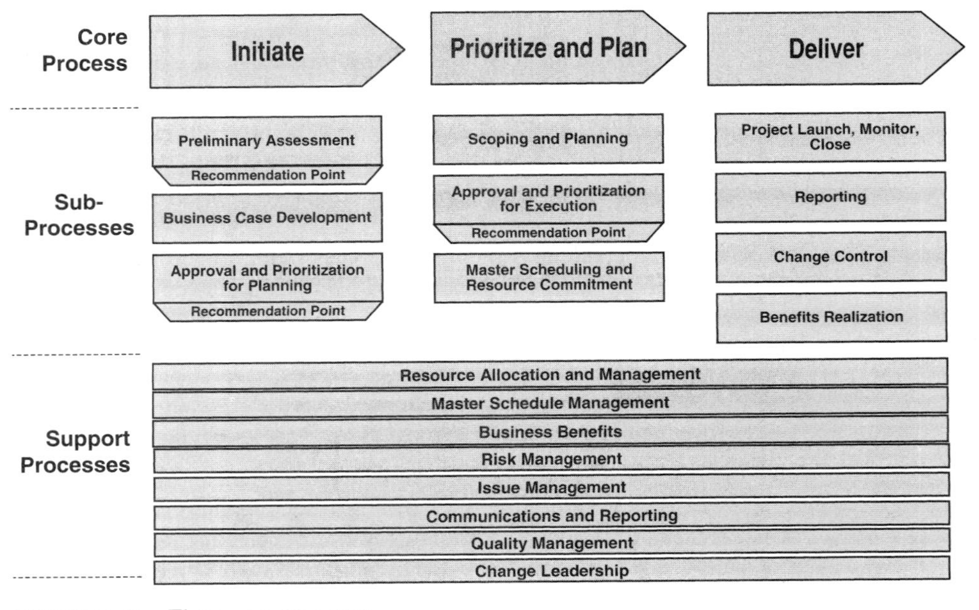

FIGURE 8–10. Three core processes.

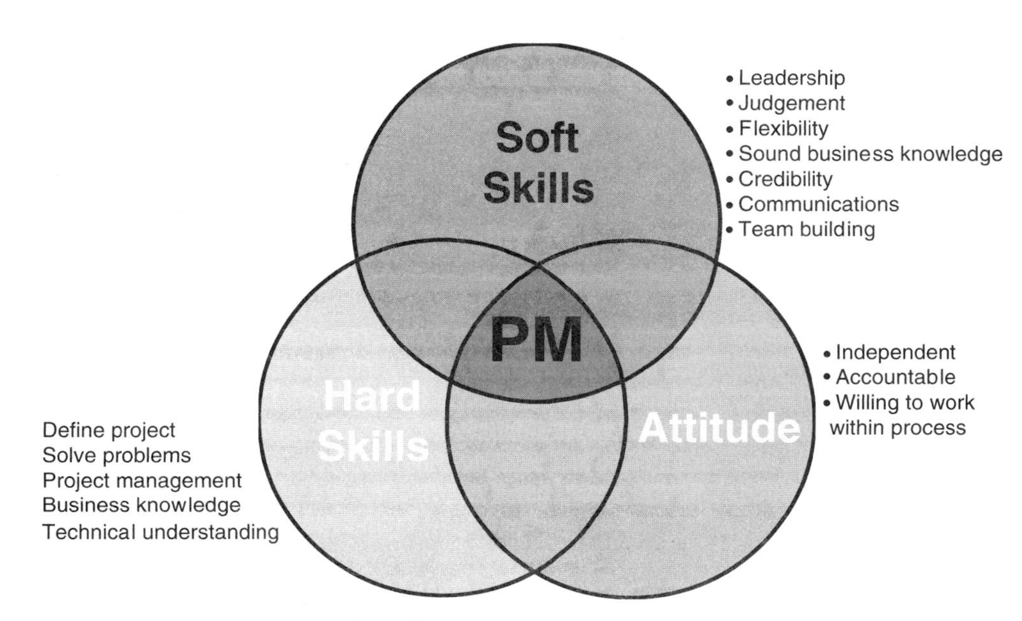

FIGURE 8–11. Project manager identification.

Three lessons learned by the FirstEnergy's I/T Program Management Office reshaped our understanding of project management:

- The first was that the cost of a dedicated project manager is more than offset by the savings achieved.
- Another was that project managers who embraced project management best practices and tools consistently perform better than their counterparts.
- The final lesson learned was that project management has a skill set that is unique. Not everybody has the skills or the interest in becoming a project manager . . . and that's OK.

The processes and procedures developed by the FirstEnergy I/T Program Management Office produced real, measurable results. Eliminating projects that didn't deliver value or were redundant or similar to other projects saved significant dollars. Applying them to projects that were consistent with corporate strategic objectives better employed these dollars. Projects that were delivered on time and on budget became the norm rather than the exception. A new, better way of doing business had been defined that objectively evaluates project performance and focused on corporate KPIs.

The FirstEnergy Program Management Office was able to improve the quality of FirstEnergy's project portfolio by reducing the size of the 2001 project portfolio by 35 percent. This is shown in Figure 8–12. The smaller portfolio produced more value per value invested.

An important component of the I/T Project Management Office is its public face, the PMO community pages of FirstEnergy's intranet. Standard reports are available to both

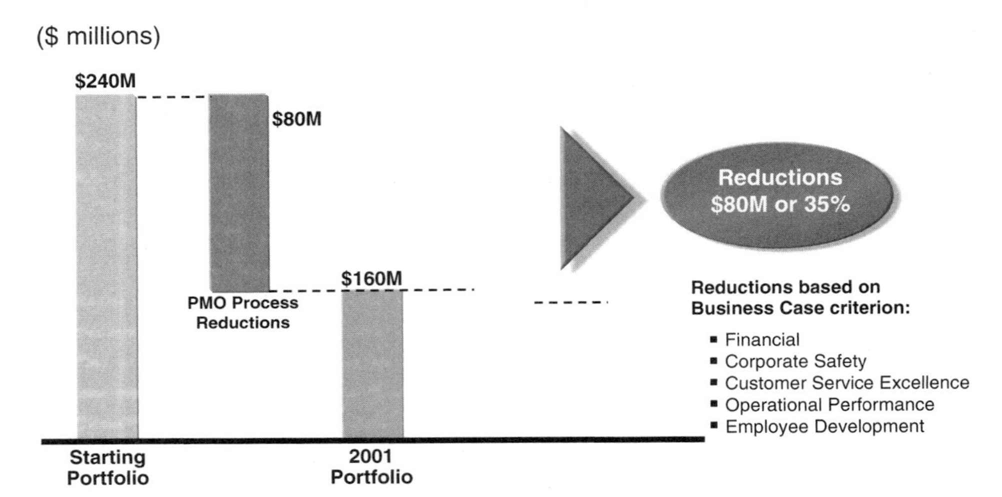

FIGURE 8–12. Project portfolio reductions.

management and the project manager. The pages offer explanations of all PMO processes as well as each of the PMO forms, instructions on their use, and examples of completed forms on the forms and reference pages. The complete Project Management Office Guide can be found online as well as links to FirstEnergy's custom Application Development and Project Management Methodologies.

8.5 PROJECT MANAGEMENT INFORMATION SYSTEMS

Given the fact that the project office is now the guardian of project management intellectual property, there must exist processes and tools for capturing this information. This information can be collected through four information systems, as shown in Figure 8–13. Each information system can be updated and managed through the company intranet.

Earned Value Measurement Information System

The earned value measurement information system is common to almost all project managers. It provides sufficient information to answer two questions:

- Where is the project today?
- Where will the project end up?

This system either captures or calculates the planned and actual value of the work, the actual costs, cost and schedule variances (in hours or dollars and in percent), the estimated cost at completion, the estimated time at completion, percent complete, and trends.

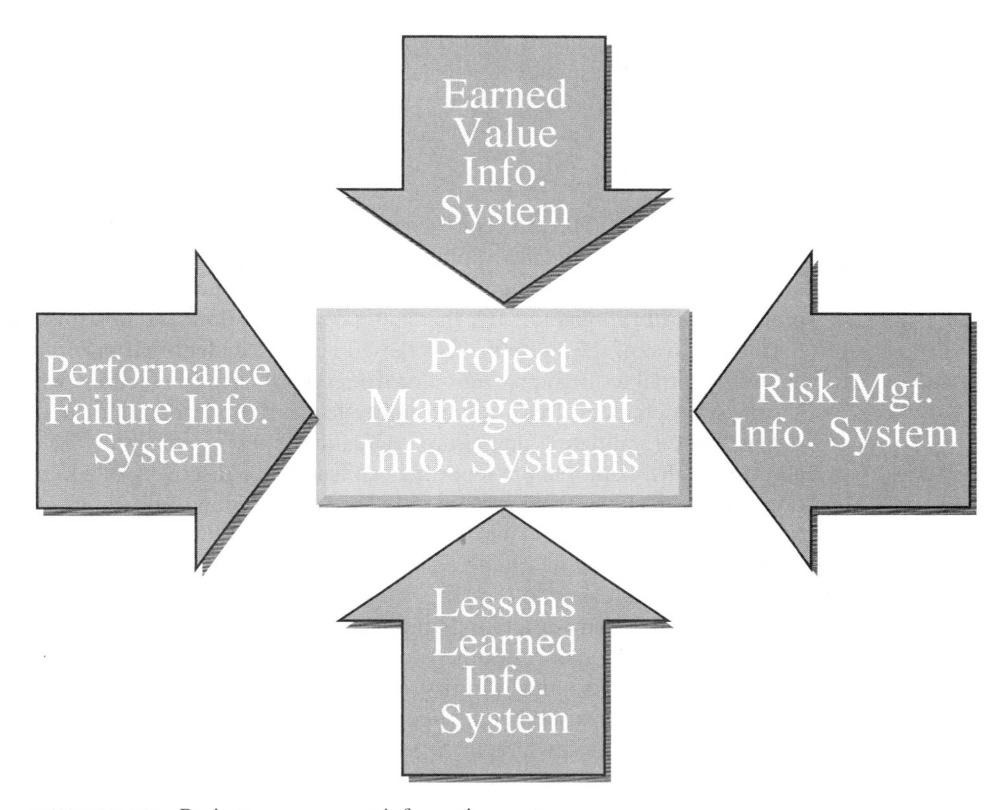

FIGURE 8–13. Project management information systems.

The Earned Value Measurement Information System is critical for a company that requires readily available information for rapid decision-making. It is easier to make small rather than large changes to a project plan. Therefore, variances from the performance management baseline must be identified quickly such that corrective action can be taken in small increments.

Risk Management Information System

The second information system provides data on risk management. The risk management information system (RMIS) stores and allows retrieval of risk-related data. It provides data for creating reports and serves as the repository for all current and historical information related to project risk. The information will include risk identification documentation (possibly by using templates), quantitative and qualitative risk assessment documents, contract deliverables if appropriate, and any other risk-related reports. The project management office will use the data from RMIS to create reports for senior management and retrieve data for the day-to-day management of projects. By using risk management templates, each project will produce a set of standard reports for periodic reporting and have the ability to create ad hoc reports in response to special queries. This information is directly related to the failure-reporting information system and the lessons-learned information system. The last two information systems are covered in more detail in the next two sections.

Performance Failure Information System

The project office may have the responsibility for maintaining the performance failure information system (PFIS). The failure could be a complete project failure or simply the failure of certain tests within the project. The PFIS must identify the cause(s) of the failure and possibly recommendations for the removal of the cause(s). The cause(s) could be identified as coming from problems entirely internal to the organization or from coordinated interactions with subcontractors.

It is the project office's responsibility to develop standards for maintaining the PFIS rather than for validating the failure. Validation is the responsibility of the team members performing the work. Failure reporting can lead to the discovery of additional and more serious problems. First, there may be resistance to reporting some failures for fear that it may reflect badly on the personnel associated with the failure, such as the project sponsors. Second, each division of a large company may have its own procedures for recording failures and may be reluctant to make the failure visible in a corporate-wide database. Third, there could exist many different definitions of what is or is not a failure. Fourth, the project office may be at the mercy of others to provide accurate, timely, and complete information.

The failure report must identify the item that failed, symptoms, conditions at the time of the failure, and any other pertinent evidence necessary for corrective action to take place. Failure analysis, which is the systematic analysis of the consequences of the failure on the project, cannot be completed until the cause(s) of the failure have been identified completely. The project office may simply function as the records keeper to standardize a single company-wide format and database for reporting the results of each project. This could be part of the lessons-learned review at the end of each project.

Consider the following example: An aerospace company had two divisions that often competed with one another through competitive bidding on government contracts. Each conducted its own R&D activities and very rarely exchanged data. One of the divisions spent six months working on an R&D project that was finally terminated and labeled as a failure. Shortly thereafter, it was learned that the sister division had worked on exactly the same project a year ago and achieved the same unproductive results. Failure information had not been exchanged, resulting in the waste of critical resources.

Everyone recognizes the necessity for a corporate-wide information system for storing failure data. But there always exists the risk that some will view this as a loss of power. Others may resist for fear that their name will be identified along with the failure. The overall risk with giving this responsibility to the project office is low to moderate.

Lessons-Learned (Postmortem Analysis) Information System

Some companies work on a vast number of projects each year, and each of these projects provides valuable information for improving standards, estimating for future bidding, and the way business is conducted. All of this information is intellectual property and must be captured for future use. Lesson-learned reviews are one way to obtain this information. Marcelino Diez, PMP, project management consultant with Edelca, describes the changes taking place at Edelca:

> Efforts are underway to normalize the process of obtaining lessons learned through our methodology from its capture, classification, storage and disposition to interested parties. Within the methodology are incorporated the guidelines related to the opportunity, participation, and recommendations.

The remarks by Mr. Diez are important because it illustrates that the current trend is to make lessons-learned reviews part of the project management methodology rather than a separate function.

Lessons-learned reviews have become commonplace. Brad Ruzicka, Senior Consulting Manager at the StoneBridge Group, believes:

> StoneBridge Group recommends a post project assessment be performed on every project. Generally, our approach is to produce a post project report. Depending on the nature of the project and the preference of the client, we may review the report with the entire team, key team members, the sponsor, and/or senior management. The key in any post project review is to keep the focus on the project and not personalities. Lessons learned are generally in the categories of sponsorship, management, scope and requirements definition, change control, and resources.

If intellectual property from projects is to be retained in a centralized location, the project office must develop expertise in how to conduct a postmortem analysis meeting. At that meeting, four critical questions must be addressed:

- What did we do right?
- What did we do wrong?
- What future recommendations can be made?
- How, when, and to whom should the information be disseminated?

Additional questions that must be asked follow the postmortem pyramid shown in Figure 8–14. The objectives for a project are established from the top of the pyramid to the bottom. However, the postmortem analysis that evaluates the project's metrics or measurements goes in reverse order from the bottom to the top. The bottom level, which is the basic level, evaluates the deliverables in terms of time, cost, quality, and scope. These constraints are often referred to as the critical success factors (CSFs) as seen through the eyes of the client.

Typical questions to consider for the critical success factors include:

- Time
    - Were the schedules realistic?
    - Was the level of detail correct?
    - Was it easy to evaluate performance from the schedule?
    - Was tracking accomplished easily?

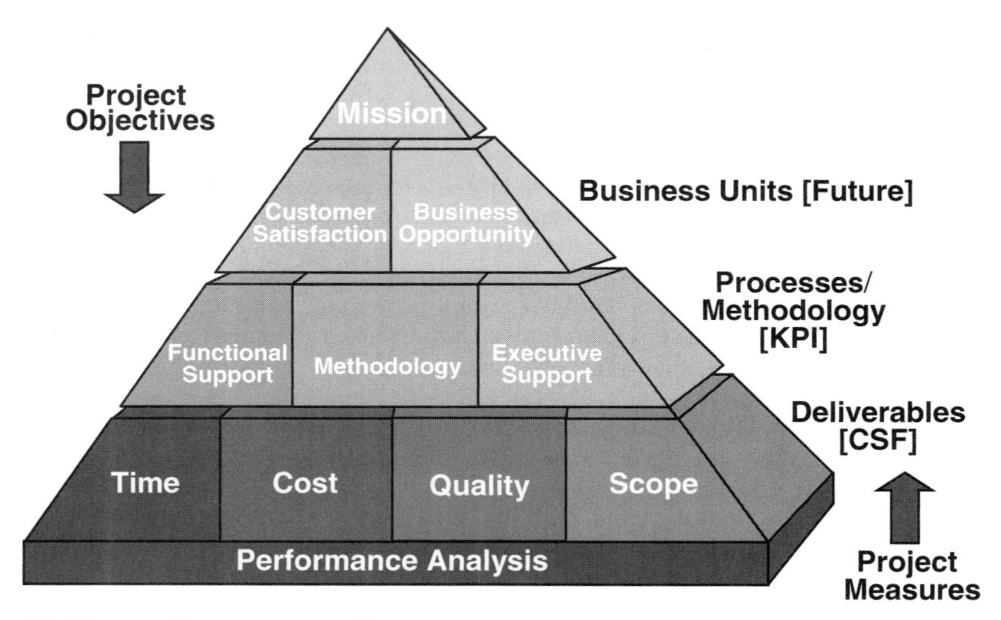

FIGURE 8–14. Postmortem pyramid.

- Cost
  - How accurate were our estimates?
  - Do our estimates need to be updated?
  - Did cost tracking follow our methodology?
  - Were there any problems with cost reporting?
- Quality
  - Did we conform to the customer's specifications?
  - Did the product perform as expected?
  - Did we evaluate durability, reliability, serviceability, and aesthetics?
- Scope
  - Was the statement of work easily understood?
  - Were the objectives clearly defined?
  - Was there proprietary technology involved?
  - If so, does the company have patent protection?
  - How difficult were the trade-offs?

The second layer in the postmortem pyramid of Figure 8–14 contains the key performance indicators (KPIs). KPIs are the internally shared learning topics that allow us to maximize what we do right and correct what we do wrong. KPIs are the internal best practices that allow us to achieve the critical success factors. Success is normally defined in terms of both CSFs and KPIs.

The KPIs can be categorized into the three areas shown in Figure 8–14. Typical questions for each KPI area might include:

- Line management support
  - Did the assigned personnel have the required expertise?
  - What was the quality of the assigned resources?
  - Did the resources possess innovative capability?
  - Was the right quantity of resources assigned?
  - Were the resources assigned in a timely manner to support the schedule?
  - Was there resource overload?
- Senior management support
  - Did senior management function as a sponsor?
  - Were they helpful?
  - Did they decentralize decision-making?
  - Did the project team have sufficient authority for the work required?
  - Was there a charter?
- Methodology
  - Did the methodology allow for quick response?
  - Was the planning performed correctly?
  - Did the methodology allow for contingency planning?
  - Were the tools to support the methodology available and state of the art?

The third layer in the postmortem pyramid of Figure 8–14 is the business unit evaluation. This evaluation focuses on two areas: customer satisfaction and future business opportunities. Typical questions for these areas include:

- Customer satisfaction
  - Was the customer pleased with the price–quality–value relationship?
  - Were the deliverables provided on time?
  - Are there value-added opportunities or is follow-on work available?
- Business opportunities
  - Were our preconceptions valid?
  - Are there additional sales opportunities other than with this client?
  - Will this project allow the organization to grow?

Computer Associates

The only true project failures are those from which nothing is learned. We can learn just as much, or even more, from project failures than from project successes. Therefore, it is imperative that both good news and bad news be recorded in the postmortem pyramid debriefing.

Computer Associates has developed an excellent process on how to conduct a post-project assessment. This process is shown in Appendix F. This assessment process is one of the ways that Computer Associates captures best practices.

Computer Associates has developed a Best Practices Library to support the way it interacts with clients. The Engagement Management Model (EMM), shown in Figure 8–15, defines the way in which CA Services carries out an engagement from end to end. This model is part of CA's Quality Management System and is ISO certified.

The Best Practices Library (BPL) maintains the methods required to provide repeatable, profitable, well-managed engagements. The Project Management method details the

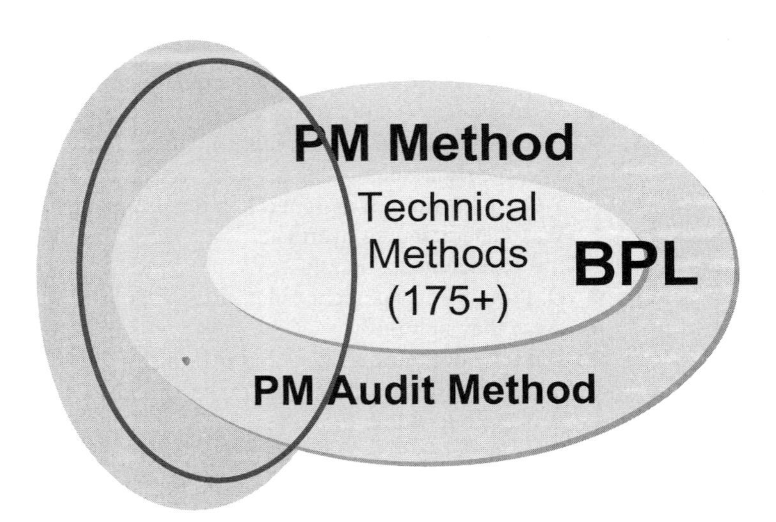

FIGURE 8–15. Engagement model.

activities and deliverables required for project management. The Project Audit method defines how to conduct an internal project audit against CA's Engagement Management Model. These internal audits help ensure that CA Services retains ISO certification. In addition to project management methods, the BPL is the repository for "technical" methods that provide detailed task descriptions for CA's consultants to deliver services.

Methods are added and revised continuously and published monthly on a BPL Web site for CA users. The number of methods documented in CA's BPL continues to grow. In February 2003, CA had 175 best practices documented.

The three major components of CA's engagement model are the:

- Best Practices Library
- Project Audit
- Project Management Method

Computer Associates maintains detailed processes for each of these components. Appendix E contains the overview for each of these three processes.

Lessons learned and best practices are intellectual property for a company. But to be effective, the best practices should be incorporated into future methods developed. How does CA's Method Management Organization (MMO) decide what methods to develop? The method development process needs to be aligned with business process. In CA's business, methods are integral with CA's services business. Method development cannot be done in a vacuum; it requires a thoughtful, well-structured collaborative process. Methods development is one of a number of important "threads" in the development and rollout of CA's service offerings. This is shown in Figure 8–16. If these threads are not connected properly, there is a risk that methods will be built that will not be used. Input is gathered from client requirements, business focus, product development, past sales, and sales forecasts. Method development priorities are determined, then subject matter experts are engaged to build new methods. Existing methods are revised as required for process improvement or product changes.

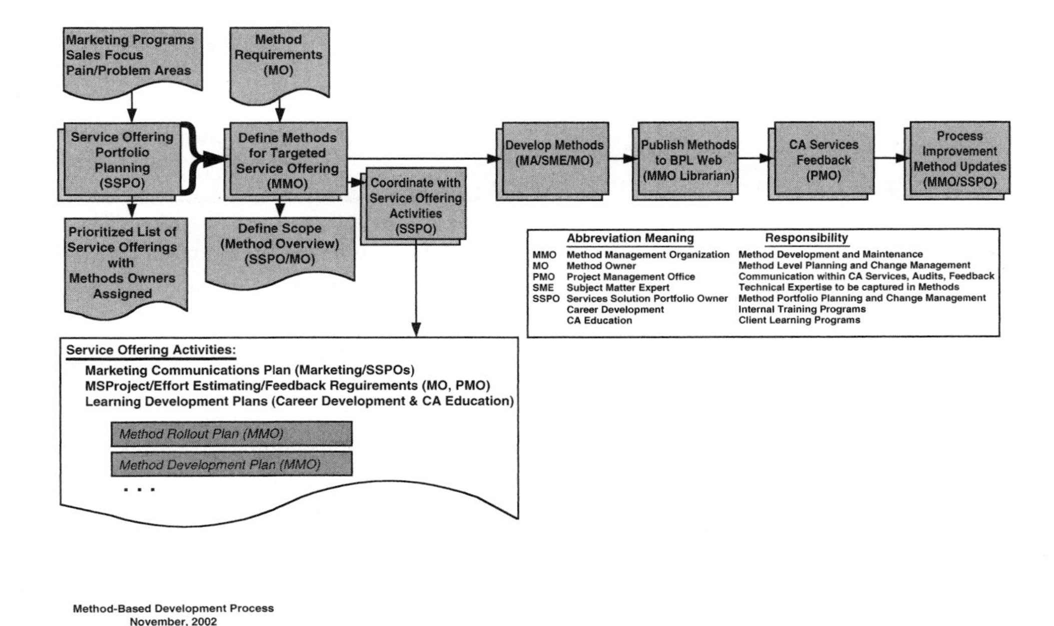

FIGURE 8–16. Method-based service development and rollout process.

Texas Instruments

Texas Instruments does an excellent job on lessons-learned reviews. According to a spokesperson at Texas Instruments:

> At Texas Instruments World Wide Facilities Central Utilities our maintenance turn-a-round teams lessons learned are one of the crucial elements in our success. Everyone from the suppliers, technicians, managers, operators, maintenance technicians, and engineers attend these powerful 2-hour sessions.
>
> We start the session inviting comments in an open format. Each comment is recorded as either a plus or a minus. To encourage participation, we go around the room, and often more than once. Then we switch gears and record necessary or recommended changes for the next year. Examples would include:
>
> - Proactive (as a plus from a supplier)
> - Safety issues were discussed in advance
> - Eliminated multiple shutdowns of equipment
> - Gained the ability to make decisions based on capacity/plan/forecast model
> - Scope needed to expand to include more equipment in the process
> - Needed increased operator involvement
> - Work with Procurement to improve/streamline the PO/invoicing process
>
> Next, a bin list is created for the next year's schedule. We list and assign action items with deliverable dates. After the meetings, minutes are sent to everyone. Some items require breakout teams. To facilitate timely completion, we found it important to maintain a regularly scheduled meeting time to address and follow-up on these issues.

Each time we have found further avenues for improvement, capital forecast items, and schedule integrations that benefit the bottom line!

8.6 DISSEMINATION OF INFORMATION

A problem facing most organizations is how to make sure that critical information, such KPIs and CSFs, are known throughout the organization. Intranet lessons-learned databases would be one way to share information. However, a better way might be for the project office to take the lead in preparing lessons-learned case studies at the end of each project. The case studies could then be used in future training programs throughout the organization and be intranet-based.

As an example, a company completed a project quite successfully, and the project team debriefed senior management at the end of the project. The company had made significant breakthroughs in various manufacturing processes used for the project, and senior management wanted to make sure that this new knowledge would be available to all other divisions.

The decision was made to dissolve the team and reassign the people to various divisions throughout the organization. After six months had passed, it became evident that very little knowledge had been passed on to the other divisions. The team was then reassembled and asked to write a lessons-learned case study to be used during project management training programs.

Although this approach worked well, there also exist detrimental consequences that make this approach difficult to implement. Another company had adopted the concept of having to prepare lessons-learned case studies. Although the end result of one of the projects was a success, several costly mistakes were made during execution of the project, due to a lack of knowledge of risk management and poor decision-making. Believing that lessons-learned case studies should include mistakes as well as successes, the project office team preparing the case study included all information.

Despite attempts to disguise the names of the workers that made the critical mistakes, everyone in the organization knew who worked on the project and was able to discover who the employees were. Several of the workers involved in the project filed a grievance with senior management over the disclosure of this information, and the case studies were then removed from training programs. It takes a strong organizational culture to learn from mistakes without retribution to the employees. The risk here may be moderate to high for the project office to administer this activity.

8.7 MENTORING

Project management mentoring is a critical project office activity. Most people seem to agree that the best way to train someone in project management is with on-the-job training. One such way would be for inexperienced project managers to work directly under the guidance of an experienced project manager, especially on large projects. This approach may become costly if the organization does not have a stream of large projects.

Perhaps the better choice would be for the project office to assume a mentoring role whereby inexperienced project managers can seek advice and guidance from the more experienced project managers, who report either "solid" or "dotted" to the project office. This approach has three major benefits. First, the line manager to whom the project manager reports administratively may not have the necessary project management knowledge or experience capable of assisting the worker in times of trouble. Second, the project manager may not wish to discuss certain problems with his or her superior for fear of retribution. Third, given the fact that the project office may have the responsibility to maintain lessons-learned files, the project mentoring program could use these files and provide the inexperienced project manager with early warning indicators that potential problems could occur.

The mentoring program could be done on a full-time basis or on an as-needed basis, which is the preferred approach. Full-time mentoring may seem like a good idea, but it includes the risk that the mentor will end up managing the project. The overall risk for project office mentoring is low.

8.8 DEVELOPMENT OF STANDARDS AND TEMPLATES

A critical component of any project office is the development of project management standards. Standards foster teamwork by creating a common language. However, developing excessive standards in the form of policies and procedures may be a mistake because it may not be possible to create policies and procedures that cover every possible situation on every possible project. In addition, the time, money, and people required to develop rigid policy and procedure standards would make project office implementation unlikely because of head-count requirements.

Forms and checklists can be prepared in a template format such that the information can be used on a multitude of projects. Templates should be custom-designed for a specific organization rather than copied from another organization that may not have similar types of projects or a similar culture. Reusable templates should be prepared *after* the organization has completed several projects, whether successfully or unsuccessfully, and where lessons-learned information can be used for the development and enhancement of the templates.

There is a danger in providing templates as a replacement for the more formalized standards. First of all, because templates serve as a guide for a general audience, they may not satisfy the needs of any particular program. Second, there is the risk that some perspective users of the templates, especially inexperienced project managers, may simply adopt the templates "as required as written" despite the fact that it does not fit his or her program.

The reason for providing templates is *not* to tell the team how to do their job, but to give the project manager and his or her staff a starting point for their own project initiation, planning, execution, control, and closure processes. Templates should stimulate proactive thinking about what has to be done and possibly some ideas on how to do it.

Templates and standards often contain significantly more information than most project managers need. However, the templates and standards should be viewed as the key to keeping things simple, and the project managers should be able to tailor the templates and standards to suit the needs of the project by focusing on the key critical areas.

Templates and standards should be updated as necessary. Since the project office is probably responsible for maintaining lessons-learned files and project postmortem analysis, it is only fitting that the project office evaluate these data to seek out key performance indicators that could dictate template enhancements. Standards and templates can be regarded as a low-risk project office activity.

Templates and standards are very important to consulting companies, not necessarily for internal use but perhaps more so for the benefit of their clients. The standards and templates must be flexible such that they can be adapted easily to the needs of different clients. Brad Ruzicka, Senior Consulting Manager at the StoneBridge Group, believes:

> StoneBridge Group stresses reusability in all phases of the project management methodology. I personally utilize a standard template for project plans, work plans and status reports. However, we also believe that, as a consulting firm, it is important for us to adapt our templates consistent with our client's standards.

8.9 PROJECT MANAGEMENT BENCHMARKING

Perhaps the most interesting and most difficult activity assigned to a project office is benchmarking. Just like mentoring, benchmarking requires the use of experienced project managers. The personnel assigned must know what to look for, what questions to ask, the ability to recognize a good fit with the company, how to evaluate the data, and what recommendations to make.

Benchmarking is directly related to strategic planning for project management and can have a pronounced effect on the corporate bottom line based on how quickly the changes are implemented. In recent years, companies have discovered that best practices can be benchmarked against organizations not necessarily in your line of business. For example, an aerospace division of a large firm had been using project management for over 30 years. During the early 1990s, the firm had been performing benchmarking studies but *only* against other aerospace firms. Complacency had set in, with the firm believing that they were in equal standing with their competitors in the aerospace field. In the late 1990s, the firm began benchmarking against firms outside their industry, specifically telecommunications, computers, electronics, and entertainment. Most of these firms had been using project management for less than five years and in that time had achieved project management performance that exceeded the aerospace firm. Now, the aerospace firm benchmarks against all industries.

Project office networking for benchmarking purposes could very well be in the near future for most firms. Project office networking could span industries and continents. In addition, it may become commonplace even for competitors to share project management knowledge. However, at present, it appears that the majority of project management benchmarking is being performed by organizations whose function is entirely benchmarking. These organizations charge a fee for their services and conduct symposiums for their membership whereby project management best practices data are shared. In addition, they offer database services against which you can compare your organization:

- Other organizations in your industry
- Other organizations in different industry sectors
- Other employee responses within your company
- Other organizations by company size
- Other organizations by project size

Some organizations have a strong resistance to benchmarking. The arguments against benchmarking include:

- It doesn't apply to our company or industry.
- It wasn't invented here.
- We're doing fine! We don't need it.
- Let's leave well enough alone.
- Why fix something that isn't broken?

Because of these concerns, benchmarking may be a high-risk activity because of the fear of recommended changes.

8.10 BUSINESS CASE DEVELOPMENT

One of the best ways for a project office to support the corporate strategic planning function is by becoming experts in business case development. More specifically, this includes expertise in feasibility studies and cost–benefit analysis. In the "Scope Management" section of the PMBOK®, one of the outputs of the Scope Initiation Process is the identification/appointment of a project manager. This is accomplished after the business case is developed. There are valid arguments for assigning the project manager after the business case is developed:

- The project manager may not be able to contribute to the business case development.
- The project might not be approved and/or funded, and it would be an added cost to have the project manager on board early.
- The project might not be defined well enough to determine at an early stage the best person to be assigned as the project manager.

Although these arguments seem to have merit, there is a more serious issue in that the project manager ultimately assigned may not have sufficient knowledge about the assumptions, constraints, and alternatives considered during the business case development.

This could lead to a less than optimal project plan. It is wishful thinking to believe that the project charter, which may have been prepared by someone completely separated from the business case development efforts, contains all of the necessary assumptions, alternatives, and constraints.

One of the axioms of project management is that the earlier the project manager is assigned, the better the plan and the greater the commitment to the project. Companies argue that the project manager's contribution is limited during business case development. The reason for this belief is because the project managers have never been trained in how to perform feasibility studies and cost–benefit analysis. These courses are virtually nonexistent in the seminar marketplace.

Business case development often results in a highly optimistic approach with little regard for the schedule and/or the budget. Pressure is then placed on the project manager to accept arguments and assumptions made during business case development. If the project fails to meet business case expectations, the blame is placed on the project manager.

The project office must develop expertise in feasibility studies, cost–benefit analysis, and business case development. This expertise lends itself quite readily to templates, forms, and checklists. The project office can then become a viable support arm to the sales force in helping them make more realistic promises to customers and possibly assist in generating additional sales. In the future, the project office might very well become the company experts in feasibility studies and cost–benefit analyses, and eventually conduct customized training for the organization on these subjects. Marketing and sales personnel who traditionally perform these activities may view this as a high risk.

8.11 CUSTOMIZED TRAINING (RELATED TO PROJECT MANAGEMENT)

For years, the training branch of the human resources group had the responsibility of working with trainers and consultants in the design of customized project management training programs. Although many of these programs were highly successful, there were many that were viewed as failures. One division of a large company recognized the need for training in project management. The training department went out for competitive bidding and selected a trainer. The training department then added in their own agendas after filtering all the information concerning the goals and deliverables sought by the division requesting the training. The trainer never communicated directly with the organization requesting the training and simply designed the course around the information presented by the training department. The training program was viewed as a failure and the consultant/trainer was never invited back. Postmortem analysis indicated the following conclusions:

- The training branch (and the requesting organization) never recognized the need to have the trainer meet directly with the requesting organization.
- The training group received input from senior management, unknown to the requesting organization, as to what information they wished to see covered, and the resulting course satisfied nobody's expectation.

- The trainer requested that certain additional information be covered while other information was considered inappropriate and should be deleted. The request fell on deaf ears.
- The training department informed the trainer that they wanted only lecture, no case studies, and minimal exercises. This was the way it was done in other courses. The participant evaluations complained about lack of exercises and case studies.

While the training group believed that their actions were in the company's best interest, the results were devastating. The trainer was also at fault for allowing this situation to exist.

Successful project management implementation has a positive effect on corporate profitability. Given that this is true, why allow nonexperts to design project management coursework? Even line managers who believe that their organization requires project management knowledge may not know what to stress and what not to stress from the PMBOK®.

The project office has the expertise in designing project management course content. The project office maintains intellectual property on lessons-learned files and project postmortem analysis giving the project office valuable insight on how to obtain the best return on investment on training dollars. This intellectual property could also be invaluable in assisting line managers in designing courses specific to their organization. This activity is a low risk for the project office.

DTE Energy

Henry Campbell, a member of the ITS Project Management Office at DTE Energy, describes how the project management office championed educational activities in project management:

> The Information Technology Services (ITS) management team at DTE Energy has recognized project management as a role, skill and asset to the organization. As part of the organizational strategy to mature this skill set, the Project Management Office created a training curriculum.
>
> Initially ad hoc courses were offered internally to introduce formal project management concepts and guide employees towards attaining Project Management Professional designation. However, to complement the re-alignment of organizational goals and the introduction of an enterprise project management information system (PMIS), a curriculum was designed to assist project managers in their growth and maturity.
>
> As part of the PMIS pilot, an assessment was conducted to identify organizational gaps in the project management knowledge areas. It was important to leverage project management training with the tool in order to increase the level of data quality. For instance, during the pilot, scheduling and earned value analysis were identified as a knowledge area gap.
>
> To address the scheduling issue, the International Institute of Learning (IIL) was contracted to conduct several courses around scheduling dynamics. As a result, the scheduling issues with the tool have been minimized since the pilot. In addition, Quentin Fleming, author of "Earned Value Project Management," was contracted to conduct Earned Value (EV) courses for project managers, supervisors and directors. Organizationally, there has been an increase in the understanding of Earned Value to the point where all projects are required to report EV metrics weekly.

Project Management professionalism and organizational continuity are important to ITS. Therefore, all employees are encouraged to attend project management training. Project Managers are encouraged to become PMI Project Management Professionals (PMP). In the fast paced changing environment of ITS, it is important to maintain core employees who understand project management and are capable of successfully leading or contributing to IT projects.

Our current training curriculum has been quantitative courses such as Project Management Professional preparation, Risk Management, Scheduling, Earned Value Analysis, and TeamPlay. A description of these courses is shown in Table 8–8. Our next step is to introduce behavioral courses into the curriculum as well as developing in-house Subject Matter Experts (SME's) to conduct internal training.

TABLE 8–8. PMO TRAINING

Project Management Curriculum

PMO 101—Managing Projects in Organizations (*Prerequisite:* None) 15 PDU's

Get a solid understanding of project management methods with this comprehensive introductory course. Gain practical experience in proven project management techniques and discover a wealth of valuable, flexible tools that you can use immediately to ensure the success of any project in any type of organization.

PMO 102—Project Management Certification Course (*Prerequisite:* PMO 101)

This advanced project management training program delivers the skills necessary for today's project professionals and incorporates all the important components of PMI®'s *A Guide to the Project Management Body of Knowledge* (PMBOK® Guide), including the five process groups across the project life cycle phases. Participants for this course must meet the PMI®'s minimum requirements for the PMP exam; please visit the PMI website for requirements.

By registering for this course, candidates verify that they have reviewed the PMP exam requirements and have the minimum qualifications to sit for the exam. The PMP exam must be taken within 90 days of successful completion of this course.

PMO 103—Project Scheduling (*Prerequisite:* PMO 102) 15 PDU's

Course Objectives

- Set up a scheduled that adheres to the highest level of professional standards of project management
- Use scheduling as an analytical tool for planning, optimizing, and managing your project
- Create schedules that reflect limits on resources, personnel, and subcontractors on your timeline
- Learn techniques for creating, estimating, and debugging your schedule, resolving negative float to reach objectives
- Avoid scheduling abuses that result in establishing unrealistic deadlines
- Learn how to use, not abuse constraints

PMO 104—Earned Value (*Prerequisite:* PMO 102) 8 PDU's

(A one-day seminar facilitated by Quentin W. Fleming, co-author of Earned Value Project Management.)

Seminar Outline

I. *An Overview of Earned Value Project Management*
   - The utility of Earned Value management
   - The essence & evolution of the Earned Value concept
   - Earned Value as contrasted with traditional project management

II. *Defining Project Scope with Use of a Work Breakdown Structure (WBS)*
   - The importance of defining the project's scope
   - The WBS and why we need it: Make or Buy decisions
   - Integration of project scope, schedule and resources with the WBS

III. *Establishing the Earned Value Project Baseline*
   - The project performance measurement baseline
   - The importance of a Control Account Plan (CAP)
   - Methods used to plan and measure earned value
   - Creating Control Account Plans

TABLE 8–8. PMO TRAINING (CONTINUED)

IV. *Monitoring Performance against the Project Baseline*
- The meaning of schedule variances in Earned Value
- Analysis of Earned Value data
- Forecasting a statistical range of estimates at completion
- Team exercise: analyze project status and report to management
- Team exercise: calculate TCP Indices, report to management

V. *Future Directions for Earned Value Project Management*
- Reengineering the process for use on all projects

PMO 105—Risk Management (*Prerequisite:* PMO 102) 22.5 PDU's

In this Risk Management course, you'll work through the proactive approach to threat and opportunity—based on a clear understanding of the powerful nature of both qualitative and quantitative approaches to risk management.

Included in the course is a multi-part case study that takes you from a risk overview at the beginning of a project through the challenges of ongoing assessment and reassessment of threats and opportunities throughout the project.

TeamPlay Curriculum

T102 TeamPlay Project Manager

TeamPlay Project Manager is a system for planning, tracking and controlling your projects. By using TeamPlay Project Manager, an organization can store and manage its projects in a central location.

T105 TeamPlay Portfolio Analyst

Provides project summary and tracking via a rich set of graphics, spreadsheets, and reports. The system provides costs, schedule, and performance rollups in a single or multi-project environment.

TeamPlay "TeamPlayer" Web-Based Training

TeamPlayer enables Web-based timekeeping and communications. Team members see all supporting information they need to coordinate and perform their work while communicating directly with the central database.

Primavision Portfolio (formerly TeamPlay ESP)

Allows users access project data via the Internet. Users can build new projects and analyze schedule and cost information for all projects using their Web browsers.

TeamPlay Resource Utilization Management

Resource Utilization Management training will provide the fundamentals, which will enable resource managers to allocate, evaluate and deploy resources to projects. This class will cover "canned" Primavera training as well as customized resource topics as required by Information Technology Services within DTE Energy.

8.12 MANAGING STAKEHOLDERS

All companies have stakeholders. Figure 8–17 depicts the broad range of stakeholders, which for simplicity sake, have been categorized as organizational, product/market, and capital markets. Apprehension may exist in the minds of some that the project office will become the ultimate project sponsor responsible for all stakeholders. Although this may happen in the future, it is highly unlikely that it will occur in the near term.

The project office focuses its attention on internal (organizational) stakeholders. It is not the intent of a project office to replace executive sponsorship. As project management matures within an organization, it is possible that not all projects will require executive sponsorship. In such situations, the project office (and perhaps middle management) may be given the added responsibility of some sponsorship activities, but probably only for internal projects.

The project office is a good starting point for building and maintaining alliances with key stakeholders. However, the project office's activities are designed to benefit the entire company, and giving the project office sponsorship responsibility may create a conflict of

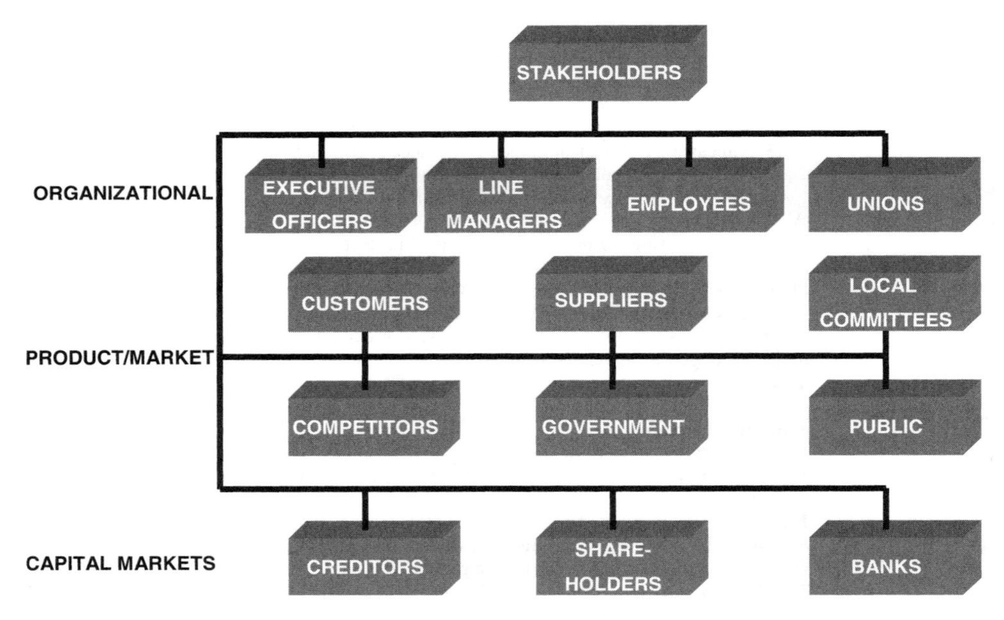

FIGURE 8–17. Project stakeholders.

interest for project office personnel. Partnerships with key stakeholders must be built and nurtured, and that requires time. Stakeholder management may rob the project office personnel of valuable time needed for other activities. The overall risk for this activity is low.

8.13 CONTINUOUS IMPROVEMENT

Given the fact that the project office is a repository of project management intellectual property, the project office may be in the best position to identify continuous improvement opportunities. The project office should not have unilateral authority for implementing the changes, but rather the ability to recommend changes. Some organizations maintain a strategic policy board or executive steering committee that, as one of its functions, evaluates project office continuous improvement opportunities.

As a starting point, continuous improvement opportunities may be classified as illustrated in Figure 8–18. Typical activities in each category might include:

- Existing process improvements
  - Integration of new or updated software
  - Easier use and application of existing tools
  - Better customer/contractor interfacing

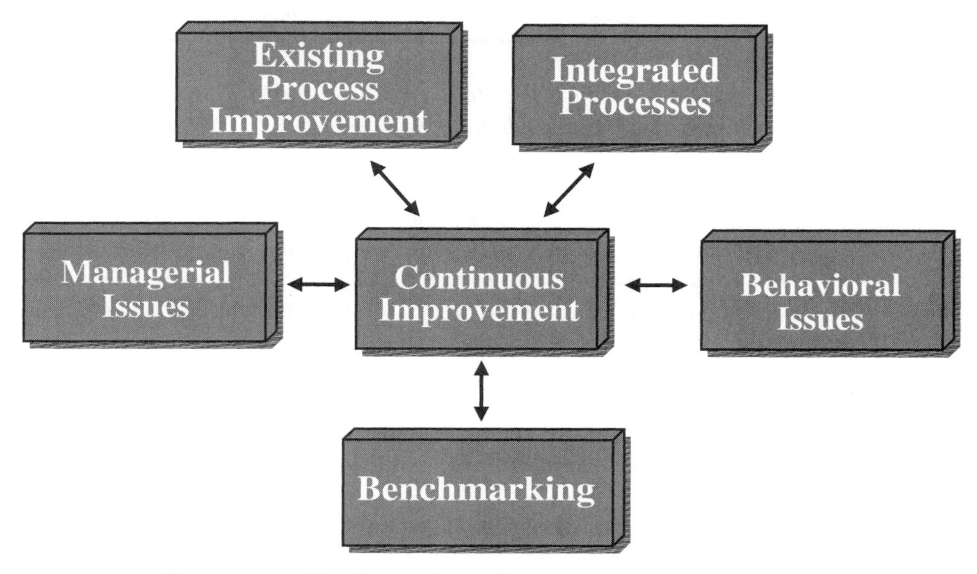

FIGURE 8–18. Factors to consider for continuous improvement.

- Convincing other internal organizations to use the project management methodology
- Integrated processes
  - Integrating other systems, such as risk and change management, into the project management system
  - Integrating other corporate databases into an integrated intranet system available to all
  - Integrating, or making more compatible, customer and contractor databases with your company's database
- Cultural issues
  - Better management of required changes in organizational behavior
  - Overcoming cultural barriers
- Benchmarking
  - Improvements in the benchmarking process
  - Increasing the number of benchmarking partners
- Managerial issues
  - Improvements in project sponsorship
  - Improvements in communications management with stakeholders
  - Projections of future resource skill levels versus existing capabilities

The overall risk is probably moderate because some may view the project office as infringing upon their turf.

8.14 CAPACITY PLANNING

Of all of the activities assigned to a project office, the most important activity in the eyes of senior management could very well be capacity planning. For executives to fulfill their responsibility as architects of the corporate strategic plan, they must know how much additional work the organization can take on, when, and where, without excessive burdening of the existing labor pool. The project office must work closely with senior management on all activities related to portfolio management and project selection. In Figure 7–1 we showed the project selection process as part of the portfolio management of projects. Strategic timing, which is the process of deciding which projects to work on and when, is a critical component of strategic planning.

Senior management could "surf" the company intranet on an as-needed basis to view the status of an individual project without requiring personal contact with the team. But to satisfy the requirements for strategic timing, all projects would need to be combined into a single database that identifies the following:

- Resources committed per time period and per functional area
- Total resource pool per functional area
- Available resources per time period per functional area

There may be some argument as to whether control of this database should fall under the administration of the project office. The author believes that this should be a project office responsibility because:

- The data would be needed by the project office to support strategic planning efforts and project portfolio management.
- The data would be needed by the project office to determine realistic timing and costs to support competitive bidding efforts.
- The project office may be delegated the responsibility to determine resource skills required to undertake additional work.
- The data will be needed by the project office for upgrades and enhancements to this database and other impacted databases.
- The data may be necessary to perform feasibility studies and cost–benefit analysis.

This activity is a high-risk effort for the project office because line managers may see this as turf infringement.

8.15 RISKS OF USING A PROJECT OFFICE

Risks and rewards go hand-in-hand. The benefits of a project office can be negated if the risks of maintaining a project office are not effectively managed. Most risks do not appear during creation of the project office, but more do so well after implementation. These risks include:

- *Head Count:* Once the organization begins recognizing the benefits of using a project office, there is a natural tendency to increase head count in the project office, in the false belief that additional benefits will be forthcoming. Although this belief may be valid in some circumstances, the most common result is diminishing returns. As more of the organization becomes knowledgeable in project management, the head count in the project office should decrease.
- *Burnout:* Employee burnout is always a risk. Using rotational or part-time assignments can minimize the risk. It is not uncommon for people working in a project office still to report "solid" to their line manager and "dotted" to the project office.
- *Excessive Paperwork:* Excessive paperwork costs millions of dollars to prepare and can waste precious time. Project activities work much better when using forms, guidelines, and checklists rather than the more rigid policies and procedures. To do this effectively requires a culture based upon trust, teamwork, cooperation, and effective communications.
- *Organizational Restructuring:* Information is power. Given the fact that the project office performs more work laterally than vertically, there can be power struggles for control of the project office, especially the project managers. Project management and a project office can work quite well within *any* organizational structure that is based upon trust, teamwork, cooperation, and effective communications.
- *Trying to Service Everyone in the Organization:* The company must establish some criteria for when the project office should be involved. Not all projects are monitored by the project office. The most common threshold limits for when to involve the project office include:
  - Dollar value of the project
  - Time duration of the project
  - Amount and complexity of cross-functionality
  - Risks to the company
  - Criticality of the project (i.e., cost reductions)

A critical question facing many executives is: How do we as executives measure the return on investment as a result of implementing a project office? The actual measurement can be described in both qualitative and quantitative terms. Qualitatively, executives can look at the number of conflicts coming up to the executive levels for resolution. With an effective project office acting as a filter, fewer conflicts should go up to the executive levels. Quantitatively, the executives can look at the following:

- *Progress Reviews:* Without a project office, there may exist multiple scheduling formats, perhaps even a different format for each project. With a project office and standardization, the reviews are quicker and more meaningful.
- *Decision-Making:* Without a project office, decisions are often delayed and emphasis is placed upon action items rather than on meaningful decisions. With a project office, meaningful decisions are possible.
- *Wasted Meetings:* Without a project office, executives can spend a great deal of time attending too many and very costly meetings. With a project office and more effective information, executives can spend less time in meetings and more time dealing with strategic rather than operational issues.

- *Quantity of Information:* Without a project office, executives can end up with too little or too much project information. This may inhibit effective decision-making. With a project office and standardization, executives find it easier to make timely decisions.

The prime responsibility of senior management is strategic planning and deployment and worrying about the future of the organization. The prime responsibility of middle- and lower-level management is to worry about operational issues. The responsibility of the project office is to act as a bridge between all the levels and to make it easier for all levels to accomplish their goals and objectives.

8.16 PROJECT OFFICE EXCELLENCE: A CASE STUDY ON JOHNSON CONTROLS, INC. (JCI) AUTOMOTIVE SYSTEMS GROUP (ASG)

When a company desires to become a world-class leader in project management, specifically the project office concept, it finds additional applications such as six sigma implementation and even the creation of a project office within a project office. Such was the case at JCI.[2]

Organization of Project Teams Using a Project Office Concept

The Organization In fall 2000, Automotive Systems Group of Johnson Controls, Inc. (North America) reorganized using a new business model. Under this model, individual projects (i.e., seats, cockpit, door panels, floor consoles, hard trim parts, etc.) were grouped together and managed as vehicle interior platforms under a platform director who was responsible for an entire vehicle (i.e., Ford F150 truck, Jeep Grand Cherokee, etc.). The platform director with his or her project managers functioned as a project office, viewing the entire vehicle interior and all interfacing components as one large project.

What Drove This Change? Johnson Controls, Inc., Automotive Systems Group, acquired two companies, Becker and Prince, the purpose being to allow JCI to become an integrated interior supplier to the automotive industry. The addition of Prince's overhead systems, door panel, and instrument panel capabilities to Becker's interior plastic trim capability and Automotive Systems Group's seating products established ASG immediately as an interior supplier. However, organizational changes were necessary to truly integrate the companies and their capabilities. The result of this was a new business model that reorganized the company into vehicle platforms and project offices.

2. Alok Kumar and Dave Kandt of the Automotive Systems Group of Johnson Controls, Inc. provided much of the information in the remainder of this section.

Acceptance There has been high acceptance of the changeover to the new business model. The expected initial resistance has dissipated, and platform director positions have been institutionalized. The customers are comfortable with this change, and quotes are now being responded to in terms of vehicle interiors rather than a collage of individual products and quotes.

How Has It Worked? This change has allowed us to function as an integrated company. The platform teams (project offices) have become true teams and feel a sense of identity based on vehicle interiors rather than doors, seats, and cockpits. Timing, profitability, and customer relationships are now viewed from the perspective of the total vehicle. The customers are happier because they have a single point of contact for their vehicle interior instead of multiple points of contact. An interesting side note is that we have been able to reduce the headcount of project managers as part of this change.

Lessons Learned For other companies that are attempting to move to (platform) project offices:

- Start with a clear organizational model. Stick with it despite concerns and contentions that "It won't work for us." Small anomalies will present themselves, but don't be deterred. Stay with a sound, overall concept.
- Change must be driven from the top. The organization must see constancy of purpose from a united leadership.
- Communication meetings must occur with all the impacted parties.
- Position descriptions within the project office must be written. Roles and responsibilities must be clear.
- The project management systems, reviews, and accountability must mirror the new organization.

Central Project Office: Organization, Roles, and Responsibilities

Johnson Controls, Inc., Automotive Systems Group, instituted a centralized project office reporting to the vice president of project management. The project office's responsibilities include:

- Methods and standards (JCI's corporate standard project management methodology is called PLUS (which stands for "Product Launch System")[3]
- Consulting/mentoring/training
- Project tracking/portfolio management/project metrics
- Project Web sites
- Personnel development/resource management/project manager availability for new projects
- Execution of the strategic plan for project management

---

3. PLUS is the Automotive Product Launch System of JCI. For further information on its development and application, see Richard Spigarelli and Carel Allen, "Implementation of an Automotive Product Launch System," *Proceedings of the Project Management Institute Annual Seminars and Symposium,* November 1–10, 2001, Nashville, Tennessee.

- Six sigma improvements of our project management systems and project execution
- Organizational structure for project management/position descriptions
- Hiring project managers
- Training and developing project managers who work for our extended enterprise partners (long-term suppliers)

The director of the project office also has the responsibility to manage all the component groups' project management departments. Managing the project office is only part of his responsibilities.

How Did We Get Here? The functions of the project office were somewhat distributed several years ago. The responsibility for our standard project management procedures was part of corporate procedures. Some engineering metrics were being led by engineering. Then approximately two years ago, a conscious decision was made to centralize these various functions under the project management organization. The key point for JCI was that the organization was ready for this change. The various groups readily gave up their resources to have them managed centrally.

The notable late additions to the project management initiatives at JCI are Six Sigma for Project Management and the PM Web sites (use of the Web for worldwide access to standard project documentation). These have been added since the centralization of the project office.

Results of Centralizing Since that time we have actually been able to reduce the head count by two in the project office (we now have four, including the director), and our output has arguably increased. Also, the six sigma initiative (applying six sigma principles to project execution) is very much enabled by having this central group. We have the opportunity to look across the entire organization for improvements and cost reductions. The project office has been able to reduce the total headcount in project management by approximately 30 percent in the last two years. This is a result of being able to manage, distribute, and redistribute resources and project management intellectual property over the entire enterprise. Also, while "pockets" of project managers developed in the past, the project office became a "final decision-maker" to determine appropriate corporate organizational structure and to simplify the organization. The project office is now involved in staffing our extended enterprise partners and finding ways to outsource project management resources to them.

Another significant improvement has been the ability to quickly reach consensus and decisions on corporate systems and processes. The project office now has the authority to make rapid calls and decisions on these issues. The most recent result was the release of a significantly simplified and therefore more effective version of PLUS. The original PLUS was developed by a committee and thus lacked a central theme and some cohesiveness. The current version has been very well received by the Automotive Systems Group. The project office played the central role in this initiative.

Obstacles to the Project Office As always, head count is reviewed very critically. Since we have been able to reduce head count (and move some project office personnel into project management positions successfully), there is little pressure at the present time to reduce further. A successful project office must be viewed as a productive, contributing group and must be on the leading edge of new initiatives, helping to reduce costs and improve project execution. Our project office has begun applying the six sigma process to improving the performance of our projects. The project office is also helping in staffing by moving project managers into black belt roles to execute six sigma projects, advanced quality engineering, purchasing, and other departments as personal development opportunities.

Future of the Project Office We are diligently working on improved metrics and evaluation of projects. At any one time, Automotive Systems Group is working on 250 projects, and there is tremendous opportunity to evaluate how the entire system is working if we look at the performance statistically. Individually, the cost overruns from a single project may not be daunting, but if we look at the waste occurring in all projects in the entire company, we see millions of dollars of opportunity. Once solutions are found, they can be applied across the company for all projects with the involvement of the central project office.

Additionally, we believe that improving the performance and capabilities of the project managers and supporting/providing them productivity enhancement tools is a permanent assignment. It must always be watched over and nurtured by this central organization.

8.17 REPORTING AND STRUCTURE

Not all companies support project management the way that Johnson Controls does. Despite the benefits that are recognizable by using a project office, disagreements still exist in many organizations as to where the project office should report and how to get the most out of a project office. However, given the responsibilities of the project office and its relationship to corporate strategic planning, capacity planning and project portfolio management, the project office must report to the executive levels of management. The shorter the distance between senior management and the project office, the quicker the benefits of project management will be recognized.

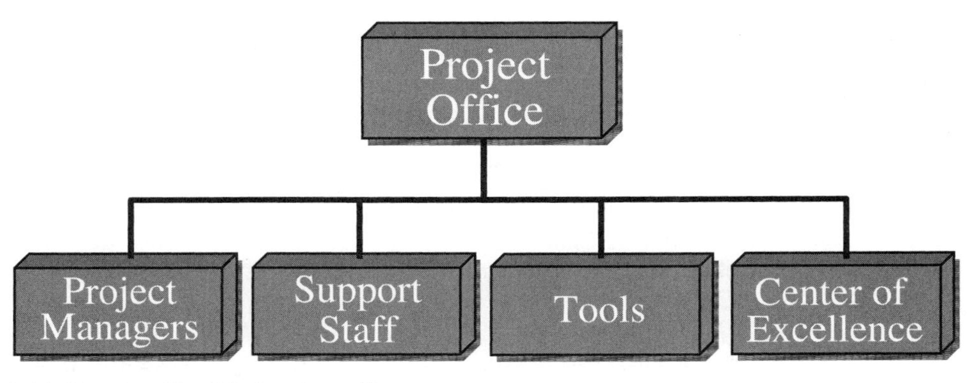

FIGURE 8–19. Simplified project office.

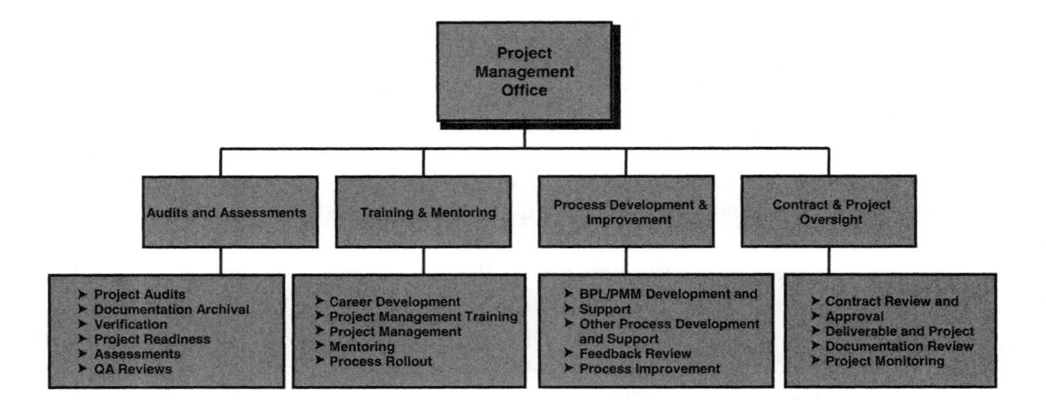

FIGURE 8–20. Computer Associates Project Management Office.

Every company can have a different structure for its project office. A typical structure for a project office appears in Figure 8–19.

The intent of the project office is to manage intellectual property on project management and, as such, does not create bureaucracy by adding layers of management. A project office does not need more than four or five people assigned, as shown in the Johnson Controls case study. In addition, individuals may be assigned part-time to a project office or could be "dotted" line, reporting to the project office while maintaining other functional responsibilities.

Because of the diversity of activities within a project office, individuals assigned to a project office could have multiple responsibilities and might serve as a backup for one another. This would reduce the head count in the project office and might make it easier to measure the return on investment of using a project office.

As companies become more experienced in project management and best practices begin to grow, additional responsibilities can be assigned to a project management office. Some companies even assign activities that are normally functional activities to a PMO if a large portion of the work is directly related to project work and the employees can be justified on a full-time basis. In Figure 8–20, we see the PMO for Computer Associates.

MULTIPLE CHOICE QUESTIONS

1. For each activity placed under the control of the project office, resistance can appear:
   A. At the executive levels
   B. At the line management levels
   C. At the worker level
   D. All of the above
2. Which type of project office would probably not have project managers assigned to it full time?

A. Functional
B. Customer groups
C. Corporate
D. All have full-time project managers

3. A company maintains a project office to handle large projects and a second project office to handle small projects. Each of these project offices probably resembles a ________ project office.
A. Functional
B. Customer groups
C. Corporate
D. B and C only

4. In the early years of project management, project offices had more people than necessary assigned full-time because:
A. The customer was willing to pay the overmanagement cost.
B. There was a requirement for the assigned personnel to have both primary and backup responsibilities.
C. Risks in the management of the project would be reduced.
D. All of the above.

5. Which type of project office probably functioned as a cost center?
A. Functional
B. Customer groups
C. Strategic
D. A and C only

6. Which type of project office probably functioned as a profit center?
A. Functional
B. Customer groups
C. Strategic
D. B and C only

7. Which of the following is generally not a responsibility of a project office?
A. Performance reviews
B. Benchmarking
C. Mentorship
D. Methods and standards

8. The prime responsibility of a project management office today appears to be:
A. Mentorship
B. Standards
C. Management of intellectual property
D. Training and education

9. Which of the following is generally regarded as a low-risk activity for a PMO?
A. Capacity planning
B. Standards and methodology development
C. Business case development
D. Continuous improvement

10. Which of the following is generally regarded as a high-risk activity for a PMO?
A. Mentorship
B. Dissemination of information
C. Lessons learned files
D. Training and education

11. Which of the following would be a specific benefit to executives as a result of establishing a PMO?
    A. Elimination or reduction of company silos
    B. Standardization of operations
    C. Company-wide prioritization of work
    D. All of the above
12. Which of the following is generally not an information system managed by the PMO?
    A. Failure reporting
    B. Compensation management
    C. Lessons learned
    D. Earned value measurement
13. Which of the following is not one of the four critical questions addressed during a postmortem analysis of a project?
    A. What went wrong?
    B. What should we do next time?
    C. Who gets what information?
    D. How should the workers be reassigned?
14. The bottom line of the postmortem pyramid evaluates:
    A. Key performance indicators
    B. Critical success factors
    C. Customer satisfaction/business opportunities
    D. Strategic mission
15. The second layer (from the bottom) in the postmortem pyramid evaluates:
    A. Key performance indicators
    B. Critical success factors
    C. Customer satisfaction/business opportunities
    D. Strategic mission

DISCUSSION QUESTIONS

1. Why is there often resistance to the implementation of a project management office?
2. What industries are the primary users of a customer group PMO?
3. What organizations are the primary users of a functional PMO?
4. When we say that the implementation of a PMO is a high or low risk, to what are we referring?
5. Why would a company want to support both a functional PMO and a corporate PMO?
6. What type of PMO is best suited to support (i.e., provide benefits to) senior management as a whole?
7. What are the four types of information systems maintained by a corporate PMO?
8. During postmortem analysis of a project, what are the differences between CSFs and KPIs?
9. Explain project manager mentorship.
10. Why is it a good idea to have the PMO act as the champion for the project management methodology?

Across

3. Type of PMO
5. _ _ _ _ _ _ value measurement
6. PMO responsibility
8. PMO responsibility
9. Type of PMIS: _ _ _ _ _ _ _ _ reporting
10. Type of PMO: _ _ _ _ _ _ _ _ _ groups
11. Executive benefit of PMO
14. PMO responsibility
16. Part of CSF
17. _ _ _ _ _ _ _ _ _ _ _ _ _ _ property
19. Implementation threat
20. Type of PMIS: _ _ _ _ management
22. Executive benefit of PMO

Down

1. Executive benefit of PMO
2. Type of PMO
4. _ _ _ _ _ _ _ _ _ _ _ _ improvement
7. PMO responsibility
12. _ _ _ _ _ _ _ _ _ _ _ _ pyramid
13. PMO responsibility
15. Implementation threat
16. _ _ _ _ _ _ _ _ _ _ planning
18. Lessons-_ _ _ _ _ _ _
21. Part of KPI

Integrated Processes

9.0 INTRODUCTION

Companies that have become extremely successful in project management have done so by performing strategic planning for project management. These companies are not happy with just matching the competition. Instead, they opt to exceed the performance of their competitors. To do this on a continuous basis requires processes and methodologies that promote continuous rather than sporadic success.

Figure 9–1 identifies the hexagon of excellence. The six components identified in the hexagon of excellence are the areas where the companies excellent in project management exceed their competitors. Each of these six areas is discussed in Chapters 9 through 14. We begin with *integrated processes.*

9.1 UNDERSTANDING INTEGRATED MANAGEMENT PROCESSES

As we discussed in Chapter 6, several new management processes since 1985 (concurrent engineering, for example) have supported the acceptance of project management. The most important complementary management processes, and the years they were introduced, are listed below:

- *1985:* Total quality management (TQM)
- *1990:* Concurrent engineering
- *1991:* Self-directed teams
- *1992:* Employee empowerment

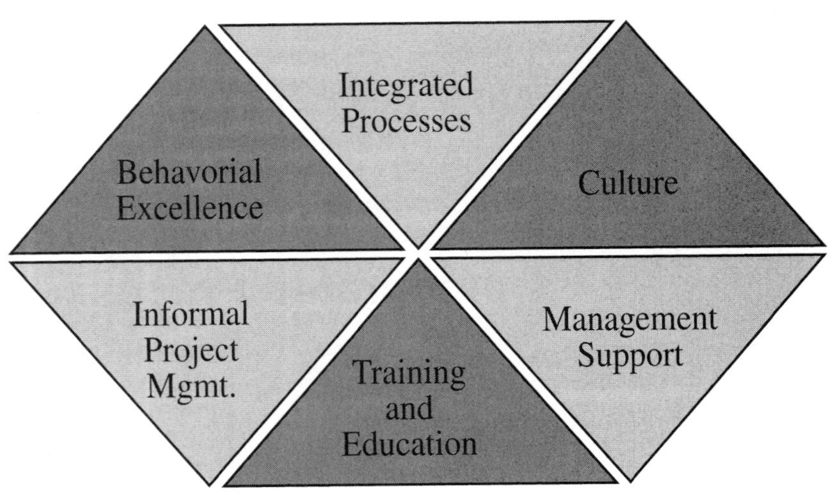

FIGURE 9–1. The six components of excellence. *Source:* Reprinted from H. Kerzner, *In Search of Excellence in Project Management.* New York: Wiley, 1998, p. 14.

- *1993:* Reengineering
- *1994:* Life cycle costing
- *1995:* Change management
- *1996:* Risk management
- *1997–1998:* Project offices and centers of excellence
- *1999:* Co-located teams
- *2000:* Multinational teams
- *2001:* Maturity models
- *2002:* Strategic planning for project management
- *2003:* Intranet status reporting
- *2004:* Capacity planning models

The *integration* of project management with these other management processes is key to achieving excellence. Not every company uses every process all the time. Companies choose the processes that work the best for them. However, whichever processes are selected, they are combined and integrated into the project management methodology. Previously we stated that companies with world-class methodologies employ a single, standard methodology based upon integrated processes.

The ability to integrate processes is based on which processes the company decides to implement. For example, if a company implemented a stage-gate model for project management, the company might find it an easy task to integrate new processes such as concurrent engineering. The only precondition would be that the new processes were not treated as independent functions but were designed from the onset to be part of a project management system already in place. For example, the four-phase model used by the General Motors Powertrain Group and the PROPS model used at Ericsson Telecom AB readily allow the assimilation of additional management processes.

This chapter discusses each of the management processes listed and how the processes enhance project management. Then we look at how some of the integrated management processes have succeeded using actual case studies.

9.2 EVOLUTION OF COMPLEMENTARY PROJECT MANAGEMENT PROCESSES

Since 1985, several new management processes have evolved parallel to project management. Of these processes, total quality management and concurrent engineering are the most relevant. Companies that reach excellence are the quickest to recognize the synergy among the many management options available today. Companies that reach maturity and excellence the quickest are those that recognize that certain processes feed on one another. As an example, consider the seven points listed below. Are these seven concepts part of a project management methodology?

- Teamwork
- Strategic integration
- Continuous improvement
- Respect for people
- Customer focus
- Management by fact
- Structured problem-solving

These seven concepts are actually the basis of Sprint's total quality management process. They could just as easily have been facets of a project management methodology.

During the 1990s, Kodak taught a course entitled, "Quality Leadership." The five principles of Kodak's Quality Leadership Program include:

Customer focus: "We will focus on our customers, both internal and external, whose inputs drive the design of products and services. The quality of our products and services is determined solely by these customers."

Management leadership: "We will demonstrate, at all levels, visible leadership in managing by these principles."

Teamwork: "We will work together, combining our ideas and skills to improve the quality of our work. We will reinforce and reward quality improvement contributions."

Analytical approach: "We will use statistical methods to control and improve our processes. Data-based analyses will direct our decisions."

Continuous improvement: "We will actively pursue quality improvement through a continuous cycle that focuses on planning, implementing, and verifying of improvements in key processes."

Had we just looked at the left-hand column we could argue that these are the principles of project management as well. More recently, in 1997, the International Organization for Standardization (ISO) in Geneva, Switzerland, developed the ISO 10006 series, which addresses quality management: guidelines to quality in project management.

Figure 9–2 shows what happens when an organization does not integrate its processes. The result is totally uncoupled processes. Companies with separate methodologies for each process may end up with duplication of effort, possibly duplication of resources, and even duplication of facilities.

As companies begin recognizing the synergistic effects of putting several of these processes under a single methodology, the first two processes to become partially coupled are project management and total quality management, as shown in Figure 9–3. As the benefits of synergy and integration become apparent, organizations choose to integrate all of these processes, as shown in Figure 9–4.

Excellent companies are able to recognize the need for new processes and integrate them quickly into existing management structures. During the early 1990s, integrating project management with total quality management and concurrent engineering was emphasized. Since the middle 1990s, two other processes have become important in addition: risk management and change management. Neither of these processes is new; it's the emphasis that's new.

Steve Gregerson, vice president for Product Development at BTR Automotive, Sealing Systems Group, describes the integrated processes in their methodology:

> Our organization has developed a standard methodology based on global best practices within our organization and on customer requirements and expectations. This methodol-

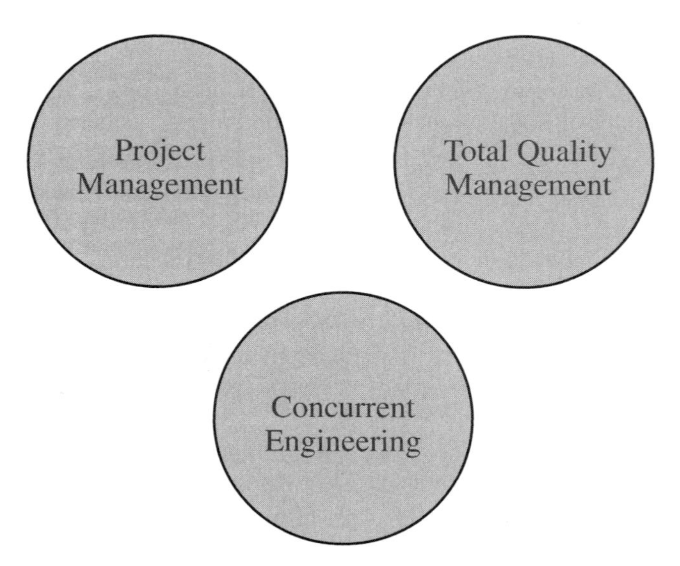

FIGURE 9–2. Totally uncoupled processes.

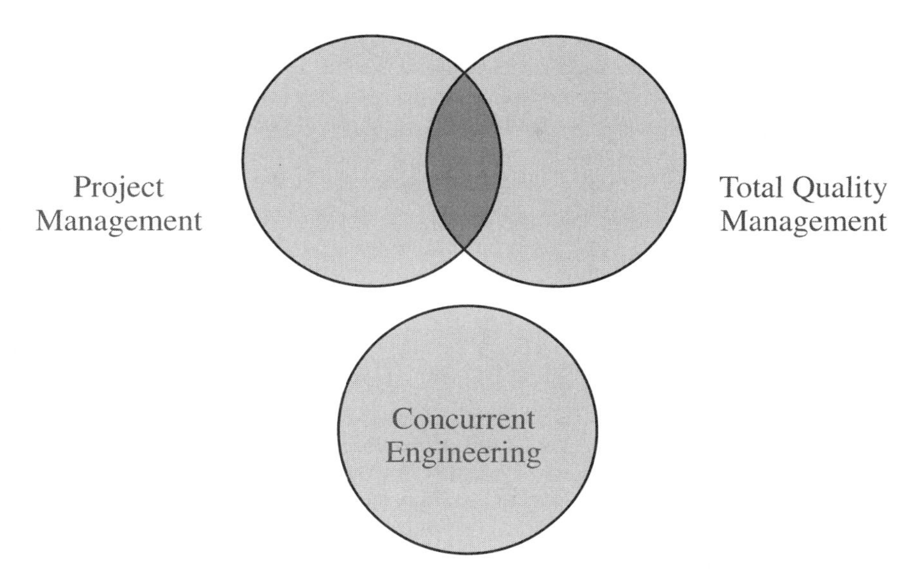

FIGURE 9–3. Partially integrated processes.

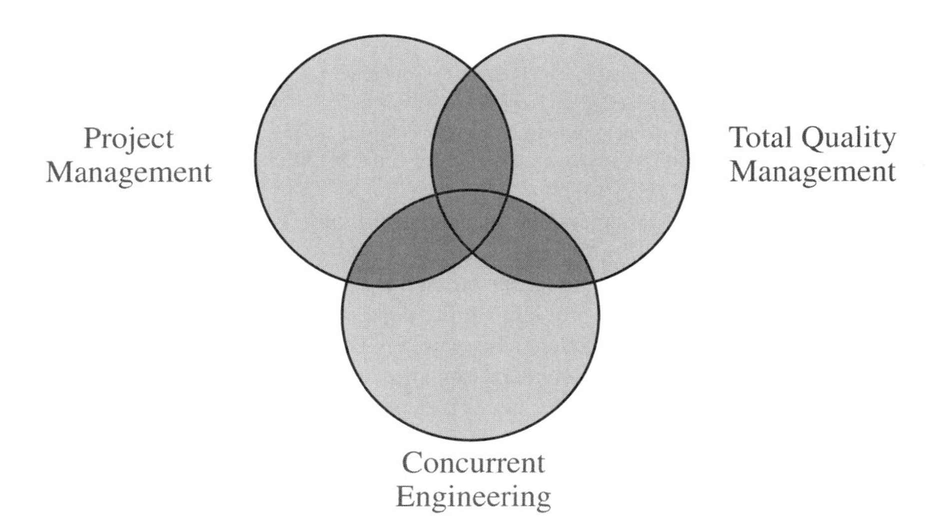

FIGURE 9–4. Totally integrated processes.

> ogy also meets the requirements of QS-9000. Our process incorporates seven gateways that require specific deliverables listed on a single sheet of paper. Some of these deliverables have a procedure and in many cases a defined format. These guidelines, checklists, forms, and procedures are the backbone of our project management structure, and also serve to capture lessons learned for the next program. This methodology is incorporated into all aspects of our business systems, including risk management, concurrent engineering, advanced quality planning, feasibility analysis, design review process, and so on."

Another example of integrated processes is the methodology employed by Nortel. Bob Mansbridge, vice president, Supply Chain Management at Nortel Networks, believes

> Nortel Networks project management is integrated with the supply chain. Project management's role in managing projects is now well understood as a series of integrated processes within the total supply chain pipeline. Total quality management (TQM) in Nortel Networks is defined by pipeline metrics. These metrics have resulted from customer and external views of "best-in-class" achievements. These metrics are layered and provide connected indicators to both the executive and the working levels. The project manager's role is to work with all areas of the supply chain and to optimize the results to the benefit of the project at hand. With a standard process implemented globally, including the monthly review of pipeline metrics by project management and business units, the implementation of "best practices" becomes more controlled, measurable, and meaningful.

The importance of risk management is finally being recognized. According to Frank T. Anbari, Professor of Project Management, George Washington University:

> By definition, projects are risky endeavors. They aim to create new and unique products, services, and processes that did not exist in the past. Therefore, careful management of project risk is imperative to repeatable success. Quantitative methods play an important role in risk management. There is no substitute for profound knowledge of these tools.

Risk management has been a primary focus among health care organizations for decades, and for obvious reasons. Today, in organizations of all kinds, risk management keeps us from pushing our problems downstream in the hope of finding an easy solution later on or of the problem simply going away by itself. Change management as a complement to project management is used to control the adverse effects of scope creep: increased costs (sometimes double or triple the original budget) and delayed schedules. With change management processes in place as part of the overall project management system, changes in the scope of the original project can be treated as separate projects or subprojects so that the objectives of the original project are not lost.

Today, excellent companies integrate five main management processes (see Figure 9–5):

- Project management
- Total quality management
- Concurrent engineering
- Risk management
- Change management

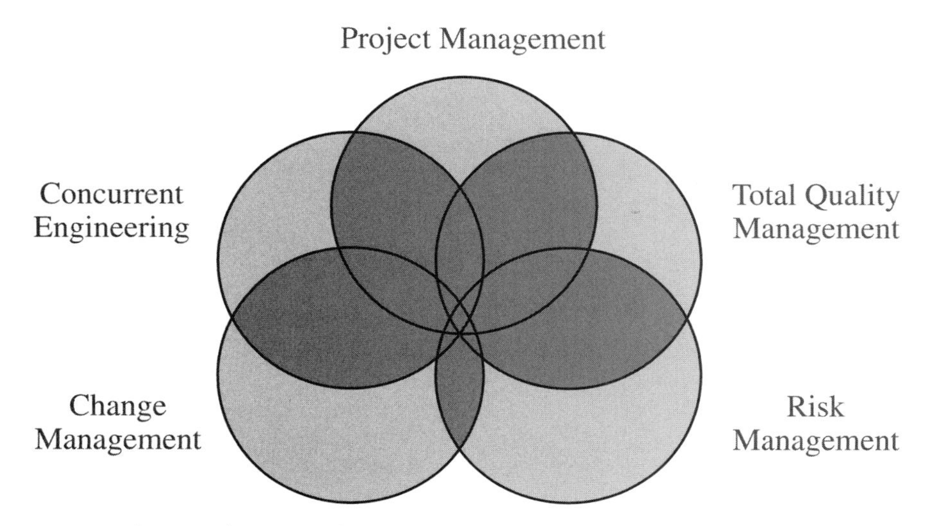

FIGURE 9–5. Integrated processes for the twenty-first century.

Self-managed work teams, employee empowerment, reengineering, and life cycle costing are also combined with project management in some companies. We briefly discuss these less widely used processes after we have discussed the more commonly used ones.

9.3 TOTAL QUALITY MANAGEMENT

During the past decade, the concept of total quality management has revolutionized the operations and manufacturing functions of many companies. Companies have learned quickly that project management principles and systems can be used to support and administer total quality management programs, and vice versa. Ultimately, excellent companies have completely integrated the two complementary systems.

The emphasis in total quality management is on addressing quality issues in total systems. Quality, however, is never an end goal. Total quality management systems run continuously and concurrently in every area in which a company does business. Their goal is to bring to market products of better and better quality and not just of the same quality as last year or the year before.

Total quality management (often referred to as TQM) was founded on the principles advocated by W. Edwards Deming, Joseph M. Juran, and Phillip B. Crosby. Deming is famous for his role in turning postwar Japan into a dominant force in the world economy. Total quality management processes are based on Deming's simple plan–do–check–act cycle (see Figure 9–6).

The cycle fits completely with project management principles. To fulfill the goals of any project, first you plan what you're going to do, then you do it. Next, you check on what

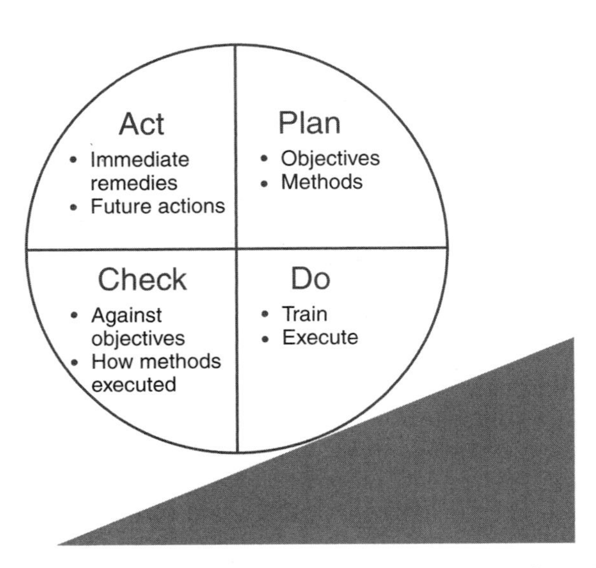

FIGURE 9–6. The Deming plan–do–check–act cycle. *Source:* Reprinted from H. Kerzner, *In Search of Excellence in Project Management.* New York: Wiley, 1998, p. 115.

you did. You fix what didn't work, then you execute what you set out to do. But the cycle doesn't end with the output. Deming's cycle works as a continuous improvement system, too. When the project is complete, you examine the lessons learned in its planning and execution. Then you incorporate those lessons into the process and begin the plan–do–check–act cycle all over again on a new project.

Many companies have achieved improvements using TQM. Here are only a sample[1]:

- *AMP:* On-time shipments improved from 65 to 95 percent, and AMP products have nationwide availability within three days or less on 50 percent of AMP sales.
- *Asea, Brown, Boveri:* Every improvement goal customers asked for—better delivery, quality responsiveness, and so on—was met.
- *Chrysler:* New vehicles are now being developed in 33 months versus as long as 60 months ten years ago.
- *Eaton:* Increased sales per employee from $65,000 in 1983 to about $100,000 in 1992.
- *Fidelity:* Handles 200,000 information calls in four telephone centers; 1,200 representatives handle 75,000 calls, and the balance is automated.
- *Ford:* Use of 7.25 person-hours of labor per vehicle versus 15 person-hours in 1980; Ford Taurus bumper uses 10 parts compared to 100 parts on similar GM cars.

1. Taken from C. Carl Pegels, *Total Quality Management.* Danvers, Massachusetts: Boyd and Fraser, 1994, p. 27.

- *General Motors:* New vehicles are now being developed in 34 months versus 48 months in the 1980s.
- *IBM Rochester:* Defect rates per million are 32 times lower than four years ago and on some products exceed six sigma (3.4 defects per million).
- *Pratt & Whitney:* Defect rate per million was cut in half; a tooling process was shortened from two months to two days: part lead times were reduced by 43 percent.
- *VF Corp:* Market response system enables 97 percent in-stock rate for retail stores, compared to 70 percent industry average.
- *NCR:* Checkout terminal was designed in 22 months versus 44 months and contained 85 percent fewer parts than its predecessor.
- *AT&T:* Redesign of telephone switch computer completed in 18 months versus 36 months; manufacturing defects reduced by 87 percent.
- *Deere & Co:* Reduced cycle time of some of its products by 60 percent, saving 30 percent of usual development costs.

Total quality management also is based on three other important elements: customer focus, process thinking, and variation reduction. Does that remind you of project management principles? It should.

One of the characteristics of companies that have won the prestigious Malcolm Baldrige Award is that each has an excellent project management system. Companies such as Motorola, Armstrong World Industries, General Motors, Kodak, Xerox, and IBM use integrated total quality management and project management systems.

9.4 CONCURRENT ENGINEERING

The need to shorten product development time has always plagued U.S. companies. Under favorable economic conditions, corporations have deployed massive amounts of resources to address the problem of long development times. During economic downturns, however, not only are resources scarce, time becomes a critical constraint. Today, the principles of concurrent engineering have been almost universally adopted as the ideal solution to the problem.

Concurrent engineering requires performing the various steps and processes in managing a project in tandem rather than in sequence. This means that engineering, research and development, production, and marketing all are involved at the beginning of a project, before any work has been done. That is not always easy, and it can create risks as the project is carried through. Superior project planning is needed to avoid increasing the level of risk later in the project. The most serious risks are delays in bringing product to market and costs when rework is needed as a result of poor planning.

Improved planning is essential to project management, so it is no surprise that excellent companies integrate concurrent engineering and project management systems. Chrysler Motors used concurrent engineering with project management to go from concept to market with the Viper sports car in less than three years. Concurrent engineering may well be the strongest driving force behind the increased acceptance of modern project management.

9.5 RISK MANAGEMENT

Risk management is an organized means of identifying and measuring risk, and developing, selecting, and managing options for handling those risks. Throughout this book, I have emphasized that tomorrow's project managers will need superior business skills in assessing and managing risk. Project managers in the past were not equipped to quantify risks, respond to risks, develop contingency plans, or keep lessons-learned records. They were forced to go to senior managers for advice on what to do when risky situations developed. Now senior managers are empowering project managers to make risk-related decisions, and that requires a project manager with solid business skills as well as technical knowledge.

Preparing a project plan is based on history. Simply stated: What have we learned from the past? Risk management encourages us to look at the future and anticipate what can go wrong, and then to develop contingency strategies to mitigate these risks.

We have performed risk management in the past, but only financial and scheduling risk management. To mitigate a financial risk, we increased the project's budget. To mitigate a scheduling risk, we added more time to the schedule. But in the 1990s, technical risks became critical. Simply adding into the plan more time and money is not the solution to mitigate technical risks. Technical risk management addresses two primary questions:

- Can we develop the technology within the imposed constraints?
- If we do develop the technology, what is the risk of obsolescence, and when might we expect it to occur?

To address these technical risks, effective risk management strategies are needed based upon technical forecasting.

On the surface, it might seem that making risk management an integral part of project planning should be relatively easy. Just identify and address risk factors before they get out of hand. Unfortunately, the reverse is likely to be the norm, at least for the foreseeable future.

For years, companies provided lip service to risk management and adopted the attitude that we should simply live with it. Very little was published on how to develop a structure risk management process. The disaster with the Space Shuttle *Challenger* in January 1986 created a great awakening on the importance of effective risk management.[2] Today, a new breed of project management experts exists with expertise in the development and application of risk management strategies for all aspects of a business. One such expert is Dr. Edmund H. Conrow CMC, CPCM, PMP, who shares with us some keys to success in risk management best practices. According to Dr. Conrow[3]:

---

2. The case study "The Space Shuttle Challenger Disaster" appears in the case study section at the end of this book.

3. This information was derived from Edmund H. Conrow, *Effective Risk Management: Some Keys to Success*, 2nd ed., American Institute of Aeronautics and Astronautics, Reston, VA, 2003. Copyright © 2003 Edmund H. Conrow. Used with permission of the author. The information provided here is selected from more than 700 lessons learned developed by Dr. Conrow from implementing risk management on a wide variety of projects over a 20+ year period. He can be reached at www.risk-services.com and (310) 374-7975.

All too often the project risk management process is either absent or poorly implemented. I will briefly describe some key aspects of a good risk management process that can be applied to a wide range of projects. At the top level, effective risk management requires both a well developed process and careful implementation. The process and implementation characteristics described below have been used on numerous projects, including high risk projects, and have been judged by customers to have greatly contributed to project success.

Some characteristics of an effective risk management process include: 1) all process steps are present, 2) the steps are in the correct order, 3) the steps are weighted equally, 4) the process is well structured, 5) the process is iterative, 6) the process is continuous, 7) the process is started early in the project phase, and 8) the process is updated for each project phase and/or any major re-baselining.

The risk management process should include planning, identification, analysis, handling, and monitoring steps, which are implemented in that order. It is surprisingly common to find one or more of these steps either missing (e.g., risk planning), weakly implemented (e.g., focus on the mitigation option of risk handling without considering the assumption, avoidance or transfer options), or out of order relative to the other steps (e.g., performing risk handling as part of risk identification or analysis). What you call a given process step is somewhat unimportant, but the functions performed for that step are very important. It is also common to find one or more process steps given considerably more attention than others. Yet, unless each step is carefully developed and implemented serious gaps may exist in the process and lead to diminished effectiveness. For example, if excessive attention is paid to risk analysis yet little thought is given to risk identification, then risk issues that should have been identified may remain undetected until they surface later in the project as problems. A well structured process includes clearly defined inputs, tools and techniques, and outputs for each process step. It also incorporates a set of procedures for each process step (developed as part of risk planning), suitable documentation and guidelines for communicating results. Because risk management is but one of many processes on many projects, it is also important to understand ahead of time how it interacts with top-level processes (e.g., project management and systems engineering), as well as same or lower level processes (e.g., cost, quality, schedule). All too often, linkages have not been properly established among the key processes and risk management is at least somewhat disconnected from the rest of the project.

The risk management process should be iterative, meaning that as the steps are performed from planning through monitoring, there is feedback from the monitoring step comparing actual vs. planned progress to the risk handling and analysis steps (and in some cases to the identification and planning steps). This feedback is in terms of cost, technical, schedule and risk-related performance and provides inputs to both evaluate and adjust results at these steps. The risk management process should also be continuous—it is not performed once or some other finite amount during the project phase, but on a regular cycle both for existing risk issues as well as to ensure that new risk issues are promptly evaluated. Risk management should be started early in the project phase else it increases the likelihood that one or more risk issues that should have been identified, analyzed and appropriately handled will "slip through the cracks" and surface as problems later in the project when it is typically much more costly to deal with them. Finally, the risk management process should be updated as needed for each succeeding project phase or as any major project re-baselining occurs. This is because not only new risk issues, but potentially new risk categories may appear that had not been given emphasis in a previous project phase or under the previous baseline. For example, early in development the architecture, then the preliminary design of an item should be explored, which often includes technology and design risk, while manufacturing and support risk should also be addressed in the trade process to develop a detailed design.

While risk management implementation will vary on a project-to-project basis, there are some considerations that apply to virtually all projects. First, there should be considerable attention paid to both organizational and behavioral implementation considerations. Even if the risk management process includes the facets mentioned above, clearly defined, nonconflicting organizational roles and responsibilities should exist and the program environment should be supportive and rewarding to permit risk management to be practiced on a daily basis.

Second, risk management should be implemented in both a "top-down" and "bottom-up" manner across the project. Key management personnel, including the project manager, should use risk management principles in decision making as well as support and encourage others to perform risk management. While the project manager (or deputy project manager) should not be the risk manager except in most cases, it is essential that he participate in risk management activities as well as using risk management principles as part of his decision making process. Without such support and active modeling, the message sent to other project personnel is that "risk management is unimportant." This can be highly detrimental towards creating a culture that both accepts risk management principles and integrates this process with other project processes.

Upper management participation must be proactive and contribute more than "don't shoot the messenger," because many organizations do not have a history of effective risk management and a suitable positive model must be developed and demonstrated to project personnel. This is very important because upper management "sets the example" for project personnel as to whether or not risk management is actually used on the project. Working-level personnel will quickly gauge whether or not upper management practices risk management, and if it is determined to be lip service, then ineffective risk management will likely result. It is equally important for working-level personnel to assimilate risk management principles into the daily job function. Finally, all project personnel should perform risk management—only the roles and responsibilities are different. This is particularly important for risk identification and if this is not the case, then it is likely that risk issues will go undetected until later in the project and surface as problems.

Third, simple but effective motivators should be developed and used, such as rewarding good risk management practice with a savings bond, dinner at a local restaurant, or as part of the personnel performance review. However, an even better approach is through intrinsic motivators. There is nothing potentially more effective to "turn on" project personnel to perform risk management than for them to see how it really can make a positive difference to the project. Those "on the fence" or even detractors often begin to use risk management principles and may even convince other people on the project perform risk management as well. Thus, successful risk management often leads to continuing success.

Fourth risk management must be tailored to each project—one size does not fit all. A common mistake is to believe that an existing, even effective risk management process can be applied directly to your project. In general I've found that there are 10 to 15 items that need to be evaluated, ranging from the obvious (e.g., contractual requirements) to the subtle (e.g., organizational behavior), before a particular risk management process can be successfully applied. More than once I've seen an effective risk management process on one project blindly applied to another project. The result was considerable inefficiency and an ineffective risk management process because project personnel did not know how to tailor the process to their project. This is made all the worse when those attempting to adapt the process have little or no "real world" experience or understanding of risk management, yet do not realize their own limitations.

While implementing the above recommendations will not guarantee a successful project, it will provide a starting point for developing effective risk management on your project.

As projects begin to grow in size and complexity, risk management requires a coordinated effort between the customer prime contractor and subcontractors. The Department of Defense (DoD) has taken the lead in developing ways to coordinating risk management efforts on large programs, especially in cultivating the risk management relationship between the government and contractors. According to the government[4]:

The prime contractor's support and assistance is required, even though the ultimate responsibility for risk management rests with the Government PM. Often, the contractor is better equipped to understand the program technical risks than the Government program office is. Both the Government and contractor need to share information, understand the risks, and develop and execute management efforts. The Government must involve the contractor early in program development so that effective risk assessment and reduction can occur.

Therefore, risk management must be a key part of the contractor's management scheme. Although the Government does not dictate how the contractor should manage risk, some characteristics of a good Government/contractor relationship include:

- Clear definition of risks and their assignment
- Flexibility for assignment of risks and risk management responsibilities among the teams
- Strong emphasis on best management and technical practices which, if followed, avoid unnecessary risks

Although risk management activities are specific and assigned to individuals rather than groups, the project manager is still ultimately responsible for risk management. The project can set up either a centralized or decentralized risk management organization. Because risk management is a responsibility shared by all team members, a decentralized risk management organization is used most frequently. An example is shown in Figure 9–7.[5]

The government has established the following guidelines applicable to all risk management organizations[6]:

- The PM is ultimately responsible for planning, allocating resources, and executing risk management. This requires the PM to oversee and participate in the risk management process.
- The PM must make optimal use of available resources; i.e., personnel, organizations, and funds. Personnel and organizational resources include the PMO, functional support offices of the host command, the prime contractor, independent risk assessors, and support contractors.

---

4. *Risk Management Guide for DoD Acquisitions,* 5th ed. Washington, DC: DoD/Defense Acquisitions University (DAU), June 2002, p. 40.

5. See note 4, p. 38.

6. See note 4, pp. 38–39.

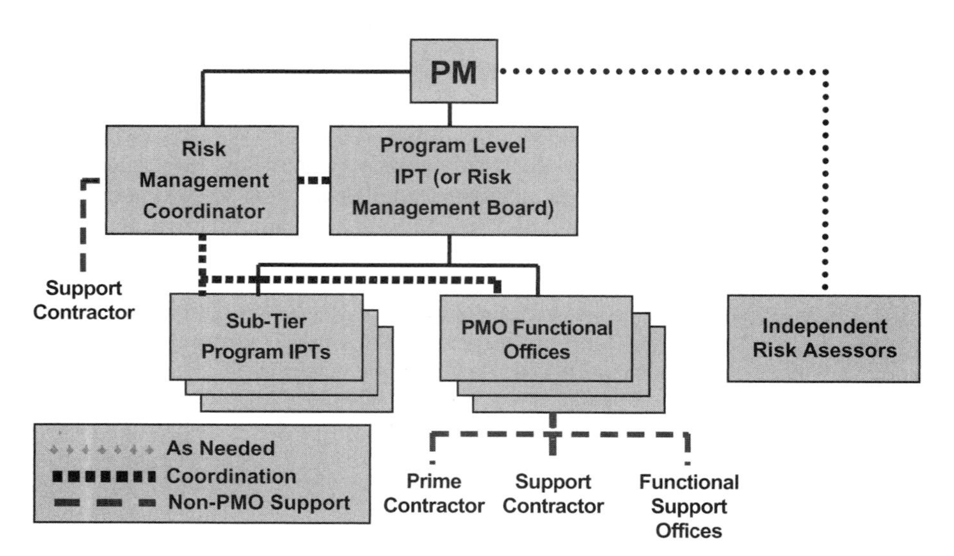

FIGURE 9–7. Decentralized risk management organization.

- Risk management is a team function. This stems from the pervasive nature of risk and the impact that risk-handling plans may have on other program plans and actions. In the aggregate, risk planning, risk assessment, risk handling, and risk monitoring affect all program activities and organizations. Any attempt to implement an aggressive forwarding-looking risk management program without the involvement of all PMO subordinate organizations could result in confusion, misdirection, and wasted resources. The only way to avoid this is through teamwork among the PMO organizations and the prime contractor. The management organizational structure can promote teamwork by requiring strong connectivity between that structure, the various PMO organizations, and the prime contractor. The teams may use independent assessments to assist them, when required.

- Figure 9–7 portrays a decentralized risk management organization. This example includes the entire PMO and selected non-PMO organizations; e.g., the prime contractor, who are members of the IPTs [Integrated Project Teams]. The figure shows that risk management is an integral part of program management and not an additional or separate function to perform. Hence, separate personnel are not designated to manage risk, but rather all individuals are required to consider risk management as a routine part of their jobs.

Another example of the relationship between risk management and process integration is provided by Lt. Col. John Driessnack. Before joining the Defense Acquisition University in 2001 as a professor, Lt. Col. Driessnack experienced the power of integrating risk management with other program managers' tools. He led the Global Broadcast System program, an effort by the Department of Defense to provide high-speed data and video capability anywhere in the world, through a rebaselining effort that considered the risks on the program and adjusted the program approach appropriately.

> We had experienced numerous failures in both hardware and software over the past two years on the program. The program was attempting to provide most of the performance in a single delivery. A different approach had to be taken. Considering the risks on the program, a spiral development approach was taken. The requirements were broken into various initial operating capability packages to be delivered on one year increments. Each increment was broken down into two six month efforts to be delivered in different parts of the world, reducing the complexity of the fielding process. A critical chain schedule, an approach that places an emphasis on getting all work done as quickly as possible, was employed. The "normal" schedule was built using standard estimates for how long each work package would take with both a success path and alternative paths for various risk mitigation strategies. Risks that could not be mitigated within the schedule were pushed off to a later release of the software. This revised level of work set the Earned Value baseline. The critical chain method then cut those standard estimates by 50% and built buffers at the end of the schedule. This pulled any slack out of the schedule and provided a sense of urgency on almost every package. Critical to the success was the integration of a six sigma approach to the software work efforts to cut down on rework. The critical chain created the buffers in the schedule that were needed to mitigate the inherent schedule risk in software development. In the first 6 month delivery, the program met the overall schedule because processes had been improved that reduced rework, but also buffers were available to absorb the realized schedule risks. Critical to the buy in by the team was setting the Earned Value baseline on the standard schedule and not on the Critical Chain schedule.

Professor Driessnack now teaches risk management to both midlevel and senior-level Department of Defense civilian and military program managers. He emphasizes in his executive risk management sessions:

> A risk process is limited if it is not integrated with cost, schedule, and performance processes and the baselines and measures used to monitor the overall program. The risk process needs to do more than add risk margins. The process needs to influence the approaches to achieve success, such as having alternative design teams on uncertain technical work packages. The additional work, often called mitigation strategies, then needs to be in the estimated costs and laid out on an alternative path on the schedule. Risk identification is the task of everyone on the project team. The team needs to consider the best approaches to mitigate both the likelihood the risk will occur and the consequence of the occurrence if it happens. A good measure of how well the risk process is working on a program is simply to ask yourself how many issues that are biting at the project team were identified risks. If less than 50% are, your management is more crisis than structured.

The Department of Defense acquisition community, which works with some of the riskiest multibillion-dollar projects, has numerous examples of success stories. For specific stories on risk and integration with other techniques and tools, Lt. Col. Driessnack suggests joining the Program Management Community of Practice (PMCOP) risk community at *www.pmcop.dau.mil.* The community of practice has grown into a large community in which practitioners exchange ideas.

Risk management today has become so important that companies are establishing separate risk management organizations within the company. However, many companies have been using risk management functional units for years, and yet this concept has gone unnoticed. The following is an overview of the program management methodology of the risk management department of an international manufacturer headquartered in Ohio. This department has been in operation for approximately 25 years.

> The Risk Management Department is part of the financial discipline of the company and ultimately reports to the Treasurer, who reports to the CFO. The overall objective of the department is to coordinate the protection of the company's assets. The primary means of meeting that objective is eliminating or reducing potential losses through Loss Prevention Programs. The department works very closely with the internal Environmental Health and Safety Department. Additionally, it utilizes outside loss control experts to assist the company's divisions in loss prevention.
>
> One method employed by the company to insure the entire corporation's involvement in the risk management process is to hold its divisions responsible for any specific losses up to a designated self-insured retention level. If there is a significant loss, the division must absorb it and its impact on their bottom line profit margin. This directly involves the divisions in both loss prevention and claims management. When a claim does occur, Risk Management maintains regular contact with division personnel to establish protocol on the claim and reserves and ultimate resolution.
>
> The company does purchase insurance above designated retention levels. As with the direct claims, the insurance premiums are allocated to its divisions. These premiums are calculated based upon sales volume and claim loss history, with the most significant percentage being allocated to claims loss history.
>
> Each of the company's locations must maintain a Business Continuity Plan for its site. This plan is reviewed by Risk Management and is audited by the Internal Audit and Environmental Health and Safety Departments.
>
> Risk management is an integral part of the corporation's operations as evidenced by its involvement in the due diligence process for acquisitions or divestitures. It is involved at the onset of the process, not at the end, and provides a detailed written report of findings as well as an oral presentation to Group Management.
>
> Customer service is part of the company's Corporate Charter. Customers served by Risk Management are the company's divisions. The department's management style with its customers is one of consensus building and not one of mandating. This is exemplified by the company's use of several workers compensation third party administrators (TPAs) in states where it is self insured. Administratively, it would be much easier to utilize one nationwide TPA. However, using strong regional TPAs with offices in states where divisions operate provides knowledgeable assistance with specific state laws to the divisions. This approach has worked very well for this company which recognizes the need for the individual state expertise.

The importance of risk management is now apparent worldwide. The principles of risk management can be applied to all aspects of a business, not just projects. Once a company begins using risk management practices, the company can always identify other applications for the risk management processes. Consider the following remarks made by Marcelino Diez, PMP, Project Management Consultant at Edelca:

> Edelca is in the process of defining and establishing structured risk management processes that will be obligatory. Today, Edelca is encouraging the development and use of processes related to the identification and mitigation of risks for all the relevant projects. These risk management processes are focused on the most critical items such as Business Objectives, Project Drivers and Objectives, Scope, Project Management Plan, Organizational and Human Resource Plans, Contracting Plan, Cost and Schedule Management Plans, Quality Management Plan, Communication Plan, Management Plan for Safety, Health and Environment, Procurement Plan, and Equipment and Materials Management.

For multinational companies that are project-driven, risk management takes on paramount importance. Not all companies, especially in undeveloped countries, have an understanding of risk management or its importance. These countries sometimes view risk management as an overmanagement expense on a project.

Consider the following scenario. As your organization gets better and better at project management, your customers begin giving you more and more work. You're now getting contracts for turnkey projects. Before, all you had to do was deliver the product on time and you were through. Now you are responsible for project installation and startup as well, sometimes even for ongoing customer service. Because the customers no longer use their own resources on the project, they worry less about how you're handling your project management system.

Alternatively, you could be working for third world clients who haven't yet developed their own systems. One hundred percent of the risk for such projects is yours, especially as projects grow more complex (see Figure 9–8). Welcome to the twenty-first century.

One subcontractor received a contract to install components in a customer's new plant. The construction of the plant would be completed by a specific date. After construction was completed, the contractor would install the equipment, perform testing, and then start up. The subcontractor would not be allowed to bill for products or services until after a successful startup. There was also a penalty clause for late delivery.

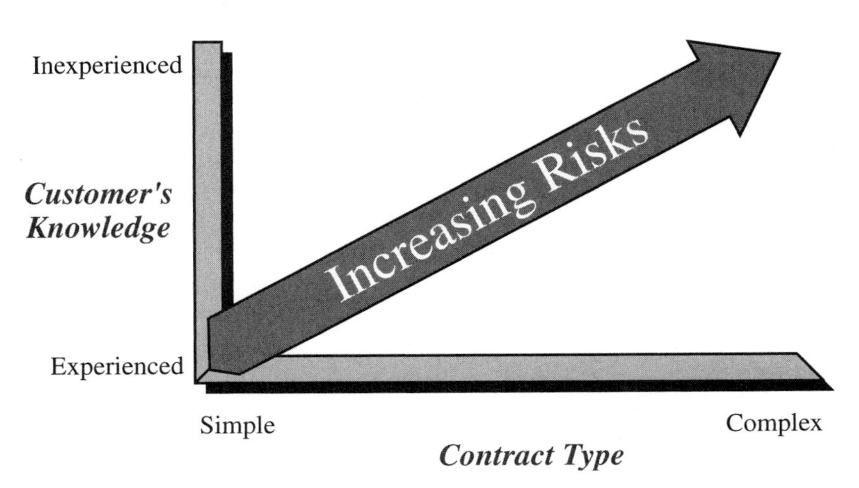

FIGURE 9–8. Future risks.

The contractor delivered the components to the customer on time, but the components were placed in a warehouse because plant construction had been delayed. The contractor now had a cash flow problem and potential penalty payments because of external dependencies that sat on the critical path. In other words, the contractor's schedule was being controlled by actions of others. Had the project manager performed business risk management rather than just technical risk management, these risks could have been reduced.

For the global project manager, risk management takes on a new dimension. What happens if the culture in the country with which you are working neither understands risk management nor has any risk management process? What happens if employees are afraid to surface bad news or identify potential problems? What happens if the project's constraints of time, cost, and quality/performance are meaningless to the local workers?

Suzanne Zale, global program manager of EDS, made the following comments on risk management:

> Risk management automatically becomes more critical for global projects. For some geographic regions, people are less likely to surface issues, especially if issues may be taken to higher level management or outside the local environment. It is very important that the risk management process be very clearly defined and people well trained in the process. This will give people a nonthreatening avenue to raise issues. It is also very important for the project manager to establish good relationships with the team members to increase the level of comfort and encourage open communications.

Even a very simple risk management process is better than having no process at all. Good companies recognize that risk management is part of the value-added chain. Steve Gregerson, vice president for product development at Metzeler Automotive Profile System, believes:

> Risk management is a major value-added function of our project management process. A simple form has been developed with the intent of predicting the likelihood of failure of a program based on its current status. The program team must complete this form for each gateway and report on the risk management in about 20 areas of the program and take countermeasures to mitigate the risk.
>
> In addition, a survey of each member of the program is taken during the gateway review, and any concern which is considered worthy of further analysis is carried either in the team open issues report or reported in the monthly management reviews, or both. I have found that a "gut feeling" of an experienced team member is as good of an early warning system as any.

Keith Rosenau, Chief Engineer and Director of Program Management for Metzeler, continues:

> The improvement process for Program Management for Monthly Management Reviews now includes senior staff review. Each senior staff member of the attending functional area is queried for program status and rating. This query requires agreement on the program rating or the lowest rating will be given to the program.

Consulting companies today are under severe pressure to assist their clients with risk management rather than either being an observer or coming on board after the risk management process is completed and not having access to information that could be critical to the success of the project. Brad Ruzicka, Senior Consulting Manager at the StoneBridge Group, explains:

> StoneBridge Group will generally recommend a risk assessment to be performed prior to undertaking projects that meet certain criteria. These criteria include project size, complexity, resource availability, client experience with projects of this type, and client expectations regarding use of new technologies, processes and/or tools. Our objective in performing the risk assessment is to identify all areas of potential high or medium risk and develop recommendations that will minimize or eliminate the risk. Generally, this approach has been very effective in lowering total project risk.

Craig Belton, vice president for Network Field Operations at Nortel Networks, believes that risk management at the work package level of the work breakdown structure is critical.

> In many areas of Nortel Networks, a risk management qualification is performed as part of the bid response. We are working toward a standard method to perform this risk analysis in order to assess the need for contingency plans. A prerequisite for the risk analysis is the Work Breakdown Structure (WBS) being proposed for the project. Each work package will be examined in terms of potential risk to the planned project time, cost, and quality. The risks are described, impacts and probabilities identified, and the mitigation cost estimated. This assessment of the risk is a quantitative way to identify the contingency plans needed for a project, and facilitates the management of these plans during the project implementation.

HP Services also uses the WBS for risk analysis. Sameh Boutros, Director Americas Engagement PMO for HP Services, states that:

> All projects must go through the SOAR (Solution Opportunity Approval and Review) process. Business management reviews projects at bid stage and at selected points within the project life cycle. A key element of each review is risk analysis and monitoring risk status. A key tool is the WBS. Risk is addressed at work package level and subjected to the risk planning process. Contingency plans are made and reserves allocated and managed. Because we are invariably competing for business which has to be profitable, risk has to be addressed in detail. If we get it wrong, we either price ourselves out of the bid or have an unprofitable project. Being able to correctly analyze and estimate risk is the key to our business.

Integrated processes are vital to the success of project-driven organizations whose entire business is projects. Jean Lehmann, manager of systems engineering methods at EDS, comments on the integrated processes included in the EDS methodology for project management:

> Product and service providers recognize the need to invest time and money in developing standard processes. EDS' Project Management methodology, PM 2, establishes a standard approach to managing projects.

PM 2 can be used as a stand-alone methodology or with other EDS or client methodologies. Process Sourcerer®, an EDS-developed tool for process definition and delivery, supports a rigorous definition of the project management discipline, and integration of that discipline with other disciplines, such as systems development and change management. Proper integration of PM 2 into any type of project provides a solid foundation that enables project managers to create project plans and effectively manage a project. Using PM 2 as a map reduces risks, enhances project communication, and improves each project manager's ability to teach others how to chart their project management paths. PM 2 provides a course to successfully navigate projects.

Project management is crucial to the delicate task of balancing people, customers, and business values. As an organization of information systems professionals, EDS is aware of the value of project management in controlling resources, reducing costs, and meeting customer expectations to make a fair profit. PM 2 guides project managers in addressing these sometimes conflicting facets of a project. PM 2 also reduces the learning curve for project managers by providing templates and examples of common work products, tips, techniques, tools, and training. PM 2, combined with leadership and management skills, provides project managers the effective process required to give EDS a powerful competitive advantage.

The EDS PM 2 methodology includes various process steps for risk management. During the start-up phase, an initial high-level risk assessment is performed to identify potential risk areas for the project. During the planning phase, formal risk management plans are developed for each risk identified. As new risks are identified throughout the project, the methodology guides the creation of a risk contingency plan for the newly identified risks. In addition, an evaluation is performed to determine the impact on other risk management plans currently in place.

One of the major advantages of a project management methodology is that it provides structure to processes such as the risk management process. Risk management templates for project management are an excellent way of providing guidance for workers on how to handle risk management. The ABB manual for project management contains an excellent section on templates for risk management, especially in the identification of the typical risk categories applicable to ABB projects. The following material has been taken from the ABB Project Management Manual and reproduced with permission of ABB:

Consider that a project may include any and all of the following risks:

- Back-to-back agreement that may include any of the risks listed below
- Contract and agreements
  - Penalty/liquidated damages
  - Customer deliberately exploits weakness in the agreement
  - Agreement contains specifications open to misinterpretation
  - General/vague contract wordings
  - Open issues, deviations—technical and/or commercial
  - Change management procedures not agreed with customer
  - Excessive reporting needs by the customer
  - Permits/licenses
  - Agents and commissions
- Responsibility and liability

   - Unclear scope or unclear limit of responsibility
   - Limits of liability
   - Cancellation caused by force majeure: seller, customers' responsibility milestones for scope of supply, noncompliance with contract, employer
- Financial
   - Payment, positive cash flow, letter of credit, procedures
   - Currency exchange rate fluctuation
   - Inflation
   - Bonds
   - Financial status—customer credit history
   - Subcontracting exposure
- Political
   - Stability of political environment
   - Political disturbances delay startup
   - Political disturbances that could inhibit our delivery performance
   - Changes in legislation
   - Export/import restrictions
   - Legal clauses of arbitration
- Warranty
   - Nonstandard/extended warranty
   - Cost of transportation to/from AAB facility (in/out warranty)
   - Repair of equipment at site
   - Ready access to equipment to allow repairs at site
- Schedule
   - Unrealistic delivery time
   - Delivery impact due to nonstandards solutions
   - Delivery commitments
   - Document exchange from all parties
   - Approval procedures
   - Site access and site readiness: power, and water, including a medium for testing (i.e., correct process conditions)
   - Work by others not finished on time
   - Compulsory interface with other suppliers/deliveries
   - Delivery impact due to peak loading of resources
- Technical and technological
   - Nonstandard application solutions
   - Process/performance guarantees
   - Environmental test simulations (i.e., heat chambers)
   - Quality assurance regulations must be observed
   - Site location and/or customer standards applies
   - Extensive documentation requirements
   - Technical interfaces: retrofit/extension of existing plant, third-party contractors/ consortia partners, responsibility for performance
   - Technical specifications: vague technical specification/function description, customer-selected technical solution or components
   - Customer acceptance criteria
- Resources
   - Personnel: skill sets needed, capacity, availability
   - Contract involves a major engineering input

    - Contract involves a large portion of external resources
    - Contract involves major demands on production equipment
    - Technical workmanship of local supply
- Supply and demand chain management
    - Delivery commitments: ABB, third party
- Customer
    - Level of experience of key personnel
    - The project will be given a low priority
    - Business counterparts/consultants/joint venture
    - Bureaucracy
    - Communications
    - Language
- Consortia
    - Sharing of penalty of consortium members
    - Third-party contractors difficult to deal with
    - Loss of payment in case of insolvency of consortium leader
    - Sharing costs in case of additional suppliers and services
- Environmental
    - Emissions (noise, dust, etc.)
    - Lack of secure storage leading to loss or damage of equipment
    - Infrastructure: harbor links (port connections), roads, telecommunications, electric power/disturbances
    - Social cultural; thefts, business ethic
- Project manager must take immediate action to manage and/or avoid any and all potential risk.
- Project manager exerts all possible influence in order to improve the economical results of the project.
- Project manager makes certain who the client is. If others are acting on the client's behalf, project manager checks the authority of the persons concerned.
- *Risk Review Report* is updated during the contract analysis in the project startup phase and again when any risk has changed.
- Project manager carries on or extends the client's contract conditions to the suppliers, as applicable.
- All possible back-charge claims are identified, and the worst-case cost is estimated.
- Project manager verifies to what extent certain risks can be covered by general company insurance.
- When a risk cannot be avoided, the related cost and influence to the schedule have to be evaluated; planned cost and schedule should be updated accordingly.

TABLE 9–1. TRAFFIC LIGHT DEFINITIONS

Symbol	Likelihood	Impact
●●●●	Certainty	Threatens success of entire project
●●●○	High	Will impact deliverables, schedule and/or budget
●●○○	Medium	May impact deliverables, schedule and/or budget
●○○○	Low	With effort, can be managed without impacting deliverables, schedule and/or budget
○○○○	Unlikely	Little or none

TABLE 9–2. SPECIFIC RISKS

Risk	Likelihood	Impact	Contingency Plan
1. List risk	●●●●	●●○○	List plan to minimize risk

Companies such as ABB truly recognize the importance of risk management. This is illustrated by the following comments made by James M. Triompo, Group Senior Vice President, Process Area Project Management at ABB:

> Risk Review is a critical part of the ABB PM process. In addition, we have a Project Risk Evaluation and Tender Procedure used for bids greater than US$ 15 million. This is an ABB Group Instruction which requires mandatory compliance. We give risk management significant attention, which has kept us out of trouble in many instances. But there is always room for improvement both in identifying and mitigating risk.

One of the signs of a company that is well positioned for the project management challenges expected in the twenty-first century is a company that recognizes that project risk management reporting is just as critical as the development of risk management templates. According to Mark Smith, Director, Global Design Excellence at Diebold Corporation:

> Of all the project management knowledge areas, risk receives less attention than time, cost, and quality management. And for development projects, especially those that involve technology, risk management can have a major impact on the success of your project. Determining the different risks, the likelihood of risk, and the impact of risk, along with a contingency plan for each risk must be defined in the planning phase of a project.
>
> We have determined risks and their impact early on a number of our development projects. We start with a brain-storming session where we list all of the different risks. Those that understand the risks determine the likelihood of the risk, the impact of the risk, and the contingency for the risk. After this is done, a follow-up meeting is held to make sure all involved agree with the overall risk assessment.
>
> Traffic lights (see Table 9–1) are used to reflect the likelihood and impact of the risk. Each specific risk is listed with the likelihood, impact, and contingency plan (Table 9–2). A summary of the risks is also provided (Table 9–3).

TABLE 9–3. RISK SUMMARY

Risk	Likelihood
Deliverables	
Scope creep or shrinkage	●●○○
Budget	
Cost overruns	●●○○
Schedule	
Delivery date slippage	●●○○

9.6 CHANGE MANAGEMENT

Companies use change management to control both internally generated changes and customer-driven changes in the scope of projects. Most companies establish a configuration control board or change control board to regulate changes. For customer-driven changes, the customer participates as a member of the configuration control board. The configuration control board addresses the following three questions at a minimum:

- What is the cost of the change?
- What is the impact of the change on project schedules?
- What added value does the change represent for the customer or end user?

The benefit of developing a change management process is that it allows you to manage your customer. When your customer initiates a change request, you must be able to predict immediately the impact of the change on schedule, safety, cost, and technical performance. This information must be transmitted to the customer immediately, especially if your methodology is such that no further changes are possible because of the life cycle phase you have entered. Educating your customer as to how your methodology works is critical in getting customer buy-in for your recommendations during the scope change process.

The importance of having a structured change control process as part of the methodology is critical. Marcelino Diez, PMP, Project Management Consultant at Edelca, states that:

> The scope change control processes in Edelca are integrated into and part of our project management methodology. At this moment, for scope change control, there are no standard or well-structured processes in place, but these controls have been managed in a discretionary form by the responsible people involved in the projects and according to the company regulations. We are developing a structured change control process that will be integrated without risk management processes.

Consulting companies must have their own templates for controlling scope changes and must also be knowledgeable with the client's change control process. Brad Ruzicka, Senior Consulting Manager of the StoneBridge Group, explains:

> As a consulting firm, the ability to control changes in scope is extremely important, as we are often asked to provide bids for our services at the outset of a project. As a result, our standard approach is to define scope, overall approach, deliverables, and budget expectations in our Statement of Work and obtain agreement from the Project Sponsor based on examples of similar projects our firm has done in the past. We then agree that any substantive changes to the above over the course of the project will be jointly agreed to by StoneBridge Group and the Sponsor, at which point we will produce a revised SOW.

Risk management and change management function together. Risks generate changes that, in turn, create new risks. For example, consider a company in which the project manager is given the responsibility for developing a new product. Management usually establishes a launch date even before the project is started. Management wants the income stream from the project to begin on a certain date to offset the development costs. Project managers view executives as their customers during new project development, but the executives view their customers as the stockholders who expect a revenue stream from the new product. When the launch date isn't met, surprises result in heads rolling, usually executive heads first.

Previously, we stated that ABB had developed excellent processes for risk management, so it is understandable that they also have structured change management processes. Figure 9–9 shows the risk management flowchart taken from the Project Management Manual at ABB and reproduced with permission of ABB. What is interesting in the flowchart is the integration of the change management process with risk management.

Benny Nyberg, Manager, Process Unit Project Management, comments on the change management process at ABB:

> The ABB Change Management process/tool is constructed to cover all aspects from a simple Query, through Changes and Claims. The ABB process stipulates: *"The Change Management process is the means by which technical issues, or issues with financial or schedule impact, are raised, and managed. This includes Queries, Changes, and Claims. Project related Faults and Non-conformances are documented through the same process."*
>
> ABB strives to establish a well-defined project baseline before a project is booked, i.e. through project people's participation in Project Sales and through a formal Handover. After the job is booked the Change Management process/tool is used to manage the scope.
>
> ABB uses a structured Change Management process of which the PM is the owner. All project team members have the responsibility to log Queries and Changes as they arise. There are no threshold limits on when to use the process. Because we have learned that many small issues over time may amount to substantial scope creep, ABB requires its Change Management process/tool to be used for all *"technical issues, or issues with financial or schedule impact."* It is up to the Project Manager whether to approach the client for a scope change.

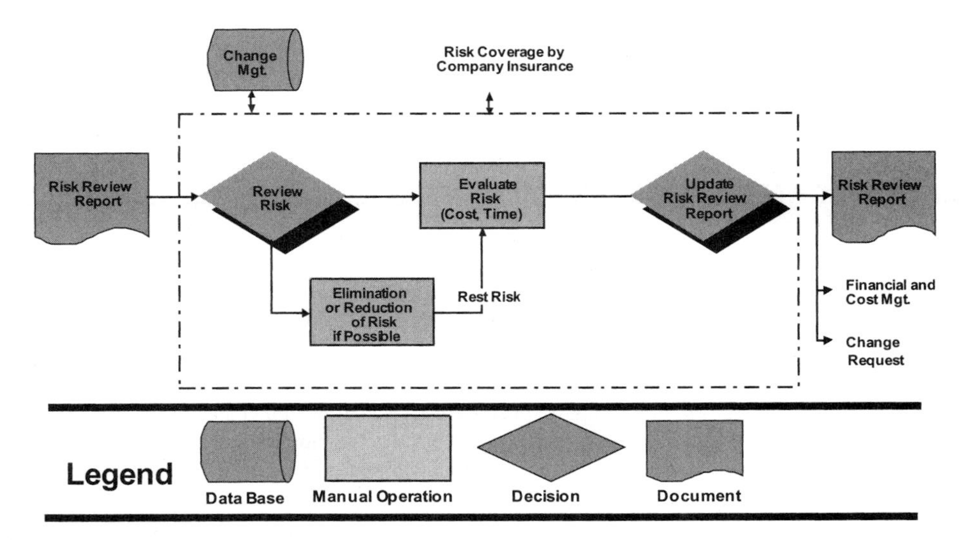

FIGURE 9–9. ABB risk management flowchart.

In companies excellent in project management, risk management and change management occur continuously throughout the life cycle of the project. The impact on product quality, cost, and timing are continuously updated and reported to management as quickly as possible. The goal is always to minimize the number and extent of surprises.

9.7 OTHER MANAGEMENT PROCESSES

Employee empowerment and self-directed work teams took the business world by storm during the early 1990s. With growing emphasis on customer satisfaction, it made sense to empower those closest to the customer—the order service people, nurses, clerks, and so on—to take action in solving customers' complaints. A logical extension of employee empowerment is the self-managed work team. A self-directed work team is a group of employees with given day-to-day responsibility for managing themselves and the work they perform. This includes the responsibility for handling resources and solving problems.

Some call empowerment a basis for the next industrial revolution, and it is true that many internationally known corporations have established self-directed work teams. Such corporations include Esso, Lockheed-Martin, Honeywell, and Weyerhauser. Time will tell whether these concepts turn out to be a trend or only a fad.

Reengineering a corporation is another term for downsizing the organization with the (often unfortunate) belief that the same amount of work can be performed with fewer people, at lower cost, and in a shorter period of time. Because project management proposes getting more done in less time with fewer people, it seems only practical to implement project management as part of reengineering. It still isn't certain that downsizing executed at the same time as the implementation of project management works, but project-driven organizations seem to consider it successful.

Life cycle costing was first used in military organizations. Simply stated, life cycle costing requires that decisions made during the research and development process be evaluated against the total life cycle cost of the system. Life cycle costs are the total cost of the organization for the ownership and acquisition of the product over its full life.

9.8 INTEGRATED PROCESSES AT WORK

In Chapter 4 we discussed the development of a project management system that include risk management and change management. It takes about two years to reach partial acceptance of a project management philosophy.

According to the manager of program management at an automotive subcontractor:

> It took [them] approximately one year [after implementation] to reach a degree of maturity in the project management process, including the structuring of procedures, calendars, pilot programs, and encompassing all programs into a regular review [plan–do–check–act] cycle. Initial programs are currently starting second-generation cycles and grandfather programs are under lessons learned review.

Embedded in these comments are two important points. First, the plan–do–check–act cycle mentioned is the one W. Edwards Deming introduced for continuous improvement as part of the total quality management process. Second, lessons learned reviews are part of ISO 9000 Certification, as well as the Malcolm Baldrige Award criteria. The manager's comments then reflect the integration of project management and total quality management. Let's look at Johnson Controls, perhaps the best example of a company that has achieved excellence through integrating management processes.

Johnson Controls

Headquartered in Plymouth, Michigan, Johnson Controls has become the worldwide leader in supplying seats for the automobile industry. The size of the product-driven company's projects ranges from $100,000 to $600 million. The company attributes its success in improving product quality and shortening project development time to the speed with which managers understood and integrated total quality management and project management. During a live video conference on the subject, "How to Achieve Maturity in Project Management," Dave Kandt, executive director of worldwide operations at Johnson Controls, commented on the reasons behind Johnson Controls' astounding success:

> We came into project management a little differently than some companies. We have combined project management and TQC (total quality control) or total quality management. Our first design and development projects in the mid 1980s led us to believe that our functional departments were working pretty well separately, but we needed to have some systems to bring them together. And, of course, a lot of what project management is about is getting work to flow horizontally through the company. What we did first was to contact Dr. Norman Feigenbaum, who is the granddaddy of TQC in North America, who helped us to establish some systems that linked together the whole company. Dr. Feigenbaum looked at quality in the broadest sense. Quality of products, quality of systems, quality of deliverables, and, of course, the quality of projects and new product launches. A key part of these systems included project management systems that addressed product introduction and the product introduction process. Integral to this was project management training, which was required to deliver these systems.
>
> We began with our executive office, and once we had explained the principles and philosophies of project management to these people, we moved to the management of plants, engineering managers, analysts, purchasing people and, of course, project managers. Only once the foundation was laid did we proceed with actual project management and with defining the role and responsibility so that the entire company would understand their role in project management once these people began to work. Just the understanding allowed us to move to a matrix organization and eventually to a stand-alone project management department. So how well did that work? Subsequently, since the mid-1980s, we have grown from two or three projects to roughly 50 in North America and Europe. We have grown from two or three project managers to 35. I don't believe it would have been possible to manage this growth or bring home this many projects without project management systems and procedures and people with understanding at the highest levels of the company.
>
> In the early 1990s we found that we were having some success in Europe, and we won our first design and development project there. And with that project, we carried to Europe not

> only project managers and engineering managers who understood these principles, but also the systems and training we incorporated in North America. So we had a company-wide integrated approach to project management. What we've learned in these last ten years that is the most important to us, I believe, is that you begin with the systems and the understanding of what you want the various people to do in the company across all functional barriers, then bring in project management training, and last implement project management.
>
> Of course, the people we selected for project management were absolutely critical, and we selected the right people. You mentioned the importance of project managers understanding business, and the people that we put in these positions are very carefully chosen. Typically, they have a technical background, a marketing background, and a business and financial background. It is very hard to find these people, but we find that they have the necessary cross-functional understanding to be able to be successful in this business.

At Johnson Controls, project management and total quality management were developed concurrently. Dave Kandt was asked during the same teleconference whether companies must have a solid total quality management culture in place before they attempt the development of a project management program. He said:

> I don't think that is necessary. The reason why I say that is that companies like Johnson Controls are more of the exception rather than the rule of implementing TQM and project management together. I know companies that were reasonably mature in project management and then ISO 9000 came along and because they had project management in place in a reasonably mature fashion, it was an easier process for them to implement ISO 9000 and TQM. There is no question that having TQM in place at the same time or even first would make it a little easier, but what we've learned during the recession is that if you want to compete in Europe and you want to follow ISO 9000 guidelines, TQM must be implemented. And using project management as the vehicle for that implementation quite often works quite well.

There is also the question of whether or not successful project management can exist within the ISO 9000 environment. According to Dave Kandt:

> Not only is project management consistent with ISO 9000, a lot of the systems that ISO 9000 require are crucial to project management's success. If you don't have a good quality system, engineering change system, and other things that ISO requires, the project manager is going to struggle in trying to accomplish and execute that project. Further, I think it's interesting that companies that are working to install and deploy ISO 9000, if they are being successful, are probably utilizing project management techniques. Each of the different elements of ISO requires training, and sometimes the creation of systems inside the company that can all be scheduled, teams can be assigned, deliverables can be established, and they can be tracked and monitored, and reports go to senior management. That's exactly how we installed TQC at Johnson Controls, and I see ISO 9000 as having a very similar thrust and intent.

When a company recognizes the benefits of integrating TQM and project management, recognition by your customers will be forthcoming. Appendix A lists the numerous quality awards received by Johnson Controls Automotive Group.

Nortel

Nortel is an international telecommunications company with business segments exclusively project-driven. Nortel recognized that it needed a consistent, global project management system. It emphasized the following factors during the development of its new project management system:

- Development of project control forms
- Training and education of employees
- Development of project management standards
- Support of project management certification for project managers

The results of the new project management system included improved financial management and risk management on projects. In addition, customer relations are being improved as a result of the new system. Education and training proved to be the most important steps in introducing new techniques to support the company's cultural evolution.

Nortel is now well positioned to manage risk in the next decade. According to Bill Marshall, former director of project management standards at Nortel during the 1990s:

> Nortel views the project manager as the risk manager. Our financial plans for projects often contain a contingency line that is identified up front along with the bid. This follows a process of risk identification and exposure. Our contracts clarify the risks borne by our company and those risks not accepted. In the case where there are risks in areas where we do not have expertise, we will pass on risks in the form of subcontracts to our partners. The risks we face in the telecom industry are with our new entrant customers, who tend to overestimate the tasks they can take on. To mitigate this risk, we sell turnkey solutions and take on some of the traditional telecom operator tasks.
>
> The other risk we face is in new area of legislation and new market entrants. The greatest rewards are for the frontier builders, for example, the cable companies' entrance into the phone business. This legislation is still unfolding, and the cable operators are oftentimes making purchases in anticipation of evolving legislation.
>
> The project manager manages the risk through tight monitoring of project deliverables (both Nortel and the customer), progress costs and payments and assessment of the operational readiness of the project. Frequent project meetings and reviews are required to identify, document, and resolve project dependencies and progress issues.

BellSouth

It's not always possible to develop a company-wide policy for risk management. Risk management may need to be customized for each new project. This is particularly true of large organizations. Just identifying the need for risk management and encouraging its use is a step in the right direction, even if the process can't be immediately standardized and consistently integrated with the company's project management system.

BellSouth is a good example of such a large organization. The telecommunications company is headquartered in Atlanta, Georgia. Its project management system during the 1990s was used for internal customers only. Projects ranged in size from $1 million to $400 million.

Ed Prieto, director of BellSouth's corporate project management office during the 1990s, commented:

> The risk management process is the responsibility of each BellSouth project manager and project team to ensure a successful project outcome. It is the project manager's responsibility to establish suitable mitigation strategies to deal with the likelihood of their occurrence. Although there is no standard procedure in place for risk management, various techniques and approaches are being reviewed by the BellSouth corporate project management office for future implementation as the standard for BellSouth. We are committed to having risk management procedures in place that will enable the project manager to identify critical risk area, take effective risk mitigation actions, and make timely decisions regarding needed resources and priorities for risk handling. Risk management in BellSouth will be a continuous proactive process, designed to develop targeted risk reduction measures with clearly identified mitigation tasks, responsibilities, and closure dates.

Armstrong World Industries Armstrong World Industries, a large company headquartered in Lancaster, Pennsylvania, uses project management for its own internal programs. The size of its internal projects ranges from $500,000 to $600 million. The company not only has achieved excellence in project management, it has also won the prestigious Malcolm Baldrige Award.

According to Stephen J. Senkowski, President and CEO of Armstrong Building Products:

> Our major risk management tool is a financial analysis model specifically developed for new product development, as well as a financial model used for major capital investments for engineering project management. In new product development, the first step in using these financial analysis tools begins when the global marketing group identifies key major development efforts that might be undertaken and evaluates these against each other on the basis of potential sales volume, resources required to develop and launch the product, strategic implications of the product, time to develop the product, and a product probability evaluation of both the potential market success and the technical and manufacturing success of the product. These are then stack ranked against one another and discussed for potential commissioning as projects by our multi-functional new product business teams, which include business leaders and middle management representatives from marketing, manufacturing, logistics, engineering, research, and product design.
>
> The financial analysis is then done on a more detailed basis once the project team is formed and a kickoff and planning meeting is held with all key constituents involved. This financial analysis is then used as a benchmark to determine if the project should be continued at various milestones during its development.
>
> Before and during the project, depending on the type of product being developed, market research is used to verify marketplace desire/acceptance so that as the product is finally developed and launched, sales positioning can be done to maximize return.
>
> It should be noted that all the financial and capital investment analyses used includes an extensive sensitivity analysis for items such as capital, product cost and product sales.
>
> To ensure a product is not going to cause a safety or environmental problem in the marketplace, a thorough product safety and design review is held with key people from qual-

ity control, product environmental, marketing, research, and legal to review all potential implications of new products. With the exception of the review of potential product categories to be developed by the global marketing team, a project manager is involved and, in most cases, coordinates the activity around these risk management tools."

Boeing Aircraft

As companies become successful in project management, risk management becomes a structured process that is performed continuously throughout the life cycle of the project. The two most common factors supporting the need for continuous risk management are how long the project lasts and how much money is at stake. For example, consider Boeing's aircraft projects. Designing and delivering a new plane might require ten years and a financial investment of more than $5 billion.

Table 9–4 shows the characteristics of risks at Boeing. (The table does not mean to imply that risks are mutually exclusive of each other.) New technologies can appease customers, but production risks increase because the learning curve is lengthened with new technology compared to accepted technology. The learning curve can be lengthened further when features are custom-designed for individual customers. In addition, the loss of suppliers over the life of a plane can affect the level of technical and production risk. The relationships among these risks require the use of risk management matrix and continued risk assessment.

TABLE 9–4. RISK CATEGORIES AT BOEING

Type of Risk	Risk Description	Risk Mitigation Strategy
Financial	Up-front funding and payback period based upon number of planes sold	• Funding by life cycle phases • Continuous financial risk management • Sharing risks with subcontractors • Risk reevaluation based upon sales commitments
Market	Forecasting customers' expectations on cost, configuration, and amenities based upon a 30- to 40-year life of a plane	• Close customer contact and input • Willingness to custom-design per customer • Development of a baseline design that allows for customization
Technical	Because of the long lifetime for a plane, must forecast technology and its impact on cost, safety, reliability, and maintainability	• A structured change management process • Use of proven technology rather than high risk technology • Parallel product improvement and new product development processes
Production	Coordination of manufacturing and assembly of a large number of subcontractors without impacting cost, schedule, quality, or safety	• Close working relationships with subcontractors • A structured change management process • Lessons learned from other new airplane programs • Use of learning curves

Motorola

Martin O'Sullivan, vice president and director of business process management at Motorola SSG, describes Motorola's risk management processes:

> The components of the risk management process are illustrated in Table 9–5. They are the risk management strategy, risk identification, risk assessment, risk management planning, and overall management and reporting.
>
> Risk identification systematically identifies the possible sources of risk and assesses the probability of occurrence and possible impact in terms of technical, schedule, and cost. Risk assessment involves the categorization of risks into risk types and areas.
>
> Project risk identification and assessment is a qualitative process, concerned with identifying significant risks, estimating their probability of occurrence, evaluating the impact in terms of time, cost, and performance. It also takes into account the causes of risk, the factors driving the magnitude of impact, and the timing of the risk. The process is regarded as the essential cornerstone of risk management.
>
> Risk management planning involved the prioritization of risks and the allocation of resources used to control risk. Overall management and reporting defines the risk management organization, the meetings that take place, and the reports that are generated and receive management review.

Gregory M. Willits, project manager process champion of Motorola SSG, provides some details on the use of the risk management model at Motorola:

> Risk management is performed on virtually all projects. Our policy is to manage all categories and classes of risk and opportunity in an organized and responsible manner by employing proven decision-making methodologies. Current business and business support risks and opportunities are assessed and managed based upon our core values of:
>
> - Customer satisfaction
> - Financial success
> - Technical superiority
> - Strategic alignment
> - Regulatory agency relations
> - Health and safety
> - Environmental protection
> - Corporate reputation
> - Employee alignment
>
> We developed a decision tool that enables project teams to assess, evaluate, prioritize, and communicate project risk and opportunity decisions consistently. The tool enables project teams to make decisions based upon the knowledge of experts and their own personal experience, but provides a universal scale upon which to make tradeoffs and decisions. The tool in no way restricts their decisions, but allows consistency in evaluating one strategy against another. This is accomplished through the use of a value model that was developed and validated to be consistent with our core values. Decisions can thus be made based upon their aggregate value (benefit) as a function of resource cost in implementing the proposed risk mitigation or opportunity capture strategy.

TABLE 9–5. FORMAL APPROACH TO RISK MANAGEMENT HELPS ENSURE PROJECT SUCCESS

Regular Updates	Risk Identification	Risk Assessment	Risk Management Planning	Overall Management and Reporting
Establish strategy for project Update as necessary	Select and brief key staff Free format questionnaires Prompt lists Expert interviews Checklists	Evaluate individual risks, using prob-impact score Identify major risk areas and types Quantify risks in terms of cost, time, and performance Highlight critical activities	Assign risk owners Prioritize risks Establish risk reduction actions based on urgency and manageability Determine fallback plans	Risk register Risk management report Risk review board meeting
Risk Evaluation and Management Information Systems (REMIS)				

Having a universal value model to use as the yardstick is a great help in both consistency and cycle time; consistency, because all assessments use the same evaluation tool, and cycle time, because the evaluation is built into the model.

MULTIPLE CHOICE QUESTIONS

1. Which of the following are facets of a project management system and are also the primary elements of Sprint's total quality management process?
   - **A.** Teamwork
   - **B.** Continuous improvement
   - **C.** Customer focus
   - **D.** Structured problem-solving
   - **E.** All of the above
2. As companies recognize the synergistic results of integrating processes, project management and __________ are usually the first two processes to become integrated.
   - **A.** Total quality management
   - **B.** Risk management
   - **C.** Scope management
   - **D.** Change management
   - **E.** Procurement management
3. Which process is used to control the adverse effects of scope creep?
   - **A.** Total quality management
   - **B.** Concurrent engineering
   - **C.** Risk management
   - **D.** Change management
   - **E.** None of the above

4. Increased project risks are best resolved by:
   - **A.** Performing risk management
   - **B.** Superior project planning
   - **C.** Scope change control
   - **D.** Performing more activities in parallel
   - **E.** None of the above
5. An effective risk management process forces project managers to:
   - **A.** Predict the future
   - **B.** Review the past
   - **C.** Bring functional employees on board earlier
   - **D.** All of the above
   - **E.** B and C only
6. Which type of project risk will probably be the most difficult risk to manage in the twenty-first century?
   - **A.** Financial risks
   - **B.** Scheduling risks
   - **C.** Quality risks
   - **D.** Technical risks
   - **E.** Customer expectations
7. If your customer's knowledge of project management is basically at an inexperienced level, the type of contract issued will probably be:
   - **A.** A simple contract
   - **B.** A complex contract
   - **C.** Some type of cost-sharing contract
   - **D.** An incentive-type contract
   - **E.** None of the above
8. Which of the following is not one of the three basic questions that must be answered during configuration control board meetings?
   - **A.** Cost of the change
   - **B.** Assignment of people requesting the change
   - **C.** Schedule impact
   - **D.** Customer value added
9. The risk management process and change management process generally function together.
   - **A.** True
   - **B.** False
10. Integrated processes are critical if an organization wants to:
    - **A.** Minimize executive sponsorship requirements
    - **B.** Provide an audit trail of all activities
    - **C.** Satisfy customer requirements
    - **D.** Create an effective, standard methodology
    - **E.** All of the above
11. Which of the following is not one of the three major processes to be integrated during the early to mid-1990s?
    - **A.** Total quality management
    - **B.** Project management
    - **C.** Risk management
    - **D.** Concurrent engineering

12. Other than the three common processes, which of the following is an integrated process that will probably be important in the next century?
    A. Risk management
    B. Customer management
    C. Scope change management
    D. All of the above
    E. A and C only
13. Improper risk management practices result in:
    A. Pushing problems downstream for resolution
    B. More frequent meetings
    C. More customer involvement
    D. More sponsorship involvement
    E. All of the above
14. Quality is an end goal.
    A. True
    B. False
15. Johnson Controls Automotive Group believes that its current success was, for the most part, due to:
    A. Their acceptance of risk management practices.
    B. Their integration of project management and total quality management
    C. The project management support received from their customers
    D. Their decision not to confuse their employees
    E. All of the above
16. Nortel's project management system was based upon:
    A. Standard forms
    B. Training and education
    C. Development of project management standards
    D. Support of project management certification
    E. All of the above
17. A company that has integrated only two of their six processes into a single methodology has:
    A. A totally uncoupled process
    B. A partially coupled process
    C. A totally coupled process
    D. A and B only
    E. None of the above
18. It is expected that most companies that have won the prestigious national Malcolm Baldrige Award are also using project management.
    A. True
    B. False
19. Which of the following is necessary to convert a standard methodology into a world-class methodology?
    A. Strong executive sponsorship
    B. Corporate-wide acceptance and use
    C. Strong line managers
    D. A project manager with a strong technical background
    E. All of the above

20. Integrated processes tend to imply that the primary skill for a project manager in the twenty-first century will be:
    A. A command of technology
    B. Knowledge of the business
    C. A strong background in interpersonal skills training
    D. A marketing orientation
    E. A production orientation
21. Based upon our knowledge today, the original companies showing excellence in project management have how many processes integrated into their world-class methodology?
    A. Two
    B. Three
    C. Four
    D. Five
    E. Six
22. Integrated processes allow for better decision-making, which may have the added benefit of:
    A. Lowering program risk
    B. Allowing for more work to be performed
    C. Reducing executive sponsorship involvement
    D. Exceeding customer specifications
    E. All of the above
23. Which of the following are characteristics of a world-class methodology?
    A. Maximum of 10 life-cycle phases
    B. Eliminating end-of-phase gate reviews
    C. Internally oriented rather than customer-oriented
    D. Overlapping life cycle phases
    E. Maximization of paperwork
24. Which of the following are benefits of a world-class methodology?
    A. Faster new product development
    B. Lower overall costs
    C. Improved communications
    D. Overall paperwork reduction
    E. All of the above
25. Control of scope changes is generally accomplished by:
    A. A Configuration Control Board (CCB)
    B. A steering committee composed of senior management
    C. A representative of the client
    D. Well-educated line managers
    E. A safety engineering team

DISCUSSION QUESTIONS

1. Why is it advantageous to have integrated processes rather than uncoupled processes?
2. Why are the principles of project management and total quality management so alike?
3. What is the relationship between risk management and concurrent engineering? Why should they be integrated together?

4. What is the relationship between risk management and scope change management? Why should they be integrated together?
5. What type of risks are the most difficult to anticipate, and why?
6. Companies such as Ericsson and Nortel expect a large portion of their income to come from third world nations in the next century. Should their need for risk management increase or decrease?
7. How can integrated processes help you to better manage your customers?
8. Johnson Controls integrated project management and TQM right from the beginning, especially through training. Explain how training related to these two methodologies.
9. Explain how risk management at Boeing might differ from risk management at other organizations.
10. Does the size of a company have an impact on which processes get integrated and how long the integration process will take?

Across

1. Type of risk
5. Type of risk
6. Integrated process: _ _ _ _ _ management
8. Type of risk
9. Benefit: _ _ _ _ _ program risk
11. Integrated process: _ _ _ _ management
12. _ _ _ _ _ _ _ _ _ _ _ _ _ _ acceptance
13. Benefit: _ _ _ _ _ _ _ _ management

Down

2. Executive _ _ _ _ _ _ _ _
3. Integrated process: _ _ _ _ _ _ management
4. Number of integrated processes
5. Integrated process: _ _ _ _ _ _ _ _ _ _ _ _ _ management (2 words)
6. _ _ _ _ _ _ _ _ _ methodology
7. Maximum number of life cycle phases
10. Type of methodology

Culture

10.0 INTRODUCTION

Perhaps the most significant characteristic of companies that are excellent in project management is their *culture.* Successful implementation of project management creates an organization and culture that can change rapidly because of the demands of each project, and yet adapt quickly to a constantly changing dynamic environment, perhaps at the same time. Successful companies have to cope with change in real time and live with the potential disorder that comes with it.

Change is inevitable in project-driven organizations. As such, excellent companies have come to the realization that competitive success can be achieved *only* if the organization has achieved a culture that promotes the necessary behavior. Corporate cultures cannot be changed overnight. Years are normally the time frame. Also, if as little as one executive refuses to support a potentially good project management culture, disaster can result.

In the early days of project management, a small aerospace company had to develop a project management culture in order to survive. The change was rapid. Unfortunately, the vice president for engineering refused to buy into the new culture. Prior to the acceptance of project management, the power base in the organization had been engineering. All decisions were either instigated by or approved by engineering. How could the organization get the vice president to buy into the new culture?

The president realized the problem but was stymied for a practical solution. Getting rid of the vice president was one alternative, but not practical because of his previous successes and technical know-how. The corporation was awarded a two-year project that was strategically important to the company. The vice president was then temporarily assigned as the project manager and removed from his position as vice president for engineering. At the completion of the project, the vice president was assigned to fill the newly created position of vice president of project management.

10.1 CREATION OF A CORPORATE CULTURE

Corporate cultures may take a long time to create and put into place, but can be torn down overnight. Corporate cultures for project management are based upon organizational behavior, *not processes*. Corporate cultures reflect the goals, beliefs, and aspirations of senior management. It may take years for the building blocks to be in place for a good culture to exist, but can be torn down quickly through the personal whims of one executive who refuses to support project management.

Project management cultures can exist within any organizational structure. The speed at which the culture matures, however, may be based upon the size of the company, the size and nature of the projects, and the type of customer, whether it be internal or external. Project management is a culture, not policies and procedures. As a result, it may not be possible to benchmark a project management culture. What works well in one company may not work equally well in another.

Good corporate cultures can also foster better relations with the customer, especially external clients. As an example, one company developed a culture of always being honest in reporting the results of testing accomplished for external customers. The customers, in turn, began treating the contractor as a partner, and routinely shared proprietary information so that the customers and the contractor could help each other out.

Within the excellent companies, the process of project management evolves into a behavioral culture based upon multiple boss reporting. The significance of multiple boss reporting cannot be understated. There is a mistaken belief that project management can be benchmarked from one company to another. Benchmarking is the process of continuously comparing and measuring against an organization anywhere in the world in order to gain information that will help your organization improve its performance and competitive position. Competitive benchmarking is where one benchmarks organizational performance against the performance of competing organizations. Process benchmarking is the benchmarking of discrete processes against organizations with performance leadership in these processes.

Since a project management culture is a behavioral culture, benchmarking works best if we benchmark best practices, which are leadership, management, or operational methods that lead to superior performance. Because of the strong behavioral influence, it is almost impossible to transpose a project management culture from one company to another. What works well in one company may not be appropriate or cost-effective in another company.

Strong cultures can form when project management is viewed as a profession. Ron Kempf, Director PM Competency and Certification at HP Services Worldwide Engagement PMO, states that:

> A project management culture is embedded in HP Services. It is recognized as a primary business differentiator. Project managers enjoy high status in the company and have clear development paths to senior levels. A PM Professional Council with members from worldwide and each regional PMO sets the direction, communicates and implements programs related to the PM Profession. The Council focuses on PM Development, knowledge sharing, and professional association leadership.
>
> HP Services has a comprehensive Project Management Development Program that covers all aspects of project management training. This world class program won Excellence

> in Practice awards in career development and organizational learning in 2002 from ASTD (American Society for Training and Development), the world's leading resource on workplace learning and performance issues. Every project manager is encouraged to become PMP (Project Management Professional) certified and PMP designation is mandatory for senior project management appointments.

Some companies use the Project Management Institute's project management body of knowledge (PMBOK®) as the basis to establish a project management culture. Nortel is attempting to do this. Peter Meldrum, vice president for service operations, Carrier Networks Europe for Nortel Networks, states:

> Nortel Networks has embraced the PMI PMBOK standards as the guide for establishing the Nortel Networks project culture. Project managers are dispersed throughout the enterprise and their business unit directors act as the project sponsors. The sponsors assist with the launch of a project through staffing and project priority decisions. Nortel Network headquarters staff for project management provides the process and tools standards support, and mentors project skills improvements.

In the two examples above, HP and Nortel recognized the importance of a good culture for project management and were able to achieve their goal. Unfortunately, not all companies are as successful. When trying to improve project management, senior management often overemphasizes the quantitative components of the project management maturity model and underemphasizes the behavioral component. These mistakes are quite common even in the best management companies.

The first step in changing a culture is to identify the key players that can influence the cultural change. In the following situation, the president identified the line managers as a critical change element.

Situation 10–1: The Automotive Supplier The Automotive Services Group (ASG) of a large corporation was floundering. Corporate management had great expectations for ASG considering the quality of their technical resources. Performance was significantly below expectation and corporate executives believed that a change in leadership was necessary for project management performance to improve.

The company hired a new divisional president for the ASG. After interviewing personnel at all levels of the organization, the president believed that the root cause of the problem was the culture at ASG. Cooperation was poor at best, and technical decision-making was dominated by the line managers. In many cases, less than optimal decisions were made and then implemented. Project management, simply stated, was not working well nor was it providing the best results.

To create a cooperative corporate culture, the new president's attention focused first on the line managers. The president believed that line managers had a much greater influence on the culture than did project managers. The president discovered that most of the line managers were technical experts who were promoted to management because they were at the top of the pay grades for their group and could not receive additional compensation without becoming a line manager. Many of these people did not want to become line managers but did so reluctantly for the additional income.

The president was granted complete autonomy by the board of directors to implement whatever changes were necessary to turn around the ASG. A four-step approach was implemented, which included:

- Creation of a new pay grade called *technical consultant* whereby management could reward through almost unlimited salary highly technical employees without forcing them to become line managers.
- Several of the existing technical line managers were given the option of relinquishing their line management position and opted for the new consultant pay scale.
- Line managers were expected to function more as administrative managers than as technical managers. Also, line managers were expected to have an understanding of technology but not necessarily a command of technology.
- The most important criteria for promotion to line manager would be a demonstrated willingness to cooperate with other line managers and project managers, as well as making decisions in the best interest of the company rather than the line function or the project.

The president focused heavily on the importance of cooperation. For more than six months, the president conducted personal interviews with employees and managers to see if a cooperative culture was being developed. Within two years the company developed a project management methodology which was being implemented more informally rather than formally because of the cooperative culture that was developed. Performance improved significantly.

Situation 10–2: The Aerospace Company

A line manager in an aerospace firm had the responsibility of providing deliverables to four high-priority projects at about the same time. These were the only projects labeled as high priority. The line manager needed at least 20 employees from his group to provide all of the deliverables on time. Unfortunately, the line manager had only 12 people available. The line manager could have simply made the decision according to the established priorities as to which projects would be serviced first. Another option would have been to elevate the problem to senior management, who probably would also follow the established priorities.

Because the organization had a cooperative culture, the line manager asked the four project managers for help. All four project managers sat around the conference table, exposed their contingency plans to one another, and did trade-offs such that none of the four project managers would be hurt significantly. The four project managers and the line manager then informed senior management of their recommendation. Senior management's response was: "Thank you! Goodbye!"

Another common mistake made by senior management is believing that project management cultures can be transplanted from one division of a business to another or from one company to another. Organizational change fails when culture is ignored. Project management cultures are based upon the size and nature of the projects, the types of customers, and the requirement of the projects. In the next situation, an executive mistakenly believed that there was very little difference between aerospace projects and consumer projects.

Situation 10–3: The Commercial Products Division

The Commercial Products Division (CPD) of a large company identified a need for project management. With very little guidance on how to change the existing culture to a project management culture, the company transferred two of their senior managers from their Aerospace Division to the Commercial Products Division. Both managers had spent at least 25 years in the Aerospace Division. The intent was to create the same culture that existed in the project-driven Aerospace Division in the non–profit-driven CPD.

The two newly transferred managers erroneously believed that cultural transfers would be easy to accomplish. Project management software that was being used on large aerospace projects was installed on the desktop computers of the project managers and line managers in the CPD. Forms, policies, procedures, and templates that were used on aerospace projects were also installed as part of the new CPD methodology for project management. The CPD project management methodology was almost identical to the aerospace division's project management methodology.

It took less than two years for the company to realize that a mistake had been made. The two managers returned to the Aerospace Division for reassignment. The company instructed the CPD to develop their own methodology and culture for project management. But the damage was done and perhaps even irrevocably. There was now significant resistance to project management. The scars left on the corporate culture would take time to repair.

When a significant change occurs to the business base of a company, there may be a necessity for a corresponding change in the corporate culture. This cultural change could be for better or for worse. When executives make a decision to change the strategic direction of the company, they should assess the potential impact on the culture. If this is not done, irrevocable damage can occur. This point is illustrated in the next two situations.

Situation 10–4: Davis Construction Company (Disguised Case)

Davis Construction Company had divisions in several large U.S. cities. Davis had a reputation for performing quality work and usually bid on only small to medium-sized jobs. Small jobs were considered to be under $10 million and medium-sized jobs were $10 to $15 million. Occasionally, Davis would bid on a large job, but most of the work then had to be subcontracted out to selected partners.

Senior management at Davis was relatively conservative. The company wanted no more than a 3 to 5 percent growth rate per year in sales. The revenue stream was fairly stable, and the firm's financial health was good.

When the U.S. economy began to derail, Wall Street analysts were predicting that gasoline prices at the pumps could exceed $2 per gallon by year end. If prices reached or exceeded $2 per gallon, the belief was that companies would invest heavily in alternative energy sources, synthetic fuel plants, oil shale research and production, and possibly a resurrection of nuclear power plant construction. These projects could easily reach $400 to $500 million in size, with nuclear power plants at about $8 billion or more.

Senor management at Davis believed that this was a once-in-a-lifetime opportunity to compete head on with the larger construction companies and to reap large profits. The decision was then made to focus heavily on the larger projects, over $100 million in size, and to give up many of the smaller projects that could tie up critical resources.

The existing project management methodology was designed to service small projects and could not effectively support megaprojects. The company considered having two separate methodologies: one for small projects and another for large projects. Since the company considered their future existence to be heavily dependent on large projects, a new methodology was created for large projects and the old methodology was discarded.

The company began bidding on the larger projects. Each time a large project would be awarded to Davis, hundreds of engineers would be hired. Within two years, the company quadrupled in size and the business was flourishing. The future looked promising.

Nobody seemed to notice or care that the culture at Davis had changed. With the old methodology, emphasis was on teamwork and cooperation. With the new methodology, emphasis was on policies and procedures. Everything was done according to policies and procedures for fear that mistakes could lead to massive cost overruns. The culture had changed at the expense of profitability.

As the economy began to improve, it became evident that gas prices would fall to approximately $1 per gallon. Companies that had prepared plans for megaprojects put everything on hold. Projects that had already been started were being canceled.

For Davis Construction, these changes were disastrous. Megaprojects were needed to support the enlarged resource base. But with fewer megaprojects available and competition becoming fierce, downsizing was inevitable. David tried vehemently to change back to the old methodology, but this was an impossible task during downsizing when people saw their jobs disappearing. Survival and internal competition took on paramount importance for the workers. Davis is no longer in business.

Situation 10–5: The Software Solutions Group (SSG)

Three graduate students who recently completed a graduate program in information systems started up SSG in the early 1990s. By 2001, the company employed over 250 people servicing exclusively U.S.-based companies. Unfortunately, the downturn in technology spending was taking its toll of SSG's profitability. SSG had to either downsize their organization or find additional work.

SSG's core business was supported by some 30 large companies that found it less expensive to outsource work to SSG than to perform the same software development efforts internally. SSG assigned their best systems programmers to these 30 clients. The working relationships with the 30 clients were on solid ground and the revenue stream from these companies was stable. Unfortunately, this revenue stream could not support the entire organization.

To maintain cash flow, the company began bidding aggressively for work outside the United States. Many of the contacts were priced out just to meet expenditures and salaries and to keep the present workforce employed. The company's business base increased, but problems began to surface.

Because many of the new global customers were in different time zones, some with as much as six hours' difference, the company went to flex-time. Workers could start and finish work whenever they wanted as long as they still put in an eight-hour day. This would alleviate some of the problems with time zone changes. This began to change the social aspect of the culture within SSG.

In order to win contracts, SSG accepted statements of work that had ill-defined requirements. Many of SSG's new clients were clueless as to what was meant by a well-defined statement of work. The number of scope changes was increasing exponentially. The project managers working on multinational projects began "stealing" resources from other projects to put out fires on their own projects. This caused havoc with their core 30 customers, who could no longer contact the resources who had been assigned to their projects previously. Maintaining the schedules on the multinational projects was being accomplished at the expense of the domestic business base. More unfortunate, the number of scope changes was expected to increase.

The president recognized that the cooperative culture that had taken 10 years to build and nurture was becoming derailed rapidly. The president immediately restructured the company on the mistaken belief that organizational structure shapes culture. To make matters worse, the president then put the project managers on salary and commission rather than just salary. The commission was a percentage of project profits.

As the number of scope changes grew, it became apparent that the multinational customers' needs were unstable. The stealing and hoarding of precious resources began to increase. The organizational culture simply could not cope with the constantly changing requirements and the new compensation system. Quality resources began leaving the company for employment elsewhere. The corporate culture had disintegrated. The president realized that he was better off with a small, focused customer base rather than marketing to the world. But the damage was done and he was unsure as to how to correct the situation quickly.

Changing a corporate culture to one of cooperation and teamwork can be viewed as a project by itself. Unfortunately, as seen in the next situation, not everyone will accept the new culture.

Situation 10–6: The Junior High School Boys' Basketball Team

During preseason practice games with other junior high schools, the coach became concerned that the "culture" of the team was so poor that a losing season was inevitable. Teamwork was simply not there. Whoever had the ball would shoot, even though there was someone closer to the basket who might have an easier shot. Each person played as an individual, looking for points rather than assists.

The coach came up with an interesting idea. During the next team practice, the coach placed boxing gloves in the middle of the basketball court. Each player had to place one boxing glove on either of his hands. The rules of the game were that if you were defending a player holding the basketball, you could "bash" him in the arm, chest, or stomach as long as he held the basketball. The intent was to force the players to pass the ball and play as a team.

Each of the players selected a boxing glove for the hand that could deliver the most powerful punch. Right-handed players put on a right-handed glove; left-handed players put on a left-handed glove. As the practice game progressed, significant bashing took place, but very few points were scored. The reason for the few points was that the strong hand that had the boxing glove was also the hand used for shooting the basketball.

At the next practice session, most of the players chose a boxing glove for the nonshooting hand. Teamwork improved somewhat, and more points were being scored. However, there were still some players who selected the boxing glove for the strong hand and preferred bashing rather than winning games and teamwork.

No matter how hard we try, we always end up with individuals who will not change or support project management. Corporate cultures can inevitably be impacted by such employees. They may be motivated only by the wage and salary administration program. Care must be taken that whatever changes are required to meet the future needs of the business do not have a major impact on the wage and salary administration program. If the impact is not fully understood and mishandled, irrevocable damage can occur.

Situation 10–7: The Electronics Manufacturer

Several years ago, a manufacturer of electronic components recognized the need to implement project management. A new department for project managers was created. Since project managers were expected to work closely with the department managers for the accomplishment of deliverables, the salary structure of the project managers was close to that of the line managers. The president believed that the working relationship between the project and line managers would function best if both parties viewed each other as equals rather than in a superior–subordinate relationship.

The corporate culture liked what the president had instituted. Over a period of several years, the culture fostered an atmosphere of trust and effective cooperation. In 2002, the president retired and a new CEO was brought in.

The new president liked the way that project management was working but believed that it could be working even better. Project managers were offered a financial incentive, similar to a bonus, based upon the profitability of their projects. Only the project managers were offered the incentives.

Project managers began to focus more on the incentives than on the working relationship with line managers. To make matters worse, several project managers were now receiving total compensation packages that were significantly larger than their functional counterparts.

The relationship between the project and line managers began to deteriorate, and this, of course, had a significant detrimental impact on the corporate culture. The president began filing vacant line management positions with project managers who had been successful in project management.

Employees began believing that the only career path to the top would be through project management. Employees and line managers began volunteering to become project managers. The technical base of the company was becoming diluted because everyone began to believe that "the grass was greener" in project management than in line management. The corporate culture for executing projects was not focusing on animosity and jealousy rather than on cooperation and teamwork.

Perhaps the greatest challenge facing senior managers in this decade will be the creation of a multinational project management culture. Cultural change requires transformations to take place, and in some countries these transformations are difficult to implement. In the next situation, the resistance to change occurred at the executive levels.

Situation 10–8: The Global Automotive Supplier

A U.S.-based automotive supplier had developed an excellent project management methodology. All of the U.S. divisions used the methodology with outstanding results. Their customers liked the methodology and the quality of the deliverables to such a degree that they began treating the company as a partner rather than as an ordinary supplier.

As business began to grow, the company began purchasing other companies around the world, to become a global automotive supplier. The original intent was to leave local management in place and to allow each division to operate autonomously as long as business was good. However, this meant that multiple project management methodologies and cultures would be in existence. Some of the newly acquired divisions had no project management in place at all. All projects were managed through line managers. Other divisions had project management but functioned without the use of a project management methodology.

Realizing that there would be rapid growth in the number of projects requiring several divisions to work together, the company needed to have all divisions managing projects in the same way. Project managers and professional trainers in project management were brought from the United States to each division to help install the project management methodology and to change the culture, if necessary.

There appeared strong resistance to project management from all levels of management, especially at the executive levels. One executive stated that as long as the customers in his country did not recognize project management as important in the U.S. auto industry, he would not support project management.

After two years of battling the resistance, corporate management in the United States began replacing the executives in these foreign divisions with American executives who could change the corporate culture to one that would accept project management. This process took almost three years to complete.

Multinational cultures have serious issues that go well beyond executive recognition of the need to change and even the desire to do so. The following are typical multinational cultural complexities that must be addressed when using the project management maturity model. For simplicity sake, they are listed according to the PMBOK® processes.

- Integration management
  - Poor understanding of the benefits of project management
  - Poor understanding of the role of the project manager
  - No project management methodology
  - Limited cross-functional access to certain groups
  - Difficulty in gaining agreement and commitment
  - Poor problem-solving capability
- Scope management
  - Improper assumptions
  - Methodology designed for a single, national project
  - Culture impacted by country's legal system

  - No definition of a charter, scope baseline or scope statement
  - Varying education and experience of planners
- Time management
  - Unrealistic estimating using an 8-hour day
  - Different interpretation about assumptions
  - Inaccurate or obsolete scheduling tools
  - Time not viewed as a critical constraint
  - Missed milestones viewed as an acceptable practice
  - Nonexistent templates
  - Varying size of resource pool
  - Differing calendars concerning holidays and vacations
- Cost management
  - Estimating impacted by currency conversion and inflation rates
  - Politics impacting the award of subcontracts
  - Government instability causing changes in or cancellation of projects
- Procurement management
  - Negotiation per local customers
  - Infrastructure delays in procurement, resulting in damaged goods
  - Specifications improperly interpreted
  - Gratuities and kickbacks perceived common practice
  - Use of resources requiring government permission
  - Source selection based upon culture and politics rather than on requirements and quality of suppliers
- Risk management
  - Ineptness and withholding of information
  - Virtually no tools for risk management
  - Too much government intervention
  - Inaccurate understanding of problems
  - Identifying risks ending one's career
- Quality management
  - Differing codes and laws
  - Inadequate skill levels
  - Differing view of quality
  - No quality policies
- Human resource management
  - Inadequate skill levels and training
  - Political instability in human resource practices
  - Decisions based upon holidays, customs and dietary considerations
  - Differing value systems
- Communications management
  - Increased barriers and filters
  - Poor understanding of each party's language
  - Time zone differences
  - Lack of trust

- Poorly understood speech etiquette
- Misinterpretation of body language

10.2 CORPORATE VALUES

An important part of the culture in excellent companies is an established set of values that all employees abide by. The values go beyond the normal "standard practice" manuals and morality and ethics in dealing with customers. Insuring that company values and project management are congruent is vital to the success of any project. In order to ensure this congruence of values, it is important that company goals, objectives, and values be well understood by all members of the project team.

Successful project management can flourish within any structure, no matter how terrible the structure looks on paper, but the culture within the organization must support the four basic values of project management:

- Cooperation
- Teamwork
- Trust
- Effective communication

The set of values, or what some people call a *corporate credo,* dictates the order of preference by which decisions are made. Humana Health Plans of Ohio has done an outstanding job in establishing this credo. When a decision has to be made at Humana Health Plans of Ohio, what is in the best interest of the customer comes first, what is in the best interest of Humana Health Plans of Ohio comes second, and what is in the best interest of the employees of Humana Health Plans of Ohio comes third. Alison Newrock Dipilla, Director of Project Management at Humana Health Plans of Ohio, comments on the use of this set of values as part of their project management culture:

> Our Project Management philosophy has incorporated Humana's corporate strategic imperatives, as well as core values or credos specific to Humana Health Plans of Ohio. Being a service industry, it is critical to think from outside in instead of inside out. Meaning, we are always trying to think like a customer and determine the best process for the customer. So, our model is fluid and enables us to meet the changing needs of the customer. Employees who participate in projects know the core values and understand the importance of building the right process or product for the customers' needs rather than the needs of specific departments within Humana Health Plans of Ohio.

When asked if this set of values actually speeds up the decision-making process, Alison Newrock Dipilla states:

> Because we have set core values that all employees embrace, including executives, we are able to avoid many of the internal disputes regarding management of projects. All

employees know their role within the scope of the project and have a base understanding that the customer is the top priority. The core values allow changes with the work environment to happen faster and more efficiently.

Johnson & Johnson

Perhaps the best measure of the effectiveness of a corporate culture is the way it reacts to a crisis project or unforeseen problems that could be damaging to the company as a whole. The damage could be financial, public image, or both. During the last week of September and the first week of October 1982, seven people died ingesting Extra-Strength Tylenol® capsules (from McNeil Consumer Products Company) laced with cyanide. Four years later, in 1986, Tylenol® product tampering occurred again, this time with the death of one person. During both crises, Johnson & Johnson set the standard on how crisis projects should be managed.

Academia has been teaching the Tylenol® case study for over 20 years as an example of morality and ethics in business and what constitutes effective corporate responsibility.[1] But with a more in-depth analysis, one finds that the secret to the success of Johnson & Johnson's handling of the Tylenol® tragedies lay in the strength of their culture and corporate credo.

The expression *crisis management* or *crisis project management* appeared in the literature as a result of the Tylenol tragedies. Crisis management focuses on the "unknown unknowns," which are tragedies without precedent. Crisis management requires a heads-up approach combined with a very quick reaction time and a concerted effort on the part of possibly all employees. In crisis management, decisions have to be made often without even partial information and perhaps before the full extent of the damage is known.

In a crisis, events happen so quickly and so unpredictably that it may be impossible to perform any kind of detailed planning. Statements of work, work breakdown structures, and detailed scheduling are nonexistent. Roles and responsibilities of key individuals may change on a daily basis. There may be very active involvement by a majority of the stakeholders. Company survival could rest entirely on how well a company reacts to and manages the crisis project.

With very little information available at that time, and very little time to act, the Tylenol® crisis project was managed using three phases. The first phase was discovery, which included the gathering of any and all information from every possible source. The full complexity of the problem had to be known as well as the associated risks. The second phase included the assessment and quantification of the risks and the containment of potential damage to the company, both financial and in image. The third phase was the establishment of a recovery plan and risk mitigation. Unlike traditional "project life cycle" phases, which could be months or years in duration, these phases would be in hours or days.

The CEO of Johnson & Johnson was James Burke, a 30-year veteran of Johnson & Johnson, who personally took charge of the crisis project. Instead of providing incomplete information or only the most critical pieces and stonewalling the media, Burke provided

---

1. A comprehensive analysis of the case can be found in "The Tylenol Tragedies: A Case Study on Managing Crisis Projects," in *Case Studies in Project Management,* by Harold Kerzner. New York: Wiley, 2003.

all information available. He quickly and honestly answered all questions from anyone. This was the first time that a corporate CEO had become so visible to the media and the public. James Burke spoke with an aura of trust.

The Tylenol® crisis quickly captured the nation's attention. Both James Burke and the culture at Johnson & Johnson were now under the nation's microscope. The world was about to see how a world-class culture would perform during a crisis.

There were several options available to Burke and the strategy committee that he chaired. Some of the options included:

- Tell the full story in the hope that the public would be sympathetic and Tylenol® could recover quickly.
- Take aggressive action in a search for the killer, placing all blame elsewhere.
- Replace the capsules with another type of product (i.e., caplets).
- Recall only those batches of Tylenol® that were contaminated.
- Recall all Extra-Strength Tylenol® capsules.
- After recall, relaunch the product under the same name but different packaging.
- After recall, relaunch the product under a different name and different packaging.

Deciding which option to select would certainly be a difficult decision. Looking over Burke's shoulder were the stakeholders who would be affected by Johnson & Johnson's decision. Among the stakeholders were stockholders, lending institutions, employees, managers, suppliers, government agencies, and consumers. How does a company prioritize stakeholders?

For the average company, there would be endless meetings and extreme difficulty in making any type of decision. But for Jim Burke and the entire strategy committee, the decision was not very difficult—just follow the corporate Credo.[2] For more than 45 years, Johnson & Johnson had a corporate Credo that clearly stated that the company's first priority is to the users of Johnson & Johnson's products and services. Everyone knew the Credo, what it stood for, and the fact that it must be followed. The corporate Credo guided the decision-making process, and everyone knew it without having to be told.

When the crisis had ended, Burke recalled that no meeting had been convened for the first critical decision: to be open with the press and put the consumer's interest first. "Every one of us knew what we had to do," Mr. Burke commented. "There was no need to meet. We had the Credo philosophy to guide us." The Credo had been in existence since 1943 and everyone knew the Credo. The way that the culture responded to the crisis, and made decisions, was dictated by the Credo.

The speed with which a company can react to a crisis is often dependent upon the corporate culture. If the culture of the firm promotes individualism and internal competition, employees will feel threatened by the crisis, become nonsupportive, and refuse to help even if it is in their own best interest. Such was not the case at Johnson & Johnson. The culture at Johnson & Johnson was one of cooperation. Employees were volunteering to assist in any way possible to help Johnson & Johnson out of the crisis.

---

2. The complete Johnson & Johnson Credo can be found on the Johnson & Johnson Web site, www.jnj.com.

On the surface, creating a set of values implies that the customer is always correct and that creeping scope will be allowed to continue forever to placate the customer. This is certainly not the case.

Virtually all of the companies excellent in project management have change management or configuration management processes to control scope changes. The two most common methods for managing scope changes are (1) allowing no changes after a specified point in the project's life cycle and (2) using enhancement projects. (Enhancement projects include subprojects and new projects developed in parallel to the original project.)

The first method is demonstrated by Roadway Express's change management system. In 1997, Kelly Baumer, director of project integration at Roadway, stated:

> At Roadway Express, the conceptual design document is presented to all management groups that will be involved in the new reengineered process/system/ product. Sign-off is obtained from the key users and line management. This helps prevent scope creep. Any changes after sign-off must be *separately* justified after implementation. This also ensures business requirements are clearly defined.

Linda Kretz of the International Institute of Learning stated what she tried to do as a project manager with US West Communications to control customer-requested scope changes:

> That [scope changes] has been a problem for us frequently. What we've done is insisted that our folks establish controls up front in the project to address the possibility of creeping scope. And so we might define in the objectives that certain changes may be allowed up to a certain point in the project, and that at a certain time we cannot accept changes without going back to the client and renegotiating exactly what the outcome is going to be because it may affect the budget or the schedule. But we do not accept the changes with the original scope of work because you are doomed to failure if you do. So, setting those controls up front really does take care of it a lot of the time.

The second method for controlling scope changes is the use of enhancement projects. The director of information services for an appliance manufacturer was given the task of preparing a uniform software package that would be acceptable for use at all of the manufacturing plants. Representatives from all of the plants met together and agreed upon the project's scope. When the project was approximately 75% complete, scope change requests were still coming in from all of the plants asserting that they could not live with the original scope and demanding that changes be made. The director of information services held her ground and stated that she would be installing the software according to the original scope. All scope changes would be clustered together into an "enhancement" project and submitted for funding and approval *after* the original project was completed. The plants were furious, but the director stood firm. After the project was installed and up and running, the plants admitted that they could indeed live with the original package and, to this date, the enhancement project has not been submitted for approval.

Walt Disney Company

Trade-offs and the management of scope changes are required on almost all projects. The ease by which this can be handled is often based upon the culture of the company. If the company has a noncooperative culture, people argue for what is in their own best interest and with little regard for the overall company. One person, with perhaps little input on alternatives and recommendations by the team, makes the decision. The time needed to make a decision may be prolonged. When a company has a cooperative culture, decision-making and trade-offs can be quick, effective, and in the best interest of the company as a whole. Such was the case at the Walt Disney Company.

The Walt Disney Company is perhaps the most famous name in family entertainment, including theme parks and recreation, motion pictures, broadcasting and commercial products. During the mid to late 1980s, I had the fortunate opportunity to provide project management training and consulting services for W.E.D. (Walt E. Disney) Enterprises in Glendale, California. W.E.D. was responsible for the project management for attractions at the theme parks.

W.E.D., as well as Disney as a whole, maintained a set of values created by Walt Disney himself and later reinforced by Michael Eisner when he took control of Disney in 1984. These values were:

- Theme park activities are attractions, not rides.
- The visitors are guests, not customers.
- Employees are treated like members of the Disney family.

These values permeated W.E.D. Enterprises. The employees referred to themselves as "imagineers" and W.E.D. was referred to as the Imagineering Division. The business cards also said *imagineering*. There was no dress code, but the employees wore clothing and jewelry with Disney characters on them. Each employee wore a badge in the shape of Mickey Mouse's face. In Mickey Mouse's eyes was the number of years that the worker was employed at Disney. One of the workers that I had interviewed had the number 46 in Mickey Mouse's eyes and was proud of the attractions that he and "Walt" had designed and built together.

There were five project management constraints on the theme park attractions:

- Time
- Cost
- Quality
- Aesthetic value
- Safety

Trade-offs often had to be made on the attraction projects. However, at no time were trade-offs ever made on safety. Walking through the Disney parks, all one sees are quality and high aesthetic appeal. With regard to the schedules, the employees were proud of the fact that only one opening day for the major attractions had been missed, and that attraction was delayed by only two days. Cost creep was occasionally allowed, to increase quality and aesthetic appeal.

At W.E.D., the cooperative culture was clearly visible. Employees were always willing to help one another. If a project manager had a problem, all workers were willing to assist without being officially asked to do so by management. Decision-making was always made for what was in the best interest of the entire company.

10.3 TYPES OF CULTURES

There are different types of project management cultures based upon the nature of the business, the amount of trust and cooperation, and the competitive environment. Typical types of cultures include:

- *Cooperative Cultures:* These are based upon trust and effective communications, not only internally but externally as well.
- *Noncooperative Cultures:* In these cultures, mistrust prevails. Employees worry more about themselves and their personal interests than what's best for the team, company, or customer.
- *Competitive Cultures:* These cultures force project teams to compete with one another for valuable corporate resources. In these cultures, project managers often demand that the employees demonstrate more loyalty to the project than to their line manager. This can be disastrous when employees are working on multiple projects at the same time.
- *Isolated Cultures:* These occur when a large organization allows functional units to develop their own project management cultures. This could also result in a culture within a culture environment. This occurs within strategic business units.
- *Fragmented Cultures:* Projects where part of the team is geographically separated from the rest of the team may result in a fragmented culture. Fragmented cultures also occur on multinational projects, where the home office or corporate team may have a strong culture for project management but the foreign team has no sustainable project management culture.

Cooperative cultures thrive on effective communications, trust, and cooperation. Decisions are made based upon the best interest of all of the stakeholders. Executive sponsorship is more passive than active, and very few problems ever go up to the executive levels for resolution. Projects are managed more informally than formally, with minimum documentation, and often with meetings held only as needed. This type of project management culture takes years to achieve and functions well during both favorable and unfavorable economic conditions.

Noncooperative cultures are reflections of senior management's inability to cooperate among themselves and possibly their inability to cooperate with the workforce. Respect is nonexistent. Noncooperative cultures can produce a good deliverable for the customer if you believe that the end justifies the means. However, this culture does not generate the number of project successes achievable with the cooperative culture.

Competitive cultures can be healthy in the short term, especially if there exists an abundance of work. Long-term effects are usually not favorable. An electronics firm would continuously bid on projects that required the cooperation of three departments. Management then implemented the unhealthy decision of allowing each of the three departments to bid on every job. Whichever department would be awarded the contract, the other two departments would be treated as subcontractors.

Management believed that this competitiveness was healthy. Unfortunately, the long-term results were disastrous. The three departments refused to talk to one another and the sharing of information stopped. In order to get the job done for the price quoted, the departments began outsourcing small amounts of work rather than using the other departments, which were more expensive. As more and more work was being outsourced, layoffs occurred. Management now realized the disadvantages of a competitive culture.

10.4 SHARED ACCOUNTABILITY AND MULTIPLE-BOSS REPORTING

As stated previously, the culture in project management is a behavioral culture dictated by multiple-boss reporting. Multiple-boss reporting requires that the organization have a clear understanding of authority, responsibility, and accountability. The relationship between them is:

$$\text{Accountability} = \text{Authority} + \text{Responsibility}$$

Assuming that the project manager and the line manager are not the same person, the majority of the resources are under the direct control of the line managers. Therefore, the line managers have the authority by virtue of their control over resources, and the project manager has the responsibility. In the excellent companies, accountability is shared between the project and line managers. However, it is a different type of accountability. Line managers are accountable for the promises made to the projects, and project managers are accountable to their executives and the customer.

With shared accountability, there is no prerequisite that project managers be selected from the technical ranks. Line managers are the technical experts. More and more project managers have business backgrounds because in the future the major background requirements for a project manager is knowledge of the business.

> Because of shared accountability, project managers should negotiate for the line manager's ability to get the job done within time, cost, and quality, rather than argue for the best resources available.

A vice president at an automotive company stated that although he knew that it was correct to negotiate for deliverables rather than specific resources, it was still a very difficult thing to get used to doing.

10.5 SHARED REWARDS

The concept of shared accountability also dictates shared rewards. There is a company in Baltimore, Maryland, that went to project management. They had a Christmas party in December 1989. At that Christmas party, the president stood up and, inviting three of the project managers to come to the podium, took out three corporate checks and handed one to each of the project managers, telling them, "You've done a fantastic job on these three projects." Everybody in the audience applauded. The following day when those project managers went back to their offices to work on the projects, the line managers shut down all three projects. After three or four months, the president called me into his office and asked, "What went wrong?" I asked him if he understood the damage that he had done in rewarding the project managers without also rewarding the line managers.

Project management advocates a team spirit, advocates team management, and advocates cooperation. We want project and line managers to view each other as peers rather than superiors and subordinates. If we put on the walls of our building that we want project management in the future to be where line management was in the past, I have to question what support the projects will get from line management. If a line manager wants to walk into a room and dig a hole six feet down and put the project manager in that hole and cover him/her up with dirt and flowers, the line manager can do that. On the other hand, the line manager wants to walk into that room and take that project manager and make him/her come out smelling like a rose, that too is possible, simply by the line manager's providing the proper resources. The single most important factor in getting project management maturity is buy-in from the line managers. You can have a relatively weak project manager and good line managers can compensate for that weakness. If you have weak line managers, however, it may be impossible for project managers, no matter how good they are, to compensate for those weaknesses in the line managers.

Project management is a team effort. As such, you either reward the team or you reward nobody. Excellent companies never reward project managers at the expense of line managers or employees. One executive commented to me several years ago,

> I totally agree with what you said about rewarding teams. That's why we reward our project managers in secret.

One can only imagine the culture that exists in this organization.

Shared accountability fosters a close working relationship between the project and line managers. A project manager in a defense contract was managing one of the company's four high-priority projects. Out of 30 projects, only the four high-priority projects had established priorities. The line manager, who shared accountability with each of the four project managers, convened a meeting to state that he did not have sufficient resources to meet the deliverable requirements that he had promised the four project managers. Twenty people were needed for the next several months, but only 12 people were available.

Senior management could have made the decision by following the ranking on the priorities and deciding "who shall live and who shall die." The line manager, knowing what the priorities were, could have made the decision also. But the line manager was smart and asked the four project managers for help.

The four project managers sat around the table and did trade-offs with one another. Each project manager took a little bit of the bite. The line manager acted strictly as an observer. At the end of the hour, the four project managers *told* the line manager how to assign his 12 people over the next four months.

This is how project management works in excellent companies. Line managers go first to the project manager when a problem arises. Likewise, project managers go first to line managers. Problems are resolved at the lowest possible levels. Executives and sponsors are brought in as a last resort. In the example above, the line manager was smart because he knew that good project managers had contingency plans. The project managers released their contingency plans to help the line manager. Shared accountability leads to behavioral excellence, which in turn creates an outstanding project management culture.

10.6 PRIORITIZATION OF WORK

Strong project management cultures minimize the necessity to prioritize work. Not all projects have to be prioritized. However, there are times where priorities are needed. A household fixture manufacturer hired an outside consultant to help identify the problems and potential solutions to the poor culture that existed in the engineering division. Every engineer interviewed blamed the problems on marketing, stating that marketing kept meddling with engineering schedules. There was unanimous agreement that the problem was marketing.

The interviews with marketing, however, painted a different picture. There were two trade shows per year where marketing would go to introduce new products. Missing a launch window at a trade show required waiting six months for another window.

On the surface, the solution appeared to be simple: just get engineering and marketing to work together. But as the consultant kept digging for information, the real problem appeared. The engineering group had a queue of over 400 projects with no established priorities. The engineers were working on whatever projects they wanted.

At first, the organization tried to solve their problem by assigning priorities to the top 100 projects. As the project management culture began to grow and mature, however, a new project management methodology was developed, which dictated that only 20 projects can be prioritized at any one time, regardless of how many projects are in the queue. Cooperative cultures minimize the necessity for prioritization and foster a "can-do" attitude among the employees.

10.7 CORPORATE CULTURES AT WORK

Cooperative cultures are based upon trust, communication, cooperation, and teamwork. As a result, the structure of the organization becomes unimportant. Restructuring a company simply to bring in project management will lead to disaster. Companies should be restructured for other reasons, such as getting closer to the customer.

> Successful project management can occur within any structure, no matter how bad the structure appears on paper, if the culture within the organization promotes teamwork, cooperation, trust, and effective communications.

Small to medium-sized companies are restructuring to "pool" management. (This is discussed later in this chapter.) Large companies are focusing on strategic business unit (SBU) project management. Figure 10–1 illustrates a Detroit automobile subcontractor's creation of three strategic business units: Ford programs, Chrysler programs, and General Motors programs. Each SBU had its own program and project managers. The functional areas of human resources, procurement, marketing, and finance were corporate functions designed to support each of the SBUs. If the SBUs were extremely large, then these corporate functions would be included under each SBU. Each of the SBU's had its own culture for project management.

Figure 10–2 shows a case where each SBU has its own program managers. The platform managers are actually line managers responsible for developing a platform and maintaining it. The SBU managers are responsible for applying the platform to the product lines within each SBU.

Boeing

In the early years of project management, the aerospace and defense contractors set up customer-focused project offices for specific customers such as the Air Force, Army, and Navy. One of the benefits of these project offices

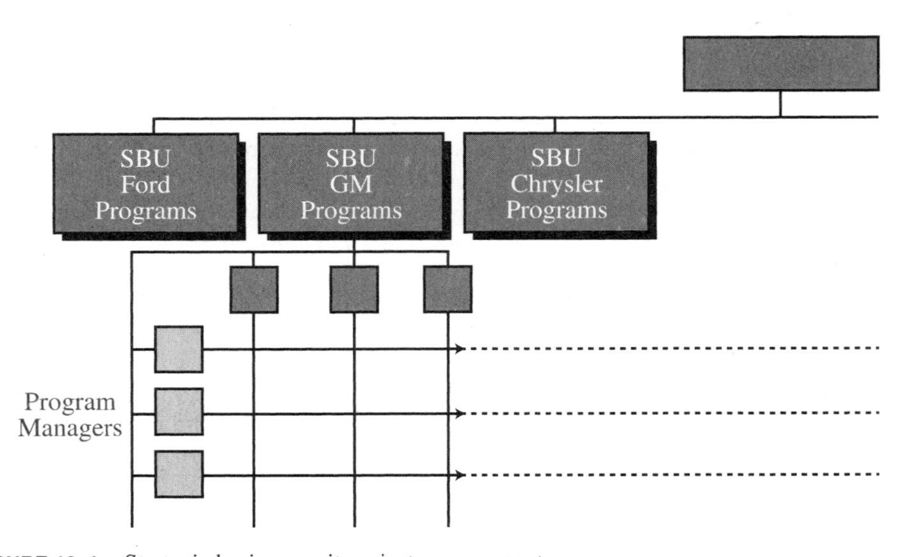

FIGURE 10–1. Strategic business unit project management.

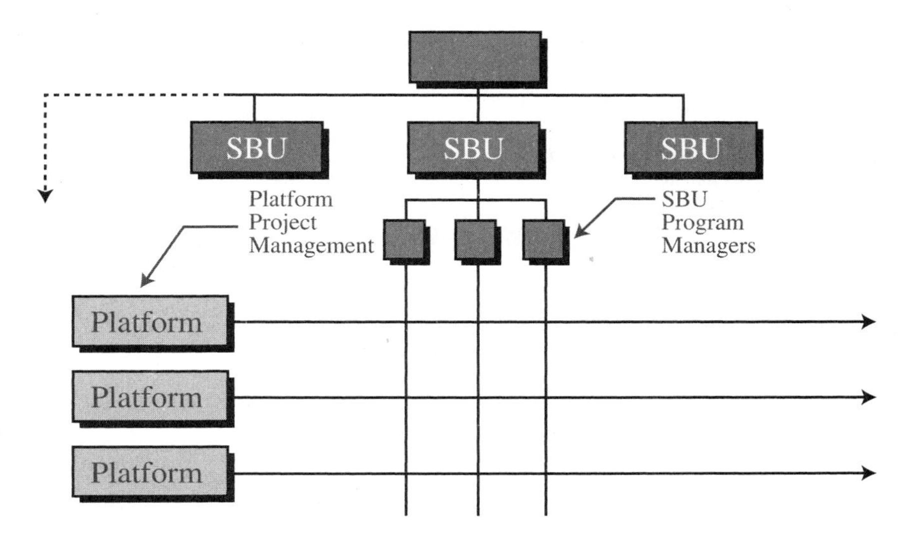

FIGURE 10–2. SBU project management using platform management.

was the ability to create a specific working relationship and culture for that customer. Developing a specific relationship or culture was justified because the projects often lasted for decades. It was like having a culture within a culture. When the projects disappeared and the project office was no longer needed, the culture within that project office might very well disappear as well.

Sometimes, one large project can require a permanent cultural change within a company. Such was the case at Boeing with the decision to design and build the Boeing 777 airplane. The Boeing 777 project would require new technology and a radical change in the way that people would be required to work together. The cultural change would permeate all levels of management, from the senior-most levels down to the workers on the shop floor.

Table 10–1 shows some of the changes that took place. At the end of this book is a Boeing 777 case study that provides additional information on the changes that took place.

Humana Health Plans of Ohio

As companies become more and more cost conscious, there is an emphasis on using nondedicated teams unless otherwise directed by the customer. Tracey Garrison, Project Manager at Humana Health Plans of Ohio, remarks on the effectiveness of using nondedicated teams at Humana Health Plans of Ohio:

> Non-dedicated teams are composed of associates who have functional responsibilities in addition to project team responsibilities. By using a large pool of associates who work in functional areas as well as on projects, we are able to maintain a fresh perspective and the associates are informed of changing business processes. This results in more creative and

TABLE 10–1. CHANGES DUE TO THE BOEING 777 NEW AIRPLANE PROJECT

Situation	Previous New Airplane Projects	Boeing 777
• Executive communications	• Secretive	• Open
• Communication flow	• Vertical	• Horizontal
• Thinking process	• Two-dimensional	• Three-dimensional
• Decision-making	• Centralized	• Decentralized
Empowerment	• Managers	• Down to factory workers
• Project managers	• Managers	• Down to nonmanagers
• Problem solving	• Individual	• Team
• Performance reviews (of managers)	• One way	• Three ways
• HR problems focus	• Weak	• Strong
• Meetings style	• Secretive	• Open
• Customer involvement	• Very low	• Very high
• Core values	• End result/quality	• Leadership/participation/ customer satisfaction
• Speed of decisions	• Slow	• Fast
• Life cycle costing	• Minimal	• Extensive
• Design flexibility	• Minimal	• Extensive

cost-effective solutions for the enterprise. It also allows employees the ability to increase their knowledge and feel like they have made a valuable contribution to the organization.

Some people argue that even the project manager can be part-time. To answer this argument, we must determine whether or not the project is for an internal or external customer. It is very difficult to convince external customers that their project is being managed correctly when the project manager is sharing his or her responsibilities on other activities.

Texas Instruments

Recognizing the importance of culture for prolonged and repeatable project management success is critical. According to a spokesperson for Texas Instruments:

> We feel *culture* at Texas Instruments is one of the most important yet most often least addressed items. The boundaries of hierarchy were removed from the beginning to allow a successful integrated team approach. With the support of upper and middle management, with a little power of persuasion, the ground work was laid to begin the process of team integration.
>
> The leaders of these teams are not supervisors generally of any of the other team members. Yet, the leaders of these teams are constantly challenged to heavily persuade and influence the team. The team members are a part of different departments. When the team all comes together, everyone from the engineer, technicians, operators, and vendors/suppliers are present. Each leverages each other to develop the best strategies: when something can be made available, when we would have the right resources, and when we would have the right parts. This coordination and cooperation is crucial for everyone to succeed. In fact, this team integration is so important that initially we page people to remind them

of meetings as well as send weekly invites, which include the latest schedule. It took a few weeks for people to let down their guard, but the synergy soared shortly thereafter. The key was in the execution stage meetings, which were a power packed 30 minutes, where every minute counted. In the planning stage, the meetings are extended to an hour.

With a central point of contact and constant, consistent participation in the meetings, the results have been shorter completion times for projects, clearer and more concise instructions for execution, reduction in maintenance costs, and a friendlier environment. We have seen warrantees' extended, prices maintained, and demands changed because of these meetings. It is not to say that these things might have still occurred, but they were the by-product of this culture and not the by-product of a specific meeting to accomplish that specific task. To sum it all up, things are done more now by consensus . . . which we view as success all in itself.

General Electric

In Chapter 6 we showed that the skills needed to be an effective project manager for the twenty-first century are knowledge of the business, integration skills, and risk management. Today, organizations have come to realize that the entire organizational staff can benefit from being trained in project management, not just the project managers. When this is done, people from anywhere in the organization can be assigned as project managers. Brian Vannoni, formerly of General Electric Plastics, states:

We have very few dedicated project managers. Our project managers might be process engineers, they might be scientists, they might be process control technicians, maintenance mechanics, degreed and nondegreed people. A short answer for GE Plastics is that anyone, any level, any function could be a project manager.

ABB

Part of the cultural change in many organizations today is the recognition that the project manager should function as a business manager rather than just as a technical subject matter expert. The cultural changes that are taking place at ABB are described by Benny Nyberg, Manager, Process Unit Project Management at ABB:

Traditionally, ABB PMs have been recruited through the engineering ranks. With the realization that project management is a management position requiring a great deal of business mind (engineering skills being beneficial but not required), we are in the process of transitioning our PMs to "Business Managers." The cultural change is on-going and begins with the senior management of the company.

General Motors Powertrain

Rose Russett, formerly with General Motors Powertrain, identified not only knowledge of the business as a training need but integration and behavioral skills as well.

In the past, we had project managers who were only responsible for the deliverables within their specific function—product engineering or manufacturing engineering, for example. We now appoint a single, focal point program manager who is responsible for the entire program from beginning to end. This individual is the integrator of all functional deliverables and must understand all of the various functions, their interrelationships, and have the ability to work within a strong matrix organization. This integration ability is a key success factor for our program managers. The ability to work in a team environment is also critical. Our organizational structure emphasizes cross-functional teams at all levels. Our executive sponsors are, by design, a small cross-functional team. The individuals that are chosen as program managers all have multifunctional backgrounds. Many have held assignments in engineering, manufacturing, quality systems, and so on. This is not an entry-level position. Their skills at the behavioral side of project management, such as managing and motivating teams, communication, and conflict resolution, become as important as the knowledge of the business itself when working in a project environment.

As project management matures and the project manager is given more and more responsibility, project managers are being given the responsibility for wage and salary administration. The excellent companies are still struggling with this new approach. The first problem is that the project manager may not be on the management pay scale in the company but is being given the right to sign performance evaluations.

The second problem is determining what method of evaluation should be used for union employees. This is probably the most serious problem, and the jury hasn't come in yet on what will and will not work. One reason why executives are a little reluctant to implement wage and salary administration that affects project management is because of the union involvement. This dramatically changes the picture, especially if a person on a project team decides that a union worker is considered to be promotable when in fact his or her line manager says, "No, that has to be based upon a union criterion." There is no black-and-white answer for the issue, and most companies have not even addressed the problem yet.

Armstrong World Industries The development of a good corporate culture for project management cannot take place overnight. Steve Senkowski, along with Matt Cook, General Manager, Global Product Development, for Armstrong Building Products, shows the chronology of events that led up to Armstrong World Industries' development of a culture for project management.

- Make it a full-time, dedicated position (1991).
- Have on-going forums so project managers could share successes, experiences, etc. . . and have presentations by various functional experts to broaden their knowledge (1992–93).
- Create a career ladder/specific position for project managers and open up the position for qualified individuals from any function (1994)
- Perform project case studies (1994).
- Create a "Best Practices" model for project manager (1996).
- Create Project Manager and NPD process database to better share best practices—be repository of methods for NPD (2000).
- Establish mentoring program for new project managers (2000).
- Establish project manager network to ensure the execution of best practice skills (2002).

Radian International

During the 1990s, Radian International was a medium-sized, international organization with corporate headquarters in Austin, Texas. It provided consulting services on project management and technical issues to external clients. Radian was a project-driven company, and its projects ranged in size between $2,000 and $2 million.

Superior corporate cultures for project management are a thing of beauty to observe. To identify the crème de la crème, one needed to go no further than Radian International. Radian had instituted the "pool" management concept. (Radian referred to it as *resource management.*) Classical management emphasizes from five to seven workers per manager. At Radian, the employee-to-manager ratio got as high as 100 to 1. What was the reason for Radian's success? Simply stated, it was Radian's corporate culture! Radian was very selective in the people it hired. All prospective employees were evaluated by selective members of the resource pool to make sure that there was a good cultural fit.

The heart of the Radian culture was its project managers. The project managers had the authority to negotiate with virtually any employee in any Radian division for assignment to a project. The employee would tell the project manager of his or her availability and then would *commit* to achieving a deliverable for the project manager within time, cost, and quality. The project managers were part of the same resource pool from which the employees came. When an employee agreed to a deliverable for the project manager, the agreement functioned as an unwritten contract. If an employee consistently missed deliverables, the project managers would no longer seek out this employee for work assignments. As a result, the employee would run out of charge numbers, and possible loss of employment may result.

At Radian, the project managers and employees viewed each other as peers. When a problem occurred, *everyone* in the pool volunteered to help that project team, quite often on their own time. Project managers had the authority to evaluate the performance of the employees at the end of each project. The project management culture at Radian was based upon teamwork and cooperation. Radian had become so good at project management that its customers were finding more work for Radian. Its customers were realizing that Radian could manage their projects at a lower cost, and even at a higher quality, than the customers could do themselves. Therefore, why should customers tie up their own resources when they could use Radian?

Perhaps the greatest strength in the Radian culture was its senior management team. Senior management at Radian had an exceptionally good knowledge of how project management should work. The project managers at Radian were empowered to make any and all decisions related to the project. While other companies provided lip service to empowerment, Radian practiced it. Because of the strong behavioral attitudes of the employees, project management excellence at Radian was strongly reinforced by the informal culture of the company. Teamwork, trust, cooperation, and communications permeated all resource pools. Project managers were not burdened with excessive paperwork.

In order for empowerment to work, there must be a great deal of trust between senior management and project management. Specifically, senior management must trust project managers to make the right decisions. This trust visibly existed at Radian. Empowerment of the project managers at Radian allowed senior managers to act as hands-off sponsors rather than hands-on supervisors. One project manager at Radian stated it best when he said:

> Once a week my executive sponsor sticks his head in my office and asks if there are any problems that he should be involved with. If I say no, he leaves.

Senior management would sponsor any training programs that the employees want or need. The only constraint was that if management were willing to pay for the training, the employees must be willing to contribute some of their own time. Most training courses were held on Friday afternoon and evening and on Saturday.

Senior management had instilled this culture at each Radian division. The employees viewed Radian as a good place to work. Employee turnover was very low. One Radian project manager summed it up best when he said:

> Before I came to Radian, I was a middle manager with another company. I spent most of my time writing useless reports to go up and down the line in order to help justify my position and that of my boss. I was a report writer rather than a manager. At Radian I am a project manager and I do, in fact, manage. It is a good feeling, and it is the best job that I have ever had.

Radian was extremely well positioned for the future, thanks to its culture, the dedication of its employees, and the quality of its senior managers.

Midwest Corporation (Disguised Company)

The larger the company, the more difficult it is to establish a uniform project management culture across the entire company. Large companies have "pockets" of project management, each of which can mature at a different rate. A large Midwest corporation had one division that was outstanding in project management. The culture was strong, and everyone supported project management. This division won awards and recognition on its ability to manage projects successfully. Yet at the same time, a sister division was approximately five years behind the excellent division in maturity. During an audit of the sister division, the following problem areas were identified:

- Continuous process changes due to new technology
- Not enough time allocated for effort
- Too much outside interference (meetings, delays, etc.)
- Schedules laid out based upon assumptions that eventually change during execution of the project
- Imbalance of workforce
- Differing objectives among groups
- Use of a process that allows for no flexibility to "freelance"
- Inability to openly discuss issues without some people taking technical criticism as personal criticism
- Lack of quality planning, scheduling, and progress tracking

- No resource tracking
- Inheriting someone else's project and finding little or no supporting documentation
- Dealing with contract or agency management
- Changing or expanding project expectations
- Constantly changing deadlines
- Last minute requirements changes
- People on projects having hidden agendas
- Scope of the project is unclear right from the beginning
- Dependence on resources without having control over them
- Finger-pointing: "It's not my problem"
- No formal cost-estimating process
- Lack of understanding of a work breakdown structure
- Little or no customer focus
- Duplication of efforts
- Poor or lack of "voice of the customer" input on needs/wants
- Limited abilities of support people
- Lack of management direction
- No product/project champion
- Poorly run meetings
- People do not cooperate easily
- People taking offense at being asked to do the job they are expected to do, while their managers seek only to develop a quality product
- Some tasks without a known duration
- People who want to be involved, but who do not have the skills needed to solve the problem
- Dependencies: making sure that when specs change, other things that depend on it also change
- Dealing with daily fires without jeopardizing the scheduled work
- Overlapping assignments (three releases at once)
- Not having the right personnel assigned to the teams
- Disappearance of management support
- Work being started in "days from due date" mode, rather than in "as soon as possible" mode
- Turf protection among nonmanagement employees
- Risk management nonexistent
- Project scope creep (incremental changes that are viewed as "small" at the time but that add up to large increments)
- Ineffective communications with overseas activities
- Vague/changing responsibilities (who is driving the bus??)

Large companies tend to favor pockets of project management rather than a companywide culture. However, there are situations where a company *must* develop a companywide culture to remain competitive. Sometimes it is simply to remain a major competitor; other times it is to become a global company. Three such companies have done an outstanding job in developing a uniform culture: BellSouth, Nortel, and United Technologies Automotive.

BellSouth

BellSouth is a large telecommunications company headquartered in Atlanta, Georgia. It is a project management hybrid corporation; that is, it does some project-driven work, but its business is based on services and products rather than projects. During the 1990s, all of its projects were performed for internal customers, and the projects ranged in size from $1 million to $400 million.

BellSouth's goal was to establish a culture to support its project management system across the organization. To accomplish that goal, the company made project management a professional discipline and instituted a continuous project management training program. BellSouth's new culture, supported by corporate sponsorship and commitment, resulted in the company's ability to meet customers' needs and even exceed their expectations.

In1997, Ed Prieto, a director in BellSouth's Strategic Management Unit, provided the following insight on the corporation's evolution of project management:

> Prior to 1994, the BellSouth project managers were responsible for the delivery of complex solutions to a diverse number of customers and project sponsors. The role of the project manager was one of project coordinator, without any authority for the project. The project manager, although skilled in the telecommunications arena, lacked the knowledge to apply modern project management methodology, tools, and techniques. In fact, the project manager was not recognized in the company and did not have a career path or a clearly defined development program. Projects were completed using various methods, tools, and reporting documentation. These tools and methodology were different from organization to organization. Confusion and inefficiency occurred due to the absence of a standard integrated methodology. Risk, quality, scope, and scheduling control management were rarely done with the adverse result of budget and schedule overruns on many projects. Since lessons learned and project metrics were not recorded, there is no history of project failure or successes available for measurement. The use of consultants was widespread. When an organization needed project management support, they usually hired outside consultants to manage the projects. At the end of the project, the consultants took their knowledge of project management with them.
>
> In 1994, a corporate project management initiative under the Reengineering Organization led the effort to implement professional project management as a discipline throughout all of BellSouth. A project management handbook was developed that combines the Project Management Institute (PMI) Body of Knowledge (PMBOK®), the Software Engineering Institute Capability Maturity Model (CMM), and BellSouth management experience.
>
> The methodology is software independent; however, Microsoft Project 98 is the most widely used project management scheduling software in BellSouth. A comprehensive continuing education program was established to build project management skills and knowledge. The professional development program offered by Educational Services Institute and The George Washington University leads to a Masters Certificate in Project Management. To date, over 100 BellSouth project managers hold the PMP certification, and hundreds more are preparing to sit for the exam this year.
>
> An interest in joining local PMI chapters grew as well. Recognizing this need, BellSouth established the PMI corporate-sponsored Membership Program to encourage

employees' involvement in local chapter activities and to provide networking opportunities with other professionals. BellSouth also sponsored the establishment of three PMI local chapters (Central Alabama, South Carolina Midlands, and East Tennessee) and will continue to support the establishment of local chapters in BellSouth's territory. Two new local chapters being sponsored by BellSouth are planning to open this year, in Louisville, Kentucky, and Nashville, Tennessee. Recognizing the value of professional project management, five levels of job descriptions for a new job family of project managers were introduced in early 1996. Realizing that to succeed in today's competitive market place, business must take an active part in developing the essential skills and career development needs of employees, a Standard Curriculum and Training Program was introduced in March, 1995. The program is a process that provides individuals with opportunities to develop and maintain the skills and knowledge necessary to create a level of proficiency in project management.

On January 1, 1997, the BellSouth Corporate Project Management Office was established with the mission to institutionalize modern project management discipline and techniques to effectively facilitate BellSouth's transition to an open, competitive environment. The goal of this organization is to standardize the BellSouth approach to project management by providing direction, developing and enhancing methodologies, selecting software tools, and development guidelines to be adopted by BellSouth for use on all projects by all project management practitioners.

Project management is viewed in BellSouth as a core competency and means to increase customer satisfaction, improve efficiency in the management of projects, and improve responsiveness to customers. Our goal is to deliver consistent, professionally managed projects throughout the company, meeting cost, functionality, quality, and schedule objectives. With the support of higher management, and through the efforts of the BellSouth Corporate Project Management Office, a truly integrated project management system will be created to deliver a common approach to meet or exceed stakeholder needs and expectations from a project. Management recognizes that project management provides the skills necessary for the twenty-first century and is creating the mind-set to bring modern project management a reality in BellSouth.

Nortel

As an international telecommunications corporation, Nortel recognized the importance of training employees in project management skills before introducing new project management tools and techniques. Perhaps the most difficult task facing large corporations today is the establishment of a global culture for project management. Bill Marshall, former director of project management standards at Nortel, comments on the evolution of project management at Nortel during the 1990s:

I am sure that Nortel is not alone in regards to our evolution. We tend to get pushed into our evolution in project management. It is only in the mid-1990s that we started acting globally in order to complement our global marketing story that we "have gone out of the box" and done some proactive evolution in project management. The competition of the market place, driven by deregulation, has been the push. There is a difference between a company that talks about global business and the company that acts globally. Nortel has broken the mold and found the formula to encourage cooperation among the project management groups globally and to implement the standards to demonstrate global project management. In doing so, Nortel has set a new standard for project management and has

> gained the high ground in serving the incumbent and the new market entrant companies in the telecom business.
>
> Up to 1990, Nortel project management was a single project based organization. Network based customers with multiple locations, often in offshore locations, have demanded a global wide view of project status. In 1993, the UK introduced a project management process that was approved by the president of the division. This drove the development and introduction of a project tool from the Mantix Corporation in the UK. This tool allowed multisite projects to be controlled through work packages. However, the work package owners were bought into the process only reluctantly, and did not feel the total ownership. This resulted in the project manager performing an abnormal amount of administration to obtain an overall status on the project. The project managers continue to evolve the process, and at this point the UK has joined force with the Nortel global efforts to refine the process.

The complexity facing Nortel is the development of a global rather than a local project management process, especially a process that is executed uniformly worldwide. Nortel UK/Europe project management evolution is described by Peter Edwards, director of supply management, AVP Projects Paris:

> The common threads of excellence in project management—work breakdown structures, financial and risk management, and so on—are being applied today across the corporation to bring us all to a common baseline of terms, tools, and processes. In Europe, we have developed a project management process document that has been in use for about five years and is now being adopted globally.
>
> We also have had a wealth of experience of managing complex projects with new telecommunications operators who are not only having to set up a network—a "relatively" straightforward project management job—but also to establish their business, business processes, service offerings, organization, staffing, and so on. The customer, therefore, is working in many dimensions of uncertainty at the same time. This puts him in a difficult position, because he has to manage his network supplier (Nortel) against his own framework, and that is in continuous evolution.
>
> To the mature supplier, this is not just an opportunity to sell further services to the customer but, more importantly, it becomes a necessity to actively work with the customer to identify areas of weakness in the customers' overall plan and to find ways to provide assistance, because a failed customer will inevitably reflect on his suppliers even though the fault may not rest with them.
>
> This requires the project manager to have additional skills, and can also change the traditional relationship between customer and supplier. The project manager, being in continual contact with the customer, needs to spend time working closely with the customer, becoming part of the team, if possible, so that he can identify areas outside the contracted scope of work where Nortel can provide assistance. He then needs to describe the work proposed and sell the concept to the customer. He essentially is working closely with the customer, becoming part of the team, if possible, so that he can identify areas outside the contracted scope of work where Nortel can provide assistance. He then needs to describe the work proposed and sell the concept to the customer. He essentially becomes a major player in the upsell process; his role is much broader than the traditional project manager in this type of contract. The project team also becomes much more of a partner in the overall success of the venture, rather than the traditional (often antagonistic) relationship. In

fact, we have taken this as far as to have totally integrated teams (customer and Nortel Networks), and to couple our payments to the success of the customer.

We are, therefore, having to ensure that our project managers are not only skilled in project management techniques, processes, and tools, but are also skilled in sales techniques, communications, customer care, and management in a supportive, constructive atmosphere.

Nortel has been very successful with this approach and has built up an excellent track record of assisting new operators with their business plans and putting them into operation through the provision of consultants or outsourcing parts of the customer operation. Perhaps the most difficult task facing large corporations today is the establishment of a global culture of project management.

Bill Marshall of Nortel continues:

Asia took the UK model one step further and introduced project management tools in many Asian countries. The architecture was such that to achieve a rollup view, the "parent" had to contain all the data of the "children." This also required an overhead of administration to coordinate the coding structure to ensure accurate data merges. This has its drawbacks.

The future tools currently being planned will be owned by the work package owners, and a very strong public relations effort will be spent up front to ensure "buy-in." All parties up and down the information "food chain" will look at the same data but will view it in their areas of accountability. This will allow the project manager an overall view of earned value (schedule variations and cost variations). The account manager views the overall progress and exceptions. The work package owner views specific work package levels of time, cost, and quality.

Financial tool evolution is seen as one of the key sources of information for the project manager. This tool evolution has been going on for some years, and Nortel is currently introducing a standard system into the various regions globally. This will continue into the next millenium.

The education tools and training must support the introduction of these new tools and culture evolution. In 1996, there were training pilots, and in 1997 many project managers will be trained in industry standards. 1998 brought training focused on new tools and the standard Nortel processes. In 1999, a standard curriculum to all project management in Nortel, globally, was introduced.

The visions and plans of the project managers in Nortel are subject to funds and resource approvals. Executive support at the highest levels in Nortel is being sought to ensure this will occur in a timely and coordinate fashion. A gate process consisting of a series of decisions ensures that the key milestones for development of tools, process, and training are brought to the team and the executives for approvals. The formation of a gate process will provide up to four checkpoints for executive approvals.

Bill Marshall of Nortel describes the impact that project management has within the corporation:

Project management is a core competency in Nortel. Centers of excellence in project management are found in the development area, internal systems support area, and the management of customer contracted services.

Chahram Bolouri, vice president of global operations, Nortel Networks, comments on the Nortel Project Management of Customer Contracts:

> Nortel is a vendor of a wide range of telecom products and services that is capable of providing city, country, and global networks. Nortel is proud of its customer satisfaction record, particularly in the area of relationships and demonstrated performance in project management. Many new market entrants are starting their telecom business spurred on by the deregulation by the WTO—World Trade Organization—and countries willing to open up their markets to competition The new entrants do not have the resources or knowledge of the incumbent telecom providers and must rely on the vendors to bring the turnkey knowledge and skills to implement complex telecom networks. Those new market entrants entering the intercountry and global telecom businesses are dependent on the vendors for standard implementation and support. Nortel project management provides a single point of contact for the customer and presents an integrated schedule and implementation plan for the total project.
>
> Nortel has attained global standards in project management and is in the process of reengineering the project management tools to ensure global visibility and control of projects. This will be evident to our customers through standard implementation of telecom networks and common practices in controlling projects.

Mark Linaugh, Nortel director of organization development, comments on the human resources side of project management:

> We are training our project managers to lead cross-functional integrated project teams. Simultaneously, we are building a supportive culture and infrastructure for project management across Nortel Networks. The project management organizations have agreed upon standards in job profiles, pay grade levels, and skills assessments. A global framework for project management organization structures and practices ensures the project manager has a career path. When he or she performs the current job up to expectations, the project manager knows another challenging job awaits the completion of the current project. Ultimately, our customers benefit from a higher, faster, and more consistent level of service delivery. One of the areas where project management in Nortel Networks exceeds other companies is the area of performance against contracted expectations and control of project deliverables.
>
> We assess project managers against a model, train them to a defined level of competency, mentor them to support the evolving project manager model, and certify them. This process is aligned to the project management body of knowledge (PMBOK®) of the Project Management Institute (PMI). By reaching agreement on the generic aspects of project management, Nortel enables project managers to move around the enterprise with transportable project management skills. In addition, this provides industry standard certification, which can make individuals more marketable in the external world as well. Project management is key to Nortel delivering better, faster, more valuable solutions to its customers.

MULTIPLE CHOICE QUESTIONS

1. Good corporate cultures are based heavily upon:
   A. Executive sponsorship
   B. Line managers relinquishing authority to the project manager
   C. How much authority the project manager exercises

D. Trust and cooperation
E. All of the above

2. Which type of organization fosters the best culture for project management?
A. Project-driven organizations
B. Non–project-driven organizations
C. Hybrid organizations
D. Classical organizations
E. Any organization can have a good project management culture

3. A common mistake made by companies is:
A. Not recognizing that project management is a culture
B. Believing that cultures can be benchmarked from company to company
C. Believing that the culture of one company can easily be duplicated into another company
D. Believing that large organizations can have a single culture for project management rather than having "pockets" of project management each with its own culture
E. All of the above

4. An automotive supplier has restructured their organization to three strategic business units (i.e., Ford, Chrysler, and General Motors) and has a separate project management culture for each SBU. This can work even if the SBUs share common resources.
A. True
B. False

5. Corporate cultures are often developed for the purpose of better interfacing with the customers.
A. True
B. False

6. Cultures where team members make decisions based upon personal interests rather than corporate interests are called __________ cultures.
A. Cooperative
B. Noncooperative
C. Competitive
D. Independent
E. Discretionary

7. Cultures where project managers aggressively compete with one another for talented resources or prestige are called __________ cultures.
A. Cooperative
B. Noncooperative
C. Competitive
D. Independent
E. Discretionary

8. A company receives a Request for Proposal (RFP) from one of its customers and allows two of its divisions (both located in the same building) to submit independent bids for the job. The resulting culture would be a __________ culture.
A. Cooperative
B. Noncooperative
C. Competitive
D. Aggressive
E. Independent

9. Cultures take a long time to build up but can be shattered quickly by:
   A. The assignment of an inexperienced project manager
   B. The assignment of the wrong sponsor
   C. A change in senior management accompanied by new personal whims
   D. All of the above
   E. A and B only
10. The culture at Humana emphasized that when a project decision has to be made, the best interest of the __________ comes first.
   A. Company
   B. Project
   C. Project manager
   D. Project sponsor
   E. Customer
11. According to the text, it took Armstrong World Industries approximately __________ years to develop the strength of its culture.
   A. 2
   B. 5
   C. 8
   D. 10
   E. 15
12. Cooperative cultures result in:
   A. Fewer scope changes
   B. Fewer team meetings
   C. More communications channels supported by large amounts of documentation
   D. Increased authority for the project manager
   E. None of the above
13. The implementation of the methodology of project management is achieved primarily through:
   A. The skills of the project manager
   B. The organizational structure
   C. The skills of the executive sponsor
   D. The skills of the team members
   E. The corporate culture
14. Corporate cultures are a reflection of:
   A. The skills of the project management
   B. The aspirations and beliefs of senior management
   C. The needs of the customers
   D. The needs of the team members
   E. None of the above
15. A good corporate culture can foster better relationships between the company and external customers.
   A. True
   B. False
16. The most complex component of a project management culture is:
   A. Multiple-boss reporting
   B. The role of the executive sponsor
   C. Management without authority

D. Customer involvement
E. Identification of risks

17. Good corporate cultures can be derailed virtually overnight through the whims of one senior executive.
A. True
B. False

18. The most commonly used definition of accountability is:
A. Authority plus responsibility
B. Power without authority
C. Power and responsibility
D. Authority and power
E. None of the above

19. A quick way to destroy a potentially good culture is by:
A. Too much reliance on authority
B. Too much involvement by the customer
C. Not sharing rewards effectively
D. Assigning an inexperienced project management
E. Too much focus on profitability

20. A good way to alleviate the pressure placed upon a corporate culture is through:
A. More employee training
B. More involved sponsorship
C. A reduction in employee workload
D. Prioritization of work
E. More customer involvement

21. A company structured according to strategic business units (SBUs) will:
A. Have an easy time developing a uniform culture across all SBUs
B. Have the same executive sponsor for all of the projects in all of the SBUs
C. Have different methodologies for each SBU but similar cultures
D. Have the same methodology for each SBU but different cultures
E. Have different methodologies and different cultures.

22. Project-driven organizations need a strong project management culture merely to survive.
A. True
B. False

23. Some people contend that the sign of a good corporate culture is when the customer treats the contractor as a partner.
A. True
B. False

24. An alternative approach to continuous scope changes is:
A. Strong executive sponsorship
B. Project termination
C. Enhancement projects
D. Life cycle change projects
E. Partial change projects

25. A complex problem facing executives in the development of a corporate culture is:
A. The amount of authority for the project manager
B. Giving the project manager the responsibility for wage and salary administration

C. Deciding the grade level of the sponsor
D. Deciding on the involvement of the customer
E. Keeping the paperwork to a minimum

DISCUSSION QUESTIONS

1. Why is it so difficult to change the culture of an organization specifically to accept project management?
2. A company develops a standard methodology for project management and a supportive culture is created. Is it possible that enhancements to the methodology will require changes to the culture? If so, would you expect the changes to the culture to be easy or difficult to implement?
3. What factors influence the speed by which a corporate culture matures?
4. Defend the statement that most people believe that corporate cultures cannot be benchmarked.
5. Why is multiple boss reporting the driving force for a behavioral culture?
6. What factors influence the decision of making continuous scope changes or using enhancement projects?
7. What is shared accountability?
8. Why is the organizational structure generally unimportant when creating a good culture for project management?
9. Why do good corporate cultures foster a minimization of prioritization?
10. What cultural changes may be necessary if the project manager is given the responsibility for wage and salary administration?

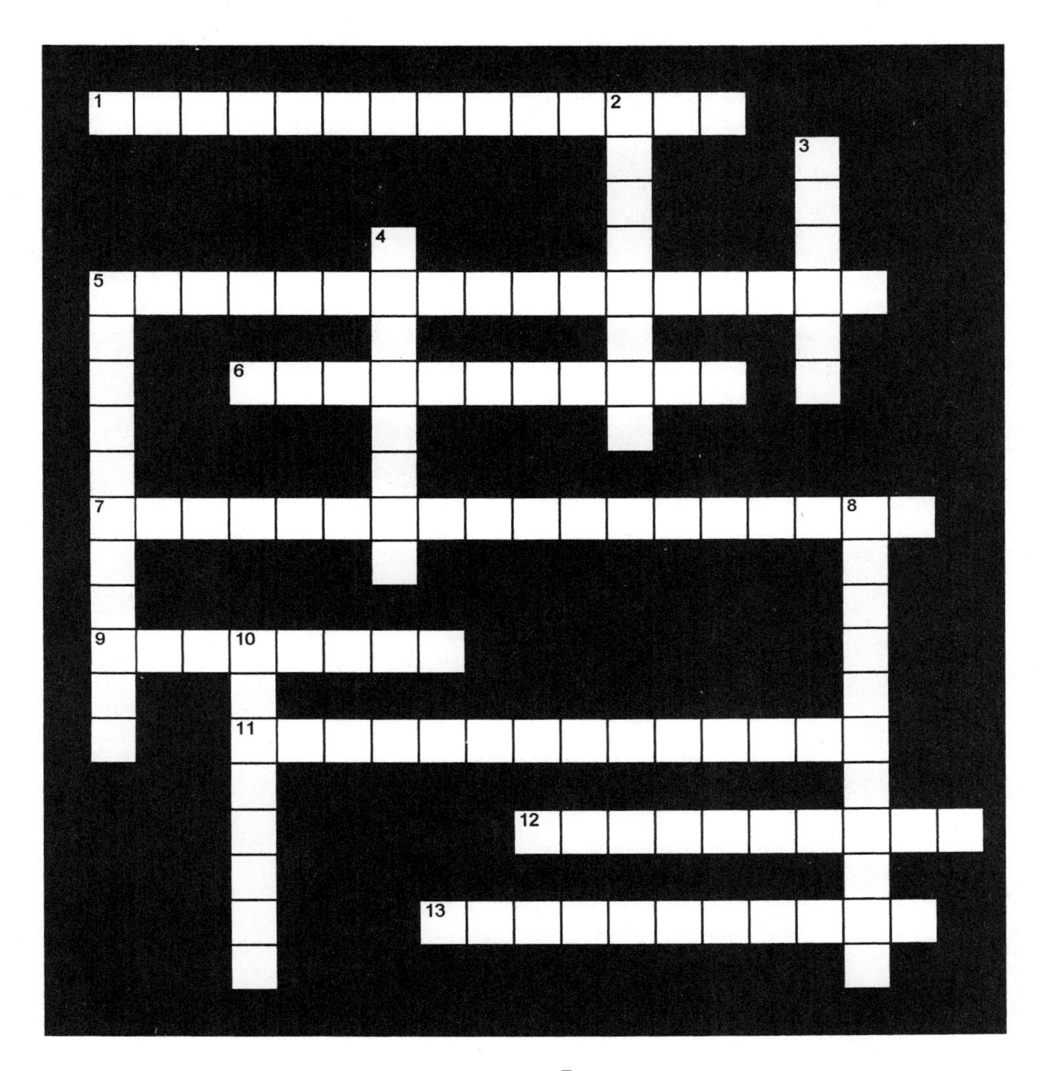

Across

1. Type of culture
5. Benefit: Better ___________________ (2 words)
6. Scope change: ____________ project
7. Cultural maturity factor (4 words)
9. Type of customer
11. Cultural maturity factor (3 words)
12. Type of culture
13. Type of culture

Down

2. Type of culture
3. Enhancement project: ______ management
4. Organizational ________
5. Cultural maturity factor (2 words)
8. Type of culture
10. Type of customer

Management Support

11.0 INTRODUCTION

As we saw in Chapter 10, senior managers are the architects of corporate culture. They are charged with making sure that their companies' cultures, once accepted, don't come apart. Visible *management support* is essential to maintaining project management culture.

This chapter examines the importance of management support in the creation and maintenance of project management cultures. Case studies illustrate the vital importance of employee empowerment and the project sponsor's role in the project management system.

11.1 VISIBLE SUPPORT FROM SENIOR MANAGERS

As project sponsors, senior managers provide support and encouragement to the project managers and the rest of the project team. Companies excellent in project management have the following characteristics:

- Senior managers maintain a hands-off approach, but they are available when problems come up.
- Senior managers expect to be supplied with concise project status reports.
- Senior managers practice empowerment.
- Senior managers decentralize project authority and decision-making.
- Senior managers expect project managers and their teams to suggest both alternatives and recommendations for solving problems, not just to identify the problems.

Robert Hershock, former vice president of 3M, said it best:

> Probably the most important thing is that they have to buy in from the top. There has to be leadership from the top, and the top has to be 100 percent supportive of this whole process. If you're a control freak, if you're someone who has high organizational skills and likes to dot all the i's and cross all the t's, this is going to be an uncomfortable process, because basically it's a messy process; you have to have a lot of fault tolerance here. But what management has to do is project the confidence that they have in the teams. They have to set the strategy and the guidelines, and then they have to give the teams the empowerment that they need in order to finish their job. The best thing that management can do after training the team is get out of the way.

To ensure their visibility, senior managers need to believe in walk-the-halls management. In this way, every employee will come to recognize the sponsor and realize that it is appropriate to approach the sponsor with questions. Walk-the-halls management also means that executive sponsors keep their doors open. It is important that everyone, including line managers and their employees, feels supported by the sponsor. Keeping an open door can occasionally lead to problems if employees attempt to go around lower-level managers by seeking a higher level of authority. But such instances are infrequent, and the sponsor can easily deflect the problems back to the appropriate manager.

11.2 PROJECT SPONSORSHIP

Executive project sponsors provide guidance for project managers and project teams. They are also responsible for making sure that the line managers who lead functional departments fulfill their commitments of resources to the projects underway. In addition, executive project sponsors maintain communication with customers.

The project sponsor usually is an upper-level manager who, in addition to his or her regular responsibilities, provides ongoing guidance to assigned projects. An executive might take on sponsorship for several concurrent projects. Sometimes, on lower priority or maintenance projects, a middle-level manager may take on the project sponsor role. One organization I know of even prefers to assign middle managers instead of executives. The company believes that avoids the common problem of lack of line manager buy-in to projects (see Figure 11–1).

In some large, diversified corporations, senior managers don't have adequate time to invest in project sponsorship. In such cases, project sponsorship falls to the level below corporate senior management or to a committee.

Some projects don't need project sponsors. Generally, sponsorship is required on large, complex projects involving a heavy commitment of resources. Large, complex projects also required a sponsor to integrate the activities of the functional lines, to dispel disruptive conflicts, and to maintain strong customer relations.

Consider one example of a project sponsor's support for a project. A project manager who was handling a project in an organization within the federal government decided that another position would be needed on his team if the project was to meet its completion deadline. He had already identified a young woman in the company who fit the qualifica-

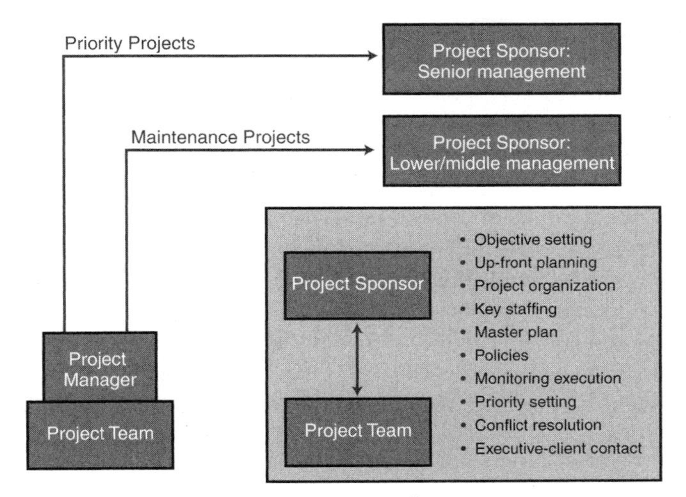

FIGURE 11–1. The roles of the project sponsor. *Source:* Reprinted from H. Kerzner, *In Search of Excellence in Project Management.* New York: Wiley, 1998, p. 159.

tions he had outlined. But adding another full-time-equivalent position seemed impossible. The size of the government project office was constrained by a unit-manning document that dictated the number of positions available.

The project manager went to the project's executive sponsor for help. The executive sponsor worked with the organization's human resources and personnel management department to add the position requested. Within 30 days, the addition of the new position was approved. Without the sponsor's intervention, it would have taken the organization's bureaucracy months to approve the position, too late to affect the deadline.

In another example, the president of a medium-size manufacturing company wanted to fill the role of sponsor on a special project. The project manager decided to use the president to the project's best advantage. He asked the president/sponsor to handle a critical situation. The president/sponsor flew to the company's headquarters and returned two days later with an authorization for a new tooling the project manager needed. The company ended up saving time on the project, and the project was completed four months earlier than originally scheduled.

Gregory M. Willits, project manager process champion of Motorola, discusses the executive-level sponsorship at Motorola's System Solution Group (SSG):

> An executive at the VP level is responsible for the project management process. This individual is the "owner" of the process. The owner has a champion who is responsible to improve the overall capability maturity in the process area. At the project level, we have project managers who act as the general manager of their program(s), and engineering project leaders who handle the day-to-day operations of their project. We use a matrixed organizational structure, biased toward the functional side. On very large programs, a projectized organizational structure is used (where all functions report to the project manager).

Sponsorship by Committee

As companies grow, it sometimes becomes impossible to assign a senior manager to every project, and so committees act in the place of individual project sponsors. In fact, the recent trend has been toward committee sponsorship in many kinds of organizations. A project sponsorship committee usually is made up of a representative from every function of the company: engineering, marketing, and production. Committees may be temporary, when a committee is brought together to sponsor one time-limited project, or permanent, when a standing committee takes on the ongoing project sponsorship of new projects.

For example, General Motors Powertrain has achieved excellence in using committee sponsorship. Two key executives, the vice president of engineering and the vice president of operations, lead the "Office of Product and Operations," a group formed to oversee the management of all product programs. This group demonstrates visible executive-level program support and commitment to the entire organization. Their roles and responsibilities are to:

- Appoint the project manager and team as part of the charter process
- Address strategic issues
- Approve the program contract and test for sufficiency
- Assure program execution through regularly scheduled reviews with program managers

EDS is also a frequent user of committee sponsorship. According to Gene Panter, a program manager with EDS,

> EDS executives, in conjunction with key customers, tend to take the role of sponsor for large-scale projects or programs. Sponsors are asked to participate in various checkpoint reviews, assure the acquisition of necessary resources, and provide support in the elimination of barriers that may exist. Large programs tend to have guidance teams in place to provide necessary support and approvals for project teams.
>
> Line managers and functional managers tend to be responsible for working with the project or program manager to identify appropriate human resources for projects. Project or program managers provide the direct management function for the project. EDS is also widely utilizing the concept of a project or program office to provide standards for project tracking, oversight, management, and control support for projects.

Phases of Project Sponsorship

The role of the project sponsor changes over the life cycle of a project. During the planning and initiation phases, the sponsor plays an active role in the following activities:

- Helping the project manager establish the objectives of the project
- Providing guidance to the project manager during the organization and staffing phases
- Explaining to the project manager what environmental or political factors might influence the project's execution

- Establishing the project's priority (working alone or with other company executives) and then informing the project manager about the project's priority in the company and the reason that priority was assigned
- Providing guidance to the project manager in establishing the policies and procedures for the project
- Functioning as the contact point for customers and clients

During the execution phase of a project, the sponsor must be very careful in deciding which problems require his or her guidance. Trying to get involved with every problem that comes up on a project will result in micromanagement. It will also undermine the project manager's authority and make it difficult for the executive to perform his or her regular responsibilities.

For short-term projects of two years or less, it's usually best that the project sponsor assignment isn't changed over the duration of the project. For long-term projects of five years, more or less, different sponsors could be assigned for every phase of the project, if necessary. Choosing sponsors from among executives at the same corporate level works best, since sponsorship at the same level creates a "level" playing field, whereas at different levels, favoritism can occur.

Project sponsors needn't come from the functional area where the majority of the project work will be completed. Some companies even go so far as assigning sponsors from line functions that have no vested interest in the project. Theoretically, this system promotes impartial decision-making.

Customer Relations

The role of executive project sponsors in customer relations depends on the type of organization (entirely project-driven or partially project-driven) and the type of customer (external or internal). Contractors working on large projects for external customers usually depend on executive project sponsors to keep the clients fully informed of progress on their projects. Customers with multimillion-dollar projects often keep an active eye on how their money is being spent. They are relieved to have an executive sponsor they can turn to for answers.

It is common practice for contractors heavily involved in competitive bidding for contracts to include both the project manager's and the executive project sponsor's résumés in proposals. All things being equal, the résumés may give one contractor a competitive advantage over another.

Customers prefer to have a direct path of communication open to their contractors' executive managers. One contractor identified the functions of the executive project sponsor as:

- Actively participating in the preliminary sales effort and contract negotiations
- Establishing and maintaining high-level client relationships
- Assisting project managers in getting the project underway (planning, staffing, and so forth)
- Maintaining current knowledge of major project activities
- Handling major contractual matters
- Interpreting company policies for project managers

- Helping project managers identify and solve significant problems
- Keeping general managers and client managers advised of significant problems with projects

Decision-Making Imagine that project management is like car racing. A yellow flag is a warning to watch out for a problem. Yellow flags require action by the project manager or the line manager. There's nothing wrong with informing an executive about a yellow-flag problem as long as the project manager is not looking for the sponsor to solve the problem. Red flags, however, usually do require the sponsor's direct involvement. Red flags indicate problems that may affect the time, cost, and performance parameters of the project. So red flags need to be taken seriously and decisions need to be made collaboratively by the project manager and the project sponsor.

Serious problems sometimes result in serious conflicts. Disagreements between project managers and line managers are not unusual, and they require the thoughtful intervention of the executive project sponsor. First, the sponsor should make sure that the disagreement can't be solved without his or her help. Second, the sponsor needs to gather information from all sides and consider the alternatives being considered. Then, the sponsor must decide whether he or she is qualified to settle the dispute. Often, disputes are of a technical nature and require someone with the appropriate knowledge base to solve them. If the sponsor is unable to solve the problem, he or she will need to identify another source of authority who has the needed technical knowledge. Ultimately, a fair and appropriate solution can be shared by everyone involved. If there were no executive sponsor on the project, the disputing parties would be forced to go up the line of authority until they found a common superior to help them. Having executive project sponsors minimizes the number of people and the amount of time required to settle work disputes.

Strategic Planning Executives are responsible for performing the company's strategic planning, and project managers are responsible for the operational planning on their assigned projects. Although the thought processes and time frames are different for the two types of planning, the strategic planning skills of executive sponsors can be useful to project managers. For projects that involve process or product development, sponsors can offer a special kind of market surveillance to identify new opportunities that might influence the long-term profitability of the organization. Furthermore, sponsors can gain a lot of strategically important knowledge from lower-level managers and employees. Who else knows better when the organization lacks the skill and knowledge base it needs to take on a new type of product? When the company needs to hire more technically skilled labor? What technical changes are likely to affect their industry?

11.3 EXCELLENCE IN PROJECT SPONSORSHIP

Many companies have achieved excellence in their application of project sponsorship. Radian International depended on single-project sponsors to empower their project managers for decision-making. General Motors proved that sponsorship by committee works.

Roadway demonstrated the vital importance of sponsorship training for both sponsorship by a single executive and sponsorship by a committee.

In excellent companies, the role of the sponsor is not to supervise the project manager but to make sure that the best interests of both the customer and the company are recognized. However, as the next two examples reveal, it's seldom possible to make executive decisions that appease everyone.

Franklin Engineering (a pseudonym) had a reputation for developing high-quality, innovative products. Unfortunately, the company paid a high price for its reputation: a large research and development (R&D) budget. Fewer than 15 percent of the projects initiated by R&D led to the full commercialization of a product and the recovery of the research costs.

The company's senior managers decided to implement a policy that mandated that all R&D project sponsors periodically perform cost-benefit analyses on their projects. When a project's cost-benefit ratio failed to reach the levels prescribed in the policy, the project was canceled for the benefit of the whole company.

Initially, R&D personnel were unhappy to see their projects canceled, but they soon realized that early cancellation was better than investing large amounts in projects that were likely to fail. Eventually, the project managers and team members came to agree that it made no sense to waste resources that could be better used on more successful projects. Within two years, the organization found itself working on more projects with a higher success rate but no addition to the R&D budget.

Another disguised case involves a California-based firm that designs and manufactures computer equipment. Let's call the company Design Solutions. The R&D group and the design group were loaded with talented individuals who believed that they could do the impossible and often did. These two powerful groups had little respect for the project managers and resented schedules because they thought schedules limited their creativity.

In June 1997, the company introduced two new products that made it onto the market barely ahead of the competition. The company had initially planned to introduce them by the end of 1996. The reason for the late releases: projects had been delayed because of the project teams' desire to exceed the specifications required and not just meet them.

To help the company avoid similar delays in the future, the company decided to assign executive sponsors to every R&D project to make sure that the project teams adhered to standard management practices in the future. Some members of the teams tried to hide their successes with the rationale that they could do better. But the sponsor threatened to dismiss the employees, and they eventually relented.

The lessons in both cases are clear. Executive sponsorship actually can improve existing project management systems to better serve the interests of the company and its customers.

11.4 EMPOWERMENT OF PROJECT MANAGERS

One of the biggest problems with assigning executive sponsors to work beside line managers and project managers is the possibility that the lower-ranking managers will feel threatened with a loss of authority. This problem is real and must be dealt with at the executive level. Frank Jackson, formerly a senior manager at MCI, believes in the idea that information is power:

> We did an audit of the teams to see if we were really making the progress that we thought or were we kidding ourselves, and we got a surprising result. When we looked at the audit, we found out that 50 percent of middle management's time was spent in filtering information up and down the organization. When we had a sponsor, the information went from the team to the sponsor to the operating committee, and this created a real crisis in our middle management area.
>
> MCI has found its solution to this problem. If there is anyone who believes that just going and dropping into a team approach environment is an easy way to move, it's definitely not. Even within the companies that I'm involved with, it's very difficult for managers to give up the authoritative responsibilities that they have had. You just have to move into it, and we've got a system where we communicate within MCI, which is MCI mail. It's an electronic mail system. What it has enabled us to do as a company is bypass levels of management. Sometimes you get bogged down in communications, but it allows you to communicate throughout the ranks without anyone holding back information.

Not only do executives have the ability to drive project management to success, they also have the ability to create an environment that leads to project failure. According to Robert Hershock, former vice president of 3M:

> Most of the experiences that I had where projects failed, they failed because of management meddling. Either management wasn't 100 percent committed to the process, or management just bogged the whole process down with reports and a lot of other innuendos. The biggest failures I've seen anytime have been really because of management. . . . Basically, there are two experiences where projects have failed to be successful. One is the management meddling where management cannot give up its decision-making capabilities, constantly going back to the team and saying you're doing this wrong or you're doing that wrong. The other side of it is when the team can't communicate its own objective. When it can't be focused, the scope continuously expands, and you get into project creep. The team just falls apart because it has lost its focus.

Project failure can often be a matter of false perceptions. Most executives believe that they have risen to the top of their organizations as solo performers. It's very difficult for them to change without feeling that they are giving up a tremendous amount of power, which traditionally is vested in the highest level of the company. To change this situation, it may be best to start small. As Frank Jackson observed:

> There are so many occasions where senior executives won't go to training and won't listen, but I think the proof is in the pudding. If you want to instill project management teams in your organizations, start small. If the company won't allow you to do it using the Nike theory of just jumping in and doing it, start small and prove to them one step at a time that they can gain success. Hold the team accountable for results—it proves itself.

It's also important for us to remember that executives can have valid reasons for micromanaging. One executive commented on why project management might not be working as planned in his company:

> We, the executives, wanted to empower the project managers and they, in turn, would empower their team members to make decisions as they relate to their project or function. Unfortunately, I do not feel that we [the executives] totally support decentralization of decision-making due to political concerns that stem from the lack of confidence we have in our project managers, who are not proactive and who have not demonstrated leadership capabilities.

In most organizations, senior managers start at a point where they trust only their fellow managers. As the project management system improves and a project management culture develops, senior managers come to trust project managers, even though they do not occupy positions high on the organizational chart. Empowerment doesn't happen overnight. It takes time and, unfortunately, a lot of companies never make it to full project manager empowerment (see Figure 11–2).

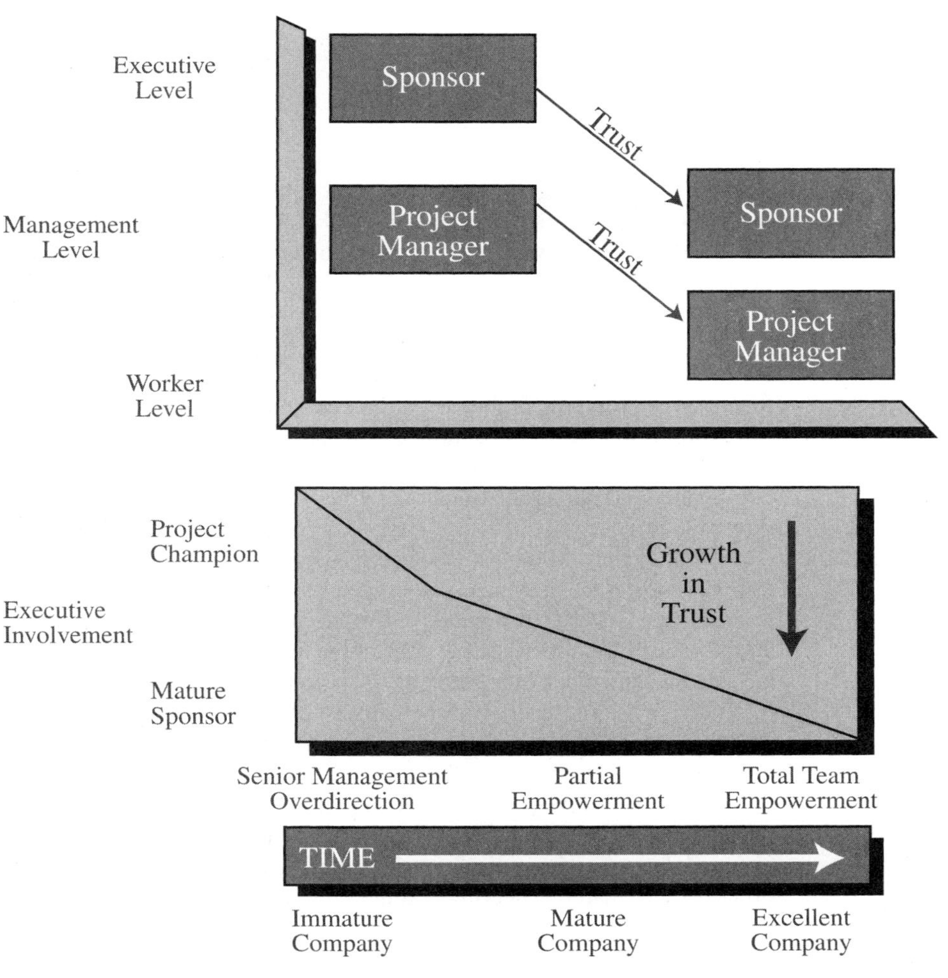

FIGURE 11–2. Time-dependent changes in trust.

Figure 11–2 shows how this can happen over time. The top part of the figure illustrates how, as trust develops, sponsorship moves down in the organization, rather than remaining at executive levels. In the figure, we see growing trust leading toward mature sponsorship. Without trust, senior management micromanages the project and acts as a champion. As executives begin to trust the team more and more, the team achieves partial empowerment, thus proving to executives the team's ability to make decisions. With total team empowerment, executives act as mature sponsors, keeping hands off except when needed.

11.5 MANAGEMENT SUPPORT AT WORK

Visible executive support is necessary for successful project management and the stability of a project management culture. But there is such a thing as too much visibility for senior managers. Take the following case example, for instance.

Midline Bank

Midline Bank (a pseudonym) is a medium-size bank doing business in a large city in the Northwest. Executives at Midline realized that growth in the banking industry in the near future would be based on mergers and acquisitions and that Midline would need to take an aggressive stance to remain competitive. Financially, Midline was well prepared to acquire other small- and middle-size banks to grow its organization.

The bank's information technology group was given the responsibility of developing an extensive and sophisticated software package to be used in evaluating the financial health of the banks targeted for acquisition. The software package required input from virtually every functional division of Midline. Coordination of the project was expected to be difficult.

Midline's culture was dominated by large, functional empires surrounded by impenetrable walls. The software project was the first in the bank's history to require cooperation and integration among the functional groups. A full-time project manager was assigned to direct the project.

Unfortunately, Midline's executives, managers, and employees knew little about the principles of project management. The executives did, however, recognize the need for executive sponsorship. A steering committee of five executives was assigned to provide support and guidance for the project manager, but none of the five understood project management. As a result, the steering committee interpreted its role as one of continuous daily direction of the project.

Each of the five executive sponsors asked for weekly personal briefings from the project manager, and each sponsor gave conflicting directions. Each executive had his or her own agenda for the project.

By the end of the project's second month, chaos took over. The project manager spent most of his time preparing status reports instead of managing the project. The executives changed the project's requirements frequently, and the organization had no change control process other than the steering committee's approval.

At the end of the fourth month, the project manager resigned and sought employment outside the company. One of the executives from the steering committee then took over the project manager's role, but only on a part-time basis. Ultimately, the project was taken over by two more project managers before it was complete, one year later than planned. The company learned a vital lesson: More sponsorship is not necessarily better than less.

Contractco

Another disguised case involves a Kentucky-based company I'll call Contractco. Contractco is in the business of nuclear fusion testing. The company was in the process of bidding on a contract with the U.S. Department of Energy. The department required that the project manager be identified as part of the company's proposal and that a list of the project manager's duties and responsibilities be included. To impress the Department of Energy, the company assigned both the executive vice president and the vice president of engineering as cosponsors.

The Department of Energy questioned the idea of dual sponsorship. It was apparent to the department that the company did not understand the concept of project sponsorship, because the roles and responsibilities of the two sponsors appeared to overlap. The department also questioned the necessity of having the executive vice president serve as a sponsor.

The contract was eventually awarded to another company. Contractco learned that a company should never underestimate the customer's knowledge of project management or project sponsorship.

Health Care Associates

Health Care Associates (another pseudonym) provides health care management services to both large and small companies in New England. The company partners with a chain of 23 hospitals in New England. More than 600 physicians are part of the professional team, and many of the physicians also serve as line managers at the company's branch offices. The physician-managers maintain their own private clinical practices as well.

It was the company's practice to use boilerplate proposals prepared by the marketing department to solicit new business. If a client was seriously interested in Health Care Associates' services, a customized proposal based on the client's needs would be prepared. Typically, the custom-design process took as long as six months or even a full year.

Health Care Associates wanted to speed up the custom-design proposal process and decided to adopt project management processes to accomplish that goal. In January 1994, the company decided that it could get a step ahead of its competition if it assigned a physician-manager as the project sponsor for every new proposal. The rationale was that the clients would be favorably impressed.

The pilot project for this approach was Sinco Energy (another pseudonym), a Boston-based company with 8,600 employees working in 12 cities in New England. Health Care Associates promised Sinco that the health care package would be ready for implementation no later than June 1994.

The project was completed almost 60 days late and substantially over budget. Health Care Associates' senior managers privately interviewed each of the employees on the Sinco project to identify the cause of the project's failure. The employees had the following observations:

- Although the physicians had been given management training, they had a great deal of difficulty applying the principles of project management. As a result, the physicians ended up playing the role of invisible sponsor instead of actively participating in the project.
- Because they were practicing physicians, the physician sponsors were not fully committed to their role as project sponsors.
- Without strong sponsorship, there was no effective process in place to control scope creep.
- The physicians had not had authority over the line managers, who supplied the resources needed to complete a project successfully.

Health Care Associates' senior managers learned two lessons. First, not every manager is qualified to act as a project sponsor. Second, the project sponsors should be assigned on the basis of their ability to drive the project to success. Impressing the customer is not everything.

MULTIPLE CHOICE QUESTIONS

1. The person or group responsible for creation and maintenance of the project management culture is:
   A. Senior management
   B. Executive sponsor
   C. Middle management
   D. Project managers
   E. Project leaders
2. Senior management visibility should be accomplished by:
   A. Executive briefings at least twice a week
   B. Attendance at all of the project team meetings
   C. Formal memos to the team members
   D. Formal memos to the project manager
   E. Walk-the-halls management
3. The most critical phase for project management support is:
   A. Project planning
   B. Project execution
   C. Project closure
   D. Gate reviews only
   E. All of the above
4. Senior management support includes:
   A. Empowerment
   B. Decentralization of authority
   C. Decentralization of decision-making
   D. A hands-off approach
   E. All of the above
5. Robert Hershock believes that the best thing senior management can do after training the team is to:
   A. Verify the level 5 WBS plan
   B. Provide guidance for cost control

C. Handle all customer communications until the project begins to wind down
D. Prepare the format desired for status reporting
E. Get out of the way

6. Who should feel supported by senior management?
A. Project managers
B. Project team
C. Line managers
D. Functional employees
E. All of the above

7. All project-related problems should go to the sponsor before being brought to line managers.
A. True
B. False

8. The prime responsibility of the sponsor in the early stage of a project is to:
A. Provide functional employee assignments
B. Run the first few team meetings
C. Plan the project
D. Establish the project objectives
E. All of the above

9. Executives should avoid acting as a sponsor for more than one project at the same time.
A. True
B. False

10. On large, complex projects, the role of the sponsor exists to:
A. Assist in integrating functional activities
B. Dispel disruptive conflicts
C. Maintain strong customer relations
D. All of the above
E. B and C only

11. Concurrent engineering activities are best handled by:
A. An executive sponsor who has a vested interest in the project
B. An executive sponsor who has no vested interest in the project
C. Sponsorship at middle management
D. Committee sponsorship
E. None of the above

12. The role of the project sponsor can change based upon which life cycle phase the project is in.
A. True
B. False

13. For a long-term project:
A. The sponsor should be the same person for the entire project
B. Senior management should not be burdened with project sponsorship
C. A different sponsor can be assigned for each life cycle phase
D. The ultimate customer will function as the sponsor
E. None of the above

14. Which of the following is a function of the sponsorship committee at General Motors Powertrain Group?
A. Approve changes to the project management process
B. Ensure compliance with the project management process

C. Promote and share lessons learned across programs
D. Review the master plan
E. All of the above

15. Continuously going to the sponsor for help can result in:
A. Micromanagement by the sponsor
B. Appointment of a new project manager
C. An invisible sponsor
D. More frequent meetings
E. All of the above

16. For external projects, an executive sponsor is needed because the customer prefers to have a direct path of communication open to the contractor's executive management.
A. True
B. False

17. Good project sponsors are careful in deciding which problems require his or her guidance.
A. True
B. False

18. Which type of problem may be inappropriate for the sponsor to resolve?
A. Safety issues
B. Technical issues
C. Financial issues
D. Scheduling issues
E. Staffing issues

19. Without project sponsorship, problems are resolved:
A. In favor of the project manager
B. In favor of the line manager
C. In favor of the customer
D. By going up the line until a common superior is identified
E. None of the above

20. Strategic planning on projects is accomplished by:
A. The project sponsor
B. The steering committee
C. The project manager
D. The project manager with input from the sponsor
E. The customer

21. Executive sponsors are expected to make decisions for what is in the best interest of:
A. The sponsor
B. The project manager
C. The project
D. The customer
E. Both the customer and the project

22. A major problem facing executive sponsorship is a:
A. Lack of technical knowledge at the sponsor level
B. Lack of desire to empower the project team
C. Lack of desire to work with lower-ranking personnel
D. Desire to run the project themselves
E. Desire to change the project management methodology

23. The key to empowerment by the sponsor is:
    A. Confidence in the ability of the customer to identify problems quickly
    B. A desire not to get involved
    C. Trust
    D. All of the above
    E. A and B only
24. When an individual goes from a project sponsor to a project champion, the result is usually:
    A. Micromanagement by the executive
    B. Mistrust by the executive
    C. Lack of faith in the use of the methodology
    D. A crisis
    E. All of the above
25. The principal reason why project management fails to reach its full potential is:
    A. Lack of training for project managers
    B. No line management support
    C. Lack of executive buy-in
    D. All of the above
    E. B and C only

DISCUSSION QUESTIONS

1. Why must management support be visible to the entire team rather than just the project manager?
2. Under what conditions would sponsorship by middle management be more appropriate than sponsorship at the executive levels?
3. How does the role of the sponsor change as we go through the project life cycle?
4. Why is sponsorship by committee becoming more common?
5. What is an "invisible" sponsor?
6. How would sponsorship differ between a cooperative culture and a noncooperative culture?
7. Would an external customer ever have a say in who is assigned as the project sponsor?
8. For internal projects, how can sponsorship be impacted by organizational trust?
9. What must exist within an organization for executives to be willing to empower the project team?
10. Can an executive sponsor drive a project to failure as well as to success?

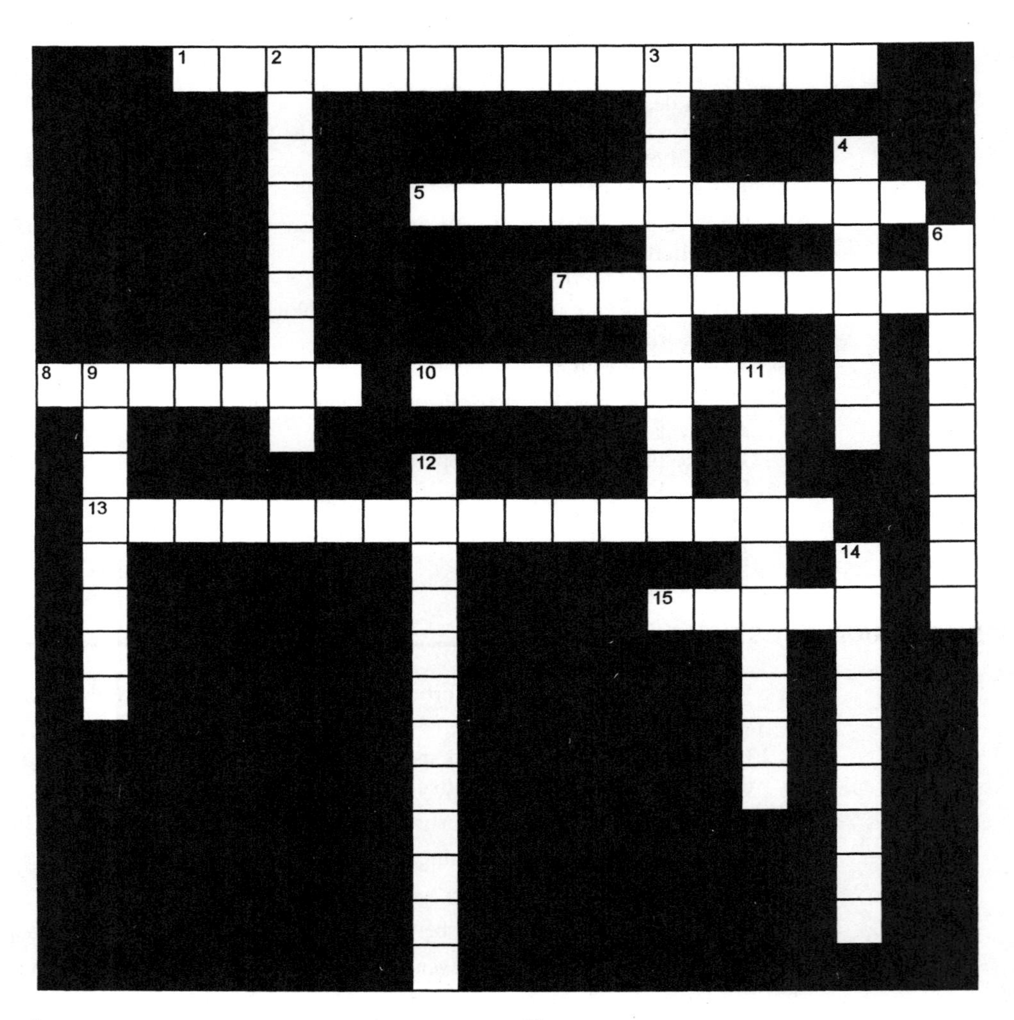

Across

1. Overinvolved
5. Executive ____________
7. Sponsorship: monitoring __________
8. Sponsorship: approving ________ to process
10. Sponsorship: _________ relations
13. ___________________ of authority
15. Cultural characteristic

Down

2. GM sponsorship
3. Form of delegation
4. ________ executive support
6. Type of sponsor: hard to find
9. Opposite of micromanagement
11. Sponsorship: conflict __________
12. Form of visibility (3 words, hyphenated)
14. Executive role: __________ planning

Training and Education

12.0 INTRODUCTION

Establishing project management *training* programs is one of the greatest challenges facing training directors because project management involves numerous complex and interrelated skills (qualitative/behavioral, organizational, and quantitative). In the early days of project management, project managers learned by their own mistakes rather than from the experience of others. Today, companies excellent in project management are offering a corporate curriculum in project management. Effective training supports project management as a *profession.*

Some large corporations offer more internal courses related to project management than most colleges and universities do. These companies include General Electric, General Motors, Kodak, the National Cryptological School, Ford Motor Company, and USAA. Such companies treat education almost as a religion. Smaller companies have more modest internal training programs and usually send their people to publicly offered training programs.

This chapter discusses processes for identifying the need for training, selecting the students who need training, designing and conducting the training, and measuring training's return on dollars invested.

Training managers around the world have now recognized the importance of project management training and education. According to Luis Bonilla, training manager for project management for the Petroleum Industry of Venezuela (PDVSA/CIED),

> All of Venezuela has come to realize the benefits of effective project management, from utilities to telecommunications and the oil industry. We at PDVSA/CIED are dedicated and committed to help lower cost, reduce schedule time at minimum risk, and maintain or improve our quality. We are accomplishing our goals through training and education. We have courses on basic and advanced project management, and in 1998 we began training our people to sit for the PMI Certification Exam.

We are currently performing strategic planning for project management education. At PDVSA/CIED, we are committed to becoming one of the world leaders in petroleum industry project management. To accomplish this, we are preparing educational needs assessments for the next five years.

Marcelino Diez, PMP, Project Management Consultant for Edelca, comments on his view for project management training:

At Edelca, the training for project management has been intensified during the last two years. There is an effort under way that all the personnel interacting with projects, direct or indirectly, be trained in the Foundations of Project Management (to be familiar with the processes described in the PMBOK® Guide) as well as with the Project Management Processes contained in the methodology developed for Edelca. Likewise, courses and workshops are offered for groups whose work is related to specific areas such as Contingency and Risks Management, Cost Estimating, Project Control, Contracting Strategies, and Value Improvement Practices, among others.

Edelca is committed to achieve stature as a national and regional leader with regard to excellence in project management. This will be accomplished by reaching progressive degrees of maturity and excellence. For this achievement to occur, Edelca is constantly in touch with all the available sources of information, such as participating and giving support to the PMI Venezuela Chapter, trying to maintain contact with the new trends in project management and with the world-class educators.

12.1 TRAINING FOR MODERN PROJECT MANAGEMENT

During the early days of project management, in the late 1950s and throughout the 1960s, training courses concentrated on the advantages and disadvantages of various organizational forms (e.g., matrix, traditional, functional). Executives learned quickly, however, that any organizational structure can be made to work effectively and efficiently when basic project management is applied. Project management skills based in trust, teamwork, cooperation, and communication can solve the worst structural problems.

Starting with the 1970s, emphasis turned away from organizational structures for project management. The old training programs were replaced with two basic programs:

- Basic project management, which stresses behavioral topics such as multiple reporting relationships, time management, leadership, conflict resolution, negotiation, team building, motivation, and basic management areas such as planning and controlling
- Advanced project management, which stresses scheduling techniques and software packages used for planning and controlling projects

Today's project management training programs include courses on behavioral as well as quantitative subjects. The most important problem facing training managers is how to achieve a workable balance between the two parts of the coursework—behavioral and quantitative (see Figure 12–1). For publicly sponsored training programs, the seminar leaders determine their own comfort levels in the "discretionary zone" between technical and behavioral subject matter. For in-house trainers, however, the balance must be preestablished by

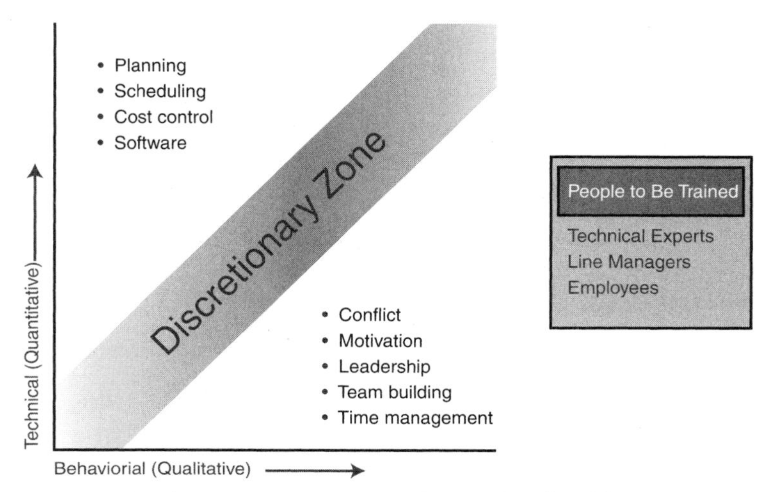

FIGURE 12–1. Types of project management training. *Source:* Reprinted from H. Kerzner, *In Search of Excellence in Project Management.* New York: Wiley, 1998, p. 174.

the training director on the basis of factors such as which students will be assigned to manage projects, types of projects, and average lengths of projects (see Table 12–1).

International Institute for Learning (IIL)

Given the importance of project management now, and in the future, there will certainly be a continuing need for high-quality project management education. E. LaVerne Johnson, President and CEO of the International Institute for Learning, comments on the growth of project management training[1]:

> Since launching International Institute for Learning over a decade ago, we have come in touch with literally thousands of companies, many of whom are leaders in their industry. Our business has become a barometer of trends in the marketplace. We have been in the unique position to witness first-hand the evolution of project management since the late 1980's. Our company has watched "project management" mature and grow in business importance. We have seen how training curriculums have evolved to keep pace with business needs. Training courses that were once sufficient a few years ago no longer adequately meet customers' needs. The global market and expanding importance of project management has dictated a whole new breed of course content and delivery mechanisms. Our barometer also allows us to anticipate some trends for the future.
>
> **The Evolutionary Years: Learning Trends**
>
> Training courses during the 1980s were mostly geared to advancing the project manager's skills. The focus of training was on the basics: the fundamental methodology and the know-how required to pass PMI's Project Management Professional (PMP) Certification

1. For more information regarding IIL, you can visit IIL Web site at *www.iil.com.*

TABLE 12–1. EMPHASES IN VARIOUS TRAINING PROGRAMS

Type of Person Assigned for PM Training (PM Source)	Training Program Emphasis	
	Quantitative/ Technology Skills	Behavior Skills
Training Needed to Function as a Project Manager		
• Technical expert on short-term projects	High	Low
• Technical expert on long-term projects	High	High
• Line manager acting as a part-time project manager	High	Low
• Line manager acting as a full-time project manager	High	Average to high
• Employees experienced in cooperative operations	High	Average to high
• Employees inexperienced in cooperative operations	High	Average to high
Training Needed for General Knowledge		
• Any employees or managers	Average	Average

Source: Reprinted from H. Kerzner, *In Search of Excellence in Project Management.* New York: Wiley, 1998, p. 175.

Exam. In response, IIL launched training courses in Project Management Fundamentals and established a comprehensive certification course that allowed individuals to prepare for and successfully pass PMI's PMP Exam. A small variety of books, traditional classroom courses and software products were made available to those individuals responsible for managing projects in their companies.

The Revolutionary Years: Marketplace Trends

In recent years, a far greater variety of companies and industries have recognized the value and business importance of managing projects more effectively. Compared to previous years, there is a revolution emerging. This has become evident in these trends:

- The volume of projects in companies is growing at a revolutionary rate. More and more companies are running their business via projects. Indeed some leading organizations undertake many hundreds of thousands of individual projects each year—some small and simple, others huge and complex.
- The ability to effectively manage projects has become critically important to business, and good project management skills have become a competitive advantage for leading companies.
- As a result of this revolutionary growth, the status and value of the PM professional has grown in importance—having this know-how allows a company to complete projects faster, at lower cost, with greater customer satisfaction and with more desirable project outcomes.
- Knowledge that was once deemed "nice to have" is now considered "mandatory."
- Today, employees with project management skills expand beyond the PM professional. Team members, middle and top management are developing expertise in the subject.
- The complexity and scope of project management methodologies grows to include new skills and new applications.

- A large number of PM-related software programs have been developed to help manage projects (such as Microsoft Project, Primavera and dozens of others).
- PM certification has become an even more valuable asset to an individual's career path. As a result, PMI membership grew to over 100,000 in 2003 and the number of registered PMP's increased to over 53,000 in that same year.
- Heretofore, the project manager's skill set has remained mostly at a technical level. But today we are seeing project managers embrace additional skills: human resources, communications, managerial, and other advanced areas of knowledge.
- Strategic planning for project management has taken on importance. Organizations are now seeking systematic ways to better align project management with business objectives.
- More and more companies are establishing project offices.
- Approaches to PM within an organization remain relatively varied and non-standardized. There is need to work toward a singular methodology in companies.

The Revolutionary Years: Learning Trends

In response to these trends, a far greater variety of courses are available to a broader and broader diversity of industries. New methods of learning have been introduced to meet a growing diversity of customer needs. Here are some examples of how IIL has responded to the revolution and established best practices in training and education for project managers:

- IIL has dozens of different course titles in "advanced" areas of knowledge to increase the scope, application and sophistication of the PM. Such courses include advanced concepts in risk management, the design and development of a project office, and how to manage multiple projects, just to name a few.
- Courses addressing the "softer" side of project management are now available to hone facilitation skills, interpersonal skills, leadership skills and other non-technical areas.
- The growth of project management software applications has demanded new training in how to effectively use these tools. IIL has a comprehensive four-level certification program for Microsoft Project 2002.
- More and more universities are recognizing project management as a part of their degree program. IIL has recently partnered with New York University to offer a Project Management Certificate Program (complete with university transcript and letter grade).
- The way we learn is changing. Employees have less time to devote to classroom study. As a result, in addition to traditional classroom training, IIL now offers innovative technology-based formats—Web-based "self-paced" training, distance learning via satellite broadcasts, "virtual" instructor-led courses, hands-on leadership simulation, CD-ROM programming and online mentoring. Even our "virtual" classroom courses are available nearly 24/7 to accommodate the busy professional's needs, budget and schedule.

A Look into the Crystal Ball: Trends and Learning Responses

It's always a challenge to try and predict the future, but there are some emerging trends that allow us to take a reasonable stab at this. For each of these trends, there will be the need to develop the appropriate learning responses.

- A key competitive factor in companies will be their ability to undertake and effectively manage many, many projects (projects to develop new products and services, to get to market faster, reduce costs, to improve customer satisfaction, to increase sales, and so on). The more well-chosen projects, the more competitive a company will become in the marketplace (particularly as it relates to projects to improve products, processes and customer satisfaction).

- We anticipate a blending of project management methodologies with other proven business strategies (such as quality management and risk management). Training in these subjects will similarly become blended.
- Project management will continue to grow in business importance—it will firmly become a strategic differentiating factor for remaining competitive.
- Senior management will become more knowledgeable and more involved in project management efforts. This will require project management training that meets the unique needs of executives.
- Strategic planning for project management will become a way of life for leading organizations. The role of the "project office" will grow in importance and its existence will become commonplace in companies. Membership will include the highest levels of executive management. Senior management will take leadership of the company's project management efforts.
- An obstacle to participation by upper managers will be their limited experience and training in project management. An essential element will be to provide upper managers with experience and training in how to manage project activities within their companies. This training must be tailored to be responsive to the unique needs and business responsibilities of upper management.
- The company's reward and recognition systems will change to stimulate and reinforce project management goals and objectives.
- Training in project management will be expanded to include all levels of the company hierarchy, including the non project management professional. Training will become responsive to the unique needs of this broad array of job functions, levels and responsibilities.
- The status of the PM professional will grow significantly, and the project manager's skills will be both technical and managerial.
- We will witness the establishment of a corporate level project management executive (Chief Project Management Officer).
- Project benchmarking and continuous project improvement will become a way-of-life in leading organizations. The project management maturity models will play an important role in this regard, as they will help companies identify their strengths, weaknesses and specific opportunities for improvement.
- The expanding importance of project management will require that more individuals are trained in project management. This in turn will necessitate the development of new and improved methods of delivery. It's likely that Web-based training will play an ever-increasingly important role.
- We will see an order of magnitude increase in the number of organizations reaching the higher levels of project management maturity.
- More and more colleges and universities will offer degree programs in project management.

The School of Project Management (SPM)

Today, project management training and education is a worldwide endeavor. Graduate-level education exists at colleges and universities throughout the world. In addition, during the past 10 years or so, many private organizations have started up with the sole objective of providing high-quality project management education suited to both undergraduate and graduate levels of education. Most of these companies start out with a dream and then cultivate the dream into a story of success. One such success story is the School of Project Management in South Africa. According to Dr. Lionel Smalley, President and CEO[2]:

> The School of Project Management (SPM) was formed in March 1989 by Dr. Lionel Smalley to promote project management training in non–project-driven enterprises primarily in South Africa. From 1980 to 1989, there were only a few project management training courses in South Africa, and these courses appeared to make three fundamental mistakes. First, the project management course was offered by companies that also promoted a multitude of other non-related courses, and very little effort was provided to insert quality into the project management course because there appeared to be some concern that project management was just a fad that would soon disappear. Second, the course addressed only the concerns of the project manager rather than the concerns of all of the team players. And third, most training programs at that time focused on how projects should be managed and controlled in mainly project-driven enterprises. The decision to direct attention to product and service enterprises, as well as project-driven enterprises, was based upon the completed research conducted by Dr. Smalley for his doctoral thesis completed in 1986 and entitled "Project Management in South Africa." The results showed that there was an urgent need for modern-day project management principles and techniques to be applied to all industries, regardless of the type, and that the training programs had to be custom-designed for all types of team players rather than just for the project managers. The research also showed that there was a growing need for graduate level education in project management and that the original belief that project management was simply a fad soon to disappear was incorrect.
>
> The SPM was started up with the objective of providing only high quality project management education for the South African community. The SPM's approach to project management training is focused on the management of projects in association with the use and application of integrated computer-based software packages such as those that would be part of a company's intranet project management system and the core for the development of a company-wide project management methodology. Training is not purely a techniques-based affair. Specialized training is provided for project managers, project team members, and project executives.
>
> For this to become a reality, the SPM had to develop a set of core competencies to satisfy the project management needs of the business community. The core competencies developed by SPM include:
>
> - *Corporate Strategy:* showing how the environment affects project performance; cost management; and trade-off analysis.
> - *Organizational Behavior:* illustrating the importance of inter-personal relationships; leadership skills; managing conflict situations: and managing cultural diversity.
> - *Project Finance:* showing how to effectively participate in feasibility studies by making decisions based upon IRR values, NPV values, profitability indices, determining payback periods, and the management of risks. Far too many project managers today still are brought on board the projects during the detailed planning stage rather than during the feasibility study phase. This is mainly due to a lack of project finance knowledge.

2. For more information regarding SPM, you can visit the SPM Web site at *www.spm.co.za*.

- *Project Negotiations:* showing how to effectively negotiate for the required resources and/or deliverables. This is extremely important at all management levels and is often downplayed in its importance.
- *Decision-Support Systems:* using linear and goal programming models to make the "right" decision in order to optimize the required outcome.

The selection of course material, whether it is a new course or enhancements to existing courses, is not a random process. The SPM is in constant communication with industry leaders and executives in both the public and private sectors to make sure that our offerings satisfy their needs. When they tell us that their needs have changed, such as the result of newly discovered best practices, we immediate include this material into our programs. The SPM also custom-designs courses to satisfy the needs of particular clients. As best practices rise to the surface, so will the changes to our offerings.

12.2 IDENTIFYING THE NEED FOR TRAINING

Identifying the need for training requires that line managers and senior managers recognize two critical factors: first, that training is one of the fastest ways to build project management knowledge in a company and, second, that training should be conducted for the benefit of the corporate bottom line through enhanced efficiency and effectiveness.

Identifying the need for training has become somewhat easier in the past ten years because of published case studies on the benefits of project management training. The benefits can be classified according to quantitative and qualitative benefits. The quantitative results include:

- Shorter product development time
- Faster, higher quality decisions
- Lower costs
- Higher profit margins
- Fewer people needed
- Reduction in paperwork
- Improved quality and reliability
- Lower turnover of personnel
- Quicker "best practices" implementation

Qualitative results include:

- Better visibility and focus on results
- Better coordination
- Higher morale
- Accelerated development of managers
- Better control
- Better customer relations

- Better functional support
- Fewer conflicts requiring senior management involvement

Companies are finally realizing that the speed by which the benefits of project management can be achieved is accelerated through proper training.

12.3 SELECTING THE STUDENTS

Selecting the people to be trained is critical. As we've already seen in a number of case studies, it's usually a mistake to train only the project managers. A thorough understanding of project management and project management skills is needed throughout the organization if project management is to be successful. For example, one automobile subcontractor invested months in training its project managers. Six months later, projects were still coming in late and over budget. The executive vice president finally realized that project management was a team effort rather than an individual responsibility. After that revelation, training was provided for all of the employees who had anything to do with the projects. Virtually overnight, project results improved.

Dave Kandt, executive director of worldwide operations at Johnson Controls, explained how his company's training plan was laid out to achieve excellence in project management:

> We began with our executive office, and once we had explained the principles and philosophies of project management to these people, we moved to the management of plants, engineering managers, cost analysts, purchasing people, and, of course, project managers. Only once the foundation was laid did we proceed with actual project management and with defining the roles and responsibilities so that the entire company would understand their role in project management once these people began to work. Just the understanding allowed us to move to a matrix organization and eventually to a stand-alone project management department.

12.4 FUNDAMENTALS OF PROJECT MANAGEMENT EDUCATION

Twenty years ago, we were somewhat limited as to availability of project management training and education. Emphasis surrounded on-the-job training in hopes that fewer mistakes would be made. Today, we have other types of programs including:

- University courses
- University seminars
- In-house seminars
- In-house curriculums
- Distance learning (E-learning)
- Computer-based training (CBT)

With the quantity of literature available today, we have numerous ways to deliver the knowledge. Typical delivery systems include:

- Lectures
- Lectures with discussion
- Exams
- Case studies on external companies
- Case studies on internal projects
- Simulation and role-playing

Training managers are currently experimenting with "when to train." The most common choices include:

- *Just-in-Time Training:* This includes training employees immediately prior to assigning them to projects.
- *Exposure Training:* This includes training employees on the core principles just to give them enough knowledge so that they will understand what is happening in project management within the firm.
- *Continuous Learning:* This is training first on basic, then on advanced, topics so that people will continue to grow and mature in project management.
- *Self-Confidence Training:* This is similar to continuous learning but on current state-of-the-art knowledge. This is to reinforce employees' belief that their skills are comparable to those in companies with excellent reputations for project management.

12.5 DESIGNING THE COURSES AND CONDUCTING THE TRAINING

Many companies have come to realize that on-the-job training may be less effective than more formal training. On-the-job training virtually forces people to make mistakes as a learning experience, but what are they learning? How to make mistakes. It seems much more efficient to train people to do their jobs the right way from the start. Table 12–2 shows an abbreviated list of the courses offered by USAA, Ford Motors, and Armstrong World Industries.

Project management has become a career path. More and more companies today allow or even require that their employees get project management certification. One company informed its employees that project management certification would be treated the same as a master's degree in the salary and career-path structure. The cost of the training behind the certification process is only 5 or 10 percent of the cost of a typical master's degree in business administration program. And certification promises a quicker return on investment for the company. Project management certification can also be useful for employees without college degrees; it gives them the opportunity for a second career path within the company.

Linda Kretz of the International Institute for Learning explained what type of project management training worked the best in her experience:

> In our experience, we have found that training them ahead of time is definitely the better route to go. We have done it the other way with people learning on the job, and that has

TABLE 12–2. TYPICAL IN-HOUSE PROJECT MANAGEMENT TRAINING PROGRAMS

USAA[a]		
Building Effective Teams	Project Management Body of Knowledge (PMBOK)® Review	Leadership Training
Risk Management	Communicating Change	Resolving Team Conflicts
The Complete Project Manager	Strategic Project Management	Negotiation
Communicating One-to-One	Managing Software Projects	Applying Project Workbench
Facilitation for Leaders	Business Orientation Assignment	Microsoft Project
Coaching for Performance	Planning, Scheduling, Cost	Business Process Reengineering
Giving a Briefing	ITAP and Investment Workshop	Seven Steps to Change
Helping Your Team Reach Consensus	Winning Support from Others	Multiproject Management
Making the Most of Team Differences	Building Collaborative Relationships	Managing Client/Server Projects
Outcome Thinking	Effective Writing	Interpersonal Skills
Advanced Project Workbench	Building a Foundation of Trust	
FORD		
Benchmarking	Models for Management	Zero Defects
Microsoft Project	Collaborative Decision Making	Achieving Win–Win Outcomes
Writing on Target	Working in a Multicultural Organization	Outstanding Meetings
ISO 9000	On-the-Job Training	Successful Work Teams
Quick Quality Function Deployment	Waste Minimization	Coaching and Counseling
Statistic Process Control	Financial Engineering	Team-Oriented Problem Solving
Communication Skills	Theory of Constraints	Working in a Matrix Organization
Effective Listening and Feedback	Failure Mode and Effects Analysis	Facilitation Skills
Shifting the Conflict Paradigm	Quality Operating Systems	Time Management
Basic Supervisory Knowledge	Reliability	
ARMSTRONG WORLD INDUSTRIES		
Project Management	Presentation Skills	Listening, Influencing and Handling Tough Situations
Mastering the Marketing Process	Facilitative Fundamentals	Leading Successful Project Team
Understanding the Total Business	Performance Appraisals	Budgeting Fundamentals
Mini Tab	Statistical Process Control	
The Enabling Leader	Technical Leadership	
	Communicating Better with Your Team	

[a]These were the courses offered during the 1990s; see H. Kerzner, *Applied Project Management: Best Practices on Implementation*. New York: Wiley, 2000, p. 261.

been a rather terrifying situation at times. When we talk about training, we are not just talking about training. We want our project managers to be certified through the Project Management Institute. We have given our people two years to certify. To that end there is quite a bit of personal study required. I do believe that training from the formal training end is great, and then you can modify that to whatever the need is in-house.

Rose Russett formerly of General Motors Powertrain, described what she believes is working well for her group:

We have a single, dedicated program manager on each of our product programs. This provides a focal point of responsibility and accountability for the entire program. These individuals are chosen for their leadership skills and ability to work within our cross-functional team structure. These program managers report to an executive sponsor committee of three directors who are committed to and fully supportive of program management. When the program manager and team are appointed, they are given training to understand the program organizational structure, roles and responsibilities of the team, and common systems and processes that will be used throughout the life of the program. This starts the entire team out with full knowledge of how work will get done and should minimize potential ambiguity and confusion that could result when working in a matrix organizational structure.

As companies master the quantitative side of project management, the emphasis in training shifts to behavioral skills. According to Brian Vannoni, formerly of General Electric Plastics,

> The behaviors that we've been driving for some time in our organizations we generally refer to as boundaryless. We are looking at driving the behaviors across the organization such that people interact and become very comfortable with matrixing with people from any of the functions. Just as the degrees of project success have grown, so also has the need for interacting with more functional and business types of situations. So, we are very broad functionally and also globally.
>
> Over the years, we have been pursuing this boundaryless behavior type of environment for each of our individuals. And over the past several years we have developed and delivered close to 50 different modules, different classes and seminars, all focused around behaviors, whether it is team skills, training and facilitation, training and process mapping, or some types based off a Zenger Miller type class. We've also based them on the Hershey Blanchard situational leadership model.

Bill Marshall, formerly director of Project Management Standards for Nortel, discussed the complexities of getting a uniform methodology in place for project management when a company wishes to become a global competitor:

> As our process evolves and the methods require a new culture, we are using training courses to bring about an understanding of this new culture. The International Institute of Learning is providing the basic project management training and the introduction into the new culture. These courses are being presented to the multiple global business segments headquartered in Richardson, Texas. Regions of the UK, Europe, Australia, and Central and Latin America are registered for courses. The project managers are also receiving updated training on Project Schedulers and receiving internal training on financial management of projects and a new contract ledger. They also receive instructions on contractual language and are receiving personal development training, preparing them for holding effective meetings and dealing with difficult situations. The training of the future will be more closely aligned with the project management tools and, in fact, may be more CBT (Computer Base Training) focused, with on-line simulations of a real project. The process driven navigation tools associated with the customized desk top for a project manager may contain training models that can be exercised in the project manager's spare time.

There is also the question of which are better: internally based or publicly held training programs. The answer depends on the nature of the individual company and how many employees need to be trained, how big the training budget is, and how deep the company's internal knowledge base is. If only a few employees at a time need training, it might be effective to send them to a publicly sponsored training course, but if large numbers of employees need training on an ongoing basis, designing and conducting a customized internal training program might be the way to go.

In general, custom-designed courses are the most effective. In excellent companies, course content surveys are conducted at all levels of management. For example, the research and development group of Babcock and Wilcox in Alliance, Ohio, needed a project management training program for 200 engineers. The head of the training department knew that she was not qualified to select core content, and so she sent questionnaires out to executive managers, line managers, and professionals in the organization. The information from the questionnaires was used to develop three separate courses for the audience. At Ford Motor Company, training was broken down into a two-hour session for executives, a three-day program for project personnel, and a half-day session for overhead personnel.

For internal training courses, choosing the right trainers and speakers is crucial. A company can use trainers currently on staff if they have a solid knowledge of project management, or the trainers can be trained by outside consultants who offer train-the-trainer programs. Either way, trainers from within the company must have the expertise the company needs. Some problems with using internal trainers include the following:

- Internal trainers may not be experienced in all the areas of project management.
- Internal trainers may not have up-to-date knowledge of the project management techniques practiced by other companies.
- Internal trainers may have other responsibilities in the company and so may not have adequate time for preparation.
- Internal trainers may not be as dedicated to project management or as skillful as external trainers.

But the knowledge base of internal trainers can be augmented by outside trainers as necessary. In fact, most companies use external speakers and trainers for their internal educational offerings. The best way to select speakers is to seek out recommendations from training directors in other companies and teachers of university-level courses in project management. Another method is contacting speakers' bureaus, but the quality of the speaker's program may not be a high as needed. The most common method for finding speakers is reviewing the brochures of publicly sponsored seminars. Of course, the brochures were created as sales materials, and so the best way to evaluate the seminars is to attend them.

After a potential speaker has been selected, the next step is to check his or her recommendations. Table 12–3 outlines many of the pitfalls involved in choosing speakers for internal training programs and how you can avoid them.

The final step is to evaluate the training materials and presentation the external trainer will use in the classes. The following questions can serve as a checklist:

- Does the speaker use a lot of slides in his or her presentation? Slides can be a problem when students don't have enough light to take notes.
- Does the instructor use transparencies? Have they been prepared professionally? Will the students be given copies of the transparencies?

TABLE 12–3. COMMON PITFALLS IN HIRING EXTERNAL TRAINERS AND SPEAKERS

Warning Sign	Preventive Step
Speaker professes to be an expert in several different areas.	Verify speaker's credentials. Very few people are experts in several areas. Talk to other companies that have used the speaker.
Speaker's résumé identifies several well-known and highly regarded client organizations.	See whether the speaker has done consulting for any of these companies more than once. Sometimes a speaker does a good job selling himself or herself the first time, but the company refuses to rehire him or her after viewing the first presentation.
Speaker makes a very dramatic first impression and sells himself or herself well. Brief classroom observation confirms your impression.	Being a dynamic speaker does not guarantee that quality information will be presented. Some speakers are so dynamic that the trainees do not realize until too late that "The guy was nice but the information was marginal."
Speaker's résumé shows 10 to 20 years or more experience as a project manager.	Ten to 20 years of experience in a specific industry or company does not mean that the speaker's knowledge is transferable to your company's specific needs or industry. Ask the speaker what types of projects he or she has managed.
Marketing personnel from the speaker's company aggressively show the quality of their company, rather than the quality of the speaker. The client list presented is the company's client list.	You are hiring the speaker, not the marketing representative. Ask to speak or meet with the speaker personally and look at the speaker's client list rather than the parent company's client list.
Speaker promises to custom-design his or her materials to your company's needs.	Demand to see the speaker's custom-designed material at least two weeks before the training program. Also verify the quality and professionalism of view graphs and other materials.

- Does the speaker make heavy use of chalkboards? Too much chalkboard work usually means too much note-taking for the trainees and not enough audiovisual preparation from the speaker.
- Does the speaker use case studies? If he or she does, are the case studies factual? It's best for the company to develop its own case studies and ask the speaker to use those so that the case will have relevance to the company's business.
- Are role playing and laboratory experiences planned? They can be valuable aids to learning, but they can also limit class size.
- Are homework and required reading a part of the class? If so, can they be completed before the seminar?

12.6 MEASURING THE RETURN ON INVESTMENT

The last area of project management training is the determination of the value earned on the dollars invested in training. It is crucial to remember that training shouldn't be performed unless there is a continuous return on dollars for the company. Keep in mind, also, that the speaker's fee is only part of the cost of training. The cost to the company of having employees away from their work during training must be included in the calculation.

Some excellent companies hire outside consultants to determine return on investment (ROI). The consultants base their evaluations on personal interviews, on-the-job assessments, and written surveys.

One company tests trainees before and after training to learn how much knowledge the trainees really gained. Another company hires outside consultants to prepare and interpret post-training surveys on the value of the specific training received.

The amount of training needed at any one company depends on two factors: whether the company is project-driven and whether it has practiced project management long enough to have developed a mature project management system. Figure 12–2 shows the amount of project management training offered (including refresher courses) against the number of years in project management. Project-driven organizations offer the most project management training courses, and organizations that have just started implementing project management offer the fewest. That's no surprise. Companies with more than 15 years of experience in applying project management principles show the most variance. (Take another look at Table 1–1 for an explanation of the types of industries.)

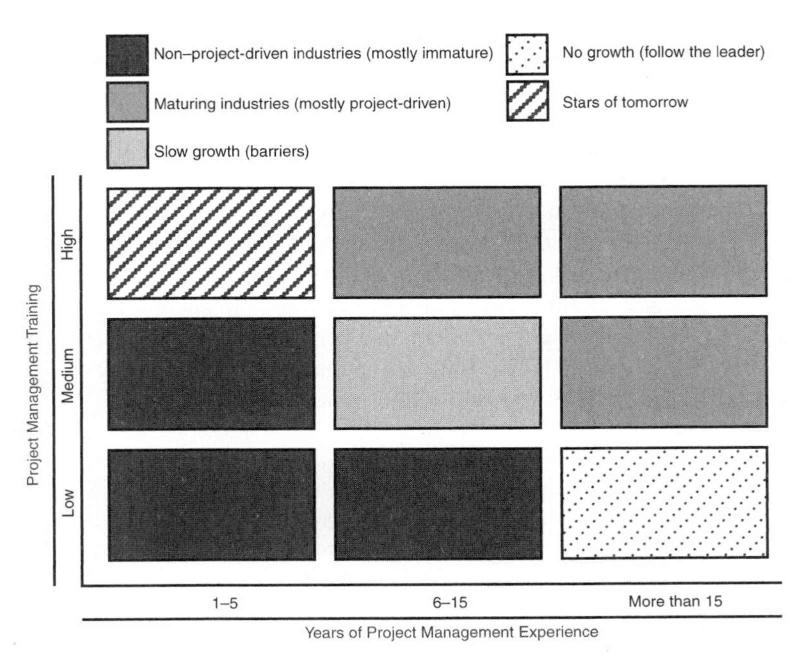

FIGURE 12–2. Amount of training by type of industry and year of project management experience. *Source:* Reprinted from H. Kerzner, *In Search of Excellence in Project Management.* New York: Wiley, 1998, p. 185.

Since measuring ROI is no easy task, companies develop their own methodologies for measurement. Randy Maxwell, formerly of Employee Organization Learning at Nortel Networks comments on his experience at Nortel during the 1990s:

> Traditionally, Nortel Networks has measured return-on-investment in training by comparing the cost benefits to the organization and training costs. Training costs include the participants' time attending courses in addition to the actual training expense. Project management training benefits include project improvements in scheduling and product costs. This measurement of return-on-investment in training is expensive and time consuming. Therefore, measures of ROI are performed on samples of learning interventions.
>
> Nortel Networks Employee and Organization Learning is now focused on strategic learning that is aligned to business unit requirements. Project Management Learning benefits are focused on the list of projects in each business unit. Projects are selected using portfolio methods linked to business strategy. The selected projects are managed with project leaders in integrated cross-functional teams that are accountable for end-to-end project implementation. Customers are involved at the front end and project charters are established. The business decision for a project is continually reviewed throughout the project life cycle. Training benefits are measured by how they contribute to organizational effectiveness, such as canceling a project when it does not provide a business impact. Training in project management is conducted frequently as it is needed in the project phases using action learning techniques.

12.7 COMPETENCY MODELS

Twenty years ago, companies prepared job descriptions for project managers to explain roles and responsibilities. Unfortunately, the job descriptions were usually abbreviated and provided very little guidance on what was required for promotion or salary increases. Ten years ago we still emphasized the job description, but it was now supported by coursework, which was often mandatory. By the late 1990s, companies began emphasizing core competency models, which clearly depicted the skills levels needed to be effective as a project manager. Training programs were instituted to support the core competency models.

Unfortunately, establishing a core competency model and the accompanying training is no easy task. Lynne Hambleton, manager for learning systems at Xerox Corporation, describes the problems and solutions at Xerox:

> Establishing project management as a core organizational competency requires more than designing a comprehensive curriculum. The learning continuum represents only a small part of the effort required for a company the size of Xerox to embrace project management. More importantly, a community around project management needs to be nurtured. This community includes both the "professional" project managers and other organizational members (e.g., management and operational or functional members).
>
> Xerox attempted to embrace project management as a key principle twice before; however, we saw room for improvement both times. The first attempt we had a strong methodology but lacked in the organization/sponsorship and learning continuum. The second attempt had a mediocre methodology and learning continuum.

> People in the workplace tend to form a "community of practice" to collectively address customer needs using best practices and a common language. Human behavior shapes work practices and procedures, creating a unique approach for each community. To improve or assimilate new information, tools, and technology into a community, we must recognize and address the way people work. Xerox's challenge is to guide, influence, and nurture the community to deliver its business goals while still promoting creativity, flexibility, and fluidity within the community.
>
> Xerox has learned from its past mistakes and is trying to build a project management community. The division president appointed an organization to champion the effort. This organization embraces learning principles that are social and collaborative. These learning principles are integrated with traditional program office responsibilities. The organization also addressed the following areas: strategy/vision, organization and sponsorship, methodology, and tools.
>
> The project management learning continuum encompassed both the organizational needs as well as four different project management positions. The breadth of the community requires a multitude of vendors (internal and external) and formats to make the learning continuum effective (e.g., CD ROM, board games, classroom lecture, workshops, certification).

Perhaps the two best competency models known to the author are Ericsson Telecom AB and Eli Lilly.

Ericsson

Ericsson has been pioneering the development of a competency model for project managers. The purpose of the competency model is to develop a set of clear expectations for both project managers and managers of project managers. The model will assist in the hiring, training, development plans, and coaching of project managers. Ericsson expects the competency model to be a valuable tool in establishing a uniform set of skills for all Ericsson's 100,000 employees in 140 countries.

The competency model consists of four levels, A, B, C, and D, each of which represents a certain level of competence. A is the lowest and D the highest. In addition, there is a level T, for training, meaning the project manager has yet not demonstrated that competence.

Level A This is the level that most project managers will begin at after having had the basic professional and technical training. The key term at this level is *able to perform.* The following descriptive sentences serve as a guideline for the performance at level A in any given competence.

The project manager:

- Is familiar with the applicable methods and procedures of the unit.
- Has participated in a project in the role of project manager.
- Is familiar with the area within which the project is or was performed.

Level B To reach this level in any given competence, the project manager has demonstrated a deeper understanding of the area. He or she has also probably functioned as a mentor in some parts for a new project manager. The key term for level B is *understand*

and share with others. This list provides some guidelines for behavior that demonstrates a level B project manager.

The project manager:

- Has been able to plan, budget for, organize, and monitor a project from start to end.
- Has performed the duties and tasks of the project manager effectively and with approved quality.
- Has kept all project stakeholders informed throughout the project.
- Has built up her or his own contact network, which contributed to the success of the project.
- Has demonstrated leadership qualities and competencies, as well as communication skills.
- Has successfully completed the project, that is, has ensured that lessons learned were shared with the project participants and stakeholders.
- Has supported new project managers in areas where the B level project manager is strong.

Level C The key term for level C is *develop.* The project manager has taken the competencies to a higher level by further developing techniques and methods that reflect a deep understanding and insight into the competencies. Examples of such behavior are instances when the project manager:

- Has been able to interpret customer requirements, both explicit and implicit, and has been able to formulate goals and objectives based on these requirements.
- Has run multiple projects and managed multiple priorities using tools and methods that may or may not have been obvious for the task, such as containing an element of development.
- Has solved problems and negotiated win-win results with the customer.
- Has developed programs to modify or improve methods, tools, and/or learning tools.

Level D Level D in any given competence is the level at which the project manager creates new processes, methods, and tools in areas where none was available. Key term for level D competency is *create.* Examples of such creativity are when the project manager:

- Has run a very large project/program of a magnitude such that there were no clear processes or methods defined.
- Has taken total financial responsibility for the program or project.
- Has a wide circle of international contacts that he or she has drawn upon to accomplish results above and beyond the normal.
- Has documented experience from several functional areas, as well as business segments (business areas) within Ericsson. This experience should, of course, have resulted in solutions outside the normal for the area in which he or she works.
- Has well-documented, exceptional leadership skills and knowledge.

The competency model is a natural follow-on to Ericsson's PROPS (PM4U) model discussed in Chapter 6. The geographical distribution of the Ericsson organization has always required some kind of project management structure. Every development and implementation of a large telecommunications system is a tremendous undertaking. The project manager has been a formally recognized function since the mid-1970s in the product development area at Ericsson. At that time, projects were spread across units in Sweden, Ireland, the United States, Australia, and many other places.

Another type of project manager was needed during the early 1980s when quality and other improvement projects were started. The concept of project management was spread into the development of other types of products beside the pure communications systems. Training development, service development, and improvement projects were all run by project managers using essentially the same skills as the "old" product development project managers. It was during this time that emphasis was put on formal project management training and methods.

The PROPS model/method was developed during the mid-1980s. This method emphasizes a common language, common decision points (tollgates), and a formal acknowledgment of project sponsors and resource owners. The model was developed as a result of a need in communications product development. This model was soon adapted to many other applications outside the pure communications systems development.

Finally, during the late 1980s and early 1990s, a focus was placed on commercial project management, where project managers were leading the implementation of Ericsson's large contracts with the customers. This has increased the demand on both formal training and competence management. Two programs have been implemented recently to ensure the quality and function of commercial project management.

The first program, which was developed in the United States during 1995–1996, is a certification program of commercial and industrialization project managers. This program has a mandatory training part and a performance part where an auditor assesses the project manager. The objective is to ensure quality projects and that the project managers work according to our processes.

The second program is a competence management initiative that started in early 1996. The idea is to identify strategically important competencies, assess our current status, and plan for actions to fill competence gaps. We developed a competence tool with which the project managers' current competencies were mapped. The managers assessed their project managers in the following 16 categories or subcompetencies: personal minimum skills to perform the assigned job position; planning; management of project; risk management; process awareness and usage; closing of project; product knowledge; leadership; communication; negotiation; conflict management; finances; language; Ericsson; customer focus; and politics. The results of this assessment are that we will need to focus competence development in the risk management area for 1997.

As a result of the preparation and groundwork laid, Ericsson has now moved into a "closed-loop" system for managing projects. Currently, Ericsson uses the PROPS (PM4U) model, Ericsson Global Career guide, competency model, and common metrics to drive projects to successful conclusion. By using the tools and processes discussed above and selecting the right project managers for the scope of work to be performed, Ericsson can assure a successful outcome by working a consistent, managed, and measured process.

Currently, Ericsson offers the following courses to support project management:

- Basic Project Management (3 days)
- PMP Mandatory Certification for Senior PM and PM Management

- PROPS (PM4U): An Introduction (1 day)
- Core Three Introduction—On Line CBT
- Leadership Skills (2 days)
- Leadership Effectiveness (5 days)
- PM Masters Programs affiliated with local colleges (GWU, UTD)
- Leadership Excellence (5 days)
- Negotiating for Win-Win (2 days)
- Presentation Skills (2 days)
- Effective Communication Skills (2 days)
- Other professional development courses

Ericsson's competency model should make it easier to pinpoint training needs for individual employees. This will accelerate Ericsson's obtaining their goal of global project management excellence by having a common language that provides a common understanding of progress and work to be done. With the developments of 2001 and 2002, the goal is in sight.

Eli Lilly

Eli Lilly has perhaps one of the most comprehensive and effective competency models in industry today. Martin D. Hynes, III, Ph.D., director, Pharmaceutical Projects Management (PPM), was the key sponsor of the initiative to develop the competency model. Thomas J. Konechnik, operations manager, Pharmaceutical Projects Management, was responsible for the implementation and integration of the competency model with other processes within the PPM group. The basis for the competency model is described below.

Lilly Research Laboratories project management competencies are classified under three major areas:

Scientific/Technical Expertise

- *Knows the Business:* Brings an understanding of the drug development process and organizational realities to bear on decisions.
- *Initiates Action:* Takes proactive steps to address needs or problems before the situation requires it.
- *Thinks Critically:* Seeks facts, data, or expert opinion to guide a decision or course of action.
- *Manages Risks:* Anticipates and allows for changes in priorities, schedules, resources, and changes due to scientific/technical issues.

Process Skills

- *Communicates Clearly:* Listens well and provides information that is easily understood and useful to others.
- *Attention to Details:* Maintains complete and detailed records of plans, meeting minutes, agreements.
- *Structures the Process:* Constructs, adapts, or follows a logical process to ensure achievement of objectives and goals.

Leadership

- *Focuses on Results:* Continually focuses own and others' attention on realistic milestones and deliverables.
- *Builds a Team:* Creates an environment of cooperation and mutual accountability within and across functions to achieve common objectives.
- *Manages Complexity:* Organizes, plans, and monitors multiple activities, people, and resources.
- *Makes Tough Decisions:* Demonstrates assurance in own abilities, judgments, and capabilities; assumes accountability for actions.
- *Builds Strategic Support:* Gets the support and level of effort needed from senior management and others to keep project on track.

We examine each of these competencies in more detail below.

1. *Knows the Business:* Brings an understanding of the drug development process and organizational realities to bear on decisions.

   Project managers/associates who demonstrate this competency will:

   - Recognize how other functions in Eli Lilly impact the success of a development effort.
   - Use knowledge of what activities are taking place in the project as a whole to establish credibility.
   - Know when team members in own and other functions will need additional support to complete an assignment/activity.
   - Generate questions based on understanding of nonobvious interactions of different parts of the project.
   - Focus attention on the issues and assumptions that have the greatest impact on the success of a particular project activity or task.
   - Understand/recognize political issues/structures of the organization.
   - Use understanding of competing functional and business priorities to reality test project plans, assumptions, time estimates, and commitments from the functions.
   - Pinpoint consequences to the project of decisions and events in other parts of the organization.
   - Recognize and respond to the different perspectives and operating realities of different parts of the organization.
   - Consider the long-term implications (pro and con) of decisions.
   - Understand the financial implications of different choices.

   Project managers/associates who do *not* demonstrate this competency will:

   - Rely on resource and time estimates from those responsible for an activity or task.
   - Make decisions based on what ideally should happen.

- Build plans and timelines by rolling up individual timelines, and so on.
- Perceive delays as conscious acts on the part of other parts of the organization.
- Assume that team members understand how their activities impact the other parts of the project.
- Focus attention on providing accurate accounts of what has happened.
- Avoid changing plans until forced to do so.
- Wait for team members to ask for assistance.

Selected consequences for projects/business of not demonstrating this competency:

- Project manager or associate may rely on senior management to resolve issues and obtain resources.
- Proposed project timelines may be significantly reworked to meet QSV guidelines.
- Attention may be focused on secondary issues rather than central business or technical issues.
- Current commitments, suppliers, and so on, may be continued regardless of reliability and value.
- Project deliverables may be compromised by changes in other parts of Lilly.
- Project plans may have adverse impact on other parts of the organization.

2. *Initiates Action:* Takes proactive steps to address needs or problems before the situation requires it.

Project managers/associates who demonstrate this competency will:

- Follow up immediately when unanticipated events occur.
- Push for immediate action to resolve issues and make choices.
- Frame decisions and options for project team, not simply facilitate discussions.
- Take on responsibility for dealing with issues for which no one else is taking responsibility.
- Formulate proposals and action plans when a need or gap is identified.
- Quickly surface and raise issues with project team and others.
- Let others know early on when issues have major implications for project.
- Take action to ensure that relevant players are included by others in critical processes or discussions.

Project managers/associates who do *not* demonstrate this competency will:

- Focus efforts on ensuring that all sides of issues are explored.
- Ask others to formulate initial responses or plans to issues or emerging events.

- Let functional areas resolve resource issues on their own.
- Raise difficult issues or potential problems after their impact is fully understood.
- Avoid interfering or intervening in areas outside own area of expertise.
- Assume team members and others will respond as soon as they can.
- Defer to more experienced team members on how to handle an issue.

Selected consequences for projects/business of not demonstrating this competency:

- Senior management may be surprised by project related events.
- Project activities may be delayed due to "miscommunications" or to waiting for functions to respond.
- Effort and resources may be wasted or underutilized.
- Multiple approaches may be pursued in parallel.
- Difficult issues may be left unresolved.

3. *Thinks Critically:* Seeks facts, data, or expert opinion to guide a decision or course of action.

   Project managers/associates who demonstrate this competency will:

   - Seek input from people with expertise or first-hand knowledge of issues, and so on.
   - Ask tough, incisive questions to clarify time estimates or to challenge assumptions, and be able to understand the answers.
   - Immerse self in project information to quickly gain a thorough understanding of a project's status and key issues.
   - Focus attention on key assumptions and root causes when problems or issues arise.
   - Quickly and succinctly summarize lengthy discussions.
   - Gather data on past projects, and so on, to help determine best future options for a project.
   - Push to get sufficient facts and data in order to make a sound judgment.
   - Assimilate large volumes of information from many different sources.
   - Use formal decision tools when appropriate to evaluate alternatives and identify risks and issues.

   Project managers/associates who do *not* demonstrate this competency will:

   - Accept traditional assumptions regarding resource requirements and time estimates.
   - Rely on team members to provide information needed.
   - Push for a new milestone without determining the reason previous milestone was missed.

- Summarize details of discussions and arguments without drawing conclusions.
- Limit inquiries to standard sources of information.
- Use procedures and tools that are readily available.
- Define role narrowly as facilitating and documenting team members' discussions.

Selected consequences for projects/business of not demonstrating this competency:

- Commitments may be made to unrealistic or untested dates.
- High risk approaches may be adopted without explicit acknowledgment.
- Projects may take longer to complete than necessary.
- New findings and results may incorporated slowly only into current Lilly practices.
- Major problems may arise unexpectedly.
- Same issues may be revisited.
- Project plan may remain unchanged despite major shifts in resources, people, and priorities.

4. *Manages Risks:* Anticipates and allows for changes in priorities, schedules, resources, and changes due to scientific/technical issues.

Project managers/associates who demonstrate this competency will:

- Double-check validity of key data and assumptions before making controversial or potentially risky decisions.
- Create a contingency plan when pursuing options that have clear risks associated with them.
- Maintain ongoing direct contact with "risky" or critical path activities to understand progress.
- Push team members to identify all the assumptions implicit in their estimates and commitments.
- Stay in regular contact with those whose decisions impact the project.
- Let management and others know early on the risks associated with a particular plan of action.
- Argue for level of resources and time estimates that allow for predictable "unexpected" events.
- Pinpoint major sources of scientific risks.

Project managers/associates who do *not* demonstrate this competency will:

- Remain optimistic regardless of progress.
- Agree to project timelines despite serious reservations.
- Value innovation and new ideas despite attendant risks.
- Accept less experienced team members in key areas.

- Give individuals freedom to explore different options.
- Accept estimates and assessments with minimal discussion.

Selected consequences for projects/business of not demonstrating this competency:

- Projects may take longer to complete than necessary.
- Project may have difficulty responding to shifts in organizational priorities.
- Major delays could occur if proposed innovative approach proves inappropriate.
- Known problem areas may remain sources of difficulties.
- Project plans may be subject to dramatic revisions.

5. *Communicates Clearly:* Listens well and provides information that is easily understood and useful to others.

Project managers/associates who demonstrate this competency will:

- Present technical and other complex issues in a concise, clear, and compelling manner.
- Target or position communication to address needs or level of understanding of recipient(s) (e.g., medical, senior management, etc.).
- Filter data to provide the most relevant information (e.g., doesn't go over all details, but knows when and how to provide an overall view).
- Keep others informed in a timely manner about decision or issues that may impact them.
- Facilitate and encourage open communication among team members.
- Set up mechanisms for regular communications with team members in remote locations.
- Accurately capture key points of complex or extended discussions.
- Spend the time necessary to prepare presentations for management.
- Effectively communicate and represent technical arguments outside own area of expertise.

Project managers/associates who do *not* demonstrate this competency will:

- Provide all the available details.
- See multiple reminders or messages as inefficient.
- Expect team members to understand technical terms of each other's specialties.
- Reuse communication and briefing materials with different audiences.
- Limit communications to periodic updates.
- Invite to meetings only those who (are presumed to) need to be there or who have something to contribute.
- Rely on technical experts to provide briefings in specialized, technical areas.

Selected consequences for projects/business of not demonstrating this competency:

- Individuals outside of the immediate team may have little understanding of the project.
- Other projects may be disrupted by "fire drills" or last minute changes in plan.
- Key decisions and discussions may be inadequately documented.
- Management briefings may be experienced as ordeals by team and management.
- Resources/effort may be wasted or misapplied.

6. *Pays Attention to Details:* Systematically documents, tracks, and organizes project details.

   Project managers/associates who demonstrate this competency will:

   - Remind individuals of due dates and other requirements.
   - Ensure that all relevant parties are informed of meetings and decisions.
   - Prepare timely, accurate, and complete minutes of meetings.
   - Continually update or adjust project documents to reflect decisions and other changes.
   - Check the validity of key assumptions in building the plan.
   - Follow up to ensure that commitments are understood.

   Project managers/associates who do *not* demonstrate this competency will:

   - Assume that others are tracking the details.
   - See formal reviews as intrusions and waste of time.
   - Choose procedures that are least demanding in terms of tracking details.
   - Only sporadically review and update or adjust project documents to reflect decisions and other changes.
   - Limit project documentation to those formally required.
   - Rely on meeting notes as adequate documentation of meetings.

   Selected consequences for projects/business of not demonstrating this competency:

   - Coordination with other parts of the organization may be lacking.
   - Documentation may be incomplete or difficult to use to review project issues.
   - Disagreements may arise as to what was committed to.
   - Project may be excessively dependent on the physical presence of project manager or associate.

7. *Structures the Process:* Constructs, adapts, or follows a logical process to ensure achievement of objectives and goals.

Project managers/associates who demonstrate this competency will:

- Choose milestones that the team can use for assessing progress.
- Structure meetings to ensure agenda items are covered.
- Identify sequence of steps needed to execute project management process.
- Maintain up-to-date documentation that maps expectations for individual team members.
- Use available planning tools to standardize procedures and structure activities.
- Create simple tools to help team members track, organize, and communicate information.
- Build a process that efficiently uses team members' time, while allowing them to participate in project decision; all team members should not attend all meetings.
- Review implications of discussion or decisions for the project plan as mechanism for summarizing and clarifying discussions.
- Keep discussions moving by noting disagreements rather than trying to resolve them there and then.
- Create and use a process to ensure priorities are established and project strategy is defined.

Project managers/associates who do *not* demonstrate this competency will:

- Trust that experienced team members know what they are doing.
- Treat complex sequences of activities as a whole.
- Share responsibility for running meetings, formulating agendas, and so on.
- Create plans and documents that are as complete and detailed as possible.
- Provide written documentation only when asked for.
- Allow team members to have their say.

Selected consequences for projects/business of not demonstrating this competency:

- Projects may receive significantly different levels of attention.
- Project may lack a single direction or focus.
- Planning documents may be incomplete or out of date.
- Presentations and briefings may require large amounts of additional work.
- Meetings may be seen as unproductive.
- Key issues may be left unresolved.
- Other parts of the organization may be unclear about what is expected and when.

8. *Focuses on Results:* Continually focuses own and others' attention on realistic project milestones and deliverables.

Project managers/associates who demonstrate this competency will:

- Stress need to keep project-related activities moving forward.
- Continually focus on ultimate deliverables (e.g., product to market, affirm/disconfirm merits of compound, value of product/program to Lilly) (Manager).

- Choose actions in terms of what needs to be accomplished rather than seeking optimal solutions or answers.
- Remind project team members of key project milestones and schedules.
- Keep key milestones visible to the team.
- Use fundamental objective of project as means of evaluating options and driving decisions in a timely fashion.
- Push team members to make explicit and public commitments to deliverables.
- Terminate projects or low-value activities in timely fashion.

Project managers/associates who do *not* demonstrate this competency will:

- Assume that team members have a clear understanding of project deliverables and milestones.
- Approach tasks and issues only when they become absolutely critical.
- Downplay or overlook negative results or outcomes.
- Keep pushing to meet original objectives, in spite of new data/major changes.
- Pursue activities unrelated to original project requirements.
- Trust that definite plans will be agreed to once team members get involved in the project.
- Allow unqualified individuals to remain on tasks.
- Make attendance at project planning meetings discretionary.

Selected consequences for projects/business of not demonstrating this competency:

- Milestones may be missed without adequate explanation.
- Functional areas may be surprised at demand for key resources.
- Commitments may be made to unreasonable or unrealistic goals or schedules.
- Projects may take longer to complete than necessary.
- Objectives and priorities may differ significantly from one team member to another.

9. *Builds a Team:* Creates an environment of cooperation and mutual accountability within and across functions to achieve common objectives.

Project managers/associates who demonstrate this competency will:

- Openly acknowledge different viewpoints and disagreements.
- Actively encourage all team members to participate regardless of their functional background or level in the organization.
- Devote time and resources explicitly to building a team identity and a set of shared objectives.
- Maintain objectivity; avoid personalizing issues and disagreements.
- Establish one-on-one relationship with team members.
- Encourage team members to contribute input in areas outside functional areas.

- Involve team members in the planning process from beginning to end.
- Recognize and tap into the experience and expertise that each team member possesses.
- Solicit input and involvement from different functions prior to their major involvement.
- Once a decision is made, insist that team accept it until additional data becomes available.
- Push for explicit commitment from team members when resolving controversial issues.

Project managers/associates who do *not* demonstrate this competency will:

- State what can and cannot be done.
- Assume that mature professionals need little support or team recognition.
- Limit contacts with team members to formal meetings and discussions.
- Treat issues that impact a team member's performance as the responsibility of functional line management.
- Help others only when explicitly asked to do so.
- Be openly critical about other team members' contributions or attitudes.
- Revisit decisions when team members resurface issues.

Selected consequences for projects/business of not demonstrating this competency:

- Team members may be unclear as to their responsibilities.
- Key individuals may move onto other projects.
- Obstacles and setbacks may undermine overall effort.
- Conflicts over priorities within project team may get escalated to senior management.
- Responsibility for project may get diffused.
- Team members may be reluctant to provide each other with support or accommodate special requests.

10. *Manages Complexity:* Organizes, plans, and monitors multiple activities, people, and resources.

Project managers/associates who demonstrate this competency will:

- Remain calm when under personal attack or extreme pressure.
- Monitor progress on frequent and consistent basis.
- Focus personal efforts on most critical tasks: apply 80–20 rule.
- Carefully document commitments and responsibilities.
- Define tasks and activities to all for monitoring and a sense of progress.
- Break activities and assignments into components that appear doable.

- Balance and optimize workloads among different groups and individuals.
- Quickly pull together special teams or use outside experts in order to address emergencies or unusual circumstances.
- Debrief to capture "best practices" and "lessons learned."

Project managers/associates who do *not* demonstrate this competency will:

- Limit the number of reviews to maximize time available to team members.
- Stay on top of all the details.
- Depend on team members to keep track of their own progress.
- Let others know how they feel about an issue or individual.
- Rely on the team to address issues.
- Assume individuals recognize and learn from their own mistakes.

Selected consequences for projects/business of not demonstrating this competency:

- Projects may receive significantly different levels of attention.
- Projects may take on a life of their own with no clear direction or attainable outcome.
- Responsibility for decisions may be diffused among team members.
- Exact status of projects may be difficult to determine.
- Major issues can become unmanageable.
- Activities of different parts of the business may be uncoordinated.
- Conflicts may continually surface between project leadership and other parts of Lilly.

11. *Makes Tough Decisions:* Demonstrates assurance in own abilities, judgments, and capabilities; assumes accountability for actions.

    Project managers/associates who demonstrate this competency will:

    - Challenge the way things are done and make decisions about how things will get done.
    - Force others to deal with the unpleasant realities of a situation.
    - Push for reassessment of controversial decisions by management when new information/data becomes available.
    - Bring issues with significant impact to the attention of others.
    - Consciously use past experience and historical data to persuade others.
    - Confront individuals who are not meeting their commitments.
    - Push line management to replace individual who fail to meet expectations.
    - Challenge continued investment in a project if data suggests it will not succeed.
    - Pursue or adopt innovative procedures that offer significant potential benefits even where limited prior experience is available.

Project managers/associates who do *not* demonstrate this competency will:

- Defer to the ideas of more experienced team members.
- Give others the benefit of the doubt around missed commitments.
- Hold off making decisions until the last possible moment.
- Pursue multiple options rather than halt work on alternative approaches.
- Wait for explicit support from others before raising difficult issues.
- Accept senior managers' decisions as "nonnegotiable."
- Rely on the team to make controversial decisions.
- Provide problematic performers with additional resources and time.

Selected consequences for projects/business of not demonstrating this competency:

- Projects may take longer to complete than necessary.
- Failing projects may be allowed to linger.
- Decisions may be delegated upward.
- Morale of team may be undermined by nonperformance of certain team members.
- "Bad news" may not be communicated until the last minute.
- Key individuals may "burn out" in effort to play catch-up.

12. *Builds Strategic Support:* Gets the support and level of effort needed from senior management and others to keep projects on track.

Project managers/associates who demonstrate this competency will:

- Assume responsibility for championing the projects while demonstrating a balance between passion and objectivity.
- Tailor arguments and presentations to address key concerns of influential decision-makers.
- Familiarize self with operational and business concerns of major functions within Lilly.
- Use network of contacts to determine best way to surface an issue or make a proposal.
- Push for active involvement of individuals with the experience and influence needed to make things happen.
- Pinpoint the distribution of influence in conflict situations.
- Presell controversial ideas or information.
- Select presenter to ensure appropriate message is sent.
- Ask senior management to help position issues with other senior managers.

Project managers/associates who do *not* demonstrate this competency will:

- Meet senior management and project sponsors only in formal settings.
- Propose major shifts in direction in group meetings.

- Make contact with key decision-makers when faced with obstacles or problems.
- Limit number of face-to-face contacts with "global" partners.
- Treat individuals as equally important.
- Avoid the appearance of "politicking."
- Depend on other team members to communicate to senior managers in unfamiliar parts of Lilly.

Selected consequences for projects/business of not demonstrating this competency:

- Viable projects may be killed without clear articulation of benefits.
- "Cultural differences" may limit success of global projects.
- Decisions may be made without the input of key individuals.
- Resistance to changes in project scope or direction may become entrenched before merits of proposals are understood.
- Key individuals/organizations may never buy in to a project's direction or scope.
- Minor conflicts may escalate and drag on.

12.8 TRAINING AND EDUCATION AT WORK

The quality of the project management training and education a company's employees receive is, along with executive buy-in, one of the most important factors in achieving success and ultimately excellence in project management. The training could be for both the employees of the company as well as for its suppliers who must interface with the customer's project management methodology. Let's look at some case examples of effective training programs.

Noveon

According to a spokesperson at Noveon:

> Noveon, a specialty chemicals company located in Brecksville, Ohio, is a project-driven operation that has used project management techniques on its capital projects for almost 20 years, including the time it was a part of The BFGoodrich Company (Noveon became an independent company in February 2001). Projects range from $500,000 to $30 million.
>
> During much of the 20 years the company has been practicing project management, the line managers from the plant have been responsible for project execution. Before 1993, projects frequently ran over budget and missed schedules because the scopes of projects were poorly defined, project controls and monitoring were weak, staffing was limited, and staff had other priorities.
>
> Additional problems existed in project cost accounting. The format, reporting, and procedures of the cost tracking system did not allow managers and executives to monitor or control project budgets. For example, international project accounting was reported in

a multi-currency format that had no single means of comparing actual spending to budget spending.

In 1994, corporate strategy changed and ultimately affected project management. The organization identified strategic goals, which included a significant expansion in product and sales, intended to support a marketing strategy based on the introduction of new products to the marketplace ahead of the competition.

It became necessary for the central support organization, including engineering, to become more responsive to the business managers than they had been in the past.

An attempt to decentralize project management had been made in 1993, but numerous problems came up. One set of problems arose because managers had multiple responsibilities at more than one plant in more than one location. Another set of problems resulted from conflicting priorities for line staff when they had project responsibilities.

The pressure to support expansion and to reduce costs led to developing a more sophisticated project management tool. Projects are now outsourced as a single, design-build contract and managed at the central engineering level. A structured process (though relatively informal) has been in place since 1993.

The improvements in the design–build project management system at this company have been accelerated primarily through education and training. The education of employees is vital, but just as critical is the training of the contractors and consultants who work with the company. Noveon has made it clear to the consultants that such training is a requirement of doing business with the company. The numerous contacts with outside contracts have resulted in common definitions, risk sharing, and clear understanding of all aspects of the project management process.

Today, the company believes that it has two special areas of strength in project management. The first is its record of success in completing projects on schedule, within budget and within specification. This allows the company to be responsive to corporate needs. The second strength is the emphasis on the project identification stage of planning. Through both the project authorization and project scope identification stage, basic goals, objectives and scope work are carefully defined.

Other strengths include a commitment to accurate budgeting with cost estimating for projects considered a priority. Another strength is the selection of contractors based on their qualifications for each project. The benefits of project management are significant and broad. It has been a crucial foundation supporting the successful corporate marketing strategy. The company estimates that project management trims down the cycle time from research and development's initiation of the project to full production to 38 months. Project management also allows improved cost control.

Hewlett-Packard

Hewlett-Packard is clearly committed to project management training and education. The following material is typically included in HP proposals. The material was provided by Ron Kempf, PMP, Director of PM Competency & Certification at HP Services Engagement:

PM Development

HP Services has a comprehensive Project Management Development Program with courses that cover all aspects of project management training. The program was established in 1995. A standard curriculum with over 25 courses is implemented throughout the world covering project leadership, management, communication, risk management, contracting, managing business performance, scheduling and cost control, and quality. The courses are based on PMI's Project Management Body of Knowledge (PMBOK®). The curriculum also encompasses specialized courses on key HP internal topics, such as the Program Methodology, as well as essential business and financial management aspects of projects. Over 4,000 students complete courses in the PM curriculum each year.

Even the most experienced project managers continue to take courses to strengthen their knowledge and skills. HP Services conducts an intensive weeklong training called Project Management University (PMU) twice a year in each major geography—Americas, Asia/Pacific, and Europe. Over 100 project managers attend each PMU. These events provide project managers with an opportunity to devote concentrated time to study and to exchange knowledge and ideas with other HP project managers from around the world.

All courses taught in HP's PM curriculum are registered in PMI's Registered Education Provider (REP) Program to ensure a consistent basis and oversight. The Project Management Development Program won Excellence in Practice awards in career development and organizational learning in 2002 from ASTD, the world's leading resource on workplace learning and performance issues.

PMP Certification

HP has a well-established program to encourage and support our project managers to achieve certification. HP Services has over 900 individuals who have earned the PMP (Project Management Professional) certification from PMI.

PMI Support

HP actively supports the Project Management Institute (PMI), a non-profit organization with more than 100,000 members. PMI has set standards for Project Management excellence that are recognized by the industry and our customers worldwide.

HP employees participate on a number of PMI boards and committees, including the Global Accreditation Center, Research Program Membership Advisory Group, development of the Certified Associate in Project Management (CAPM), development of Certificates of Added Qualification (CAQ) in IT Systems, IT Networks, and Project Management Office, PMBOK® 2000 review, and PMBOK® 2004 Update Team. Many HP employees hold leadership positions in PMI Chapters and SIGs throughout the world.

HP sponsored the Monday Keynote Speaker at the annual PMI Symposiums in October 2002 and was one of the top companies represented at the Symposium, with 8 papers presented by HP employees.

Exel

Training and education can accelerate the project management maturity process especially if project management training is accompanied by training on the corporate project management methodology. This approach has been successfully implemented at Exel. According to Francena D. Gargaro, PMP, Director Project and Resource Management, Americas:

> Project Management training is currently not mandatory, however, all sectors are aware of the Project Management Methodology and the training provided.
>
> The Enterprise Project Management group is responsible for scheduling and conducting the project management training courses. To date, over 1,500 associates have been trained in Exel's Project Management Methodology.

Project Management 101 (PM101); Exel's introductory PM course, is a 2-day, interactive course consisting of lecture, exercises, interactive team activities, and a comprehensive case study. An outside training consultant, who is intimately familiar with Exel's business and a certified PMI training provider, delivers the course. Members from the EPM group supplement the training with instruction on Exel PM tools, real-world examples, and facilitation of team exercises.

PM101 participants include members of all sectors, company-wide who have involvement in projects as team members, project managers, sponsors, or business development.

Presently, the PM training programs are trending toward a curriculum-based format. Originally, PM101 was the only, optional project management training offered in-house. Over the past two years, we have expanded the program to include an on-line Project Management Primer, which serves as a precursor to the PM101 course. The PM Primer is a 30-minute, online tutorial on the basic concepts of project management and Exel's project management methodology.

Additionally, Exel is in the process of developing the next step in the PM training program—PM201 (Advanced Project Management). It is geared more toward project managers, specifically targeting topics such as managing critical path, risk management, conflict management, customer relationship management, presentation skills, management/leadership skills, and cost/schedule management.

The PM training curriculum can also lead to certification training, as described by Todd Daily, PMP, Project Manager, Enterprise Project Management, Americas:

The PM training curriculum does lead to certification training, in some cases. Not everyone who attends PM training or performs the role of project manager pursues formal certification. It is, however, becoming more prominent and recognized within the organization. Today we have 16 certified project management professionals (PMP) in the U.S. We incorporate information about certification in the PM training courses and our course material follows the PMBOK® approach.

The EPM group provides assistance to candidates within Exel who are interested in pursuing formal certification in project management. We have material and processes to guide potential candidates in their pursuit. We also have begun developing assistance for the Certified Associate in Project Management (CAPM) program and have candidates currently pursuing this certification.

Internationally, we have candidates pursuing certification in UK/Europe, Mexico, and Brazil.

Exel Course Listings

- Project Management—On-line primer
- Project Management 101
- Project Management 201 (in development)

Measuring a return on investment in training dollars is not always easy. Sometimes, the returns are more easily identified from a qualitative rather than a quantitative perspective. Francena D. Gargaro, PMP, Director Project and Resource Management, Americas for Exel, believes that:

Current measures are more qualitative than quantitative. Course costs are generally covered by each participant's cost center, along with any associated travel expenses.

All course participants, at the end of the class, complete a course and instructor evaluation. Feedback is reviewed, archived and used to improve course material curriculum. Additionally, participants of the course are paired with a "buddy" to incorporate one new project management idea, learned from the course, into their day-to-day job activities and share experiences with their "buddy."

Future considerations include a periodic post-training review with participant's managers/supervisors to evaluate improvements in job performance following the training. That information would then be communicated to the Enterprise Project Management group for tracking and evaluation. This approach is currently being considered, but has yet to be implemented.

ABB

Best results from project management training and education usually occur when the course is custom-designed for the company's specific problems and integrated with the company's project management methodology. Customization and methodology integration need not be highly complex. For example, ABB has achieved excellent success with three project management courses custom-designed for their particular use. The following is a brief description of these three courses taken from the ABB Project Management Development Program catalog:

Project Management 1

Project Management 1 provides proven strategies and practical tools for planning, executing and controlling complex projects. Like Project Management Overview, the course covers the entire project management life cycle, but offers more detailed and sophisticated instruction in the critical areas of scheduling, cost control and risk management.

Students improve their ability to define the scope of a project and manage within that definition. They learn how best to identify and sequence tasks, estimate duration of tasks, schedule events and activities, control variances, manage costs and utilize resources. They also learn qualitative and quantitative techniques for identifying, analyzing and mitigating risks, as well as the best ways and times to apply these techniques in the project environment. Students also get knowledge about customer satisfaction management and get introduced to the ABB Project Management Process.

Individual and small group learning exercises help students develop these skills. Students master key theories, concepts and practices, then put this knowledge to work in the classroom through case studies and other practical learning activities. Students must pass an objective course exam.

This course is one of the entry level courses in the ABB Project Management Development Program.

Syllabus

- Project Definition and Selection
- Resource Allocation
- Risk Management
- Customer Satisfaction Management
- Project Economy
- Network Logic and Scheduling
- Project Control
- Project Completion
- ABB Project Management Process

Project Management 2

Project Management 2 is based on a week-long case study and uses the knowledge gained in Project Management 1. The course synthesizes and reinforces the knowledge and skills that are essential in managing complex projects. The course also teaches advance approaches that build on competencies gained in Project Management Overview, Project Management 1 and Leadership 1. The course emphasizes the importance of active project control by risk management based on sound planning and management in the tendering and negotiation stages, as well as the execution and closeout stages of a project.

Students benefit from participating in this advanced simulation. Working in teams, they complete an extensive and realistic project in which they propose, plan, carry out, and close out a project under normal job conditions. The simulation spans all phases in the project life cycle. Each team must resolve issues of performance, scheduling, and control, and at the same time address leadership and other management concerns. Students take turns acting as the project manager who must lead the team in defining and producing the deliverables needed to complete the project successfully.

Syllabus

- Team Building
- Pre-proposal Analysis and Planning
- Proposal Kick-off and Preparation
- Planning for Project Execution
- Negotiating and Finalizing the Contract
- Project Execution
- Project Completion and Closeout

Project Management 3

Project Management 3 is developed for experienced project managers facing strategic, global challenges in integrating the goals, objectives and policies of the project and the company. It provides a deeper understanding of and insight into the role of the project manager as the team leader, as the expert on current issues and trends in project management, and as the principal force in mitigating cross-cultural variations in the project environment.

The purpose of this course is twofold: to improve the capability of experienced project managers in managing an international project globally, and to increase the chance of successful completion of international business projects. These objectives are achieved by considering and systematically analyzing the impact of significant factors, such as organizational relations, culture, conflicts, and stakeholders that affect the successful completion of those international business projects.

Syllabus

- Organizational Relationships
- Culture
- Conflict Management
- Opportunities and Risks in the Strategic Context
- Stakeholder Management
- What Is Strategic for Execution of a Business Project?
- Preparing Strategies for Project Execution
- Project Management—The Future

MULTIPLE CHOICE QUESTIONS

1. In the 1950s and 1960s, training courses on project management concentrated on:
   A. Organizational structures
   B. Scope change management
   C. Scheduling techniques
   D. Cost control
   E. Behavioral issues
2. During the 1970s and 1980s, training courses in project management concentrated on:
   A. Organizational structures
   B. Scope change management
   C. Scheduling techniques
   D. Cost control
   E. Behavioral issues
3. During the late 1980s and early 1990s, training courses on project management concentrated on:
   A. Organizational structures
   B. Scope change management
   C. Scheduling techniques
   D. Cost control
   E. Behavioral issues
4. A common mistake made by companies is to train:
   A. Everyone
   B. Only the project managers
   C. Only the line people
   D. Only the project office personnel
   E. Only the project sponsor
5. Effective project management and TQM are similar when it comes to training. Both processes require that training begin with:
   A. Employees
   B. Project managers
   C. Functional employees
   D. Line managers
   E. Executives
6. The need for project management training has increased during the past decade because:
   A. More training funds are available
   B. Training guarantees project success
   C. Project management is now recognized as a profession
   D. Corporations are mandating that each employee attend 80 hours of training per year
   E. All of the above
7. If funding is not an issue, most companies would prefer to learn project management through:
   A. University seminars
   B. University courses
   C. Self-study courses
   D. Customized internal training programs
   E. Computer-based training

8. Which of the following problems is common when using internal project management trainers?
- **A.** Limited project management knowledge areas
- **B.** Limited up-to-date knowledge of principles
- **C.** Other responsibilities above and beyond training
- **D.** Lack of dedication to project management
- **E.** All of the above

9. Which of the following is a common pitfall in hiring external speakers and trainers?
- **A.** Not verifying credentials
- **B.** Not verifying quality of information before hand
- **C.** Not requesting a custom-designed program
- **D.** Working with the speaker's marketing representative rather than the speaker
- **E.** All of the above

10. Evaluation of training materials requires an examination of:
- **A.** Use of slides
- **B.** Use of transparencies
- **C.** Use of chalkboard
- **D.** Use of case studies
- **E.** All of the above

11. Which of the following is usually the largest part of the cost of training?
- **A.** Speaker fees
- **B.** Materials
- **C.** Facility rental
- **D.** Meals/lodgings
- **E.** Employee release time

12. Training should not be performed unless there is a return on investment on training dollars.
- **A.** True
- **B.** False

13. The amount of employee training needed is dependent upon _____ and ______.
- **A.** Type of company (i.e., project-driven); experience in project management
- **B.** Age of employees; funding available
- **C.** Availability of qualified trainers; skills required
- **D.** Company size; union demands
- **E.** None of the above

14. General Electric Plastics emphasizes _____ training for their project personnel.
- **A.** Scheduling techniques
- **B.** Cost control
- **C.** Boundarylessness behavior
- **D.** Diversity management
- **E.** Sponsorship management

15. Which of the following is a quantitative benefit of project management training?
- **A.** Higher morale
- **B.** Better visibility and focus on results
- **C.** Better control
- **D.** Faster, higher quality decisions
- **E.** All of the above

16. Which of the following is a qualitative benefit of training?
    A. A better visibility and focus on results
    B. Reduction in paperwork
    C. Lower costs
    D. Lower turnover of personnel
    E. Improved quality and reliability

17. Project management training can accelerate the benefits of project management.
    A. True
    B. False

18. Which of the following is the fastest growing type of project management training program?
    A. University courses
    B. In-house seminars
    C. University seminars
    D. On-the-job
    E. Computer-based training

19. The most common decision on when to train is:
    A. Just-in-time
    B. Exposure
    C. Continuous learning
    D. Self-confidence
    E. All of the above

20. A trainer tells you that he/she uses all of the following methods for project management training. Which of the following is the most difficult to evaluate before hand?
    A. Lectures/discussions
    B. Case studies (external firms)
    C. Case studies (internal projects)
    D. Simulation/role playing
    E. Exams

21. One of the best ways to decide what information should be included in the training program for project managers is by:
    A. Performing a core competency analysis
    B. Asking other companies
    C. Asking project management trainers
    D. Surveying executives
    E. All of the above

22. The complexity in measuring the return on investment for training dollars is to determine:
    A. When to evaluate
    B. What to evaluate
    C. How to evaluate
    D. How to verify achievement of the objectives
    E. All of the above

23. A company spends $20,000 training a project management team immediately before they work on the project. The project is completed $40,000 below the original budget. One could argue that the return on investment on training dollars is:
    A. 50 percent
    B. 100 percent
    C. 150 percent

D. 200 percent
E. Cannot be determined

24. Inviting your customers and contractors to attend your project management training programs can be just as important as training your own people.
A. True
B. False

25. Some large companies offer more internal courses related to project management than most colleges and universities do.
A. True
B. False

DISCUSSION QUESTIONS

1. Describe the evolution of project management training from the 1950s to the twenty-first century.
2. Why is it so difficult to decide who to train in project management?
3. Has the acceptance of project management as a career path affected training?
4. Why are custom-designed internal training courses normally the most effective?
5. What are some of the typical costs associated with an internal training program?
6. What is core competency analysis?
7. When hiring a project management speaker, why is it good practice to evaluate, in advance, the training material to be used?
8. What factors must be considered when performing a return on investment on training dollars for project management?
9. A company believes in just-in-time project management training. A team of 20 employees are about to begin working on a new project and training is set for the week immediately before startup of the project. Is this good or bad?
10. To bring project management into an organization, should training be bottom up or top down?

Across

1. 1950s training
4. Training program: _ _ _ _ _ _ _ _ _ learning
5. ROI on _ _ _ _ _ _ _ _ _ dollars
6. Behavioral training at GE
8. Benefit: _ _ _ _ _ _ morale
9. Type of training: university _ _ _ _ _ _ _
11. Way to train
12. Core _ _ _ _ _ _ _ _ _ _ _ analysis

Down

2. Benefit: lower _ _ _ _ _ _ _ _ _ of personnel
3. Type of training: university _ _ _ _ _ _ _ _
6. 1990s training
7. 1970s training
10. Benefit: lower _ _ _ _ _

Informal Project Management

13.0 INTRODUCTION

Over the past 20 years, the most significant change in project management has been the idea that *informal project management* does work. In the 1950s and 1960s, the aerospace, defense, and large construction industries were the primary users of project management techniques and tools. Because project management was a relatively new management process, customers of the contractors and subcontractors wanted evidence that the system worked. Documentation of the policies and procedures to be used became part of the written proposal. Formal project management, supported by hundreds of policies, procedures, and forms, became the norm. After all, why would a potential customer be willing to sign a $10 million contract for a project to be managed informally?

This chapter clarifies the difference between informal and formal project management, then discusses the four critical elements of informal project management.

13.1 INFORMAL VERSUS FORMAL PROJECT MANAGEMENT

Formal project management has always been expensive. In the early years, the time and resources spent on preparing written policies and procedures had a purpose: They placated the customer. As project management became established, formal documentation was created mostly for the customer. Contractors began managing more informally, while the customer was still paying for formal project management documentation. Table 13–1 shows

TABLE 13–1. FORMAL VERSUS INFORMAL PROJECT MANAGEMENT

Factor	Formal Project Management	Informal Project Management
Project manager's level	High	Low to middle
Project manager's authority	Documented	Implied
Paperwork	Exorbitant	Minimal

the major differences between formal and informal project management. As you can see, the most relevant difference is the amount of paperwork.

Paperwork is expensive. Even a routine handout for a team meeting can cost $500 to $2,000 per page to prepare. Executives in excellent companies know that paperwork is expensive. They encourage project teams to communicate without excessive amounts of paper. However, some people are still operating under the mistake believe that ISO 9000 certification requires massive paperwork.

Figure 13–1 shows the changes in paperwork requirements in project management. The early 1980s marked the heyday for lovers of paper documentation. At that time, the average policies and procedures manual probably cost between $3 million and $5 million to prepare initially and $1 million to $2 million to update yearly over the lifetime of the project. Project managers were buried in forms to complete to the extent that they had very little time left for actually managing the projects. Customers began to complain about the high cost of subcontracting, and the paperwork boom started to fade.

Real cost savings didn't materialize until the early 1990s with the growth of concurrent engineering. Concurrent engineering shortened product development times by taking activities that had been done in series and performing them in parallel instead. This change increased the level of risk in each project, which required that project management back away from some of its previous practices. Formal guidelines were replaced by less detailed and more generic checklists.

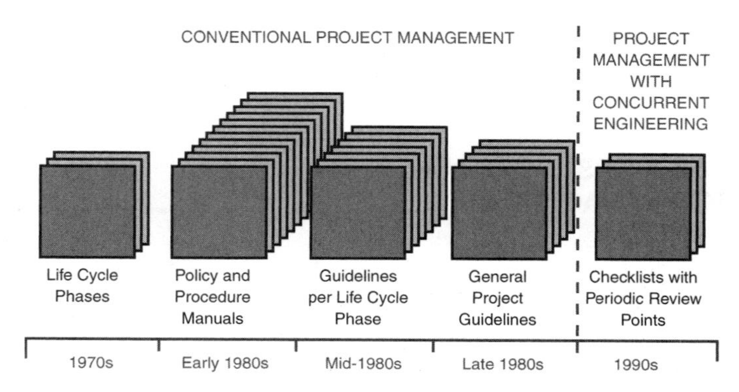

FIGURE 13–1. Evolution of policies, procedures, and guidelines. *Source:* Reprinted from H. Kerzner, *In Search of Excellence in Project Management.* New York: Wiley, 1998, p. 196.

Policies and procedures represent formality. Checklists represent informality. But informality doesn't eliminate project paperwork altogether. It reduces paperwork requirements to minimally acceptable levels. To move from formality to informality demands a change in organizational culture (see Figure 13–2). The four basic elements of an informal culture are these:

- Trust
- Communication
- Cooperation
- Teamwork

Large companies quite often cannot manage projects on an informal basis although they want to. The larger the company, the greater the tendency for formal project management to take hold. Patty Goyette, vice president IOC sales operations and customer service at Nortel Networks, believes that:

> The introduction of enterprise-wide project process and tools standards in Nortel Networks and the use of pipeline metrics (customer defined, industry standard measures) provides a framework for formal project management. This is necessary given the complexity of telecom projects we undertake and the need for an integrated solution in a short time frame. The Nortel Networks project manager crosses many organizational boundaries to achieve the results demanded by customers in a dynamic environment.

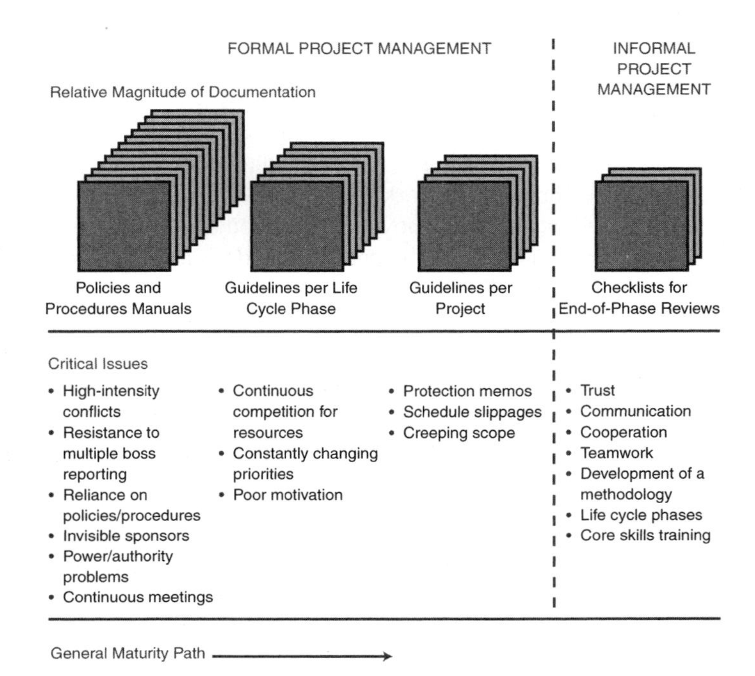

FIGURE 13–2. Evolution of paperwork and change of formality levels. *Source:* Reprinted from H. Kerzner, *In Search of Excellence in Project Management.* New York: Wiley, 1998, p. 198.

Most companies manage either formally or informally. However, if your company is project-driven and has a very strong culture for project management, you may have to manage either formally or informally based upon the needs of your customers. Carl Isenberg, director of EDS project management consulting, describes this dual approach at EDS:

> Based on the size and scope of a project, formal or informal project management is utilized. Most organizations within EDS have defined thresholds that identify the necessary project management structure. The larger the project (in cost, schedule, and resource requirements), the more strictly the methodology is followed. Smaller projects (such as a modification to a report) require a more cursory approach to the management of that project.
>
> Larger projects require full development of a start-up plan (i.e., the plan for the plan), full-scale planning, execution of the plan and crucial closedown activities to ensure proper project completion and documentation of lessons learned. These types of projects have a formal project workbook in place (typically in electronic format) to archive project documentation and deliverables. These archives are generally available to be leveraged on future projects that are similar in nature.
>
> The execution of EDS project management methodology is also sometimes dependent on the customer environment and relationship. The contractual nature and/or relationship EDS has with various customers sometimes affect whether project management is done formally or informally. The relationship also affects whose methodology will be used.

13.2 TRUST

Trusting everyone involved in executing a project is critical. You wake up in the morning, get dressed, and climb into your car to go to work. On a typical morning, you operate the foot pedal for your brakes maybe 50 times. You've never met the people who designed the brakes, manufactured the brakes, or installed the brakes. Yet you still give no thought to whether the brakes will work when you need them. No one broadsides you on the way to work. You don't run over anyone. Then you arrive at work and push the button for the elevator. You've never met the people who designed the elevator, manufactured it, installed it, or inspected it. But again you feel perfectly comfortable riding the elevator up to your floor. By the time you get to your office at 8 A.M., you have trusted your life to uncounted numbers of people whom you've never even met. Still, you sit down in your office and refuse to trust the person in the next office to make a $50 decision.

Trust is the key to the successful implementation of informal project management. Without it, project managers and project sponsors would need all that paperwork just to make sure that everyone working on their projects was doing the work just as he or she had been instructed. And trust is also key in building a successful relationship between the contractor/subcontractor and the client. Let's look at an example:

Perhaps the best application of informal project management that I have seen is the Heavy Vehicle Systems Group of Bendix Corporation. Bendix hired a consultant to conduct a three-day training program. The program was custom-designed, and during the design phase the consultant asked the vice president and general manager of the division whether he wanted to be trained in formal or informal project management. The vice pres-

TABLE 13–2. BENEFITS OF TRUST IN CUSTOMER–CONTRACTOR WORKING RELATIONSHIPS

Without Trust	With Trust
Continuous competitive bidding	Long-term contracts, repeat business, and sole-source contracts
Massive documentation	Minimal documentation
Excessive customer-contractor team meetings	Minimal number of team meetings
Team meetings with documentation	Team meetings without documentation
Sponsorship at executive levels	Sponsorship at middle-management levels

ident opted for informal project management. The reason for his decision? The culture of the division was already based on trust. Line managers were not hired solely based on technical expertise. Hiring and promotions were based on how well the new manager would communicate and cooperate with the other line managers and project managers in making decisions in the best interests of both the company and the project.

When the relationship between a customer and a contractor is based on trust, numerous benefits accrue to both parties. The benefits are apparent in companies such as Hewlett-Packard, Computer Associates, and various automobile subcontractors. Table 13–2 shows the benefits.

13.3 COMMUNICATION

In traditional, formal organizations, employees usually claim that communication is poor. Senior managers, however, usually think that communication in their company is just fine. Why the disparity? In most companies, executives are inundated with information communicated to them through frequent meetings and dozens of weekly status reports coming from every functional area of the business. The quality and frequency of information moving down the organizational chart is less consistent, especially in more formal companies. But whether it's a problem with the information flowing up to the executive level or down to the staff, the problem usually originates somewhere upstairs. Senior managers are the usual suspects when it comes to requiring reports and meetings. And many of those reports and meetings are unnecessary and redundant.

Most project managers prefer to communicate verbally and informally. The cost of formal communication can be high. Project communication includes dispensing information on decisions made, work authorizations, negotiations, and project reports. Project managers in excellent companies believe that they spend as much as 90 percent of their time on internal interpersonal communication with their teams. Figure 13–3 illustrates the communication channels used by a typical project manager. In project-driven organizations, project managers may spend most of their time communicating externally to customers and regulatory agencies.

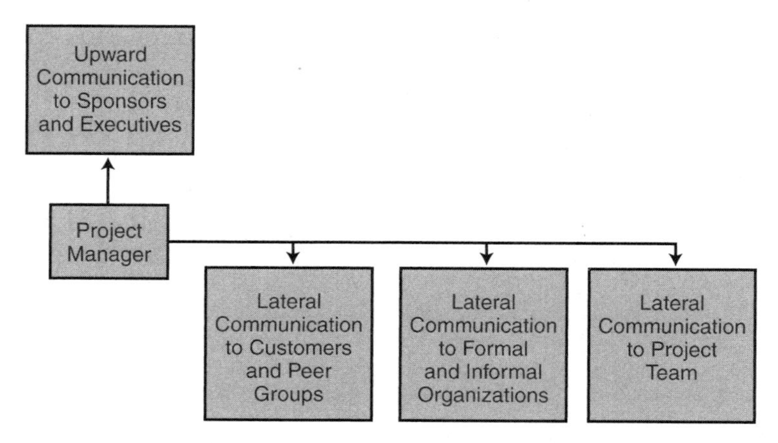

FIGURE 13–3. Internal and external communication channels for project management. *Source:* Reprinted from H. Kerzner, *In Search of Excellence in Project Management.* New York: Wiley, 1998, p. 200.

Good project management methodologies not only promote informal project management, but also promote effective communications laterally as well as vertically. The methodology itself functions as a channel of communication. A senior executive at a large financial institution commented on his organization's project management methodology, called Project Management Standards (PMS):

> The PMS guides the project manager through every step of the project. The PMS not only controls the reporting structure, but also sets the guidelines for who should be involved in the project itself and the various levels of review. This creates an excellent communication flow between the right people. The communication of a project is one of the most important factors for success. A great plan can only go so far if it is not communicated well.

Most companies believe that a good project management methodology will lead to effective communications, which will allow the firm to manage more informally than formally. The question, of course, is how long it will take to achieve effective communications. With all employees housed under a single roof, the time frame can be short. For global projects, geographical dispersion and cultural differences may mandate decades before effective communication will occur. Even then, there is no guarantee that global projects will ever be managed informally.

Suzanne Zale, global program manager of EDS, emphasized:

> With any global project, communications becomes more complex. It will require much more planning up front. All constituents for buy-in need to be identified early on. In order to leverage existing subject matter, experts conversant with local culture, and suppliers, the need for virtual teams becomes more obvious. This increases the difficulty for effective communications.

> The mechanism for communication may also change drastically. Face-to-face conversations or meetings will become more difficult. We tend to rely heavily on electronic communications, such as video and telephone conferencing and electronic mail. The format for communications needs to be standardized and understood up front so that information can be sent out quickly. Communications will also take longer and require more effort because of cultural and time differences.

One of the implied assumptions for informal project management to exist is that the employees understand their organizational structure and their roles and responsibilities within both the organizational and project structure. Forms such as the linear responsibility chart and the responsibility assignment matrix are helpful. Communication tools are not used today with the same frequency as in the 1970s and 1980s.

For multinational projects, the organizational structure, roles, and responsibilities must be clearly delineated. Effective communications is of paramount importance and probably must be accomplished more formally than informally.

Suzanne Zale, global program manager of EDS, stated:

> For any global project, the organizational structure must be clearly defined to minimize any potential misunderstandings. It is best to have a clear-cut definition of the organizational chart and roles and responsibilities. Any motivation incentives must also contemplate cultural differences. The drivers and values for different cultures can vary substantially.

The two major communication obstacles that must be overcome when a company truly wants to cultivate an informal culture are what I like to call hernia reports and forensic meetings. Hernia reports result from senior management's belief that that which has not been written has not been said. Although there is some truth to such a belief, the written word comes with a high price tag. We need to consider more than just the time consumed in the preparation of reports and formal memos. There's all the time that recipients spend reading them, as well as all the support time taken up in processing, copying, distributing, and filing them.

Status reports written for management are too long if they need a staple or a paper clip. Project reports greater than five or ten pages often aren't even read. In companies excellent in project management, internal project reports answer these three questions as simply as possible:

- Where are we today?
- Where will we end up?
- Are there any problems that require management's involvement?

All of these questions can be answered on one sheet of paper.

In 1997, Steve Sauer, director of the corporate project management office at BellSouth, pointed out the reasons for paperwork reduction at BellSouth:

> Further evidence of the evolution of project management at BellSouth is revealed in the areas of project metrics and reporting. BellSouth's officers have requested that standard report formats be developed for projects that focus on the key metrics by which projects

> should ideally be tracked and measured. This will allow them to review the same information across all projects and facilitate like-for-like comparisons during progress reviews and prioritization sessions.
>
> The objective of this type of "dashboard" report is to have a standardized, one-page format that focuses on key project deliverables in the areas of cost, schedule, functionality, and quality. This allows the project to be evaluated based on earned values analysis, which has become widely accepted as standard of project analysis. The report would also list key items in the areas of risk management, issues management, and change control. Plans are to tie output from the financial reporting systems directly into the project management databases so that status reports can be generated automatically as information is updated.

The second obstacle is the forensic team meeting. A forensic team meeting is a meeting scheduled to last 30 minutes that actually lasts for more than three hours. Forensic meetings are created when senior managers meddle in routine work activities. Even project managers fall into this trap when they present information to management that management should not be dealing with. Such situations are an invitation to disaster.

When it is implemented and used correctly, project management software can reduce the number and length of meetings. For example, the electrical components division of a Fortune 500 company installed project management software to help team members communicate among one another and with managers. Considering just one department within the company, these simple calculations can be made:

- Typical cost per meeting = $550
- Meetings eliminated per week (as a result of using the new software) = 4
- Total annual savings = ($550/meeting) × (4 meetings/week) × 52 weeks/year) = $114,400 per year

When company executives realized the potential cost savings, they looked at the 204 projects within that division that might be submitted for approval over the next two years. Approximately 75 percent of the projects would go through the approval process over the next 18 months. The company executives called for a 50 percent reduction in the number of team meetings necessary to achieve schedule approvals.

- Current average number of meetings required per project schedule approval = 10
- Expected average number after organization begins using project management software = 5
- Average number of attendees per meeting = 10
- Average duration of meetings = 1.25 hours
- Fully burdened labor rate = $70 per hour
- Total annual savings = (204 projects) × (5 meetings/project) × (10 people) × (1.25 hours/meeting) × ($70/hour) = $892,500

Software alone did not save the organization millions of dollars. The company also has a culture based on trust, teamwork, communication, and cooperation. But the company did show that minimizing the number of team meetings can save millions of dollars and improve productivity and efficiency.

13.4 COOPERATION

Cooperation is the willingness of individuals to work with others for the benefit of all. It includes the voluntary actions of a team working together toward a favorable result. In companies excellent in project management, cooperation is the norm and takes place without the formal intervention of authority. The team members know the right thing to do, and they do it.

In the average company (or the average group of any kind, for that matter), people learn to cooperate as they get to know each other. That takes time, something usually in short supply for project teams. But companies such as Ericsson Telecom AB, the General Motors Powertrain Group, and Hewlett-Packard create cultures that promote cooperation to the benefit of everyone.

13.5 TEAMWORK

Teamwork is the work performed by people acting together with a spirit of cooperation under the limits of coordination. Some people confuse teamwork with morale, but morale has more to do with attitudes toward work than it has to do with the work itself. Obviously, however, good morale is beneficial to teamwork.

In excellent companies, teamwork has these characteristics:

- Employees and managers share ideas with each other and establish high levels of innovation and creativity in work groups.
- Employees and managers trust each other and are loyal to each other and the company.
- Employees and managers are committed to the work they do and the promises they make.
- Employees and managers share information freely.
- Employees and managers are consistently open and honest with each other.

Making people feel that they are part of a team doesn't necessarily require a great deal of effort. Consider the situation at the Engineering and Construction Services Division of Dow Chemical Corporation several years ago. Dow Chemical had requested a trainer to develop a project management training course. The trainer interviewed several of the seminar participants before the training program to identify potential problem areas. The biggest problem appeared to be a lack of teamwork. This shortcoming was particularly evident in the drafting department. The drafting department personnel complained that too many changes were being made to the drawings. They simply couldn't understand the reasons behind all the changes.

The second problem identified, and perhaps the more critical one, was the project managers didn't communicate with the drafting department once the drawings were complete. The drafting people had no idea of the status of the projects they were working on, and they didn't feel as though they were part of the project team.

During the training program, one of the project managers, who was responsible for constructing a large chemical plant, was asked to explain why so many changes were being made to the drawings on his project. He said, "There are three reasons for the changes. First, the customers don't always know what they want up front. Second, once we have the preliminary drawings to work with, we build a plastic model of the plant. The model often shows us that equipment needs to be moved for maintenance or safety reasons. Third, sometimes we have to rush into construction well before we have final approval from the Environmental Protection Agency. When the agency finally gives its approval, that approval is often made contingent on making major structural changes to the work already complete." One veteran employee at Dow commented that in his 15 years with the company, no one had ever before explained the reasons behind drafting changes.

The solution to the problem of insufficient communication was also easy to repair once it was out in the open. The project managers promised to take monthly snapshots of the progress on building projects and share them with the drafting department. The drafting personnel were delighted and felt more like a part of the project team.

13.6 COLOR-CODED STATUS REPORTING

The use of colors for status reporting, whether it be for printed reports or intranet-based visual presentations, has grown significantly. Color-coded reports encourage informal project management to take place. Colors can reduce risks by alerting management quickly that a potential problem exists. One company prepared complex status reports but color-

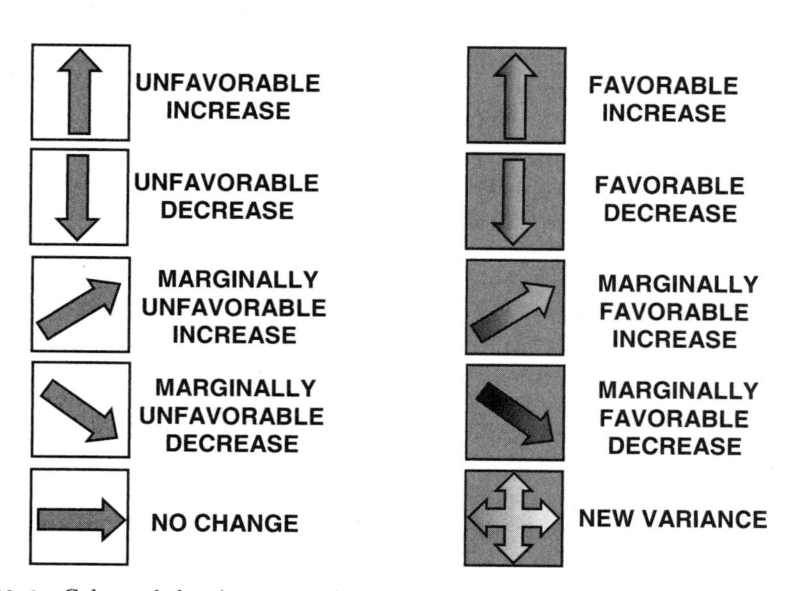

FIGURE 13–4. Color-coded variance reporting.

TABLE 13–3. RISK REPORTING

Symbol	Likelihood	Impact
●●●●	Certainty	Threat to total project success
●●●○	High	Will impact constraints/results
●●○○	Medium	May impact constraints/results
●○○○	Low	With effort can be managed
○○○○	Unlikely	Little impact

coded the right-hand margins of each page designed for specific audiences and levels of management. One executive commented that he now reads only those pages that are color-coded for him specifically rather than having to search through the entire report. In another company, senior management discovered that color-coded intranet status reporting allowed senior management to review more information in a timely manner just by focusing on those colors that indicated potential problems. Colors can be used to indicate:

- Status has not been addressed.
- Status is addressed, but no problems exist.
- Project is on course.
- A potential problem might exist in the future.
- A problem definitely exists and is critical.
- No action is to be taken on this problem.
- Activity has been completed.
- Activity is still active and completion date has passed.

Historically, we used the colors red, yellow, and green to represent status. Today, we use many more colors and shading techniques, such as shown in Figure 13–4. This type of display makes it easy to determine status or to focus specific information for specific audiences.

Another type of color-coded reporting is being used in the area of risk management as shown in Tables 13–3 and 13–4. This system has the advantage of allowing for degrees of risk but may suffer from the subjectivity of the decision-maker who controls the color of the circles.

Sometimes, executives and managers prefer to compare the numbers other than in a graphical format. In Figure 13–5, the actual total project expenditure to date is $1.58 million, whereas the position of the arrow indicates that the planned expenditure to date was

TABLE 13–4. RISK REPORTING

Risk	Likelihood	Impact	Contingency Plan
Late delivery of some raw materials	●●●○	●●○○	Secondary source suppliers qualified and available
A critical resource not available for two weeks	●●●○	●○○○	Use of overtime when resource is available

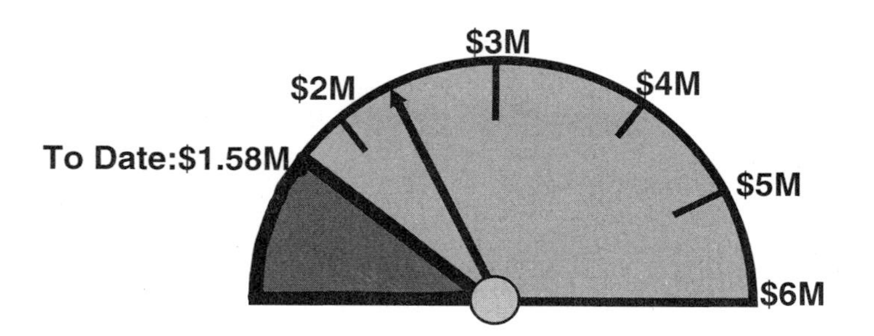

FIGURE 13–5. Project expenditures.

$2.4 million. In Figure 13–6 we see the timing variance, which shows that we predict that we will be about two weeks beyond our target end date.

13.7 INFORMAL PROJECT MANAGEMENT AT WORK

Let's review two case studies that illustrate informal project management in action.

Polk Lighting

Polk Lighting (a pseudonym) is a $35 million company located in Jacksonville, Florida. The company manufactures lamps, flashlights, and a variety of other lighting instruments. Its business is entirely based in products and services, and the company doesn't take on contract projects from outside customers. The majority of the company's stock is publicly traded. The president of Polk Lighting has held his position since the company's start-up in 1985.

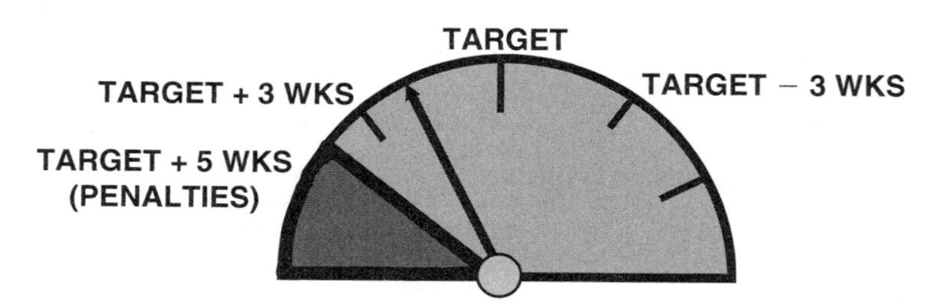

FIGURE 13–6. Customer launch: timing variance.

In 1994, activities at Polk centered on the research and development group, which the president oversaw personally, refusing to hire an R&D director. The president believed in informal management for all aspects of the business, but he had a hidden agenda for wanting to use informal project management. Most companies use informal project management to keep costs down as far as possible, but the president of Polk favored informal project management so that he could maintain control of the R&D group. However, if the company were to grow, the president would need to add more management structure, establish tight project budgets, and possibly make project management more formal than it had been. Also, the president would probably be forced to hire an R&D director.

Pressure from the company's stockholders eventually forced the president to allow the company to grow. When growth made it necessary for the president to take on heavier administrative duties, he finally hired a vice president of research and development.

Within a few years, the company's sales doubled, but informal project management was still in place. Although budgets and schedules were established as the company grew, the actual management of the projects and the way teams worked together remained informal.

The president learned two important lessons. First, the success of informal project management depends more on the culture of the organization than it does on budgets and schedules. Second, growth does not necessarily destroy an effective informal project management system.

Boeing Aerospace

Boeing was the prime contractor for the U.S. Air Force's new short-range attack missile (SRAM) and awarded the subcontract for developing the missile's propulsion system to the Thiokol Corporation.

It's generally assumed that communication between large customers and contractors must be formal because of the potential for distrust when contracts are complex and involve billions of dollars. The use of on-site representatives, however, can change a potentially contentious relationship into one of trust and cooperation when informality is introduced into the relationship.

Two employees from Boeing were carefully chosen to be on-site representatives at the Thiokol Corporation to supervise the development of the SRAM's propulsion system. The working relationship between Thiokol's project management office and Boeing's on-site representatives quickly developed into shared trust. Team meetings were held without the exchange of excessive documentation. And each party agreed to cooperate with the other. The Thiokol project manager trusted Boeing's representatives well enough to give them raw data from test results even before Thiokol's engineers could formulate their own opinions on the data. Boeing's representatives in return promised that they wouldn't relay the raw data to Boeing until Thiokol's engineers were ready to share their results with their own executive sponsors.

The Thiokol–Boeing relationship on this project clearly indicates that informal project management can work between customers and contractors. Large construction contractors have had the same positive results in using informal project management and on-site representatives to rebuild trust and cooperation.

MULTIPLE CHOICE QUESTIONS

1. In the early days of project management, formality was required because the customers wanted:
   - **A.** Proof that project management worked
   - **B.** Documentation
   - **C.** Well-defined communication channels
   - **D.** All of the above
   - **E.** A and B only
2. Aerospace and defense contractors were mandated by their customers during the early days of project management to:
   - **A.** Prepare formal documentation on policies and procedures
   - **B.** Include the policies and procedures in written proposals
   - **C.** Provide verification that the policies and procedures are being used
   - **D.** All of the above
   - **E.** A and C only
3. Which of the following is not a major difference between formal and informal project management?
   - **A.** Reporting level
   - **B.** Authority
   - **C.** Responsibility
   - **D.** Paperwork
   - **E.** None of the above
4. A typical cost per page for customer handouts is:
   - **A.** $10
   - **B.** $25
   - **C.** $50 to $100
   - **D.** $100 to $500
   - **E.** $50 to $2,000
5. Today, the execution of the project management methodology is heavily biased toward:
   - **A.** Policy manuals
   - **B.** Procedure manuals
   - **C.** Guidelines per life cycle phase
   - **D.** Guidelines per project
   - **E.** Checklists
6. Which of the following is *not* one of the four basic elements of an informal culture?
   - **A.** Authority
   - **B.** Trust
   - **C.** Cooperation
   - **D.** Effective communication
   - **E.** Teamwork
7. Formal policies and procedures are needed when which of the following exists?
   - **A.** High-intensity conflicts
   - **B.** Resistance to multiple boss reporting
   - **C.** Invisible sponsors
   - **D.** Power and authority problems
   - **E.** All of the above

8. Long-term contracts, repeat business, and sole-source contracts are a sign of:
A. Mistrust
B. Trust
C. Empowerment
D. Low cost labor, products, or services
E. None of the above

9. The actions of as little as one executive can cause an organization to go from informal to formal project management.
A. True
B. False

10. Historically, companies must go through formal project management before achieving informal project management.
A. True
B. False

11. Sponsorship at the executive levels implies mistrust.
A. True
B. False
C. Depends on other factors as well

12. Trust results in more team meetings with more documentation.
A. True
B. False

13. When an organization accepts the concept of informal project management, care must be taken that all new employees hired into management position must support the concept.
A. True
B. False

14. In the traditional organization, there is usually a disparity between employees and executives as to whether the communication process is working well.
A. True
B. False

15. Project managers spend the majority of their time communicating:
A. Upward to sponsors
B. Upward to executives
C. Laterally
D. Downward
E. Cannot be determined

16. Which of the following can be eliminated through effective communication and trust?
A. Hernia reports
B. Forensic meetings
C. Periodic team meetings
D. All of the above
E. A and B only

17. Informal project management practices allow organizations to develop standard report formats.
A. True
B. False

18. Reasonable status reports should not have staples or paper clips.
A. True
B. False

19. Status reporting must provide answers to:
A. Where are we today?
B. Where will we end up?
C. Special problems
D. All of the above
E. A and B only

20. BellSouth's "dashboard" report contains how many pages?
A. 1
B. 2
C. 5
D. 10
E. No more than 15

21. Project management software can reduce the frequency and number of meetings.
A. True
B. False

22. Cooperation can be defined as:
A. Trust
B. Effective communications
C. Effective leadership
D. Willingness to work with others
E. Teamwork

23. Work performed by people acting together with a spirit of cooperation is called:
A. Trust
B. Effective communications
C. Effective leadership
D. Willingness to work
E. Teamwork

24. Some people confuse teamwork with:
A. Morale
B. Trust
C. Cooperation
D. Effective leadership
E. Effective communication

25. __________ has more to do with attitudes than with the work itself.
A. Morale
B. Trust
C. Cooperation
D. Effective leadership
E. Effective communication

DISCUSSION QUESTIONS

1. From a quantitative perspective, what are the differences between formal and informal project management?

2. From a qualitative perspective, what are the differences between formal and informal project management?
3. Can documentation increase with informal project management? If so, under what circumstances?
4. Why is formal project management more expensive than informal project management?
5. Does informal project management eliminate paperwork?
6. Does a standard project management methodology automatically imply that a company is using informal project management?
7. Can an informal project management culture actually increase the number of team meetings with the customer?
8. Some people contend that project managers communicate vertically upward to sponsors and executives, but vertically downward to the project team. Why is it "politically incorrect" to say this?
9. Why do advocates of project management argue in favor of a one-page status report?
10. What is the difference between cooperation, teamwork, and morale?

Across

4. Willingness to work together
6. Formal sponsorship level
7. People acting together
8. Necessity for informality
9. Necessity for informality
10. Direction of communication
12. Informality benefit (2 words)
13. Communication direction
14. Formal documentation

Down

1. Benefit of informality (2 words)
2. Informal documentation
3. Necessity for informality
5. Type of project management
7. Faith
8. Necessity for informality
11. Informal benefit: _ _ _ _ _ _ _ documentation

Behavioral Excellence

14.0 INTRODUCTION

In Chapter 12 we saw that companies excellent in project management strongly emphasize training for behavioral skills. In the past it was thought that project failures were due primarily to poor planning, inaccurate estimating, inefficient scheduling, and lack of cost control. Today, excellent companies realize that project failures have more to do with behavioral shortcomings—poor employee morale, negative human relations, low productivity, and lack of commitment.

This chapter discusses these human factors in the context of situational leadership and conflict resolution. It also provides information on staffing issues in project management. Finally, the chapter offers advice on how to achieve *behavioral excellence.*

14.1 SITUATIONAL LEADERSHIP

As project management has begun to emphasize behavioral management over technical management, situational leadership has also received more attention. The average size of projects has grown, and so has the size of project teams. Process integration and effective interpersonal relations have also taken on more importance as project teams have gotten larger. Project managers now need to be able to talk with many different functions and departments. There's a contemporary project management proverb that goes something like this: "When researcher talks to researcher, there is 100 percent understanding. When researcher talks to manufacturing, there is 50 percent understanding. When researcher talks to sales, there is 0 percent understanding. But the project manager talks to all of them."

Randy Coleman, former senior vice president of the Federal Reserve Bank of Cleveland, emphasizes the importance of tolerance:

> The single most important characteristic necessary in successful project management is tolerance: tolerance of external events and tolerance of people's personalities. . . . Generally, there are two groups here at the Fed-lifers and drifters. You have to handle the two groups differently, but at the same time you have to treat them similarly. You have to bend somewhat for the independents [younger drifters] who have good creative ideas and whom you want to keep—those who take risks. You have to acknowledge that you have some trade-offs to deal with.

A senior project manager in an international accounting firm states how his own leadership style has changed from a traditional to a situational leadership style since becoming a project manager:

> I used to think that there was a certain approach that was best for leadership, but experience has taught me that leadership and personality go together. What works for one person won't work for others. So you must understand enough about the structure of projects and people, and then adopt a leadership style that suits your personality so that it comes across as being natural and genuine. It's a blending of a person's experience and personality with his or her style of leadership.

Many companies start applying project management without understanding the fundamental behavioral differences between project managers and line managers. If we assume that the line manager is not also functioning as the project manager, here are the behavioral differences:

- Project managers have to deal with multiple reporting relationships. Line managers report up a single chain of command.
- Project managers have very little real authority. Line managers hold a great deal of authority by virtue of their titles.
- Project managers often provide no input into employee performance reviews. Line managers provide formal input into the performance reviews of their direct reports.
- Project managers are not always on the management compensation ladder. Line managers always are.
- The project manager's position may be temporary. The line manager's position is permanent.
- Project managers sometimes are a lower grade level than the project team members. Line managers usually are paid at a higher grade level than their subordinates.

Several years ago, when Ohio Bell was still a subsidiary of American Telephone and Telegraph, a trainer was hired to conduct a three-day course on project management. During the customization process, the trainer was asked to emphasize planning, scheduling, and controlling, and not to bother with the behavioral aspects of project management. At that time, AT&T offered a course on how to become a line supervisor that all of the seminar participants had already taken. In the discussion that followed between the trainer and the course-content designers, it became apparent that leadership, motivation, and conflict resolutions were being taught from a superior-to-subordinate point of view in AT&T's course. When the course-content designers realized from the discussion that project managers provide leadership, motivation, and conflict resolution to employees who do not report directly to them, the trainer was allowed to include project management–related behavioral topics in the seminar.

Organizations must recognize the importance of behavioral factors in working relationships. When they do, they come to understand that project managers should be hired for their overall project management competency, not for their technical knowledge alone. Brian Vannoni, formerly site training manager and principal process engineer at GE Plastics, described his organization's approach to selecting project managers:

> The selection process for getting people involved as project managers is based primarily on their behavioral skills and their skills and abilities as leaders with regard to the other aspects of project management. Some of the professional and full-time project managers have taken senior engineers under their wing, coached and mentored them, so that they learn and pick up the other aspects of project management. But the primary skills that we are looking for are, in fact, the leadership skills.

Project managers who have strong behavioral skills are more likely to involve their teams in decision-making, and shared decision-making is one of the hallmarks of successful project management. Today, project managers are more managers of people than they are managers of technology. According to Robert Hershock, former vice president at 3M:

> The trust, respect, and especially the communications are very, very important. But I think one thing that we have to keep in mind is that a team leader isn't managing technology; he or she is managing people. If you manage the people correctly, the people will manage the technology.

In addition, behaviorally oriented project managers are more likely to delegate responsibility to team members than technically strong project managers. In 1996, Frank Jackson, formerly a senior manager at MCI, said that:

> . . . team leaders need to have a focus and a commitment to an ultimate objective. You definitely have to have accountability for your team and the outcome of your team. You've got to be able to share the decision-making. You can't single out yourself as the exclusive holder of the right to make decisions. You have got to be able to share that. And lastly again, just to harp on it one more time, is communications. Clear and concise communications throughout the team and both up and down a chain of command is very, very important.

Some organizations prefer to have a project manager with strong behavioral skills acting as the project manager, with technical expertise residing with the project engineer. Other organizations have found the reverse to be effective. Rose Russett, the program management process manager for General Motors Powertrain, states:

> We usually appoint an individual with a technical background as the program manager and an individual with a business and/or systems background as the program administrator. This combination of skills seems to complement one another. The various line managers

are ultimately responsible for the technical portions of the program, while the key responsibility of the program manager is to provide the integration of all functional deliverables to achieve the objectives of the program. With that in mind, it helps for the program manager to understand the technical issues, but they add their value not by solving specific technical problems, but by leading the team through a process that will result in the best solutions for the overall program, not just for the specific functional area. The program administrator, with input from all team members, develops the program plans, identifies the critical path, and regularly communicates this information to the team throughout the life of the program. This information is used to assist with problem solving, decision-making, and risk management.

14.2 CONFLICT RESOLUTION

Opponents of project management claim that the primary reason why some companies avoid changing over to a project management culture is that they fear the conflicts that inevitably accompany change. Conflicts are a way of life in companies with project management cultures. Conflict can occur on any level of the organization, and conflict is usually the result of conflicting objectives.

The project manager is a conflict manager. In many organizations, the project managers continually fight fires and handle crises arising from interpersonal and interdepartmental conflicts. They are so busy handling conflicts that they delegate the day-to-day responsibility for running their projects to the project teams. Although this arrangement is not the most effective, it is sometimes necessary, especially after organizational restructuring or after a new project demanding new resources has been initiated.

The ability to handle conflicts requires an understanding of why conflicts occur. We can ask four questions, the answers to which are usually helpful in handling, and possibly preventing, conflicts in a project management environment:

- Do the project's objectives conflict with the objectives of other projects currently in development?
- Why do conflicts occur?
- How can we resolve conflicts?
- Is there anything we can do to anticipate and resolve conflicts before they become serious?

Although conflicts are inevitable, they can be planned for. For example, conflicts can easily develop in a team in which the members don't understand each other's roles and responsibilities. Responsibility charts can be drawn to map out graphically who is responsible for doing what on the project. With the ambiguity of roles and responsibilities gone, the conflict is resolved, or future conflict averted.

Resolution means collaboration, and collaboration means that people are willing to rely on each other. Without collaboration, mistrust prevails and progress documentation increases.

The most common types of conflict involve the following:

- Manpower resources
- Equipment and facilities
- Capital expenditures
- Costs
- Technical opinions and trade-offs
- Priorities
- Administrative procedures
- Schedules
- Responsibilities
- Personality clashes

Each of these types of conflict can vary in intensity over the life of the project. The relative intensity can vary as a function of:

- Getting closer to project constraints
- Having met only two constraints instead of three (for example, time and performance but not cost)
- The project life cycle itself
- The individuals who are in conflict

Conflict can be meaningful in that it results in beneficial outcomes. These meaningful conflicts should be allowed to continue as long as project constraints are not violated and beneficial results accrue. An example of a meaningful conflict might be two technical specialists arguing that each has a better way of solving a problem. The beneficial result would be that each tries to find additional information to support his or her hypothesis.

Some conflicts are inevitable and occur over and over again. For example, consider a raw material and finished goods inventory. Manufacturing wants the largest possible inventory of raw materials on hand to avoid possible production shutdowns. Sales and marketing wants the largest finished goods inventory so that the books look favorable and no cash flow problems are possible.

Let's consider five methods that project managers can use to resolve conflicts:

- Confrontation
- Compromise
- Facilitation (or smoothing)
- Force (or forcing)
- Withdrawal

Confrontation is probably the most common method used by project managers to resolve conflict. Using confrontation, the project manager faces the conflict directly. With the help of the project manager, the parties in disagreement attempt to persuade one another that their solution to the problem is the most appropriate.

When confrontation doesn't work, the next approach project managers usually try is compromise. In compromise, each of the parties in conflict agrees to tradeoffs or makes concessions until a solution is arrived at that everyone involved can live with. This give-and-take approach can easily lead to a win-win solution to the conflict.

The third approach to conflict resolution is facilitation. Using facilitation skills, the project manager emphasizes areas of agreement and deemphasizes areas of disagreement. For example, suppose that a project manager said, "We've been arguing about five points, and so far we've reached agreement on the first three. There's no reason why we can't agree on the last two points, is there?" Facilitation of a disagreement doesn't resolve the conflict. Facilitation downplays the emotional context in which conflicts occur.

Force is also a method of conflict resolution. A project manager uses force when he or she tries to resolve a disagreement by exerting his or her own opinion at the expense of the other people involved. Often, forcing a solution onto the parties in conflict results in a win–lose outcome. Calling in the project sponsor to resolve a conflict is another form of force project managers sometimes use.

The least used, and least effective, mode of conflict resolution is withdrawal. A project director can simply withdraw from the conflict and leave the situation unresolved. When this method is used, the conflict does not go away and is likely to recur later.

Personality conflicts might well be the most difficult conflicts to resolve. Personality conflicts can occur at any time, with anyone, and over anything. Furthermore, they can seem almost impossible to anticipate and plan for.

Let's look at how one company found a way to anticipate and avoid personality conflicts on one of its projects. Foster Defense Group (a pseudonym) was the government contract branch of a Fortune 500 company. The company understood the potentially detrimental effects of personality clashes on its project teams, but it didn't like the idea of getting the whole team together to air its dirty laundry. The company found a better solution. The project manager put the names of the project team members on a list. Then he interviewed each of the team members one-on-one and asked each to identify the names on the list that he or she had had a personality conflict with in the past. The information remained confidential, and the project manager was able to avoid potential conflicts by separating clashing personalities.

If at all possible, conflict resolution should be handled by the project manager. When the project manager is unable to defuse the conflict, then and only then should the project sponsor be brought in to help solve the problem. Even then, the sponsor should not come in and force a resolution to the conflict. Instead, the sponsor should facilitate further discussion between the project managers and the team members in conflict.

14.3 STAFFING FOR EXCELLENCE

Project manager selection is always an executive-level decision. In excellent companies, however, executives go beyond simply selecting the project manager. They use the selection process to accomplish the following:

- Project managers are brought on board early in the life of the project to assist in outlining the project, setting its objectives, and even planning for marketing and sales. The project manager's role in customer relations becomes increasingly important.

- Executives assign project managers for the life of the project and project termination. Sponsorship can change over the life cycle of the project, but not the project manager.
- Project management is given its own career ladder.
- Project managers given a role in customer relations are also expected to help sell future project management services long before the current project is complete.
- Executives realize that project scope changes are inevitable. The project manager is viewed as a manager of change.

Companies excellent in project management are prepared for crises. Both the project manager and the line managers are encouraged to bring problems to the surface as quickly as possible so that there is time for contingency planning and problem solving. Replacing the project manager is no longer the first solution for problems on a project. Project managers are replaced only when they try to bury problems.

A defense contractor was behind schedule on a project, and the manufacturing team was asked to work extensive overtime to catch up. Two of the manufacturing people, both union employees, used the wrong lot of raw materials to produce a $65,000 piece of equipment needed for the project. The customer was unhappy because of the missed schedules and cost overruns that resulted from having to replace the useless equipment. An inquisition-like meeting was convened and attended by senior executives from both the customer and the contractor, the project manager, and the two manufacturing employees. When the customer's representative asked for an explanation of what had happened, the project manager stood up and said, "I take full responsibility for what happened. Expecting people to work extensive overtime leads to mistakes. I should have been more careful." The meeting was adjourned with no one being blamed. When word spread through the company about what the project manager did to protect the two union employees, everyone pitched in to get the project back on schedule, even working uncompensated overtime.

Human behavior is also a consideration in assigning staff to project teams. Team members should not be assigned to a project solely on the basis of technical knowledge. It has to be recognized that some people simply can't work effectively in a team environment. For example, the director of research and development at a New England company had an employee, a 50-year-old engineer, who held two master's degrees in engineering disciplines. He had worked for the previous 20 years on one-person projects. The director reluctantly assigned the engineer to a project team. After years of working alone, the engineer trusted no one's results but his own. He refused to work cooperatively with the other members of the team. He even went so far as redoing all the calculations passed on to him from other engineers on the team.

To solve the problem, the director assigned the engineer to another project on which he supervised two other engineers with less experience. Again, the older engineer tried to do all of the work himself, even if it meant overtime for him and no work for the others.

Ultimately, the director had to admit that some people are not able to work cooperatively on team projects. The director went back to assigning the engineer to one-person projects on which the engineer's technical abilities would be useful.

Robert Hershock, former vice president at 3M, once observed:

> There are certain people whom you just don't want to put on teams. They are not team players, and they will be disruptive on teams. I think that we have to recognize that and make sure that those people are not part of a team or team members. If you need their expertise, you can bring them in as a consultant to the team but you never, never put people like that on the team.
>
> I think the other thing is that I would never, ever eliminate the possibility of anybody being a team member no matter what the management level is. I think if they are properly trained, these people at any level can be participators in a team concept.

In 1996, Frank Jackson, formerly a senior manager at MCI, believed that it was possible to find a team where any individual can contribute:

> People should not be singled out as not being team players. Everyone has got the ability to be on a team and to contribute to a team based on the skills and the personal experiences that they have had. If you move into the team environment one other thing that is very important is that you not hinder communications. Communications is the key to the success of any team and any objective that a team tries to achieve.

One of the critical arguments still being waged in the project management community is whether an employee (even a project manager) should have the right to refuse an assignment. At Minnesota Power and Light, an open project manager position was posted, but nobody applied for the job. The company recognized that the employees probably didn't understand what the position's responsibilities were. After more than 80 people were trained in the fundamentals of project management, there were numerous applications for the open position.

> It's the kiss of death to assign someone to a project manager's job if that person is not dedicated to the project management process and the accountability it demands.

14.4 INTEGRATED PRODUCT/PROJECT TEAMS

In recent years there has been an effort to substantially improve the formation and makeup of teams required to develop a new product or to implement a new practice. These teams have membership from across the entire organization and are called *integrated product/project teams* (IPTs).

Since the concept of an IPT is well suited to large long-term projects, it is no wonder that the Department of Defense has been researching best practices for an IPT.[1] The gov-

1. *DOD Teaming Practices Not Achieving Potential Results,* Best Practices Series (GAO-01-510). Government Accounting Office, April 2001; much of the material in this section has been taken from this study.

ernment looked at four projects in both the public and private sectors that were highly successful using the IPT approach and four government projects that had less than acceptable results. The four successful projects using IPTs were:

- DaimlerChrysler, an automobile manufacturer located in Auburn Hills, Michigan. DaimlerChrysler's Minivan Platform team was responsible for the design, development, and production of new minivans. Team Epic, part of the minivan platform team, designed and developed electric vehicles.
- 3M, a manufacturer of a variety of industrial and consumer products located in St. Paul, Minnesota. 3M's Pluto team was responsible for the development of a new dental material.
- Hewlett-Packard, a high-technology electronic products manufacturer located in Palo Alto, California. Hewlett-Packard's Snakes Program was responsible for developing new computer workstations.
- The government's Advanced Amphibious Assault Vehicle Program.

The four projects will be discussed in the remainder of this section, together with a comparison of government and commercial IPT practices.

Background

Product development, whether for commercial or defense application, is a complex undertaking. The process begins with a concept or idea for meeting a customer's need, the idea is converted to detailed design drawings, and the design is translated into articles or prototypes that can be tested. During the product's development, the processes for manufacturing the product must also be identified and tested. The development process is characterized by a tension between competing demands on the product. These demands include the desire for the highest performance and the most features, the lowest cost and shortest time to market, and the ease with which the product can be produced in both quantity and quality. Trade-offs between these demands must be made to provide the customer with a desirable product quickly and at a reasonable price. If performance features are allowed to dominate, the product may become too expensive. If costs are cut too much, the product's quality may suffer. A product design that ignores the limits of manufacturing processes may never make it into the hands of the customer.

Taking a product from idea to delivery requires expertise from a number of different professions or functions, which can vary depending on the type of product. To illustrate, designing a product's features may require the collaboration of people with expertise in areas such as mechanical, electrical, materials, and software engineering. People with a financial management background are needed to accurately estimate the cost of the product and to keep track of the budget. People expert in test and evaluation are needed to objectively assess the performance of product prototypes. Production engineers make sure that the design lends itself to proven manufacturing processes, even developing new processes when necessary. Quality assurance experts ensure that defects are kept out of the product design and manufacturing processes. Yet another group of people is responsible for understanding and representing the customer's needs, often part of the marketing function in commercial industry.

In the 1980s, companies began to look for better ways to bring the knowledge of the people in different functions together in the design phase of a new product to reduce rework and shorten cycle times. They organized teams made up of a cross section of the different functional disciplines and gave them responsibility for developing an entire product. These efforts evolved into the IPT approach as it is known today. In the 1990s, Boeing received acclaim for the success of its 777 aircraft, developed by using design/build teams that were IPTs.

Under the IPT approach, each team possesses the knowledge to collaboratively identify problems and propose solutions, minimizing the amount of rework that has to be done. When this knowledge is accompanied by the authority to make key product decisions, IPTs can make trade-offs between competing demands and more quickly make design changes, if necessary.

Two elements are essential to determining whether a team is, in fact, an integrated product team: the knowledge and authority needed to recognize problems and make cross-cutting decisions expeditiously. Knowledge is sufficient when the team has the right mix of expertise to master the different facets of product development. Authority is present when the team is responsible for making both day-to-day decisions and delivering the product. In the programs experiencing problems, the teams either did not have the authority or the right mix of expertise to be considered integrated product teams. If a team lacks expertise, it will miss opportunities to recognize potential problems early; without authority, it can do little about them. Although these teams were called integrated product teams, by and large they were not.

Leading commercial firms took steps to create an environment more conducive to the integrated product team approach. They committed to making the approach integral to the product development process and backed up the commitment through actions to ensure that implementation was not left to chance. Importantly, the pressures of competing in the commercial market meshed well with the decision-making advantages of integrated product teams.

Reducing Cost and Cycle Time

Officials at leading commercial firms and the Advanced Amphibious Assault Vehicle Program attribute their successful product outcomes directly to their IPTs. Specifically, DaimlerChrysler officials attributed reduced cycle time, improved product performance, and better market success to their switch to IPTs. Hewlett-Packard officials stated that the company's teaming approach resulted in higher product quality, better design results, and improved system integration. A Hewlett-Packard official stated that the Snakes Program team simultaneously developed three computer workstations in nine months, half the time normally required, with four times the performance of existing workstations. A Hewlett-Packard IPT developing printer equipment increased productivity sixfold, despite using one-fourth fewer employees, and reduced the product defect rate to 2 percent, of which the majority were cosmetic defects. In another example, company officials said that in the past, test equipment was developed at a cost between $25,000 and $70,000 and required up to four years to develop—which was well behind the performance of their competitors. Now, their IPT approach enables the company to develop a higher-quality product in two-thirds less time and with a price of $10,000 to $25,000.

A 3M official in the dental products division stated that the Pluto IPT created a revolutionary dental material that surpasses similar products on the market. The team leader reported that members developed a material that shrinks 50 percent less than current materials and can withstand 80 percent more stress. In addition, team members filed five patents, of which four have been issued—a valuable benefit to the company. The team leader attributes the IPTs decision-making, including the trade-off between product performance and schedule, with shortening the product development time as much as 18 months. In addition, the team leader believes that the IPT will outperform the competition because 3M's patents make it difficult for other companies to bring a product to market in a similar technology area. Finally, Advanced Amphibious Assault Vehicle Program officials believe their IPT approach as critical to the program's ability to meet or exceed its cost, schedule, and performance objectives since it began in 1995—atypical for large DoD programs.

The successful IPT projects are shown in Table 14–1. The unsuccessful IPT projects are shown in Table 14–2. In analyzing the data, the government came up with the results shown in Figure 14–1. Each vertical line in Figure 14–1 is a situation where the IPT must go outside its own domain to seek information and approvals. Each time this happens, it is referred to as a "hit." The government research indicated that the greater the number of hits, the more likely it is that the time, cost, and performance constraints will not be achieved. The research confirmed that if the IPT has the knowledge necessary, and the authority, to make decisions, the desired performance would be achieved. Hits will delay decisions and cause schedule slippages.

The Makeup of an IPT

The IPT consists of a sponsor, program (or project) manager, and the core team. For the most part, members of the core team are assigned full-time to the team but may not be on the team for the duration of the entire project.

The skills needed to be a member of the core team include:

- Self-starter ability
- Work without supervision
- Good communication skills

TABLE 14–1. EFFECTIVE IPTs

Program	Cost Status	Schedule Status	Performance Status
DaimlerChrysler	Product cost was lowered	Decreased development cycle months by 50 percent	Improved vehicle designs
Hewlett-Packard	Lowered cost by over 60 percent	Shortened development schedule by over 60 percent	Improved system integration and product design
3M	Outperformed cost goals	Product deliveries shortened by 12 to 18 months	Improved performance by 80 percent
Advanced Amphibious Assault Vehicle	Product unit cost lower than original estimate	Ahead of original development schedule	Demonstrated five-fold increase in speed

TABLE 14–2. INEFFECTIVE IPTs

Program	Cost Status	Schedule Status	Performance Status
CH-60S Helicopter	Increased cost but due to additional purchases	Schedule delayed	Software and structural difficulties
Extended Range Guided Munitions	Increases in development costs	Schedule slipped three years	Redesigning due to technical difficulties
Global Broadcast Service	Experiencing cost growth	Schedule slipped 1.5 years	Software and hardware design shortfalls
Land Warrior	Cost increase of about 50 percent	Schedule delayed four years	Overweight equipment, inadequate battery power and design

- Cooperative
- Technical understanding
- Willing to learn backup skills
- Able to perform feasibility studies and cost–benefit analyses
- Able to perform or assist in market research studies
- Able to evaluate asset utilization
- Decision-maker
- Knowledgeable in risk management
- Understand the need for continuous validation

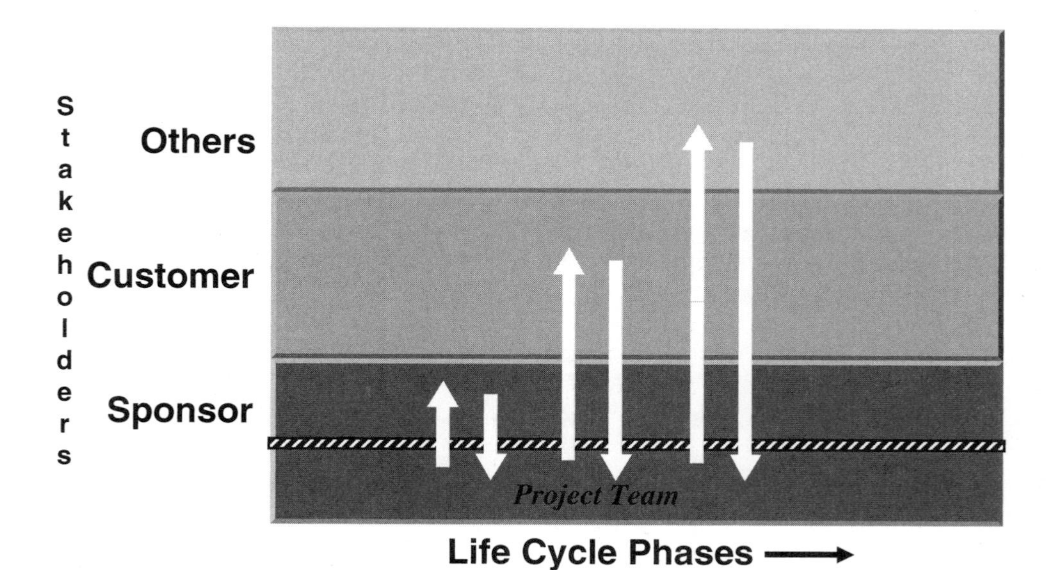

FIGURE 14–1. Knowledge and authority.

Each IPT is given a project charter that identifies the project's mission and identifies the assigned project manager. However, unlike traditional charters, the IPT charter can also identify the key members of the IPT by name or job responsibility.

Unlike traditional project teams, the IPT thrives on sharing information across the team and collective decision-making. IPTs eventually develop their own culture and, as such, can function in either a formal or informal capacity.

IPTs at leading commercial firms are given control over selecting members. The firms believe that it is very important that a team have the right expertise on the team. As a result, team leaders are hand-selected by upper management based on reputation, knowledge, and/or expertise. In turn, team leaders select team members they believe have the expertise and the interpersonal skills that would match the team's needs.

At 3M, when a new team is formed, an announcement is sent to employees notifying them of the team's purpose, time frame, and skills needed. Employees are allowed to volunteer for teams. Team leaders select team members from the pool of volunteers. According to company officials, the self-nomination process allows staff to demonstrate commitment and alignment with the team goals and ensures that the team members share a common purpose.

Authority and Knowledge

Effective IPTs possess the knowledge and authority essential to the kind of decision-making that is their hallmark. Knowledge is sufficient when the team has the right mix of expertise to master the different facets of product development. Authority is present when the team is responsible for making day-to-day decisions and delivering the product. These two elements are essential to determining whether a team is in fact an IPT. Other factors significantly enhance an IPTs effectiveness. For the programs we studied, effective IPTs had key members physically collocated where possible to facilitate the communication, interaction, and overall operations. When physical collocation was not possible, resources were provided to connect members through virtual means, such as shred software. Effective IPTs were also given control over selecting members, and changes in membership were driven by the team's need for different knowledge of skills.

In the programs experiencing product development problems, the teams either did not have responsibility for product development or were missing key areas of expertise. Although called IPTs, in reality they were not. If a team is missing either the knowledge or the authority to recognize and make difficult decisions, it is ill equipped to carry out the role expected from an IPT. Some of these programs had separate DoD and defense contractor teams, which further dispersed knowledge and authority. Moreover, DoD did not routinely collocate team members. Less effective teams also did not have control over their composition. Team membership fluctuated often but did not appear to be tied directly to the needs of the project; members left and joined the team due to personnel rotation policies or for other reasons.

Research shows that product development responsibility and cross-functional membership are fundamental IPT elements. If a team lacks expertise, it will miss opportunities to recognize potential problems early; without authority, it can do little about them. IPTs in leading commercial firms and the Advanced Amphibious Assault Vehicle Program had the right cross section of functional disciplines to develop new products. Their IPTs were responsible for developing and delivering the product and making day-to-day decisions on cost, design, performance, quality, test, and manufacturing issues. The combination of product responsibility and expertise equipped the IPTs with the information needed to tackle crucial issues—such as trade-offs—without having to rely heavily on organizations outside the IPT. Once so-equipped, the collocation of team members and control over the selection of members made the IPTs even better.

Along with being responsible for developing a complex new dental material, 3M's Pluto IPT had the authority to conduct research, select material attributes based on customer needs, determine the delivery schedule, estimate the cost of the material, and perform and evaluate the scientific experiments to create the material. To meet these expectations, the team possessed all key areas of expertise.

Hewlett-Packard's Snakes IPT consisted of representatives from research and development, marketing, quality, leadership, finance, and manufacturing. Collectively, the IPT is responsible for designing, developing, and building new computer workstations. Company officials noted that the breadth of knowledge on the IPT not only speeded up the pace of development but the amount of innovation as well. They also stated that IPTs may also include customers and suppliers.

Similarly, DaimlerChrysler's minivan platform team were comprised of design engineers and representatives from planning, finance, marketing, procurement, and manufacturing. They were vested with full authority to design, develop, and produce new vehicle lines. Given the complexity of developing a vehicle, smaller IPTs concentrated on developing component parts, such as the door. Even the door IPT included specialists for sheet metal, glass, hardware, wiring, electrical switches, customer liaison, and manufacturing. This IPT addressed day-to-day issues on designing door features, determining performance characteristics, and constructing the door. Equally important, the IPT was responsible for ensuring that the entire door is ready when production of the vehicle starts. If it were not, the IPT could delay the entire delivery schedule.

Similarly, the Firepower IPT on the Advanced Amphibious Assault Vehicle program had responsibility for designing, developing, prototyping, and testing the gun system, including the barrel, ammunition feeder, and the gunner's station. The IPT had members from engineering, testing, logistics, cost estimating, manufacturing, and modeling and simulation. Importantly, these members were drawn from the Marine Corps acquisition workforce, weapon system operators, and the defense contractor and subcontractors responsible for building the system.

Commercial versus Government Organizations

Differences in the environment in which teams operate can have a significant effect on implementing the IPT approach successfully. The study found that leading commercial firms provided a more supportive foundation for IPTs. Company leaders committed to the IPT approach and backed up that commitment through actions designed to ensure that implementation was not left to chance. In short, they created a different, more conducive environment for IPTs. Although the DoD endorsed the IPT approach and had issued policies and other guidance, it had not taken steps to ensure that IPTs were implemented at the program execution level. In essence, the IPT approach had been left to germinate in an unchanged environment that is not necessarily conducive to IPTs. Successful implementation was thus more dependent on the ingenuity of people working on programs.

Differences in how commercial firms and DoD managers measure success and in the pressures they face in starting programs significantly affect the environment for integrated product teams. Commercial products' success is measured in terms of the customer's acceptance of the final product and cycle times short enough to beat the competition. These conditions create incentives for gaining knowledge early, forming realistic goals and estimates, and holding teams accountable for delivering the product—all of which favor an IPT approach. In DoD, the pressures to launch new programs successfully and protect their funding, coupled with long cycle times, create incentives to be overly optimistic in seeing program goals and to focus on process—versus product—concerns. DoD's necessary reliance on defense contractors introduces another complication for IPTs because two major organizations (DoD and defense contractors) are responsible for the product, and they do not necessarily share the same incentives. Notably, the amphibious vehicle program has overcome these obstacles and made IPTs work in the DoD environment.

DaimlerChrysler, 3M, and Hewlett-Packard all provided an environment that supported the IPT approach to product development. Corporate leaders not only embraced the IPT approach, but demonstrated their commitment by reorganizing to better align their structure with IPTs and making targeted investments in physical assets, training, and other forms of help. The firms delegated considerable power to IPTs, such as in setting product development goals, but held the teams accountable for delivering on those goals. In addition, the pressures of competing successfully in the marketplace—that foster realism, short cycle times, and satisfying the customer—play well to the strengths of the IPT approach.

Although DaimlerChrysler and 3M did not plan to restructure their organizations when they decided to implement IPTs, they found that their former organizations were at odds with the IPT approach. For example, in the 1980s, DaimlerChrysler (then Chrysler) had separate organizations for key functions, such as engineering, finance, and manufacturing. Moreover, engineers were organized around the types of components—such as climate control—rather than product types. This organization made it difficult even for all the engineers working on a particular vehicle to talk with one another, let alone interact with functions other than engineering, such as finance. DaimlerChrysler realized that IPTs could not simply be patched across such organizations. This realization was followed by a corporate reorganization along platform lines—classes of vehicles—to reinforce the emphasis on products rather than functions or components.

The companies took other steps to reinforce their commitment to IPTs. For example, DaimlerChrysler officials noted that some employees were resistant to the IPT approach. To encourage employee acceptance and ensure that organizational and product goals were achieved, a two-pronged performance appraisal process was instituted that solicited input from both an IPT member's immediate supervisor and other members and from organizations with which the member interfaced. The study found that at 3M and Hewlett-Packard, the IPT leader either prepared the members' performance evaluations or provided significant input to it. Officials noted that capturing an individual's performance on an IPT was a driving factor in garnering acceptance of the IPT approach.

In addition to the physical infrastructure investments made to collocate and integrate the workplace, the companies invested other resources to ensure that IPTs were implemented successfully at the product development level. In an earlier report on best training practices, researchers noted that leading firms focus on a few key initiatives at any one time and deliver targeted, hands-on training to ensure that implementation is successful at the product development level.[2] DaimlerChrysler, 3M, and Hewlett-Packard took the same approach. These companies offered extensive front-end planning assistance. For example, Hewlett-Packard helps new teams plan and define their priorities and track their progress. A company official believed that this help could reduce a project's time by 10 to 20 percent. At 3M, the company provided team sponsors who were top managers who established the IPT and assisted the team with leadership and high-level decision-making. In some cases, the companies provided facilitators who assisted IPTs with hands-on guidance to enhance their daily performance.

14.5 VIRTUAL PROJECT TEAMS

Historically, project management was a face-to-face environment where team meetings involved all players meeting together in one room. Today, because of the size and complexity of projects, it is impossible to find all team members located under one roof. Duarte and Snyder define seven types of virtual teams.[3] These are shown in Table 14–3.

Culture and technology can have a major impact on the performance of virtual teams. Duarte and Snyder have identified some of these relationships in Table 14–4.

The importance of culture cannot be understated. Duarte and Snyder identify four important points to remember concerning the impact of culture on virtual teams. The four points are[4]:

1. There are national cultures, organizational cultures, functional cultures, and team cultures. They can be sources of competitive advantages for virtual teams that know how to use cultural differences to create synergy. Team leaders and members who understand and are sensitive to cultural differences can create more robust outcomes than can members of homogeneous teams with members who think and act alike. Cultural differences can create distinctive advantages for teams if they are understood and used in positive ways.

---

2. *DOD Training Can Do More to Help Weapon Systems Programs Implement Best Practices,* Best Practices Series *(GAO/NSIAD-99-206),* August 16, 1999.

3. Deborah L. Duarte and Nancy Tennant Snyder, *Mastering Virtual Teams.* San Francisco: Jossey-Bass, an imprint of John Wiley & Sons, 2001, p. 10; reproduced by permission of John Wiley & Sons.

4. See note 3, p. 70.

TABLE 14–3. TYPES OF VIRTUAL TEAMS

Type of Team	Description
Network	Team membership is diffuse and fluid; members come and go as needed. Team lacks clear boundaries with the organization.
Parallel	Team has clear boundaries and distinct membership. Team works in the short term to develop recommendations for an improvement in a process or system.
Project or product development	Team has fluid membership, clear boundaries, and a defined customer base, technical requirement, and output. Longer-term team task is nonroutine, and the team has decision-making authority.
Work or production	Team has distinct membership and clear boundaries. Members perform regular and outgoing work, usually in one functional area.
Service	Team has distinct membership and supports ongoing customer network activity.
Management	Team has distinct membership and works on a regular basis to lead corporate activities.
Action	Team deals with immediate action, usually in an emergency situation. Membership may be fluid or distinct.

2. The most important aspect of understanding and working with cultural differences is to create a team culture in which problems can be surfaced and differences can be discussed in a productive, respectful manner.
3. It is essential to distinguish between problems that result from cultural differences and problems that are performance based.
4. Business practices and business ethics vary in different parts of the world. Virtual teams need to clearly articulate approaches to these that every member understands and abides by.

TABLE 14–4. TECHNOLOGY AND CULTURE

Cultural Factor	Technological Considerations
Power distance	Members from high-power-distance cultures may participate more freely with technologies that are asynchronous and allow anonymous input. These cultures sometimes use technology to indicate status differences between team members.
Uncertainty avoidance	People from cultures with high uncertainty avoidance may be slower adopters of technology. They may also prefer technology that is able to produce more permanent records of discussions and decisions.
Individualism–collectivism	Members from highly collectivistic cultures may prefer face-to-face interactions.
Masculinity–femininity	People from cultures with more "feminine" orientations are more prone to use technology in a nurturing way, especially during team startups.
Context	People from high-context cultures may prefer more information-rich technologies, as well as those that offer opportunities for the feeling of social presence. They may resist using technologies with low social presence to communicate with people they have never met. People from low-context cultures may prefer more asynchronous communications.

Source: Deborah L Duarte and Nancy Tennant Snyder, *Mastering Virtual Teams*. San Francisco: Jossey-Bass, an imprint of John Wiley & Sons, 2001, p. 60.

14.6 REWARDING PROJECT TEAMS

Today, most companies are using project teams. However, there still exist challenges in how to reward project teams for successful performance. The importance of how teams are rewarded is identified by Parker, McAdams, and Zielinski[5]:

> Some organizations are fond of saying, "We're all part of the team,' but too often it is merely management-speak. This is especially common in conventional hierarchical organizations; they say the words but don't follow up with significant action. Their employees may read the articles and attend the conferences and come to believe that many companies have turned collaborative. Actually, though, few organizations today are genuinely team-based.
>
> Others who want to quibble point to how they reward or recognize teams with splashy bonuses or profit-sharing plans. But these do not by themselves represent a commitment to teams; they're more like a gift from a rich uncle. If top management believes that only money and a few recognition programs ('team of year' and that sort of thing) reinforce teamwork, they are wrong. These alone do not cause fundamental change in the way people and teams are managed.
>
> But in a few organizations, teaming is a key component of the corporate strategy, involvement with teams is second nature, and collaboration happens without great thought or fanfare. There are natural work groups (teams of people who do the same or similar work in the same location), permanent cross-functional teams, ad hoc project teams, process improvement teams, and real management teams. Involvement just happens.

Why is it so difficult to reward project teams? To answer this question, we must understand what a team is and is not. According to Parker et al.[6]:

> Consider this statement: an organizational unit can act like a team, but a team is not necessarily an organizational unit, at least for describing reward plans.
>
> An organizational unit is just that, a group of employees organized into an identifiable business unit that appears on the organizational chart. They may behave in a spirit of teamwork, but for the purposes of developing reward plans they are not a "team." The organizational unit may be a whole company, a strategic business unit, a division, a department, or a work group.
>
> A "team" is a small group of people allied by a common project and sharing performance objectives. They generally have complementary skills or knowledge and an interdependence that requires that they work together to accomplish their project's objective. Team members hold themselves mutually accountable for their results. These teams are not found on an organization chart.

Incentives are difficult to apply because project teams may not appear on an organizational chart. Figure 14–2 shows the reinforcement model for employees.[7] For project teams, the emphasis is the three arrows on the right-hand side of Figure 14–2.

---

5. G. Parker, J. McAdams, and D. Zielinski, *Rewarding Teams.* San Francisco: Jossey-Bass, an imprint of John Wiley & Sons, 2000, p. 17; reproduced by permission of John Wiley & Sons.

6. See note 5, p. 17.

7. See note 5, p. 29.

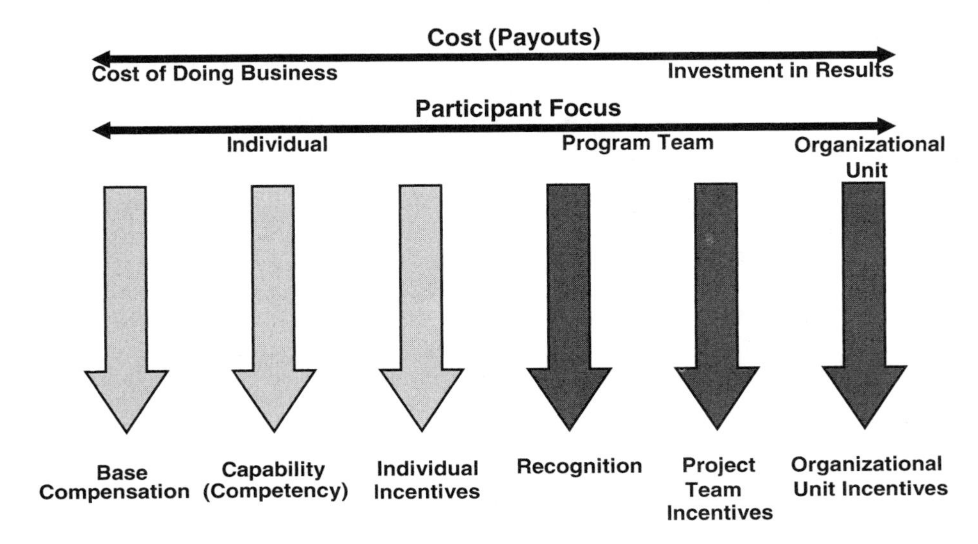

FIGURE 14–2. Reinforcement model.

Project team incentives are important because team members expect appropriate rewards and recognition. According to Parker et. al.[8]:

> Project teams are usually, but not always, formed by management to tackle specific projects or challenges with a defined time frame—reviewing processes for efficiency or cost-savings recommendations, launching a new software product, or implementing enterprise resource planning systems are just a few examples. In other cases, teams self-form around specific issues or as part of continuous improvement initiatives such as team-based suggestion systems.
>
> Project teams can have cross-functional membership or simply be a subset of an existing organizational unit. The person who sponsors the team—its "champion"—typically creates an incentive plan with specific objective measures and an award schedule tied to achieving those measures. To qualify as an incentive, the plan must include preannounced goals, with a "do this, get that" guarantee for teams. The incentive usually varies with the value added by the project.
>
> Project team incentive plans usually have some combination of these basic measures:
>
> - *Project Milestones:* Hit a milestone, on budget and on time, and all team members earn a defined amount. Although sound in theory, there are inherent problems in tying financial incentives to hitting milestones. Milestones often change for good reason (technological advances, market shifts, other developments) and you don't want the team and management to get into a negotiation on slipping dates to trigger the incentive. Unless milestones are set in stone and reaching them is simply a function of the team

8. See note 5, pp. 38–39.

doing its normal, everyday job, it's generally best to use recognition—after-the-fact celebration of reaching milestones—rather than typing financial incentives to it.

Rewards need not always be time-based, such that when the team hits a milestone by a certain date it earns a reward. If, for example, a product development team debugs a new piece of software on time, that's not necessarily a reason to reward it. But if it discovers and solves an unsuspected problem or writes better code before a delivery date, rewards are due.

- *Project Completion:* All team members earn a defined amount when they complete the project on budget and on time (or to the team champion's quality standards).
- *Value Added:* This award is a function of the value added by a project, and depends largely on the ability of the organization to create and track objective measures. Examples include reduced turnaround time on customer requests, improved cycle times for product development, cost savings due to new process efficiencies, or incremental profit or market share created by the product or service developed or implemented by the project team.

One warning about project incentive plans: they can be very effective in helping teams stay focused, accomplish goals, and feel like they are rewarded for their hard work, but they tend to be exclusionary. Not everyone can be on a project team. Some employees (team members) will have an opportunity to earn an incentive that others (non-team members) do not. There is a lack of internal equity. One way to address this is to reward core team members with incentives for reaching team goals, and to recognize peripheral players who supported the team, either by offering advice, resources, or a pair of hands, or by covering for project team members back at their regular job.

Some projects are of such strategic importance that you can live with these internal equity problems and non-team members' grousing about exclusionary incentives. Bottom line, though, is this tool should be used cautiously.

Some organizations focus only on cash awards. However, Parker et al. have concluded from their research that noncash awards can work equally well, if not better, than cash awards[9]:

Many of our case organizations use non-cash awards because of their staying power. Everyone loves money, but cash payments can lose their motivational impact over time. However, non-cash awards carry trophy value that has great staying power because each time you look at that television set or plaque you are reminded of what you or your team did to earn it. Each of the plans encourages awards that are coveted by the recipients and, therefore, will be memorable.

If you ask employees what they want, they will invariably say cash. But providing it can be difficult if the budget is small or the targeted earnings in an incentive plan are modest. If you pay out more often than annually and take taxes out, the net amount may look pretty small, even cheap. Non-cash awards tend to be more dependent on their symbolic value than their financial value.

Non-cash awards come in all forms: a simple thank-you, a letter of congratulations, time off with pay, a trophy, company merchandise, a plaque, gift certificates, special services, a

9. See note 5, pp. 190–191.

dinner for two, a free lunch, a credit to a card issued by the company for purchases at local stores, specific items or merchandise, merchandise from an extensive catalogue, travel for business or a vacation with the family, and stock options. Only the creativity and imagination of the plan creators limit the choices.

14.7 KEYS TO BEHAVIORAL EXCELLENCE

There are some distinguishing actions that project managers can take to ensure the successful completion of their projects. These include:

- Insisting on the right to select key project team
- Negotiating for key team members with proven track records in their fields
- Developing commitment and a sense of mission from the outset
- Seeking sufficient authority from the sponsor
- Coordinating and maintaining a good relationship with the client, parent, and team
- Seeking to enhance the public's image of the project
- Having key team members assist in decision-making and problem-solving
- Developing realistic cost, schedule, and performance estimates and goals
- Maintaining backup strategies (contingency plans) in anticipation of potential problems
- Providing a team structure that is appropriate, yet flexible and flat
- Going beyond formal authority to maximize their influence over people and key decisions
- Employing a workable set of project planning and control tools
- Avoiding overreliance on one type of control tool
- Stressing the importance of meeting cost, schedule, and performance goals
- Giving priority to achieving the mission or function of the end-item
- Keeping changes under control
- Seeking ways to assure job security for effective project team members

Earlier in this book, I claimed that a project cannot be successful unless it is recognized as a project and gains the support of top-level management. Top-level management must be willing to commit company resources and provide the necessary administrative support so that the project becomes part of the company's day-to-day routine of doing business. In addition, the parent organization must develop an atmosphere conducive to good working relationships among the project manager, parent organization, and client organization.

There are actions that top-level management should take to ensure that the organization as a whole supports individual projects and project teams, as well as the overall project management system:

- Showing a willingness to coordinate efforts
- Demonstrating a willingness to maintain structural flexibility

- Showing a willingness to adapt to change
- Performing effective strategic planning
- Maintaining rapport
- Putting proper emphasis on past experience
- Providing external buffering
- Communicating promptly and accurately
- Exhibiting enthusiasm
- Recognizing that projects do, in fact, contribute to the capabilities of the whole company

Executive sponsors can take the following actions to make project success more likely:

- Selecting a project manager at an early point in the project who has a proven track record in behavioral skills and technical skills
- Developing clear and workable guidelines for the project manager
- Delegating sufficient authority to the project manager so that she or he can make decisions in conjunction with the project team members
- Demonstrating enthusiasm for and commitment to the project and the project team
- Developing and maintaining short and informal lines of communication
- Avoiding excessive pressure on the project manager to win contracts
- Avoiding arbitrarily slashing or ballooning the project team's cost estimate
- Avoiding "buy-ins"
- Developing close, not meddlesome, working relationships with the principal client contact and the project manager

The client organization can exert a great deal of influence on the behavioral aspects of a project by minimizing team meetings, rapidly responding to requests for information, and simply allowing the contractor to conduct business without interference. The positive actions of client organizations also include:

- Showing a willingness to coordinate efforts
- Maintaining rapport
- Establishing reasonable and specific goals and criteria for success
- Establishing procedures for making changes.
- Communicating promptly and accurately
- Committing client resources as needed
- Minimizing red tape
- Providing sufficient authority to the client's representative, especially in decision-making

With these actions as the basic foundation, it should be possible to achieve behavioral success, which includes:

- Encouraging openness and honesty from the start from all participants
- Creating an atmosphere that encourages healthy competition, but not cutthroat situations or liar's contests
- Planning for adequate funding to complete the entire project

- Developing a clear understanding of the relative importance of cost, schedule, and technical performance goals
- Developing short and informal lines of communication and a flat organizational structure
- Delegating sufficient authority to the principal client contact and allowing prompt approval or rejection of important project decisions
- Rejecting "buy-ins"
- Making prompt decisions regarding contract okays or go-aheads
- Developing close working relationships with project participants
- Avoiding arm's-length relationships
- Avoiding excessive reporting schemes
- Making prompt decisions on changes

Companies that are excellent in project management have gone beyond the standard actions as listed previously. These additional actions for excellence include the following:

- The outstanding project manager has these demonstrable qualities:
  - Understands and demonstrates competency as a project manager
  - Works creatively and innovatively in a nontraditional sense only when necessary; does not look for trouble
  - Demonstrates high levels of self-motivation from the start
  - Has a high level of integrity; goes above and beyond politics and gamesmanship
  - Is dedicated to the company and not just the project; is never self-serving
  - Demonstrates humility in leadership
  - Demonstrates strong behavioral integration skills both internally and externally
  - Thinks proactively rather than reactively
  - Is willing to assume a great deal of risk and will spend the appropriate time needed to prepare contingency plans
  - Knows when to handle complexity and when to cut through it; demonstrates tenaciousness and perseverance
  - Is willing to help people realize their full potential; tries to bring out the best in people
  - Communicates in a timely manner and with confidence rather than despair
- The project manager maintains high standards of performance for self and team, as shown by these approaches:
  - Stresses managerial, operational, and product integrity
  - Conforms to moral codes and acts ethically in dealing with people internally and externally
  - Never withholds information
  - Is quality conscious and cost conscious
  - Discourages politics and gamesmanship; stresses justice and equity
  - Strives for continuous improvement but in a cost-conscious manner
- The outstanding project manager organizes and executes the project in a sound and efficient manner:
  - Informs employees at the project kickoff meeting how they will be evaluated
  - Prefers a flat project organizational structure over a bureaucratic one

- Develops a project process for handling crises and emergencies quickly and effectively
- Keeps the project team informed in a timely manner
- Does not require excessive reporting; creates an atmosphere of trust
- Defines roles, responsibilities, and accountabilities up front
- Establishes a change management process that involves the customer

- The outstanding project manager knows how to motivate:
  - Always uses two-way communication
  - Is empathetic with the team and a good listener
  - Involves team members in decision-making; always seeks ideas and solutions; never judges an employee's idea hastily
  - Never dictates
  - Gives credit where credit is due
  - Provides constructive criticism rather than making personal attacks
  - Publicly acknowledges credit when credit is due but delivers criticism privately
  - Makes sure that team members know that they will be held accountable and responsible for their assignments
  - Always maintains an open-door policy; is readily accessible, even for employees with personal problems
  - Takes action quickly on employee grievances; is sensitive to employees' feelings and opinions
  - Allows employees to meet the customers
  - Tries to determine each team member's capabilities and aspirations; always looks for a good match; is concerned about what happens to the employees when the project is over
  - Tries to act as a buffer between the team and administrative/operational problems
- The project manager is ultimately responsible for turning the team into a cohesive and productive group for an open and creative environment. If the project manager succeeds, the team will exhibit the following behaviors:
  - Demonstrates innovation
  - Exchanges information freely
  - Is willing to accept risk and invest in new ideas
  - Is provided with the necessary tools and processes to execute the project
  - Dares to be different, is not satisfied with simply meeting the competition
  - Understands the business and the economics of the project
  - Tries to make sound business decisions rather than just sound project decisions

MULTIPLE CHOICE QUESTIONS

1. In the past, we believed that project failure was due to:
   A. Poor scheduling
   B. Poor estimating
   C. Poor behavior
   D. All of the above
   E. A and B only
2. Today, we believe that project failure is due to:

- **A.** Poor employee morale
- **B.** Negative human relations
- **C.** Lack of commitment
- **D.** Low productivity
- **E.** All of the above

3. Failure occurs when companies apply project management without first:
- **A.** Understanding the differences between project and line management
- **B.** Assigning an executive sponsor
- **C.** Preparing job descriptions for project managers
- **D.** Making project management a career path position
- **E.** Providing the project manager with authority

4. Project management is closely aligned with __________ leadership.
- **A.** Project manager-centered
- **B.** Sponsor-centered
- **C.** Situational
- **D.** Authoritative
- **E.** Facilitative

5. The most critical behavioral problem in project management is:
- **A.** Assigning authority to the project manager
- **B.** Roles and responsibilities
- **C.** Executive sponsorship
- **D.** Customer interface
- **E.** Multiple boss reporting

6. In general, which of the following is a behavioral difference between project and line management?
- **A.** Amount of authority
- **B.** Wage and salary administration for employees
- **C.** Salary structure
- **D.** Reporting structure
- **E.** All of the above

7. Randy Coleman, former senior vice president of the Federal Reserve Bank of Cleveland, believes that the most important behavioral skill in project management is:
- **A.** Honesty
- **B.** Decision-making
- **C.** Tolerance
- **D.** Respect for people
- **E.** Persistence

8. Project managers today are selected more for their __________ skills than for their __________ skills.
- **A.** Decision-making; technical
- **B.** Risk management; scheduling
- **C.** Financial; technical
- **D.** Risk management; organizational
- **E.** Behavioral; technical

9. Resistance to project management implementation exists because people today believe that conflicts inevitably accompany change.
- **A.** True
- **B.** False

10. Conflict is usually the result of:
- **A.** Poor scheduling
- **B.** Assigning the wrong people
- **C.** Conflicting objectives
- **D.** Personality clashes
- **E.** Poor project administration

11. The current view of conflict is:
- **A.** Conflict is bad and should be avoided
- **B.** Conflict management has no place in project management
- **C.** The project sponsor is responsible for conflict resolution
- **D.** Conflict can be good if managed correctly
- **E.** None of the above

12. Although conflicts are inevitable, they can be planned for.
- **A.** True
- **B.** False

13. The intensity of a conflict can vary as a function of:
- **A.** Approaching the constraints
- **B.** Having fewer constraints
- **C.** The project life cycle phase
- **D.** The individuals who are in conflict
- **E.** All of the above

14. Conflicts that result in beneficial outcomes are called __________ conflicts.
- **A.** Low-level
- **B.** Low-tolerance
- **C.** Inconsequential
- **D.** Meaningful
- **E.** Strategic

15. The most common conflict resolution mode used by project managers is:
- **A.** Confrontation
- **B.** Compromise
- **C.** Smoothing
- **D.** Forcing
- **E.** Withdrawal

16. Which conflict resolution mode is generally a give-and-take or win-win solution?
- **A.** Confrontation
- **B.** Compromise
- **C.** Smoothing
- **D.** Forcing
- **E.** Withdrawal

17. Most project managers seem to agree that the least used conflict resolution mode is:
- **A.** Confrontation
- **B.** Compromise
- **C.** Smoothing
- **D.** Forcing
- **E.** Withdrawal

18. The most difficult conflicts to resolve might very well be:
- **A.** Scheduling

B. Costs
C. Manpower
D. Technical opinions
E. Personalities

19. Project sponsors should be brought in to resolve a conflict only as a last resort.
A. True
B. False

20. Project manager selection is a(an) __________ decision.
A. Executive-level
B. Steering committee
C. Middle-management
D. Customer-focused
E. None of the above

21. Project __________ should be on board for the life of the project, whereas project __________ can change over the life of the project.
A. Team members; sponsors
B. Team members; managers
C. Managers; team members
D. Managers; sponsors
E. None of the above are valid

22. It is not the responsibility of the project manager to sell future project management services to a customer.
A. True
B. False

23. A sign of a good project management culture is when both the project and line managers bring problems to the surface for resolution as quickly as possible.
A. True
B. False

24. Functional team members should be assigned to a project team solely based upon their technical knowledge rather than on their ability to work effectively in a team environment.
A. True
B. False

25. The chances of project management success are relatively poor if the individual assigned as project leader is not dedicated to the project management process.
A. True
B. False

DISCUSSION QUESTIONS

1. What factors have caused the need for behavioral excellence in project management to grow?
2. Why do some people believe that the best project managers are the people who utilize a management style based upon situational leadership?
3. Who selects the project manager? Project office personnel? Functional team members?
4. Do project managers today manage people or manage technology?

5. Which type of conflicts can exist at any time and in any life cycle phase?
6. Can the lack of an organizational career path ladder for project management be a detriment to behavioral excellence?
7. What is meant by a project manager "going beyond formal authority": to maximize his or her influence over people and key decisions?
8. Why is confrontation regarded as the best conflict resolution mode for project management?
9. How does a project sponsor develop a close, but not meddlesome, working relationship with the project manager?
10. What are the relationships between behavioral excellence, culture, and implementing a project management methodology?

Across

3. Type of excellence
6. Conflict resolution mode
7. Conflict resolution mode
8. Type of leadership
9. Type of conflict
11. Conflict resolution mode
13. Type of conflict
14. Conflict resolution mode
15. Type of conflict

Down

1. Type of conflict
2. Conflict resolution mode
4. Wage and salary
5. Type of conflict
10. Type of conflict
12. _ _ _ _ _ _ _ _ _ boss reporting

The Effect of Mergers and Acquisitions on Project Management

15.0 INTRODUCTION

All companies strive for growth. Strategic plans are prepared identifying new products and services to be developed and new markets to be penetrated. Many of these plans require mergers and acquisitions to obtain the strategic goals and objectives. Yet even the best-prepared strategic plans often fail. Too many executives view strategic planning as planning only, often with little consideration given to implementation. Implementation success is vital during the merger and acquisition process.

15.1 PLANNING FOR GROWTH

Companies can grow in two ways: internally and externally. With internal growth, companies cultivate their resources from within and may spend years attaining their strategic targets and marketplace positioning. Since time may be an unavailable luxury, meticulous care must be given to make sure that all new developments fit the corporate project management methodology and culture.

External growth is significantly more complex. External growth can be obtained through mergers, acquisitions, and joint ventures. Companies can purchase the expertise they need very quickly through mergers and acquisitions. Some companies execute occasional acquisitions, whereas other companies have sufficient access to capital such that they can perform continuous acquisitions. However, once again, companies often neglect to consider the impact on project management. Best practices in project management may not be transferable from one company to another. The impact on project management systems resulting from mergers and acquisitions is often irreversible, whereas joint ventures can be terminated.

This chapter focuses on the impact on project management resulting from mergers and acquisitions. Mergers and acquisitions allow companies to achieve strategic targets at a speed not easily achievable through internal growth, provided that the sharing of assets and capabilities can be done quickly and effectively. This synergistic effect can produce opportunities that a firm might be hard-pressed to develop themselves.

Mergers and acquisitions focus on two components: preacquisition decision-making and postacquisition integration of processes. Wall Street and financial institutions appear to be interested more in the near-term financial impact of the acquisition rather than the long-term value that can be achieved through better project management and integrated processes. During the mid-1990s, companies rushed into acquisitions in less time than the company required for a capital expenditure approval. Virtually no consideration was given to the impact on project management and whether or not the expected best practices would be transferable. The result appears to have been more failures than successes.

When a firm rushes into an acquisition, very little time and effort appears to be spent on postacquisition integration. Yet, this is where the real impact of best practices is felt. Immediately after an acquisition, each firm markets and sells products to each other's customers. This may appease the stockholders, but only in the short term. In the long term, new products and services will need to be developed to satisfy both markets. Without an integrated project management system where both parties can share the same best practices, this may be difficult to achieve.

When sufficient time is spent on preacquisition decision-making, both firms look at combining processes, sharing resources, transferring intellectual property, and the overall management of combined operations. If these issues are not addressed in the preacquisition phase, unrealistic expectations may occur during the postacquisition integration phase.

15.2 THE PROJECT MANAGEMENT VALUE-ADDED CHAIN

Mergers and acquisitions are expected to add value to the firm and increase its overall competitiveness. Some people define value as the ability to maintain a certain revenue stream. A better definition of value might be defined as the competitive advantages that a firm possesses as a result of customer satisfaction, lower cost, efficiencies, improved quality, effective utilization of personnel, or the implementation of best practices. True value occurs *only* in the postacquisition integration phase, well after the actual acquisition itself.

Value can be analyzed by looking at the value chain: the stream of activities from upstream suppliers to downstream customers. Each component in the value chain can provide a competitive advantage and enhance the final deliverable or service. Every company has a value chain, as illustrated in Figure 15–1. When a firm acquires a supplier, the value chains are combined and expected to create a superior competitive position. Similarly, the same result is expected when a firm acquires a downstream company. But it may not be possible to integrate the best practices.

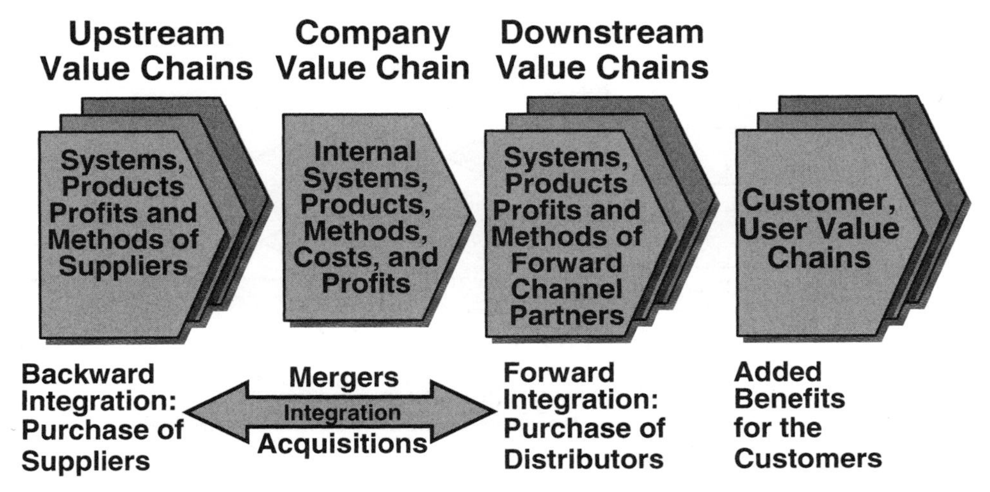

FIGURE 15–1. Generic value-added chain.

Historically, value chain analysis was used to look at a business as a whole.[1] However, for the remainder of this chapter, the sole focus will be the project management value-added chain and the impact of mergers and acquisitions on the performance of the chain.

Figure 15–2 shows the project management value-added chain. The primary activities are those efforts needed for the physical creation of a product or service. The primary activities can be considered to be the five major process areas of project management: project initiation, planning, execution, control, and closure.

The support activities are those company-required efforts needed for the primary activities to take place. At an absolute minimum, the support activities must include:

- *Procurement Management:* The quality of the suppliers and the products and services they provide to the firm.
- *Technology Development:* The quality of the intellectual property controlled by the firm and the ability to apply it to products and services both offensively (new product development) or defensively (product enhancements).
- *Human Resource Management:* The ability to recruit, hire, train, develop, and retain project managers. This includes the retention of project management intellectual property.
- *Supportive Infrastructure:* The quality of the project management systems necessary to integrate, collate, and respond to queries on project performance. Included within the supportive infrastructure are the project management methodology, project management information systems, total quality management system, and any

1. Michael E. Porter, *Competitive Advantage.* New York: Free Press, 1985, Chap. 2.

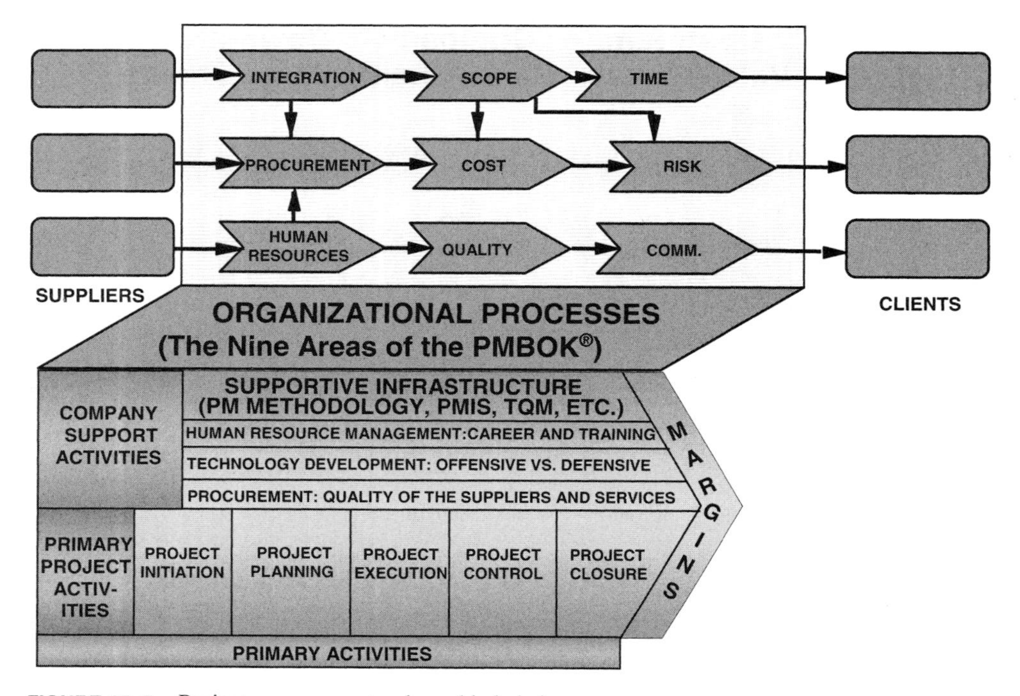

FIGURE 15–2. Project management value-added chain.

other supportive systems. Since customer interfacing is essential, the supportive infrastructure can also include processes for effective supplier–customer interfacing.

These four support activities can be further subdivided into the nine process areas of the PMBOK®. The arrows connecting the nine PMBOK® areas indicate their interrelatedness. The exact interrelationships may vary for each project, deliverable, and customer.

Each of these primary and support activities, together with the nine process areas, is required to convert material received from your suppliers into deliverables for your customers. In theory, Figure 15–2 represents a work breakdown structure for a project management value-added chain:

- *Level 1:* value chain
- *Level 2:* primary activities
- *Level 3:* support activities
- *Level 4:* nine PMBOK® process areas

The project management value-added chain allows a firm to identify critical weaknesses where improvements must take place. This could include better control of scope changes, the need for improved quality, more timely status reporting, better customer relations, or better project execution. The value-added chain can also be useful for supply chain management. The project management value-added chain is a vital tool for continuous improvement efforts and can easily lead to the identification of best practices.

Executives regard project costing as a critical, if not the most critical component of project management. The project management value chain is a tool for understanding a project's cost structure and the cost control portion of the project management methodology. In most firms, this is regarded as a best practice. Actions to eliminate or reduce a cost or schedule disadvantage need to be linked to the location in the value chain where the cost or schedule differences originated.

The "glue" that ties together elements within the project management chain is the project management methodology. A project management methodology is a grouping of forms, guidelines, checklists, policies, and procedures necessary to integrate the elements within the project management value-added chain. A methodology can exist for an individual process such as project execution or for a combination of processes. A firm can also design its project management methodology for better interfacing with upstream or downstream organizations that interface with the value-added chain. Ineffective integration at supplier–customer interface points can have a serious impact on supply chain management and future business.

15.3 PREACQUISITION DECISION-MAKING

The reason for most acquisitions is to satisfy a strategic and/or financial objective. Table 15–1 shows the six most common reasons for an acquisition and the most likely strategic and financial objectives. The strategic objectives are somewhat longer-term than the financial objectives that are under pressure from stockholders and creditors for quick returns.

The long-term benefits of mergers and acquisitions include:

- Economies of combined operations
- Assured supply or demand for products and services
- Additional intellectual property, which may have been impossible to obtain otherwise

TABLE 15–1. TYPES OF OBJECTIVES

Reason for Acquisition	Strategic Objective	Financial Objective
Increase customer base	Bigger market share	Bigger cash plow
Increase capabilities	Provide solutions	Wider profit margins
Increase competitiveness	Eliminate costly steps	Stable earnings
Decrease time-to-market (new products)	Market leadership	Earnings growth
Decrease time-to-market (enhancements)	Broad product lines	Stable earnings
Closer to customers	Better price-quality-service mix	Sole-source procurement

- Direct control over cost, quality, and schedule rather than being at the mercy of a supplier or distributor
- Creation of new products and services
- Pressure on competitors through the creation of synergies
- Cost cutting by eliminating duplicated steps

Each of these can generate a multitude of best practices.

The essential purpose of any merger or acquisition is to create lasting value that becomes possible when two firms are combined, and value that would not exist separately. The achievement of these benefits, as well as attaining the strategic and financial objectives, could rest on how well the project management value-added chains of both firms integrate, especially the methodologies within their chains. Unless the methodologies and cultures of both firms can be integrated, and reasonably quickly, the objectives may not be achieved as planned.

Project management integration failures occur after the acquisition happens. Typical failures are shown in Figure 15–3. These common failures result because mergers and acquisitions simply cannot occur without organizational and cultural changes that are often disruptive in nature. Best practices can be lost. It is unfortunate that companies often rush into mergers and acquisitions with lightning speed but with little regard for how the project management value-added chains will be combined. Planning for better project management should be of paramount importance, but unfortunately is often lacking.

The first common problem area in Figure 15–3 is the inability to combine project management methodologies within the project management value–added chains. This occurs because of:

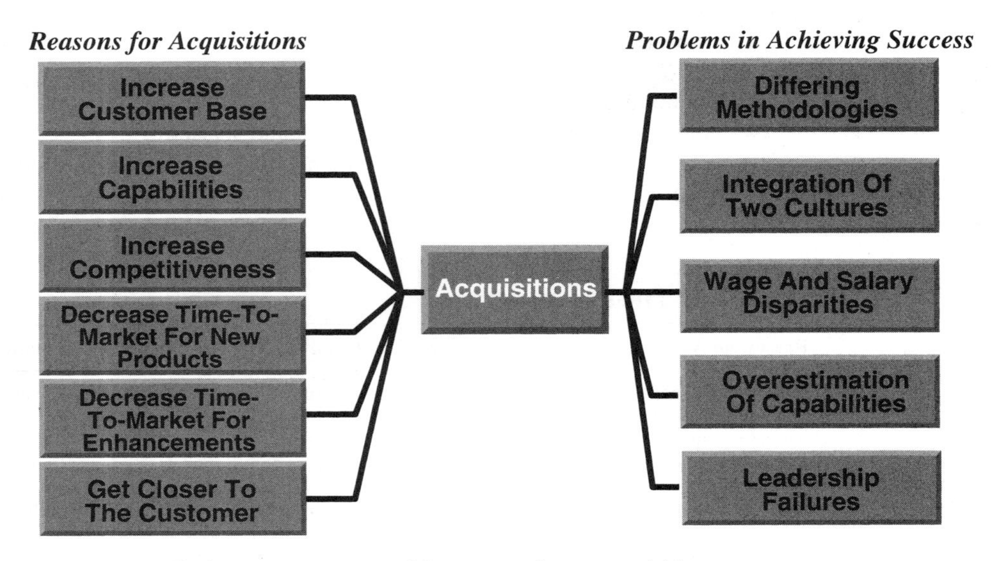

FIGURE 15–3. Project management problem areas after an acquisition.

- A poor understanding of each other's project management practices prior to the acquisition
- No clear direction during the preacquisition phase on how the integration will take place
- Unproven project management leadership in one or both of the firms
- A persistent attitude of "we–them"

Some methodologies may be so complex that a great amount of time is needed for integration to occur, especially if each organization has a different set of clients and different types of projects. As an example, a company developed a project management methodology to provide products and services for large publicly held companies. The company then acquired a small firm that sold exclusively to government agencies. The company realized too late that integration of the methodologies would be almost impossible because of requirements imposed by the government agencies for doing business with the government. The methodologies were never integrated and the firm servicing government clients was allowed to function as a subsidiary, with its own specialized products and services. The expected synergy never took place.

Some methodologies simply cannot be integrated. It may be more prudent to allow the organizations to function separately than to miss windows of opportunity in the marketplace. In such cases, "pockets" of project management may exist as separate entities throughout a large corporation.

The second major problem area in Figure 15–3 is the existence of differing cultures. Although project management can be viewed as a series of related processes, it is the working culture of the organization that must eventually execute these processes. Resistance by the corporate culture to support project management effectively can cause the best plans to fail. Sources for the problems with differing cultures include:

- A culture in one or both firms that has limited project management expertise (i.e., missing competencies)
- A culture that is resistant to change
- A culture that is resistant to technology transfer
- A culture that is resistant to transfer of any type of intellectual property
- A culture that will not allow for a reduction in cycle time
- A culture that will not allow for the elimination of costly steps
- A culture that must "reinvent the wheel"
- A culture in which project criticism is viewed as personal criticism

Integrating two cultures can be equally difficult during both favorable and unfavorable economic times. People may resist any changes in their work habits or comfort zones, even though they recognize that the company will benefit by the changes.

Multinational mergers and acquisitions are equally difficult to integrate because of cultural differences. Several years ago, a U.S. automotive supplier acquired a European firm. The American company supported project management vigorously and encouraged its employees to become certified in project management. The European firm provided very little support for project management and discouraged their workers from becoming certified, using the argument that their European clients do not regard project management in such high esteem as do General Motors, Ford, and Chrysler. The European subsidiary saw no need for project management. Unable to combine the methodologies, the U.S. parent company slowly replaced the European executives with American executives to drive home the need for a singular project management approach across all divisions. It took almost five years for the complete transformation to take place. The parent company believed that the resistance in the European division was more of a fear of change in their comfort zone than a lack of interest by their European customers.

During the past 40 years, Philip Morris has systematically pursued a diversification strategy to reduce their dependence on cigarettes. In addition to their cigarette business, they now own Clark Chewing Gum, Kraft General Foods, Miller, Miller Lite, Lowenbrau, Jello-O, Oscar Meyer, Maxwell House, Sealtest, Orowheat Baked Goods, and Louis Kemp Seafood. The strategy of Philip Morris was to acquire well-established industry leaders. Each organization acquired had a distinctive corporate culture. Yet even though all of the organizations were successful, some of the expected synergies were not realized because of cultural dissimilarities. Each acquisition was then treated as a stand-alone organization. Although it could be argued that there was no reason to merge these firms into one culture, it does show that even highly successful firms are often resistant to change.

Sometimes there are clear indications that the merging of two cultures will be difficult. When Federal Express acquired Flying Tiger, the strategy was to merge the two into one smoothly operating organization. At the time of the merger, Federal Express employed a younger workforce, many of whom were part-time. Flying Tiger had full-time, older, longer-tenured employees. Federal Express focused on formalized policies and procedures and a strict dress code. Flying Tiger had no dress code and management conducted business according to the chain of command, where someone with authority could bend the rules. Federal Express focused on a quality goal of 100 percent on-time delivery, whereas Flying Tiger seemed complacent with a 95 to 96 percent target. Combining these two cultures had to be a monumental task for Federal Express. In this case, even with these potential integration problems, Federal Express could not allow Flying Tiger to function as a separate subsidiary. Integration was mandatory. Federal Express had to address quickly those tasks that involved organizational or cultural differences.

Planning for cultural integration can also produce favorable results. Most banks grow through mergers and acquisitions. The general belief in the banking industry is to grow or be acquired. National City Corporation of Cleveland, Ohio, recognized this and developed project management systems that allowed National City to acquire other banks and integrate the acquired banks into National City's culture in less time than other banks allowed for mergers and acquisitions. National City views project management as an asset that has a very positive effect on the corporate bottom line. Many banks today have manuals for managing merger and acquisition projects.

The third problem area in Figure 15–3 is the impact on the wage and salary administration program. The common causes of the problems with wage and salary administration include:

- Fear of downsizing
- Disparity in salaries

- Disparity in responsibilities
- Disparity in career path opportunities
- Differing policies and procedures
- Differing evaluation mechanisms

When a company is acquired and integration of methodologies is necessary, the impact on the wage and salary administration program can be profound. When an acquisition takes place, people want to know how they will benefit individually, even though they know that the acquisition is in the best interest of the company.

The company being acquired often has the greatest apprehension about being lured into a false sense of security. Acquired organizations can become resentful to the point of physically trying to subvert the acquirer. This will result in value destruction, where self-preservation becomes of paramount importance to the workers, often at the expense of the project management systems.

Consider the following situation. Company A decides to acquire Company B. Company A has a relatively poor project management system in which project management is a part-time activity and not regarded as a profession. Company B, on the other hand, promotes project management certification and recognizes the project manager as a full-time, dedicated position. The salary structure for the project managers in Company B was significantly higher than for their counterparts in Company A. The workers in Company B expressed concern that "We don't want to be like them," and self-preservation led to value destruction.

Because of the wage and salary problems, Company A tried to treat Company B as a separate subsidiary. But when the differences became apparent, project managers in Company A tried to migrate to Company B for better recognition and higher pay. Eventually, the pay scale for project managers in Company B became the norm for the integrated organization.

When people are concerned with self-preservation, the short-term impact on the combined value-added project management chain could be severe. Project management employees must have at least the same, if not better, opportunities after acquisition integration as they did prior to the acquisition.

The fourth problem area in Figure 15–3 is the overestimation of capabilities after acquisition integration. Included in this category are:

- Missing technical competencies
- Inability to innovate
- Speed of innovation
- Lack of synergy
- Existence of excessive capability
- Inability to integrate best practices

Project managers and those individuals actively involved in the project management value-added chain rarely participate in preacquisition decision-making. As a result, decisions are made by managers who may be far removed from the project management value-added chain and whose estimates of postacquisition synergy are overly optimistic.

The president of a relatively large company held a news conference announcing that his company was about to acquire another firm. To appease the financial analysts attending the news conference, he meticulously identified the synergies expected from the combined operations and provided a time line for new products to appear on the marketplace. This announcement did not sit well with the workforce, who knew that the capabilities were overestimated and that the dates were unrealistic. When the product launch dates were missed, the stock price plunged and blame was placed erroneously on the failure of the integrated project management value-added chain.

The fifth problem area in Figure 15–3 is leadership failure during postacquisition integration. Included in this category are:

- Leadership failure in managing change
- Leadership failure in combining methodologies
- Leadership failure in project sponsorship
- Overall leadership failure
- Invisible leadership
- Micromanagement leadership
- Believing that mergers and acquisitions must be accompanied by major restructuring

Managed change works significantly better than unmanaged change. Managed change requires strong leadership, especially with personnel experienced in managing change during acquisitions.

Company A acquires Company B. Company B has a reasonably good project management system but with significant differences from Company A. Company A then decides that "We should manage them like us," and nothing should change. Company A then replaces several Company B managers with experienced Company A managers. This was put into place with little regard for the project management value-added chain in Company B. Employees within the chain in Company B were receiving calls from different people, most of who were unknown to them and were not provided with guidance on who to contact when problems arose.

As the leadership problem grew, Company A kept transferring managers back and forth. This resulted in smothering the project management value-added chain with bureaucracy. As expected, performance was diminished rather than enhanced.

Transferring managers back and forth to enhance vertical interactions is an acceptable practice after an acquisition. However, it should be restricted to the vertical chain of command. In the project management value-added chain, the main communication flow is lateral, not vertical. Adding layers of bureaucracy and replacing experienced chain managers with personnel inexperienced in lateral communications can create severe roadblocks in the performance of the chain.

Any of the problem areas, either individually or in combination with other problem areas, can cause the chain to have diminished performance, such as:

- Poor deliverables
- Inability to maintain schedules

- Lack of faith in the chain
- Poor morale
- Trial by fire for all new personnel
- High employee turnover
- No transfer of project management intellectual property

15.4 LANDLORDS AND TENANTS

Previously, it was shown how important it is to assess the value chain, specifically the project management methodology, during the preacquisition phase. No two companies have the same value chain for project management as well as the same best practices. Some chains function well; others perform poorly.

For simplicity sake, the "landlord" will be the acquirer and the "tenant" will be the firm being acquired. Table 15–2 identifies potential high-level problems with the landlord–tenant relationship as identified in the preacquisition phase. Table 15–3 shows possible postacquisition integration outcomes.

The best scenario occurs when both parties have good methodologies and, most important, are flexible enough to recognize that the other party's methodology may have desirable features. Good integration here can produce a market leadership position.

If the landlord's approach is good and the tenant's approach is poor, the landlord may have to force a solution upon the tenant. The tenant must be willing to accept criticism, see the light at the end of the tunnel, and make the necessary changes. The changes, and the reasons for the changes, must be articulated carefully to the tenant to avoid cultural shock.

Quite often a company with a poor project management methodology will acquire an organization with a good approach. In such cases, the transfer of project management intellectual property must occur quickly. Unless the landlord recognizes the achievements of the tenant, the tenant's value-added chain can diminish in performance and there may be a loss of key employees.

The worst-case scenario occurs when neither the landlord nor the tenant have good project management systems. In this case, all systems must be developed anew. This could be a blessing in disguise because there may be no hidden bias by either party.

TABLE 15–2. POTENTIAL PROBLEMS WITH COMBINING METHODOLOGIES BEFORE ACQUISITIONS

Landlord	Tenant
Good methodology	Good methodology
Good methodology	Poor methodology
Poor methodology	Good methodology
Poor methodology	Poor methodology

TABLE 15–3. POSSIBLE INTEGRATION OUTCOMES

Methodology		
Landlord	**Tenant**	**Possible Results**
Good	Good	Based upon flexibility, Good Synergy achievable; market leadership possible at a low cost.
Good	Poor	Tenant must recognize weaknesses and be willing to change; possible cultural shock.
Poor	Good	Landlord must see present and future benefits; strong leadership essential for quick response.
Poor	Poor	Chances of success limited; good methodology may take years to get.

15.5 BEST PRACTICES: A CASE STUDY ON JOHNSON CONTROLS, INC.[2]

The Automotive Systems Group (ASG) of Johnson Controls, Inc. (JCI) is one of the best-managed organizations in the world, with outstanding expertise in the management of the project management value-added chain. ASG is a global supplier of automotive interior systems and batteries.

During the 1990s, JCI automotive business expanded more than tenfold, with some of the growth coming from strategic acquisitions. AGS was increasingly working on development projects that required multiple teams in multiple locations developing products for a single customer. It was clear that AGS needed to have one common global project management methodology or process that would allow *all* AGS employees and teams to communicate better and improve the efficiency of the development process. Without this, the results could have been devastating. AGS was no longer supplying simply products. They were now providing complete solutions for their customers: namely, the interior of the vehicle.

Johnson Controls, Inc. Automotive Systems Group Integration with Prince and Becker Corporation

JCI purchased both Prince and Becker approximately five years ago, with the intent of becoming an integrated interior supplier to the automotive industry. The addition of Prince's overhead systems, door panel, and instrument panel capabilities to Becker's interior plastic trim capability and ASG seating products, established ASG immediately as an interior supplier. Organizational changes were necessary to truly integrate the companies and their capabilities. The result was a new business model from which the company would be reorganized into vehicle platforms so as to provide complete solutions for their customers rather than merely components.

All three companies had different project management systems, project management position descriptions, and organizational structures. For example, Becker in Europe had

2. Dave Kandt and Alok Kumar of the Automotive Systems Group of Johnson Controls, Inc. provided much of the information in the remainder of this section.

project accounters, which were combined project managers and sales personnel. Prince, on the other hand, in addition to project managers, had project coordinators who handled the administrative functions for the project managers. JCI followed a traditional approach, with a well-defined matrix organization and project managers who were fully responsible for all aspects of project execution and success. JCI decided quickly that the three project management systems had to be integrated and commonized. This proved to be a difficult but productive task.

Integrating the Methodologies Johnson Controls' Automotive Systems Group had a 15-year history of project management with a project management methodology that was well integrated with total quality control. Between 1995 and 2000, ASG had received more than 164 quality awards from their customers and other organizations, and much of the credit was given to the way in which projects were managed. This system had nine life cycle phases. Prince's system was referred to as New Product Development/Product Development Process and had seven life cycle phases. Becker's system had six life cycle phases and was called the Projekt Management Handbuch (Project Management Handbook). Not only were the systems different, the languages were different. Each company thought that their system was superior and should be adopted by the entire company. Reaching agreement just on position titles required extensive discussion.

Corporate Cultures In addition to different organizations and systems, the three companies had different values and all were understandably proud of their own ways of operating. Prince Corporation had a very well integrated culture, with a strong focus on the leadership that had been provided by the founder, Ed Prince. Prince Corp. (located in Holland, Michigan) had never had any real need to consider European interests, especially those of Becker, which had traditionally been a competitor or supplier. JCI had a very strong presence in Europe, with its central office in Burscheid, Germany. The Burscheid office followed North America's lead for project management systems, but had strong opinions on how these systems should function in the European culture, and followed a much more laissez-faire approach to managing projects. In keeping with European culture and traditional separation of European countries, the various Automotive Systems Group development centers located in different countries in Europe operated pretty much on their own, with little centralized influence. The end result was a lot of opinions on how to integrate the project management systems, with some common principles and some widely varying values.

Product Focus In addition to the differences listed above, the purchase of Prince and Becker was intended to bring JCI into an automotive total interior systems position with the vehicle companies. This meant that not only differences in culture, organization, and systems had to be overcome, but also differences in product, equipment, and core manufacturing processes. ASG had to find a way to commonize. In addition, the Automotive Systems Group had to position the new systems to allow the development and launch of total vehicle interiors, a fundamentally different scope for all three of the companies involved. The newly developed integrated project management value-added chain would be a completely new approach for all three companies.

The Integration A team was created to integrate the various systems, organization, and values of the three companies. Project managers from all three companies, including North America and Europe, were appointed. Also, representatives from all functional departments were represented. Representatives from Quality, Engineering, Manufacturing, and Finance were all part of the team and were able to influence the direction.

The team was challenged to create a project management methodology (and multinational project management value-added chain) that would achieve the following goals:

- Combine best practices from all existing project management methodologies and project management value-added chains
- Create a methodology that encompasses the entire project management value added chain from suppliers to customers
- Meet the industry standards established by the Automotive Industry Action Group (AIAG), Project Management Institute (PMI), and International Organization for Standardization (ISO)
- Share best practices among all ASG global locations
- Achieve the corporate launch goals of timing, cost, quality, and efficiency
- Accommodate all ASG automotive products
- Optimize procedures, deliverables, roles, and responsibilities
- Provide clear and useful documentation

The system that evolved, called PLUS (Product Launch System), incorporated all three companies' ideas and best practices.[3] PLUS had five phases (plus an initial phase called phase 0 for Ideation, which was Prince's Product Creation Phase). All three companies were able to identify with this new system, and provided the basic project management structure for JCI to pursue its strategic target of providing vehicle interiors. PLUS repackaged and improved the three existing systems into a new process with optimized procedures, roles, responsibilities, and deliverables. The new system was also painstakingly mapped to the Quality System Requirements QS9000 (AIAG, March 1998), Advanced Product Quality Planning (APQP) and Control Plan (AIAG, June 1994), the Automotive Project Management Guide (AIAG, 1997), and PMBOK® (2000).

There were still some problems that had to be resolved. This became evident when teams with more than one year of development time were required to convert to PLUS. PLUS was introduced to the workers through an online PLUS Overview Course that was not sufficient to prepare the teams for using PLUS. Additional training programs were then put in place with customization for particular audiences of workers. The training program provided guidance on why PLUS was developed, how to apply it, and procedures for suggested improvements.

The Outcome PLUS was well received by all three companies. More than 350 programs were using PLUS. The team that created the new system was given the Chairman's Award for Customer Satisfaction. AGS now had a basic system that allowed everyone in the com-

---

3. The complexities in the development and roll-out of PLUS, and the mapping of PLUS against PMBOK® can be found in Richard Spigarelli and Carel Allen, "Implementation of an Automotive Product Launch System," *Proceedings of the Project Management Institute Annual Seminars and Symposiums,* November 1–10, 2001, Nashville, Tennessee.

pany to utilize a common language for product development and launch. In fact, this development paved the way for an organizational change that came one year later, referred to as the New Business Model, which reorganized the company into vehicle platform teams, emphasizing JCI's strategic direction to become a total vehicle interior supplier.

This change allowed ASG to function as an integrated company. The platform teams (project offices) had become true teams and felt a sense of identity based on the entire vehicle rather than on doors, seats, and cockpits. The major metrics, such as timing, profitability, and customer relations, were now viewed from the perspective of the entire vehicle. The customers were happier because they now had a single point of contact for their vehicle instead of multiple points of contact.

Follow-on Work The benefits of PLUS were clear: integrated system, common roles and responsibilities, and the ability to manage projects as integrated platforms. The downside was that PLUS was designed by a committee. Because everyone had a vote, the system was somewhat more bulky and less elegant than it could be. The next step was obvious—PLUS needed to be simplified and streamlined. Out of respect for the strong opinions and cultures of the three companies, PLUS was allowed to remain as it was for approximately $1\frac{1}{2}$ years. After that, a central group of project management directors took a minimalist approach to PLUS, removed some unnecessary content, focused on the critical deliverables, and made it much simpler and user-friendlier. The revised system, launched in October 2001, focused on the critical deliverables and on a one deliverable, one owner principle. The results were well accepted in Europe and North America.

The system is revised twice a year at an interval of six months. The focus continues on making it intuitive and keeping it flexible, driven by critical deliverables and having clear accountability. Proper integration of methodologies, along with possibly the entire project management value-added chain, can provide excellent benefits. At JCI, the following benefits were found:

- Common terminology across the entire organization
- Unification of all companies
- Common forms and reports
- Guidelines for less experienced project managers and team members
- Clearer definition of roles and responsibilities
- Reduction in the number of procedures and forms
- No duplication in reporting
- Reduction in the number of time line items from 184 to 110

Recommendations for Other Companies The following recommendations can be made:

- *Use a common written system for managing programs.* If new companies are acquired, bring them into the basic system as quickly as is reasonable.
- *Respect all parties.* You cannot force one company to accept another company's systems. There has to be selling, consensus, and modifications.
- *It takes time to allow different corporate cultures to come together.* Pushing too hard will simply alienate people. Steady emphasis and pushing by management is ultimately the best way to achieve integration of systems and cultures.

- *Sharing management personnel among the merging companies helps to bring the systems and people together quickly* .
- *There must be a common "process owner" for the project management system.* A person on the vice-presidental level would be appropriate.

15.6 INTEGRATION RESULTS

The best prepared plans do not necessarily guarantee success. Reevaluation is always necessary. Evaluating the integrated project management value added after acquisition and integration is completed can be done using the modified Boston Consulting Group Model (BCG), shown in Figure 15–4. The two critical parameters are the perceived value to the company and the perceived value to customers.

If the final chain has a low perceived value to both the company and the customers, it can be regarded as a *dog*. The characteristics of a dog include:

- There is a lack of internal cooperation, possibly throughout the entire value-added chain.
- The value chain does not interface well with the customers.

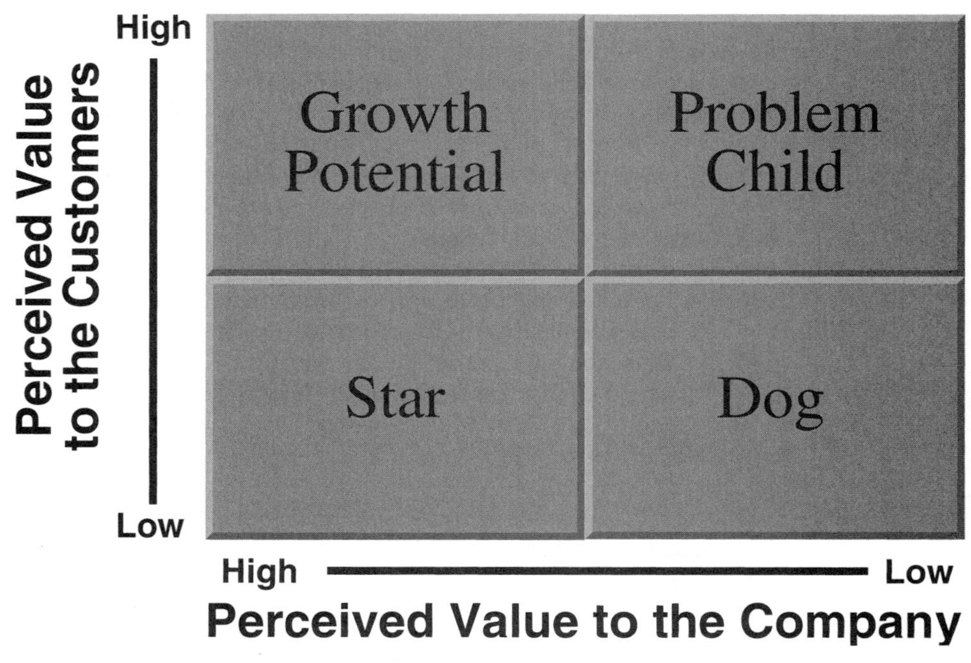

FIGURE 15–4. Project management system after acquisition.

- The customer has no faith in the company's ability to provide the required deliverables.
- The value-added chain processes are overburdened with excessive conflicts
- Preacquisition expectations were not achieved, and the business may be shrinking.

Possible strategies to use with a dog include:

- Downsize, descope, or abandon the project management value-added chain.
- Restructure the company to either a projectized or departmentalized project management organization.
- Allow the business to shrink and focus on selected projects and clients.
- Accept the position of a market follower rather than a market leader.

The *problem child* quadrant in Figure 15–4 represents a value-added chain which has a high perceived value to the company but is held in low esteem by customers. The characteristics of a problem child include:

- The customer has some faith in the company's ability to deliver but no faith in the project management value-added chain.
- Incompatible systems may exist within the value-added chain.
- Employees are still skeptical about the capability of the integrated project management value-added chain.
- Projects are completed more on a trial-by-fire basis rather than on a structured approach.
- Fragmented pockets of project management may still exist in both the landlord and the tenant.

Possible strategies for a problem child value chain include:

- Invest heavily in training and education to obtain a cooperative culture.
- Carefully monitor cross-functional interfacing across the entire chain.
- Seek out visible project management allies in both the landlord and the tenant.
- Use of small breakthrough projects may be appropriate.

The *growth-potential* quadrant in Figure 15–4 has the potential to achieve preacquisition decision-making expectations. This value-added chain is perceived highly by both the company and its clients. The characteristics of a growth potential value-added chain include:

- Limited, successful projects are using the chain.
- The culture within the chain is based upon trust.
- Visible and effective sponsorship exists.
- Both the landlord and the tenant regard project management as a profession.

Possible strategies for a growth potential project management value-added chain include:

- Maintain slow growth leading to larger and more complex projects.
- Invest in methodology enhancements.

- Begin selling complete solutions to customers rather than simply products or services.
- Focus on improved customer relations using the project management value added chain.

In the final quadrant in Figure 15–4 the value chain is viewed as a *star.* This has a high perceived value to the company but a low perceived value to the customer. The reason for the customer's low perceived value is because you have already convinced the customer of the ability of your chain to deliver, and your customers now focus on the deliverables rather than the methodology.

The characteristics of a star project management value-added chain include:

- A highly cooperative culture exists.
- The triple constraint is satisfied.
- Your customers treat you as a partner rather than as a contractor.

Potential strategies for a star value-added chain include:

- Invest heavily in state-of-the-art supportive subsystems for the chain.
- Integrate your PMIS into the customer's information systems.
- Allow for customer input into enhancements for your chain.

15.7 VALUE CHAIN STRATEGIES

At the beginning of this chapter the focus was on the strategic and financial objectives established during preacquisition decision-making. However, to achieve these objectives, the company must understand its competitive advantage and competitive market after acquisition integration. Four generic strategies for a project management value-added chain are shown in Figure 15–5. The company must address two fundamental questions concerning postacquisition integration:

- Will the organization now compete on cost or uniqueness of products and services?
- Will the postacquisition marketplace be broad or narrow?

The answer to these two questions often dictates the types of projects that are ideal for the value-added chain project management methodology. This is shown in Figure 15–6. Low-risk projects require noncomplex methodologies, whereas high-risk projects require complex methodologies. The complexity of the methodology can have an impact on the time needed for postacquisition integration. The longest integration time occurs when a company wants a project management value-added chain to provide complete solution project management, which includes product and service development, installation, and

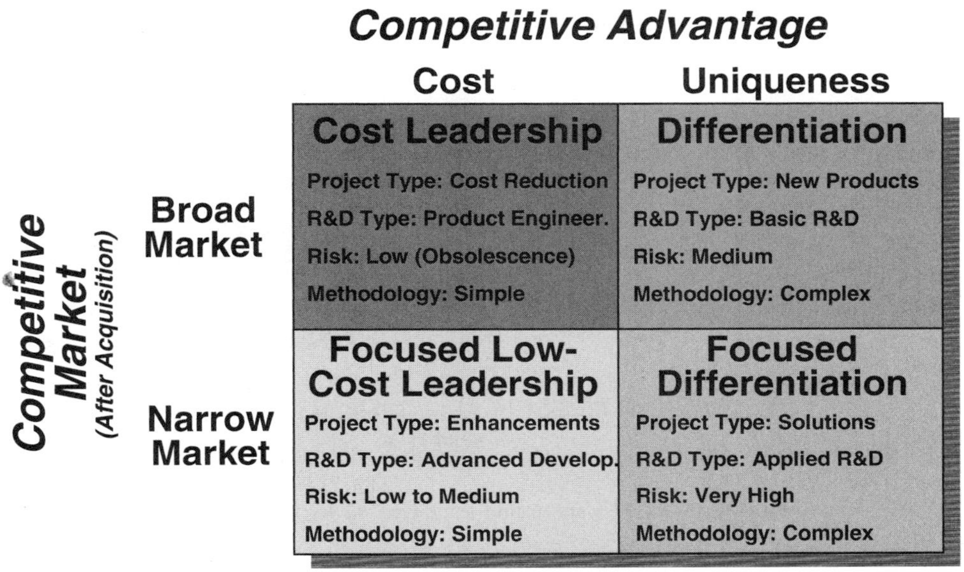

Competitive Advantage
Cost
Uniqueness
Competitive Market
(After Acquisition)
Broad Market
Narrow Market
Cost Leadership
Project Type: Cost Reduction
R&D Type: Product Engineer.
Risk: Low (Obsolescence)
Methodology: Simple
Differentiation
Project Type: New Products
R&D Type: Basic R&D
Risk: Medium
Methodology: Complex
Focused Low-Cost Leadership
Project Type: Enhancements
R&D Type: Advanced Develop.
Risk: Low to Medium
Methodology: Simple
Focused Differentiation
Project Type: Solutions
R&D Type: Applied R&D
Risk: Very High
Methodology: Complex

FIGURE 15–5. Four generic strategies for project management.

High Risk
Low Risk
Solutions
New Products
Complex Project Management Methodologies
Enhanced Products
Similar Products
Noncomplex Project Management Methodologies

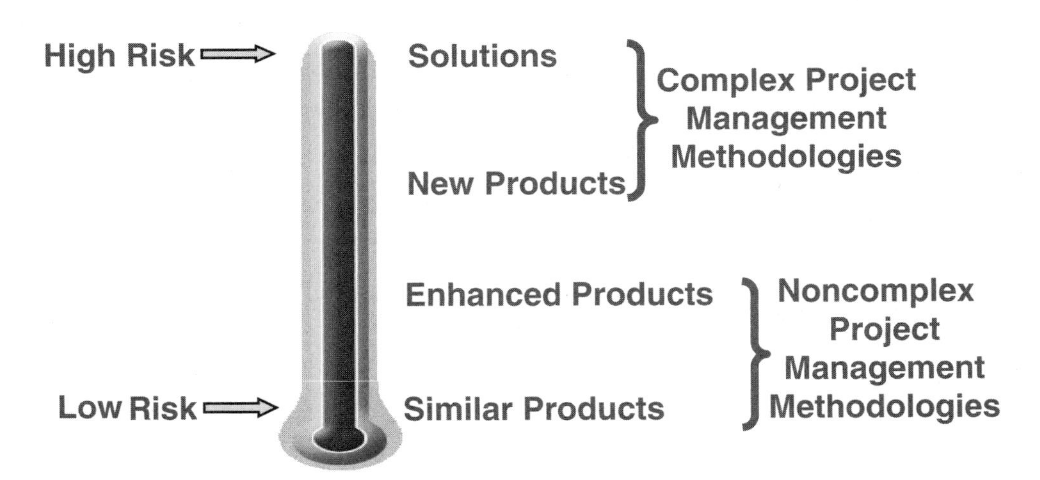

FIGURE 15–6. Risk spectrum for type of project.

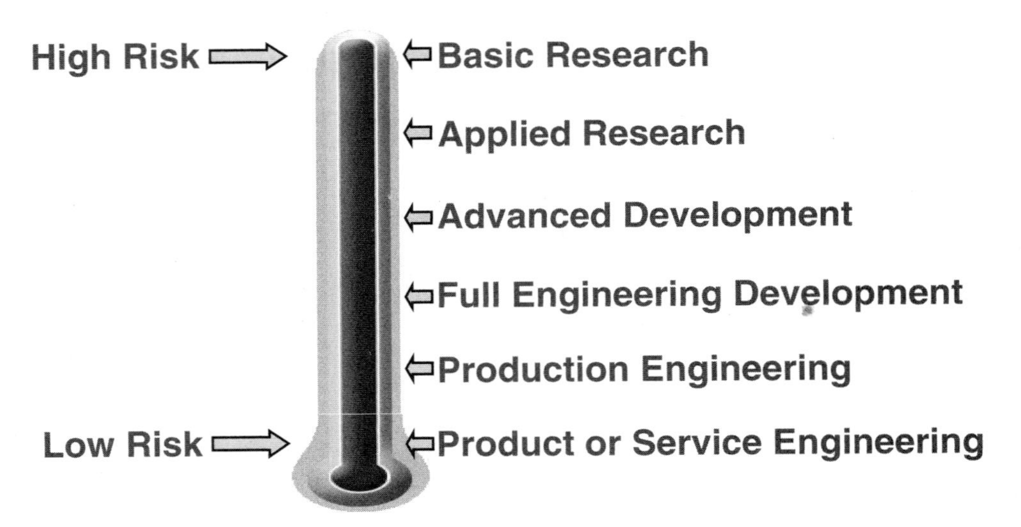

FIGURE 15–7. Risk spectrum for the types of R&D projects.

follow-up. It can also include platform project management, as was the case with Johnson Controls, Inc. Emphasis is on customer satisfaction, trust, and follow-on work.

Project management methodologies are often a reflection of a company's tolerance for risk. As shown in Figure 15–7, companies with a high tolerance for risk develop project management value-added chains capable of handling complex R&D projects and become market leaders. At the other end of the spectrum are enhancement projects that focus on maintaining market share and becoming a follower rather than a market leader.

15.8 FAILURE AND RESTRUCTURING

Great expectations often lead to great failures. When integrated project management value-added chains fail, the company has three viable but undesirable alternatives:

- Downsize the company.
- Downsize the number of projects and compress the value-added chain.
- Focus on a selected customer business base.

The short- and long-term outcomes for these alternatives are shown in Figure 15–8.

Failure often occurs because the preacquisition decision-making phase was based on illusions rather than fact. Typical illusions include:

- Integrating project management methodologies will automatically reduce or eliminate duplicated steps in the value-added chain.

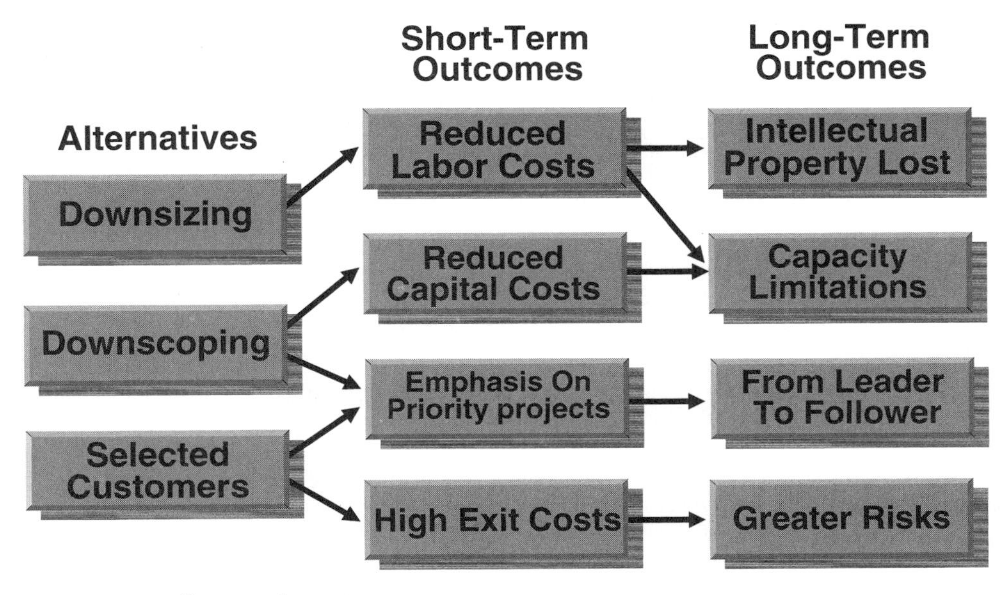

FIGURE 15–8. Restructuring outcomes.

- Expertise in one part of the project management value-added chain could be directly applicable to upstream or downstream activities in the chain.
- A landlord with a strong methodology in part of their value-added chain could effectively force a change on a tenant with a weaker methodology.
- The synergy of combined operations can be achieved overnight.
- Postacquisition integration is a guarantee that technology and intellectual property will be transferred.
- Postacquisition integration is a guarantee that all project managers will be equal in authority and decision-making.

Mergers and acquisitions will continue to take place regardless of whether the economy is weak or strong. Hopefully, companies will now pay more attention to postacquisition integration and recognize the potential benefits.

MULTIPLE CHOICE QUESTIONS

1. The two most common ways for companies to grow are:
  A. Mergers and acquisitions
  B. Internally and mergers
  C. Internally and externally
  D. Internally and acquisitions
  E. Internally and joint ventures

2. When firms rush into acquisitions, very little time is spent on:
   - **A.** Post-acquisition integration
   - **B.** Resource evaluation
   - **C.** Financial analysis
   - **D.** Evaluation of senior management's capabilities
   - **E.** All of the above
3. When sufficient time is spent on preacquisition decision-making, both firms look at:
   - **A.** Combining processes
   - **B.** Sharing resources
   - **C.** Combining operations
   - **D.** Transferring intellectual property
   - **E.** All of the above
4. The stream of activities from upstream suppliers to downstream customers is called the:
   - **A.** Performance chain
   - **B.** Profit chain
   - **C.** Value (or value-added) chain
   - **D.** Integration chain
   - **E.** Competitive chain
5. Backward integration is the purchasing of or merging with:
   - **A.** Suppliers
   - **B.** Distributors
   - **C.** Customers
   - **D.** Internal manufacturing skills
   - **E.** ENone of the above
6. Which of the following is not part of the project management value-added chain?
   - **A.** PMBOK® processes
   - **B.** Company primary activities
   - **C.** Company secondary activities
   - **D.** Company support activities
   - **E.** Supportive infrastructure
7. The long-term benefits of mergers and acquisitions include:
   - **A.** Economies of combined operations
   - **B.** Direct control over cost, quality, and schedule
   - **C.** Elimination of duplicated steps
   - **D.** Creation of new products
   - **E.** All of the above
8. The essential purpose of any merger or acquisition is to:
   - **A.** Increase profits
   - **B.** Reduce cost
   - **C.** Improve quality
   - **D.** Increase sales
   - **E.** Create lasting value
9. Merger and acquisition failures can result from:
   - **A.** Differing methodologies
   - **B.** Poor integration of two cultures
   - **C.** Wage and salary disparities
   - **D.** Leadership failures
   - **E.** All of the above

10. Which of the following can cause wage and salary administration problems after an acquisition?
- **A.** Fear of downsizing
- **B.** Disparity in salaries
- **C.** Differing evaluation mechanisms
- **D.** Disparity in responsibilities
- **E.** All of the above

11. The worse-case scenario during a merger or acquisition is when the "landlord" has a __________ methodology and the "tenant" has a __________ methodology.
- **A.** Good; good
- **B.** Good; poor
- **C.** Poor; good
- **D.** Poor; poor
- **E.** All of the above are bad

12. Which of the following is not one of the quadrants in the modified BCG model in Figure 15–4?
- **A.** Star
- **B.** Limited potential
- **C.** Growth potential
- **D.** Problem child
- **E.** Dog

13. Possible strategies to use with a "dog" include:
- **A.** Invest heavily in the growth of the project management value-added chain
- **B.** Restructure the company to a projectized organizational form
- **C.** nvest heavily to become a market leader rather than a follower
- **D.** Focus on a selected customer base
- **E.** Invest heavily in marketing and R&D

14. When your customers treat you as a partner rather than as a contractor, the project management value-added chain is most likely a:
- **A.** Star
- **B.** Problem child
- **C.** Dog
- **D.** Growth potential
- **E.** Limited growth potential

15. Which of the following is not a typical type of project in the risk spectrum of projects?
- **A.** Solutions
- **B.** New products
- **C.** Enhanced products
- **D.** Similar products
- **E.** All of above are project types

DISCUSSION QUESTIONS

1. Why does external growth have a greater potential negative impact on project management?
2. From a project management perspective, what are the two critical components of mergers and acquisitions? Which one has the greatest impact on project management?

3. Looking at the generic value-added chain in Figure 15–1, is a project management methodology more important to a company for backward or forward integration?
4. Where does the project management methodology fit in the project management value-added chain?
5. Why is there often a difference between the strategic objectives and the financial objectives during mergers and acquisitions? Which objective does project management usually have as their primary focus?
6. Why do project management integration failures occur?
7. What would probably happen if during postacquisition integration the project management methodologies could not be integrated successfully?
8. Can a "tenant" force its methodology upon the "landlord," assuming that the tenant has the better methodology?
9. Is the integration of two diverse methodologies more important to a company that wants a competitive advantage in cost or in uniqueness?
10. If integration of methodologies cannot take place, what are the most likely potential outcomes?

Across

3. Restructuring alternative: selected __________
5. Supportive ________________
7. Type of combined system (2 words)
9. Restructuring alternative
10. External growth
12. Acquisition failure: wage and salary ____________
13. Type of objective
15. Value-added _____
16. Landords and ________
18. Type of combined system (2 words)
19. Type of combined system
20. Restructuring alternative
21. Type of combined system
22. Type of integration

Down

1. External growth
2. Acquisition problem: ___________ failure
4. Chain "Glue"
5. Post-acquistion ____________
6. Type of objective
7. ________________ decision-making or planning
8. Acquistion problems: differing ______________
11. Type of growth
14. Type of growth
17. Type of integration

Rising Stars and Future Directions

16.0 INTRODUCTION

Some companies are simply complacent, doing what they've done and maintaining their market share, no matter how small. Other companies, however, aren't satisfied with the status quo. They want to exceed their customers' expectations and not just meet them. Such companies thrive on continuous improvement in everything they do. Their cultures are second to none. And what's truly amazing is that companies such as Lear, Computer Associates, Texas Instruments, Sun Microsystems, Johnson Controls, and Roadway Express have accomplished their excellence in project management in less than five years. This chapter looks at these stars of the present and the future.

16.1 COMPUTER ASSOCIATES[1]

Today, most companies recognize that that they possess some type of best practices in project management. The best practices are then recorded in a repository for best practices. But what happens next separates the winners from the losers. The losers regard their best practices as proprietary knowledge and often withhold some of the results even from their own employees. The winners, on the other hand, maximize the results of their best practices through continuous improvement efforts, integrating the best practices into their other

1. © 2002 Computer Associates International, Inc. (CA). All trademarks, trade names, service marks, and logos referenced herein belong to their respective companies.

processes, developing support systems for their best practices, and most important, help their customers benefit from these best practices. Computer Associates belongs in the winner category.

Islandia, New York–based Computer Associates International, Inc. (CA) provides solutions to help its customers establish effective business environments—the capabilities required to build and deploy applications, exploit the knowledge contained within them, and manage the systems and databases that comprise their infrastructures.

Three years ago, CA acquired a variety of consulting service companies to ensure the availability of the talent, skills, and expertise required to undertake the breadth and depth of projects that are necessary to align business and IT organizations. Each company had its preferred methods for delivering a range of project engagements, including developing business applications, deploying data warehouses, and implementing systems management infrastructures. It was critical to ensure that these methods were captured and properly integrated so that CA Services℠ project managers and consultants throughout the world delivered engagements in consistent, repeatable ways.

A Methods Management Organization (MMO) led by Vice President Robert Caravella (shown in Figure 16–1) was established with the mission of deploying a corporate-wide Best Practices Library (BPL). Since its inception in August 1998, the MMO has successfully implemented a world-class BPL with more than 150 methods covering a range of activity from strategy definition and transition planning to architecture planning and technology implementation. The MMO Method Development Process is shown in Figure 16–2.

Caravella selected the AllFusion™ Process Management Suite as the vehicle for rolling out the BPL. "I have been in the methods game a long time, and the AllFusion Process Management Suite was the only one capable of building and managing a library of this magnitude—leading to the automated generation of project schedules," said Caravella. "By capturing and leveraging institutional knowledge, we created a 'corporate insurance policy.' As a repository of our process knowledge, the BPL alleviates the prob-

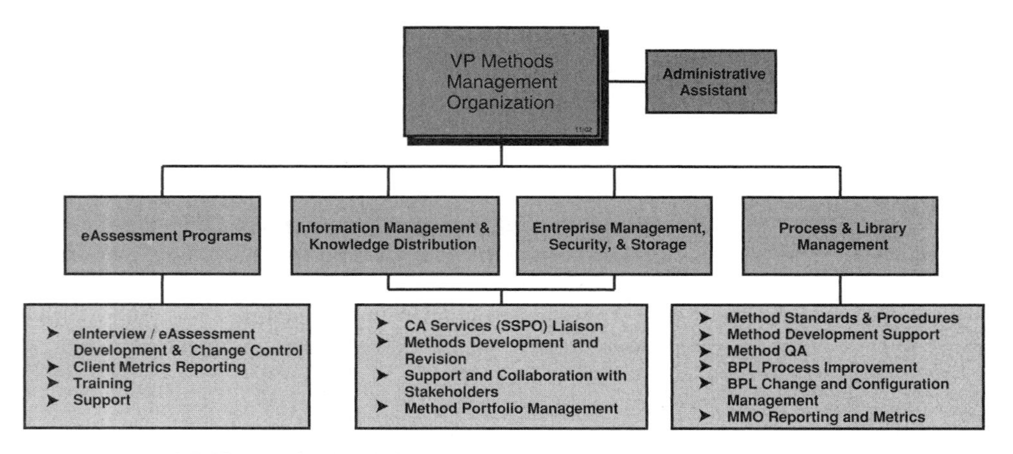

FIGURE 16–1. MMO organizational chart.

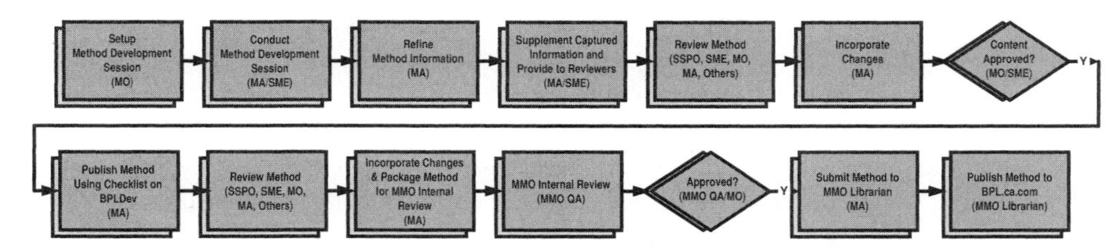

FIGURE 16–2. MMO Method Development Process.

lem of having expertise locked in the heads of a few subject matter experts and minimizes the risk of losing corporate knowledge if these key people decide to pursue other endeavors." The AllFusion Process Management Suite is illustrated in Figure 16–3.

Integrated Process Management Capabilities

The AllFusion Process Management Suite manages all critical factors for project success, including Best Practices, projects, resources, and deliverables. Gary Starkey, a CA Senior Vice President and General Manager, stated: "With the investment in the MMO and the Best Practices Library it supports, CA is ensuring the optimal success for every services engagement." He

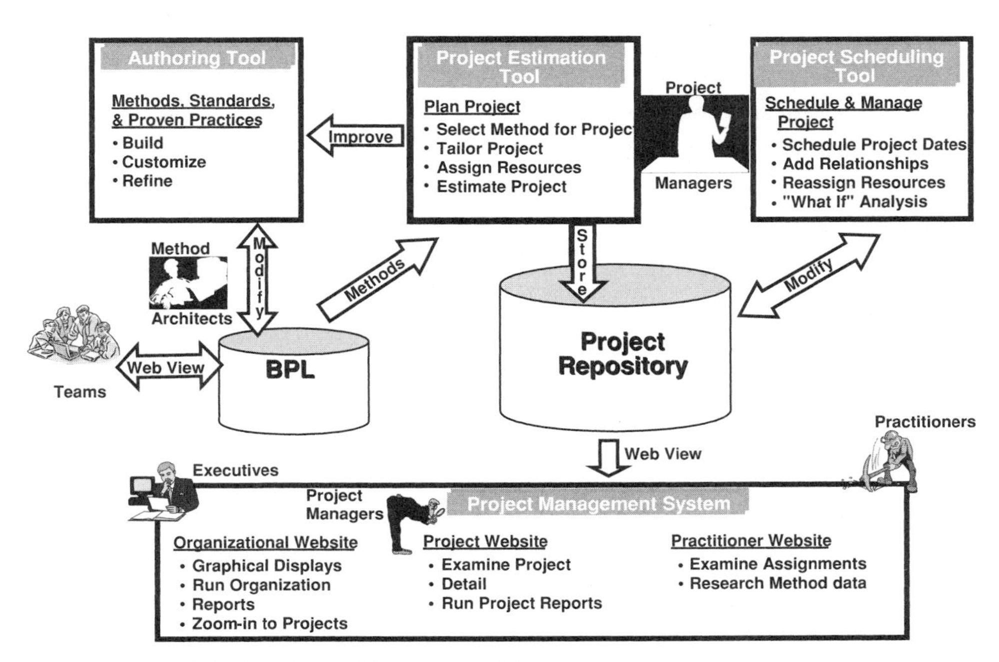

FIGURE 16–3. AllFusion Process Management Suite.

added: "When consultants are able to take advantage of Best Practices from subject matter experts across the organization, the result is the best possible solution delivered to our customers in a consistent manner."

Starkey believes that the decision to establish a CA Best Practices Program was a fairly simple one from a return-on-investment (ROI) standpoint. Starkey continued: "We have observed firsthand how CA customers using the AllFusion Process Management Suite have increased productivity and improved the quality of software development and services provided. It just made good business sense for us to roll out the AllFusion Process Management Suite internally."

One specific product in the AllFusion Process Management Suite that has been particularly helpful to Caravella is AllFusion™ Process Engineer™. This "process-authoring tool" is used to create and store methods in the BPL. "The process of authoring a method requires the collaboration of method architects with subject matter experts," said Caravella. "In addition to being stored in the library for use by project managers in generating project schedules, methods are published to a website where they are available to practitioners for reference. A consultant goes to the website to get a clear understanding of tasks assigned, preferred techniques to use and even what sample work products look like."

Best Practices Library in Use

According to Caravella, use of methods in the BPL leads to an improvement in the quality of CA's project engagements. With proven success from past engagements, CA can leverage its proven method templates for future implementations, which results in higher customer satisfaction. "I started using the BPL before the AT&T Broadband project started," explained Harry Johns, a Senior Project Manager in Englewood, Colorado. "I was looking for information and ideas on how to manage the projects I was assigned more efficiently. I was also looking for any tools I could find to make my job easier and more effective. I felt that an organized project would provide the client with a positive perspective of the CA Services team."

The AT&T Broadband project began in the spring of 2000. One of the first tasks for Johns was to develop a Project Schedule and Project Charter. He used the Project Charter template on the BPL site, which served as the foundation for the first Project Charter for Phase One of the AT&T Broadband project. "Both the Project Schedule and Project Charter were living documents that were used throughout the project," said Johns. "Future phases of the AT&T Broadband project used updated versions of the Project Charter from the BPL." The BPL also provided Johns with a variety of forms, including Change Request, Risk, Project Issues Log, Project Progress Report, Milestone Acceptance Form and others to help him effectively manage the project. Finally, the BPL gave Johns fresh ideas for kick-off and closeout meetings with the client; it also provided guidelines on risk management and Change Control considerations. "AT&T was very pleased with how well the project was managed and with the professionalism of the CA Services team," said Johns.

Adding Value to the Best Practices Library

Caravella insists that CA's future business success will be defined not only by the quality and relevance of its products, but also by its ability to "scale" its services business globally. "Treating institutional knowledge as a corporate asset is critical. The AllFusion Process Management Suite helps us capture our process knowledge in a reusable form and get it to the right people at the right time. Our customers value the BPL because they can be assured of getting quality results based on industry-proven practices. CA Services consultants and project managers appreciate it because it helps them learn the best ways to do a project and be more productive. Finally, management is keen on our approach because it empowers effective process management—a centerpiece of our ISO 9001:2000 certification program. It's a win–win situation for all involved," he said.

Three Steps toward an Effective eBusiness Environment

Companies around the world are positioning themselves for an eBusiness future despite the recent decline of the dot-coms. Projects for business-to-business applications, eMarketplaces, ePurchasing, knowledge management, and customer relationship management are on the rise.

IT organizations are searching for the most effective way to establish their eBusiness environment—the processes and tools that will allow them to build and deploy applications, exploit the knowledge contained within them, and manage the systems and databases that comprise their infrastructure.

Customers under increased pressure to respond to the eBusiness challenge approach Computer Associates International, Inc. for solutions that will enable them to establish and maintain an effective eBusiness environment. At the forefront of the eBusiness challenge, customers are asking themselves the following questions:

- How healthy is my current eBusiness environment?
- Is there a faster, more systematic way to establish an eBusiness environment?
- What are the pitfalls that other IT organizations have faced? And what are the lessons learned?

CA has developed a systematic approach supported by proven practices. In this section we share some of the lessons and best practices that we have learned and suggest some ideas that may help organizations prepare.

Specifically, CA recommends three key practical steps toward building and maintaining an eBusiness environment:

- Establish a manageable eBusiness environment.
- Create a systematic approach with proven practices.
- Define metrics to set priorities.

Step 1: Make the eBusiness environment manageable. Define the eBusiness challenge as several discrete implementation "programs." At first glance, establishing an effective eBusiness environment appears to be a daunting task. Before you set out to accomplish this task, it is essential that your management team share a common view of the scope of what lies ahead. The eBusiness challenge should be defined and categorized into separate programs that can be easily managed.

Establishing an effective eBusiness environment requires that an organization successfully implement the following:

- Build and deploy eBusiness applications.
- (Application Management).

- Exploit knowledge (Knowledge Management).
- Manage systems and databases (Systems Management).

Effective eBusiness Application Management requires attention to the following programs:

- *Project and Resource Management:* Allocate and control project resources to meet schedule, budget, and user "time-to-market" expectations.
- *Requirements Management:* Establish a common, precise understanding of eBusiness application system requirements among end users and the application development project team.
- *Application Development:* Provide the "design time environment" and process by which applications are developed, integrated, and rolled out into production.
- *Application Integration:* Provide the "runtime environment" to enable integration and interoperability of applications built upon multivendor platforms and database environments.
- *Change and Configuration Management:* Establish an environment to maintain the integrity of eBusiness application software throughout its life cycle.
- *Quality Management:* Minimize software defects before the application reaches the production state.

Similarly, the area of knowledge management can be categorized into the following focused, manageable programs:

- *Process Management:* Ensure understanding of current business processes to identify information needs.
- *Metadata Management:* Define and manage the attributes of enterprise data assets.
- *Data Transformation and Warehousing:* Transform structured data into information and organize unstructured data for decision-making.
- *Query, Analysis, and Reporting:* Provide query, analysis, and reporting facilities to knowledge workers.
- *Information Distribution and Collaboration (Portal):* Offer personalized graphical user interface and access to enterprise knowledge resources and applications. Provide an effective, efficient environment to encourage interchange of ideas and to promote innovation within the enterprise and with partners, suppliers, and customers.

Systems management encompasses managing systems and databases, including:

- *Storage Management:* Ensure the availability and integrity of enterprise data assets.
- *Security:* Ensure secure, productive access to resources and information.
- *Operations Management:* Provide 24 × 7 availability and reliability of systems and networks and process workloads.
- *Customer Service:* Provide proactive customer service.
- *Asset and Resource Management:* Monitor and track the physical location, financial value, and performance of IT resources.

Figure 16–4 will serve as a useful starting point to help your organization determine the scope of your eBusiness environment by defining (1) management areas (application, knowledge, and systems) and (2) discrete programs that fall within each area.

It is possible to break down each program further into lower-level management functions to achieve individual objectives. For example, in Figure 16–5 we have taken the systems management area and further defined the storage management program into backup/restore, archiving, data integrity management, and disaster recovery.

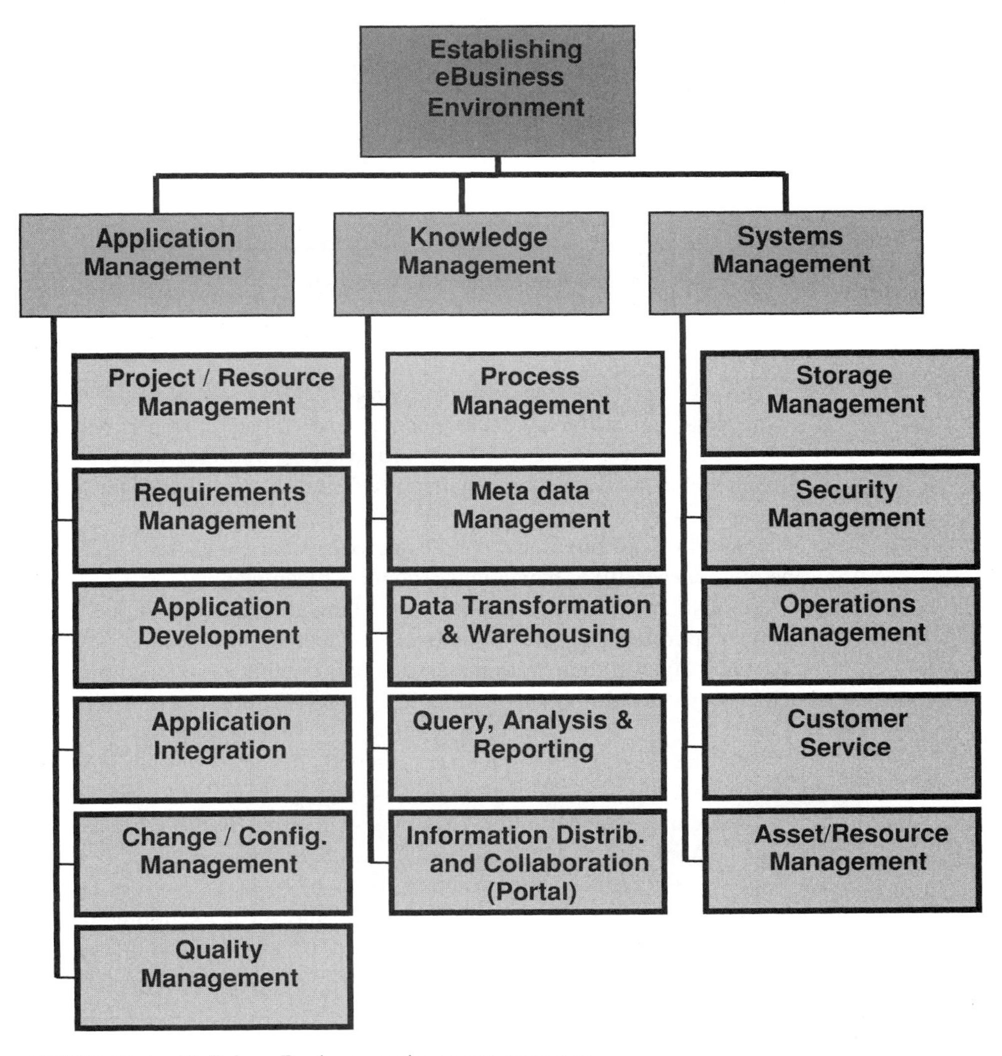

FIGURE 16–4. Defining eBusiness environment programs.

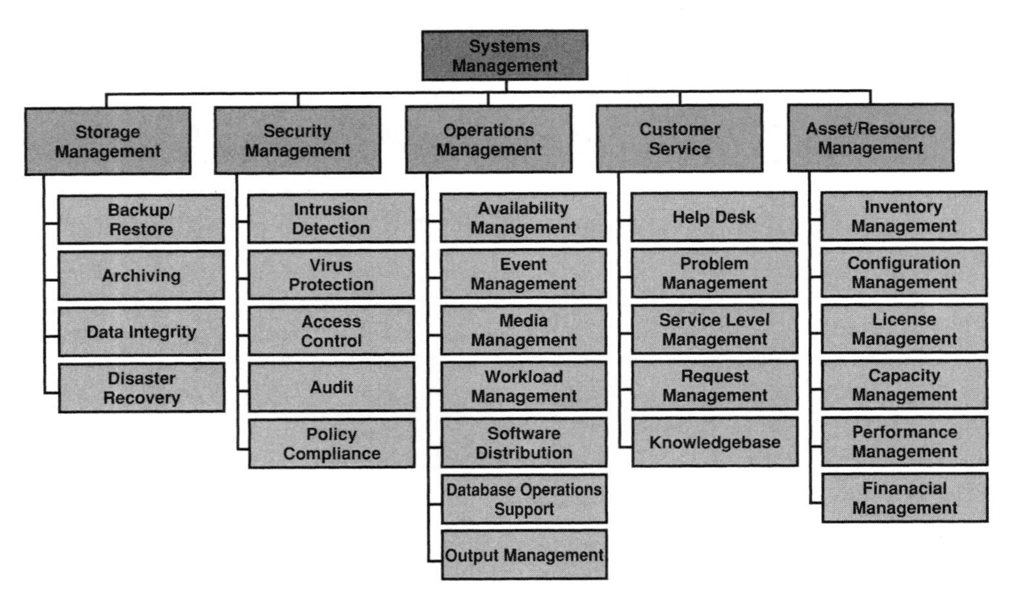

FIGURE 16–5. Programs break down into functions.

Step 2: Leverage a systematic approach and proven practices. Your approach should ensure that a shared vision is defined and implemented using effective methods.

With a more precise definition of the scope, you can now focus on a systematic approach for addressing these challenges. Three key management areas are suggested in Figure 16–4.

Based on our broad and solid experience with a wide range of customers, we believe the most successful organizations recognize that a planned, "layered" approach is required to establish an effective eBusiness environment. Figure 16–6 provides a visualization of such an approach, entitled "CA's Best Practices Framework."

Each layer in the pyramid suggests that organizations need one or more methods[2] to achieve successful implementation. Here are some observations on best practices:

- The *Strategy Layer* requires methods designed to gain buy-in to a shared vision and direction. For example, this may translate into a method to facilitate and articulate the definition of strategy in each of the three areas referenced above (Applications Management, Knowledge Management, and Systems Management). It also suggests the need to have methods to assess your current situation. In the Applications Management area, this may mean examining the maturity of your application development environment or assessing your organization's ability to

2. A *method* is a collection of information that provides guidance on how to plan for, manage, and carry out a particular effort to achieve a well-defined set of outcomes.

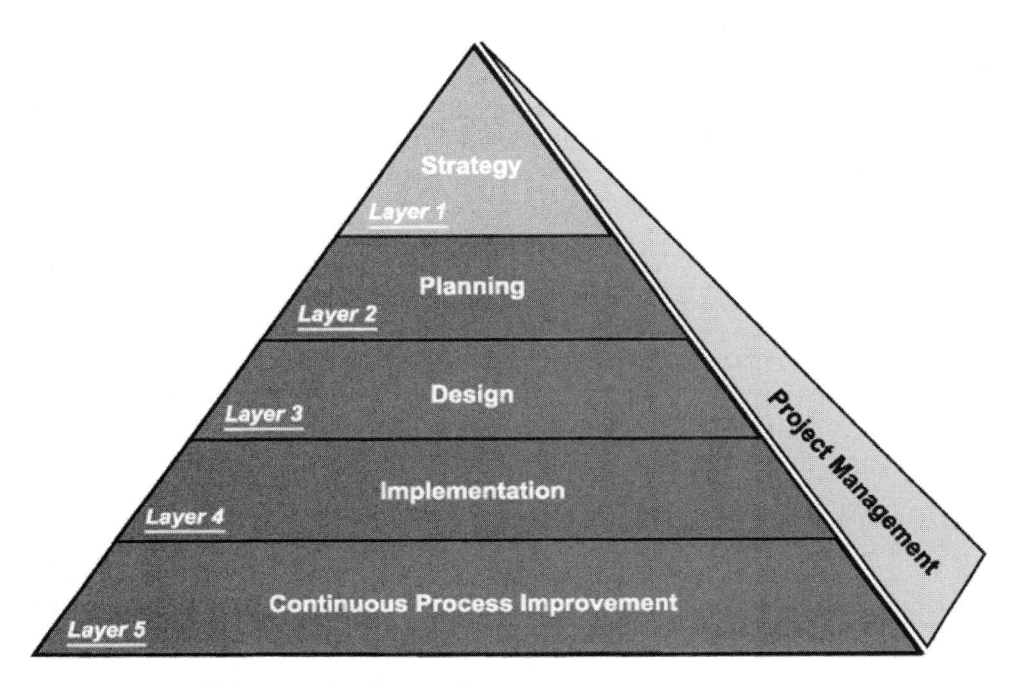

FIGURE 16–6. CA's best practices framework.

build and reuse components in your development process. In the Systems Management area, it requires articulating a strategy for integrating the major Systems Management programs identified in Figure 16–4.

- The *Planning Layer* requires methods that clearly define the nature of the management areas and programs identified in Figure 16–4. These methods should assess your "as is" program condition, characterize a desired "to be" state, and define appropriate transition states. The desired result is a clear definition of the projects required within a given program to achieve its goals.
- The *Design Layer* requires methods such as process definition and architecture planning. At this layer, sufficient information is compiled to lead to "build versus buy" product decisions.
- The *Implementation Layer* contains the methods required to implement products and build/integrate custom applications. This may include methods for configuring and implementing various technologies (e.g., portal implementation, application runtime environment configuration, and data warehouse implementation). Your library may also include a variety of application development methods ranging from object-oriented approaches to more traditional development processes.
- The *Continuous Process Improvement Layer* is a reminder that ongoing assessment methods are required to help manage and improve the "steady-state" eBusiness environment.

- The *Project Management* dimension of the pyramid indicates that regardless of management area, a common approach to managing projects is essential to ensure consistent, repeatable, and quality results. Thus, whether you are implementing systems management infrastructure components, building a new business application, or establishing a data warehouse, a standard template for project management will help ensure success.

Using this framework, you and your team can identify the methods required to ensure that your eBusiness environment is implemented and supported properly. Table 16–1 provides some examples of the types of methods required to set up an effective eBusiness environment. Ideally, these methods should be authored in a rigorous, consistent manner and stored in a repository for use by authorized staff within your organization.[3]

Step 3: Use metrics to set priorities. Each organization faces different business scenarios and occupies a different point on the eBusiness readiness scale. Regardless of the specific model your organization adopts as an eBusiness plan, you will still be faced with time and resource constraints. IT organizations continue to struggle to get ahead of the power curve. So where should you begin?

Assessments are great tools to help your team quickly focus on the right areas and programs. Metrics-based assessments can help pinpoint critical areas that your organization should focus on for early payback. Begin with a set of self-assessments aligned with the eBusiness environment programs shown in Figure 16–4. A good "starter set" might include the following eight assessments:

- Executive Level
- Application Management
- Knowledge Management
- Customer Service
- Asset/Resource Management
- Security Management
- Storage Management
- Operations Management

The Executive Level assessment is intended for senior IT executives who have knowledge of their organization's Application Management, Knowledge Management, and Systems Management areas. The Applications Management and Knowledge Management surveys are intended for individuals with more in-depth knowledge of these areas. The remaining assessments are critical building blocks that span the entire Systems Management area and are intended for individuals with specific knowledge and interest in Customer Service, Asset/Resource Management, Security Management, Storage Management, and Operations Management.

---

3. At a minimum, we suggest that each method in your library contain (1) a work breakdown structure of the activities involved, (2) descriptions of the techniques used in the activities, (3) estimating parameters that enable the creation of project schedules, (4) role definitions, and (5) samples of the deliverables created by the method. In this way, use of the method ensures consistency, repeatability, and predictability of quality results.

TABLE 16–1. EXAMPLES OF METHODS REQUIRED TO ESTABLISH AN eBUSINESS

Layer	Applications Management	Knowledge Management	Systems Management
Strategy	• Component-Based Development (CBD) Assessment • CBD Strategy • Process Improvement Strategy • Application Portfolio Strategy	• Knowledge Management Strategy • Data Warehousing Strategy • Metadata Management Strategy	• Enterprise Systems Management Strategy • Strategy Risk Assessment
Planning	• Application Solution Scoping • Feasibility Study	• Metadata Planning • Extract, Transform, and Load Planning	Program Definition and Transition Planning for: • Customer Service • Security • Asset/Resource Management • Storage Management • Operations Management
Design	• Solution Definition Using Object-Oriented Techniques • Solution Definition Using Traditional Techniques	• Portal Requirements Definition and Design • Data Warehouse Architecture Planning	• Customer Service Process Help Desk Architecture Planning • Security Architecture Planning
Implementation	• Solution Construction Using Object-Oriented Techniques • Rule-Based Application Development • Configuration Management Implementation • Installation and Configuration of Design Time and Runtime Environment	• Enterprise Information Portal Implementation • Data Warehouse Implementation • Repository Implementation • Data Migration	• Systems Management Infrastructure Deployment • Help Desk Implementation • Application Instrumentation Implementation • Intrusion Detection Implementation • Single Sign-on Implementation
Continuous process implementation	• Capability Maturity Model Assessment	• Knowledge Management Assessment	• Enterprise Systems Management Readiness Assessment • Security Penetration Testing and Analysis

Each self-assessment should be designed to reveal the *importance* of program attributes as compared with your organization's *capability* in this area. Ideally, the attributes that you evaluate should reflect proven practices. For example, consider the Customer Service Program identified in Figure 16–5 (one of five programs listed in the Systems Management area). CA's Best Practices suggest that effective Customer Service requires the integration of five functions:

- *Help Desk:* Establishes, implements, and maintains the policies, procedures, and tools required to receive, route, track, respond to, and report on incidents and information requests that are received by Help Desks.

- *Problem Management:* Undertakes actions to identify, record, diagnose, and correct problems as they are brought to the attention of the system through either events or calls. Problems are tracked, reported, and resolved within specific time frames.
- *Service-Level Management:* Manages business relationships with customers of IT services. Areas of focus include (1) working with customers to understand business requirements, (2) helping customers plan for the application of information technology, (3) negotiating service level agreements with customers (including workload processing requirements and associated service level and cost expectations), (4) managing customer (future) expectations, and (5) conducting regular reviews with customers.
- *Request Management:* Initiates, processes, and approves user requests for the acquisition, modification, transfer, or disposal of hardware and software products.
- *Knowledge Base:* Establishes the policies, procedures, and tools used to acquire, store, retrieve, and build upon expert information (knowledge) that is used for research and problem-solving.

It stands to reason that the maturity of your Customer Service Program is related to how well these functions are implemented and integrated. Thus, your self-assessment survey should feature a number of "best practices" statements about each of these functions. Table 16–2 provides an example of the statements you might consider for your "Help Desk" function.

Assessing other functions associated with the Customer Service program in a similar way provides important insight into the organization. The resulting analysis can be repre-

TABLE 16–2. SAMPLE ASSESSMENT STATEMENTS FOR HELP DESK

	Importance					Capability				
	Low				High	Low				High
Proven Practices—Help Desk	**1**	**2**	**3**	**4**	**5**	**1**	**2**	**3**	**4**	**5**
Out IT customers rarely experience any delays in contacting a help desk support person.				×		×				
Our help desk is able to answer the majority of customer requests on the initial call.			×				×			
We maintain accurate, up-to-date information on the status of all customer-reported incidents.				×			×			
We have a well-defined method for prioritizing incidents as they are received.				×				×		
We have a well-defined process for escalating incidents that remain unresolved.					×		×			
We have an accurate history of past incidents and their causes.			×			×				
We do an excellent job of keeping our customers apprised of the status of any incidents they have reported.					×		×			

sented graphically showing the "gap" between *Importance* and *Capability* for each function, as shown in Figure 16–7.

Be prepared to find the results of your team's "perception comparison" enlightening. Very often, there will be substantial differences in what members of the same team consider "important." You may also be surprised by the variance in opinion on the state of your capabilities. Nevertheless, it helps to understand these perceptions at the outset. Setting up an effective eBusiness environment is a sufficiently challenging assignment and becomes far more difficult when your team cannot agree on the state of your strengths and weaknesses.

Business Benefits

By following the course of action described here, you can gain the following business benefits:

- *Consensus on a Shared Vision:* The self-assessment approach is a provocative way to quickly bring your management team together in agreement on the scope of the eBusiness challenge and priorities for improvement.
- *Consistency, Repeatability, and Predictability:* Use of a standard project management methodology in concert with proven technology implementation methods will help ensure a more timely, cost-effective rollout of your company's eBusiness environment.
- *eBusiness Time-to-Market Expectations:* An effective application development environment, supported by an appropriate systems management infrastructure, will

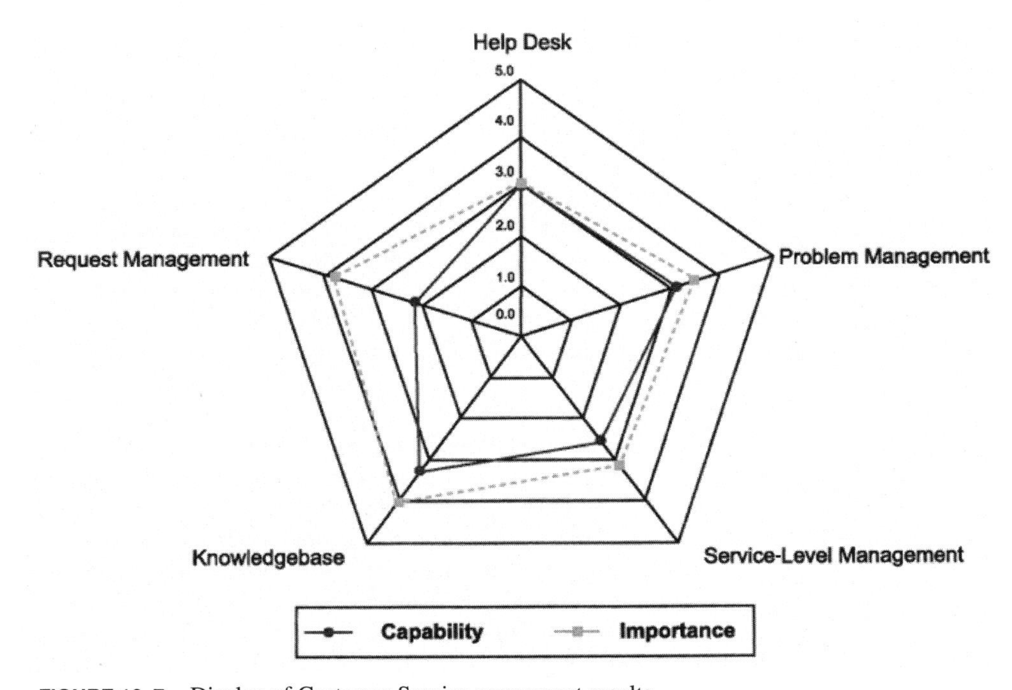

FIGURE 16–7. Display of Customer Service assessment results.

greatly enhance your ability to meet "time to market" and application availability/performance expectations.

- *Lessen Your "Skill Shortage Woes":* Dr. Gary Hamel, author and chairman of Strategos, states that in the future, "businesses will not be by what they do, but what they know. Strengths will be defined by deep core competencies and what unique knowledge they possess." Many IT organizations have difficulty getting sufficient talent to address the eBusiness challenge. Moreover, you are at risk of key players leaving the company and taking their knowledge with them. The approach suggested in this chapter ensures that critical IT management and technical knowledge is captured in your Best Practices Library. This not only creates a learning environment that helps attract and retain the best people but also minimizes the possibility of your staff leaving the organization without documented procedures or information.

Conclusion

The trillion-dollar eBusiness investment levels and payback expectations cited by industry analysts are difficult to comprehend. Understandably, demands on IT organizations to deliver have never been higher. IT executives must move quickly to establish the capabilities to build, deploy, and manage the rush of technologies that pervade eBusiness applications: wireless, peer-to-peer, instant messaging, and more.

A systematic approach supported by proven practices will help you set up the kind of eBusiness environment required to deal with rapidly rising business expectations. It will also help you establish the type of work environment that attracts and retains talented people.

You're Not Alone

Computer Associates' Best Practices have led to a number of offerings to assist clients in setting up and maintaining an eBusiness environment:

- *eInterviews:* Computer Associates has devised an easy-to-use online tool that may help clients quickly focus on the right issues. eInterviews provide the client with a metrics-based assessment approach to target priorities and problem areas.
- *CA's Best Practices Library:* Whenever a client uses CA Services℠, they are leveraging our experience and insight gained over countless customer engagements and captured into CA's Best Practices Library (BPL) in order to ensure success. The Library consists of IT business management and technology-specific methods, as well as CA's Project Management methodology.
- *AllFusion™ Process Management Suite:* This is CA's tool set used to build and deploy the BPL. This suite of process management tools enables customers to author processes and to use them to generate project schedules automatically—ensuring consistency, repeatability and predictability across a wide range of project types (such as application development, data warehousing, business intelligence, help desk implementation, and many more). The client may be interested in exploring its use as the basis for Best Practices and project management methods.
- Finally, CA offers a wide range of software products to support eBusiness heterogeneous platforms. These include tools that will help establish an application "design time" environment; set up an effective "runtime" environment to link and in-

tegrate legacy systems with Internet-based applications; mine and exploit knowledge through data warehousing and portal technology; and manage performance and secure the systems and databases that comprise their infrastructure.

We know that eBusiness needs are mission-critical and that clients do not necessarily have the time, resources, or tools required to set up and maintain their own environment both quickly and thoroughly. CA has made a considerable investment to build and maintain Best Practices, and to invite clients to benefit from our processes, lessons learned, and extensive set of tools.

16.2 LEAR

Lear Corporation, a Fortune 150 company headquartered in Southfield, Michigan, focuses on integrating complete automotive interiors, including seat systems, interior trim, and electrical systems. With annual net sales of $13.6 billion in 2001, Lear ranks as the world's largest automotive interior supplier and the world's fifth-largest automotive supplier. Lear's world-class products are designed, engineered, and manufactured by over 115,000 employees in more than 300 facilities located in 33 countries.

Like many large companies, acquisitions played a large role in the growth of Lear. With each acquisition came many new business practices. In 2001, Lear established a corporate-wide Program Management Office to develop a common, world-class, project management infrastructure that would be deployed globally throughout the enterprise.

Eighteen months later, Lear is managing its entire portfolio of projects utilizing the Lear Program Management Process (LPMP) and Web-based project management software (ProFile) developed in-house, where over 5000 associates collaborate and manage over 700 programs. The program management processes and tools deployed have led to improved communication, clarification of roles and responsibilities, early identification and mitigation of risk, increased accountability, and more, which ultimately facilitates better products and service to Lear's customers.[4]

Lear is a customer-focused organization and our program managers are responsible for organizing all the activities within Lear to assure our customers are getting the best possible products and services. Our program management office was created with the mission to develop a business infrastructure focused on making our program managers more successful.

We will always be working to improve, but we feel we have achieved some good successes already. Some of the key factors in our success are described below.

Global Program Management Office

To lead fundamental changes in how we managed product development and effectively drive those changes through divisional, political, and cultural barriers, it was important that a single entity have the

4. Jim Sistek, Director of Global Business Practices at Lear, who is responsible for Lear's program management office, provided the text for the remainder of this section.

responsibility and authority to develop the vision and manage the activities to achieve that vision. For Lear our PMO is that entity. The program management office will inevitably be trying to change the way program team members do their jobs; therefore, the PMO should report into the senior executive ultimately responsible for the program team members. At Lear that is the Chief Operating Officer of the company. Other success factors for the PMO include the following.

- *Global Authority:* Business practices tended to be specific to divisions within the organization. To create "One Lear" we had to establish a neutral PMO at the corporate level that could manage across divisional lines worldwide.
- *Lean:* Keep the PMO small. The Global PMO at Lear is staffed with approximately five people. The development of complex business processes, like a program management system, is not achieved by consensus. We consult with subject matter experts in various areas of the organization and quickly make decisions and deploy. If we miss the mark the teams are quick to let us know and we make adjustments. We don't want to get bogged down in bureaucracy.
- *Divisional Champions:* The PMO staff is supplemented with program management champions from each division. The champions report to their respective divisions so there is ownership and accountability to the success of the program management initiatives. The champions provide input on future directions, drive methodologies into their organizations, and serve as the front-line support to users in their organization. This is not a separate, full-time job. Typically, the division champions are people in fairly senior management positions who already have some direct responsibility for program management performance.

Standardized Program Management Methodology

Our customers all have their own set of requirements and nomenclature, and internally we had different methods of managing one product versus another. The different combinations of products and customers lead to a complex situation to manage. Developing a standardized program management methodology gave our company a common language we could use to communication during product development and a process we could monitor and improve.

We started by identifying the key things we do from idea to production. We linked these requirements together into a process that supports all products and all customers. Where required, forms and processes were developed and included. The entire process was documented and released to the organization through our intranet as the Lear Program Management Process (LPMP). The core process uses a gate-phase approach. We have five phases, each of which has a number of required deliverables, and cross-functional review processes are used to evaluate the quality of the deliverables and address risks and overall performance. In each phase we perform some combination of the following reviews:

- *Design Reviews (DRs):* Address compliance to specifications, fulfillment of customer expectations, manufacturability ease of installation, service, etc.
- *Supplier Reviews (SRs):* Ensure that the supply base clearly understands the latest design, timing, and shipping requirements and assesses their readiness to provide quality parts on time.

- *Plant Reviews (PRs):* Cross-functional teams go to our manufacturing plants to ensure that all requirements are clearly understood, evaluate readiness for the upcoming product build, and evaluate initial product and processes.
- *Management Gateway Reviews (MGRs):* Program teams present overall program status to senior executives, including financial assessment, evidence of completed phase requirements, and top risks with mitigation plans.

The process is depicted in Figure 16–8. Key success factors of the Lear Program Management Process include:

- *Web-Based Manual:* There are no paper copies of manual, forms, etc. Once the process was developed, it was deployed via an interactive Web site on the company's intranet. This allows us to make continuous improvements and deploy them immediately to the organization. From a document control perspective we know that everyone is always working to the latest version.
- *Deliverable Focused:* For every item in LPMP there is a clearly defined and tangible output. Where it makes sense, standard forms and templates are provided and actual filled-out examples are made available.

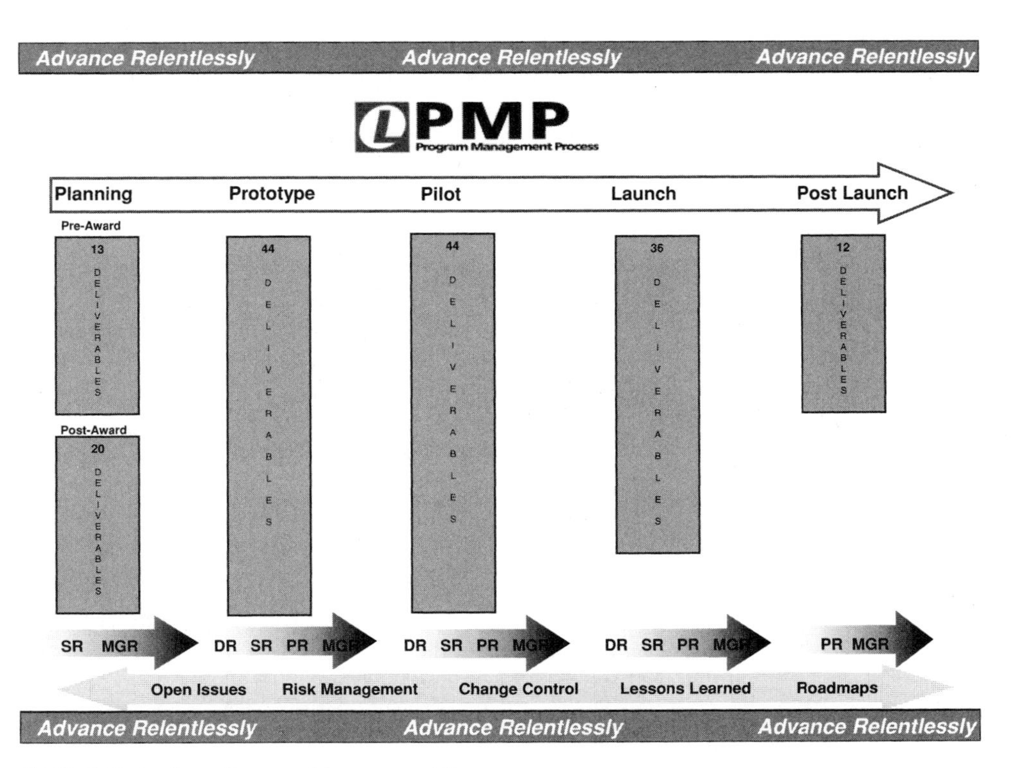

FIGURE 16–8. Lear Program Management Process.

- *Best-Practice Oriented:* Our Web site provides users the ability to view best practice examples of how some other teams completed their deliverables. It is not practical to make everyone do every deliverable exactly the same way. Where we need consistency, as in our financial assessments, everyone is required to do them in the same way. Where format is not critical, we provide examples of what are considered the best practices we have seen from some of our teams. This helps keep everyone from having to reinvent the wheel but also gives them the freedom to customize and continuously improve how we complete the deliverables.
- *Enterprise-Wide Solution:* LPMP was created generically enough that it can be used for all products and for all customers. In essence it provides the core requirements for success, and those are then supplemented by the program teams, who add unique product or customer requirements to their project plan.
- *Cradle to Grave:* Initially, we focused on developing a process that started with the activities required to prepare a quote for new business and ended with a post-launch phase which ensured that the product and processes met all expected targets. Once that was in place, we extended the process to provide a methodology for managing advanced product development activities. Lear now has a single methodology for the development of an idea into a marketable product concept, selling it, and transitioning to full-scale production.
- *Industry Requirements and Standards Built In:* By building industry requirements for product development into LPMP, we were able to provide a way to make sure that all programs were conducting the necessary activities to comply. Program managers and their teams do not have to do separate activities or create special documents just to comply. If they're doing LPMP, compliance is automatic. Cross-referencing to required standards is provided so that teams can demonstrate compliance during audits.
- *Design for Six Sigma:* Design for six sigma is about applying six sigma methodologies to the product development process throughout the entire product development life cycle. Design for six sigma deliverables are built into LPMP to facilitate product designs that can be manufactured at six sigma quality levels.
- *Flexible Process:* There are situations where it does not add value to accomplish some of the deliverables. Teams are given the flexibility to add and delete deliverables from the process as required. However, our project management software automatically flags items that are deleted and drives them into our gateway reviews, where concurrence by senior management must be obtained. This was a critical success factor. People don't want to be treated like robots. Our focus was to provide solid building blocks that teams could use to piece together a plan that makes sense for their program. Making teams do something just to satisfy a process is a waste and it takes the sense of ownership away from the teams.
- *Baseline for Continuous Improvement:* LPMP serves as a road map for continuous improvement throughout the organization. It documents *what* we need to do. Now that we have the *what* defined, we are using six sigma to develop robust processes for *how* to accomplish each deliverable and deploying standard processes across the entire organization.

Enterprise Project Management Software

Having a documented process such as LPMP gave the organization a common framework to guide our teams to success; however, effectively carrying out the process, sharing results, and monitoring status were still difficult. Project plans were typically maintained by the PM on their computer, with copies occasionally distributed to the team; documents produced during the process were stored by one team member in a drawer or binder; and key issues were tracked on spreadsheets by the program manager. Generally speaking, information was held by a few and was not readily available to the organization. To help teams execute LPMP more effectively and provide valuable project information to the entire organization, Lear developed a Web-based program management software solution called ProFile. ProFile is a method for managing project timing plans, financial information, customer information, open issues, documentation, supplier status, manufacturing plant status, profit improvement roadmaps, tooling and equipment plans, and team member information.

According to Spencer Gill, Manager Lear PMO, "ProFile has been instrumental in the evolution of our program management capabilities." An enterprise-wide management system has a number of advantages:

- *Information Access for All Lear Associates:* With a few clicks on the computer, associates can access program information needed to do their job better. Team members can log on from anywhere in the world and update their issues and actions and share documents as well.
- *Increased Visibility and Accountability:* Without requiring special meetings, presentations, and the time necessary to prepare for them, management can view the real-time status of every program in Lear. They can quickly see if a program is behind schedule, experiencing critical issues, not meeting financial objectives, etc. This instant access to information provides the visibility necessary to flag management when intervention is required and raises the accountability of team members. This leads to more focus on preventing issues and on early identification and resolution when issues do occur, meaning less cost to the organization and better service to our customers.
- *Disseminated Workload:* In ProFile all team members are responsible for maintaining their own actions. Previously, the program manager served as the team administrator, getting bogged down in keeping spreadsheets up to date with all the team's inputs.
- *Smooth Employee Transitions:* In today's work environment, team members can move about frequently. With ProFile the program information is not held by an individual's knowledge. If new program managers are assigned, they instantly have the entire program history available to them as well as the current plans. By simply replacing one team member in the system, a new team member automatically picks up the assignments of the previous team member.
- *Improved Performance of Dispersed Teams:* Nearly all of the programs that Lear has require the collaboration of team members from several different countries. Making sure that everyone is on the same page and has accurate information to do their job is a significant challenge. With enterprise program management tools like ProFile, teams can better communicate and know they are working with the latest

information. Being able to tie teams together regardless of their physical location enables a company to leverage resources in low-cost regions for product development.

- *On-Line Program Reviews:* To the extent possible, program managers are no longer asked to prepare special reports and/or presentations to meet management's information needs. These presentations often took significant time for the teams to prepare. Because ProFile is real time, information reports can be generated right out of the system with no special preparation. Weekly team meetings, monthly divisional reviews, and even Management Gateway Reviews are conducted by projecting a system-generated report up on a screen. As questions are asked and more information is needed, it can immediately be accessed in the system. According to Gill, the technical challenge of creating reports is not as big as the cultural challenge of getting everyone to give up their special reports. It took a commitment from the top down. Senior management still gets the information they need, but program managers are no longer burdened with continual reports and presentations.

ProFile was developed in-house at Lear. "We just didn't see a tool on the market that could deliver all that we wanted. We didn't want to have to run our business to satisfy some software, we wanted the software to support the way we wanted to run the business," stated Sistek.

Other Success Factors

Other factors cited by Sistek and Gill include:

- *Co-location by Customer:* Communication is a key to our success. Co-locating the program teams as much as possible improves communication, provides faster response to issues, reduces errors, and gives us a more efficient utilization of resources. All of these support our objectives as a customer focused organization.
- *Global Business Practices Group:* Lear established a corporate group to build an efficient product development infrastructure throughout the organization. The PMO is a part of this group. The PMO helps make the overall management of projects more successful, but many times the success relies on team members having good tools to get their work done. The business practices group is working to give team members from all disciplines the tools to perform their work better. The program manager can work with an engineer to be sure a drawing gets released on time and that tooling will be available on time, etc., but if that drawing has errors, we will still have problems. The business practices group is focused on tools that would eliminate those errors, thus contributing to the overall success of our programs.
- *Rapid Deployment/Relentless Improvement:* From the beginning our focus has been to move forward by taking small but quick steps. Whether it is a process, a form, software functionality, organizational modifications, etc., as soon as we thought we had something developed enough to add value, we rolled it out to the organization. A few people in a program management office are not going to be able to build the perfect program management system alone. Our focus was to build infrastructure in a flexible way so that we could react to feedback quickly.

Our product teams know best what works and what doesn't work. The PMO threw out a starting point, and feedback from the product teams has shaped our current systems to maximize value to the teams. We have to react quickly or our teams will get disillusioned with the system. It won't work if they don't believe it.

- *Six Sigma:* Six sigma is used extensively throughout Lear. Much as manufacturing processes can be improved with six sigma, so can business processes. LPMP provided a high-level business process model that we continually improved with the accomplishment of six sigma projects managed by our certified six sigma champions.

Bill Pumphrey, President of Lear's DaimlerChrysler Division, summarizes Lear's situation in this way. "The days of flexibility and 'I'll scratch your back, you scratch mine' in the automotive industry are long gone. Cost competitiveness and reduced timing have driven precision and accuracy into product development processes. Without disciplined, requirements driven program management, supported by global communication tools, suppliers will not survive. The cost of mismanaged scope changes, poor risk assessment and planning, and weak task fulfillment will all drive intolerable financial results and an inability to compete. Survival of the fittest takes on a whole new meaning in the next decade. Lear has advanced relentlessly at establishing best in class program management and we expect to win."

16.3 TEXAS INSTRUMENTS

Companies such as Texas Instruments have best practices in project management throughout their organization. In previous chapters we discussed several of these best practices. In this chapter we discuss the best practices in project management related to capital projects.[5]

Two-in-a-Box

Until very recently our process was as is currently typical with most construction firms, to build the entire plant and then turn it over to the system owner or sustaining team (the team that has to move in after construction and maintain/operate the building or process). Many times there is a single person or small contingent (two or three people) representing the customer, to interface with the builder. Typically, this team is more concerned with architectural issues rather than the more important engineering design concerns. Unfortunately, at "handoff" you must accept what you get, assuming that the construction specifications were met. Typically, the budget has already been expanded past the comfort level for upper management. When the sustaining team moves in to begin operating the equipment, many times their reaction is: "That's not the system I would have installed in this process." Management response is often: "Can you make it work until we can get this place producing and can afford to replace it?" It all depends on the management style of the senior management team.

5. A spokesperson at Texas Instruments provided the information for the remainder of this section.

In reality, what has happened is that a customer representative team (that typically does not look for engineering concerns) reviewed the original construction plans and approved them. Then during the construction, when it appeared that the project was going to overrun the budget, the same customer representative team agreed with the general contractor on several areas that had opportunities for cost savings (referred to at TI as *value engineering*). Many of these ideas are valid and would make sense to even the most technical. For example: The plant may be designed for immediate ramp to 100 percent production. In an effort to save money, the management team may decide to stagger the ramp from 50 percent production the first year to 100 percent over the next three years. In this case it might make sense that you only install the number of systems necessary for the first production ramp plan. Where this gets tricky, what systems can be cut back? Many of these decisions require a qualified engineering staff.

Two-in-a-box is how we get the engineering staff working directly with the construction staff throughout the life of the project. Two-in-a-box is a project management process that we developed in the late 1990s to overcome the "hot potato," "new project," and "run out of money, project over" construction completion syndromes.

- The "hot potato" syndrome is, of course, a reaction by the project manager to drop (close) the project as soon as possible, many times prior to completion of the planned scope. Most of the time, the completion of the project is handed off to the sustaining team and may or may not include the necessary budget to finish the project. Typically, this is due to manpower, budget, or schedule conflict problems.
- The "new project" syndrome is, in our case, the most popular. It is where a project manager, toward the end of a project, is assigned a new project to start as soon as he or she is near completion with the current one. The management concern is that the project manager will become less productive toward the end of a project and not having new goals (projects), will squander time on nonproductive issues. On the project management side, they have spent many days/months/years on the current project and look forward to something new and more interesting. As the project manager's attention moves from the current project to the new one, things such as documentation, punch lists, and commissioning lose their value, and less than perfect work is accepted to bring the project to a close as quickly as possible.
- The "run out of money, project over" syndrome is where the project team hands off a project that is not complete because the project budget has been expended, with the expectation that the sustaining team will complete the project with its financial resources. This is typical in the case where scope-creep takes over toward the end of a project; which happens because the sustaining management team did not have a good grasp on the design or scope and the project manager is too agreeable and pleasing. This problem is more typical and can be more extreme on government projects. Government projects have a tendency to use low-bid contractors, who make a living on the change notices that can run a project out of money fast. Government projects also have a problem with unrelated people controlling the "purse strings" and can be halted in midproject, due to funding changes made by high-level planning committees.

Our two-in-a-box process is designed to overcome these project deficiencies. The basis of the process is that the project manager and sustaining manager (or system owner) work together from the inception of the project. Both are equally responsible for the project scope, budget, and schedule. This allows for the future system owner to be involved in and thus responsible for the system/building/process that they will accept and maintain when the project is complete. The responsibility for the product comes from participating in and agreeing to the project designs, installation quality, and commissioning.

The first question that you might have is: How is the work divided? It is obviously not efficient for two people to be doing the same job, and I can assure you that this process would not be allowed unless it added value to the project. TI management is a proponent of the Texas Ranger slogan, "One riot, one Ranger." In other words, it should take only one person to resolve one problem/issue/project. However, looking at what the two-in-a-box process is expected to accomplish, you can begin to see the lines of responsibility being drawn.

The system owner contributes more heavily to the systems design and scope creation, Quality control (QC) during the installation/construction phase, providing engineering support for changes and value engineering, reviews submittals and ensures that commissioning and documentation occur, as specifications require prior to project close. The project manager contributes more heavily to writing the CSI format scope specifications document, ensuring that the scope matches the budget, budget tracking, issuing purchase orders (POs) and processing invoices and change notices, ensuring that contractors stay on schedule and managing the ever-present scope-creep.

The second question you might have is that since this would appear to be a competitive environment, and with the aggressive personalities required for either one of these positions to be successful, how do you get these two positions to work together for this endeavor to be successful? The key is that both are accountable for all project-related decisions made and will want to help each other be successful. That means that the project manager must want to ensure that the system owner has the best design, quality installation, and that all parts of the system are tested (commissioned) and documented prior to completion and project handoff. The system owner must also want to ensure the success of the project manager considering scope, budget, and schedule; in other words, they must be able to communicate and work in a team environment. To the right project manager and system owner, it is obvious that this team arrangement allows both of the participants to do their jobs more efficiently and pay more attention to the details of their individual responsibilities. These communication and teaming traits must be a consideration when interviewing for these positions.

This process is basically a system of checks and balances. The expectation is that the project manager, with his or her scope, schedule, and budget concerns, will offset the system owner's need for the best system that money can buy. They both have to agree on project priorities. A good example is an air-handling unit in the administration wing that should have a commercial-grade valve and actuator ($300) to control the chill water flow to the cooling coils; whereas an industrial-grade valve and actuator ($5,000) are necessary for deionized water to support the production environment. The key point in this example is reliability; one can fail without effect to production and the other will have a substantial effect on production. To make this work, both the project manager and the system owner must have similar goals and incentives. These goals must support the premise that priority 1 is quality work, 2 is system reliability, 3 is budget, and 4 is schedule. Sometimes, of course, schedule is a higher priority than budget, but typically not. The bottom line is that the system should function as designed, meet all relevant codes, require little to no unscheduled maintenance in the first few years of service, and provide quality service through the depreciation period. It is the responsibility of both the project manager and system owner to make this a successful project.

Checkbook Process

I have taken several project management courses and none of them covered how to track project costs or change notices, document purchase orders issued or signature authority level, how to track Invoices to POs, or even the significance of these and other financial issues. Each project manager either designs his or her own financial tracking process or edit someone else's to fit their particular needs. In the early 1990s the Worldwide Construction Team of Texas Instruments, Inc. (TI) created a "checkbook" for project finances. It is designed to handle the construction of a semiconductor manufacturing plant. I have simplified the version here so that it can more easily accommodate a simple one-contractor project or can be expanded to a multicontractor project.

There are many things that must be included in a project financial tracking process. These include but are not limited to:

- Project name
- Contractor name
- Dates for all transactions
- Project budgets
- Purchase order numbers and amounts
- Change notice numbers and amounts
- Invoice numbers and amounts
- Open commitments remaining on project budgets
- Release of liens (lien wavers)
- Retainage

Each company may have other bits of important information that must be tracked, such as accruals, taxable and nontaxable items, subcontracts, contract management teams, early project completion incentives, penalties for missed schedule milestones, and many others. The matrix I have included can easily be edited to accommodate these additional tracking categories. It is best to do this tracking from as few files as possible. I prefer a single Excel file for each major project with several worksheets. I would recommend electronic tracking between worksheets so that there are fewer transcription errors. Electronic tracking between worksheets also helps with number crunching, which can very quickly point out errors in invoices or budget balances. Data entry should be limited to only two or three of the worksheets, and these data can be rolled up to populate the other worksheets that provide the management-required project reports.

Open commitment is an area I pay close attention to. The amount invoiced should correspond with the scope completed and that remaining to be completed. Every project manager looks at this area of financial tracking in different ways and may base the approval of invoices on many different criteria. For example:

- Equipment delivered to the project
- Percent of scope completed
- Estimated time/schedule remaining on the project, perceived volume of work remaining on the project, lien wavers accumulated to date
- Subcontract invoices accumulated
- Time and material for work completed
- Project milestones

Whichever method you choose to approve invoices should be covered very clearly in the contract terms and conditions. The key here is not to lose track of the idea that pay for performance and project balance is an incentive to complete the project in a timely, high-quality manner.

Retainage is the only saving grace for some methods of payment. Here at TI we allow for progress billing on equipment delivered to the project and prior to payment of invoices require signed letters of transmittal for the equipment and lien wavers from the companies supplying labor and equipment. Standard progress billing is based on percent of scope completed. We require that a punch list be established and a substantial completion form be signed by all parties prior to final invoice, and we hold retainage from all project invoices until the punch list is complete and the final completion form is signed by all parties.

Every company has a different legal staff and therefore has different contractual requirements for contractor financial agreements. The key is that all financial transactions be documented and retained in a central repository. Basically, what I am suggesting here is that instead of putting the invoice in the contractor folder in the file cabinet, these transactions should be entered into a financial tracking matrix that can provide immediate updates to the financial status of each purchase order and the project as a whole.

Change Notice Process The change notice process is typically the downfall of most project managers. To be a success, you must be able to react to an interruption in the flow of the project immediately, or the schedule and budget can suffer severely. The project manager must be able to make a decision within hours to rectify a problem or interruption, document that decision and the scope of the change, and assure payment or schedule compensation for the change to the contractor or team providing the service. The main premise here is that the project manager has the authority to make the decisions and authorize the work. Unfortunately, in many cases the project manager has only the responsibility to bring the project in on schedule and on budget, but lacks the authority to make financial decisions. In these cases the project is destined, at a minimum, to miss the budget and schedule plan: budget because it cost more for rework and downtime than the original uninterrupted scope of work, and schedule because you have people (sometimes multiple levels of subcontractors) waiting for decisions and the team in front to finish before they can start.

Below is the basis for what we call our ABC Change Notice Form. It is easy to see why we call it the "ABC" form:

A. It provides for the documentation of the change in scope.
B. It provides a notice to proceed with the work described.
C. It provides at some later date, based on previously agreed to unit pricing for time and material work, a final cost for the work performed.

"ABC" can obviously apply to its simplicity as well. We agree on a scope of work, we agree to pay for the changes, and we both have agreed on unit pricing for time and material work in the original contract.

16.4 SUN MICROSYSTEMS

Established as a distinct group within Sun in July 1993, Sun Services is the global service and support business of Sun Microsystems, Inc., a worldwide leader in enterprise network computing and technical network computing and the home of the worldwide Java™ technology paradigm. Sun Services provides the capability to plan, implement, and manage enterprise network computing environments on a global scale.[6]

With deep technology expertise, broad service offerings, and global experience serving enterprises of all types and sizes, Sun Professional Services (SunPS) has a proven record of being the best choice for companies looking to reduce the time, cost, and risk of transforming their business with technology. Our expert consultants can help you with every phase of developing and deploying the innovative services that will give sustained business advantage: maximizing return on investment while minimizing total cost of ownership. Project management services from Sun Professional Services are designed to help customers manage a project throughout the IT life cycle. Our project managers adhere to industry standard and Sun-specific methodologies and principles to develop complete project planning that addresses quality assurance and risk control. This includes our use of the SunTone℠ Architecture Methodology, PRINCE2 project methodologies, and the principles defined by the Project Management Institute (PMI). Sun project managers can become both partners and mentors in delivering successful projects. Through the exchange of information during workshops, leveraging industry best practices, and the presence of project managers, Sun can help not only guide a project but also manage the transfer of project management knowledge. The Sun BluePrints™ programs are the leading source for in-depth technical information on best practices using Sun Solutions. The Sun BluePrints program publications, synonymous with Best Practices, are applied methodologies, derived from practical experience, that result in solutions for a well-defined problem.

A crucial feature of the project management service is an emphasis on risk management. Before our project managers begin work on a project, we conduct a risk workshop

6. Sun Microsystems provided the information for the remainder of this section.

with the customer and partners to develop the foundation for risk planning and administration. Project management services can help:

- Deliver the project on time. One can start reaping business benefits as early as possible.
- Provide the expertise to help deliver projects within budget while meeting customers' expectations. This can mean lower total cost of ownership and improved cost-effectiveness.
- Improve decision-making processes and more efficiently leverage technical resources.
- Increase advantages by linking Sun technology and the "best of breed" solutions provided by Sun partners.

Sun's project management professionals are specifically trained and often certified in the techniques and skills that can help customers deliver projects on time, within budget, and that meet customer expectations of work. Our project managers must continually demonstrate excellence through quality reviews and certification training by the Project Management Institute and PRINCE2 organizations. Projects can benefit from our expertise in these specialized areas:

Risk Management

- Delivers a plan that helps resolve current objectives, identify uncertainties and refine project scope
- Provides a foundation for tactical and strategic planning sessions
- Analyzes each identified risk to assess the impact on schedule, costs, and quality

Risk Workshop

- Brings together the Sun project management team, members of the team, and any partners to discuss identified risks
- Sets the foundation for risk planning and administration

Quality Assurance

- Begins at project conception
- Develops a plan that incorporates customer standards, organization, and PRINCE2 methodology
- Evaluates project products against this plan

Quality Review

- Provides a quality review (quality coaching/quality delivery oversight) service checkpoint
- Documents findings and appropriate corrective actions

Program Management Office

- Coordinates the development of a Program Management Office and Architecture Review Board
- Reviews operations to optimize IT systems and processes on an ongoing basis
- Promotes the use of best practices during the IT life cycle

- Helps identify and mitigate project elements that may affect key service-level agreements

Integrated Processes

With 35,000 employees in 100 countries, Sun Microsystems is carrying out project management on a truly global scale. Sun is challenged with defining project management "best practices;" leveraging the work of hundreds of project management experts around the world and in every part of the organization, and developing processes for ensuring repeatable, measurable project management that adds value to what we do.

Within Sun's Services organization, the Professional Services arm (SunPS) has responsibility for developing internal project management expertise for a network of close to 500 project management leaders. SunPS coordinates the development of best practices and executes continuous process improvement using Sun Sigma tools and methodologies. This rigorous approach keeps customer needs at the forefront of decision-making. Sun Sigma is Sun's adaptation of the six sigma practices pioneered by Motorola in the 1970s. Sun continues to improve these practices and has recently undertaken major initiatives in the following areas.

Sun CAP

When solving a problem or implementing an initiative, a good-quality solution that is widely accepted will have more endurance than will a higher-quality solution that is not accepted. The two main factors to the success of any solution, quality (Q) and acceptance (A), are both necessary for the initiative to be effective and may be formulated as

$$Q \times A = E \text{ (effectiveness)}$$

Although historically, Sun has put more emphasis on the quality part of a solution, it is now realized that there can be greater return on time and energy by increasing the acceptance of a good solution rather than by trying to improve the quality even more. And often, working on the acceptance will lead to a higher-quality solution.

The Sun Change Acceleration Program (Sun CAP) helps people who are working on solutions to focus on the acceptance as well. Sun CAP provides a change acceptance methodology to accelerate the acceptance of change in the business. Training is provided in these Change Acceleration methodologies and tools, with particular emphasis on stakeholder analysis and ongoing stakeholder management. Facilitation and applications of these tools are discussed and practiced in detail.

Sun Sigma

The Sun Sigma methodologies and tools are being applied throughout our business to achieve customer satisfaction and profitability. Efforts are under way to:

- Identify core/enabling cross-functional/cross-business processes within Sun
- Determine the key steps within these processes

- Establish unambiguous ownership for each process or subprocess
- Develop baseline process metrics from the customer's perspective
- Institutionalize an ongoing monitoring and corrective action mechanism for these processes

Products, services, and processes meet customer requirements [Voice of the Customer (VOC)] which are Critical to Quality (CTQ), such as reliability, on-time delivery, repeatability, and so on; and investment trade-off decisions (ranging from design requirements to resource allocation to merger and acquisition) are sound, well understood, and meet company, business, and market objectives.

Sun Sigma improvement and design outcomes guide us in allocating our resources rationally to best meet Sun customer and shareholder objectives. The Sun Sigma road map will ensure that all employees will be trained to the appropriate degree, so that Sun Sigma methodologies are ingrained in everyday business operations at Sun, from product design and manufacturing to marketing, human resources, workplace resources, and all other responsibilities.

Set for Success

Sun's Engagement Methodology is called "Set for Success." It is scalable to accommodate the different types of services that Sun Professional Services sells and delivers to the customer. The engagement tool kit consists of documents and templates organized so that business processes and supporting documentation fit the size and nature of each engagement. Within that context:

- Every engagement follows the SunPS Customer Engagement Life Cycle.
- Every engagement fits into one or more of three service categories:
  - *Placement Services:* The customer contracts for a specific skill set for a predetermined time period. Generally used to augment customer staff or to transfer expert knowledge on Sun products and the associated implementation.
  - *Product Services:* The customer contracts for a third-party service through Sun, usually as a portion of a product/services bundle.
  - *Project Services:* Customers contract for a custom-deliverable technical service that meets their requirements.

Every service category fits into one of three opportunity types, based on the profile of risk characteristics for the engagement. These opportunity types determine the level of project autonomy, signature authorities, and documentation requirements:

- *Type I:* Mainly low-price packaged services with simple deliverables and small teams.
- *Type II:* Includes large fixed-price services and the smaller custom Project Services.
- *Type III:* Custom consulting engagements, usually spanning several months and necessitating a larger delivery team.

Every opportunity type follows the core Set for Success guidelines and may require additional standards, depending on the time zone and country that the project is delivered in.

Work is currently under way to enhance the Engagement Methodology in two significant areas:

- A project has been created to improve the processes and procedures around winning worldwide, large corporate engagements called the Deal Management Guidelines.
- "Stretching the envelope" and going for larger custom engagements in order to grow the business. This is likely to have higher levels of risk than Type III and consequently, more senior involvement. This Opportunity Type is called Type X.

PRINCE2

PRINCE (PRojects In Controlled Environments) is a UK government-sponsored project management methodology that was established in 1989 by CCTA in the UK. The later version of PRINCE, called PRINCE2, is now the recognized method of managing projects in many industry sectors. It provides a structure to identify the project deliverables and ensures delivery on time and within budget, honoring quality and the customer's business drivers. PRINCE2 remains in the public domain and the Crown retains copyright.

Sun Professional Services UK began the PRINCE2 push several years ago, and this effort continues to gain momentum and support:

- More than two-thirds of Sun's project management consultants are PRINCE2 Foundation certified, including over 90 percent of the UK-based staff.
- About a third of the project management consultants have achieved the more rigorous PRINCE2 Practitioner designation, including 80 percent of the UK-based staff.

SunTone Architecture Methodology

Sun recognized early on that PRINCE2 needed to be supplemented with another methodology that specifically addressed technical issues. In response to this need, Sun drew upon concepts of the highly successful Rational Unified Process (RUP) and developed the SunTone Architecture Methodology (SunTone AM). This new methodology incorporates the best practices identified in RUP with the industry-leading architectural design standard, 3-DM. SunTone AM enhances the meaning of "architecture centric" by incorporating pattern-based reasoning based on systemic qualities. In addition, SunTone AM keeps the overall process description smaller and more manageable.

SunTone AM builds on the concepts of phases and work flows from RUP. A phase provides focus and priority on key issues. The partitioning of the project time line into phases serves to clarify and emphasize these priorities both internally and externally to the project. Each phase is defined by the products that constitute the deliverables, which in turn drive the activities that must occur within that phase. The SunTone AM is Sun's technical methodology, concentrating on Quality of Service across the entire customer architecture.

SunPS has devoted considerable resources to developing a set of global standards and best practices for consistent field delivery of consulting services. This complex task has involved members from Sun organizations worldwide. These groups have been focused on making the way we do business as rock-solid as Sun's technologies. The mission continues to be focused on implementing the methodology, tools, communication, and leadership that support the SunTone quality standards.

While there are multiple components of the SunTone program, SunPS focuses on a few key investments within the initiative:

- SunPS certifies its own services to ensure the same quality that we encourage our customers and partners to practice.
- SunPS consultants use SunTone AM to define the architectural principles and best practices employed to address a customer's infrastructure across the development life cycle: namely, Architect, Implement, and Manage.
- SunTone Architecture Framework is a tool that defines our services for sales and partners. This framework brings together the concepts identified in the architecture stack with those of an enterprise life cycle.

Culture

Focus, dedication, innovation, leadership, experience, expertise, and bringing together multiple teams to address a common goal—customer satisfaction—has propelled Sun into its position as a leading provider of industrial-strength hardware, software, and services that make the Net work. At the cornerstone of Sun's success are global corporate initiatives to actively involve our partners and our customers to shape the way that technology unfolds. Innovation reigns supreme, and most important, ensure, that the Net works.

Sun believes that innovation is the hallmark of a great company. Sun's cutting-edge technology developments have set Sun apart from the pack and will continue to do so. Sun's ongoing commitment to innovation and technology development is based on our belief that innovation is a core asset. It is fundamental to Sun's long-term success; our ability to develop exciting, useful products and services; and our mission of making our customers successful.

Everything we do goes back to what we call the "Voice of the Customer," which has to be the voice that resonates loudest within Sun. The Project Management Performance Model is tied closely to quality and customer loyalty metrics, with numerous points for measurement and "course correction" to assure that we meet these metrics. The performance model, managed on a company-wide basis by our 500-plus PM Leaders, is a powerful tool for global PM at Sun Microsystems.

At Sun we believe that we should motivate our project management team by focusing on their strengths. By investing more time in making use of these strengths rather than putting in place training plans trying to iron out the weaknesses, PMs are motivated to develop their strengths even more. Rewards for the project management team are often given in the form of corporate announcements such as "Win of the Week" or by being nominated to attend corporate events by their peers and team members. Monetary rewards are usually given in the form of stock option incentives.

Management Support

Project management expertise within SunPS is organized around four primary competencies:

- Project Management Performance Model
- Project Quality Activities
- CLI and Sun Sigma Activities
- Quality Feedback Loop

Project Management Performance Model The Project Management Performance Model is our tool for applying appropriate performance metrics to the PM practice. We are continually refining the model in response to customer and stakeholder feedback and organizational changes and to take advantage of improved PM practices. We are also building Sun Sigma quality processes into the model.

- *People:* Everything we do must meet the customer's business needs while developing our project management professionals. Processes, best practices, and tools exist only as a means to this end.
- *Processes and Best Practices:* Our project managers adhere to industry standard and Sun-specific PM methodologies and principles. All of the methodologies we employ are driven by three principles: customer business case, quality, and risk management.
- *Tools:* Our project managers use a variety of the "usual suspect" tools to enhance their performance. The latest additions to our tool chest is the implementation of Oracle Project Accounting (OPA) and an automatically generated PM Practices Survey, which triggers our customer loyalty survey.

Project Quality Activities SunPS is in the process of rolling out and perfecting our project quality activities based on the recent implementation of OPA. The customer feedback component has been online for several years, but the capability to generate the PM Practices Survey as a component of this is a huge step. Here is how it works.

- *Startup:* During project startup a Sun project manager is selected based on the skill sets required for the project. To identify the correct project manager, two tools are used, the Resource Management System and Skilltrax.
- *Delivery:* Quality assurance begins at project conception. The project manager works with the project team (which includes the customer) to develop a plan to measure the success of the project. During the course of delivery, the team evaluates project products as well as the project delivery against this plan.

The quality review team provides a Quality Review Service. These checkpoint reviews include document reviews and interviews with the team and the customer. Findings are documented and appropriate corrective actions are taken based on the quality plan. (This information also becomes part of the quality feedback loop to the project manager, PM expertise, technical team, and engineering.)

Project Closure When the project is closed, a PM Practices Survey is generated and the project manager answers questions about the project and best practices used on the project. The completion of this survey triggers what we call a Voice of the Customer survey, which measures the Customer Loyalty Index (CLI). This process is managed by an independent third-party firm. About 250 CLI surveys are generated per quarter. The CLI survey program is supported by a Web-based application which offers survey design and maintenance, and analysis and reporting of CLI results.

Other project closure activities for the project manager include creating a lessons-learned report and harvesting project management IP for cleansing and reuse in our knowledge management system, the Field Automated Service Tool. This information also provides a component of our quality feedback loop.

Once the project closure activities are completed, data from the PM Practices Survey, CLI responses, and other data points are analyzed by the Project Management Expertise Center and SunPS World Wide Quality Office. The outcome of this analysis may trigger a Sun Sigma quality initiative.

Quality Feedback Loop The CLI and Sun Sigma activities provide improvement information to the PM Expertise Center. This is used to improve PM best practices in project delivery as well as in PM development.

SunPS Organization

The program managers in each region are responsible for a "slice" of the business (could be by vertical, location, or customer). New business opportunities are owned by business managers, who interface directly to general sales. In some countries the business manager and program manager roles are executed by the same person, often called the engagement manager.

Line management is by function—all SunPS people are professionally qualified in their field:

- Solution delivery, which owns project managers, technical project managers, and project cordinators
- Consultants
- Engineers
- Managed services practitioners

Each of the groups has a professional development manager, who staff members can use as required. Communications are through Transfer of Information (TOI), where groups of like-minded people get together for an update, regular corporate focus groups, email, and webcasts.

Engagements run as projects using the Customer Engagement Life Cycle as defined in Set for Success, which uses a seven-stage model covering the following phases: Prospect, Qualify, Propose, Negotiate, Initiate, Deliver, and Close.

- The business manager is responsible for Prospect, Qualify, Propose, and Negotiate phases.
- The program manager is responsible for Initiate, Deliver, and Close phases.

All projects use an internal project board, and some have both internal and external project boards, depending on the size of engagement and the customer's ways of working. Project board members represent the business (program manager), supplier (industry lead consultant or technical design authority), and customer (lead account manager or business manager). The program manager is the most likely supplier representative of the customer's project board.

Project controls are at the following points in the project:

- Initial Opportunity Review
- Opportunity Review
- Bid Review
- Contract Review
- Project Initiation Document Review
- Initiation and Kick-off
- End Stage Review
- Closure Review

Control mechanisms include:

- Risk Typing Form to determine project Type (I, II, III, or X)
- Detailed plans (some of which are autogenerated in the case of packaged service engagements)
- Monthly Highlight Reports to project board
- Exception Reports

Any investment required in developing the offerings is made with reference to the local steering group and the regional project review board, who consider the business case and decide whether or not to invest. The development projects may be delivered using PRINCE2; however, they are more likely to be run using the Sun Sigma methodology, sponsored by a senior member of the management team called the Champion.

Training and Education

The ownership for the training and education for the project management professional lies with the individual in SunPS. This is a conscious and considered strategy that complements and emphasizes the strong empowered and autonomous message of the company. To this end there is a dedicated support to the self ownership and management of the project management professional in the person of the professional development manager. This role supports the dual purpose of a dynamic, empowered self-directed learning environment married with a strategically focused professional capability development. The emphasis is placed strongly on the development of individual strengths and their further development. The role can be seen as the enabling tension between individual aspiration and business imperative. Considering the fact that education and training candidates are self-selecting, this volunteerism has the consequential benefit of strengthening the community's learning bias and emphasizing the autonomous professional behavior of the project management professional.

The professional development framework has two major project management threads: the behavioral and the technical (qualitative and quantitative). SunPS leverages the offering of professional bodies (PMI and APM), partners (WPM, etc.), and in-house expertise to offer a wide ranging formal curriculum to the professional. The capability stages utilized within this process are from awareness, understanding, and practice, to expertise. The training course input is largely focused on the awareness and understanding levels of this capability grading. Practice and expertise are firmly rooted in experiential learning, which is achieved through use of the community of practice infrastructure described below.

The professional development framework is enhanced by the community of practice infrastructure that provides a wide-ranging and networked capability that allows for the creation of knowledge, the development of expertise, and the dissemination of information in a dynamic and just-in-time manner. Initiatives that support this infrastructure are email interest threads, SIGs with Web pages, virtual meetings and actual meetings, community-wide transfer of information and workshops on a regular basis, practice-specific road shows, and development-sensitive client engagement matching. The development and fostering of the informal communities that straddle all professionals in SunPS are a great source of learning that complements the knowledge acquired formally. This also reinforces the behaviors required of a top-class project management professional.

Self-managed learning is developed in a supportive organizational climate, and client engagement opportunities offer further development of outstanding practices. The professional development framework and infrastructure allow SunPS to share experiences, and drive front-line expertise that focuses on client results and employee development.

Informal Project Management

SunPS encourages:

- Trust
- Communication
- Teamwork
- Empowerment

Although SunPS does have guidelines for cycle phases and projects and has developed various methods, the project manager is still empowered to make decisions that will be supported by senior management. It is expected that any significant deviation from the guidelines is recorded along with the rationale for the change.

Organizational Boundaries

Sun is a global company having its primary organization through sales in each country. SunPS is managed by a matrix of two organizations, its country manager (for sales) and its functional manager in the United States. A number of large projects for corporate clients span many countries; the project manager appointed will be responsible for the total solution across all organizational boundaries, providing a single interface to the customer.

Communications

Formal communications within the project are defined by the project; there is no laid-down process for doing this, however, by following the Change Acceleration Process (adopted by Sun as part of the Sun Sigma rollout), a communications plan will emerge for the project. The tools used during this process with the project team are:

- Change Acceleration Process Profile
- Stakeholder Action Plan
- Threats, opportunities, proof, and implications
- Vision affinity
- Transition map
- Elevator pitch
- Risk assessment

Information about the status of the project is provided by the project manger on a monthly basis to the project board in a prescribed format which is less than two pages long. Exception reports are written and communicated to the project board when the project is predicted to go outside the tolerance limits set by the project board. General communications are through Transfer of Information (TOI), where groups of like-minded people get together for an update, regular corporate focus groups, email, and webcasts.

Behavior Excellence

The behaviors exhibited by the empowered PM professionals in SunPS are the route by which Sun achieves the successful delivery of projects to both our customers and to the business. There is a sense of pride in the project management community that regulates the expected high standards of behavior within the community. The professional development infrastructure and framework described earlier can be seen as the backdrop to the behavioral excellence pursued and expected by the community.

The PM professional will thrive by utilizing all of those behaviors that we know assist in the successful delivery of projects in the internal SunPS interactions. Soundness of judgment, appropriate use of others, team working, honesty, trustworthiness, commitment, resolution of conflict, communication, and utilization of the internal network are all useful in client interactions. The emergent experts in the community, those who negotiate with great effectiveness or who can manage conflict superbly, are recognized by the community. They are sought out by their peers and either formally or informally mentor their peers. This dynamic and emergent capability further reinforces the supportive, team-working framework and translates into more effective client interactions. The emphasis is on the community and the network. Self-management and empowered working reinforce the behaviors that we wish to focus on and allow for the vision of business leadership to be interpreted appropriately by the professional.

The PM community aligns behind a vision that allows for the professional judgment of the individual and the ethos of the community. This vision stresses the unity of direction, the primacy of customer focus, and the power of teamwork, and complements the skills and strengths of others. Through "living" these behaviors, professionals can translate this highly successful model into their own practices on customer engagements.

Meeting customer expectations as defined by the contract begins at project conception. At this early stage, the project manager develops plans that incorporate customer standards, organization, and the PRINCE2 and SunTone Methodology into a process that enables the routine, periodic evaluation of work in progress. In addition, work is also overseen by the Quality Delivery Oversight (QDO) service, an independent body not involved in the engagement. QDO can also provide a quality checklist and engagement reports, which document findings and recommend corrective actions.

The purpose of the quality review is to help make sure that:

- The appropriate steps in the quality assurance process have been carried through.
- All project members have the correct version of the project control documents for the project.
- The specific responsibility and authority assigned during the different phases of the project have been applied.

The program management office, which maintains high service levels while developing and deploying new systems and services, requires careful coordination. Sun project managers and architects can assist organizations create a program management office and architecture review board to help ensure that best practices are followed on this and any subsequent projects. Sun has been successful in instituting a strong post-project measurement program. To improve project quality, Sun is currently focusing efforts on managing quality and risk. Tools have been developed for both of these to provide the project manager with a checklist, report template, planning guide, plan overview, and project plan. It is recognized that applying these tools to daily activities have far-reaching results:

- Product delivery is reliable and within reasonable tolerances of time and budget.
- Products meet or exceed the expectations of the recipients and users.
- Critical steps in the process are not overlooked or ignored.
- Preparation and cross-check steps promote proactivity rather than reactivity.
- History and consistency allow accurate estimates.
- Appropriate people participate in the process, and expectations are actively managed.
- Continuous improvement is promoted.

16.5 MOTOROLA SYSTEM SOLUTION GROUP

The concept of continuous improvement is critical to those companies that are excellent in project management. According to Martin O'Sullivan, vice president and director of business process management at Motorola's System Solution Group (SSG),

> Motorola is constantly striving to improve our processes. What worked well last year, or this year, may not meet our needs in the future. We have goals to improve the effectiveness of project performance and product development. We believe that how effectively we

> implement the project management process is a key to achieving our goal. We intend to be the best in class at project management and are investing resources to this end.

Martin O'Sullivan describes why project management is important to Motorola, what Motorola is doing to advance its maturity level, and how serious Motorola is about advancing its level of maturity:

> First, timeliness in new product and system development is critical. Over 77 percent of companies responding to a recent survey believe in the value of being first to market. We agree. PM offers a disciplined approach to meeting time commitments. Second, customers expect not only timeliness in their projects, but also service, support, and insight into their needs on the part of the project provider. A major element of PM is to address this pervasive customer desire for overall project quality. Third, customers today seek system solutions that require contributions from many parts of the company and our suppliers to produce effective results. Project management fulfills this role as an integrating management discipline. And fourth, Motorola believes in the power of teamwork in achieving best results. PM plays an important role in orchestrating team efforts for optimum results.
>
> Our current focus is on improving project management for new product and system development and deployment. We are working at two levels in the company, the corporate level and within each business unit, leveraging the lessons of the company's quality improvement experiences.
>
> Motorola has a Corporate Council devoted to advancing maturity in product development and project management. The Council is composed of project management and engineering champions from 20 of Motorola's business units. The Council has agreed on goals and metrics for timeliness of projects. It has enhanced its longstanding maturity standards with additional project management criteria. The Council and Motorola University developed a comprehensive education program to foster professionalism in project management. Conferences are held that focus on best practices in project management. There are also publications and a Web site to promote information sharing and communication. Each business unit project management champion has a team laying specific, measurable implementation plans to advance project management maturity. The plans support standards and measures agreed to by the Council and create an environment for continuous improvement in project management maturity. Most importantly, the senior management of the company have assumed a leadership role by sponsoring this initiative and by devoting time to seeing that it is successfully executed.
>
> Based on Motorola's experience in quality advancements, we appreciate that a journey toward continuous improvement requires very serious intent. That's why some business units intend to allocate a portion of executive compensation to reward those managers who contribute to advancing maturity and timeliness of execution. In addition, the progress of business units in reaching maturity goals is reviewed by the Corporate Council and the senior management of the company.
>
> Our quality experience taught us that having a clear definition of maturity is essential to being able to advance. So the company has had a maturity standard for new product and system development for a number of years. Earlier this year we decided to "raise the bar" by updating and strengthening the standard.
>
> We used a two-pronged approach. First, we sought an external standard for an updated definition of the generic process areas comprising project management. Fortunately, the Project Management Institute published its Body of Knowledge last year, from which we

> were able to leverage. We found that its emphasis on the integrating role of project management, risk management, and overall change control and project quality management supported Motorola's values. Second, we considered best practices observed in benchmarking both within Motorola and within a number of excellent companies, including some companies mentioned in your new book. These two sources were useful in enhancing the existing internal standards. To reach the top of the maturity rating, a business unit will have to be best in class not just in a few attributes, but in all ten attributes covered by the standard. Continuous progress toward the highest levels of maturity is the goal. And it's a never-ending task, considering that what is mature today tomorrow will continue to advance.
>
> Organizational assessments to the enhanced standard will be conducted beginning in 1998 and be repeated on a periodic basis within the business units. Assessments are done by a team of trained assessors from within and outside each business. Assessments require several days to complete. There are interviews with management and project management practitioners, as well as reviews of policies, processes, training, and tools. Each assessment will culminate in a numeric rating, as well as briefing to the management of the organization on the observed areas of strength and the opportunities for improvement. Business unit management will respond with an action plan to capitalize on the opportunities. Areas of strength are candidates for sharing across the company to help us move up the maturity curve.
>
> As our Council has shared experiences across the company, there developed a distinct correlation between indicators of advanced maturity in product and system development, including project management, and bottom line results. These results include measures such as shortened development time, meeting commitments on time, and at improved levels of customer satisfaction. The correlation is strong and unmistakable.

Martin O'Sullivan then continues by describing the challenges for project management that Motorola sees now and in the future:

> In my judgment, customers will drive the direction of project management in the future. Customers continue to seek total solutions to their problems, solutions that go well beyond simply buying a company's products. They look to the project provider to be aware of their business and operational constraints and just take care of them. This will require project managers and teams to have more insight into the customers' and users' needs than ever before. So the ability of the project team to walk in the customers' and users' shoes will be vitally important.
>
> There is also a trend for projects to be increasingly international in scope, requiring the resources of large, disbursed industrial teams to take them on. So the ability to bring up a project quickly, among organizations that may not have worked together before, is important. Project management will be challenged to bridge the gaps between organizations, time zones, and cultures.

16.6 NORTEL NETWORKS

The project management achievements at Nortel over the past six years have been remarkable. With over 250 sales offices spread out over more than 100 countries worldwide, Nortel is developing and implementing a standard methodology for the project-driven areas in Nortel, regardless of the location. Bill Marshall, director for project management standards at Nortel Networks, describes the development process:

> Nortel Networks is developing and implementing a reengineering methodology that provides a forum for debate and consensus before tool solutions are sought. This is an area in which we perform extremely well. This methodology tends to get diverse groups on the "same page" and takes the passion and focus out of tools selection. Project management tool selection is notorious for people jumping to the solution before the problem is well understood. In Nortel Networks, the definition of the problem and the solutions are examined through the use of a business process flow. The business flow is first detailed from the macro to the micro level with all of the characteristics such as mega level interfaces to other business processes and the data elements defined to support the business flow. The business flow incorporates process roles, current day benchmark timings, resource consumption, performance metrics and desired business results. Next global consensus is sought and then agreed upon through a voting procedure. The process is "locked down" and placed under change control at this point. It does not mean that changes in the process cannot occur, but rather that these changes are brought forth to global forums and debated. If and when they are agreed upon by all, the previous reference model is adjusted to reflect the new business process changes. It is also recognized that there may be regional reasons why an authorized variation from the standard process must be approved. Once the enterprise has reached this level of consensus on the standard business process, the selection of a supporting tool is more objective and less passionate. For a large corporation that works to a common business process, it means that "lessons learned" in one part of the corporation can be shared with all the parts and a corporate learning process kicks in. It has been long recognized in Nortel Networks that no one area of the corporation has the perfect management methods for projects, but through the reengineering process and sharing of "lessons learned" mentioned above, the maturity in project management throughout the enterprise will improve.

16.7 BATTELLE MEMORIAL INSTITUTE

Another organization that has become a rising star in project management is the Pacific Northwest National Laboratory (PNNL), which is operated by Battelle for the U.S. Department of Energy (DoE). Project management discipline has been recognized as important by PNNL and its primary customers for many years. On large multidisciplinary projects, project management concepts have been applied for approximately 10–15 years. On smaller traditional research projects, these concepts continue to be introduced. The William R. Wiley Environmental Molecular Sciences Laboratory (EMSL) Project, a $230 million facility construction and equipment acquisition effort funded by DoE and managed by Battelle staff, was a significant driver that forced PNNL to improve its project management practices.

The EMSL was chartered as a DoE line-item project in 1989. PNNL formally endorsed project management concepts in a professional sense to enable delivery of the EMSL scope, and the application of these concepts was expanded to a new Battelle-wide business model that is currently being implemented. Prior to 1989, PNNL used an informal project management process to deliver approximately 1000 projects per year that were undertaken within the laboratory's major research programs.

In the new Battelle business model, the basic business unit is the research project. Traditionally, Battelle performs projects that range from basic scientific research to enhanced engineering activities. Projects with a more basic science orientation generally are funded by grants with very little project management discipline required by the customer. These projects typically are funded for less than $100,000. Larger, multidisciplinary projects, where coordination and collaboration are necessary, usually have a greater application of project management discipline. As the marketplace has come to demand more accountability for adherence to scope, schedule, and budget, project management concepts have begun to play a more important role in PNNL's management activities.

Prior to 1989, PNNL was an inconsistent project management practitioner on large projects. They could deliver the smaller $50,000 to $100,000 projects with regularity, but the larger projects were frequently overbudget and/or late. PNNL consistently scored high marks technically, but lost credibility and, therefore, opportunity due to poor baseline project performance. The opportunity to deliver EMSL in 1989 improved PNNL's project management focus. Professional project managers were hired to deliver the EMSL scope, and the skill set brought by those managers has spread and continued to grow in effectiveness and application within the research environment.

Over the last seven years, PNNL has made improving its project management systems a high priority, and the momentum of this effort continues. Maturity in an individual project can be achieved quickly; however, it takes many years of new systems and requirements to become accepted and routinely used and for the full value of the changes to be realized.

Asked what the PNNL did to drive the maturity process forward, key project management practitioners said:

- "Clarified roles, responsibilities, accountabilities, and authorities of line and project managers."
- "Published and widely distributed the project management requirements of our customer(s)."
- "Implemented a work plan authorization process for major task elements that is derived from the project WBS."

The PNNL now performs the following elements of project management very well:

- "Project management application in the research environment, which blends the need for research flexibility with project management control."
- "Integrated project scheduling and earned value management on large projects."
- "Extended project management principles into the operations environment as part of the maturity process, the Battelle business model, and the future of PNNL and Battelle."

Battelle and the PNNL have achieved these benefits:

- "Five-year noncompetitive extensions of the operating contract for PNNL based primarily on outstanding performance."

- "Better 'on-time' and 'on-budget' delivery of contracted work."
- "Improved creditability re: project management decisions = competitive edge."
- "Better ability to *define* the cost of a project and, therefore, make rational decisions about descoping or impacts due to change = better cost control and customer interaction."

Project managers at the PNNL have a vision of the changes they expect to see during the next ten years.

> The national laboratory system is facing tremendous challenges. The lab system must 1) deal with significantly reduced federal research budgets, 2) define a distinctive mission or capability for each lab, and 3) integrate distinctive capabilities or skills within individual labs or among several labs to solve larger problems facing DoE and the nation. The application of established project management principles in managing complex, integrated research projects and laboratory operations can help us to effectively meet these challenges and make a real contribution to the success of PNNL. We also expect to see:
>
> - More informality in the project management career path.
> - More formal qualification requirements for project managers.
> - Enhanced and broader support specialists and systems.
> - Greater use of information from clients as the basis for continuously improving performance.

16.8 JOHNSON CONTROLS

Any list of the stars of tomorrow in project management must include the Automotive Systems Group of Johnson Controls. Earlier in the book, we looked at how the Plymouth, Michigan, company grew by integrating project management with TQM and concurrent engineering. Richard J. Crayne, manager of engineering operations in the Automotive Systems Group, provided a chronology of how the company achieved excellence in project management:

> We have used project management for manufacturing start-ups on "white-box" programs for over 12 years. As we have grown into a full-service, Tier 1 supplier, we have integrated PM with our Total Quality Control (TQC) and Simultaneous Development Team (SDT) methodology. We have been using PM intensely for over ten years.
>
> We needed tools to help us meet aggressive timing requirements on product development projects. We could not be late for program milestones. We developed a "phased" approach with gateways that we still use today.
>
> We gained responsibility for management of complete seat programs from initial concepts to production including supply-base management and sourcing decisions. At the time, most projects included building a new plant and launching all new products. As we were working toward using SDT methods, we knew that we needed a structured framework for all projects.
>
> We brought the process to maturity as part of developing our division-wide TQC system. This took about one year to develop and two years to incorporate into all projects. The

process continued in the engineering areas where we added two more levels of subprojects. We are applying this approach now to our manufacturing areas to build a multiproject program management system.

We appointed a TQC director and a director of program management. Their duties included developing and implementing a mature process. Additional levels were added after we formed a Project Office for Engineering. This group now develops, implements, and supports several software tools for quoting, creating timelines from templates, using standard documents and document control. We developed and implemented a comprehensive training program to support SDTs in the use of these tools and other aspects of PM.

We excel at managing timing and overcoming timing crisis situations due to our customers' focus on meeting milestone commitments. We are now controlling project costs via our custom quoting software. We are improving our quality of execution by standardizing and auditing our key subprocesses. We are applying this expertise toward development of quotation packages for seat system proposals to acquire new business.

Excellent companies recognize the need for continuous improvement. Crayne believes that the following improvements will occur over the next several years at Johnson Controls:

We focus on continuous improvement. All parts of the company feel an obligation to make things run more smoothly. Team members fresh from a launch are full of ideas on how to improve the next program. The trick is to channel these energies into a knowledge base that everyone can access. Our methodologies are under constant scrutiny for improvement. Team members expect the methodologies, procedures, and tools to help them improve performance. Project management methods are followed consistently because people believe it will help.

We are integrating our systems for PM, project data management, product change, and core technology use. We are integrating our top to bottom levels for PM, engineering, and manufacturing to better monitor and manage these relationships and impacts. We are working toward resource requirements and leveling at the task level across multiple levels and disciplines.

We have standard methodologies, and they are based upon policies, procedures, and guidelines. We have published standards for creation of timelines, project reporting, phase exits, and these are audited monthly by the Project Office. Standards are available through our intranet to everyone in the company.

Our methodology divides the process into phases, and we have procedures for gate reviews at the end of each phase.

Even two years after the acquisition of Prince, the organization is still struggling a bit with integrating the respective product development processes. One organization (JCI) was largely process-driven; the other (Prince) was largely procedure-driven. However, the management team recognized that integrating product development and project management processes could help to unify the organizations. A team is now working on developing a revised product development process that integrates all of the points in question from both organizations; a standard methodology driven by a process *and* procedures. The methodology is very simple in terms of the number of process steps (approximately 150 tasks). However, it thoroughly integrates the various efforts that are required to develop new products, procedures, guidelines, standard forms, customer quality standards (QS-9000, APQP, etc.), training courses, functional roles, and so on.

Our methodology is integrated with total quality management, change management, risk management, and concurrent engineering. It has been accepted as the basis for our QS-9000 certification for Design Control. Our next release will be fully integrated via our intranet with our TQC systems.

Scope change management and product change management are integrated. We have procedures for both that are connected to the methodology.

Concurrent engineering has been part of our methodology since its inception in the mid-1980's. We have expanded this to a full team approach and pulled manufacturing in as early as possible.

Risk management is used throughout the process. Teams are encouraged to do up-front risk management and to report using standardized methods on risk assessment and mitigation strategies from start to finish on all projects.

Risk management is used throughout the process. Teams are encouraged to identify risks at the beginning of every project. We have developed courses on how to leverage our knowledge of things gone wrong (and right) from previous projects. This is combined with a simulation to test the team's mitigation strategies.

We also have standardized reports for risk assessment and mitigation strategies from start to finish on all projects. We have reports for design, manufacturing, phase exits, and the total project. These roll up to a business unit level report and are presented to regional management. This is operating successfully on four continents.

The benefits seen by Johnson Controls include:

We have reduced our product development time and improved our ability to maintain our "best methods" for many customers' timing requirements. We have increased our customers' confidence in our ability to deliver. We are using these systems as part of our "solution provider" approach. We have used this to help win new business. By controlling our processes, we are in a much better position to do proper value analysis and value engineering (VA/VE) analysis at the right times. Future integration of our core technology is expected to yield large project and product cost reductions.

Richard Crayne provides the following list of additional accomplishments for Johnson Controls Automotive Group:

- We have dedicated project managers in all departments for all projects.
- Our engineering budgets are created with a custom software program called ProGen.
- We hold monthly project reviews in all areas of the company at the general manager/VP level.
- We have implemented standardized project reports that are sent to the president of the company every month, based on the nine knowledge areas of PMBOK.
- We have standard in-house training in project management.
- We promote external training (PMI [Project Management Institute] Certification).
- We have implemented a set of standard project management tools—PIP/EPM.
- We have established phase exits, drawing/prototype/test reports, monthly project reviews with standard reporting, and a center of excellence in project management.
- We formally monitor project launch success based on 0 PPMs, 100 percent labor efficiency, and 100 percent financial performance during the first 30 days of production (0–100–30).

- We apply our standard project management methodology (PIP/EPM) to our supply base through the Internet for standard monthly reporting.
- Few have trained our supply base in our methodology (PIP/EPM).

Johnson Controls has been able to quantify the benefits of project management:

- Our engineering costs have been reduced from 4.5 to 2 percent of sales over the last five years.
- Our company has grown from $200 million in sales to $16 billion in 15 years, partly due to our reputation for executing projects successfully.
- We have expanded into Europe, Asia, Australia, China, and South America, with a standard project management methodology everywhere.
- Growth of projects managed successfully is from approximately six in 1986 to over 300 today.

16.9 METZELER AUTOMOTIVE PROFILE SYSTEM

Before the 1990s, companies handled lessons learned through team debriefings. With that approach only the people on the project team benefit from the lessons learned. For example, one company achieved tremendous success on a particular project. The division that ran the project had made improvements to the manufacturing process that shortened product development time. To spread the new process to other divisions, the project team was disbanded, and the team members were assigned to other divisions. Somehow, however, the team members' new knowledge was not transplanted to the sister divisions. The vice president then reconvened the team and asked them to write a lessons learned case study. The resulting case study included seed questions and a teacher's guide on how to use the case study. The case study then became required reading in almost all training programs. Finally, the word about the manufacturing changes spread quickly.

There are two ways to learn: from successes (let's hope they're yours) and failures (let's pray they're someone else's). Both ways provide valuable information. Some people contend that knowledge learned from failure may be more valuable than knowledge learned from success. Other people contend, however, that learning to do the right things right the first time (success) is the way to go. In any case, documented lessons learned case studies will likely be emphasized in the future. Each project team will be required to prepare a documented lessons learned case study, which will then be used by training personnel.

Not all companies favor the idea of preparing case studies, especially on project failures. These companies make two points: (1) No team is going to want to prepare a case study highlighting their own mistakes and then sign off on it. (2) Even if the case study were disguised, colleagues would know who worked on the project. For documented lessons learned case studies to be effective, senior managers must avoid promoting the idea that there's someone to blame for every mistake.

Lessons learned can also apply to the senior level of management. Senior managers are, and always will be, the architects of corporate cultures. Steve Gregerson, vice president for product development at the Metzeler Automotive Profile System in Madison

Heights, Michigan, was asked three questions to gain his extensive experience in project/ program management:

- What did you do at your previous employers to help create excellence in project management?
- What lessons were learned in evolving this system of excellence in program management?
- What are you now planning to do at Metzeler to accelerate program management maturity and excellence?

In response to the first question, Gregerson described four activities that led to excellence in program management:

- *Customer Focused, Colocated Teams:* Teams were established that had complete responsibility for all aspects of satisfying customer requirements and gaining new business. These teams were linked via a solid line reporting structure that connected all of the appropriate company activities. They were also loosely matrixed with "Centers of Excellence" that had the responsibility to develop and implement best practices across the teams. These teams evolved into "Global Network Teams," which are virtually colocated as customer requirements required multiple country participation.
- *Program Office and Program Planners:* As the need for more cumbersome program management techniques became apparent, I recognized the need to both develop these techniques within our company and to give the program manager relief from the dedicated tasks of program planning and tracking. The establishment of a program office, linked via a matrixed solid-line relationship to each team program planner allowed a mechanism to develop consistent program management practices in the teams, while shifting many of the technical aspects to the program planner.
- *Team Business Operating System (BOS) Metrics:* The teams developed a set of about 20 key metrics, which were linked to our customer satisfaction objectives. This system complied with standard BOS approaches and provided a way to improve team focus, awareness, and improvements in its performance.
- *Gateway Process:* A basic sequencing of "the right way" of developing new products was developed and published as a Product Development and Launch manual for the company. It is an enhanced version of APQP, the Automotive Industry's Action Group Advanced Products Quality Planning process. We also created new Gateways (seven in all) through which each team needed to pass in order to proceed. We then used Microsoft Project to create the ideal model for sequencing a program including events, deliverables, relationships, durations, responsibilities, and so on. These master models then served as the starting point for program plans within all the teams. This allows the teams to coordinate customer milestones and our Gateway requirements.

Regardless of how great the final result looks, lessons are always learned along the way, according to Gregerson:

- *Too Much Focus on Compliance, Not Value-Added Activity:* We did move the organization into doing more, earlier. A lesson learned was we also needed to improve the value of accomplishing a given deliverable. Example: Process Failure Mode and Effects Analysis (PFMEA).

- *The Plan Was Great Until the Customer Blew It Apart:* Great focus was put on doing things the right way. This worked well as long as the plan was somewhat on track. What occurred too often was a significant program change (timing, design, scope, etc.) that was not within our capabilities to respond. Under increasingly greater pressure from our customer, we were too often forced to do something which put us in a high-risk position, which eventually resulted in failure. Since last minute change is a way of life, the lesson learned is that all our planning, systems, and capabilities have to be more flexible.
- *Planning Model Was Too Complex:* We had a cross-functional, cross-team group put together a master model of the "right way" to execute a program. This model incorporated our WBS [work breakdown structure], including relationships, durations, responsibilities, and so on. The final model incorporated over 600 separate tasks using Microsoft Project. The resulting model was simply too complex to serve as a value-added tool.
- *Metrics Work!* The BOS metric system provided a mechanism to focus and guide our activity.

Metzeler is committed and accelerating the program management process toward excellence:

- *Training:* Effective program management must focus on much more than simply the techniques to plan, execute, track, and control programs. Equally important to a program's success are the "softer" skill sets, such as personnel mastery, leadership, and team building. Leadership in particular is one of the most important, yet less trained, aspect of project management. As a starting point of evolving my new organization, we focus on training key associates using the best programs available anywhere in both the hard and soft skill sets. To further continuous improvement, communication training is now directed to understanding the full value chain of how each event is affecting the program and the company.
- *Direct Global Linkages:* The revised solid-line matrix organization incorporated three basic changes versus my prior organization: (1) Centers of Excellence were directly matrixed with the Customer Focused Teams (CFTs); (2) the COEs are directly aligned to our global organization; and (3) the CFTs are linked globally to ensure global customer focus, coordination, and support.
- *Planning Simplified Value-Added Events:* We left the complex models behind and instead focused on centralized planning of only significant events and key deliverables. We reduced those tracked by the program team from over 600 to about 30. This significantly improved the flexibility and user friendliness of the model. We also increased the number of significant "value-added events" in the same line. These included senior management reviews (often named Gateway), fresh eyes design reviews, design reviews, and process reviews. Detailed planning within the work breakdown structure is the responsibility of each function entity.
- *Communication:* No single element is more important than effective communications. All too often teams believe they do not have time for yet another meeting. We have implemented clockwork team meetings, where the team is brought together once a week (same day, same time, same location) via virtual colocation tools such as video conferencing and teleconferencing. The clockwork scheduling drastically improves the team ability to maintain a high level of participation.

Keith Rosenau, Chief Engineer and Director of Program Management for Metzeler, states:

> The use of communication technology at Metzeler allows us to transfer information faster than ever before. E-mail allows questions to be sent in a time zone 6 or more hours different

and have the answer the next day. Custom developed databases on company servers allow a multi-location team to 24 hours a day/7 days a week information on open issues, procedures, forms and standards. This reduces the amount of time spent waiting for information or feedback from other team members. Open communication and data allow full disclosure to the team of any items needing attention.

When an organization is managed poorly, change is very slow and tedious. But when an organization is managed well, change can be swift. Consider the following sequence of events at Metzeler Automotive Profile System:

- *January 1996:* Gateway process established.
- *August 1996:* The reorganization was accompanied by training and added discipline.
- *August–December 1996:* PMI Certification was required for all project managers.
- *November 1996:* The program office was created.
- *November 1996:* The Global Metzeler Program Management Team was created.
- *December 1996:* The BOS metrics for program management and customer satisfaction were finalized.

The program management structure, systems, and organization were essentially established during 1996. During this year, a matrix organization was created linking all functional activities via Customer Focused Teams. Systems, procedures, checklists, training, and metrics were put in place. Over the last two years, the company's product development and launch capabilities continued to improve as the result of experience applying the process to several major programs, and as the result of the established lessons learned process. Two highlights of this improvement were:

- *Creation of Monthly Management Review Meetings:* The normal gateway process operated effectively in monitoring the progress of programs through the development cycle; however, a weakness was identified in handling significant risks identified during the process. The Monthly Management Review meetings were created, where senior management were brought together on the third Friday of each month to discuss specific issues identified in the gateway process and to assist the team in identifying corrective actions and providing needed resources. Programs would continue in a red or yellow status until all issues were closed. The status depended on whether plans and resources had adequately mitigated the identified risk.

Keith Rosenau, Chief Engineer and Director of Program Management for Metzeler, states:

Improvement of the Management Reviews continues. Original intent of the reviews was only for yellow or red status programs. It became apparent that program issues were being delayed until the formal Gateway meetings. Due to timing and coordination of the 7 Gateways this could be months or even a year between reviews. Metzeler's updated Management Review now requires all active programs to be reviewed monthly. This prevents problems and issues from lingering and not being resolved. The monthly review has the benefit of continuously involving and updating senior management on program status.

- *Establishment of a Design Freeze Process:* Even the thought of establishing a design freeze in the automotive industry has been the source of laughter in more than one gathering. All too often, this mood is sobered when a supplier is forced into making a last minute change that elevates the risk to the point of failure. We first needed to change this paradigm with our customer and within our own organization—this we're still working on! The second step has been to establish a design freeze process that carefully evaluates every change that occurs after the design freeze date relative to its impact on the organization, so that a course of action can be determined. This typically involves a "negotiated" change in terms of product, timing, and cost. Until the process is integrated into our culture, all program changes require approval by engineering, sales, program management, manufacturing, and the customer. Final approval is required from senior management and the director of the manufacturing facility.

Metzeler has defined Design Freeze as follows:

A milestone in a program schedule, after which, scope or product changes can adversely affect cost, quality or delivery.

Historically, we could not develop and launch a major new product without significant cost overruns and customer dissatisfaction. Over the last several years, we have successfully launched new products equal to half our total turnover and have been cited by two different customers as being one of the best suppliers in the industry relative to new product development. What we do well is to continuously improve by quickly identifying strengths and weaknesses and locking them into future program development processes via our program management procedures.

Benefits of our improvements are clearly identified in the costs spent in launching a similar program in 1995 compared to a launch in 2000. In the automotive industry, a program must be qualified and delivered to the automotive assembly plant slightly in advance of the initial production date. Failure to accomplish this can cost a supplier millions of dollars as the entire new vehicle launch process is stopped or completely disrupted. Our performance has drastically improved over a two-year period where launch related expenses have been reduced 20 to 40 percent compared to similar programs only three years ago.

Our organization has developed a standard methodology based on global best practices within our organization and on customer requirements and expectations. This methodology also meets the requirement of QS-9000. Our process incorporates seven gateways, which require specific deliverables listed on a single sheet of paper. Some of these deliverables have a procedure and in many cases a defined format. These guidelines, checklists, forms, and procedures are the backbone of our PM structure and also serve to capture lessons learned for the next program. The methodology is incorporated into all aspects of our business systems, including risk management, concurrent engineering, advanced quality planning, feasibility analysis, design review process, and so on.

16.10 EDS

When a company is non–project-driven, the pressure to accept project management principles may be a slow and tedious process. However, when a company is project-driven and correctly recognizes that its entire business is projects, change takes place quickly, and improvement is done on a continuous basis, rather than sporadically. EDS is a prime example of such an organization, and its successes are well documented. Carl Isenberg, director of EDS Project Management Consulting, describes the growth of project management at EDS:

> EDS, which was founded in 1962, has been following project management principles for most of its history. In the early years, however, project management tended to be ad hoc at best. In the late 1980s, EDS formed a Project Management Consulting Group. Today, this group has offices worldwide and provides consulting support to EDS clients and EDS project teams. In the early 1990s, numerous separate methods and training were beginning to be combined at a corporate level (including international).
>
> In August of 1993, EDS released the first corporate project management methodology based on the Project Management Institute Body of Knowledge and best practices throughout EDS. The second major release of this methodology occurred in November of 1995, incorporating Software Engineering Institute's Capability Maturity Model (SEI CMM) and ISO [International Organization for Standardization] project management requirements. The EDS Project Management methodology version 2 (PM2) has undergone continuous improvement since its release. To date, four additional updates to this methodology have been published based on continuous improvement suggestions and lessons learned.
>
> EDS introduced a formal project management career path in 1994. The job family focused on individuals with experience in project management and a solid understanding of project management disciplines. This job family had three levels: project planning specialist, senior project planning specialist, and project planning specialist advisor. In 1997, EDS made two additional changes to its job families that significantly addressed the importance of project management in all we do for our customers. The change completely modified the technical job families within EDS to include core competencies in basic project management at all levels. This change also created specialty areas in project management concepts (project planning, project cost management, etc.). Acquiring and demonstrating skills in the various specialty areas qualifies individuals to progress up the job family.
>
> At this time, EDS recognizes the need to separate the traditional role of the functional manager from the role of a project manager. Under the revised leadership job families, there is now a career path in program/project management. This new program/project management job family has various levels of progression based on skill development level and experience, including the pursuit of formal professional certification as a project management professional.

Gene Panter, program manager with EDS, added:

> EDS has also spent the past eight years focusing on implementing operational changes to achieve higher levels of maturity under the Software Engineering Institute (SEI) Capability Maturity Model (CMM). In the Defined Level (Level 2) of this model, the majority of the focus is on project management practices. A significant portion of EDS has attained Level 2 or higher of the SEI CMM. A large number of EDS organizations have also achieved ISO certification.
>
> EDS has a proven track record of success in support of its customers. To be successful, EDS has followed some form of project management. The success of EDS organizations that achieve SEI CMM assessments was based on their use of the EDS PM2 methodology, in conjunction with the other corporate and organizational methodologies and processes.

The PM2 methodology was developed based on the key practice areas identified in SEI CMM Levels 2 and 3, the Project Management Institute's Project Management Body of Knowledge (PMBOK), and best practices throughout EDS. Given these proven inputs, the methodology is very comprehensive.

16.11 USAA

United Services Automobile Association (USAA) is a leading financial services company based in San Antonio, Texas. Founded in 1922, USAA seeks to be the provider of choice for the military community. To obtain homeowners' or auto insurance products through USAA, one must have a military connection, either personally or through a family member. A Fortune 500 company, USAA has more than 4.9 million members and 21,000 employees.

Getting to World-Class Status

In 1997, USAA's project management capabilities were considered one of the world's best. The company achieved above-average success with on-time and on-budget project completions. Realizing that effective project management is absolutely necessary for any business to succeed, USAA made a commitment to project management and sought to develop it as one of the company's core competencies.

Business Project Management Process (BPMP) was born. A very simple book was prepared that outlined the process and defined all the phases and steps associated with a project. This book described the "tollgates" through which every project would travel for budget and schedule approvals. In the new environment, work was leveraged through a single project manager accountable for all aspects of the project throughout the entire company.

With a process in place, the next step was to determine governance and project oversight. Realizing that it was inefficient to have multiple business units spending separate budgets on projects, an integration steering committee was created in 1999. Eliminating silos and focusing on the greater good of the organization opened the door for USAA to deliver world-class project management. The committee eliminated duplication of efforts and allowed specific business units to leverage common business drivers.

Coming Together

USAA's current CEO, Bob Davis, supported the push for effective project management. He recognized the value of disciplined project management in ensuring efficient and effective execution of projects with potential for bottom-line impact. His leadership led to the creation of a project management center, where all project managers worked together.

Today, an objective team of executives reviews all proposed projects against a defined set of criteria, including potential for service improvement, automation of manual processes, return on investment, and ease of doing business for the customer. Under these criteria, all projects compete for limited resources and must build a business case for their development and implementation. There is only one project management tool used for scheduling and tracking, and individuals working on projects track both planned and actual time. Precise budgets are created with the information and tracked on a weekly basis. Resources across the entire company are planned over a rolling six-month period, allowing for proper resource allocation.

Teamwork

Putting a new process in place requires the ultimate in teamwork. By the end of 2002, USAA teams reached unprecedented levels of success. Sound business decisions accounted for less than 8 percent of projects being stopped, and more than 80 percent of projects finished on time *and* on budget. Utilizing a center of excellence model, project management overhead has been reduced on a per-project basis, bringing the average overhead ratio to well under 10 percent of the total budget.

Every new process faces challenges. It takes leadership, commitment, and a willingness to change to overcome them. Through teamwork, USAA was able to make this transition a success.

"Necessity Is the Mother of Invention"

"Benjamin Franklin said, 'Necessity is the mother of invention,'" recalls Tim Handren, Executive Vice-President of USAA's Enterprise Business Operations, the group in charge of USAA Project Management. Never was this more true than when USAA adopted BPMP.

According to Handren, "When there are limited budgets and limited resources, a project manager must get very creative. We found that people were making smarter business decisions that led to a projects successful completion." Handren added that job satisfaction went through the roof because people believed that they were making a difference.

Because of BPMP, USAA project managers utilize resources more effectively, make better use of personnel, improve efficiency, and reduce costs. "The bottom line," says Handren, "is that our members reap the rewards of BPMP, and our association becomes more competitive and fiscally sound through good use of its resources and sound business decisions."

16.12 CHANGING TIMES

Once the executives recognize the need for flexibility, project management practices can begin the evolution toward excellence. Table 16–3 summarizes how most of the excellent companies view project management present and future. Emphasis will be placed on the project manager's knowledge of the business. Shared authority and nondedicated teams will be the norm everywhere except in very large organizations that can afford dedicated teams.

Many of these changes did not occur by themselves. The evolution of project management processes was accelerated by the acceptance of concurrent engineering and total

TABLE 16–3. CHANGING TIMES FOR PROJECT MANAGEMENT

Factor	Past View	Present View	Future View
Definition of success	Technical terms only	Time, cost, technology, and customer acceptance	Time, cost, technology, and customer acceptance; minimum scope changes; no business disturbance
Project manager's background	Technical	Technical or non-technical	Must understand the business
Organization	Dedicated teams	Partially dedicated teams	Nondedicated teams
Authority	Project manager has maximum authority	Project and line managers share authority	Shared authority with team empowerment
Human resources	Negotiate for best people	Negotiate for best team	Negotiate for results
Team building	Sensitivity sessions	Selected coursework	Certification training and curriculum development

Source: Reprinted from H. Kerzner, *In Search of Excellence in Project Management.* New York: Wiley, 1998, p. 246.

quality management. Table 16–4 shows the impact that concurrent engineering has had. Project managers are now dedicated to one and only one project because of risk management. In order to perform risk management, especially business risks, the project manager *must* understand the business. Strong integration skill may be a necessity in order to mitigate business risks.

TABLE 16–4. CONCURRENT ENGINEERING AND CHANGE

	Present Organization		New Organization
Critical Issues	**Project-Driven**	**Non–Project-Driven**	**Concurrent Engineering**
Number of hats for the project manager	1	2	1
Availability	Full-time	Part-time	Full-time
Primary skill required	Understand technology, understand people	Technical expert, understand people	Knowledge of business, risk management, integration skills
Career path	Line manager to project manager to executive	Project manager to line manager to executive	Multiple
Promotion ladders	Management, technical, project management	Management, technical	Management, technical, project management
Project management department	Yes	No	Yes
Certification required in the near future	Probably	No	Highly probable

Source: Reprinted from H. Kerzner, *In Search of Excellence in Project Management.* New York: Wiley, 1998, p. 247.

To add professionalism to project management, companies are creating project management departments. This sometimes provides a competitive advantage when submitting proposals to customers. Project management departments are usually followed by a triple ladder career path: a management ladder, a technical ladder, and a project management ladder. The idea behind the project management ladder is quite simple. If an employee becomes experienced in project management and really likes the assignment, why ask her or him to change jobs and/or ladders for advancement? It simply makes no sense!

Combining Tables 16–3 and 16–4, we can create Table 16–5, which gives us a glimpse into the future. Executive recognition of the need for flexibility has allowed corporations to restructure into strategic business units. The project managers within the SBUs are more business managers than technical managers. The line managers who have the ultimate responsibility for the technical quality of the project or product must now share accountability with the project manager. Since the project manager is more a business manager, the project manager, therefore, manages at the management levels of the WBS (levels 1–3) and allows the line managers, with whom accountability is shared, to manage at the technical levels of the WBS (levels 4–6).

The future of project management will be driven by both the customer and the contractor. Contractors who recognize the need for change will be the stars of the twenty-first century. As an example, the automotive industry is moving away from continental vehicle platforms to a global vehicle platform. The future will be team management on a global basis. The stars of tomorrow have already planned for this.

Linda Kretz, a senior consultant with the International Institute for Learning, has provided us with her prediction of the project manager's role in the future:

> Project management as a professional discipline is undergoing a metamorphosis. Corporations use the term "project management" to describe a variety of functions that can be characterized as expediting or brute force coordination. The difference between project management and these other functions lies mainly in the expectations of the client or sponsor and when the project manager is assigned to the project.

TABLE 16–5. PROJECT MANAGEMENT'S EVOLUTIONARY PROCESS

Time Frame	WBS Level at Which PM Executes	PM Educational Background	Organizational Structure	Accountability
Early 1960s	Technical levels of WBS	Engineering	Traditional	With line managers
Late 1960s	Technical levels of WBS	Mostly engineering, some business	Strong matrix	With PMs
1970s–1980s	Management levels of WBS	Mostly business, some engineering	Weak matrix	Partially shared
1990s	Management levels of WBS	Mostly business, some engineering	SBU project management	Totally shared

Source: Reprinted from H. Kerzner, *In Search of Excellence in Project Management.* New York: Wiley, 1998, p. 248.

Most "project managers" today are brought into the project during the implementation or execution phase. They are told what the budget is, what the contractual constraints are and to "go forth and do." They haven't been privy to market analysis, economic evaluation, or whether the project aligns with corporate goals. In fact, up-front planning, if it takes place at all, is typically done by others. Yet, project managers are told that they must be accountable for the result. The question begs to be asked, "How accountable can you be for someone else's plan?" In fact, oftentimes the project manager has no control of the budget whatsoever. Figures are produced after the fact with no analysis during the process, which might change the complexion of the outcome.

Project management in the future will be a multifaceted discipline that recognizes project managers for the value they bring to the corporate bottom line instead of a cost center perceived in the same way internal audits are viewed. Killing the messenger will be a thing of the past, because the message will meet the expectations of the stakeholders. Instead of finding out at the end of the fiscal year that corporate financial margins have not been met, project managers will have the capability to avert disaster by proactively managing the process rather than reacting to ongoing risk events.

Project managers in the future will be catalysts for corporate change, quality improvement efforts and be pivotal in their ability to help meet corporate fiscal objectives. No matter what senior managers are talking about, it's always about money. And when the money isn't there, the first thing they do is downsize, reengineer, or restructure. Why are they downsizing? Because profits aren't what they should be. Why aren't the profits coming in as expected? Because projects have not been proactively managed. The costs far exceed the expected revenue as we reacted to serious risk events. No wonder the fabric of American business is being rewoven! Of course, profits aren't coming in. How can they when project managers have all the accountability but no authority to manage the purse strings? Senior management believes the only answer is to cut back, and the quickest way to do that is to downsize. And even that is not well planned. This is evidenced by the increasing number of businesses who let their key people go and hire them back as consultants!

Today, downsizing and restructuring are done rather regularly without the type of up-front planning necessary to keep these changes transparent to customers. Morale is declining as downsized workers fall back down Maslow's pyramid from self-esteem to security or even physiological needs. Corporate change is good. Downsizing is necessary. But it can be done with a systematic project management approach that accommodates the needs of the corporation as well as the employee.

In summary then, project management will be viewed as a value-added entity. Project managers will be considered professionals, capable of enhancing the bottom line. They will be proactive instead of reactive. They will be the catalyst for corporate change. They will be senior management's staunchest ally. They will be viewed as having the ability to participate in economic justification for projects, in feasibility studies and have the authority and empowerment to manage the project budget.

In the face of emerging technology, project management will catapult successful businesses into the twenty-first century. Those companies who prefer the status quo may find themselves without customers as well as diminished work force.

APPENDIX A

Quality Awards at Johnson Controls Automotive Systems Group

Alagon, Spain	ISO 9002
Almusafes, Spain	Ford Q101, 1995 Q1, 1996; ISO 9002, 1996
Anderlecht, Belgium	ISO 9002
Athens, TN	Ford Q1 Rating, 1986 (maintained)
Bardstown, KY	Toyota Excellent VE/VA Award, 1994 Toyota Excellent Value Improvement Award, 1995, 1996
Belcamp, MD	GM Mark of Excellence, 1988 Venture Award, 1988 GM Zero Follow-up Award, 1988 GM Supplier Self Certification, 1989
Bochum, Germany	GM Supplier of the Year ISO 9002; QS 9000
Burton-Upon-Trent, UK	Toyota VE/VA Award, 1994 ISO 9002
Cadiz, KY	Ford Q1 Rating, 1987 (maintained) GM Supplier Self Certification ISO 9002
Ceska Lipa, Czech Rep	ISO 9002; VW “A” Supplier
Cuautitlan, Mexico (Remomesa)	Ford Q1
Dagenham, UK	ISO 9002; QS 9000

Espelkamp, Germany	Ford Q1 VW Formel Q ISO 9002
FoaMech, Georgetown, KY	Shingo Prize for Manufacturing Excellence, 1996
Friedensdorf, Germany	VW "A" Supplier ISO 9002
Geel, Belgium	GM Supplier of the Year Ford Q1 ISO 9002; QS-9000
Georgetown, KY	Toyota Excellent Delivery Award 1989, 1990, 1991, 1995, 1996 Internal Partnership, 1992 (from EPD) Toyota Excellent Quality Award, 1994 QS-9000 Shingo Prize for Manufacturing Excellence, 1997
Greenfield, OH	Ford Q1, 1988 (maintained) GM Gold Medal Achievement ISO 9002; QS-9000
Jefferson City, MO	Ford Q1, 1988 (maintained) GM Mark of Excellence Shingo Prize for Manufacturing Excellence, 1997
John's Creek, GA	GS 9000
Kansas City, MO	Ford Q1, 1995
King's Norton, UK	ISO 9002
Lapeer, MI	Ford Q1, 1994 Chrysler Gold Pentastar
Lewisburg, TN	Saturn 100% Delivery JD Power Highest Quality Seat Supplier, 1995
Lexington, TN	Ford Q1, 1986 (maintained) Chrysler Preferred Supplier Award, 1990 Chrysler QE, 1990, 1991 Ford Q101, 1991 Shingo Prize for Manufacturing Excellence, 1998
Linden, TN	Ford Q1, 1986 (maintained) Ford Preferred Supplier Chrysler Self Certified Shingo Prize for Manufacturing Excellence, 1997
Livermore, CA	NUMMI Partnership Award, 1989, 1990, 1991, 1992, 1993, 1994, 1995 NUMMI Excellence Award for Cost, 1995 NUMMI Excellence Award for Delivery, 1995
Mallersdorf, Germany	ISO 9001
Mandling, Austria	Ford Q1 ISO 9001
Mansfield, UK	ISO 9002

Martorell, Spain	VW Best Supplier, 1994 VW Value to the Customer Award, 1995 ISO 9002
Mlada Bolkeslav, Czech Republic	Skoda Quality Award,m 1996
Mt. Clemens, MI	Chrysler QE, 1992 Chrysler Gold Pentastar, 1994, 1995
Murfreesboro, TN	Nissan Quality Achievement Award, 1986 Nissan Quality Masters Award, 1987–1990 Nissan First Team Supplier Award, 1993, 1994 Shingo Prize for Manufacturing Excellence, 1995
Naucalpan, Mexico (Autoseat)	Chrysler Pentastar Nissan Zero Defects, 1994 Nissan Master Supplier, 1995
Nelas, Portugal	ISO 9002
Orangeville, ON	Toyota Pinnacle Award, 1992, 1994, 1995 Chrysler QE, 1993 Chrysler Gold Pentastar, 1994, 1995 Shingo Prize for Manufacturing Excellence, 1996
Ossian, IN	GM Mark of Excellence, 1988 GM T&B Self Certification QS-9000
Pretoria, South Africa	Q101
Pulaski, TN	Ford Q1, 1991 Shingo Prize for Manufacturing Excellence, 1997
Roudnice, Czech Republic	ISO 9002
Schwalbach, Germany	ISO 9002
Shelbyville, KY	Ford Q1
Shreveport, OA	GM Mark of Excellence, 1988 GM Zero Follow-up Award, 1989 GM Supplier Self Certification
Silloth, UK	ISO 9002
Speke, UK	ISO 9002; QS-9000
St. Mary's OH	Honda 100% Delivery Award Honda Quality Performance Award, 1995 Honda Productivity Improvement Award, 1997
Stratford, ON	Chrysler QE
Strasbourg, France	ISO 9001
Straz Podraiskam, Czech Republic	ISO 9002
Strongsville, OH	Ford Q1, 1991, 1993, 1994
Sunderland, UK	ISO 9001
Sycamore, IL	Chrysler Gold Pentastar, 1988, 1989, 1991, 1993, 1994, 1995, 1996 Chrysler Quality Excellence Award, 1988, 1989, 1990, 1991, 1992 Chrysler Self Certification

Taylor, MI	Chrysler Gold Pentastar, 1994
Telford, UK	Ford Q1, 1991 ISO 9001; QS-9000; ISO 9002 Rover Sterling Award, 1991 Preferred Supplier Award Rover Supplier Excellence, 1994
Tillsonburg, ON	Ford Q1, 1989 (maintained) Chrysler Quality Excellence Award, 1985, 1989
Tiazala, Mexico	Nissan Master of Quality Award, 1993, 1994, 1995 Nissan Award of Excellence, 1994 Nissan Zero Defects, 1995 Nissan Master Supplier, 1995
Tlazcala, Mexico	Chrysler Gold Pentastar VW "A" Supplier
Uitenhage, South Africa	VW Supplier Award
Waghausel, Germany	ISO 9002
Wednesbury, UK	ISO 9002
Zwickau, Germany	ISO 9002

APPENDIX B

Project Management Maturity Questionnaire

On the next several pages you will find 20 questions concerning how mature you believe your organization to be. Beside each question you will circle the number that corresponds to your opinion. In the example below, your choice would have been "Slightly Agree."

−3 Strongly Disagree
−2 Disagree
−1 Slightly Disagree
0 No Opinion
(+1) Slightly Agree
+2 Agree
+3 Strongly Agree

Example: (−3, −2, −1, 0, (+1), +2, +3)

The row of numbers from −3 to +3 will be used later for evaluating the results. After answering Question 20, you will grade the exercise.

The following 20 questions involve maturity. Please answer each question as honestly as possible. Circle the answer you feel is correct, not the answer you think the instructor is looking for.

1. My company recognizes the need for project management. This need is recognized at all levels of management, including senior management. −3 −2 −1 0 +1 +2 +3

2. My company has a system in place to manage both cost and schedule. The

system requires charge numbers and cost account codes. The system reports variances from planned targets.	−3	−2	−1	0	+1	+2	+3
3. My company has recognized the benefits that are possible from implementing project management. These benefits have been recognized at all levels of management, including senior management.	−3	−2	−1	0	+1	+2	+3
4. My company (or division) has a well-definable project management methodology using life cycle phases.	−3	−2	−1	0	+1	+2	+3
5. Our executives visibly support project management through executive presentations, correspondence, and by occasionally attending project team meetings/briefings.	−3	−2	−1	0	+1	+2	+3
6. My company is committed to quality up-front planning. We try to do the best we can at planning.	−3	−2	−1	0	+1	+2	+3
7. Our lower- and middle-level line managers totally and visibly support the project management process.	−3	−2	−1	0	+1	+2	+3
8. My company is doing everything possible to minimize “creeping” scope (i.e., scope changes) on our projects.	−3	−2	−1	0	+1	+2	+3
9. Our line managers are committed not only to project management, but also to the promises made to project managers for deliverables.	−3	−2	−1	0	+1	+2	+3
10. The executives in my organization have a good understanding of the principles of project management.	−3	−2	−1	0	+1	+2	+3
11. My company has selected one or more project management software packages to be used as the project tracking system.	−3	−2	−1	0	+1	+2	+3
12. Our lower- and middle-level line managers have been trained and educated in project management.	−3	−2	−1	0	+1	+2	+3
13. Our executives both understand project sponsorship and serve as project sponsors on selected projects.	−3	−2	−1	0	+1	+2	+3

14. Our executives have recognized or identified the applications of project management to various parts of our business.	−3	−2	−1	0	+1	+2	+3
15. My company has successfully integrated cost and schedule control together for both managing projects and reporting status.	−3	−2	−1	0	+1	+2	+3
16. My company has developed a project management curriculum (i.e., more than one or two courses) to enhance the project management skills of our employees.	−3	−2	−1	0	+1	+2	+3
17. Our executives have recognized what must be done in order to achieve maturity in project management.	−3	−2	−1	0	+1	+2	+3
18. My company views and treats project management as a profession rather than a part-time assignment.	−3	−2	−1	0	+1	+2	+3
19. Our lower- and middle-level line managers are willing to release their employees for project management training.	−3	−2	−1	0	+1	+2	+3
20. Our executives have demonstrated a willingness to change our way of doing business in order to mature in project management.	−3	−2	−1	0	+1	+2	+3

SCORING SHEET

Each response you circled in Questions 1–20 had a column value between −3 and +3. In the appropriate spaces below, place the circled value (between −3 and +3) beside each question.

Embryonic	**Executive**	**Line Management**
1. ______	5. ______	7. ______
3. ______	10. ______	9. ______
14. ______	13. ______	12. ______
17. ______	20. ______	19. ______
Total ______	Total ______	Total ______

Growth	**Maturity**
4. ______	2. ______
6. ______	15. ______
8. ______	16. ______
11. ______	18. ______
Total ______	Total ______

Grand Total ______

Transpose your total score in each category to the table below by placing an "X" in the appropriate area.

Points													
Stages	−12	−10	−8	−6	−4	−2	0	+2	+4	+6	+8	+10	+12
Maturity													
Growth													
Line Management													
Executive													
Embryonic													

Grading System

High scores (usually +6 or greater) indicate that these evolutionary stages of maturity have been achieved or at least you are now in this stage. Stages with very low numbers have not been achieved yet.

Consider the following scores:

Embryonic:	+ 8
Executive:	+10
Line Management:	+ 8
Growth:	+ 3
Maturity:	− 4

This indicates that you have probably completed the first three stages and are now entering the Growth Stage. Keep in mind that the answers are not always this simple because companies can achieve portions of one stage in parallel with portions of a second or third stage.

APPENDIX C
Project Management Excellence Questionnaire

On the next several pages are 42 multiple choice questions that will allow you to compare your organization against those companies that are discussed in this text. After you complete Question 42, a grading system is provided. You can then "benchmark" your organization against some of the best.

1. Your company *actively* uses the following processes:
   A. Total quality management (TQM) only
   B. Concurrent engineering (shortening deliverable development time) only
   C. TQM and concurrent engineering only
   D. Risk management only
   E. Risk management and concurrent engineering only
   F. Risk management, concurrent engineering, and TQM
2. On what percent of your projects do you use the principles of total quality management?
   A. 0 percent
   B. 5–10 percent
   C. 10–25 percent
   D. 25–50 percent
   E. 50–75 percent
   F. 75–100 percent
3. On what percent of your projects do you use the principles of risk management?
   A. 0 percent
   B. 5–10 percent
   C. 10–25 percent
   D. 25–50 percent
   E. 50–75 percent
   F. 75–100 percent

4. On what percent of your projects do you try to compress product/deliverable schedules by performing work in parallel rather than in series?
   A. 0 percent
   B. 5–10 percent
   C. 10–25 percent
   D. 25–50 percent
   E. 50–75 percent
   F. 75–100 percent
5. Your company's risk management process is based upon:
   A. You do not use risk management
   B. Financial risks only
   C. Technical risks only
   D. Scheduling risks only
   E. A combination of financial, technical, and scheduling risks based upon the project
6. The risk management methodology in your company is:
   A. Nonexistent
   B. More informal than formal
   C. Based upon a structured methodology supported by policies and procedures
   D. Based upon a structured methodology supported by policies, procedures, and standardized forms to be completed
7. How many different project management methodologies exist in your organization (i.e., consider a systems development methodology for MIS projects different than a product development project management methodology)?
   A. You have no methodologies
   B. 1
   C. 2–3
   D. 4–5
   E. More than 5
8. With regard to benchmarking:
   A. Your company has never tried to use benchmarking
   B. Your company has performed benchmarking and implemented changes but not for project management
   C. Your company has performed project management benchmarking but no changes were made
   D. Your company has performed project management benchmarking and changes were made
9. Which of the following best describes your corporate culture?
   A. Single-boss reporting
   B. Multiple-boss reporting
   C. Dedicated teams without empowerment
   D. Nondedicated teams without empowerment
   E. Dedicated teams with empowerment
   F. Nondedicated teams with empowerment
10. With regard to morals and ethics, your company believes that:
    A. The customer is always right
    B. Decisions should be made in the following sequence: best interest of the customer first, then the company, then the employees
    C. Decisions should be made in the following sequence: best interest of company first, customer second, and the employees last
    D. Your company has no such written policy or set of standards

11. Your company conducts internal training courses on:
    A. Morality and ethics within the company
    B. Morality and ethics in dealing with customers
    C. Good business practices
    D. All of the above
    E. None of the above
    F. At least two of the first three

12. With regard to scope creep or scope changes, your culture:
    A. Discourages changes after project initiation
    B. Allows changes only up to a certain point in the project's life cycle using a formal change control process
    C. Allows changes anywhere in the project life cycle using a formal change control process
    D. Allows changes but without any formal control process

13. Your culture seems to be based upon:
    A. Policies
    B. Procedures (including forms to be filled out)
    C. Policies and procedures
    D. Guidelines
    E. Policies, procedures, and guidelines

14. Cultures are either quantitative (policies, procedures, forms, and guidelines), behavioral, or a compromise. The culture in your company is probably _________ percent behavioral.
    A. 10–25
    B. 25–50
    C. 50–60
    D. 60–75
    E. Greater than 75

15. Your organizational structure is:
    A. Traditional (predominantly vertical)
    B. A strong matrix (i.e., project manager provides most of the technical direction)
    C. A weak matrix (i.e., line managers provide most of the technical direction)
    D. We use co-located teams
    E. I don't know what the structure is; management changes it on a daily basis

16. When assigned as a project leader, your project manager obtains resources by:
    A. "Fighting" for the best people available
    B. Negotiating with line managers for the best people available
    C. Negotiating for deliverables rather than people
    D. Using senior management to help get the appropriate people
    E. Taking whatever he or she can get, no questions asked

17. Your line managers:
    A. Accept total accountability for the work in their line
    B. Ask the project managers to accept total accountability
    C. Try to share accountability with the project managers
    D. Hold the assigned employees accountable
    E. Don't know the meaning of the word *accountability;* it is not part of your company's vocabulary.

18. In the culture within your company, the person most likely to be held accountable for the ultimate technical integrity of the final deliverable is/are:
    A. The assigned employees
    B. The project manager
    C. The line manager
    D. The project sponsor
    E. The whole team
19. In your company, the project manager's authority comes from:
    A. Within himself/herself, whatever he/she can get away with
    B. The immediate superior to the project manager
    C. Documented job descriptions
    D. Informally through the project sponsor in the form of a project charter or appointment letter
20. After project go-ahead, your project sponsors tend to:
    A. Become invisible, even when needed
    B. Micromanage
    C. Expect summary-level briefings once a week
    D. Expect summary-level briefings once every two weeks
    E. Get involved only when a critical problem occurs or at the request of the project manager or line managers
21. What percentage of your projects have sponsors who are at the director level or above?
    A. 0–10 percent
    B. 10–25 percent
    C. 25–50 percent
    D. 50–75 percent
    E. More than 75 percent
22. Your company offers approximately how many different *internal* training courses for the employees (courses that can be regarded as project-related)?
    A. Less than 5
    B. 6–10
    C. 11–20
    D. 21–30
    E. More than 30
23. With regard to your previous answer, what percentage of the courses are more behavioral than quantitative?
    A. Less than 10 percent
    B. 10–25 percent
    C. 25–50 percent
    D. 50–75 percent
    E. More than 75 percent
24. Your company believes that:
    A. Project management is a part-time job
    B. Project management is a profession
    C. Project management is a profession and employees should become certified as project management professionals but at their own expense
    D. Project management is a profession and the company pays for employees to become certified as project management professionals
    E. There are no project managers in your company

25. Your company believes that training should be:
    A. Performed at the request of employees
    B. Performed to satisfy a short-term need
    C. Performed to satisfy both long- and short-term needs
    D. Should be performed only if there exists a return on investment on training dollars.

26. Your company believes that the content of training courses is best determined by the:
    A. Instructor
    B. Human Resources Department
    C. Management
    D. Employees who will receive the training
    E. Customization after an audit of the employees and managers

27. What percentage of the training courses in project management contain *documented* lessons learned case studies from other projects within your company?
    A. None
    B. Less than 10 percent
    C. 10–25 percent
    D. 25–50 percent
    E. More than 50 percent

28. What percentage of the executives in your functional (not corporate) organization have attended training programs or executive briefings specifically designed to show executives what they can do to help project management mature?
    A. None! The executives know everything
    B. Less than 25 percent
    C. 25–50 percent
    D. 50–75 percent
    E. More than 75 percent

29. In your company, employees are promoted to management because:
    A. They are technical experts
    B. They demonstrate the administrative skills of a professional manager
    C. They know how to make sound business decisions
    D. They are at the top of their pay grade
    E. There is no place else to put them

30. A report must be written and presented to the customer. Neglecting the cost to accumulate the information, the approximate cost per page for a typical report is:
    A. You have no idea
    B. $100–200 per page
    C. $200–500 per page
    D. Greater than $500 per page
    E. Free; exempt employees in our company prepare the reports at home on their own time

31. Which of the following best describes the culture within your organization?
    A. Informal project management based upon trust, communication, and cooperation
    B. Formality based upon policies and procedures for everything
    C. Project management thrives on formal authority relationships
    D. Executive meddling, which forces an overabundance of documentation
    E. Nobody trusts the decisions of our project managers

32. What percentage of the project manager's time each week is spent preparing reports?
    A. 5–10 percent
    B. 10–20 percent
    C. 20–40 percent
    D. 40–60 percent
    E. Greater than 60 percent

33. During project *planning,* most of your activities are accomplished using:
    A. Policies
    B. Procedures
    C. Guidelines
    D. Checklists
    E. None of the above

34. The typical time duration for a project status review meeting with senior management is:
    A. Less than 30 minutes
    B. 30–60 minutes
    C. 60–90 minutes
    D. 90 minutes–2 hours
    E. Greater than 2 hours

35. Your customers mandate that you manage your projects:
    A. Informally
    B. Formally, but without customer meddling
    C. Formally, but with customer meddling
    D. It is your choice as long as the deliverables are met

36. You company believes that less competent employees:
    A. Should never be assigned to teams
    B. Once assigned to a team, are the responsibility of the project manager for supervision
    C. Once assigned to a team, are the responsibility of their line manager for supervision
    D. Can be effective if assigned to the right team
    E. Should be promoted into management

37. Employees who are assigned to a project team (either full-time or part-time) have a performance evaluation conducted by:
    A. Their line manager only
    B. The project manager only
    C. Both the project and line managers
    D. Both the project and line managers, together with a review by the sponsor

38. Which pair of skills is probably the most important for project managers of your company into the twenty-first century?
    A. Technical knowledge and leadership
    B. Risk management and knowledge of the business
    C. Integration skills and risk management
    D. Integration skills and knowledge of the business
    E. Communication skills and technical understanding

39. In your organization, the people assigned as project leaders are usually:
    A. First-line managers

B. First- or second-line managers
C. Any level of management
D. Usually nonmanagement employees
E. Anyone in the company

40. The project managers in your organization have undergone at least some degree of training in:
    A. Feasibility studies
    B. Cost-benefit analyses
    C. Both A and B
    D. Your project managers are brought on board after project approval/award.

41. Your project managers are encouraged to:
    A. Take risks
    B. Take risks upon approval by senior management
    C. Take risks upon approval of project sponsors
    D. Avoid risks

42. Consider the following statement: Your project managers have a sincere interest in what happens to each team member *after* the project is scheduled to be completed.
    A. Strongly agree
    B. Agree
    C. Not sure
    D. Disagree
    E. Strongly disagree

The assignment of the points is as follows:

Integrated Processes

Question	Points					
1.	A. 2	B. 2	C. 4	D. 2	E. 4	F. 5
2.	A. 0	B. 0	C. 1	D. 3	E. 4	F. 5
3.	A. 0	B. 0	C. 3	D. 4	E. 5	F. 5
4.	A. 0	B. 1	C. 3	D. 4	E. 5	F. 5
5.	A. 0	B. 2	C. 2	D. 2	E. 5	
6.	A. 0	B. 2	C. 4	D. 5		
7.	A. 0	B. 5	C. 4	D. 2	E. 0	

Culture

Question	Points					
8.	A. 0	B. 2	C. 3	D. 5		
9.	A. 1	B. 3	C. 4	D. 4	E. 5	F. 5
10.	A. 1	B. 5	C. 4	D. 0		
11.	A. 3	B. 3	C. 3	D. 5	E. 0	F. 4

Question	Points				
12.	A. 1	B. 5	C. 5	D. 3	
13.	A. 2	B. 3	C. 4	D. 5	E. 4
14.	A. 2	B. 3	C. 4	D. 5	E. 5

Management Support

Question	Points				
15.	A. 1	B. 5	C. 5	D. 5	E. 0
16.	A. 2	B. 3	C. 5	D. 0	E. 2
17.	A. 4	B. 2	C. 5	D. 1	E. 0
18.	A. 2	B. 3	C. 5	D. 0	E. 3
19.	A. 1	B. 2	C. 2	D. 4	E. 5
20.	A. 1	B. 1	C. 3	D. 4	E. 5
21.	A. 1	B. 3	C. 5	D. 4	E. 4

Training and Education

Question	Points				
22.	A. 1	B. 3	C. 5	D. 5	E. 5
23.	A. 0	B. 2	C. 4	D. 5	E. 5
24.	A. 0	B. 3	C. 4	D. 5	E. 0
25.	A. 2	B. 3	C. 4	D. 5	
26.	A. 2	B. 1	C. 2	D. 3	E. 5
27.	A. 0	B. 1	C. 3	D. 5	E. 5
28.	A. 0	B. 1	C. 3	D. 4	E. 5

Informal Project Management

Question	Points				
29.	A. 2	B. 4	C. 5	D. 1	E. 0
30.	A. 0	B. 3	C. 4	D. 5	E. 0
31.	A. 5	B. 2	C. 3	D. 1	E. 0
32.	A. 3	B. 5	C. 4	D. 2	E. 1
33.	A. 2	B. 3	C. 4	D. 5	E. 0
34.	A. 4	B. 5	C. 3	D. 1	E. 0
35.	A. 3	B. 4	C. 3	D. 5	

Behavioral Excellence

Question	Points				
36.	A. 1	B. 2	C. 4	D. 5	E. 0
37.	A. 3	B. 1	C. 5	D. 2	E. 0
38.	A. 3	B. 5	C. 5	D. 5	E. 4
39.	A. 2	B. 2	C. 2	D. 5	E. 3
40.	A. 3	B. 3	C. 5	D. 1	
41.	A. 5	B. 3	C. 4	D. 1	
42.	A. 5	B. 4	C. 2	D. 1	E. 1

Determine your points for each of the questions and complete the following:

A. Points for Integrated Processes (Questions 1–7): ________

B. Points for Culture (Questions 8–14): ________

C. Points for Management Support (Questions 15–21): ________

D. Points for Training and Education (Questions 22–28): ________

E. Points for Informal Project Management (Questions 29–35): ________

F. Points for Behavioral Excellence (Questions 36–42): ________

Grand Total: ________

Each of the six areas are components of the Hexagon of Excellence discussed in Chapters 9–14. The total points can be interpreted as follows:

Points	Interpretation
169–210	Your company compares very well to the companies discussed in this text. You are on the right track for excellence, assuming that you have not achieved it yet. Continuous improvement will occur.
147–168	Your company is going in the right direction, but more work is still needed. Project management is not totally perceived as a profession. It is also possible that your organization simply does not fully understand project management. Emphasis is probably more toward being non–project-driven than project-driven.
80–146	The company is probably just providing lip service to project management. Support is minimal. The company believes that it is the right thing to do, but has not figured out the true benefits or what they, the executives, should be doing. The company is still a functional organization.
Below 79	Perhaps you should change jobs or seek another profession. The company has no understanding of project management, nor does it appear that the company wishes to change. Line managers want to maintain their existing power base and may feel threatened by project management.

APPENDIX D

Software Development Methodology at Computer Associates[1]

1.0 PURPOSE

The purpose of this document is to describe the software development methodology at Computer Associates.

2.0 SCOPE

This methodology applies to CA Branded Products.
Note: The implementation for Service Packs is described in Section 7.0 of this document.

3.0 REFERENCES

None.

---

1. © 2003 by Computer Associates International, Inc., all rights reserved; reproduced by permission of Computer Associates International, Inc

4.0 DEFINITIONS

- *Beta Ready Product:* A functionally complete product, as defined in the SRS, has been through a complete QA cycle, and has documentation for features defined in the SRS.
- *CA:* Computer Associates.
- *Certification:* The ability of the English version of the product to work on a non-English based operating system.
- *DDS:* Detailed Design Specification.
- *GA Product:* Generally Available. A product with no known severity one or two issues, a tested license mechanism, incorporated published fixes for the prior release, and a set of documentation as defined in the TPS.
- *Localization:* CA's products are produced in English and are translated to run in local languages. Localization includes QA engineering.
- *LRS:* Localization Requirements Specification.
- *PCQ:* Product Carrier Quality.
- *PGC:* Product Group Carrier.
- *PRS:* Product Requirements Specification.
- *PS:* Porting Specification.
- *Service Pack:* A collection of published fixes to an existing product, which can be delivered as fixes only or as part of a new build (i.e. genlevel). It may contain a new feature for the product as long as the feature has minimal impact on the overall product and has the approval of a Senior Vice President in development.
- *SRS:* Software Requirements Specification.
- *STS:* Support Turnover Specification.
- *TPS:* The Technical Publications Specification.

5.0 RESPONSIBILITIES

- Updating and maintaining this procedure is the responsibility of the Manager—Corporate QA.

6.0 PROCEDURE

Product Planning

The first phase in the methodology is Product Planning. During this phase, the PGC and brand manager work together to define the project at a high-level, and receive executive approval for the project to proceed. Once approval has been received, a project manager will be assigned to the project by the PGC to lead the project and to conduct cross-functional meetings. Existing members of the project team may also take on the roles and responsibilities of the project manager if required. All projects are required to be added and maintained by the PGC, or their assigned editors, within the Development Project List.

Product Planning phase documentation deliverables include:

- *Product Requirements Specification (PRS):* A PRS is a business justification of the project produced jointly by the PGC/technology owner and the brand manager. Inputs to the PRS include customer feedback, market needs and technology requirements. Approval of the PRS signifies that CA executive management has authorized resources to start development on a given product/project idea. Once approved, the PRS is updated when there is change in the GA date, or a major function is added or deleted from the product. Either event will require re-approval of the PRS by executive management.

- *Software Requirements Specification (SRS):* An SRS describes the desired features of the product and provides a target schedule. The creation of the SRS is the responsibility of the PGC/technology owner and input for the target schedule is received form the development manager, PCQ/QA manager and technical writing manager.

Communication at this phase requires that the SRS be distributed to groups involved who may provide feedback to the PGC. Cross-functional meetings will be scheduled during this phase and groups will be made aware of when they are held.

Exit criteria for this phase are an approved PRS and an SRS. The PRS is distributed with the consent of the PGC and the SRS is distributed to all managers involved in the project.

Product Design

The Product Design phase defines how the product is to be implemented including coding, testing and documentation structure. The majority of the project documentation is created during this phase. Management will review the documents for accuracy and completeness. During this phase, the project's master schedule is detailed to reflect time, resources and dependencies.

Product Design phase documentation deliverables include:

- *Detailed Design Specifications (DDS):* A DDS is written for each major feature or component that is defined in the SRS. It describes the high-level organization of the software's external interfaces (including screens, commands, and on-line help) and the organization and behavior of the product's internal processes that support the external interfaces. It is the responsibility of the development manager to ensure that all DDS' are created and distributed for review. The DDS provides the details needed to create the remaining documentation deliverables in this phase.
- *Support Turnover Specification (STS):* One STS is written for every project. This document describes the procedures necessary to successfully build the product and helps to ensure a successful turnover to Level 2 Support after the product is GA. The STS is the responsibility of the development manager.
- *Localization Requirements Specification (LRS):* The LRS defines the level of the localization for the project. If the level of localization for the project requires either certification or localization, an LRS is provided. The localization manager is responsible for creation and distribution of the LRS with the development manager being responsible for the content of the document.
- *Porting Specification (PS):* Software porting is the engineering process of transforming an existing application so that the resulting software will execute properly on a new platform. The Porting Specifications (PS) template is used to detail the porting project. A Porting Specification is only needed if the product being developed is ported to other platforms. The Development Manager is responsible for the creation of the PS.
- *QA Project Plan:* The QA Project Plan is the quality plan for the project. It includes the QA schedule, risks, and requirements for the entire project. It is the responsibility of the QA manager to create this document and distribute it to the project team members.

- *Technical Publications Specification (TPS):* The TPS identifies what documentation will be required to support the external interfaces and the internal supporting processes, the technical writing resources assigned to the project, and includes a target schedule for documentation review. The technical writing manager is responsible for the creation of the TPS and for its distribution.
- *Master Schedule:* The master schedule contains major milestones for the project and is the responsibility of the project manager. All groups involved may have input to the master schedule. To help with the creation of the project master schedule, the Project Schedule for SDM template may be used. However, if a project has created a master schedule within a project schedule tool or on a project website, that is the schedule that will be used for the project documentation.

Communication at this phase requires:

- *Cross-Functional Meetings (Project Meetings):* The project manager will run the meetings and the attendees will include representatives from the following groups: Development, QA, Services (Support and Education), Technical Writing, Product Analyst and Marketing. Each group is to be represented at each meeting and be prepared to:
    - Discuss the status of the project from their groups viewpoint.
    - Raise any product issues that need group discussion.
    - Provide updates to the project schedule.
- *Document Review Meetings:* Once an SDM document has been created and is ready to be reviewed, the reviewers should be notified via email. This is done automatically when using a tool such as SAM (SDM Asset Manager). The cover page of each SDM project document provides a suggested list of reviewers, however, the author of the document may add to this list to ensure adequate review coverage.

The review process should then proceed as follows:

- A default review period (usually no longer than one week) with completion date should be agreed upon prior to document distribution and detailed in the meeting minutes, otherwise review deadlines should be specified in the email notification.
- The purpose of Document Review is two fold; to ensure that the project team understands the document and to expose and update necessary changes.
- Cross-functional meetings are held as needed, which is usually every 1–2 weeks. Feedback from document reviews should be discussed at these meetings or take place in an equivalent meeting involving all required reviewers.
- Information exchanged in these meetings should be captured in the meeting minutes and then staged on the project's central server along with all SDM documentation.

Some probable outcomes of the review meeting are as follows:

- Acceptance of the SDM project document without any changes
- Acceptance of the SDM project document subject to change agreed upon and noted in the review meeting minutes
- Agreement that the document must be revised and another review meeting schedule

- *Meeting Minutes:* Minutes are taken and distributed at cross-functional meetings and document review meetings. It is the responsibility of the project manager to ensure that the minutes are complied and distributed to all project team members and management. The document titled Meeting Minutes Template contains a template for these minutes.

- *Project Documentation Updates:* The review and project meetings may result in changes to the project documentation. All documents created must be updated continuously throughout the project life cycle. At the end of the Product Design phase, the SRS and other project documentation is updated to reflect the required changes. Features determined to be too costly at this time are deferred to a future release of the product, and all schedules are re-evaluated.

Exit criteria for this phase requires that project documentation be created, reviewed and updated as necessary, and the master project schedule is created.

Product Implementation

During this phase, the product or enhancement is developed, built and tested and the product documentation is created. For each build created, unit testing and build certification testing are required to occur before QA can begin their testing. Unit tests, building-certification tests, and verified results of these tests are provided to QA by Development. QA Test Plans are developed and executed. Developers review the QA Test Plans and answer questions as they arise on the product's functionality. This is an iterative process with coding and testing occurring simultaneously.

The development manager and the QA manager work with the project manager to define what the product builds will contain and what the expectations are for each group during this phase. These managers define what the product builds will contain and what the expectations are for each group during this phase.

For example, addition of maintenance fixes from a previous release should not be left to the end of this phase. Application of these fixes late in the cycle can destabilize the code and have serious impact on the product's quality. To minimize the destabilization of the product, the milestone would be: "all fixes written as of a certain date are incorporated in the second build given to QA." The expectation would be that development would integrate the fixes and perform build integration testing to test for product destabilization and QA would test to make sure the old problems do not re-occur and perform regression test on the product's functionality.

Technical Writing will use the DDS documents to write the product documentation and developers will answer questions as they arise on the product's functionality. QA, Development and Support assist Technical Writing with the documentation review cycles as needed.

This phase will have much iteration prior to moving onto the next phase. It is expected that a project can be in the Product Design and Product Implementation phases simultaneously. As components are developed, the design may need to be revised, resulting in changes to the SRS, the DDS and the QA Projects Plan. The project manager will ensure that all project documentation is updated appropriately and the master schedule reflects any changes required. At this time, the product should be Beta Ready and is sent to the Release QA group.

Product Implementation phase documentation deliverables include:

- *QA Test Plan:* QA Test Plans address the testing implications and structures that will be needed to test each major feature of a project. In addition, they list the actual test cases that will be run to test the major features of the product. If a QA group is using an automated test system or a project website to store the test cases, then the QA Test Plan may reference the location of the test cases. There will be at least one QA Test

Plans, one for each features described in the project SRS. It is the responsibility of the QA Manager to ensure that all Test Plans required are created and distributed to Development and QA for review.

- *QA Turnover Form:* The QA Turnover Form lets QA know what features are being turned over to them and what testing has been done to ensure their completeness. The development manager is responsible for completing this form and for each build scheduled for delivery to QA as defined by the project's milestones.

Communication at this phase requires:

- *Cross-Functional Meetings:* Cross-functional meetings will be held during this time. Depending on the project, the team may decide that the meetings need to be held weekly. During these meetings the agenda will include, but is not limited to a:
  - Review of QA issues
  - Project status update from each group
  - Review of project checklist for action items
  - Status of Localization effort if applicable
- *Meeting Minutes:* Minutes must be taken and distributed at all cross-functional meetings. The minutes are especially crucial at this time as they are the primary means of communication between the project team and management. It is the responsibility of the project manager to ensure that the minutes are compiled and distributed to all project team members and management.

Exit Criterion for this phase is a Beta-ready product, including documentation that has gone through a complete QA cycle. In addition, the project team must agree that the product is ready for customer usage.

Product Verification

During this phase, the QA Test Plans are executed for a final time against the Beta-candidate software, product documentation is finalized and the Product Analyst makes the product available to the Beta sites and Release QA. This is all performed to validate the final product against the project requirements.

Regular Beta status meetings are held with the customers and their status is reported during the project's cross-functional meetings. As the product approaches the end of the Beta cycle, final code changes are made to fix issues discovered by QA and the Beta customers deemed necessary for a successful product launch. QA performs a final test cycle and quality results are made available to management. Internal training is provided to the Services (Support, Education and FSG) groups. The media master is also produced and the final product is packaged. The product analyst is responsible for updating all internal systems, creating the Bill of Materials (BOM) for the product and distributing the master media.

Product Verification phase documentation deliverables include:

- *Updates to All Existing Project Documentation:* Owners of all documents are to ensure the latest updates are in each of the documents. The project manager is responsible for ensuring that the original owners of the documents are making the updates as needed.
- *Beta Readiness Checklist:* This checklist covers the tasks to be completed to consider a product Beta-ready. The project manager is responsible for ensuring that the tasks have been completed. This checklist is an optional component of the SDM.

- *Release QA Entrance Criteria Checklist:* All products need to be sent to the Release QA group when they are Beta-ready. Release QA provides a list of tasks that need to be completed in order for them to accept the product. It is the responsibility of the PCQ/QA manager and the project manager to ensure that the tasks are complete.
- *Release QA Turnover Form:* Release QA has a turnover form that provides essential information about the product being turned over. It is the responsibility of the PCQ/QA manager of the project to complete this form. The Release QA Turn Over Form is a form that provides essential information about the product being turned over to Release QA.

Communication at this phase requires:

- *Cross-Functional Meetings:* Cross-functional meetings will be held during this phase. During these meetings the agenda will include:
  - Review of QA Issues
  - Status update from each Beta site
  - Review of project checklist for action items
- *Meeting Minutes:* Minutes must be taken and distributed at all cross-functional meetings. These minutes will alert management to any problems found during the Beta testing that will impact the project schedule. It is the responsibility of the project manager to ensure that the minutes are compiled and distributed to all project team members and management.

Exit criteria for this phase are a successful Beta cycle and a product ready to be announced Generally Available GA. The PGC determines if the Beta cycle was successful.

Product Rollout

The purpose of the Product Rollout phase is to make the GA product available to the client base, to turn over the responsibility of the product to Support and to review the project's life cycle to see how the process could be improved for the next project.

Product Rollout phase documentation deliverables include:

- *GA Readiness Checklist:* This checklist covers the tasks to be completed to consider a product GA-ready. The project manager is responsible for completing the GA Readiness Checklist. This checklist is an optional component of the SDM.
- *Support Turnover Specification:* This document which is first created during the design phase, provides Support the information needed to take responsibility of the GA product. The document is completed during the Product Rollout Phase. The development manager is responsible for completing the Support Turnover Specification.

Communication at this phase requires:

- *Post-mortem Meeting:* The Post Mortem Meeting gives the project manager and team members the opportunity to review the project life cycle and implement process improvements. The project manager is responsible for scheduling the meeting and distributing the minutes from the meeting. Refer to post Mortem Standards document, available at the Corporate QA website for information on how to run a Post Mortem meeting.

- *Cross-Functional Meetings:* Cross-functional meetings will be held during this phase. During these meetings the agenda will include:
  - Review of GA Readiness Checklist
  - Review of the Support Turnover Specification

Exit criterion for this phase is a generally available product on the shelf ready for customer orders.

7.0 SERVICE PACK PROCEDURE

A Service Pack is not a CA Research and Development project. Each service pack will contain all published fixes and updates from any previous service packs for the product release number. Each service pack requires the completion of a Service Pack Specification (SPS) and are required to be added and maintained by the PGC, or their assigned editors, within the Development Project List.

The SPS encompasses all the project documentation required by:

- Product Planning
- Product Design
- Product Implementation
- Product Rollout

Since there is no Beta cycle and service packs do not go through Release QA, product verification is not required.

When service packs require new features or major source code changes, cross functional meetings must be held to ensure that all groups involved understand the modifications being made to the product by means of the service pack.

The project manager is responsible for the creation of the service pack. This will be the Level 2 Support Manager for distributed products and the level 2 support manager, the QA manager or the development manager for mainframe products. While it is the projects manager's responsibility to create the SPS, all other groups involved in the service pack creation will be responsible for providing input to the SPS.

8.0 RESEARCH & DEVELOPMENT PROTOTYPES

A Research & Development Prototype is an R&D internal task that commissions pure research work into a particular area or technology. This type of work is typically performed prior to the creation of a PRS or SPS.

The purpose of a Research & Development Prototype is to:

- Provide a vehicle to represent and account for conceptual research work that may or may not form part of an official Development project in the future
- Enable basic prototypes to be created to prove a concept, or

- Enable feasibility-study work to be performed to determine whether or not a particular concept or development goal is possible, thus allowing an educated decision on whether or not actual implementation (via an official Development project) is feasible

Under no circumstances will models, code or components developed during a Research & Development Prototype be supplied to a customer or be distributed outside of the immediate CA Research & Development organization unless it subsequently becomes part of an approved SDM project in which case it must be referenced in an PRS/SPS and subject to the normal reviews and approvals defined in the SDM.

There are no mandatory procedures or documentation that applies for prototype work.

9.0 DOCUMENTATION

All project documentation must be stored on a common server and accessible by all members of the project team.

- Beta Readiness Checklist (Optional)
- Detailed Design Specification (DDS)
- GA Readiness Checklist (Optional)
- Localization Requirements Specification (LRS)
- Meeting Minutes Template
- Porting Specification (PS)
- Post Mortem
- Product Requirements Specification (PRS)
- Project Schedule
- Quality Assurance (QA)
- QA Turnover Form
- Release QA Entrance Criteria Checklist

APPENDIX E[1]

Best Practices Library Development at Computer Associates

- Library Planning Method Overview
- Project Audit Method Overview
- Project Management Method Overview

CUSTOM BEST PRACTICES LIBRARY DEVELOPMENT: METHOD OVERVIEW

Project Scoping Information

Short Description The Custom Best Practices Library (BPL) Development method documents how to quickly create and implement a BPL customized to the requirements of a Client by using CA's AllFusion Process Engineer tool. The method uses a 'build as you go' approach and mentors the Client's own staff on basic Process Development concepts.

The method describes how to capture the knowledge of Subject Matter Experts (SMEs) into a repository of information for project schedule creation and for process improvement by practitioners.

The following tasks are completed as part of the engagement:

- Identification of key organizational processes that provide a business benefit to the Client when documented for reuse
- Documentation of selected processes into a custom Process Engineer library
- Publication of these processes to the Client's intranet for access by authorized users

---

1. © 2003 Computer Associates International, Inc., all rights reserved; reproduced by permission of Computer Associates International, Inc.

Client processes are selected based on an initial Process Portfolio discussion held with the Client Project Sponsor. Methods in the CA Best Practices Library will be used as models for the custom BPL development effort. An alternative to this approach is to conduct a formal Portfolio Development Workshop. This is documented in the Best Practices Library Planning method.

Up to three Process Architects (as identified by the Client) will be mentored in process development with the Process Engineer tool. Up to three custom processes will be captured as 'best practices' and documented in the Client's Custom BPL.

A custom 'Home Page' will be created to make the Client's BPL available for on-line viewing along with the CA Process Libraries. Knowledge transfer to a Client-identified Process Librarian will also be done to facilitate the continued development and upgrading of processes.

The method is comprised of three stages:

- Project Planning
- BPL Development
- Project Documentation Delivery

The main Client deliverables from the engagement are:

- A viewable on-line library containing the Client's custom processes
- A custom 'Home Page' from which the Client's custom library as well as the CA Process Libraries can be accessed
- Policies and procedures for process development
- Procedures for the Client's Librarian to follow for each library update cycle
- An Installation Summary report
- A Custom BPL Development Summary report

Client Problem Addressed Today, most organizations maintain process and best practice information in a variety of ways, which are often not logically organized or available when needed. Usually, Project Managers obtain data from various sources in order to generate a project schedule. Technical people spend many hours searching for answers even though they are readily available from subject matter experts within the enterprise. These organizations need to improve access to knowledge by capturing intellectual capital from experts and making it available for reuse via a BPL.

CA's AllFusion Process Management Suite provides all the tools needed to build and access a business's processes and intellectual capital in an efficient, productive manner. It is, however, generally recognized that Process Engineering tools, as a class, require a change to user behavior and work practices. Implementation of these tools therefore requires that the implementation be performed in small enough pieces for the Client to easily incorporate the change into their culture.

This implementation approach is based on CA's experiences in implementing and maintaining best practices internally. Thus, it is a proven, practical approach and provides documented procedures in addition to the tools

When the engagement ends, Client personnel are able to use, maintain, and build on the process base that has been established. This is a first step towards a custom corporate wide process library.

This engagement responds to these Client challenges by:

- Publishing information on Best Practices to the Client's intranet
- Managing project deliverables by providing a central store of standardized project documents

- Supporting the Software Engineering Institute (SEI) Capability Maturity Model (CMM)
- Increasing overall competency by capturing skills and expertise as Best Practices for consistent application across all areas
- Forming the basis for continuous improvement of Client Best Practices
- Capturing project experience and continually improve performance

Service Offering Objectives The main objectives of this service are:

- Install any purchased CA Process Libraries to a single NT server and ensure browsing capability via the Client's intranet browser.
- Install the Process Engineer and Project Engineer software to the same NT server.
- Conduct a process portfolio planning session with the Client Project Sponsor to identify the processes to be developed and the personnel who will participate in the custom BPL development effort.
- Install Process Engineer and Project Engineer locally on the PCs of those persons identified by the Client as the participants in the custom BPL development effort.
- Conduct Interactive Development Sessions with the Client's Process Architects and Process Owners.
- Conduct additional mentoring sessions with the Client's Process Architects to refine the baseline processes.
- Conduct mentoring sessions with the Client's Librarian.
- Develop effort estimates for the processes designed to enable creating project schedules using AllFusion Project Engineer and the Client's available scheduling tool.
- Develop a QA procedure for use before publishing a process to the BPL.
- Publish the processes to the Client's intranet.
- Establish procedures and structures for continued process development.

Client Obligations This offering is for a new or existing Process Continuum Client.

The Client must provide software for web sharing via their intranet.

The Client must provide Server Administrator expertise for the server installation activities.

The Client must provide an individual to serve as the discussion leader for the initial Process Development sessions.

The Client must provide a working environment properly configured for intranet access, with project management scheduling software already installed, and with sufficient hard disk space on the workstations for installation of the Process Engineer and Project Engineer software (about 100 MB).

The Client must provide Project Management software that is one of the following:

- Microsoft Project 2000, 2000 SR-1
- Microsoft Project 98 SR-1
- Niku Project Workbench 4.0.x
- Niku Project Workbench 5.0-5.20 (Repository version only, no support for desktop tool)
- Primavera Project Planner 2.0/3.0

Scope Limitations The following limitations apply:

- A maximum of 3 initial Process Development sessions will be held.
- A maximum of 3 custom processes will be developed and consolidated into a single Custom Client BPL. Each process will not exceed three levels of detail (work

breakdown structure) and no more than 100 total activities included in each process. Additional process development and process improvement is not within the scope of this engagement.

- This service does not cover maintenance procedures for the CA Best Practices libraries or for customization to components in those libraries.
- Version control or configuration management techniques are not addressed.
- Project Engineer will be used only to create project schedules for retrieval from the Client intranet. No other mentoring is provided in the use of this tool. Full implementation of Project Engineer requires training in the tool.
- Client personnel will be provided with enough background and tool skills to build and maintain straightforward processes.
- This service does not include communication and marketing of the processes developed. Plans to do this must be developed by the Client as part of the overall process of implementation. CA can assist in these efforts under a separate engagement.
- Training in advanced Methodology concepts is not provided as part of this engagement.

Estimating Guidelines The offering is provided on a time and material basis and should be completed in 35–40 days by a single CA Senior Architect or Senior Consultant. The overall duration of the effort is spread over a three-month period to allow adequate time for Client development and QA efforts. Care should be taken to produce the required deliverables within the project time frames and budget. If the complexity of the Client's environment exceeds the limitations described above, the CA Services Project Manager should utilize the Change Control process to revise the time and budget for the engagement.

The initial Process Development Workshops are timeboxed for 1 day each.

Required CA Resources A CA Consultant or Architect knowledgeable in implementing Process Engineer 9.0 and Project Engineer 9.0 is required. They must also have previous customer experience doing custom process development using Process Engineer. The Consultant must also have knowledge of how to develop effort estimates. Experience with editing HTML files or with FrontPage is needed.

The Consultant will assist the CA Project Manager to manage project activities according to the method, plan and budget.

The Consultant must have strong interpersonal and analytical skills.

Mentoring, report writing, and presentation skills are required.

A CA Project Manager is assigned. This ensures that EMM standards are met and that project activities are managed according to the method, plan, and budget.

Additional Information

Installation Requirements For complete information on the most recent guidelines for hardware and software requirements, refer to the appropriate product's Installation Guidelines.

Additional Training For further training in Process Engineer, the suggested course is CT210—AllFusion Process Engineer: CA Certified Professional (CACP). Another suggested course is PP330—AllFusion Process Engineer Introduction.

For further training in Project Engineer, the suggested course is CT200—AllFusion Project Engineer CA Certified Professional (CACP).

Stage Diagram See Figure E–1.

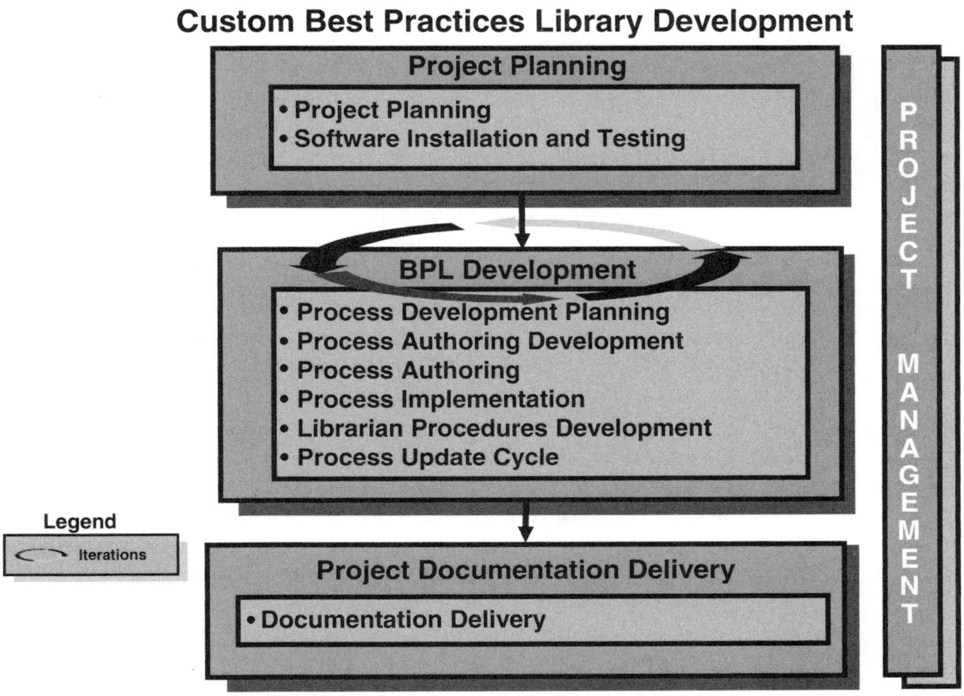

Custom Best Practices Library Development

Project Planning
- Project Planning
- Software Installation and Testing

BPL Development
- Process Development Planning
- Process Authoring Development
- Process Authoring
- Process Implementation
- Librarian Procedures Development
- Process Update Cycle

Project Documentation Delivery
- Documentation Delivery

PROJECT MANAGEMENT

Legend
Iterations

FIGURE E–1

PROJECT AUDIT: METHOD OVERVIEW

Project Scoping Information

Short Description The primary purpose of the Project Audit method is to carry out systematic inspections of CA Services projects in order to measure the degree to which they comply with CA's established standards for project management. Examining the project management deliverables and verifying that the correct deliverables have been produced and that their content is appropriate to the project and consistent with CA standards do this.

Client Problem Addressed Customers for CA Services projects must have confidence that the project will be delivered successfully and that future projects for which they engage CA Services will be managed in the same way and with the same degree of success.

IT Services projects have traditionally been difficult to manage and difficult to deliver successfully. But to ensure customer satisfaction, CA Services must become capable of managing these projects with consistent success.

Correct project management is encapsulated in the current version of CA's Project Management Methodology, supported by the techniques, work products, and tools in the Best Practices Library (BPL). In order to ensure that CA Services projects are being managed in the best way possible and to ensure that all CA projects are run in a consistent way, CA Services needs to be able to measure how well project management on its projects is following this established best practice.

In addition to this, CA Services needs to know how effective its methods are in practice, and to be able to identify areas for process improvement.

The Project Audit method allows CA to measure how well the project is being managed and to capture information about the success of its methodology; thus providing quality assurance to the customer and, at the same time, ensuring that the basic data is available to CA to effect future improvement to the methodology itself.

Method Objectives Repeated application of the Project Audit method will lead to increased consistency by project managers in applying best practices to the projects they manage and to future process improvement by CA Services. This will, in turn, lead to better project success for CA Services' customers and will increase customer satisfaction with CA for having helped them to achieve their own business objectives.

Client Obligations The following assumptions apply:

- The project must be managed by a CA Services Project Manager.
- The project must be run using CA's Project Management methodology. Where a project is being run using a customer's own methodology, there will be limited value in carrying out a project audit, although it will assist CA to identify and catalogue those projects where risk was incurred on the project as a result of not following CA's own methods.

Scope Limitations There are no major limitations on the use of the method. Projects may usefully be audited at any stage in the project life cycle.

Estimating Guidelines For the vast majority of cases, a single Auditor should be able to perform the audit in two days—a half day for preparation time (to notify the Project Manager of the impending audit via a conference call and to verify the schedule), two days to carry out the audit, and a half-day to complete the final audit report.

Other factors that might affect the effort could be:

- Difficulty in arranging the audit, for political or other reasons.
- The need to travel a long distance to a site (e.g., for access to paper documents).
- The number of project files that must be examined, since most of them will have to be read. An Auditor should be able to review about 10 documents or items per hour. This is, however, contingent on the availability of the documents and the number of pages per document.
- A lengthy report to write.

All of the above should be considered when preparing the project schedule.

Required CA Resources A CA Project Manager or PMO consultant with project management accreditation and experience should carry out the audit.

Very large projects may require two Auditors to collaborate in performing the audit.

No specific client resources are required for the audit.

Additional Information

Method Deliverables The major deliverable from the Project Audit method is the Project Audit Report. This report has several potential audiences. To be useful to all of them, the report presents:

- For senior management, a concise summary of how well the project is being run and what corrective actions are required.

Audit Preparation

- Audit Notification
- On-Site Preparation Activities

↓

Audit Execution

- Project Planning and Monitoring
- Risk Management
- Change Management
- Human Resources Management
- Financial Management
- Quality Management
- Document Management

↓

Audit Reporting

FIGURE E–2

- For the project and its immediate management, an analysis of the strengths and weaknesses leading to detailed corrective actions.
- For PMO and MMO, suggested changes to current standards for consideration.

The suggested table of contents for the audit report is:

- Summary
- History of the Project
- Strengths
- Weaknesses
- Corrective actions for this project
- Recommendations for future projects
- Answers to audit questions

Stage Diagram See Figure E–2.

PROJECT MANAGEMENT: METHOD OVERVIEW

Project Scoping Information

Short Description The Project Management Method described here is the CA Services adaptation of the Project Management Institute's recommended Project Management approach. It provides the framework and procedures for the successful management of the various issues common to all projects, including Scope Management, Communication Management, Human Resources Management, Risk Management, Change Management, Schedule Management, Procurement Management, Configuration Management, Financial Management and Quality Management.

This Method is a fundamental component of the CA Services Engagement Management Model—which documents the overall Engagement lifecycle, the Policies, Procedures and Tools involved in conducting our business. These relate not only to the activities of the Project Manager within the project, but also to the way the engagement opportunity is qualified and contracted, the CA Services management and administrative processes, the policies which govern our activities and tools which support them. Local variations to global policies and procedures are also documented in the Engagement Management Model.

CA Services recognizes that while all projects will involve the management issues listed above, the degree to which they may need to be addressed will vary due to the duration and complexity of the project. CA Project Managers must be familiar with this method and know how to apply it in principle to all projects. For shorter projects (less than 30 days effort) however, it may be appropriate to streamline some of the procedures or waive some of the required forms. For more detail on what is required for different types of engagements, refer to the Engagement Management Model.

Project management is a key factor in CA's software development and systems integration projects. The completion of projects on schedule and within budget can only be achieved with the application of sound project management principles and practices.

Project management is also an essential element of CA's Quality Management System (QMS). The Project Management Method will be followed in the development of Project Charters to ensure that technology solutions delivered to Clients conform to an agreed level of quality.

The Project Management Method is unique in the BPL in that, in most cases, this method is not exercised individually, but rather in conjunction with methodology for one or more technical disciplines. It is used for all services projects regardless of industry vertical or technical discipline.

The tasks associated with Project Management are grouped into multiple steps, which make up three major stages:

- *Project Setup and Planning:* The creation and maintenance of a set of plans to define the scope, development schedules, resource requirements, deliverables, cost, and quality criteria for the project
- *Project Monitoring and Controlling:* The execution of Project Charter for the ongoing management of the project, and the measurement of project progress and quality
- *Project Closure:* Ensuring that CA has fulfilled its contractual obligations and the formal closure of the project occurs

Client Problem Addressed Recent studies have found that the majority of IT projects are not successful. This has been traced to a variety of issues, such as:

- Lack of management support
- Lack of user involvement
- Unclear requirements and unrealistic expectations
- Lack of planning
- Lack of appropriately skilled and focused staff

The CA Project Management Methodology is designed to overcome those items that are within our control, and direct client focus and commitment to the items that are solely within their control.

The Project Charter process, conducted as part of the Setup and Planning stage of the project, provides the foundation for a successful project by confirming the requirements, expectations and scope of the project, as well as establishing the procedures that will be employed for the ongoing management of the project. Depending on the effort associated with the engagement, different options for completion of the Project Charter requirement are available to the Project Manager.

By breaking the project up into multiple stages and steps, it becomes easier for both CA and the Client to track project progress. Requiring review of all deliverables and formal acceptance of one stage before progressing to the next also assures that the requirements are being met. Structured communication and project reporting has proven value in setting expectations and keeping management and project team involvement at required levels.

Method Objectives Even the best technical expertise and solid implementation methods require skilled project management to keep projects on track, on schedule and on budget. This method is intended to apply proven project management practices to any of the Service offerings provided by Computer Associates.

Client Obligations The following assumptions apply:

- One Client staff member is required to be the principal client contact for this effort.
- An open exchange of information between the parties is necessary for the success of the project. Pre-scheduled regular meetings to be held at least weekly will be established

with the client that will be attended by key onsite project personnel from both CA and the client to discuss status, issues and any outstanding problems. Adequate notice shall be given in the event that either party cannot attend. These regularly scheduled meetings are not meant to replace normal and necessary communication exchange on a daily or hourly basis as issues arise.

- Both the Client and CA Services will mutually agree upon acceptance of deliverables, and agreement shall not be unreasonably withheld. The Client, prior to the commencement of work on any subsequent phase, must accept the deliverables for the prior stage in writing. Criteria and timelines for acceptance will be defined during the initial stage of the project.
- In the initial stage of the project, the Client and CA Services will complete a deliverable acceptance schedule. This schedule will become the basis for test review and acceptance.

Scope Limitations This method does not apply if the customer prefers to engage CA consultants in a staff augmentation mode. In that case, the client is managing the project and the consultant simply performs the technical tasks as assigned and their time is charged on a time and materials basis.

Estimating Guidelines The price of this service offering is based on the estimated effort identified for the technical discipline method, and the current Project/Program Manager charge rate. Typically, along with the specific project management tasks, additional time is added for

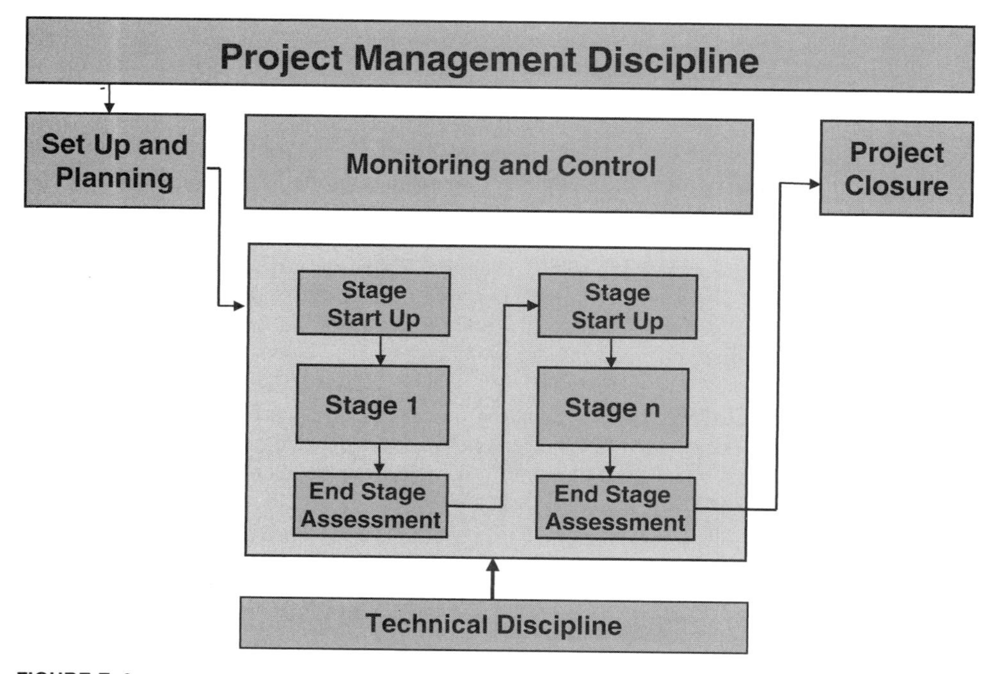

FIGURE E–3

overseeing consultant activity, at 20% of the consultant effort. For fixed price projects, an additional 30% contingency is required.

The duration of the engagement is dependent on the Technical nature of the Project and estimating information can be determined by referring to the specific technical method in the Best Practices Library.

Required CA Resources Project Managers or Program Managers who have received training on the Computer Associates Project Management Methodology and Best Practices Library are required.

Additional Information

Process Flow See Figure E–3.

Determining the Charter Format Before beginning work on the various sections of a Project Charter, the Project Manager must determine which format of the Project Charter is appropriate for this project.

The use of a Project Charter is a critical element of the CA Project Management Methodology. The Charter serves to confirm the project scope, identify any assumptions or

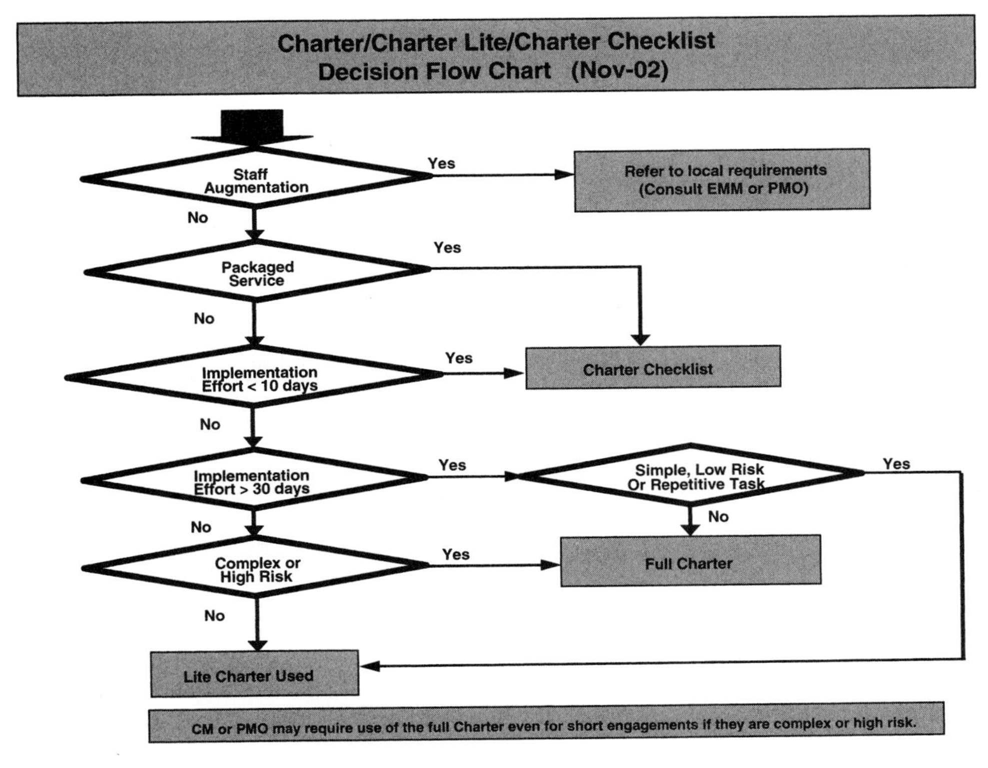

FIGURE E–4

constraints, and describe the processes that will be used to manage the project. All of these items must be considered at the beginning of each project; however, it is understood that the detail to which they must be documented will vary depending on the scope and nature of the project.

There are three different templates available in this method for producing the required Project Charter: the Project Charter Full Content template, the Project Charter 'Lite,' and the Project Charter Checklist.

The selection process is shown in Figure E–4.

APPENDIX F

Post Project Assessment Process at Computer Associates[1]

This template provides a format for documenting the findings of the Post Project Assessment. The report should be completed by the PM, sometimes in consultation with PMO Consultant if they have been involved in the Post Project Assessment. The final report must be submitted to the PMO Consultant for review and acceptance in order for the project to be closed. It is not necessary that every section be completed for all projects. In areas where everything went as expected, there is no need to state that. But if something went very well, or very badly, then it must be recorded.

The PM or CM must still make use of the CA Services Feedback site to report issues related to specific tools or methods.

1. PREFACE

Post Project Assessment: Process and Purpose At the completion of any CA Services Project a Post Project Assessment must be conducted. It is a critical component of our Project Management Methodology to review how successfully a project met its objectives, identify any issues with methods, tools or resources, document any lessons learned and provide feedback for areas of improvement.

1. © 2003 Computer Associates International, Inc., all rights reserved; reproduced by permission of Computer Associates International, Inc.

This Post Project Assessment Report summarizes the project and documents the findings of the Assessment. It is intended for internal use only and is not distributed to the client.

Referenced Documents

Enter the creation date of each document referenced by, this Report.

Document Name	Date of document
Proposal	Data entry reqd
Letter of Engagement	Data entry reqd
Statement of Work	Data entry reqd
Change Request #nnn	

Distribution List

Enter the names of the primary recipients of this Report

Name & Title		Purpose
Data entry reqd	(Consulting Manager)	Review
Data entry reqd	(PMO Consultant)	Review and Acceptance
Data entry reqd	(CRM)	Information
Data entry reqd	etc.	Information

2. PROJECT SUMMARY

Provide a summary of the project. This would normally include:

- a short description/list of the client's objectives
- a summary of the project scope
- a high level description of the project outcome

Mention if the project had any particular significance for CA Services (high visibility, new technology, dramatically improved client's impression of CA, etc. If there was any significant feedback received from the client during project closure, include that here as well.

3. CONTRACTUAL PERFORMANCE

Indicate what type of contract was agreed (T&M or Fixed Price) and if there were any special obligations. Provide a list of agreed deliverables if any. Then in one or two sentences, indicate if CA provided all the required deliverable and met the contractual obligations. State if Client acceptance was received for all project milestones and/or products, including project closure.

4. PROJECT ADMINISTRATION

Use this section to report how the project complied with the Communication Management, Change Management, Risk Management and Configuration Management Plans.

First, indicate what format of Project Charter was used, and whether the client took an active role in its development. For each of the plans listed above, state whether any exceptions were agreed to the CA standard approach. For each plan, state how the Project complied, e.g.,

Change Management Plan

— Four (4) Change Requests were raised during the project and recorded in the Project Change Request Register.

Risk Management Plan

— A Risk Assessment Spreadsheet was compiled at the beginning of the project and used as the basis for review at scheduled project review meetings.

5. PERFORMANCE AGAINST SCOPE, OBJECTIVE, SCHEDULE, BUDGET AND QUALITY

Use this section to report how the project complied with the Scope Management, Financial Management, Schedule Management and Quality Management Plans.

For each of the plans listed above, state whether any exceptions were agreed to the CA standard approach, then for each plan, state how the Project complied. If a detailed design or functional specification was agreed as part of the project, include a description of that process under Scope Management.

Scope Management Plan

— Scope of technical services and key project activities completed as described.
— Project deliverables were as documented.
— Project constraints and exclusions were as documented.
— Tasks and responsibilities of CA were as documented.

Detail Design Document Review

- Documentation of USVD customizations were included in the Detail Design Documentation and updated at the end of the engagement to reflect the details of all final system configurations (screen names, script files, etc.)

Quality Management Plan

— Acceptance Testing was performed during the project to ensure the quality/deliverables of the solution, as documented in the Scope Management Plan.
— A PMO Audit of the project records confirmed that the project was managed according to the CA Project Methodology.

Financial Management Plan

— The approved project budget for CA consulting services was $30,000, and project costs were tracked through the project schedule, and reported in informal project communications.
— The final project cost for CA consulting services was $25,630. This included two consulting days at no charge as per current CA/AGSM contracts (one day of free consulting per month). The ‘chargeable incurred’ billing was $28,106.
— The work effort for all activities was within accepted ranges of schedule cost estimates. The only exception to this was for the ‘Install USVD Web Components’ task (refer to 1.3 Project Schedule review). Task issues were discussed and minuted during Project Progress meetings, including AGSM approval for both the extra work effort required to complete this task, and associated cost.

6. ORGANIZATIONAL CONCERNS

In this section, document any issues or irregularities related to the organization of the project. This could include Client or CA organization changes during the course of the project, resource issues, etc.

7. PROJECT MANAGEMENT CONSIDERATIONS

Use this section to document any special Project Management issues. These could include, but are not limited to project budget, schedule problems, special client requirements, etc.

8. AUDIT FINDINGS AND FOLLOWUP

Use this section to document any findings or recommendations made during Project Audits—either by PMO or External Auditors, along with what was done to address them.

9. LESSONS LEARNED

Use this section to document any lessons learned from the project.

These might include suggestions for updates to a method in the BPL, techniques that worked well and might be used in future projects, feedback on internal CA Services process, or special considerations for future dealings with this client or similar organizations.

Specific feedback on CA Services tools, procedures and applications should still be submitted to the CA Services Feedback site. However, this report provides the opportunity to capture the benefits of the experience gained on the project and communicate it to the Consulting Manager, and also make it available to others in CA Services. The CM may in turn identify issues that should be escalated or areas for improvement or future training.

Case 1

Clark Faucet Company

Background

By 1999, Clark Faucet Company had grown into the third largest supplier of faucets for both commercial and home use. Competition was fierce. Consumers would evaluate faucets on artistic design and quality. Each faucet had to be available in at least 25 different colors. Commercial buyers seemed more interested in the cost than the average consumer, who viewed the faucet as an object of art, irrespective of price.

Clark Faucet Company did not spend a great deal of money advertising on the radio or on television. Some money was allocated for ads in professional journals. Most of Clark's advertising and marketing funds were allocated to the two semiannual home and garden trade shows and the annual builders trade show. One large builder could purchase more than 5,000 components for the furnishing of one newly constructed hotel or one apartment complex. Missing an opportunity to display the new products at these trade shows could easily result in a 6 to 12 month window of lost revenue.

Culture

Clark Faucet had a noncooperative culture. Marketing and engineering would never talk to one another. Engineering wanted the freedom to design new products, whereas marketing wanted final approval to make sure that what was designed could be sold.

The conflict between marketing and engineering became so fierce that early attempts to implement project management failed. Nobody wanted to be the project manager. Functional team members refused to attend team meetings and spent most of their time working on their own 'pet' projects rather than the required work. Their line managers also showed little interest in supporting project management.

Project management became so disliked that the procurement manager refused to assign any of his employees to project teams. Instead, he mandated that all project work come through him. He eventually built up a large brick wall around his employees. He claimed that this would protect them from the continuous conflicts between engineering and marketing.

The Executive Decision

The executive council mandated that another attempt to implement good project management practices must occur quickly. Project management would be needed not only for new product development but also for specialty products and enhancements. The vice presidents for marketing and engineering reluctantly agreed to try and patch up their differences, but did not appear confident that any changes would take place.

Strange as it may seem, nobody could identify the initial cause of the conflicts or how the trouble actually began. Senior management hired an external consultant to identify the problems, provide recommendations and alternatives, and act as a mediator. The consultant's process would have to begin with interviews.

Engineering Interviews

The following comments were made during engineering interviews:

- "We are loaded down with work. If marketing would stay out of engineering, we could get our job done."
- "Marketing doesn't understand that there's more work for us to do other than just new product development."
- "Marketing personnel should spend their time at the country club and in bar rooms. This will allow us in engineering to finish our work uninterrupted!"
- "Marketing expects everyone in engineering to stop what they are doing in order to put out marketing fires. I believe that most of the time the problem is that marketing doesn't know what they want up front. This leads to change after change. Why can't we get a good definition at the beginning of each project?"

Marketing Interviews

- "Our livelihood rests on income generated from trade shows. Since new product development is 4–6 months in duration, we have to beat up on engineering to make sure that our marketing schedules are met. Why can't engineering understand the importance of these trade shows?"
- "Because of the time required to develop new products [4–6 months], we sometimes have to rush into projects without having a good definition of what is required. When a customer at a trade show gives us an idea for a new product, we rush to get the project underway for introduction at the next trade show. We then go back to the customer and ask for more clarification and/or specifications. Sometimes we must ork with the customer for months to get the information we need. I know that this is a problem for engineering, but it cannot be helped."

The consultant wrestled with the comments but was still somewhat perplexed. "Why doesn't engineering understand marketing's problems?" pondered the consultant. In a follow-up interview with an engineering manager, the following comment was made:

> We are currently working on 375 different projects in engineering, and that includes those which marketing requested. Why can't marketing understand our problems?

Questions

1. What is the critical issue?
2. What can be done about it?

3. Can excellence in project management still be achieved and, if so, how? What steps would you recommend?
4. Given the current noncooperative culture, how long will it take to achieve a good cooperative project management culture, and even excellence?

Case 2

Photolite Corporation (A)

Photolite Corporation is engaged in the sale and manufacture of cameras and photographic accessories. The company was founded in Baltimore in 1980 by John Benet. After a few rough years, the company began to flourish, with the majority of its sales coming from the military. By 1985, sales had risen to $5 million.

By 1995, sales had increased to almost $55 million. However, in 1996 competition from larger manufacturers and from some Japanese and German imports made itself felt on Photolite's sales. The company did what it could to improve its product line, but due to lack of funds, it could not meet the competition head-on. The company was slowly losing its market share and was approached by several larger manufacturers as to the possibility of a merger or acquisition. Each offer was turned down.

During this time period, several meetings took place with department heads and product managers regarding the financial health of Photolite. At one of the more recent meetings, John Benet expressed his feelings in this manner:

> I have been offered some very attractive buyouts, but frankly the companies that want to acquire us are just after our patents and processes. We have a good business, even though we are experiencing some tough times. I want our new camera lens project intensified. The new lens is just about complete, and I want it in full-scale production as soon as possible! Harry Munson will be in charge of this project as of today, and I expect everyone's full cooperation. This may be our last change for survival.

With that, the meeting was adjourned.

Project Information

The new lens project was an innovation that was sure to succeed if followed through properly. The innovation was a lens that could be used in connection with sophisticated camera equipment. It was more intense than the wide-angle lens and had no distortion. The lens was to be manufactured in three different sizes, enabling the lens to be used with the top selling cameras already on the market. The lens would not only be operable with the camera equipment manufactured by Photolite, but also that of their competitors.

Management was certain that if the manufactured lens proved to be as precise as the prototypes, the CIA and possibly government satellite manufacturers would be their largest potential customers.

The Project Office

Harry Munson was a young project manager, 29 years of age, who had both sales and engineering experience, in addition to an MBA degree. He had handled relatively small projects in the past and realized that this was the most critical, not only to his career but also for the company's future.

Project management was still relatively new at Photolite, having been initiated only 15 months earlier. Some of the older department heads were very much against letting go of their subordinates for any length of time, even though it was only a sharing arrangement. This was especially true of Herb Wallace, head of the manufacturing division. He felt his division would suffer in the long run if any of his people were to spend much time on projects and reporting to another manager or project leader.

Harry Munson went directly to the personnel office to review the personnel files of available people from the manufacturing division. There were nine folders available for review. Harry had expected to see at least 20 folders, but decided to make the best of the situation. Harry was afraid that it was Herb Wallace's influence that had reduced the number of files down to nine.

Harry Munson had several decisions to make before looking at the folders. He felt that it was important to have a manufacturing project engineer assigned full-time to the project, rather than having to negotiate for part-time specialists who would have to be shared with other projects. The ideal manufacturing project engineer would have to coordinate activity in production scheduling, quality control, manufacturing engineering, procurement, and inventory control. Because project management had only recently been adopted, there were no individuals qualified for this position. This project would have to become the training ground for development of a manufacturing project engineer.

Due to the critical nature of the project, Harry realized that he must have the most competent people on his team. He could always obtain specialists on a part-time basis, but his choice for the project engineering slot would have to be not only the best person available, but someone who would be willing to give as much extra time as the project demanded for at least the next 18 months. After all, the project engineer would also be the assistant project manager since only the project manager and project engineer would be working full-time on the project. Now, Harry Munson was faced with the problem of trying to select the individual who would be best qualified for this slot. Harry decided to interview each of the potential candidates, in addition to analyzing their personnel files.

Questions

1. What would be the ideal qualifications for the project engineering slot?
2. What information should Harry look for in the personnel files?
3. Harry decided to interview potential candidates after reviewing the files. This is usually a good idea, because the files may not address all of Harry's concerns. What questions should Harry ask during the interviews? Why is Harry interviewing candidates? What critical information may not appear in the personnel files?

Case 3

Photolite Corporation (B)

On October 3, 1998, a meeting was held between Jesse Jaimeson, the director of personnel, and Ronald Ward, the wage and salary administrator. The purpose of the meeting was to discuss the grievances by the functional employees that Photolite's present employee evaluation procedures are inadequate for an organization that supports a project management structure.

Jesse Jaimeson: Ron, we're having a lot of trouble with our functional employees over their evaluation procedures. The majority of the complaints stem from situations where the functional employee works closely with the project manager. If the functional manager does not track the work of this employee closely, then the functional manager must rely heavily upon the project manager for information during employee evaluation.

Ron Ward: There aren't enough hours in a day for a functional manager to keep close tabs on all of his or her people, especially if those people are working in a project environment. Therefore, the functional manager will ask the project manager for evaluation information. This poses several problems. First, there are always situations where functional and project management disagree as to either direction or quality of work. The functional employee has a tendency of bending toward the individual who signs his or her promotion and evaluation form. This can alienate the project manager into recommending a poor evaluation regardless of how well the functional employee performs.

In the second situation, the functional employee will spend most of this time working by her or himself, with very little contact with the project manager. In this case, the project manager tends to give an average evaluation, even if the employee's performance is superb. This could result from a situation where the employee has perhaps only a one to two week effort on a given project. This doesn't give that employee enough time to get to know anybody.

In the third situation, the project manager allows personal feelings to influence his or her decision. A project manager who knows an employee personally might be tempted to give a strong or weak recommendation, regardless of the performance. When personalities influence the evaluation procedure, chaos usually results.

Jaimeson: There's also a problem if the project manager makes an overly good recommendation to a functional manager. If the employee knows that he or she has received a good appraisal for work done on a given project, that employee feels that he or she should be given an above average pay increase or possibly a promotion. Many times this puts severe pressure upon the functional manager. We have one functional manager here at Photolite who gives only average salary increases to employees who work a great deal of time on one project, perhaps away from view of the functional manager. In this case, the functional manager claims that he cannot give the individual an above average evaluation because he hasn't seen him enough. Of course, this is the responsibility of the functional manager.

We have another manager who refuses to give employees adequate compensation if they are attached to a project that could eventually grow into a product line. His rationale is that if the project grows big enough to become a product line, then the project will have its own cost center account and the employee will then be transferred to the new cost center. The functional manager thus reserves the best salary increases for those employees who he feels will stay in his department and make him look good.

Ward: Last year we had a major confrontation on the Coral Project. The Coral Project Manager took a grade 5 employee and gave him the responsibilities of a grade 7 employee. The grade 5 employee did an outstanding job and naturally expected a large salary increase or even a promotion. Unfortunately, the functional manager gave the employee an average evaluation and argued that the project manager had no right to give the employee this added responsibility without first checking with the functional manager. We're still trying to work this problem out. It could very easily happen again.

Jaimeson: Ron, we have to develop a good procedure for evaluating our employees. I'm not sure if our present evaluation form is sufficient. Can we develop multiple evaluation forms, one for project personnel and another one for nonproject personnel?

Ward: That might really get us in trouble. Suppose we let each project manager fill out a project evaluation form for each functional employee who works more than, say, 60 hours on a given project. The forms are then given to the functional manager. Should the project manager fill out these forms at project termination or when the employee is up for evaluation?

Jaimeson: It would have to be at project termination. If the evaluation were made when the employee is up for promotion and the employee is not promoted, then that employee might slack off on the job if he or she felt that the project manager rated him or her down. Of course, we could always show the employee the project evaluation sheets, but I'm not sure that this would be the wise thing to do. This could easily lead into a situation where every project manager would want to see these forms before staffing a project. Perhaps these forms should be solely for the functional manager's use.

Ward: There are several problems with this form of evaluation. First, some of our functional employees work on three or four projects at the same time. This could be a problem if some of the evaluations are good while others are not. Some functional people are working on departmental projects and, therefore, would receive only one type of evaluation. And, of course, we have the people who charge to our overhead structure. They also would have one evaluation form.

Jaimeson: You know, Ron, we have both exempt and nonexempt people charging to our projects. Should we have different evaluation forms for these people?

Ward: Probably so. Unfortunately, we're now using just one form for our exempt, nonexempt, technical, and managerial personnel. We're definitely going to have to change. The question is how to do it without disrupting the organization.

Jaimeson: I'm dumping this problem into your lap, Ron. I want you to develop an equitable way of evaluating our people here at Photolite Corporation, and I want you to develop the appropriate evaluation forms. Just remember one thing—I do not want to open Pandora's Box. We're having enough personnel problems as it is.

Questions

1. Can a company effectively utilize multiple performance evaluation forms within an organization? What are the advantages and disadvantages?
2. If we use only one form, what information should be evaluated so as to be equitable to everyone?
3. If multiple evaluation forms are used, what information should go into the form filled out by the project manager?
4. What information can and cannot a project manager effectively evaluate? Could it depend upon the project manager's educational background and experience?

Case 4

Photolite Corporation (C)

On December 11, 1998, after more than two months of effort, Ron Ward (the wage and salary administrator for Photolite Corporation) was ready to present his findings on the most equitable means of evaluating personnel who are required to perform in a project management organizational structure. Jesse Jaimeson (the director of personnel) was eagerly awaiting the results.

Ron Ward: Well, Jesse, after two months of research and analysis we've come to some reasonable possibilities. My staff looked at the nine basic performance appraisal techniques. They are:

1. Essay appraisal
2. Graphic rating scale
3. Field review
4. Forced choice rating
5. Critical incident appraisal
6. Management by objectives
7. Work-standards approach
8. Ranking methods
9. Assessment centers

(Exhibit I contains a brief description of each technique.)

We tried to look at each technique objectively. Unfortunately, many of my people are not familiar with project management and, therefore, had some difficulties. We had no so-called "standards of performance" against which we could evaluate each technique. We, therefore, listed the advantages and disadvantages that each technique would have if utilized in a project management structure.

Jesse Jaimeson: I'm not sure of what value your results are in this case because they might not directly apply to our project management organization.

Ward: In order to select the technique most applicable to a project management structure, I met with several functional and project managers as to the establishment of a selection criteria. The functional managers felt that conflicts were predominant in a project organization, and that these conflicts could be used as a comparison. I, therefore, decided to compare each of the appraisal techniques to the seven most commonly mentioned conflicts that exist in project management organizational forms. The comparison is shown in Exhibit II.

Analysis of Exhibit II shows the management by objectives technique to be the most applicable system. Factors supporting this conclusion are as follows:

Essay Appraisal: This technique appears in most performance appraisals and is characterized by a lack of standards. As a result, it tend to be subjective and inconsistent.

Graphic Rating Scale: This technique is marked by checking boxes and does not have the flexibility required by the constantly changing dynamic structure required in project management.

Field Review: This system would probably account for the majority of performance appraisal problems. However, it is costly and provides for another management overlay, as well as an additional cost (time) factor.

Forced-Choice Rating: This technique has the same problems as the essay technique with the added problem of being inflexible.

Critical Incident Appraisal: This technique centers on the individual's performance and does not take into account decisions made by one's superiors or the problems beyond the individual's control. Again, it is time-consuming.

Management by Objectives (MBO): This technique allows all parties, the project manager, the functional manager, and the employee, to share and to participate in the appraisal. It epitomizes the systems approach since it allows for objectives modification without undue or undeserved penalty to the employee. Finally, it uses objective data and downplays subjective data.

Work-Standards Approach: This technique lends itself easily to technical projects. Though not usually recognized formally, it is probably the most common project management performance appraisal technique. However, it is not flexible and downplays the effect of personality conflicts with little employee input.

Ranking Method: This method allows for little individual input. Most conflict possibilities are maximized with this technique.

Assessment Centers: This method is not utilizable on site and is very costly. It is probably most applicable (if not the best technique) for selecting project management human resources.

In summary, MBO appears to be the best technique for performance appraisal in a project management organization.

Jaimeson: Your conclusions lead me to believe that the MBO appraisal technique is applicable to all project management appraisal situations and should be recommended. However, I do have a few reservations. A key point is that the MBO approach does not eliminate, or even minimize, the problems inherent in project and matrix management organizations. MBO provides the technique through which human resources can be fairly appraised (and, of course, rewarded and punished). MBO has the weakness that it prohibits individual input and systems that employ poorly trained appraisers and faulty follow-up techniques. Of course, such weaknesses would kill any performance appraisal system. The MBO technique most exemplifies the systems approach and, even with its inherent weaknesses, should be considered when the systems approach to management is being employed.

Ward: There is another major weakness that you have omitted. What about those situations where the employee has no say in setting the objectives? I'm sure we have project managers, as well as functional managers, who will do all of the objective-setting themselves.

Jaimeson: I'm sure this situation either exists now or will eventually exist. But that's not what worries me. If we go to an MBO approach, how will it affect our present evaluation forms? We began this study to determine the best appraisal method for our organization. I've yet to see any kind of MBO evaluation form that can be used in a project management environment. This should be our next milestone.

Exhibit I. Basic appraisal techniques

Essay Appraisal

This technique asks raters to write a short statement covering a particular employee's strengths, weaknesses, areas for improvement, potential, and so on. This method is often used in the selection of employees when written recommendations are solicited from former employers, teachers, or supervisors. The major problem with this type of appraisal is the extreme variability in length and content, which makes comparisons difficult.

Graphic Rating Scale

A typical graphic rating scale assesses a person on the quality and quantity of his or her work and on a variety of other factors that vary with the specific job. Usually included are personal traits such as flexibility, cooperation, level of self-motivation, and organizational ability. The graphic rating scale results in more consistent and quantifiable data, though it does not provide the depth of the essay appraisal.

Field Review

As a check on reliability of the standards used among raters, a systematic review process may be utilized. A member of the personnel or central administrative staff meets with small groups of raters from each supervisory unit to go over ratings for each employee to identify areas of dispute and to arrive at an agreement on the standards to be utilized. This group judgment technique tends to be more fair and valid than individual ratings, but is considerably more time-consuming.

Forced-Choice Rating

There are many variations of this method, but the most common version asks raters to choose from among groups of statements those that best fit the person being evaluated and those that least fit. The statements are then weighted and scored in much the same way psychological tests are scored. The theory behind this type of appraisal is that since the rater does not know what the scoring weight of each statement is, he or she cannot play favorites.

Critical Incident Appraisal

Supervisors are asked to keep a record on each employee and to record actual incidents of positive and negative behavior. While this method is beneficial in that it deals with actual behavior rather than abstractions, it is time-consuming for the supervisor, and the standards of recording are set by the supervisor.

Management by Objectives

In this approach, employees are asked to set, or help set, their own performance goals. This approach has considerable merit in its involvement of the individual in setting the standards by which he or she will be judged, and the emphasis on results rather than on abstract personality characteristics.

Work-Standards Approach

Instead of asking each employee to set his or her own performance standards, many organizations set measured daily work standards. The work-standards technique establishes work and staffing targets aimed at increasing productivity. When realistically used and when standards are fair and visible, it can be an effective type of performance appraisal. The most serious problem is that of comparability. With different standards for different people, it is difficult to make comparisons for the purposes of promotion.

Ranking Methods

For purposes of comparing people in different units, the best approach appears to be a ranking technique involving pooled judgment. The two most effective ranking methods include alternation-ranking and paired-comparison ranking. Essentially, supervisors are asked to rank who is "most valuable."

Assessment Centers

Assessment centers are coming into use more for the prediction and assessment of future potential. Typically, individuals from different areas are brought together to spend two or three days working on individual and group assignments. The pooled judgment of observers leads to an order-of-merit ranking of participants. The greatest drawback to this system is that it is very time-consuming and costly.

Exhibit II. Rating evaluation techniques against types of conflict

	Rating Evaluation Technique								
Type of Conflict	Essay Appraisal	Graphic Rating Scale	Field Review	Forced-Choice Review	Critical Incident Appraisal	Management by Objectives	Work Standards Approach	Ranking Methods	Assessment Center
Conflict over schedules	●	●		●	●		●	●	
Conflict over priorities	●	●		●	●		●	●	
Conflict over technical issues	●			●			●		
Conflict over administration	●	●	●	●			●	●	●
Personality conflict	●	●		●			●		
Conflict over cost	●		●	●	●		●	●	●

Note: Shaded circles indicate areas of difficulty.

Questions

1. Do you agree with the results in Exhibit II? Why or why not? Defend your answers.
2. Are there any other techniques that may be better?

Case 5

Photolite Corporation (D)

On June 12, 1999, Ron Ward (the wage and salary administrator for photolite corporation) met with Jesse Jaimeson (the director of personnel) to discuss their presentation to senior management for new evaluation techniques in the recently established matrix organization.

Jesse Jaimeson: I've read your handout on what you're planning to present to senior management, and I feel a brief introduction should also be included (see Exhibit I). Some of these guys have been divorced from lower-level appraisals for over 20 years. How do you propose to convince these guys?

Ron Ward: We do have guidelines for employee evaluation and appraisal. These include:

A. To record an individual's *specific* accomplishments for a given period of time.
B. To formally communicate to the individual on four basic issues:
   1. What is expected of him/her (in specifics).
   2. How he/she is performing (in specifics).
   3. What his/her manager thinks of his/her performance (in specifics).
   4. Where he/she could progress within the present framework.
C. To improve performance.
D. To serve as a basis for salary determination.
E. To provide a constructive channel for upward communication.

Linked to the objectives of the performance appraisal, we must also consider some of the possible negative influences impacting on a manager involved in this process. Some of these factors could be:

- A manager's inability to control the work climate.
- A normal dislike to criticize a subordinate

- A lack of communication skills needed to handle the employee interview.
- A dislike for the general mode in the operation of the business.
- A mistrust of the validity of the appraisal instrument.

To determine the magnitude of management problems inherent in the appraisal of employees working under the matrix concept, the above-mentioned factors could be increased four or five times, the multiplier effect being caused by the fact that an employee working under the project/matrix concept could be working on as many as four or five projects during the appraisal period, thereby requiring all the project managers and the functional manager to input their evaluation regarding a subordinate's performance and the appraisal system itself.

Jaimeson: Of course, managers cannot escape making judgments about subordinates. Without these evaluations, Photolite would be unable to adequately administer its promotion and salary policies. But in no instance can a performance appraisal be a simple accept or reject concept involving individuals. Unlike the quality appraisal systems used in accepting or rejecting manufactured units, our personnel appraisal systems must include a human factor. This human factor must take us beyond the scope of job objectives into the values of an individual's worth, human personality, and dignity. It is in this vein that any effective personnel appraisal system must allow the subordinate to participate fully in the appraisal activities.

Ward: Prior to 1998, this was a major problem within Photolite. Up to that time, all appraisals were based on the manager or managers assessing an individual's progress toward goals that had been established and passed on to subordinates. Although an employee meeting was held to discuss the outcome of an employee's appraisal, in many instances it was one-sided, without meaningful participation by the person being reviewed. Because of such a system, many employees began to view the appraisal concept as inconsistent and without true concern for the development of the individual. This also led many to believe that promotions and salary increases were based on favoritism rather than merit.

Problems inherent in these situations are compounded in the matrix organization when an individual is assigned to several projects with varying degrees of importance placed on each project, but knowing that each project manager will contribute to the performance appraisal based on the success of their individual projects. Such dilemmas can only be overcome when the individual is considered as the primary participating party in the appraisal process and the functional manager coordinates and places prime responsibility of the subordinate contributor in the project for which prime interest has been focused by the company. Other project contributions are then considered, but on a secondary basis.

Jaimeson: Although we have discussed problems that are inherent in a matrix organization and can be compounded by the multiple performance determination, a number of positives can also be drawn from such a work environment. It is obvious, based on its design, that a project/matrix organization demands new attitudes, behavior, knowledge, and skills. This in turn has substantial implications for employee selection, development, and career progression. The ultimate success of the individual and the project depends largely on the ability of the organization to help people learn how to function in new ways.

The matrix organization provides an opportunity for people to develop and grow in ways and rates not normally possible in the more traditional functional organizational setting. Although the project/matrix organization is considered to be high tension in nature, it places greater demands on people but offers greater development and career opportunities than does the functional organization.

Because of the interdependencies of projects in a matrix, increased communications and contact between people is necessary. This does not mean that in a functional organization interdependency and communication are not necessary. What it does say, however, is that in a functional setting, roles are structured so that individuals can usually resolve conflicting demands by talking to their functional manager. In a matrix, such differences would be resolved by people from different functions who have different attitudes and orientations.

Ward: From the very outset, organizations such as Photolite ran into conflict between projects involving such items as:

- Assignment of personnel to projects
- Manpower costs
- Project priority
- Project management status (as related to functional managers)
- Overlap of authority and power in the matrix

If not adequately planned for in advance, these factors could be significant factors in the performance appraisal of matrix/project members. However, where procedures exist to resolve authority and evaluation conflicts, a more equitable performance appraisal climate exists. Unfortunately, such a climate rarely exists in any functioning organization.

With the hope of alleviating such problems, my group has redefined its approach to Exempt Performance Appraisals (see Exhibits I and II). This approach is based on the management by objectives technique. This approach allows both management and employees to work together in establishing performance goals. Beyond this point of involvement, employees also perform a self-evaluation of their performance, which is considered a vital portion of the performance appraisal. Utilization of this system also opens up communication between management and the employee, thereby allowing two-way communication to become a natural item. Although it is hoped that differences can be reconciled, if this cannot occur, the parties involved have at least established firm grounds on which to disagree. These grounds are not hidden to either and the employee knows exactly how his/her performance appraisal was determined.

Jaimeson: O.K. I'm convinced we're talking the same language. We won't have any problem convincing these people of what we're trying to do.

Exhibit I. Recommended approach

I. Prework
- Employee and manager record work to be done using goals, work plans, position guide.
- Employee and manager record measurements to be used.

Note: This may not be possible at this time since we are in the middle of a cycle. For 1999 only, the process will start with the employees submitting a list of their key tasks (i.e., job description) as they see it. Manager will review that list with the employee.

II. Self-Appraisal
- Employee submits self-appraisal for key tasks.
- It becomes part of the record.

III. Managerial Appraisal
- Manager evaluates each task.
- Manager evaluates total effort.
- Skills displayed are recorded.
- Development effort required is identified.

Note: Appraisals should describe what happened, both good and bad.

IV. Objective Review
- Employee relations reviews the appraisal.
  - Assure consistent application of ratings.
  - Assist in preparation, if needed.
  - Be a sounding board.

V. One-over-One Review
- Managerial perspective is obtained.
- A consistent point of view should be presented.

VI. Appraisal Discussion
- Discussion should be participative.
- Differences should be reconciled. If this is not possible, participants must agree to disagree.
- Work plans are recycled.
- Career discussion is teed-up.
- Employee and manager commit to development actions.

VII. Follow-up
- Checkpoints on development plan allow for this follow-up.

Exhibit II. Performance summary

When writing the overall statement of performance:
- Consider the degree of difficulty of the work package undertaken in addition to the actual results.
- Reinforce performance outcomes that you would like to see in the future by highlighting them here.
- Communicate importance of missed targets by listing them here.
- Let employees know the direction that performance is taking so that they can make decisions about effort levels, skill training emphasis, future placement possibilities, and so on.

When determining the overall rating number:
- Choose the paragraph that best describes performance in total, then choose the number that shades the direction it leans.
- Use the individual task measurements plus some weighting factor—realistically some projects are worth more than others and should carry more weight.
- Again, consider the degree of difficulty of the work package undertaken.

Strong points are:
- Demonstrated in the accomplishment of the work.

- Found in the completion of more than one project.
- Relevant—avoid trivia.
- Usually not heard well by employees.
- Good subjects for sharpening and growing.

Areas requiring improvement usually:

- Show up in more than one project.
- Are known by subordinate.
- Limit employee effectiveness.
- Can be improved to some degree.

Areas of disagreement:

- Can be manager or subordinate initiated.
- Need not be prepared in advance.
- Require some effort on both parts before recording.
- Are designed to keep problems from hiding beneath the surface.

Your review of the self-appraisal may surface some disagreement. Discuss this with the employee before formally committing it to writing.

Questions

1. If you were an executive attending this briefing, how would you react?
2. Are there any additional questions that need to be addressed?

Case 6

Continental Computer Corporation

"We have a unique situation here at Continental," remarked Ed White, Vice President for Engineering.

> We have three divisions within throwing distance of one another, and each one operates differently. This poses a problem for us at corporate headquarters because career opportunities and administrative policies are different in each division. Now that we are looking at project management as a profession, how do we establish uniform career path opportunities across all divisions?

Continental Computer Corporation (CCC) was a $9 billion a year corporation with worldwide operations encompassing just about every aspect of the computer field. The growth rate of CCC had exceeded 13 percent per year for the last eight years, primarily due to the advanced technology developed by their Eton Division, which produces disk drives. Continental is considered one of the "giants" in computer technology development, and supplies equipment to other computer manufacturers.

World headquarters for CCC is in Concord, Illinois, a large suburb northwest of Chicago in the heart of Illinoisís technology center. In addition to corporate headquarters, there are three other divisions: the Eton Division, which manufactures disk drives, the Lampco Division, which is responsible for Department of Defense (DoD) contracts such as for military application, satellites, and so on, and the Ridge Division, which is the primary research center for peripherals and terminals.

According to Ed White:

> Our major problems first began to surface during the early nineties. When we restructured our organization, we assumed that each division would operate as a separate entity (i.e., strategic business unit) without having to communicate with one another except through corporate headquarters. Therefore, we permitted each of our division vice presidents and general managers to

set up whatever organizational structure they so desired in order to get the work accomplished. Unfortunately, we hadn't considered the problem of coordinating efforts between sister divisions because some of our large projects demanded this.

The Lampco Division is by far the oldest, having been formed in 1989. The Lampco Division produces about $2 billion worth of revenue each year from DoD funding. Lampco utilizes a pure matrix structure. Our reason for permitting the divisions to operate independently was cost reporting. In the Lampco Division, we must keep two sets of books: one for government usage and one for internal control. This was a necessity because of DoD's requirement for earned value reporting on our large, cost-reimbursable contracts. It has taken us about five years or so to get used to this idea of multiple information systems, but now we have it well under control.

We have never had to lay people off in the Lampco Division. Yet, our computer engineers still feel that a reduction in DoD spending may cause massive layoffs here. Personally, I'm not worried. We've been through lean and fat times without having to terminate people.

The big problem with the Lampco Division is that because of the technology developed in some of our other divisions, Lampco must subcontract out a good portion of the work (to our other divisions). Not that Lampco can't do it themselves, but we do have outstanding R&D specialists in our other divisions.

We have been somewhat limited in the salary structure that we can provide to our engineers. Our computer engineers in the Lampco Division used to consider themselves as aerospace engineers, not computer engineers, and were thankful for employment and reasonable salaries. But now the Lampco engineers are communicating more readily with our other divisions and think that the grass is greener in these other divisions. Frankly, they're right. We've tried to institute the same wage and salary program corporate-wide, but came up with problems. Our engineers, especially the younger ones who have been with us five or six years, are looking for management positions. Almost all of our management positions in engineering are filled with people between 35 and 40 years of age. This poses a problem in that there is no place for these younger engineers to go. So, they seek employment elsewhere.

We've recently developed a technical performance ladder that is compatible to our management ladder. At the top of the technical ladder we have our consultant grade. Here our engineers can earn just about any salary based, of course, on their performance. The consultant position came about because of a problem in our Eton Division. I would venture to say that in the entire computer world, the most difficult job is designing disk drives. These people are specialists in a world of their own. There are probably only 25 people in the world who possess this expertise. We have five of them here at Continental. If one of our competitors would come in here and lure away just two of these guys, we would literally have to close down the Eton Division. So we've developed a consultant category. Now the word has spread and all of our engineers are applying for transfer to the Eton Division so as to become eligible for this new pay grade. In the Lampco Division alone I have had over 50 requests for transfer from engineers who now consider themselves as computer engineers. To make matters worse, the job market in computer technology is so good today that these people could easily leave us for more money elsewhere.

We've been lucky in the Lampco Division. Most of our contracts are large, and we can afford to maintain a project office staffed with three or four project engineers. These project engineers consider themselves as managers, not engineers. Actually they're right in doing so because theoretically they are engineering managers, not doers. Many of our people in Lamco are title-oriented and would prefer to be a project engineer as opposed to any other position. Good project engineers have been promoted, or laterally transferred, to project management so that we can pay them more. Actually, they do the same work.

In our Eton Division, we have a somewhat weird project management structure. We're organized on a product form rather than a project form of management. The engineers are considered to be strictly support for the business development function, and are not permitted to speak to the customers except under special circumstances. Business development manages both the product lines and R&D projects going on at one time. The project leader is selected by the director of engineering and can be a functional manager or just a functional employee. The project leader reports to his normal supervisor. The project leader must also report informally to one of the business development managers who is also tracking this project. This poses a problem in that when a conflict occurs, we sometimes have to take it up two or three levels before it can be resolved. Some conflicts have been so intense that they've had to be resolved at the corporate level.

The Eton Division happens to be our biggest money maker. We're turning out disk drives at an incredible rate and are backlogged with orders for at least six months. Many of our top R&D engineers are working in production support capacities because we cannot get qualified people fast enough. Furthermore, we have a yearly turnover rate in excess of 10 percent among our engineers below 30 years of age. We have several engineers who are earning more than their department managers. We also have five consultant engineers who are earning more than the department managers. We also have four consultant engineers who are earning as much as division managers.

We've had the greatest amount of problems in this division. Conflicts continuously arise due to interdependencies and misunderstandings. Our product line managers are the only people permitted to see the customers. This often alienates our engineering and manufacturing people, who are often called upon to respond to customer requests.

Planning is another major problem that we're trying to improve upon. We have trouble getting our functional mangers to make commitments. Perhaps this is a result of our inability to develop a uniform procedure for starting up a program. We always argue about when to anchor down the work. Our new, younger employees want to anchor everything down at once, whereas the poor project managers say not to anchor down anything. We, therefore, operate at all levels of the spectrum.

We can carry this problem one step further. How do we get an adequate set of objectives defined initially? We failed several times before because we couldn't get corporate agreement or understanding. We're trying to establish a policy for development of an architectural design document that will give good front-end definition.

Generally we're O.K. if we're simply modifying an existing product line. But with new product lines we have a problem in convincing people, especially our old customers.

The Ridge Division was originally developed to handle all corporate R&D activities. Unfortunately, our growth rate became so large and diversified that this became impractical. We, therefore, had to decentralize the R&D activities. This meant that each division could do their own R&D work. Corporate then had the responsibility for resolving conflicts, establishing priorities, and ensuring that all division are well-informed of the total R&D picture. Corporate must develop good communication channels between the divisions so that duplication of effort does not occur.

Almost all of our technical specialists have advanced degrees in engineering disciplines. This poses a severe problem for us, especially since we have a pure traditional structure. When a new project comes up, the project is assigned to the functional department that has the majority of the responsibility. One of the functional employees is then designated as the project manager. We realize that the new project manager has no authority to control resources that are assigned to other departments. Fortunately, our department managers realize this also, and usually put forth a concerted effort to provide whatever resources are needed. Most of the conflicts that do occur are resolved at the department manager level.

When a project is completed, the project manager returns to his or her former position as an engineering member of a functional organization. We've been quite concerned about these people that continuously go back and forth between project management and functional project engineering. This type of relationship is a must in our environment because our project managers must have a command of technology. We continuously hold in-house seminars on project management so as to provide our people with training in management skills, cost control, planning, and scheduling. We feel that we've been successful in this regard. We are always afraid that if we continue to grow, we'll have to change our structure and throw the company into chaos. Last time when we began to grow, corporate reassigned some of our R&D activities to other divisions. I often wonder what would have happened if this had not been done.

For R&D projects that are funded out of house, we generally have no major management problems for our project managers or project engineers. For corporate funded projects, however, life becomes more complex mainly because we have a tough time distinguishing when to kill a project or to pour money into it. Our project managers always argue that with just a little more corporate funding they can solve the world's greatest problems.

From the point of view of R&D, our biggest problems are in "grass roots projects." Let me explain what I mean by this. An engineer comes up with an idea and wants some money to pursue it. Unfortunately, our division managers are not budgeted for "seed monies" whenever an employee comes up with an idea for research or new product development. Each person must have a charge number to bill his time against. I know of virtually no project manager who would out-and-out permit someone to do independent research on a budgeted project.

So the engineer comes to us at corporate looking for seed money. Occasionally, we at corporate provide up to $50,000 for short-term seed money. That $50,000 might last for three to four months if the engineer is lucky. Unfortunately, obtaining the money is the lesser of the guy's problems. If the engineer needs support from another department, he's not going to get it because his project is just an informal "grass roots" effort, whereas everything else is a clearly definable, well-established project. People are reluctant to attach themselves to a "grass roots" effort because history has shown that the majority of them will be failures.

The researcher now has the difficult job of trying to convince people to give him support while continuously competing with other projects that are clearly defined and have established priorities. If the guy is persistent, however, he has a good chance to succeed. If he succeeds, he gets a good evaluation. But if he fails, he's at the mercy of his functional manager. If the functional manager felt that this guy could have been of more value to the company on a project basis, the he's liable to grade him down. But even with these risks, we still have several "seed money" requests each month by employees looking for glory.

Everyone sat around the gable listening to Ed White' comments. What had started out as a meeting to professionalize project management as a career path position, uniformly applied across all divisions seemed to have turned into a complaint session. The problems identified by Ed White now left people with the notion that there may be more pressing problems.

Questions

1. Is it common for companies to maintain two or more sets of books for cost accounting?
2. Is the matrix structure well suited for the solution to the above question?
3. Why do most project management structures find the necessity for a dual ladder system?
4. Should companies with several different types of projects have a uniform procedure for planning projects?
5. Is it beneficial to have to take conflicts up two or three levels for resolution?
6. Should project managers be permitted to talk to the customer even if the project is in support of a product line?
7. Should corporate R&D be decentralized?
8. What is meant by seed money?
9. How does control of seed money differ in a decentralized versus a centralized R&D environment?
10. Should the failure of a "grass roots" project affect an employee's opportunity for promotion?
11. If you were the vice president of either engineering or R&D, would you prefer centralized or decentralized control?
12. In either case, how would you handle each of the previously defined problems?

Case 7

Goshe Corporation

"I've called this meeting to try to find out why we're having a difficult time upgrading our EDP [Electronic Data Processing] Department to an MIS [Managment Information Systems] Division," remarked Herb Banyon, executive vice president of Goshe Corporation.

> Last year we decided to give the EDP Department a chance to show that they could contribute to corporate profits by removing the department from under the control of the Finance Division and establishing an MIS Division. The MIS Division should be a project-driven division using a project management methodology. I expected great results. I continuously get reports stating that we're having major conflicts and personality clashes among the departments involved in these MIS projects and that we're between one month to three months behind on almost all projects. If we don't resolve this problem right now, the MIS Division will be demoted to a department and once again find itself under the jurisdiction of the finance director.

Background

In June 1987, Herb Banyon announced that Goshe Corporation would be giving salary increases amounting to an average of 7 percent company-wide, with the percent distribution as shown in Exhibit I. The EDP Department, especially the scientific programmers, were furious because this was the third straight year they had received below-average salary increases. The scientific programmers felt that they were performing engineering-type work and, therefore, should be paid according to the engineering pay scale. In addition, the software that was developed by the scientific programs was shortening schedules and lowering manufacturing costs. The scientific programmers were contributing to corporate profitability.

The year before, the scientific programmers had tried to convince management that engineering needed its own computer and that there should be established a separate engineering computer programming department within the Engineering Division. This suggestion had strong support form the engineering community because they would benefit by having complete control of their own computer. Unfortunately, management rejected the idea, fearing that competition and conflict would develop by having two data processing units, and that one centralized unit was the only viable solution.

As a result of management's decision to keep the EDP Department intact and not give them a chance to demonstrate that they can and do contribute to profits, the EDP personnel created a closed shop environment and developed a very hostile attitude toward all other departments, even those within their own Finance Division.

The Meeting of the Minds

In January 1988, Banyon announced the organizational restructuring that would upgrade the EDP Department. Al Grandy, the EDP Department manager, was given a promotion to division manager, provided that he could adequately manage the MIS project activities. By December 1988, it became apparent that something must be done to remedy the deteriorating relationship between the functional departments and the MIS personnel. Banyon called a meeting of all functional and divisional managers in hopes that some of the problems could be identified and worked out.

Herb Banyon: For the past ten months I've watched you people continuously arguing back and forth about the MIS problems, with both sides always giving me the BS about how we'll work it out. Now, before it's too late, let's try to get at the root cause of the problem. Anyone want to start the ball rolling?

Cost Accounting Manager: The major problem, as I see it, is the lack of interpersonal skills employed by the MIS people. Our MIS personnel have received only on-the-job training. The Human Resources Department has never provided us with any project management training, especially in the behavioral areas of project management. Our organization here is, or should I say has been up to now, purely traditional, with each person reporting to and working for and with one manager. Now we have horizontal projects in which the MIS project leaders must work with several functional managers, all of whom have different management styles, different personalities, and different dispositions. The MIS group just can't turn around in one or two weeks and develop these necessary skills. It takes time and training.

Training Manager: I agree with your comments. There are two types of situations that literally demand immediate personnel development training. The first situation is when personnel are required to perform in an organizational structure that has gone from the relatively simple, pure structure to a complex, partial matrix structure. This is what has happened to us. The second situation is when the task changes from simple to complex.

With either situation by itself, there is usually some slack time. But when both occur almost instantaneously, as is our case, immediate training should be undertaken. I told this to Grandy several times, but it was like talking to deaf ears. All he kept saying was that we don't have time now because we're loaded down with priority projects.

Al Grandy: I can see from the start that we're headed for a rake-Grandy-over-the-coals meeting. So let me defend each accusation as it comes up. The day Banyon announced the organizational change, I was handed a list of 15 MIS projects that had to be completed within unrealistic time schedules. I performed a manpower requirements projection and found that we were understaffed by 35 percent. Now I'm not stupid. I understand the importance of training my people. But how am I supposed to release my people for these training sessions when I have been given specific instructions that each of these 15 projects had a high priority? I can just see myself walking into your office, Herb, telling you that I want to utilize my people only half-time so that they can undergo professional development training.

Banyon: Somehow I feel that the buck just got passed back to me. Those schedules that I gave you appeared totally realistic to me. I just can't imagine any simple computer program requiring more time than my original estimates. And had you come to me with a request for training, I would have checked with personnel and then probably would give you the time to train your people.

Engineering Manager: I wish to make a comment or two about schedules. I'm not happy when an MIS guy walks into my office and tells me, or should I say demands, that certain resources be given to him so that he can meet a schedule or milestone date that I've had no input into establishing. My people are just not going to become pawns in the power struggle for MIS supremacy. My people become very defensive if they're not permitted to participate in the planning activities, and I have to agree with them.

Manufacturing Manager: The Manufacturing Division has a project with the MIS group for purchasing a hardware system that will satisfy our scheduling and material handling system requirements. My people wanted to be involved in the hardware selection process. Instead, the MIS group came to us with proposal in hand identifying a system that was not a practical extension of the state of the art and that did not fall within our cost and time constraints.

We in manufacturing, being nice guys, modified our schedules to be compatible with the MIS project leaders' proposal. We then tried to provide more detailed information for the MIS team so that . . .

Grandy: Just a minute here! Your use of the word "we" is somewhat misleading. Project management is designed and structured so that sufficient definition of work to be performed can be obtained in order that a more uniform implementation can result. My people requested a lot of detailed information from your staff and were told to do the work ourselves and find our own information. "After all," as one of the functional employees put it, "if we're going to pass all of the responsibility over to you guys in project management; you people can just do it all."

Therefore, because my people had insufficient data, between us we ended up creating a problem, which was further intensified by a lack of formal communication between the MIS group and the functional departments, as well as between the functional departments themselves. I hold functional management responsible for this problem because some of the managers did not seem to have understood that they are responsible for the project work under their cognizance. Furthermore, I consider you, the manufacturing manager, as being remiss in your duties by not reviewing the performance of our personnel assigned to the project.

Manufacturing Manager: Your people designed a system that was way too complex for our needs. Your people consider this project as a chance for glory. It is going to take us ten years to grow into this complex system you've created.

Grandy: Let me make a few comments about our delays in the schedule. One of our projects was a six month effort. After the third month, there was a new department manager assigned in the department that was to be the prime user of this project. We were then given a change in user requirements and incurred additional delays in waiting for new user authorization.

Of course, people problems always affect schedules. One of my most experienced people became sick and had to be replaced by a rookie. In addition, I've tried to be a "good guy" by letting my people help out some of the functional managers when non-MIS problems occur. This other work ended up encroaching on staff time to a degree where it impacted the schedules.

Even though the MIS group regulates computer activities, we have no control over computer downtime or slow turnabout time. Turnabout time is directly proportional to our priority lists, and we all know that these lists are established from above.

And last, we have to consider both company and project politics. All the MIS group wanted to do was to show that we can contribute to company profits. Top management consistently tries to give us unwanted direction and functional management tries to sabotage our projects for fear that if we're successful, then it will be less money for their departments during promotion time.

Banyon: Well, I guess we've identified the major problem areas. The question remaining is: What are we going to do about it?

Exhibit I. Goshe organizational chart. Note: Percentages indicate 1987 salary increases

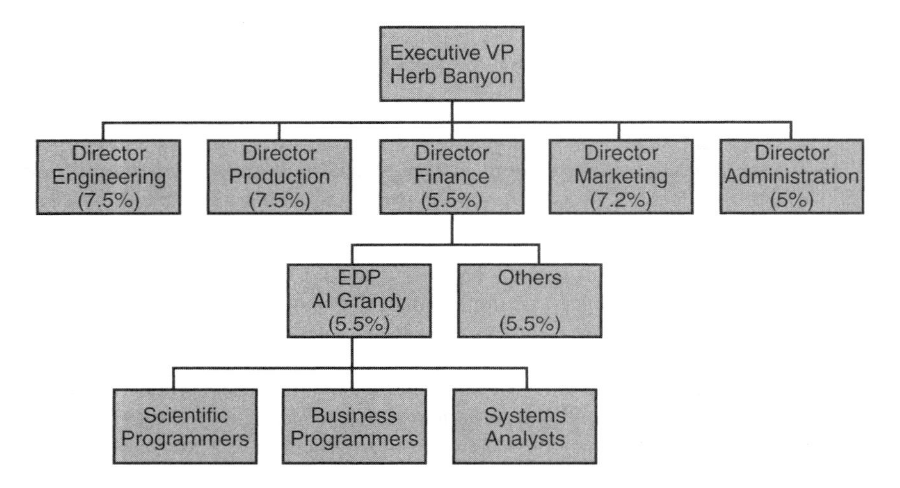

Questions

1. What are the major problems in the case study?
2. What are the user group's perceptions of the problem?
3. Was the company committed to project management?
4. Was project management forced upon the organization?
5. Did Goshe jump blindly into project management or was there a gradual introduction?
6. Did the company consider the problems that could manifest themselves with the implementation of change (i.e., morale)?
7. Did the company have a good definition of project management?
8. Should there have been a new set of company policies and procedures when the MIS group was developed?
9. How were project deadlines established?
10. Who established responsibilities for resource management?
11. Was there an integrated planning and control system?

12. Was there any training for division or project managers
13. Should Grandy have been promoted to his current position or should someone have been brought in from outside?
14. Can Grandy function effectively as both a project manager and a division manager?
15. Do you feel that Banyon understands computer programming?
16. Did anyone consider employee performance evaluations?
17. Did the company have good vertical communications?
18. Can a company without good vertical communications still have (or develop) good horizontal communications?
19. With the development of the MIS group, should each division be given 7 percent in the future?
20. What are the alternatives that are available?
21. What additional recommendations would you make?

Case 8

Hyten Corporation

On June 5, 1998, a meeting was held at Hyten Corporation, between Bill Knapp, director of sales, and John Rich, director of engineering. The purpose of the meeting was to discuss the development of a new product for a special customer application. The requirements included a very difficult, tight-time schedule. The key to the success of the project would depend on timely completion of individual tasks by various departments.

Bill Knapp: The Business Development Department was established to provide coordination between departments, but they have not really helped. They just stick their nose in when things are going good and mess everything up. They have been out to see several customers, giving them information and delivery dates that we can't possibly meet.

John Rich: I have several engineers who have MBA degrees and are pushing hard for better positions within engineering or management. They keep talking that formal project management is what we should have at Hyten. The informal approach we use just doesn't work all the time. But I'm not sure that just any type of project management will work in our division.

Knapp: Well, I wonder who Business Development will tap to coordinate this project? It would be better to get the manager from inside the organization instead of hiring someone from outside.

Company Background

Hyten Company was founded in 1982 as a manufacturer of automotive components. During the Gulf War, the company began manufacturing electronic components for the military. After the war, Hyten continued to prosper.

Hyten became one of the major component suppliers for the Space Program, but did not allow itself to become specialized. When the Space Program declined, Hyten developed other product lines, including energy management, building products, and machine tools, to complement their automotive components and electronics fields.

Hyten has been a leader in the development of new products and processes. Annual sales are in excess of $600 million. The Automotive Components Division is one of Hyten's rapidly expanding business areas (see the organizational chart in Exhibit I).

The Automotive Components Division

The management of both the Automotive Components Division and the Corporation itself is young and involved. Hyten has enjoyed a period of continuous growth over the past 15 years as a result of careful planning and having the right people in the right positions at the right time. This is emphasized by the fact that within five years of joining Hyten, every major manager and division head has been promoted to more responsibility within the corporation. The management staff of the Automotive Components Division has an average age of 40 and no one is over 50. Most of the middle managers have MBA degrees and a few have Ph.D.s. Currently, the Automotive Components Division has three manufacturing plants at various locations throughout the country. Central offices and most of the nonproduction functions are located at the main plant. There has been some effort by past presidents to give each separate plant some minimal level of purchasing, quality, manufacturing engineering and personnel functions.

Informal Project Management at Hyten Corporation

The Automotive Components Division of Hyten Corporation has an informal system of project management. It revolves around each department handling their own functional area of a given product development or project. Projects have been frequent enough that a sequence of operations has been developed to take a new product from concept to market. Each department knows its responsibilities and what it must contribute to a project.

A manager within the Business Development Department assumes informal project coordination responsibility and calls periodic meetings of the department heads involved. These meetings keep everyone advised of work status, changes to the project, and any problem areas. Budgeting of the project is based on the cost analysis developed after the initial design, while funding is allocated to each functional department based on the degree of its involvement. Funding for the initial design phase is controlled through business development. The customer has very little control over the funding, manpower, or work to be done. The customer, however, dictates when the new product design must be available for integration into the vehicle design, and when the product must be available in production quantities.

The Business Development Department

The Business Development Department, separate from Marketing/Sales, functions as a steering group for deciding which new products or customer requests are to be pursued and which are to be dropped. Factors which they consider in making these decisions are: (1) the company's long- and short-term business plans, (2) current sales forecasts, (3) economic and industry indicators, (4) profit potential, (5) internal capabilities (both volume and technology), and (6) what the customer is willing to pay versus estimated cost.

The duties of Business Development also include the coordination of a project or new product from initial design through market availability. In this capacity, they have no formal authority over either functional managers or functional employees. They act strictly on an informal basis to keep the project moving, give status reports, and report on potential problems. They are also responsible for the selection of the plant that will be used to manufacture the product.

The functions of Business Development were formerly handled as a joint staff function where all the directors would periodically meet to formulate short-range plans and solve problems associated with new products. The department was formally organized three years ago by the then 38 year old president as a recognition of the need for project management within the Automotive Components Division.

Manpower for the Business Development Department was taken from both outside the company and from within the division. This was done to honor the Corporation's commitment to hire people from the outside only after it was determined that there were no qualified people internally (an area that for years has been a sore spot to the younger managers and engineers).

When the Business Development Department was organized, its level of authority and responsibility was limited. However, the Department's authority and responsibility have subsequently expanded, though at a slow rate. This was done so as not to alienate the functional managers who were concerned that project management would undermine their "empire."

Introduction of Formal Project Management at Hyten Corporation

On July 10, 1998, Wilbur Donley was hired into the Business Development Department to direct new product development efforts. Prior to joining Hyten, he worked as project manager with a company that supplied aircraft hardware to the government. He had worked both as an assistant project manager and as a project manager for five years prior to joining Hyten.

Shortly after his arrival, he convinced upper management to examine the idea of expanding the Business Development group and giving them responsibility for formal project management. An outside consulting firm was hired to give an in-depth seminar on project management to all management and supervisor employees in the Division.

Prior to the seminar, Donley talked to Frank Harrel, manager of quality and reliability, and George Hub, manager of manufacturing engineering, about their problems and what they thought of project management.

Frank Harrel is 37 years old, has an MBA degree, and has been with Hyten for five years. He was hired as an industrial engineer and three years ago was promoted to manager of quality and reliability. George Hub is 45 years old and has been with Hyten for 12 years as manager of manufacturing engineering.

Wilbur Donley: Well, Frank, what do you see as potential problems to the timely completion of projects within the Automotive Components Division?

Frank Harrel: The usual material movement problems we always have. We monitor all incoming materials in samples and production quantities, as well as in-process checking of production and finished goods on a sampling basis. We then move to 100 percent inspection if any discrepancies are found. Marketing and Manufacturing people don't realize how much time is required to inspect for either internal or customer deviations. Our current manpower requires that schedules be juggled to accommodate 100 percent inspection levels on "hot items." We seem to be getting more and more items at the last minute that must be done on overtime.

Donley: What are you suggesting? A coordination of effort with marketing, purchasing, production scheduling, and the manufacturing function to allow your department to perform their routine work and still be able to accommodate a limited amount of high-level work on "hot" jobs?

Harrel: Precisely, but we have no formal contact with these people. More open lines of communication would be of benefit to everyone.

Donley: We are going to introduce a more formal type of project management than has been used in the past so that all departments who are involved will actively participate in the planning cycle of the project. That way we they will remain aware of how they affect the function of other departments and prevent overlapping of work. We should be able to stay on schedule and get better cooperation.

Harrel: Good, I'll be looking forward to the departure from the usual method of handling a new project. Hopefully, it will work much better and result in fewer problems.

Donley: How do you feel, George, about improving the coordination of work among various departments through a formal project manager?

George Hub: Frankly, if it improves communication between departments, I'm all in favor of the change. Under our present system, I am asked to make estimates of cost and lead times to implement a new product. When the project begins, the Product Design group starts making changes that require new cost figures and lead times. These changes result in cost overruns and in not meeting schedule dates. Typically, these changes continue right up to the production start date. Manufacturing appears to be the bad guy for not meeting the scheduled start date. We need someone to coordinate the work of various departments to prevent this continuous redoing of various jobs. We will at least have a chance at meeting the schedule, reducing cost, and improving the attitude of my people.

Personnel Department's View of Project Management

After the seminar on project management, a discussion was held between Sue Lyons, director of personnel, and Jason Finney, assistant director of personnel. The discussion was about changing the organization structure from informal project management to formal project management.

Sue Lyons: Changing over would not be an easy road. There are several matters to be taken under consideration.

Jason Finney: I think we should stop going to outside sources for competent people to manage new projects that are established within Business Development. There are several competent people at Hyten who have MBA's in Systems/Project Management. With that background and their familiarity with company operations, it would be to the company's advantage if we selected personnel from within our organization.

Lyons: Problems will develop whether we choose someone form inside the company or from an outside source.

Finney: However, if the company continues to hire outsiders into Business Development to head new projects, competent people at Hyten are going to start filtering to places of new employment.

Lyons: You are right about the filtration. Whoever is chosen to be a project manager must have qualifications that will get the job done. He or she should not only know the technical aspect behind the project, but should also be able to work with people and understand their needs. Project managers have to show concern for team members and provide them with work challenge. Project managers must work in a dynamic environment. This often requires the implementation of change. Project managers must be able to live with change and provide necessary leadership to implement the change. It is the project manager's responsibility to develop an atmosphere to allow people to adapt to the changing work environment.

In our department alone, the changes to be made will be very crucial to the happiness of the employees and the success of projects. They must feel they are being given a square deal, especially in the evaluation procedure. Who will do the evaluation? Will the functional manager be solely responsible for the evaluation when, in fact, he or she might never see the functional employee for the duration of a project? A functional manager cannot possibly keep tabs on all the functional employees who are working on different projects.

Finney: Then the functional manager will have to ask the project managers for evaluation information.

Lyons: I can see how that could result in many unwanted situations. To begin with, say the project manager and the functional manager don't see eye to eye on things. Granted, both should be at the same grade level and neither one has authority over the other, but let's say there is a situation where the two of them disagree as to either direction or quality of work. That puts the functional employee in an awkward position. Any employee will have the tendency of bending toward the individual who signs his or her promotion and evaluation form. This can influence the project manager into recommending an evaluation below par regardless of how the functional employee performs. There is also the situation where the employee is on the project for only a couple of weeks, and spends most of his or her time working alone, never getting a chance to know the project manager. The project manager will probably give the functional employee an average rating, even though the employee has done an excellent job. This results from very little contact. Then what do you do when the project manager allows personal feelings to influence his or her evaluation of a functional employee? A project manager who knows the functional employee personally might be tempted to give a strong or weak recommendation, regardless of performance.

Finney: You seem to be aware of many difficulties that project management might bring.

Lyons: Not really, but I've been doing a lot of homework since I attended that seminar on project management. It was a good seminar, and since there is not much written on the topic, I've been making a few phone calls to other colleagues for their opinions on project management.

Finney: What have you learned from these phone calls?

Lyons: That there are more personnel problems involved. What do you do in this situation? The project manager makes an excellent recommendation to the functional manager. The functional employee is aware of the appraisal and feels he or she should be given an above average pay increase to match the excellent job appraisal, but the functional manager fails to do so. One personnel manager from another company incorporating project management ran into problems when the project manager gave an employee of one grade level responsibilities of a higher grade level. The employee did an outstanding job taking on the responsibilities of a higher grade level and expected a large salary increase or a promotion.

Finney: Well, that's fair, isn't it?

Lyons: Yes, it seems fair enough, but that's not what happened. The functional manager gave an average evaluation and argued that the project manager had no business giving the functional employee added responsibility without first checking with him. So, then what you have is a disgruntled employee ready to seek employment elsewhere. Also, there are some functional managers who will only give above-average pay increases to those employees who stay in the functional department and make that manager look good.

Lyons: Right now I can see several changes that would need to take place. The first major change would have to be attitudes toward formal project management and hiring procedures. We do have project management here at Hyten but on an informal basis. If we could administer it formally, I feel we could do the company a great service. If we seek project managers from within, we could save on time and money. I could devote more time and effort on wage and salary grades and job descriptions. We would need to revise our evaluation forms—presently they are not adequate. Maybe we should develop more than one evaluation form: one for the project manager to fill out and give to the functional manager, and a second form to be completed by the functional manager for submission to Personnel.

Finney: That might cause new problems. Should the project manager fill out his or her evaluation during or after project completion?

Lyons: It would have go be after project completion. That way an employee who felt unfairly evaluated would not feel tempted to screw up the project. If an employee felt the work wasn't justly evaluated, that employee might decide not to show up for a few days—these few days of absence could be most crucial for timely project completion.

Finney: How will you handle evaluation of employees who work on several projects at the same time? This could be a problem if employees are really enthusiastic about one project over another. They could do a terrific job on the project they are interested in and slack off on other projects. You could also have functional people working on departmental jobs but charging their time to the project overhead. Don't we have exempt and nonexempt people charging to projects?

Lyons: See what I mean? We can't just jump into project management and expect a bed of roses. There will have to be changes. We can't put the cart before the horse.

Finney: I realize that, Sue, but we do have several MBA people working here at Hyten who have been exposed to project management. I think that if we start putting our heads together and take a systematic approach to this matter, we will be able to pull this project together nicely.

Lyons: Well, Jason, I'm glad to see that you are for formal project management. We will have to approach top management on the topic. I would like you to help coordinate an equitable way of evaluating our people and to help develop the appropriate evaluation forms.

Project Management as Seen by the Various Departments

The general manager arranged through the personnel department to interview various managers on a confidential basis. The purpose of the interview was to evaluate the overall acceptance of the concept of formal project management. The answers to the question, "How will project management affect your department?" were as follows:

Frank Harrel, quality and reliability manager

> Project management is the actual coordination of the resources of functional departments to achieve the time, cost, and performance goals of the project. As a consequence, personnel interfacing is an important component toward the success of the project. In terms of quality control, it means less of the attitude of the structured workplace where quality is viewed as having the function of finding defects and, as a result, is looked upon as a hindrance to production. It means that the attitude toward quality control will change to one of interacting with other departments to minimize manufacturing problems. Project management reduces suboptimization among functional areas and induces cooperation. Both company and department goals can be achieved. It puts an end to the "can't see the forest for the trees" syndrome.

Harold Grimes, plant manager

I think that formal project management will give us more work than long-term benefits. History indicates that we hire more outside people for new positions than we promote from within. Who will be hired into these new project management jobs? We are experiencing a lot of backlash from people who are required to teach new people the ropes. In my opinion, we should assign inside MBA graduates with project management training to head up projects and not hire an outsider as a formal project manager. Our present system would work fine if inside people were made the new managers in the Business Development Department.

Herman Hall, director of MIS

I have no objections to the implementation of formal project management in our company. I do not believe, however, that it will be possible to provide the reports needed by this management structure for several years. This is due to the fact that most of my staff are deeply involved in current projects. We are currently working on the installation of minicomputers and on-line terminals throughout the plant. These projects have been delayed by the late arrival of new equipment, employee sabotage, and various start-up problems. As a result of these problems, one group admits to being six months behind schedule and the other group, although on schedule, is 18 months from their scheduled completion date. The rest of the staff currently assigned to maintenance projects consists of two systems analysts who are nearing retirement and two relatively inexperienced programmers. So, as you can readily see, unless we break up the current project teams and let those projects fall further behind schedule, it will be difficult at this time to put together another project team.

The second problem is that even if I could put together a staff for the project, it might take up to two years to complete an adequate information system. Problems arise from the fact that it will take time to design a system that will draw data from all the functional areas. This design work will have to be done before the actual programming and testing could be accomplished. Finally, there would be a debugging period when we receive feedback from the user on any flaws in the system or enhancements that might be needed. We could not provide computer support to an "overnight" change to project management.

Bob Gustwell, scheduling manager

I am happy with the idea of formal project management, but I do see some problems implementing it. Some people around here like the way we do things now. It is a natural reaction for employees to fight against any changes in management style.

But don't worry about the scheduling department. My people will like the change to formal project management. I see this form of management as a way to minimize, of not eliminate, schedule changes. Better planning on the part of both department and project managers will be required, and the priorities will be set at corporate level. You can count on our support because I'm tired of being caught between production and sales.

John Rich, director of engineering

It seems to me that project management will only mess things up. We now have a good flowing chain of command in our organization. This new matrix will only create problems. The engineering department, being very technical, just can't take direction from anyone outside the department. The project office will start to skimp on specifications just to save time and dollars. Our products are too technical to allow schedules and project costs to affect engineering results.

Bringing in someone from the outside to be the project manager will make things worse. I feel that formal project management should not be implemented at Hyten. Engineering has always directed the projects, and we should keep it that way. We shouldn't change a winning combination.

Fred Kuncl, plant engineering

I've thought about the trade-offs involved in implementing formal project management at Hyten and feel that plant engineering cannot live with them. Our departmental activities are centered around highly unpredictable circumstances, which sometimes involve rapidly changing priorities related to the production function. We in plant engineering must be able to respond quickly and appropriately to maintenance activities directly related to manufacturing activities. Plant engineering is also responsible for carrying out critical preventive maintenance and plant construction projects.

Project management would hinder our activities because project management responsibilities would burden our manpower with additional tasks. I am against project management because I feel that it is not in the best interest of Hyten. Project management would weaken our department's functional specialization because it would require cross-utilization of resources, manpower, and negotiation for the services critical to plant engineering.

Bill Knapp, director of marketing

I feel that the seminar on formal project management was a good one. Formal project management could benefit Hyten. Our organization needs to focus in more than one direction at all times. In order to be successful in today's market, we must concentrate on giving all our products sharp focus. Formal project management could be a good way of placing individual emphasis on each of the products of our company. Project management would be especially advantageous to us because of our highly diversified product lines. The organization needs to efficiently allocate resources to projects, products, and markets. We cannot afford to have expensive resources sitting idle. Cross-utilization and the consequent need for negotiation ensures that resources are used efficiently and in the organization's best overall interest.

We can't afford to continue to carry on informal project management in our business. We are so diversified that all of our products can't be treated alike. Each product has different needs. Besides, the nature of a team effort would strengthen our organization.

Stanley Grant, comptroller

In my opinion, formal project management can be profitably applied in our organization. Management should not, however, expect that project management would gain instant acceptance by the functional managers and functional employees, including the finance department personnel.

The implementation of formal project management in our organization would have an impact on our cost control system and internal control system, as well.

In the area of cost control, project cost control techniques have to be formalized and installed. This would require the accounting staff to: (1) beak comprehensive cost summaries into work packages, (2) prepare commitment reports for "technical decision makers," (3) approximate report data and (4) concentrate talent on major problems and opportunities. In project management, cost commitments on a project are made when various functional departments, such as engineering, manufacturing and marketing, make technical decisions to take some kind of action. Conventional accounting reports do not show the cost effects of these technical decisions until it is too late to reconsider. We would need to provide the project manager with cost commitment reports at each decision state to enable him or her to judge when costs are getting out of control. Only by receiving such timely cost commitment reports, could the project manager take needed corrective actions and be able to approximate the cost effect of each technical decision. Providing all these reports, however, would require additional personnel and expertise in our department.

In addition, I feel that the implementation of formal project management would increase our responsibilities in finance department. We would need to conduct project audits, prepare periodic comparisons of actual versus projected costs and actual versus programmed manpower allocation, update projection reports and funding schedules, and sponsor cost improvement programs.

> In the area of internal control, we will need to review and modify our existing internal control system to effectively meet our organization's goals related to project management. A careful and proper study and evaluation of existing internal control procedures should be conducted to determine the extent of the tests to which our internal auditing procedures are to be restricted. A thorough understanding of each project we undertake must be required at all times.
>
> I'm all in favor of formal project management, provided management would allocate more resources to our department so we could maintain the personnel necessary to perform the added duties, responsibilities, and expertise required.

After the interviews, Sue Lyons talked to Wilbur Donley about the possibility of adopting formal project management. As she put it,

> You realize that regardless of how much support there is for formal project management, the general manager will probably not allow us to implement it for fear it will affect the performance of the Automotive Components Division.

Exhibit I. Organizational chart of the automotive division, Hyten Corporation

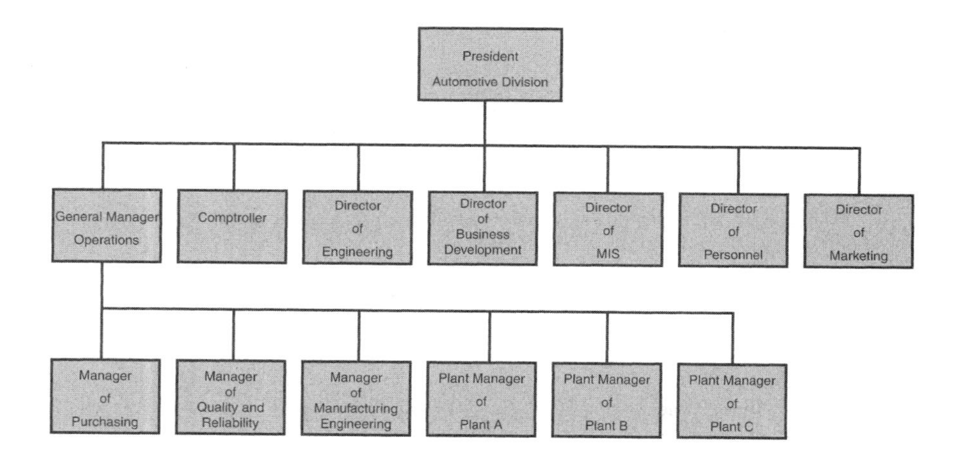

Questions

1. What are some of the major problems facing the management of Hyten in accepting formalized project management? (Include attitude problems/personality problems.)
2. Do any of the managers appear to have valid arguments for their beliefs as to why formal project management should not be considered?
3. Are there any good reasons why Hyten should go to formal project management?
4. Has Hyten taken a reasonable approach toward implementing formal project management?
5. Has Hyten done anything wrong?
6. Should formal project management give employees more room for personal growth?
7. Will formalized project management make it appear as though business development has taken power away from other groups?
8. Were the MBAs exposed to project management?

9. Were the organizational personnel focusing more on the problems (disadvantages) or advantages of project management?
10. What basic fears do employees have in considering organizational change to formal project management?
11. Must management be sold on project management prior to implementation?
12. Is it possible that some of the support groups cannot give immediate attention to such an organizational change?
13. Do functional managers risk a loss of employee loyalty with the new change?
14. What recommendations would you make to Hyten Corporation?

Case 9

Acorn Industries

Acorn Industries, prior to July of 1996, was a relatively small mid-western corporation dealing with a single product line. The company dealt solely with commercial contracts and rarely, if ever, considered submitting proposals for government contracts. The corporation at that time functioned under a traditional form of organizational structure, although it did possess a somewhat decentralized managerial philosophy within each division. In 1993, upper management decided that the direction of the company must change. To compete with other manufacturers, the company initiated a strong acquisition program whereby smaller firms were bought out and brought into the organization. The company believed that an intensive acquisition program would solidify future growth and development. Furthermore, due to their reputation for possessing a superior technical product and strong marketing department, the acquisition of other companies would allow them to diversify into other fields, especially within the area of government contracts. However, the company did acknowledge one shortcoming that possibly could hurt their efforts—they had never fully adopted, nor implemented, any form of project management.

In July of 1996, the company was awarded a major defense contract after four years of research and development and intensive competition from a major defense organization. The company once again relied on their superior technological capabilities, combined with strong marketing efforts, to obtain the contract. According to Chris Banks, the current marketing manager at Acorn Industries, the successful proposal for the government contract was submitted solely through the efforts of the marketing division. Acorn's successful marketing strategy relied on three factors when submitting a proposal:

1. Know exactly what the customer wants.
2. Know exactly what the market will bear.
3. Know exactly what the competition is doing and where they are going.

The contract awarded in July 1996 led to subsequent successful government contracts and, in fact, eight more were awarded amounting to $80 million each. These contracts were to last anywhere from seven to ten years, taking the company into early 2009 before expiration would occur. Due to their extensive growth, especially with the area of government contracts as they pertained to weapon systems, the company was forced in 1997 to change general managers. The company brought in an individual who had an extensive background in program management and who previously had been heavily involved in research and development.

Problems Facing the General Manager

The problems facing the new general manager were numerous. Prior to his arrival, the company was virtually a decentralized manufacturing organization. Each division within the company was somewhat autonomous, and the functional managers operated under a Key Management Incentive Program (KMIP). The prior general manager had left it up to each division manager to do what was required. Performance had been measured against attainment of goals. If the annual objective was met under the KMIP program, each division manager could expect to receive a year-end bonus. These bonuses were computed on a percentage of the manager's base pay, and were directly correlated to the ability to exceed the annual objective. Accordingly, future planning within each division was somewhat stagnant, and most managers did not concern themselves with any aspect of organizational growth other than what was required by the annual objective.

Because the company had previously dealt with a single product line and interacted solely with commercial contractors, little, if any, production planning had occurred. Interactions between research and development and the production engineering departments were virtually nonexistent. Research and Development was either way behind or way ahead of the other departments at any particular time. Due to the effects of the KMIP program, this aspect was likely to continue.

Change within the Organizational Structure

To compound the aforementioned problems, the general manager faced the unique task of changing corporate philosophy. Previously, corporate management was concerned with a single product with a short term production cycle. Now, however, the corporation was faced with long-term government contracts, long cycles, and diversified products. Add to this the fact that the company was almost void of any individuals who had operated under any aspect of program management, and the tasks appeared insurmountable.

The prime motivating factor for the new general manager during the period from 1997 to 1999 was to retain profitability and maximize return on investment. In order to do this, the general manager decided to maintain the company's commercial product line, operating it at full capacity. This decision was made because the company was based in solid financial management and the commercial product line had been extremely profitable. According to the general manager, Ken Hawks,

> The concept of keeping both commercial and government contracts separate was a necessity. The commercial product line was highly competitive and maintained a good market share. If the adventure into weaponry failed, the company could always fall back on the commercial products. At any rate, the company at this time could not solely rely on the success of government contracts, which were due to expire.

In 1996, Acorn reorganized its organizational structure and created a project management office under the direct auspices of the general manager (see Exhibit I).

Expansion and Growth

In late 1996, Acorn initiated a major expansion and reorganization within its various divisions. In fact, during the period between 1996 and 1997, the government contracts resulted in the acquiring of three new companies and possibly the acquisition of a fourth. As before, the expertise of the marketing department was heavily relied upon. Growth objectives for each division were set by corporate headquarters with the advice and feedback of the division managers. Up to 1996, Acorn's divisions had not had a program director. The program management functions for all divisions were performed by one program manager whose expertise was entirely within the commercial field. This particular program manager was concerned only with profitability and did not closely interact with the various customers. According to Chris Banks,

> The program manager's philosophy was to meet the minimum level of performance required by the contract. To attain this, he required only adequate performance. As Acorn began to become more involved with government contracts, the position remained that given a choice between high technology and low reliability, the company would always select an acquisition with low technology and high reliability. If we remain somewhere in between, future government contracts should be assured.

At the same time, Acorn established a Chicago office headed by a group executive. The office was mainly for monitoring for government contracts. Concurrently, an office was established in Washington to monitor the trends within the Department of Defense and to further act as a lobbyist for government contracts. A position of director of marketing was established to interact with the program office on contract proposals. Prior to the establishment of a director of program management position in 1997, the marketing division had been responsible for contract proposals. Acorn believed that marketing would always, as in the past, set the tone for the company. However, in 1997, and then again in 1998 (see Exhibits II and III), Acorn underwent further organizational changes. A full-time director of project management was appointed, and a program management office was set up, with further subdivisions of project managers responsible for the various government contracts. It was at this time that Acorn realized the necessity of involving the program manager more extensively in contract proposals. One faction within corporate management wanted to keep marketing responsible for contract proposals. Another decided that a combination between the marketing input and the expertise of the program director must be utilized. According to Chris Banks,

> We began to realize that marketing no longer could exclude other factors within the organization when preparing contract proposals. As project management became a reality, we realized that the project manager must be included in all phases of contract proposals.

Prior to 1996, the marketing department controlled most aspects of contract proposals. With the establishment of the program office, interface between the marketing department and the program office began to increase.

Responsibilities of the Project Manager

In 1997, Acorn, for the first time, identified a director of project management. This individual reported directly to the general manager and had under his control:

1. The project managers
2. The operations group
3. The contracts group

Under this reorganization, the director of project management, along with the project managers, possessed greater responsibility relative to contract proposals. These new responsibilities included:

1. Research and development
2. Preparation of contract proposals
3. Interaction with marketing on submittal of proposals
4. Responsibility for all government contracts
   a. Trade-off analysis
   b. Cost analysis
5. Interface with engineering department to insure satisfaction of customer's desires

Due to the expansion of government contracts, Acorn was now faced with the problem of bringing in new talent to direct ongoing projects. The previous project manager had had virtual autonomy over operations and maintained a singular philosophy. Under his tenure, many bright individuals left Acorn because future growth and career patterns were questionable. Now that the company is diversifying into other product lines, the need for young talent is crucial. Project management is still in the infancy stage.

Acorn's approach to selecting a project manager was dependent upon the size of the contract. If the particular contract was between $2 and $3 billion, the company would go with the most experienced individual. Smaller contracts would be assigned to whoever was available.

Interaction with Functional Departments

Due to the relative newness of project management, little data was available to the company to fully assess whether operations were successful. The project managers were required to negotiate with the functional departments for talent. This aspect has presented some problems due to the long-term cycle of most government contracts. Young talent within the organization saw involvement with projects as an opportunity to move up within the organization. Functional managers, on the other hand, apparently did not want to let go of young talent and were extremely reluctant to lose any form of autonomy.

Performance of individuals assigned to projects was mutually discussed between the project manager and the functional manager. Problems arose, however, due to length of projects. In some instances, if an individual had been assigned longer to the project manager than to the functional manager, the final evaluation of performance rested with the project manager. Further problems thus occurred when performance evaluations were submitted. In some instances, adequate performance was rated high in order to maintain an individual within the project scheme. According to some project managers, this aspect was a reality that must be faced, due to the shortage of abundant talent.

Current Status

In early 1998, Acorn began to realize that a production shortage relative to government contracts would possibly occur in late 2001 or early 2003. Acorn initiated a three-pronged attack to fill an apparent void:

1. Do what you do best.
2. Look for similar product lines.
3. Look for products that do not require extensive R&D.

To facilitate these objectives, each division within the corporation established its own separate marketing department. The prime objective was to seek more federal funds through successful contract proposals and utilize these funds to increase investment into R&D. The company had finally realized that the success of the corporation was primarily attributed to the selection of the proper general manager. However, this had been accomplished at the exclusion of proper control over R&D efforts. A more lasting problem still existed, however. Program management was still less developed than in most other corporations.

Exhibit I. 1996 organizational structure

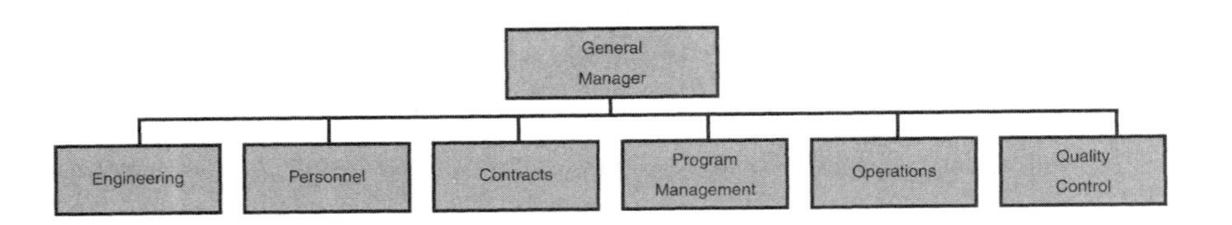

Exhibit II. 1997 organizational structure

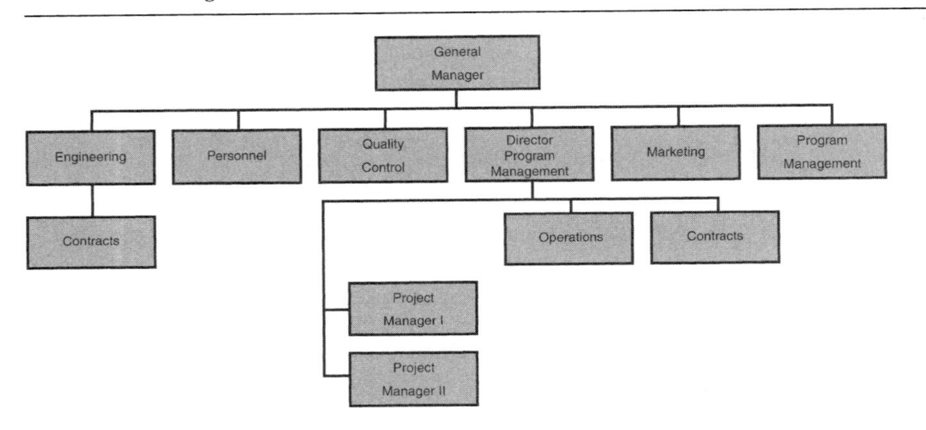

Exhibit III. 1998 organizational structure (10/1/98)

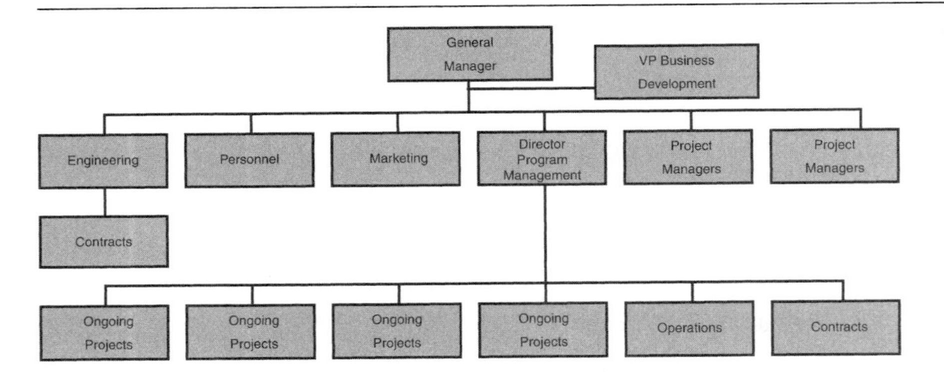

Questions

1. What are the strengths of Acorn?
2. What are the weaknesses of Acorn?
3. What are your recommendations?
4. Additional questions:
   A. Why was project management so slow in getting off the ground?
   B. Can marketing continue to prepare proposals without functional input?
   C. What should be the working relationship between the product manager and the proposal?
   D. Does KMIP benefit project management?
   E. Should KMIP be eliminated?

Case 10

Mohawk National Bank

"You're really going to have your work cut out for you, Randy," remarked Pat Coleman, vice president for operations. "It's not going to be easy establishing a project management organizational structure on top of our traditional structure. We're going to have to absorb the lumps and bruises and literally 'force' the system to work."

Background

Between 1978 and 1988, Mohawk National matured into one of Maine's largest full-service banks, employing a full-time staff of some 1,200 employees. Of the 1,200 employees, approximately 700 were located in the main offices in downtown Augusta.

Mohawk matured along with other banks in the establishment of computerized information processing and decision-making. Mohawk leased the most up-to-date computer equipment in order to satisfy customer demands. By 1984, almost all departments were utilizing the computer.

By 1985, the bureaucracy of the traditional management structure was creating severe administrative problems. Mohawk's management had established many complex projects to be pursued, each one requiring the involvement of several departments. Each department manager was setting his or her own priorities for the work that had to be performed. The traditional organization was too weak structurally to handle problems that required integration across multiple departments. Work from department to department could not be tracked because there was no project manager who could act as focal point for the integration of work.

Understanding the Changeover Problem

It was a difficult decision for Mohawk National to consider a new organizational structure, such as a matrix. Randy Gardner, director of personnel, commented on the decision:

> Banks, in general, thrive on traditionalism and regimentation. When a person accepts a position in our bank, he or she understands the strict rules, policies, and procedures that have been established during the last 30 years.
>
> We know that it's not going to be easy. We've tried to anticipate the problems that we're going to have. I've spent a great deal of time with our vice president of operations and two consultants trying to predict the actions of our employees.
>
> The first major problem we see is with our department managers. In most traditional organizations, the biggest functional department emerges as the strongest. In a matrix organization, or almost any other project form for that matter, there is a shift in the balance of power. Some managers become more important in their new roles and others not so important. We think our department managers are good workers and that they will be able to adapt.
>
> Our biggest concern is with the functional employees. Many of our functional people have been with us between 20 and 30 years. They're seasoned veterans. You must know that they're going to resist change. These people will fight us all the way. They won't accept the new system until they see it work. That'll be our biggest challenge: to convince the functional team members that the system will work.

Pat Coleman, the vice president for operations, commented on the problems that he would be facing with the new structure:

> Under the new structure, all project managers will be reporting to me. To be truthful, I'm a little scared. This changeover is like a project in itself. As with any project, the beginning is the most important phase. If the project starts out on the right track, people might give it a chance. But if we have trouble, people will be quick to revert back to the old system. Our people hate change. We cannot wait one and a half to two years for people to get familiar with the new system. We have to hit them all at once and then go all out to convince them of the possibilities that can be achieved.
>
> This presents a problem in that the first group of project managers must be highly capable individuals with the ability to motivate the functional team members. I'm still not sure whether we should promote from within or hire from the outside. Hiring from the outside may cause severe problems in that our employees like to work with people they know and trust. Outside people may not know our people. If they make a mistake and aggravate our people, the system will be doomed to failure.
>
> Promoting from within is the only logical way to go, as long as we can find qualified personnel. I would prefer to take the qualified individuals and give them a lateral promotion to a project management position. These people would be on trial for about six months. If they perform well, they will be promoted and permanently assigned to project management. If they can't perform or have trouble enduring the pressure, they'll be returned to their former functional positions. I sure hope we don't have any inter- or intramatrix power struggles.
>
> Implementation of the new organizational form will require good communications channels. We must provide all of our people with complete and timely information. I plan on holding weekly meetings with all of the project and functional managers. Good communications channels must be established between all resource managers. These team meetings will give people a chance to see each other's mistakes. They should be able to resolve their own problems and conflicts. I'll be there if they need me. I do anticipate several conflicts because our functional managers are not going to be happy in the role of a support group for a project manager. That's the balance of power problem I mentioned previously.

> I have asked Randy Gardner to identify from within our ranks the four most likely individuals who would make good project managers and drive the projects to success. I expect Randy's report to be quite positive. His report will be available next week.

Two weeks later, Randy Gardner presented his report to Pat Coleman and made the following observations.

> I have interviewed the four most competent employees who would be suitable for project management. The following results were found:
>
> Andrew Medina, department manager for cost accounting, stated that he would refuse a promotion to project management. He has been in cost accounting for 20 years and does not want to make a change into a new career field.
>
> Larry Foster, special assistant to the vice president of commercial loans, stated that he enjoyed the people he was working with and was afraid that a new job in project management would cause him to lose his contacts with upper level management. Larry considers his present position more powerful than any project management position.
>
> Chuck Folson, personal loan officer stated that in the 15 years he's been with Mohawk National, he has built up strong interpersonal ties with many members of the bank. He enjoys being an active member of the informal organization and does not believe in the applications of project management for our bank.
>
> Jane Pauley, assistant credit manager, stated that she would like the position, but would need time to study up on project management. She feels a little unsure about herself. She's worried about the cost of failure.

Now Pat Coleman had a problem. Should he look for other bank employees who might be suitable to staff the project management functions or should he look externally to other industries for consultants and experienced project managers?

Questions

1. How do you implement change in a bank?
2. What are some of the major reasons why employees do not want to become project managers?
3. Should the first group of project managers be laterally assigned?
4. Should the need for project management first be identified from within the organization?
5. Can project management be forced upon an organization?
6. Does the bank appear to understand project management?
7. Should you start out with permanent or temporary project management positions?
8. Should the first group of project managers be found from within the organization?
9. Will people be inclined to support the matrix if they see that the project managers are promoted from within?
10. Suppose that the bank goes to a matrix, but without the support of top management. Will the system fail?
11. How do you feel about in-house workshops to soften the impact of project management?

Case 11

First Security Bank of Cleveland

The growth rate of First Security of Cleveland had caused several executives to do some serious thinking about whether the present organizational structure was adequate for future operations. The big question was whether the banking community could adapt to a project management structure.

Tom Hood had been the president of First Security for the past ten years. He had been a pioneer in bringing computer technology into the banking industry. Unfortunately, the size and complexity of the new computer project created severe integration problems, problems with which the present traditional organization was unable to cope. What was needed was a project manager who could drive the project to success and handle the integration of work across functional lines.

Tom Hood met with Ray Dallas, one of the bankís vice presidents, to discuss possible organizational restructuring:

Tom Hood: I've looked at the size and complexity of some 20 projects that First Security did last year. Over 50 percent of these projects required interaction between four or more departments.

Ray Dallas: What's wrong with that? We're growing and our problems are likewise becoming more complex.

Hood: It's the other 50 percent that worry me. We can change our organizational structure to adapt to complex problem-solving and integration. But what happens when we have a project that stays in one functional department? Who's going to drive it home? I don't see how we can tell a functional manager that he or she is a support group in one organizational form and a project manager in the other and have both organizational forms going on at the same time.

We can have either large, complex projects or small ones. The small ones will be the problem. They can exist in one department or be special projects assigned to one person or a task force team. This means that if we incorporate project management, we'll have to live with a variety of structures. This can become a bad situation. I'm not sure that our people will be able to adapt to this changing environment.

Dallas: I don't think it will be as bad as you make it. As long as we clearly define each person's authority and responsibility, we'll be all right. Other industries have done this successfully. Why can't we?

Hood: There are several questions that need answering. Should each project head be called a project manager, even if the project requires only one person? I can see our people suddenly becoming title-oriented. Should all project managers report to the same boss, even if one manager has 30 people working on the project and the other manager has none? This could lead to power struggles. I want to avoid that because it can easily disrupt our organization.

Dallas: The problem you mentioned earlier concerns me. If we have a project that belongs in one functional department, the ideal solution is to let the department manager wear two hats, the second one being project manager. Disregarding for the moment the problem that this manager will have in determining priorities, to whom should he or she report to as to the status of the work? Obviously, not to the director of project management.

Hood: I think the solution must be that all project managers report to one person. Therefore, even if the project stays in one functional department, we'll still have to assign a project manager Under project management organizational forms, functional managers become synonymous with resource managers. It is very dangerous to permit a resource manager to act also as a project manager. The resource manager might consider the project as being so important that he or she will commit all the department's best people to it and make it into a success at the expense of all the department's other work. That would be like winning a battle but losing the war.

Dallas: You realize that we'll need to revamp our wage and salary administration program if we go to project management. Evaluating project managers might prove difficult. Regardless of what policies we establish, there are still going to be project managers who try to build empires, thinking that their progress is dependent upon the number of people they control. Project management will definitely give some people the opportunity to build a empire. We'll have to watch that closely.

Hood: Ray, I'm a little worried that we might not be able to get good project managers. We can't compete with the salaries the project managers get in other industries such as engineering, construction, or computers. Project management cannot be successful unless we have good managers at the controls. What's your feeling on this?

Dallas: We'll have to promote from within. That's the only viable solution. If we try to make project management salaries overly attractive, we'll end up throwing the organization into chaos. We must maintain an adequate salary structure so that people feel that they have the same opportunities in both project management and the functional organization. Of course, we'll still have some people who will be more title-oriented than money-oriented, but at least each person will have the same opportunity for salary advancement.

Hood: See if you can get some information from our personnel people on how we could modify our salary structure and what salary levels we can pay our project managers. Also, check with other banks and see what they're paying their project managers. I don't want to go into this blind and then find out that we're setting the trend for project management salaries. Everyone would hate us. I'd rather be a follower than a leader in this regard.

Questions

1. What are the major problems identified in the case?
2. What are your solutions to the above question and problems?

Case 12

Como Tool and Die (A)*

Como Tool and Die was a second-tier component supplier to the auto industry. Their largest customer was Ford Motor Company. Como had a reputation for delivering a quality product. During the 1980s and the early 1990s, Como's business grew because of their commitment to quality. Emphasis was on manufacturing operations, and few attempts were made to use project management. All work was controlled by line managers who, more often than not, were overburdened with work.

The culture at Como underwent a rude awakening in 1996. In the summer of 1996, Ford Motor Company established four product development objectives for both tier one and tier two suppliers:

- Lead time: 25–35 percent reduction
- Internal resources: 30–40 percent reduction
- Prototypes: 30–35 percent reduction (time and cost)
- Continuous process improvement and cost reductions

The objectives were aimed at consolidation of the supply base with larger commitments to tier one suppliers, who would now have greater responsibility in vehicle development, launch, process improvement, and cost reduction. Ford had established a time frame of 24 months for achievement of the objectives. The ultimate goal for Ford would be the creation of one global, decentralized vehicle development system that would benefit from the efficiency and technical capabilities of the original equipment manufacturers (OEMs) and the subsupplier infrastructure.

---

*Fictitious case

Strategic Redirection: 1996

Como realized that it could no longer compete on quality alone. The marketplace had changed. The strategic plan for Como was now based upon maintaining an industry leadership position well into the twenty-first century. The four basic elements of the strategic plan included:

- First to market (faster development and tooling of the right products)
- Flexible processes (quickly adaptable to model changes)
- Flexible products (multiple niche products from shared platforms and a quick-to-change methodology)
- Lean manufacturing (low cost, high quality, speed, and global economies of scale)

The implementation of the strategy mandated superior project management performance, but changing a 60-year culture to support project management would not be an easy task.

The president of the company established a task force to identify the cultural issues of converting over to an informal project management system. The president believed that project management would eventually become the culture and, therefore, that the cultural issues must be addressed first. The following list of cultural issues was identified by the task force:

- Existing technical, functional departments currently do not adequately support the systemic nature of projects as departmental and individual objectives are not consistent with those of the project and the customer.
- Senior management must acknowledge the movement away from traditional, "over the fence," management and openly endorse the significance of project management, teamwork, and delegation of authority as the future.
- The company must establish a system of project sponsorship to support project managers by trusting them with the responsibility and then empowering them to be successful.
- The company must educate managers in project and risk management and the cultural changes of cross-functional project support; it is in the manager's self interest to support the project manager by providing necessary resources and negotiating for adequate time to complete the work.
- The company must enhance information systems to provide cost and schedule performance information for decision-making and problem resolution.
- Existing informal culture can be maintained while utilizing project management to monitor progress and review costs. Bureaucracy, red tape, and lost time must be eliminated through project management's enhanced communications, standard practices, and goal congruence.

The task force, as a whole, supported the idea of informal project management and believed that all of the cultural issues could be overcome. The task force identified four critical risks and the method of resolution:

- Trusting others and the system.
  - *Resolution:* Training in the process of project management and understanding of the benefits. Interpersonal training to learn to trust in each other and in keeping commitments will begin the cultural change.
- Transform 60 years of tradition in vertical reporting into horizontal project management.
  - *Resolution:* Senior management sponsor the implementation program, participate in training, and fully support efforts to implement project management across functional lines with encouragement and patience as new organizational relationships are forged.

- Capacity constraints and competition for resources.
  - *Resolution:* Work with managers to understand constraints and to develop alternative plans for success. Develop alternative external capacity to support projects.
- Inconsistency in application after introduction.
  - *Resolution:* Set the clear expectation that project management is the operational culture and the responsibility of each manager. Set the implementation of project management as a key measurable for management incentive plans. Establish a model project and recognize the efforts and successes as they occur.

The president realized that project management and strategic planning were related. The president wondered what would happen if the business base would grow as anticipated. Could project management excellence enhance the business base even further? To answer this question, the president prepared a list of competitive advantages that could be achieved through superior project management performance:

- Project management techniques and skills must be enhanced, especially for the larger, complex projects.
- Development of broader component and tooling supply bases would provide for additional capacity.
- Enhanced profitability would be possible through economies of scale to utilize project managers and skilled trades resources more efficiently through balanced workloads and level production.
- Greater purchasing leverage would be possible through larger purchasing volume and sourcing opportunities.
- Disciplined coordination, reporting of project status and proactive project management problem-solving must exist to meet timing schedules, budgets, and customer expectations.
- Effective project management of multitiered supply base will support sales growth beyond existing, capital intensive, internal tooling, and production capacities.

The wheels were set in motion. The president and his senior staff met with all of the employees of Como Tool and Die to discuss the implementation of project management. The president made it clear that he wanted a mature project management system in place within 36 months.

Questions

1. Does Como have a choice in whether or not to accept project management as a culture?
2. How much influence should a customer be able to exert on how the contractors manage projects?
3. Was Como correct in attacking the cultural issues first?
4. Does the time frame of 36 months seem practical?
5. What chance of success do you give Como?
6. What dangers exist when your customers are more knowledgeable than you are concerning project management?

Case 13

Como Tool and Die (B)*

By 1997, Como had achieved partial success in implementing project management. Lead times were reduced by 10 percent rather than the target of 25–35 percent. Internal resources were reduced by only 5 percent. The reduction in prototype time and cost was 15 percent rather than the expected 30–35 percent.

Como's automotive customers were not pleased with the slow progress and relatively immature performance of Como's project management system. Change was taking place, but not fast enough to placate the customers. Como was on target according to its 36 month schedule to achieve some degree of excellence in project management, but would its customers be willing to wait another two years for completion, or should Como try to accelerate the schedule?

Ford Introduces "Chunk" Management

In the summer of 1997, Ford announced to its suppliers that it was establishing a "chunk" management system. All new vehicle metal structures would be divided into three or four major portions with each chosen supplier (i.e., chunk manager) responsible for all components within that portion of the vehicle. To reduce lead time at Ford and to gain supplier commitment, Ford announced that advanced placement of new work (i.e., chunk managers) would take place without competitive bidding. Target agreements on piece price, tooling cost, and lead time would be established and equitably negotiated later with value engineering work acknowledged.

Chunk managers would be selected based upon superior project management capability, including program management skills, coordination responsibility, design feasibility, prototypes, tooling, testing, process sampling, and start of production for components and subassemblies. Chunk managers would function as the second tier component suppliers and coordinate vehicle build for multiple, different vehicle projects at varied stages in the development–tool–launch process.

*Fictitious case.

Strategic Redirection: 1997

Ford Motor Company stated that the selection of the chunk managers would not take place for another year. Unfortunately, Como's plan to achieve excellence would not have been completed by then and its chances to be awarded a chunk management slot were slim.

The automotive division of Como was now at a critical junction. Como's management believed that the company could survive as a low-level supplier of parts, but its growth potential would be questionable. Chunk managers might find it cost-effective to become vertically integrated and produce for themselves the same components that Como manufactured. This could have devastating results for Como. This alternative was unacceptable.

The second alternative required that Como make it clear to Ford Motor Company that Como wished to be considered for a chunk manager contract. If Como were to be selected, then Como's project management systems would have to:

- Provide greater coordination activities than previously anticipated
- Integrate concurrent engineering practices into the company's existing methodology for project management
- Decentralize the organization so as to enhance the working relationship with the customers
- Plan for better resource allocation so as to achieve a higher level of efficiency
- Force proactive planning and decision-making
- Drive out waste and lower cost while improving on-time delivery

There were also serious risks if Como were to become a chunk manager. The company would be under substantially more pressure to meet cost and delivery targets. Most of its resources would have to be committed to complex coordination activities rather than new product development. Therefore, value-added activities for its customers would be diminished. Finally, if Como failed to live up to its customers' expectations as a chunk manager, it might end up losing all automotive work.

The decision was made to inform Ford of Como's interest in chunk management. Now Como realized that its original three-year plan for excellence in project management would have to be completed in 18 months. The question on everyone's mind was: "How?"

Questions

1. What was the driving force for excellence before the announcement of chunk management, and what is it now?
2. How can Como accelerate the learning process to achieve excellence in project management? What steps should management take based upon their learning so far?
3. What are their chances for success? Justify your answer.
4. Should Como compete to become a chunk manager?

Case 14

Apache Metals, Inc.

Apache Metals is an original equipment manufacturer of metal working equipment. The majority of Apache's business is as a supplier to the automotive, appliance, and building products industries. Each production line is custom-designed according to application, industry, and customer requirements.

Project managers are assigned to each purchase order only after the sales department has a signed contract. The project managers can come from anywhere within the company. Basically, anyone can be assigned as a project leader. The assigned project leaders can be responsible for as many as ten purchase orders at one time.

In the past, there has not been enough emphasis on project management. At one time, Apache even assigned trainees to perform project coordination. All failed miserably. At one point, sales dropped to an all-time low, and cost overruns averaged 20–25 percent per production line.

In January 1997, the Board of Directors appointed a new senior management team that would drive the organization to excellence in project management. Project managers were added through recruitment efforts and a close examination of existing personnel. Emphasis was on individuals with good people and communication skills.

The following steps were implemented to improve the quality and effectiveness of the project management system:

- Outside formal training for project managers
- Development of an apprenticeship program for future project managers
- Modification of the current methodology to put the project manager at the focal point
- Involvement of project managers to a greater extent with the customer

Questions

1. What problems can you see in the way project managers were assigned in the past?
2. Will the new approach taken in 1997 put the company on a path to excellence in project management?
3. What skill set would be ideal for the future project managers at Apache Metals?
4. What overall cultural issues must be considered in striving for excellence in project management?
5. What time frame would be appropriate to achieve excellence in project management? What assumptions must be made?

Case 15

Cordova Research Group

Cordova Research Group spent more than 30 years conducting pure and applied research for a variety of external customers. With the reduction, however, in R&D funding, Cordova decided that the survival of the firm would be based upon becoming a manufacturing firm as well as performing R&D. The R&D culture was close to informal project management with the majority of the personnel holding advanced degrees in technical disciplines. To enter the manufacturing arena would require hiring hundreds of new employees, mostly nondegreed.

Questions

1. What strategic problems must be solved?
2. What project management problems must be solved?
3. What time frame is reasonable?
4. If excellence can be achieved, would it occur most likely using formal or informal project management?

Case 16
Cortez Plastics

Cortez Plastics was having growing pains. As the business base of the company began to increase, more and more paperwork began to flow through the organization. The "informal" project management culture that had worked so well in the past was beginning to deteriorate and was being replaced by a more formal project management approach. Recognizing the cost implications of a more formal project management approach, senior management at Cortez Plastics decided to take some action.

Questions

1. How can a company maintain informal project management during periods of corporate growth?
2. If the organization persists in creeping toward formal project management, what can be done to return to a more informal approach?
3. How would you handle a situation where only a few managers or employees are promoting the more formal approach?

Case 17

Haller Specialty Manufacturing

For the past several years, Haller has been marginally successful as a specialty manufacturer of metal components. Sales would quote a price to the customer. Upon contract award, engineering would design the product. Manufacturing had the responsibility to produce the product as well as shipping the product to the customer. Manufacturing often changed the engineering design package to fit manufacturing capabilities.

The vice president of manufacturing was perhaps the most powerful position in the company next to the president. Manufacturing was considered to be the main contributor to corporate profits. Strategic planning was dominated by manufacturing.

To get closer to the customer, Haller implemented project management. Unfortunately, the vice president for manufacturing would not support project management for fear of a loss of power and authority.

Questions

1. If the vice president for manufacturing is a hindrance to excellence, how should this situation be handled?
2. Would your answer to the above question be different if the resistance came from middle or lower level management?

Case 18

Macon, Inc.

Macon was a 50-year-old company in the business of developing test equipment for the tire industry. The company had a history of segregated departments with very focused functional line managers. The company had two major technical departments: mechanical engineering and electrical engineering. Both departments reported to a vice president for engineering, whose background was always mechanical engineering. For this reason, the company focused all projects from a mechanical engineering perspective. The significance of the test equipment's electrical control system was often minimized when, in reality, the electrical control systems were what made Macon's equipment outperform that of the competition.

Because of the strong autonomy of the departments, internal competition existed. Line managers were frequently competing with one another rather than focusing on the best interest of Macon. Each would hope the other would be the cause for project delays instead of working together to avoid project delays altogether. Once dates slipped, fingers were pointed and the problem would worsen over time.

One of Macon's customers had a service department that always blamed engineering for all of their problems. If the machine was not assembled correctly, it was engineering's fault for not documenting it clearly enough. If a component failed, it was engineering's fault for not designing it correctly. No matter what problem occurred in the field, customer service would always put the blame on engineering.

As might be expected, engineering would blame most problems on production claiming that production did not assemble the equipment correctly and did not maintain the proper level of quality. Engineering would design a product and then throw it over the fence to production without ever going down to the manufacturing floor to help with its assembly. Errors or suggestions reported from production to engineering were being ignored. Engineers often perceived the assemblers as incapable of improving the design.

Production ultimately assembled the product and shipped it out to the customer. Oftentimes during assembly the production people would change the design as they saw fit without involving engineering. This would cause severe problems with documentation. Customer service would later inform engineering that the documentation was incorrect, once again causing conflict among all departments.

The president of Macon was a strong believer in project management. Unfortunately, his preaching fell upon deaf ears. The culture was just too strong. Projects were failing miserably. Some failures were attributed to the lack of sponsorship or commitment from line managers. One project failed as the result of a project leader who failed to control scope. Each day the project would fall further behind because work was being added with very little regard for the project's completion date. Project estimates were based upon a "gut feel" rather than upon sound quantitative data.

The delay in shipping dates was creating more and more frustration for the customers. The customers began assigning their own project managers as "watchdogs" to look out for their companies' best interests. The primary function of these "watchdog" project managers was to ensure that the equipment purchased would be delivered on time and complete. This involvement by the customers was becoming more prominent than ever before.

The president decided that action was needed to achieve some degree of excellence in project management. The question was what action to take, and when.

Questions

1. Where will the greatest resistance for excellence in project management come from?
2. What plan should be developed for achieving excellence in project management?
3. How long will it take to achieve some degree of excellence?
4. Explain the potential risks to Macon if the customer's experience with project management increases while Macon's knowledge remains stagnant.

Case 19

Jones and Shephard Accountants*

By 1970, Jones and Shephard Accountants, Inc. (J&S) was ranked eighteenth in size by the American Association of Accountants. In order to compete with the larger firms, J&S formed an Information Services Division (ISO), designed primarily for studies and analyses. By 1975, the ISD had 15 employees.

In 1977, the ISD purchased three minicomputers. With this increased capacity, J&S expanded its services to help satisfy the needs of outside customers. By September 1978, the internal and external workloads had increased to a point where the ISD employed over 50 people.

The director of the division was very disappointed in the way that activities were being handled. There was no single person assigned to push through a project, and outside customers did not know whom to call to get answers regarding project status. The director found that most of his time was being spent on day-to-day activities, such as conflict resolution, instead of on strategic planning and policy formulation.

The biggest problems facing the director were the two continuous internal projects (called Project X and Project Y, for simplicity) that required month-end data collation and reporting. The director felt that these two projects were important enough to require a full-time project manager on each effort.

In October 1978, corporate management announced that the ISD director would be reassigned on February 1, 1979, and that the announcement of his replacement would not be made until the middle of January. The same week that the announcement was made, two individuals were hired from outside the company to take charge of Project X and Project Y. Exhibit I shows the organizational structure of the ISD.

---

*Reprinted from H. Kerzner, *Project Management: A Systems Approach to Planning, Scheduling, and Controlling,* 8th ed. New York: Wiley, 2003, pp. 136–138.

Exhibit I. ISD organizational chart

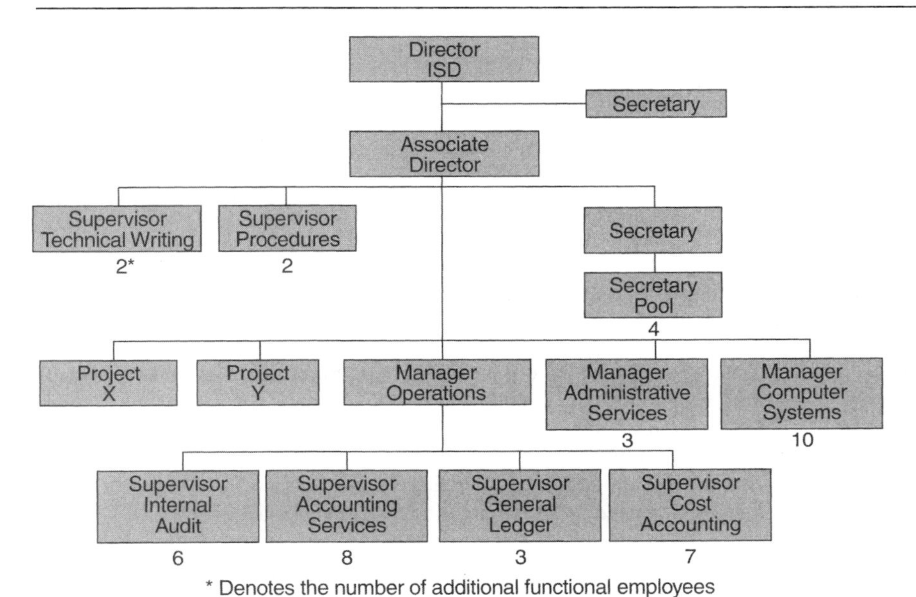

Within the next 30 days, rumors spread throughout the organization about who would become the new director. Most people felt that the position would be filled from within the division and that the most likely candidates were the two new project managers. In addition, the associate director was due to retire in December, thus creating two openings.

On January 3, 1979, a confidential meeting was held between the ISD director and the systems manager.

ISD Director: Corporate has approved my request to promote you to division director. Unfortunately, your job will not be an easy one. You're going to have to restructure the organization somehow so that our employees will not have as many conflicts as they are now faced with. My secretary is typing up a confidential memo for you explaining my observations on the problems within our division.

Remember, your promotion should be held in the strictest confidence until the final announcement later this month. I'm telling you this now so that you can begin planning the restructuring. My memo should help you. (See Exhibit II for the memo.)

The systems manager read the memo and, after due consideration, decided that some form of matrix structure would be best. To help him structure the organization properly, an outside consultant was hired to help identify the potential problems with changing over to a matrix. The following problem areas were identified by the consultant:

1. The operations manager controls more than 50 percent of the people resources. You might want to break up his empire. This will have to be done very carefully.
2. The secretary pool is placed too high in the organization.

3. The supervisors who now report to the associate director will have to be reassigned lower in the organization if the associate director's position is abolished.
4. One of the major problem areas will be trying to convince corporate management that this change will be beneficial. You'll have to convince them that this change can be accomplished without having to increase division manpower.
5. You might wish to set up a separate department or a separate project for customer relations.
6. Introducing your employees to the matrix will be a problem. Each employee will look at the change differently. Most people have the tendency of looking first at the shift in the balance of power—have I gained or have I lost power and status?

The systems manager evaluated the consultant's comments and then prepared a list of questions to ask the consultant at their next meeting:

1. What should the new organizational structure look like? Where should I put each person, specifically the managers?
2. When should I announce the new organizational change? Should it be at the same time as my appointment or at a later date?
3. Should I invite any of my people to provide input to the organizational restructuring? Can this be used as a technique to ease power plays?
4. Should I provide inside or outside seminars to train my people for the new organizational structure? How soon should they be held?

Exhibit II. Confidential memo

From: ISD Director
To: Systems Manager
Date: January 3, 1979

Congratulations on your promotion to division director. I sincerely hope that your tenure will be productive both personally and for corporate. I have prepared a short list of the major obstacles that you will have to consider when you take over the controls.

1. Both Project X and Project Y managers are highly competent individuals. In the last four or five days, however, they have appeared to create more conflicts for us than we had previously. This could be my fault for not delegating them sufficient authority, or could be a result of the fact that several of our people consider these two individuals as prime candidates for my position. In addition, the operations manager does not like other managers coming into his "empire" and giving direction.
2. I'm not sure that we even need an associate director. That decision will be up to you.
3. Corporate has been very displeased with our inability to work with outside customers. You must consider this problem with any organizational structure you choose.
4. The corporate strategic plan for our division contains an increased emphasis on special, internal MIS projects. Corporate wants to limit our external activities for a while until we get our internal affairs in order.
5. I made the mistake of changing our organizational structure on a day-to-day basis. Perhaps it would have been better to design a structure that could satisfy advanced needs, especially one that we can grow into.

Case 20

The Trophy Project*

The ill-fated Trophy Project as in trouble right from the start. Reichart, who had been an assistant project manager, was involved with the project from its conception. When the Trophy Project was accepted by the company, Reichart was assigned as the project manager. The program schedules started to slip from day one, and expenditures were excessive. Reichart found that the functional managers were charging direct labor time to his project but working on their own "pet" projects. When Reichart complained of this, he was told not to meddle in the functional manager's allocation of resources and budgeted expenditures. After approximately six months, Reichart was requested to make a progress report directly to corporate and division staffs.

Reichart took this opportunity to bare his soul. The report substantiated that the project was forecasted to be one complete year behind schedule. Reichart's staff, as supplied by the line managers, was inadequate to stay at the required pace, let alone make up any time that had already been lost. The estimated cost at completion at this interval showed a cost overrun of at least 20 percent. This was Reichart's first opportunity to tell his story to people who were in a position to correct the situation. The result of Reichart's frank, candid evaluation of the Trophy Project was very predictable. Nonbelievers finally saw the light, and the line managers realized that they had a role to play in the completion of the project. Most of the problems were now out in the open and could be corrected by providing adequate staffing and resources. Corporate staff ordered immediate remedial action and staff support to provide Reichart a chance to bail out his program.

The results were not at all what Reichart had expected. He no longer reported to the project office; he now reported directly to the operations manager. Corporate staff's interest in the

*Reprinted from H. Kerzner, *Project Management: A Systems Approach to Planning, Scheduling, and Controlling,* 8th ed. New York: Wiley, 2003, pp. 253–254.

project became very intense, requiring a 7:00 A.M. meeting every Monday morning for complete review of the project status and plans for recovery. Reichart found himself spending more time preparing paperwork, reports, and projections for his Monday morning meetings than he did administering the Trophy Project. The main concern of corporate was to get the project back on schedule. Reichart spent many hours preparing the recovery plan and establishing manpower requirements to bring the program back onto the original schedule.

Group staff, in order to closely track the progress of the Trophy Project, assigned an assistant program manager. The assistant program manager determined that a sure cure for the Trophy Project would be to computerize the various problems and track the progress through a very complex computer program. Corporate provided Reichart with 12 additional staff members to work on the computer program. In the meantime, nothing changed. The functional managers still did not provide adequate staff for recovery, assuming that the additional manpower Reichart had received from corporate would accomplish that task.

After approximately $50,000 was spent on the computer program to track the problems, it was found that the program objectives could not be handled by the computer. Reichart discussed this problem with a computer supplier and found that $15,000 more was required for programming and additional storage capacity. It would take two months for installation of the additional storage capacity and the completion of the programming. At this point, the decision was made to abandon the computer program.

Reichart was now a year and a half into the program with no prototype units completed. The program was still nine months behind schedule with the overrun projected at 40 percent of budget. The customer had been receiving his reports on a timely basis and was well aware of the fact that the Trophy Project was behind schedule. Reichart had spent a great deal of time with the customer explaining the problems and the plan for recovery. Another problem that Reichart had to contend with was that the vendors who were supplying components for the project were also running behind schedule.

One Sunday morning, while Reichart was in his office putting together a report for the client, a corporate vice president came into his office. "Reichart," he said, "in any project I look at the top sheet of paper and the man whose name appears at the top of the sheet is the one I hold responsible. For this project your name appears at the top of the sheet. If you cannot bail this thing out, you are in serious trouble in this corporation." Reichart did not know which way to turn or what to say. He had no control over the functional managers who were creating the problems, but he was the person who was being held responsible.

After another three months the customer, becoming impatient, realized that the Trophy Project was in serious trouble and requested that the division general manager and his entire staff visit the customer's plant to give a progress and "get well" report within a week. The division general manager called Reichart into his office and said, "Reichart, go visit our customer. Take three or four functional line people with you and try to placate him with whatever you feel is necessary." Reichart and four functional line people visited the customer and gave a four-and-a-half-hour presentation defining the problems and the progress to that point. The customer was very polite and even commented that it was an excellent presentation, but the content was totally unacceptable. The program was still six to eight months late, and the customer demanded progress reports on a weekly basis. The customer made arrangements to assign a representative in Reichart's department to be "on-site" at the project on a daily basis and to interface with Reichart and his staff as required. After this turn of events, the program became very hectic.

The customer representative demanded constant updates and problem identification and then became involved in attempting to solve these problems. This involvement created many changes in the program and the product in order to eliminate some of the problems. Reichart had trouble with the customer and did not agree with the changes in the program. He expressed his disagreement vocally when, in many cases, the customer felt the changes were at no cost. This caused a deterioration of the relationship between client and producer.

One morning Reichart was called into the division general manager's office and introduced to Mr. "Red" Baron. Reichart was told to turn over the reins of the Trophy Project to Red immediately. "Reichart, you will be temporarily reassigned to some other division within the corporation. I suggest you start looking outside the company for another job." Reichart looked at Red and asked, "Who did this? Who shot me down?"

Red was program manager on the Trophy Project for approximately six months, after which, by mutual agreement, he was replaced by a third project manager. The customer reassigned his local program manager to another project. With the new team the Trophy Project was finally completed one year behind schedule and at a 40 percent cost overrun.

Questions

1. Did the project appear to be planned correctly?
2. Did functional management seem to be committed to the project?
3. Did senior management appear supportive and committed?

Case 21

The Blue Spider Project*

"This is impossible! Just totally impossible! Ten months ago I was sitting on top of the world. Upper-level management considered me one of the best, if not the best, engineer in the plant. Now look at me! I have bags under my eyes, I haven't slept soundly in the last six months, and here I am, cleaning out my desk. I'm sure glad they gave me back my old job in engineering. I guess I could have saved myself a lot of grief and aggravation had I not accepted the promotion to project manager."

History

Gary Anderson had accepted a position with Parks Corporation right out of college. With a Ph.D. in mechanical engineering, Gary was ready to solve the world's most traumatic problems. At first, Parks Corporation offered Gary little opportunity to do the pure research that he eagerly wanted to undertake. However, things soon changed. Parks grew into a major electronics and structural design corporation during the big boom of the late 1950s and early 1960s when Department of Defense (DoD) contracts were plentiful.

Parks Corporation grew from a handful of engineers to a major DoD contractor, employing some 6,500 people. During the recession of the late 1960s, money became scarce and major layoffs resulted in lowering the employment level to 2,200 employees. At that time, Parks decided to get out of the R&D business and compete as a low-cost production facility while maintaining an engineering organization solely to support production requirements.

After attempts at virtually every project management organizational structure, Parks Corporation selected the matrix form. Each project had a program manager who reported to the director of program management. Each project also maintained an assistant project manager—normally a project engineer—who reported directly to the project manager and indirectly to the

*Reprinted from H. Kerzner, *Project Management: A Systems Approach to Planning, Scheduling, and Controlling,* 6th ed. New York: Wiley, 1998, pp. 494–505.

director of engineering. The program manager spent most of his time worrying about cost and time, whereas the assistant program manager worried more about technical performance.

With the poor job market for engineers, Gary and his colleagues began taking coursework toward MBA degrees in case the job market deteriorated further.

In 1975, with the upturn in DoD spending, Parks had to change its corporate strategy. Parks had spent the last seven years bidding on the production phase of large programs. Now, however, with the new evaluation criteria set forth for contract awards, those companies winning the R&D and qualification phases had a definite edge on being awarded the production contract. The production contract was where the big profits could be found. In keeping with this new strategy, Parks began to beef up its R&D engineering staff. By 1978, Parks had increased in size to 2,700 employees. The increase was mostly in engineering. Experienced R&D personnel were difficult to find for the salaries that Parks was offering. Parks was, however, able to lure some employees away from the competitors, but relied mostly upon the younger, inexperienced engineers fresh out of college.

With the adoption of this corporate strategy, Parks Corporation administered a new wage and salary program that included job upgrading. Gary was promoted to senior scientist, responsible for all R&D activities performed in the mechanical engineering department. Gary had distinguished himself as an outstanding production engineer during the past several years, and management felt that his contribution could be extended to R&D as well.

In January 1978, Parks Corporation decided to compete for Phase I of the Blue Spider Project, an R&D effort that, if successful, could lead into a $500 million program spread out over 20 years. The Blue Spider Project was an attempt to improve the structural capabilities of the Spartan missile, a short-range tactical missile used by the Army. The Spartan missile was exhibiting fatigue failure after six years in the field. This was three years less than what the original design specifications called for. The Army wanted new materials that could result in a longer life for the Spartan missile.

Lord Industries was the prime contractor for the Army's Spartan Program. Parks Corporation would be a subcontractor to Lord if they could successfully bid and win the project. The criteria for subcontractor selection were based not only on low bid, but also on technical expertise as well as management performance on other projects. Park's management felt that they had a distinct advantage over most of the other competitors because they had successfully worked on other projects for Lord Industries.

The Blue Spider Project Kickoff

On November 3, 1977, Henry Gable, the director of engineering, called Gary Anderson into his office.

Henry Gable: Gary, I've just been notified through the grapevine that Lord will be issuing the RFP for the Blue Spider Project by the end of this month, with a 30-day response period. I've been waiting a long time for a project like this to come along so that I can experiment with some new ideas that I have. This project is going to be my baby all the way! I want you to head up the proposal team. I think it must be an engineer. I'll make sure that you get a good proposal manager to help you. If we start working now, we can get close to two months of research in before proposal submittal. That will give us a one-month's edge on our competitors.

Gary was pleased to be involved in such an effort. He had absolutely no trouble in getting functional support for the R&D effort necessary to put together a technical proposal. All of the functional managers continually remarked to Gary, "This must be a biggy. The director of engineering has thrown all of his support behind you."

On December 2, the RFP was received. The only trouble area that Gary could see was that the technical specifications stated that all components must be able to operate normally and successfully through a temperature range of −65°F to 145°F. Current testing indicated the Parks Corporation's design would not function above 130°F. An intensive R&D effort was conducted over the next three weeks. Everywhere Gary looked, it appeared that the entire organization was working on his technical proposal.

A week before the final proposal was to be submitted, Gary and Henry Gable met to develop a company position concerning the inability of the preliminary design material to be operated above 130°F.

Gary Anderson: Henry, I don't think it is going to be possible to meet specification requirements unless we change our design material or incorporate new materials. Everything I've tried indicates we're in trouble.

Gable: We're in trouble only if the customer knows about it. Let the proposal state that we expect our design to be operative up to 155°F. That'll please the customer.

Anderson: That seems unethical to me. Why don't we just tell them the truth?

Gable: The truth doesn't always win proposals. I picked you to head up this effort because I thought that you'd understand. I could have just as easily selected one of our many moral project managers. I'm considering you for program manager after we win the program. If you're going to pull this conscientious crap on me like the other project managers do, I'll find someone else. Look at it this way; later we can convince the customer to change the specifications. After all, we'll be so far downstream that he'll have no choice.

After two solid months of 16-hour days for Gary, the proposal was submitted. On February 10, 1978, Lord Industries announced that Parks Corporation would be awarded the Blue Spider Project. The contract called for a ten-month effort, negotiated at $2.2 million at a firm-fixed price.

Selecting the Project Manager

Following contract award, Henry Gable called Gary in for a conference.

Gable: Congratulations, Gary! You did a fine job. The Blue Spider Project has great potential for ongoing business over the next ten years, provided that we perform well during the R&D phase. Obviously you're the most qualified person in the plant to head up the project. How would you feel about a transfer to program management?

Anderson: I think it would be a real challenge. I could make maximum use of the MBA degree I earned last year. I've always wanted to be in program management.

Gable: Having several masters' degrees, or even doctorates for that matter, does not guarantee that you'll be a successful project manager. There are three requirements for effective program management: You must be able to communicate both in writing and orally; you must know how to motivate people; and you must be willing to give up your car pool. The last one is extremely important in that program managers must be totally committed and dedicated to the program, regardless of how much time is involved.

But this is not the reason why I asked you to come here. Going from project engineer to program management is a big step. There are only two places you can go from program management—up the organization or out the door. I know of very, very few engineers who failed in program management and were permitted to return.

Anderson: Why is that? If I'm considered to be the best engineer in the plant, why can't I return to engineering?

Gable: Program management is a world of its own. It has its own formal and informal organizational ties. Program managers are outsiders. You'll find out. You might not be able to keep the strong personal ties you now have with your fellow employees. You'll have to force even your best friends to comply with your standards. Program managers can go from program to program, but functional departments remain intact.

I'm telling you all this for a reason. We've worked well together the past several years. But if I sign the release so that you can work for Grey in program management, you'll be on your own, like hiring into a new company. I've already signed the release. You still have some time to think about it.

Anderson: One thing I don't understand. With all of the good program managers we have here, why am I given this opportunity?

Gable: Almost all of our program managers are over forty-five years old. This resulted from our massive layoffs several years ago when we were forced to lay off the younger, inexperienced program managers. You were selected because of your age and because all of our other program managers have worked only on production-type programs. We need someone at the reins who knows R&D. Your counterpart at Lord Industries will be an R&D type. You have to fight fire with fire.

I have an ulterior reason for wanting you to accept this position. Because of the division of authority between program management and project engineering, I need someone in program management whom I can communicate with concerning R&D work. The program managers we have now are interested only in time and cost. We need a manager who will bend over backwards to get performance also. I think you're that man. You know the commitment we made to Lord when we submitted that proposal. You have to try to achieve that. Remember, this program is my baby. You'll get all the support you need. I'm tied up on another project now. But when it's over, I'll be following your work like a hawk. We'll have to get together occasionally and discuss new techniques.

Take a day or two to think it over. If you want the position, make an appointment to see Elliot Grey, the director of program management. He'll give you the same speech I did. I'll assign Paul Evans to you as chief project engineer. He's a seasoned veteran and you should have no trouble working with him. He'll give you good advice. He's a good man.

The Work Begins

Gary accepted the new challenge. His first major hurdle occurred in staffing the project. The top priority given to him to bid the program did not follow through for staffing. The survival of Parks Corporation depended on the profits received from the production programs. In keeping with this philosophy Gary found that engineering managers (even his former boss) were reluctant to give up their key people to the Blue Spider Program. However, with a little support from Henry Gable, Gary formed an adequate staff for the program.

Right from the start Gary was worried that the test matrix called out in the technical volume of the proposal would not produce results that could satisfy specifications. Gary had 90 days after go-ahead during which to identify the raw materials that could satisfy specification requirements. Gary and Paul Evans held a meeting to map out their strategy for the first few months.

Anderson: Well, Paul, we're starting out with our backs against the wall on this one. Any recommendations?

Paul Evans: I also have my doubts about the validity of this test matrix. Fortunately, I've been through this before. Gable thinks this is his project and he'll sure as hell try to manipulate us. I have to report to him every morning at 7:30 A.M. with the raw data results of the previous day's testing. He wants to see it before you do. He also stated that he wants to meet with me alone.

Lord will be the big problem. If the test matrix proves to be a failure, we're going to have to change the scope of effort. Remember, this is an FFP contract. If we change the scope of work and do additional work in the earlier phases of the program, then we should prepare a trade-off analysis to see what we can delete downstream so as to not overrun the budget.

Anderson: I'm going to let the other project office personnel handle the administrating work. You and I are going to live in the research labs until we get some results. We'll let the other project office personnel run the weekly team meetings.

For the next three weeks Gary and Paul spent virtually 12 hours per day, 7 days a week, in the research and development lab. None of the results showed any promise. Gary kept trying to set up a meeting with Henry Gable but always found him unavailable.

During the fourth week, Gary, Paul, and the key functional department managers met to develop an alternate test matrix. The new test matrix looked good. Gary and his team worked frantically to develop a new workable schedule that would not have impact on the second milestone, which was to occur at the end of 180 days. The second milestone was the final acceptance of the raw materials and preparation of production runs of the raw materials to verify that there would be no scale-up differences between lab development and full-scale production.

Gary personally prepared all of the technical handouts for the interchange meeting. After all, he would be the one presenting all of the data. The technical interchange meeting was scheduled for two days. On the first day, Gary presented all of the data, including test results, and the new test matrix. The customer appeared displeased with the progress to date and decided to have its own in-house caucus that evening to go over the material that was presented.

The following morning the customer stated its position: "First of all, Gary, we're quite pleased to have a project manager who has such a command of technology. That's good. But every time we've tried to contact you last month, you were unavailable or had to be paged in the research laboratories. You did an acceptable job presenting the technical data, but the administrative data was presented by your project office personnel. We, at Lord, do not think that you're maintaining the proper balance between your technical and administrative responsibilities. We prefer that you personally give the administrative data and your chief project engineer present the technical data.

"We did not receive any agenda. Our people like to know what will be discussed, and when. We also want a copy of all handouts to be presented at least three days in advance. We need time to scrutinize the data. You can't expect us to walk in here blind and make decisions after seeing the data for ten minutes.

"To be frank, we feel that the data to date is totally unacceptable. If the data does not improve, we will have no choice but to issue a work stoppage order and look for a new contractor. The new test matrix looks good, especially since this is a firm-fixed-price contract. Your company will bear the burden of all costs for the additional work. A trade-off with later work may be possible, but this will depend on the results presented at the second design review meeting, 90 days from now.

"We have decided to establish a customer office at Parks to follow your work more closely. Our people feel that monthly meetings are insufficient during R&D activities. We would like our customer representative to have daily verbal meetings with you or your staff. He will then keep us posted. Obviously, we had expected to review much more experimental data than you have given us.

"Many of our top-quality engineers would like to talk directly to your engineering community, without having to continually waste time by having to go through the project office. We must insist on this last point. Remember, your effort may be only $2.2 million, but our total package is $100 million. We have a lot more at stake than you people do. Our engineers do not like to get information that has been filtered by the project office. They want to help you.

"And last, don't forget that you people have a contractual requirement to prepare complete minutes for all interchange meetings. Send us the original for signature before going to publication."

Although Gary was unhappy with the first team meeting, especially with the requests made by Lord Industries, he felt that they had sufficient justification for their comments. Following the team meeting, Gary personally prepared the complete minutes. "This is absurd," thought Gary. "I've wasted almost one entire week doing nothing more than administrative paperwork. Why do we need such detailed minutes? Can't a rough summary suffice? Why is it that customers want everything documented? That's like an indication of fear. We've been completely cooperative with them. There has been no hostility between us. If we've gotten this much paperwork to do now, I hate to imagine what it will be like if we get into trouble."

A New Role

Gary completed and distributed the minutes to the customer as well as to all key team members.

For the next five weeks testing went according to plan, or at least Gary thought that it had. The results were still poor. Gary was so caught up in administrative paperwork that he hadn't found time to visit the research labs in over a month. On a Wednesday morning, Gary entered the lab to observe the morning testing. Upon arriving in the lab, Gary found Paul Evans, Henry Gable, and two technicians testing a new material, JXB-3.

Gable: Gary, your problems will soon be over. This new material, JXB-3, will permit you to satisfy specification requirements. Paul and I have been testing it for two weeks. We wanted to let you know, but were afraid that if the word leaked out to the customer that we were spending their money for testing materials that were not called out in the program plan, they would probably go crazy and might cancel the contract. Look at these results. They're super!

Anderson: Am I supposed to be the one to tell the customer now? This could cause a big wave.

Gable: There won't be any wave. Just tell them that we did it with our own IR&D funds. That'll please them because they'll think we're spending our own money to support their program.

Before presenting the information to Lord, Gary called a team meeting to present the new data to the project personnel. At the team meeting, one functional manager spoke out: "This is a hell of a way to run a program. I like to be kept informed about everything that's happening here at Parks. How can the project office expect to get support out of the functional departments if we're kept in the dark until the very last minute? My people have been working with the existing materials for the last two months and you're telling us that it was all for nothing. Now you're giving us a material that's so new that we have no information on it whatsoever. We're now going to have to play catch-up, and that's going to cost you plenty."

One week before the 180-day milestone meeting, Gary submitted the handout package to Lord Industries for preliminary review. An hour later the phone rang.

Customer: We've just read your handout. Where did this new material come from? How come we were not informed that this work was going on? You know, of course, that our customer, the Army, will be at this meeting. How can we explain this to them? We're postponing the review meeting until all of our people have analyzed the data and are prepared to make a decision.

The purpose of a review or interchange meeting is to exchange information when *both* parties have familiarity with the topic. Normally, we (Lord Industries) require almost weekly interchange meetings with our other customers because we don't trust them. We disregard this policy with Parks Corporation based on past working relationships. But with the new state of developments, you have forced us to revert to our previous position, since we now question Parks Corporation's integrity in communicating with us. At first we believed this was due to an inexperienced program manager. Now, we're not sure.

Anderson: I wonder if the real reason we have these interchange meetings isn't to show our people that Lord Industries doesn't trust us. You're creating a hell of a lot of work for us, you know.

Customer: You people put yourself in this position. Now you have to live with it.

Two weeks later Lord reluctantly agreed that the new material offered the greatest promise. Three weeks later the design review meeting was held. The Army was definitely not pleased with the prime contractor's recommendation to put a new, untested material into a multimillion-dollar effort.

The Communications Breakdown

During the week following the design review meeting Gary planned to make the first verification mix in order to establish final specifications for selection of the raw materials. Unfortunately, the manufacturing plans were a week behind schedule, primarily because of Gary, since he had decided to reduce costs by accepting the responsibility for developing the bill of materials himself.

A meeting was called by Gary to consider rescheduling of the mix.

Anderson: As you know we're about a week to ten days behind schedule. We'll have to reschedule the verification mix for late next week.

Production Manager: Our resources are committed until a month from now. You can't expect to simply call a meeting and have everything reshuffled for the Blue Spider Program. We should have been notified earlier. Engineering has the responsibility for preparing the bill of materials. Why aren't they ready?

Engineering Integration: We were never asked to prepare the bill of materials. But I'm sure that we could get it out if we work our people overtime for the next two days.

Anderson: When can we remake the mix?

Production Manager: We have to redo at least 500 sheets of paper every time we reschedule mixes. Not only that, we have to reschedule people on all three shifts. If we are to reschedule your mix, it will have to be performed on overtime. That's going to increase your costs. If that's agreeable with you, we'll try it. But this will be the first and last time that production will bail you out. There are procedures that have to be followed.

Testing Engineer: I've been coming to these meetings since we kicked off this program. I think I speak for the entire engineering division when I say that the role that the director of engineering is playing in this program is suppressing individuality among our highly competent personnel. In new projects, especially those involving R&D, our people are not apt to stick their necks out. Now our people are becoming ostriches. If they're impeded from contributing, even in their own slight way, then you'll probably lose them before the project gets completed. Right now I feel that I'm wasting my time here. All I need are minutes of the team meetings and I'll be happy. Then I won't have to come to these pretend meetings anymore.

The purpose of the verification mix was to make a full-scale production run of the material to verify that there would be no material property changes in scale-up from the small mixes made in the R&D laboratories. After testing, it became obvious that the wrong lots of raw materials were used in the production verification mix.

A meeting was called by Lord Industries for an explanation of why the mistake had occurred and what the alternatives were.

Lord: Why did the problem occur?

Anderson: Well, we had a problem with the bill of materials. The result was that the mix had to be made on overtime. And when you work people on overtime, you have to be willing to accept mistakes as being a way of life. The energy cycles of our people are slow during the overtime hours.

Lord: The ultimate responsibility has to be with you, the program manager. We, at Lord, think that you're spending too much time doing and not enough time managing. As the prime contractor, we have a hell of a lot more at stake than you do. From now on we want documented weekly technical interchange meetings and closer interaction by our quality control section with yours.

Anderson: These additional team meetings are going to tie up our key people. I can't spare people to prepare handouts for weekly meetings with your people.

Lord: Team meetings are a management responsibility. If Parks does not want the Blue Spider Program, I'm sure we can find another subcontractor. All you (Gary) have to do is give up taking the material vendors to lunch and you'll have plenty of time for handout preparation.

Gary left the meeting feeling as though he had just gotten raked over the coals. For the next two months, Gary worked sixteen hours a day, almost every day. Gary did not want to burden his staff with the responsibility of the handouts, so he began preparing them himself. He could have hired additional staff, but with such a tight budget, and having to remake verification mix, cost overruns appeared inevitable.

As the end of the seventh month approached, Gary was feeling pressure from within Parks Corporation. The decision-making process appeared to be slowing down and Gary found it more and more difficult to motivate his people. In fact, the grapevine was referring to the Blue

Spider Project as a loser, and some of his key people acted as though they were on a sinking ship.

By the time the eighth month rolled around, the budget had nearly been expended. Gary was tired of doing everything himself. "Perhaps I should have stayed an engineer," thought Gary. Elliot Grey and Gary Anderson had a meeting to see what could be salvaged. Grey agreed to get Gary additional corporate funding to complete the project. "But performance must be met, since there is a lot riding on the Blue Spider Project," asserted Grey. He called a team meeting to identify the program status.

Anderson: It's time to map out our strategy for the remainder of the program. Can engineering and production adhere to the schedule that I have laid out before you?

Team Member, Engineering: This is the first time that I've seen this schedule. You can't expect me to make a decision in the next ten minutes and commit the resources of my department. We're getting a little unhappy being kept in the dark until the last minute. What happened to effective planning?

Anderson: We still have effective planning. We must adhere to the original schedule, or at least try to adhere to it. This revised schedule will do that.

Team Member, Engineering: Look, Gary! When a project gets in trouble it is usually the functional departments that come to the rescue. But if we're kept in the dark, then how can you expect us to come to your rescue? My boss wants to know, well in advance, every decision that you're contemplating with regard to our departmental resources. Right now, we . . .

Anderson: Granted, we may have had a communications problem. But now we're in trouble and have to unite forces. What is your impression as to whether your department can meet the new schedule?

Team Member, Engineering: When the Blue Spider Program first got in trouble, my boss exercised his authority to make all departmental decisions regarding the program himself. I'm just a puppet. I have to check with him on everything.

Team Member, Production: I'm in the same boat, Gary. You know we're not happy having to reschedule our facilities and people. We went through this once before. I also have to check with my boss before giving you an answer about the new schedule.

The following week the verification mix was made. Testing proceeded according to the revised schedule, and it looked as though the total schedule milestones could be met, provided that specifications could be adhered to.

Because of the revised schedule, some of the testing had to be performed on holidays. Gary wasn't pleased with asking people to work on Sundays and holidays, but he had no choice, since the test matrix called for testing to be accomplished at specific times after end-of-mix.

A team meeting was called on Wednesday to resolve the problem of who would work on the holiday, which would occur on Friday, as well as staffing Saturday and Sunday. During the team meeting Gary became quite disappointed. Phil Rodgers, who had been Gary's test engineer since the project started, was assigned to a new project that the grapevine called Gable's new adventure. His replacement was a relatively new man, only eight months with the company. For an hour and a half, the team members argued about the little problems and continually avoided the major question, stating that they would first have to coordinate commitments with their bosses. It was obvious to Gary that his team members were afraid to make major decisions and therefore "ate up" a lot of time on trivial problems.

On the following day, Thursday, Gary went to see the department manager responsible for testing, in hopes that he could use Phil Rodgers this weekend.

Department Manager: I have specific instructions from the boss (director of engineering) to use Phil Rodgers on the new project. You'll have to see the boss if you want him back

Anderson: But we have testing that must be accomplished this weekend. Where's the new man you assigned yesterday?

Department Manager: Nobody told me you had testing scheduled for this weekend. Half of my department is already on an extended weekend vacation, including Phil Rodgers and the new man. How come I'm always the last to know when we have a problem?

Anderson: The customer is flying down his best people to observe this weekend's tests. It's too late to change anything. You and I can do the testing.

Department Manager: Not on your life. I'm staying as far away as possible from the Blue Spider Project. I'll get you someone, but it won't be me. That's for sure!

The weekend's testing went according to schedule. The raw data was made available to the customer under the stipulation that the final company position would be announced at the end of the next month, after the functional departments had a chance to analyze it.

Final testing was completed during the second week of the ninth month. The initial results looked excellent. The materials were within contract specifications, and although they were new, both Gary and Lord's management felt that there would be little difficulty in convincing the Army that this was the way to go. Henry Gable visited Gary and congratulated him on a job well done.

All that now remained was the making of four additional full-scale verification mixes in order to determine how much deviation there would be in material properties between full-sized production-run mixes. Gary tried to get the customer to concur (as part of the original trade-off analysis) that two of the four production runs could be deleted. Lord's management refused, insisting that contractual requirements must be met at the expense of the contractor.

The following week, Elliot Grey called Gary in for an emergency meeting concerning expenditures to date.

Elliot Grey: Gary, I just received a copy of the financial planning report for last quarter in which you stated that both the cost and performance of the Blue Spider Project were 75 percent complete. I don't think you realize what you've done. The target profit on the program was $200,000. Your memo authorized the vice president and general manager to book 75 percent of that, or $150,000, for corporate profit spending for stockholders. I was planning on using all $200,000 together with the additional $300,000 I personally requested from corporate headquarters to bail you out. Now I have to go back to the vice president and general manager and tell them that we've made a mistake and that we'll need an additional $150,000.

Anderson: Perhaps I should go with you and explain my error. Obviously, I take all responsibility.

Grey: No, Gary. It's our error, not yours. I really don't think you want to be around the general manager when he sees red at the bottom of the page. It takes an act of God to get money back once corporate books it as profit. Perhaps you should reconsider project engineering as a career instead of program management. Your performance hasn't exactly been sparkling, you know.

Gary returned to his office quite disappointed. No matter how hard he worked, the bureaucratic red tape of project management seemed always to do him in. But late that afternoon, Gary's disposition improved. Lord Industries called to say that, after consultation with the Army, Parks Corporation would be awarded a sole-source contract for qualification and production of Spartan missile components using the new longer-life raw materials. Both Lord and the Army felt that the sole-source contract was justified, provided that continued testing showed the same results, since Parks Corporation had all of the technical experience with the new materials.

Gary received a letter of congratulations from corporate headquarters, but no additional pay increase. The grapevine said that a substantial bonus was given to the director of engineering.

During the tenth month, results were coming back from the accelerated aging tests performed on the new materials. The results indicated that although the new materials would meet specifications, the age life would probably be less than five years. These numbers came as a shock to Gary. Gary and Paul Evans had a conference to determine the best strategy to follow.

Anderson: Well, I guess we're now in the fire instead of the frying pan. Obviously, we can't tell Lord Industries about these tests. We ran them on our own. Could the results be wrong?

Evans: Sure, but I doubt it. There's always margin for error when you perform accelerated aging tests on new materials. There can be reactions taking place that we know nothing about. Furthermore, the accelerated aging tests may not even correlate well with actual aging. We must form a company position on this as soon as possible.

Anderson: I'm not going to tell anyone about this, especially Henry Gable. You and I will handle this. It will be my throat if word of this leaks out. Let's wait until we have the production contract in hand.

Evans: That's dangerous. This has to be a company position, not a project office position. We had better let them know upstairs.

Anderson: I can't do that. I'll take all responsibility. Are you with me on this?

Evans: I'll go along. I'm sure I can find employment elsewhere when we open Pandora's box. You had better tell the department managers to be quiet also.

Two weeks later, as the program was winding down into the testing for the final verification mix and final report development, Gary received an urgent phone call asking him to report immediately to Henry Gable's office.

Gable: When this project is over, you're through. You'll never hack it as a program manager, or possibly a good project engineer. We can't run projects around here without honesty and open communications. How the hell do you expect top management to support you when you start censoring bad news to the top? I don't like surprises. I like to get the bad news from the program manager and project engineers, not secondhand from the customer. And of course, we cannot forget the cost overrun. Why didn't you take some precautionary measures?

Anderson: How could I when you were asking our people to do work such as accelerated aging tests that would be charged to my project and was not part of program plan? I don't think I'm totally to blame for what's happened.

Gable: Gary, I don't think it's necessary to argue the point any further. I'm willing to give you back your old job, in engineering. I hope you didn't lose too many friends while working in program management. Finish up final testing and the program report. Then I'll reassign you.

Gary returned to his office and put his feet up on the desk. "Well," thought Gary, "perhaps I'm better off in engineering. At least I can see my wife and kids once in a while." As Gary began writing the final report, the phone rang:

Functional Manager: Hello, Gary. I just thought I'd call to find out what charge number you want us to use for experimenting with this new procedure to determine accelerated age life.

Anderson: Don't call me! Call Gable. After all, the Blue Spider Project is his baby.

Questions

1. If you were Gary Anderson, would you have accepted this position after the director stated that this project would be his baby all the way?
2. Do engineers with MBA degrees aspire to high positions in management?
3. Was Gary qualified to be a project manager?
4. What are the moral and ethical issues facing Gary?
5. What authority does Gary Anderson have and to whom does he report?
6. Is it true when you enter project management, you either go up the organization or out the door?
7. Is it possible for an executive to take too much of an interest in an R&D project?
8. Should Paul Evans have been permitted to report information to Gable before reporting it to the project manager?
9. Is it customary for the project manager to prepare all of the handouts for a customer interchange meeting?
10. What happens when a situation of mistrust occurs between the customer and contractor?
11. Should functional employees of the customer and contractor be permitted to communicate with one another without going through the project office?
12. Did Gary demonstrate effective time management?
13. Did Gary understand production operations?
14. Are functional employees authorized to make project decisions?
15. On R&D projects, should profits be booked periodically or at project termination?
16. Should a project manager ever censor bad news?

Case 22

Corwin Corporation*

By June 1983, Corwin Corporation had grown into a $150 million per year corporation with an international reputation for manufacturing low-cost, high-quality rubber components. Corwin maintained more than a dozen different product lines, all of which were sold as off-the-shelf items in department stores, hardware stores, and automotive parts distributors. The name "Corwin" was now synonymous with "quality." This provided management with the luxury of having products that maintained extremely long life cycles.

Organizationally, Corwin had maintained the same structure for more than 15 years (see Exhibit I). The top management of Corwin Corporation was highly conservative and believed in using a marketing approach to find new markets for existing product lines rather than exploring for new products. Under this philosophy, Corwin maintained a small R&D group whose mission was simply to evaluate state-of-the-art technology and its application to existing product lines.

Corwin's reputation was so good that they continually received inquiries about the manufacturing of specialty products. Unfortunately, the conservative nature of Corwin's management created a "do not rock the boat" atmosphere opposed to taking any type of risks. A management policy was established to evaluate all specialty-product requests. The policy required answering yes to the following questions:

- Will the specialty product provide the same profit margin (20 percent) as existing product lines?
- What is the total projected profitability to the company in terms of follow-on contracts?
- Can the specialty product be developed into a product line?

---

*Reprinted from H. Kerzner, *Project Management: A Systems Approach to Planning, Scheduling, and Controlling,* 8th ed. New York: Wiley, 2003, pp. 368–375.

Exhibit I. Organizational chart for Corwin Corporation

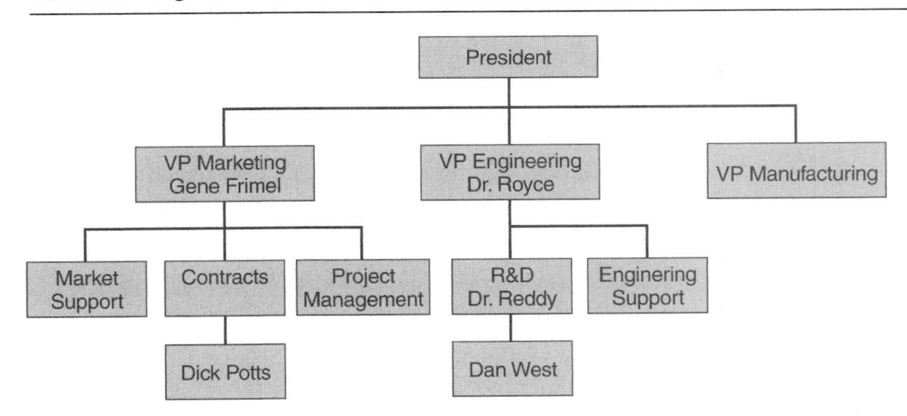

- Can the specialty product be produced with minimum disruption to existing product lines and manufacturing operations?

These stringent requirements forced Corwin to no-bid more than 90 percent of all specialty-product inquiries.

Corwin Corporation was a marketing-driven organization, although manufacturing often had different ideas. Almost all decisions were made by marketing with the exception of product pricing and estimating, which was a joint undertaking between manufacturing and marketing. Engineering was considered as merely a support group to marketing and manufacturing.

For specialty products, the project managers would always come out of marketing even during the R&D phase of development. The company's approach was that if the specialty product should mature into a full product line, then there should be a product line manager assigned right at the onset.

The Peters Company Project

In 1980, Corwin accepted a specialty-product assignment from Peters Company because of the potential for follow-on work. In 1981 and 1982, and again in 1983, profitable follow-on contracts were received, and a good working relationship developed, despite Peters' reputation for being a difficult customer to work with.

On December 7, 1982, Gene Frimel, the vice president of marketing at Corwin, received a rather unusual phone call from Dr. Frank Delia, the marketing vice president at Peters Company.

Frank Delia: Gene, I have a rather strange problem on my hands. Our R&D group has $250,000 committed for research toward development of a new rubber product material, and we simply do not have the available personnel or talent to undertake the project. We have to go outside. We'd like your company to do the work. Our testing and R&D facilities are already overburdened.

Gene Frimel: Well, as you know, Frank, we are not a research group even though we've done this once before for you. And furthermore, I would never be able to sell our management on such an undertaking. Let some other company do the R&D work and then we'll take over on the production end.

Delia: Let me explain our position on this. We've been burned several times in the past. Projects like this generate several patents, and the R&D company almost always requires that our contracts give them royalties or first refusal for manufacturing rights.

Frimel: I understand your problem, but it's not within our capabilities. This project, if undertaken, could disrupt parts of our organization. We're already operating lean in engineering.

Delia: Look, Gene! The bottom line is this: We have complete confidence in your manufacturing ability to such a point that we're willing to commit to a five-year production contract if the product can be developed. That makes it extremely profitable for you.

Frimel: You've just gotten me interested. What additional details can you give me?

Delia: All I can give you is a rough set of performance specifications that we'd like to meet. Obviously, some trade-offs are possible.

Frimel: When can you get the specification sheet to me?

Delia: You'll have it tomorrow morning. I'll ship it overnight express.

Frimel: Good! I'll have my people look at it, but we won't be able to get you an answer until after the first of the year. As you know, our plant is closed down for the last two weeks in December, and most of our people have already left for extended vacations.

Delia: That's not acceptable! My management wants a signed, sealed, and delivered contract by the end of this month. If this is not done, corporate will reduce our budget for 1983 by $250,000, thinking that we've bitten off more than we can chew. Actually, I need your answer within 48 hours so that I'll have some time to find another source.

Frimel: You know, Frank, today is December 7, Pearl Harbor Day. Why do I feel as though the sky is about to fall in?

Delia: Don't worry, Gene! I'm not going to drop any bombs on you. Just remember, all that we have available is $250,000, and the contract must be a firm-fixed-price effort. We anticipate a six-month project with $125,000 paid on contract signing and the balance at project termination.

Frimel: I still have that ominous feeling, but I'll talk to my people. You'll hear from us with a go or no-go decision within 48 hours. I'm scheduled to go on a cruise in the Caribbean, and my wife and I are leaving this evening. One of my people will get back to you on this matter.

Gene Frimel had a problem. All bid and no-bid decisions were made by a four-man committee composed of the president and the three vice presidents. The president and the vice president for manufacturing were on vacation. Frimel met with Dr. Royce, the vice president of engineering, and explained the situation.

Royce: You know, Gene, I totally support projects like this because it would help our technical people grow intellectually. Unfortunately, my vote never appears to carry any weight.

Frimel: The profitability potential as well as the development of good customer relations makes this attractive, but I'm not sure we want to accept such a risk. A failure could easily destroy our good working relationship with Peters Company.

Royce: I'd have to look at the specification sheets before assessing the risks, but I would like to give it a shot.

Frimel: I'll try to reach our president by phone.

By late afternoon, Frimel was fortunate enough to be able to contact the president and received a reluctant authorization to proceed. The problem now was how to prepare a proposal within the next two or three days and be prepared to make an oral presentation to Peters Company.

Frimel: The Boss gave his blessing, Royce, and the ball is in your hands. I'm leaving for vacation, and you'll have total responsibility for the proposal and presentation. Delia wants the presentation this weekend. You should have his specification sheets tomorrow morning.

Royce: Our R&D director, Dr. Reddy, left for vacation this morning. I wish he were here to help me price out the work and select the project manager. I assume that, in this case, the project manager will come out of engineering rather than marketing.

Frimel: Yes, I agree. Marketing should not have any role in this effort. It's your baby all the way. And as for the pricing effort, you know our bid will be for $250,000. Just work backwards to justify the numbers. I'll assign one of our contracting people to assist you in the pricing. I hope I can find someone who has experience in this type of effort. I'll call Delia and tell him we'll bid it with an unsolicited proposal.

Royce selected Dan West, one of the R&D scientists, to act as the project leader. Royce had severe reservations about doing this without the R&D director, Dr. Reddy, being actively involved. With Reddy on vacation, Royce had to make an immediate decision.

On the following morning, the specification sheets arrived and Royce, West, and Dick Potts, a contracts man, began preparing the proposal. West prepared the direct labor man-hours, and Royce provided the costing data and pricing rates. Potts, being completely unfamiliar with this type of effort, simply acted as an observer and provided legal advice when necessary. Potts allowed Royce to make all decisions even though the contracts man was considered the official representative of the president.

Finally completed two days later, the proposal was actually a ten-page letter that simply contained the cost summaries (see Exhibit II) and the engineering intent. West estimated that 30 tests would be required. The test matrix described the test conditions only for the first five tests. The remaining 25 test conditions would be determined at a later date, jointly by Peters and Corwin personnel.

On Sunday morning, a meeting was held at Peters Company, and the proposal was accepted. Delia gave Royce a letter of intent authorizing Corwin Corporation to begin working on the project immediately. The final contract would not be available for signing until late January, and the letter of intent simply stated that Peters Company would assume all costs until such time that the contract was signed or the effort terminated.

West was truly excited about being selected as the project manager and being able to interface with the customer, a luxury that was usually given only to the marketing personnel.

Exhibit II. Proposal cost summaries

Direct labor and support	$ 30,000
Testing (30 tests at $2,000 each)	60,000
Overhead at 100%	90,000
Materials	30,000
G&A (general and administrative, 10%)	21,000
Total	$231,000
Profit	19,000
Total	$250,000

Although Corwin Corporation was closed for two weeks over Christmas, West still went into the office to prepare the project schedules and to identify the support he would need in the other areas, thinking that if he presented this information to management on the first day back to work, they would be convinced that he had everything under control.

The Work Begins . . .

On the first working day in January 1983, a meeting was held with the three vice presidents and Dr. Reddy to discuss the support needed for the project. (West was not in attendance at this meeting, although all participants had a copy of his memo.)

Reddy: I think we're heading for trouble in accepting this project. I've worked with Peters Company previously on R&D efforts, and they're tough to get along with. West is a good man, but I would never have assigned him as the project leader. His expertise is in managing internal rather than external projects. But, no matter what happens, I'll support West the best I can.

Royce: You're too pessimistic. You have good people in your group and I'm sure you'll be able to give him the support he needs. I'll try to look in on the project every so often. West will still be reporting to you for this project. Try not to burden him too much with other work. This project is important to the company.

West spent the first few days after vacation soliciting the support that he needed from the other line groups. Many of the other groups were upset that they had not been informed earlier and were unsure as to what support they could provide. West met with Reddy to discuss the final schedules.

Reddy: Your schedules look pretty good, Dan. I think you have a good grasp on the problem. You won't need very much help from me. I have a lot of work to do on other activities, so I'm just going to be in the background on this project. Just drop me a note every once in a while telling me what's going on. I don't need anything formal. Just a paragraph or two will suffice.

By the end of the third week, all of the raw materials had been purchased, and initial formulations and testing were ready to begin. In addition, the contract was ready for signature. The contract contained a clause specifying that Peters Company had the right to send an in-house representative into Corwin Corporation for the duration of the project. Peters Company informed Corwin that Patrick Ray would be the in-house representative, reporting to Delia, and would assume his responsibilities on or about February 15.

By the time Pat Ray appeared at Corwin Corporation, West had completed the first three tests. The results were not what was expected, but gave promise that Corwin was heading in the right direction. Pat Ray's interpretation of the tests was completely opposite to that of West. Ray thought that Corwin was "way off base," and that redirection was needed.

Pat Ray: Look, Dan! We have only six months to do this effort and we shouldn't waste our time on marginally acceptable data. These are the next five tests I'd like to see performed.

Dan West: Let me look over your request and review it with my people. That will take a couple of days, and, in the meanwhile, I'm going to run the other two tests as planned.

Ray's arrogant attitude bothered West. However, West decided that the project was too important to "knock heads" with Ray and simply decided to cater to Ray the best he could. This was not exactly the working relationship that West expected to have with the in-house representative.

West reviewed the test data and the new test matrix with engineering personnel, who felt that the test data was inconclusive as yet and preferred to withhold their opinion until the results of the fourth and fifth tests were made available. Although this displeased Ray, he agreed to wait a few more days if it meant getting Corwin Corporation on the right track.

The fourth and fifth tests appeared to be marginally acceptable just as the first three had been. Corwin's engineering people analyzed the data and made their recommendations.

West: Pat, my people feel that we're going in the right direction and that our path has greater promise than your test matrix.

Ray: As long as we're paying the bills, we're going to have a say in what tests are conducted. Your proposal stated that we would work together in developing the other test conditions. Let's go with my test matrix. I've already reported back to my boss that the first five tests were failures and that we're changing the direction of the project.

West: I've already purchased $30,000 worth of raw materials. Your matrix uses other materials and will require additional expenditures of $12,000.

Ray: That's your problem. Perhaps you shouldn't have purchased all of the raw materials until we agreed on the complete test matrix.

During the month of February, West conducted 15 tests, all under Ray's direction. The tests were scattered over such a wide range that no valid conclusions could be drawn. Ray continued sending reports back to Delia confirming that Corwin was not producing beneficial results and there was no indication that the situation would reverse itself. Delia ordered Ray to take any steps necessary to ensure a successful completion of the project.

Ray and West met again as they had done for each of the past 45 days to discuss the status and direction of the project.

Ray: Dan, my boss is putting tremendous pressure on me for results, and thus far I've given him nothing. I'm up for promotion in a couple of months and I can't let this project stand in my way. It's time to completely redirect the project.

West: Your redirection of the activities is playing havoc with my scheduling. I have people in other departments who just cannot commit to this continual rescheduling. They blame me for not communicating with them when, in fact, I'm embarrassed to.

Ray: Everybody has their problems. We'll get this problem solved. I spent this morning working with some of your lab people in designing the next 15 tests. Here are the test conditions.

West: I certainly would have liked to be involved with this. After all, I thought I was the project manager. Shouldn't I have been at the meeting?

Ray: Look, Dan! I really like you, but I'm not sure that you can handle this project. We need some good results immediately, or my neck will be stuck out for the next four months. I don't want that. Just have your lab personnel start on these tests, and we'll get along fine. Also, I'm planning on spending a great deal of time in your lab area. I want to observe the testing personally and talk to your lab personnel.

West: We've already conducted 20 tests, and you're scheduling another 15 tests. I priced out only 30 tests in the proposal. We're heading for a cost overrun condition.

Ray: Our contract is a firm-fixed-price effort. Therefore, the cost overrun is your problem.

West met with Dr. Reddy to discuss the new direction of the project and potential cost overruns. West brought along a memo projecting the costs through the end of the third month of the project (see Exhibit III).

Reddy: I'm already overburdened on other projects and won't be able to help you out. Royce picked you to be the project manager because he felt that you could do the job. Now, don't let him down. Send me a brief memo next month explaining the situation, and I'll see what I can do. Perhaps the situation will correct itself.

During the month of March, the third month of the project, West received almost daily phone calls from the people in the lab stating that Pat Ray was interfering with their job. In fact, one phone call stated that Ray had changed the test conditions from what was agreed on in the latest test matrix. When West confronted Ray on his meddling, Ray asserted that Corwin personnel were very unprofessional in their attitude and that he thought this was being carried down to the testing as well. Furthermore, Ray demanded that one of the functional employees be removed immediately from the project because of incompetence. West stated that he would talk to the employee's department manager. Ray, however, felt that this would be useless and said, "Remove him or else!" The functional employee was removed from the project.

By the end of the third month, most Corwin employees were becoming disenchanted with the project and were looking for other assignments. West attributed this to Ray's harassment of the employees. To aggravate the situation even further, Ray met with Royce and Reddy, and demanded that West be removed and a new project manager be assigned.

Royce refused to remove West as project manager, and ordered Reddy to take charge and help West get the project back on track.

Reddy: You've kept me in the dark concerning this project, West. If you want me to help you, as Royce requested, I'll need all the information tomorrow, especially the cost data. I'll expect you in my office tomorrow morning at 8:00 A.M. I'll bail you out of this mess.

Exhibit III. Projected cost summary at the end of the third month

	Original Proposal Cost Summary for Six-Month Project	Total Project Costs Projected at End of Third Month
Direct labor/support	$ 30,000	$ 15,000
Testing	60,000 (30 tests)	70,000 (35 tests)
Overhead	90,000 (100%)	92,000 (120%)[a]
Materials	30,000	50,000
G&A	21,000 (10%)	22,700 (10%)
Totals	$231,000	$249,700

[a]Total engineering overhead was estimated at 100 percent, whereas the R&D overhead was 120 percent.

West prepared the projected cost data for the remainder of the work and presented the results to Dr. Reddy (see Exhibit IV). Both West and Reddy agreed that the project was now out of control, and severe measures would be required to correct the situation, in addition to more than $250,000 in corporate funding.

Reddy: Dan, I've called a meeting for 10:00 A.M. with several of our R&D people to completely construct a new test matrix. This is what we should have done right from the start.

West: Shouldn't we invite Ray to attend this meeting? I'm sure he'd want to be involved in designing the new test matrix.

Reddy: I'm running this show now, not Ray!! Tell Ray that I'm instituting new policies and procedures for in-house representatives. He's no longer authorized to visit the labs at his own discretion. He must be accompanied by either you or me. If he doesn't like these rules, he can get out. I'm not going to allow that guy to disrupt our organization. We're spending our money now, not his.

West met with Ray and informed him of the new test matrix as well as the new policies and procedures for in-house representatives. Ray was furious over the new turn of events and stated that he was returning to Peters Company for a meeting with Delia.

On the following Monday, Frimel received a letter from Delia stating that Peters Company was officially canceling the contract. The reasons given by Delia were as follows:

1. Corwin had produced absolutely no data that looked promising.
2. Corwin continually changed the direction of the project and did not appear to have a systematic plan of attack.
3. Corwin did not provide a project manager capable of handling such a project.
4. Corwin did not provide sufficient support for the in-house representative.
5. Corwin's top management did not appear to be sincerely interested in the project and did not provide sufficient executive-level support.

Royce and Frimel met to decide on a course of action in order to sustain good working relations with Peters Company. Frimel wrote a strong letter refuting all of the accusations in the Peters letter, but to no avail. Even the fact that Corwin was willing to spend $250,000 of their own funds had no bearing on Delia's decision. The damage was done. Frimel was now thoroughly convinced that a contract should not be accepted on "Pearl Harbor Day."

Exhibit IV. Estimate of total project completion costs

Direct labor/support	$ 47,000[a]
Testing (60 tests)	120,000
Overhead (120%)	200,000
Materials	103,000
G&A	47,000
	$517,000
Peters contract	250,000
Overrun	$267,000

[a]Includes Dr. Reddy.

Questions

1. What were the major mistakes made by Corwin?
2. Should Corwin have accepted the assignment?
3. Should companies risk bidding on projects based upon rough draft specifications?
4. Should the shortness of the proposal preparation time have required more active top management involvement before the proposal went out-of-house?
5. Are there any risks in not having the vice president for manufacturing available during the go or no-go bidding decision?
6. Explain the attitude of Dick Potts during the proposal activities.
7. None of the executives expressed concern when Dr. Reddy said, "I would never have assigned him (West) as project leader." How do you account for the executives' lack of concern?
8. How important is it to inform line managers of proposal activities even if the line managers are not required to provide proposal support?
9. Explain Dr. Reddy's attitude after go-ahead.
10. How should West have handled the situation where Pat Ray's opinion of the test data was contrary to that of Corwin's engineering personnel?
11. How should West have reacted to the remarks made by Ray that he informed Delia that the first five tests were failures?
12. Is immediate procurement of all materials a mistake?
13. Should Pat Ray have been given the freedom to visit laboratory personnel at any time?
14. Should an in-house representative have the right to remove a functional employee from the project?
15. Financially, how should the extra tests have been handled?
16. Explain Dr. Reddy's attitude when told to assume control of the project.
17. Delia's letter, stating the five reasons for canceling the project, was refuted by Frimel, but with no success. Could Frimel's early involvement as a project sponsor have prevented this?
18. In retrospect, would it have been better to assign a marketing person as project manager?

Case 23

Denver International Airport*

Background How does one convert a $1.2 billion project into a $5.0 billion project? It's easy. Just build a new airport in Denver. The decision to replace Denver's Stapleton Airport with Denver International Airport (DIA) was made by well-intentioned city officials. The city of Denver would need a new airport eventually, and it seemed like the right time to build an airport that would satisfy Denver's needs for at least 50–60 years. DIA could become the benchmark for other airports to follow.

A summary of the critical events is listed below:

1985: Denver Mayor Federico Pena and Adams County officials agree to build a replacement for Stapleton International Airport.
Project estimate: $1.2 billion

1986: Peat Marwick, a consulting firm, is hired to perform a feasibility study including projected traffic. Their results indicate that, depending on the season, as many as 50 percent of the passengers would change planes. The new airport would have to handle this smoothly. United and Continental object to the idea of building a new airport, fearing the added cost burden.

May 1989: Denver voters pass an airport referendum.
Project estimate: $1.7 billion

March 1993: Denver Mayor Wellington Webb announces the first delay. Opening day would be postponed from October, 1993 to December 1993. (Federico Pena becomes Secretary

---

*Reprinted from H. Kerzner, *Project Management: A Systems Approach to Planning, Scheduling, and Controlling*, 6th ed. New York: Wiley, 1998, pp. 607–640.

of Transportation under Clinton).
Project estimate: $2.7 billion

October 1993: Opening day is to be delayed to March 1994. There are problems with the fire and security systems in addition to the inoperable baggage handling system.
Project estimate: $3.1 billion

December 1993: The airport is ready to open, but without an operational baggage handling system. Another delay is announced.

February 1994: Opening day is to be delayed to May 15, 1994 because of baggage handling system.

May 1994: Airport misses the fourth deadline.

August 1994: DIA finances a backup baggage handling system. Opening day is delayed indefinitely.
Project estimate: $4 billion plus.

December 1994: Denver announces that DIA was built on top of an old Native American burial ground. An agreement is reached to lift the curse.

Airports and Airline Deregulation

Prior to the Airline Deregulation Act of 1978, airline routes and airfare were established by the Civil Aeronautics Board (CAB). Airlines were allowed to charge whatever they wanted for airfare, based upon CAB approval. The cost of additional aircraft was eventually passed on to the consumer. Initially, the high cost for airfare restricted travel to the businessperson and the elite who could afford it.

Increases in passenger travel were moderate. Most airports were already underutilized and growth was achieved by adding terminals or runways on existing airport sites. The need for new airports was not deemed critical for the near term.

Following deregulation, the airline industry had to prepare for open market competition. This meant that airfares were expected to decrease dramatically. Airlines began purchasing hoards of planes, and most routes were "free game." Airlines had to purchase more planes and fly more routes in order to remain profitable. The increase in passenger traffic was expected to come from the average person who could finally afford air travel.

Deregulation made it clear that airport expansion would be necessary. While airport management conducted feasibility studies, the recession of 1979–1983 occurred. Several airlines, such as Braniff, filed for bankruptcy protection under Chapter 11 and the airline industry headed for consolidation through mergers and leveraged buyouts.

Cities took a wait-and-see attitude rather than risk billions in new airport development. Noise abatement policies, environmental protection acts, and land acquisition were viewed as headaches. The only major airport built in the last 20 years was Dallas–Ft. Worth, which was completed in 1974.

Does Denver Need a New Airport?

In 1974, even prior to deregulation, Denver's Stapleton Airport was experiencing such rapid growth that Denver's Regional Council of Governments concluded that Stapleton would not be able to handle the necessary traffic expected by the year 2000. Modernization of Stapleton could have extended the inevitable problem to 2005. But were the headaches with Stapleton better cured through modernization or by building a new airport? There was no question that insufficient airport capacity would cause Denver to lose valuable business. Being 500 miles from other major cities placed enormous pressure upon the need for air travel in and out of Denver.

In 1988, Denver's Stapleton International Airport ranked as the fifth busiest in the country, with 30 million passengers. The busiest airports were Chicago, Atlanta, Los Angeles, and Dallas–Ft. Worth. By the year 2000, Denver anticipated 66 million passengers, just below Dallas–Ft. Worth's 70 million and Chicago's 83 million estimates.

Delays at Denver's Stapleton Airport caused major delays at all other airports. By one estimate, bad weather in Denver caused up to $100 million in lost income to the airlines each year because of delays, rerouting, canceled flights, putting travelers into hotels overnight, employee overtime pay, and passengers switching to other airlines. Denver's United Airlines and Continental comprised 80 percent of all flights in and out of Denver. Exhibit I shows the service characteristics of United and Continental between December 1993 and April 1994. Exhibit II shows all of the airlines serving Denver as of June 1994. Exhibit III shows the cities that are serviced from Denver. It should be obvious that delays in Denver could cause delays in each of these cities. Exhibit IV shows the top ten domestic passenger origin-destination markets from Denver Stapleton.

Stapleton was ranked as one of the ten worst air traffic bottlenecks in the United States. Even low clouds at Denver Stapleton could bring delays of 30 to 60 minutes.

Stapleton has two parallel north-south runways that are close together. During bad weather where instrument landing conditions exist, the two runways are considered as only one. This drastically reduces the takeoffs and landings each hour.

Exhibit I. Current service characteristics: United Airlines and Continental Airlines, December 1993 and April 1994

	Enplaned Passengers[a]	Scheduled Seats[b]	Boarding Load Factor	Scheduled Departures[b]	Average Seats per Departure
December, 1993					
United Airlines	641,209	1,080,210	59%	7,734	140
United Express	57,867	108,554	53%	3,582	30
Continental Airlines	355,667	624,325	57%	4,376	143
Continental Express	52,680	105,800	50%	3,190	33
Other	236,751	357,214	66%	2,851	125
Total	1,344,174	2,276,103	59%	21,733	105
April 1994					
United Airlines	717,093	1,049,613	68%	7,743	136
United Express	44,451	92,880	48%	3,395	27
Continental Airlines	275,948	461,168	60%	3,127	147
Continental Express	24,809	92,733	27%	2,838	33
Other	234,091	354,950	66%	2,833	125
Total	1,296,392	2,051,344	63%	19,936	103

[a] Airport management records.
[b] Official Airline Guides, Inc. (on-line database), for periods noted.

Exhibit II. Airlines serving Denver, June 1994

Major/National Airlines

America West Airlines
American Airlines
Continental Airlines
Delta Air Lines
Markair
Midway Airlines
Morris Air[a]
Northwest Airlines
TransWorld Airlines
United Airlines
USAir

Charter Airlines

Aero Mexico
American Trans Air
Casino Express
Express One
Great American
Private Jet
Sun Country Airlines

Foreign Flag Airlines (scheduled)

Martinair Holland
Mexicana de Aviacion

Regional/Commuter Airlines

Air Wisconsin (United Express)[b]
Continental Express
GP Express Airlines
Great Lakes Aviation (United Express)
Mesa Airlines (United Express)
Midwest Express[b]

Cargo Airlines

Airborne Express
Air Vantage
Alpine Air
American International Airways
Ameriflight
Bighorn Airways
Burlington Air Express
Casper Air
Corporate Air
DHL Worldwide Express
Emery Worldwide
Evergreen International Airlines
EWW Airline/Air Train
Federal Express
Kitty Hawk
Majestic Airlines
Reliant Airlines
United Parcel Service
Western Aviators

[a] Morris Air was purchased by Southwest Airlines in December 1993. The airline announced that it would no longer serve Denver as of October 3, 1994.
[b] Air Wisconsin and Midwest Express have both achieved the level of operating revenues needed to qualify as a national airline as defined by the FAA. However, for purposes of this report, these airlines are referred to as regional airlines.
Source: Airport Management, June 1994.

The new airport would have three north-south runways initially with a master plan calling for eight eventually. This would triple or quadruple instrument flights occurring at the same time to 104 aircraft per hour. Currently, Stapleton can handle only 30 landings per hour under instrument conditions with a *maximum* of 80 aircraft per hour during clear weather.

The runway master plan called for ten 12,000 foot and two 16,000 foot runways. By opening day, three north-south and one east-west 12,000 foot runways would be in operation and one of the 16,000 foot north-south runways would be operational shortly thereafter.

The airfield facilities also included a 327-foot FAA air traffic control tower (the nation's tallest) and base building structures. The tower's height allowed controllers to visually monitor runway thresholds as much as three miles away. The runway/taxiway lighting system, with lights imbedded in the concrete pavement to form centerlines and stopbars at intersections, would allow air traffic controllers to signal pilots to wait on taxiways and cross active runways, and to lead them through the airfield in poor visibility.

Due to shifting winds, runway operations were shifted from one direction to another. At the new airport, the changeover would require four minutes as opposed to the 45 minutes at Stapleton.

Exhibit III. U.S. airports served nonstop from Denver

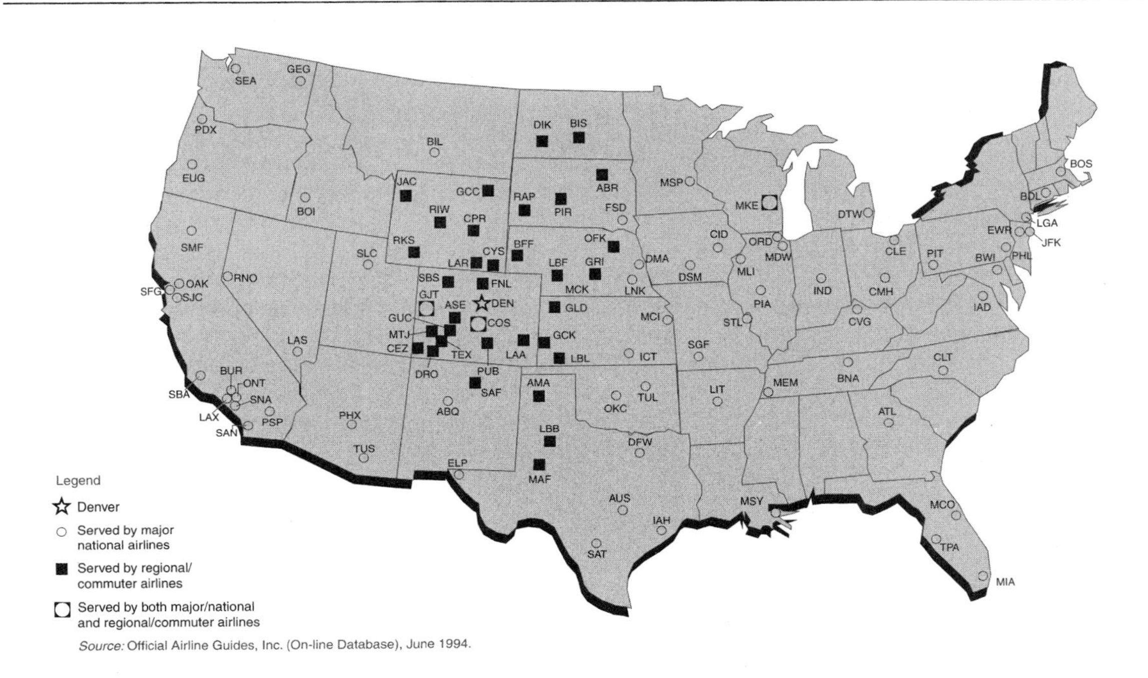

Source: Official Airline Guides, Inc. (On-line Database), June 1994.

Sufficient spacing was provided for in the concourse design such that two FAA Class 6 aircraft (i.e. 747-XX) could operate back-to-back without impeding each other. Even when two aircraft (one from each concourse) have pushed back at the same time, there could still exist room for a third FAA Class 6 aircraft to pass between them.

City officials believed that Denver's location, being equidistant from Japan and Germany, would allow twin-engine, extended range transports to reach both countries nonstop. The international opportunities were there. Between late 1990 and early 1991, Denver was entertaining four groups of leaders per month from Pacific Rim countries to look at DIA's planned capabilities.

In the long term, Denver saw the new airport as a potential hub for Northwest or USAir. This would certainly bring more business to Denver. Very few airports in the world can boast of multiple hubs.

The Enplaned Passenger Market

Perhaps the most critical parameter that illustrates the necessity for a new airport is the enplaned passenger market. (An enplaned passenger is one who gets on a flight, either an origination flight or connecting flight.)

Exhibit V identifies the enplaned passengers for individual airlines servicing Denver Stapleton for 1992 and 1993.

Connecting passengers were forecast to decrease about 1 million between 1993 and 1995 before returning to a steady 3.0 percent per year growth, totaling 8,285,500 in 2000. As a result, the number of connecting passengers is forecast to represent a smaller share (46 percent) of total

Exhibit IV. Top ten domestic passenger origin-destination markets and airline service, Stapleton International Airport (for the 12 months ended September 30, 1993)

City of Orgin or Destination[a]	Air Miles from Denver	Percentage of Certificated Airline Passengers	Average Daily Nonstop Departures[b]
1. Los Angeles[c]	849	6.8	34
2. New York[d]	1,630	6.2	19
3. Chicago[e]	908	5.6	26
4. San Francisco[f]	957	5.6	29
5. Washington, D.C.[g]	1,476	4.9	12
6. Dallas–Forth Worth	644	3.5	26
7. Houston[h]	864	3.2	15
8. Phoenix	589	3.1	19
9. Seattle	1,019	2.6	14
10. Minneapolis	693	2.3	16
Cities listed		43.8	210
All others		56.2	241
Total		100.0	451

[a] Top ten cities based on total inbound and outbound passengers (on large certificated airlines) at Stapleton International Airport in 10 percent sample for the 12 months ended September 30, 1993.
[b] Official Airline Guides, Inc.(on-line database), April 1994. Includes domestic flights operated at least four days per week by major/national airlines and excludes the activity of foreign-flag and commuter/regional airlines.
[c] Los Angeles International, Burbank–Glendale–Pasadena, John Wayne (Orange County), Ontario International, and Long Beach Municipal Airports.
[d] John F. Kennedy International, LaGuardia, and Newark International Airports.
[e] Chicago-O'Hare International and Midway Airports.
[f] San Franciscio, Metropolitan Oakland, and San Jose International Airports.
[g] Washington Dulles International, Washington National, and Baltimore/Washington International Airports.
[h] Houston Intercontinental and William P. Hobby Airports.
Sources: U.S. Department of Transportation/Air Transport Association of America, "Origin-Destination Survey of Airline Passenger Traffic, Domestic," third quarter 1993, except as noted.

enplaned passengers at the Airport in 2000 than in 1993 (50 percent). Total enplaned passengers at Denver are forecast to increase from 16,320,472 in 1993 to 18,161,000 in 2000—an average increase of 1.5 percent per year (decreasing slightly from 1993 through 1995, then increasing 2.7 percent per year after 1995).

The increase in enplaned passengers will necessitate an increase in the number of aircraft departures. Since landing fees are based upon aircraft landed weight, more parrivals and departures will generate more landing fee revenue. Since airport revenue is derived from cargo operations as well as passenger activities, it is important to recognize that enplaned cargo is also expected to increase.

Land Selection[1]

The site selected was a 53-square-mile area 18 miles northeast of Denver's business district. The site would be larger than the Chicago O'Hare and Dallas–Ft. Worth airports combined.

1. Adapted from David A. Brown, "Denver Aims for Global Hub Status with New Airport under Construction," *Aviation Week and Space Technology,* March 11, 1991, p. 44.

Exhibit V. Enplaned passengers by airline, 1992–1993, Stapleton International Airport

Enplaned Passengers	1992	1993
United	6,887,936	7,793,246
United Express[a]	470,841	578,619
	7,358,777	8,371,865
Continental	5,162,812	4,870,861
Continental Express	514,293	532,046
	5,677,105	5,402,907
American Airlines	599,705	563,119
America West Airlines	176,963	156,032
Delta Air Lines	643,644	634,341
MarkAir	2,739	93,648
Northwest Airlines	317,507	320,527
TransWorld Airlines	203,096	182,502
USAir	201,949	197,095
Other	256,226	398,436
	2,401,829	2,545,700
Total	15,437,711	16,320,472

[a] Includes Mesa Airlines, Air Wisconsin, Great Lakes Aviation, and Westair Airlines.
Source: Department of Aviation management records.

Unfortunately, a state law took effect prohibiting political entities from annexing land without the consent of its residents. The land was in Adams County. Before the vote was taken, Adams County and Denver negotiated an agreement limiting noise and requiring the creation of a buffer zone to protect surrounding residents. The agreement also included continuous noise monitoring, as well as limits on such businesses as airport hotels that could be in direct competition with existing services provided in Adams County. The final part of the agreement limited DIA to such businesses as airline maintenance, cargo, small package delivery, and other such airport-related activities.

With those agreements in place, Denver annexed 45 square miles and purchased an additional 8 square miles for noise buffer zones. Denver rezoned the buffer area to prohibit residential development within a 65 LDN (Level Day/Night) noise level. LDN is a weighted noise measurement intended to determine perceived noise in both day and night conditions. Adams County enacted even stiffer zoning regulations, calling for no residential development with an LDN noise level of 60.

Most of the airport land embodied two ranches. About 550 people were relocated. The site had overhead power lines and gas wells, which were relocated or abandoned. The site lacked infrastructure development and there were no facilities for providing water, power, sewage disposal, or other such services.

Front Range Airport

Located 2.5 miles southeast of DIA is Front Range Airport, which had been developed to relieve Denver's Stapleton Airport of most nonairline traffic operations. As a satellite airport to DIA, Front Range Airport had been offering six aviation business services by 1991:

- Air cargo and air freight, including small package services. (This is direct competition for DIA.)

- Aircraft manufacturing.
- Aircraft repair. (This is direct competition for DIA.)
- Fixed base operators to service general (and corporate) aviation.
- Flight training.
- Military maintenance and training.

The airport was located on a 4,800-acre site and was surrounded by a 12,000-acre industrial park. The airport was owned and operated by Adams County, which had completely different ownership than DIA. By 1991, Front Range Airport had two east-west runways: a 700-foot runway for general aviation use and an 8,000-foot runway to be extended to 10,000 feet. By 1992, the general plans called for two more runways to be built, both north-south. The first runway would be 10,000 feet initially with expansion capability to 16,000 feet to support wide body aircraft. The second runway would be 7,000 feet to service general aviation.

Opponents of DIA contended that Front Range Airport could be enlarged significantly, thus reducing pressure on Denver's Stapleton Airport, and that DIA would not be necessary at that time. Proponents of DIA argued that Front Range should be used to relieve pressure on DIA if and when DIA became a major international airport as all expected. Both sides were in agreement that initially, Front Range Airport would be a competitor to DIA.

Airport Design

The Denver International Airport was based upon a "Home-on-the-Range" design. The city wanted a wide open entry point for visitors. In spring of 1991, the city began soliciting bids.

To maintain a distinctive look that would be easily identified by travelers, a translucent tent-like roof was selected. The roof was made of two thicknesses of translucent, Teflon-coated glass fiber material suspended from steel cables hanging from the structural supports. The original plans for the roof called for a conventional design using 800,000 tons of structural steel. The glass fiber roof would require only 30,000 tons of structural steel, thus providing substantial savings on construction costs. The entire roof would permit about 10 percent of the sunlight to shine through, thus providing an open, outdoors-like atmosphere.

The master plan for the airport called for four concourses, each with a maximum of 60 gates. However, only three concourses would be built initially, and none would be full size. The first, Concourse A, would have 32 airline gates and 6 commuter gates. This concourse would be shared by Continental and any future international carriers. Continental had agreed to give up certain gate positions if requested to do so in order to accommodate future international operations. Continental was the only long-haul international carrier, with one daily flight to London. Shorter international flights were to Canada and Mexico.

Concourses B and C would each have 20 gates initially for airline use plus 6 commuter gates. Concourse B would be the United Concourse. Concourse C would be for all carriers other than Continental or United.

All three concourses would provide a total of 72 airline gates and 18 commuter gates. This would be substantially less than what the original master plan called for.

Although the master plan identified 60 departure gates for each concourse, cost became an issue. The first set of plans identified 106 departure gates (not counting commuter gates) and was then scaled down to 72 gates. United Airlines originally wanted 45 departure gates, but settled for 20. The recession was having its effect.

The original plans called for a train running through a tunnel beneath the terminal building and the concourses. The train would carry 6,000 passengers per hour. Road construction on and adjacent to the airport was planned to take one year. Runway construction was planned to take one year but was deliberately scheduled for two years in order to save on construction costs.

The principal benefits of the new airport compared to Stapleton were:

- A significantly *improved airfield configuration* that allowed for triple simultaneous instrument landings in all weather conditions, improved efficiency and safety of airfield operations, and reduced taxiway congestion
- *Improved efficiency in the operation of the regional airspace,* which, coupled with the increased capacity of the airfield, was supposed to significantly reduce aircraft delays and airline operating costs both at Denver and system-wide
- *Reduced noise impacts* resulting from a large site that was situated in a relatively unpopulated area
- *A more efficient terminal/concourse/apron layout* that minimized passenger walking distance, maximized the exposure of concessions to passenger flows, provided significantly greater curbside capacity, and allowed for the efficient maneuvering of aircraft in and out of gates
- *Improved international facilities* including longer runway lengths for improved stage length capability for international flights and larger Federal Inspection Services (FIS) facilities for greater passenger processing capability
- *Significant expansion capability* of each major functional element of the airport
- *Enhanced efficiency of airline operations* as a result of new baggage handling, communications, deicing, fueling, mail sorting, and other specialty systems

One of the problems with the airport design related to the high wind shears that would exist where the runways were placed. This could eventually become a serious issue.

Project Management

The city of Denver selected two companies to assist in the project management process. The first was Greiner Engineering, an engineering, architecture, and airport planning firm. The second company was Morrison-Knudsen Engineering (MKE) which is a design-construct firm. The city of Denver and Greiner/MKE would function as the project management team (PMT) responsible for schedule coordination, cost control, information management, and administration of approximately 100 design contracts, 160 general contractors, and more than 2000 subcontractors.

In the selection of architects, it became obvious that there would be a split between those who would operate the airport and the city's aspirations. Airport personnel were more interested in an "easy-to-clean" airport and convinced the city to hire a New Orleans-based architectural firm with whom Stapleton personnel had worked previously. The city wanted a "thing of beauty" rather than an easy-to-clean venture.

In an unusual split of responsibilities, the New Orleans firm was contracted to create standards that would unify the entire airport and to take the design of the main terminal only through schematics and design development, at which point it would be handed off to another firm. This sharing of the wealth with several firms would later prove more detrimental than beneficial.

The New Orleans architectural firm complained that the direction given by airport personnel focused on operational issues rather than aesthetic values. Furthermore, almost all decisions seemed to be made in reaction to maintenance or technical issues. This created a problem for the design team because the project's requirements specified that the design reflect a signature image for the airport, one that would capture the uniqueness of Denver and Colorado.

The New Orleans team designed a stepped-roof profile supported by an exposed truss system over a large central atrium, thus resembling the structure of train sheds. The intent was to bring the image of railroading, which was responsible for Denver's early growth, into the jet age.

The mayor, city council, and others were concerned that the design did not express a $2 billion project. A blue-ribbon commission was formed to study the matter. The city council eventually approved the design.

Financial analysis of the terminal indicated that the roof design would increase the cost of the project by $48 million and would push the project off schedule. A second architectural firm was hired. The final design was a peaked roof with Teflon-coated fabric designed to bring out the image of the Rocky Mountains. The second architectural firm had the additional responsibility to take the project from design development through to construction. The cost savings from the new design was so substantial that the city upgraded the floor finish in the terminal and doubled the size of the parking structure to 12,000 spaces.

The effectiveness of the project management team was being questioned. The PMT failed to sort out the differences between the city's aspirations and the maintenance orientation of the operators. It failed to detect the cost and constructability issues with the first design even though both PMT partners had vast in-house expertise. The burden of responsibility was falling on the shoulders of the architects. The PMT also did not appear to be aware that the first design may not have met the project's standards.

Throughout the design battle, no one heard from the airlines. Continental and United controlled 80 percent of the flights at Stapleton. Yet the airlines refused to participate in the design effort, hoping the project would be canceled. The city ordered the design teams to proceed for bids without any formal input from the users.

With a recession looming in the wings and Contentinal fighting for survival, the city needed the airlines to sign on. To entice the airlines to participate, the city agreed to a stunning range of design changes while assuring the bond rating agencies that the 1993 opening date would be kept. Continental convinced Denver to move the international gates away from the north side of the main terminal to terminal A, and to build a bridge from the main terminal to terminal A. This duplicated the function of a below-ground people-mover system. A basement was added the full length of the concourses. Service cores, located between gates, received a second level.

United's changes were more significant. It widened concourse B by 8 feet to accommodate two moving walkways in each direction. It added a second level of service cores, and had the roof redesigned to provide a clerestory of natural light. Most important, United wanted a destination-coded vehicle (DCV) baggage handling system where bags could be transferred between gates in less than 10 minutes, thus supporting short turnaround times. The DCV was to be on Concourse B (United) only. Within a few weeks thereafter, DIA proposed that the baggage handling system be extended to the entire airport. Yet even with these changes in place, United and Continental *still* did not sign a firm agreement with DIA, thus keeping bond interest expense at a higher than anticipated level. Some people contended that United and Continental were holding DIA hostage.

From a project management perspective, there was no question that disaster was on the horizon. Nobody knew what to do about the DCV system. The risks were unknown. Nobody realized the complexity of the system, especially the software requirements. By one account, the launch date should have been delayed by at least two years. The contract for DCV hadn't been awarded yet, and terminal construction was already under way. Everyone wanted to know why the design (and construction) was not delayed until after the airlines had signed on. How could DIA install and maintain the terminal's baseline design without having a design for the baggage handling system? Everyone felt that what they were now building would have to be ripped apart.

There were going to be massive scope changes. DIA management persisted in its belief that the airport would open on time. Work in process was now $130 million per month.

Acceleration costs, because of the scope changes, would be $30–$40 million. Three shifts were running at DIA with massive overtime. People were getting burned out to the point where they couldn't continue.

To reduce paperwork and maintain the schedule, architects became heavily involved during the construction phase, which was highly unusual. The PMT seemed to be abdicating control to the architects who would be responsible for coordination. The trust that had developed during the early phases began evaporating.

Even the car rental companies got into the act. They balked at the fees for their in-terminal location and said that servicing within the parking structures was inconvenient. They demanded and finally received a separate campus. Passengers would now be forced to take shuttle buses out of the terminal complex to rent or return vehicles.

The Baggage Handling System

DIA's $200 million baggage handling system was designed to be state of the art. Conventional baggage handling systems are manual. Each airline operates its own system. DIA opted to buy a single system and lease it back to the airlines. In effect, it would be a one-baggage-system-fits-all configuration.

The system would contain 100 computers, 56 laser scanners, conveyor belts, and thousands of motors. As designed, the system would contain 400 fiberglass carts, each carrying a single suitcase through 22 miles of steel tracks. Operating at 20 miles per hour, the system could deliver 60,000 bags per hour from dozens of gates. United was worried that passengers would have to wait for luggage since several of their gates were more than a mile from the main terminal. The system design was for the luggage to go from the plane to the carousel in 8–10 minutes. The luggage would reach the carousel before the passengers.

The baggage handling system would be centered on track-mounted cars propelled by linear induction motors. The cars slow down, but don't stop, as a conveyor ejects bags onto their platform. During the induction process, a scanner reads the bar-coded label and transmits the data through a programmable logic controller to a radio frequency identification tag on a passing car. At this point, the car knows the destination of the bag it is carrying, as does the computer software that routes the car to its destination. To illustrate the complexity of the situation, consider 4,000 taxicabs in a major city, all without drivers, being controlled by a computer through the streets of a city.

1989

Construction began without a signed agreement from Continental and United.

Early Risk Analysis

By March 1991, the bidding process was in full swing for the main terminal, concourses, and tunnel. Preliminary risk analysis involved three areas: cost, human resources, and weather.

- *Cost:* The grading of the terminal area was completed at about $5 million under budget and the grading of the first runway was completed at about $1.8 million under budget. This led management to believe that the original construction cost estimates were accurate. Also, many of the construction bids being received were below the city's own estimates.
- *Human Resources:* The economic recession hit Denver a lot harder than the rest of the nation. DIA was at that time employing about 500 construction workers. By late 1992, it was anticipated that 6000 construction workers would be needed. Although more than 3000 applications were on file, there remained the question of available, qualified

labor. If the recession were to be prolonged, then the lack of qualified suppliers could be an issue as well.

- *Bad Weather:* Bad weather, particularly in the winter, was considered as the greatest risk to the schedule. Fortunately, the winters of 1989–1990 and 1990–1991 were relatively mild, which gave promise to future mild winters. Actually, more time was lost due to bad weather in the summer of 1990 than in either of the two previous winters.

March 1991 By early March 1991, Denver had already issued more than $900 million in bonds to begin construction of the new airport. Denver planned to issue another $500 million in bonds the following month. Standard & Poor's Corporation lowered the rating on the DIA bonds from BBB to BBB−, just a notch above the junk grade rating. This could prove to be extremely costly to DIA because any downgrading in bond quality ratings would force DIA to offer higher yields on their new bond offerings, thus increasing their yearly interest expense.

Denver was in the midst of an upcoming mayoral race. Candidates were calling for the postponement of the construction, not only because of the lower ratings, but also because Denver *still* did not have a firm agreement with either Continental or United Airlines that they would use the new airport. The situation became more intense because three months earlier, in December of 1990, Continental had filed for bankruptcy protection under Chapter 11. Fears existed that Continental might drastically reduce the size of its hub at DIA or even pull out altogether.

Denver estimated that cancelation or postponement of the new airport would be costly. The city had $521 million in contracts that could not be canceled. Approximately $22 million had been spent in debt service for the land, and $38 million in interest on the $470 million in bond money was already spent. The city would have to default on more than $900 million in bonds if it could not collect landing fees from the new airport. The study also showed that a two year delay would increase the total cost by $2 billion to $3 billion and increase debt service to $340 million per year. It now appeared that the point of no return was at hand.

Fortunately for DIA, Moody's Investors Service, Inc. did *not* lower their rating on the $1 billion outstanding of airport bonds. Moody's confirmed their conditional Baa1 rating, which was slightly higher than the S & P rating of BBB−. Moody's believed that the DIA effort was a strong one and that even at depressed airline traffic levels, DIA would be able to service its debt for the scaled-back airport. Had both Moody's and S & P lowered their ratings together, DIA's future might have been in jeopardy.

April 1991 Denver issued $500 million in serial revenue bonds with a maximum yield of 9.185 percent for bonds maturing in 2023. A report by Fitch Investors Service estimated that the airport was ahead of schedule and 7 percent below budget. The concerns of the investor community seemed to have been tempered despite the bankruptcy filing of Continental Airlines. However, there was still concern that no formal agreement existed between DIA and either United Airlines or Continental Airlines.

May 1991 The city of Denver and United Airlines finally reached a tentative agreement. United would use 45 of the potential 90–100 gates at Concourse B. This would be a substantial increase from the 26 gates DIA had originally thought that United would require. The 50 percent increase in gates would also add 2,000 reservations jobs. United also expressed an interest in building a $1 billion maintenance facility at DIA employing 6,000 people.

United stated later that the agreement did not constitute a firm commitment but was contingent upon legislative approval of a tax incentive package of $360 million over 30 years plus $185 million in financing and $23 million in tax exemptions. United would decide by the summer in which city the maintenance facility would be located. United reserved the right to renegotiate the hub agreement if DIA was not chosen as the site for the maintenance facility.

Some people believed that United had delayed signing a formal agreement until it was in a strong bargaining position. With Continental in bankruptcy and DIA beyond the point of no return, United was in a favorable position to demand tax incentives of $200 million in order to keep its hub in Denver and build a maintenance facility. The state legislature would have to be involved in approving the incentives. United Airlines ultimately located the $1 billion maintenance facility at the Indianapolis Airport.

August 1991 Hotel developers expressed concern about building at DIA, which is 26 miles from downtown compared to 8 miles from Stapleton to downtown Denver. DIA officials initially planned for a 1,000-room hotel attached to the airport terminal, with another 300–500 rooms adjacent to the terminal. The 1,000-room hotel had been scaled back to 500–700 rooms and was not likely to be ready when the airport was scheduled to open in October 1993. Developers had expressed resistance to building close to DIA unless industrial and office parks were also built near the airport. Even though ample land existed, developers were putting hotel development on the back burner until after 1993.

November 1991 Federal Express and United Parcel Service (UPS) planned to move cargo operations to the smaller Front Range Airport rather than to DIA. The master plan for DIA called for cargo operations to be at the northern edge of DIA, thus increasing the time and cost for deliveries to Denver. Shifting operations to Front Range Airport would certainly have been closer to Denver but would have alienated northern Adams County cities that counted on an economic boost in their areas. Moving cargo operations would have been in violation of the original agreement between Adams County and Denver for the annexation of the land for DIA.

The cost of renting at DIA was estimated at $0.75 per square foot, compared to $0.25 per square foot at Front Range. DIA would have higher landing fees of $2.68 per 1000 pounds compared to $2.15 for Front Range. UPS demanded a cap on landing fees at DIA if another carrier were to go out of business. Under the UPS proposal, area landholders and businesses would set up a fund to compensate DIA if landing fees were to exceed the cap. Cargo carriers at Stapleton were currently paying $2 million in landing fees and rental of facilities per year.

As the "dog fight" over cargo operations continued, the Federal Aviation Administration (FAA) issued a report calling for cargo operations to be collocated with passenger operations at the busier metropolitan airports. This included both full cargo carriers as well as passenger cargo (i.e., "belly cargo") carriers. Proponents of Front Range argued that the report didn't preclude the use of Front Range because of its proximity to DIA.

December 1991 United Airlines formally agreed to a 30-year lease for 45 gates at Concourse B. With the firm agreement in place, the DIA revenue bonds shot up in price almost $30 per $1000 bond. Earlier in the year, Continental signed a five-year lease agreement.

Other airlines also agreed to service DIA. Exhibit VI sets forth the airlines that either executed use and lease agreements for, or indicated an interest in leasing, the 20 gates on Concourse C on a first-preferential-use basis.

Exhibit VI. Airline agreements

Airline	Term (Years)	Number of Gates
American Airlines	5	3
Delta Air Lines[a]	5	4
Frontier Airlines	10	2
MarkAir	10	5
Northwest Airlines	10	2
TransWorld Airlines	10	2
USAir[a]	5	2
Total		20

[a] The City has entered into Use and Lease Agreements with these airlines. The USAir lease is for one gate on Concourse C and USAir has indicated its interest in leasing a second gate on Concourse C.

January 1992

BAE was selected to design and build the baggage handling system. The airport had been under construction for three years before BAE was brought on board. BAE agreed to do eight years of work in two years to meet the October, 1993 opening date.

June 1992

DIA officials awarded a $24.4 million conract for the new airport's telephone services to U.S. West Communication Services. The officials of DIA had considered controlling its own operations through shared tenant service, which would allow the airport to act as its own telephone company. All calls would be routed through an airport-owned computer switch. By grouping tenants together into a single shared entity, the airport would be in a position to negotiate discounts with long distance providers, thus enabling cost savings to be passed on to the tenants.

By one estimate, the city would generate $3 million to $8 million annually in new, nontax net revenue by owning and operating its own telecommunication network. Unfortunately, DIA officials did not feel that sufficient time existed for them to operate their own system. The city of Denver was unhappy over this lost income.

September 1992

By September 1992, the city had received $501 million in Federal Aviation Administration grants and $2.3 billion in bonds with interest rates of 9.0–9.5 percent in the first issue to 6 percent in the latest issue. The decrease in interest rates due to the recession was helpful to DIA. The rating agencies also increased the city's bond rating one notch.

The FAA permitted Denver to charge a $3 departure tax at Stapleton with the income earmarked for construction of DIA. Denver officials estimated that over 34 years, the tax would generate $2.3 billion.

The cities bordering the northern edge of DIA (where the cargo operations were to be located) teamed up with Adams County to file lawsuits against DIA in its attempt to relocate cargo operations to the southern perimeter of DIA. This relocation would appease the cargo carriers and hopefully end the year-long battle with Front Range Airport. The Adams County Commissioner contended that relocation would violate the Clean Air Act and the National Environmental Policy Act and would be a major deviation from the original airport plan approved by the FAA.

October 1992

The city issued $261 million of Airport Revenue Bonds for the construction of facilities for United Airlines. (See Exhibit A at the end of this case.)

March 1993

The city of Denver announced that the launch date for DIA would be pushed back to December 18 rather than the original October 30 date in order to install and test all of the new equipment. The city wanted to delay the opening until late in the first quarter of 1994 but deemed it too costly because the airport's debt would have to be paid without an adequate stream of revenue. The interest on the bond debt was now at $500,000 per day.

The delay to December 18 angered the cargo carriers. This would be their busiest time of the year, usually twice their normal cargo levels, and a complete revamping of their delivery service would be needed. The Washington-based Air Freight Association urged the city to allow the cargo carriers to fly out of Stapleton through the holiday period.

By March 1993, Federal Express, Airborne Express, and UPS (reluctantly) had agreed to house operations at DIA after the city pledged to build facilities for them at the south end of the airport. Negotiations were also underway with Emery Worldwide and Burlington Air Express. The "belly" carriers, Continental and United, had already signed on.

UPS had wanted to create a hub at Front Range Airport. If Front Range Airport were a cargo-only facility, it would free up UPS from competing with passenger traffic for runway access even though both Front Range and DIA were in the same air traffic control pattern. UPS stated that it would not locate a regional hub at DIA. This would mean the loss of a major development project that would have attracted other businesses that relied on UPS delivery.

For UPS to build a regional hub at Front Range would have required the construction of a control tower and enlargement of the runways, both requiring federal funds. The FAA refused to free up funds for Front Range, largely due to a lawsuit by United Airlines and environmental groups.

United's lawsuit had an ulterior motive. Adams County officials repeatedly stated that they had no intention of building passenger terminals at Front Range. However, once federal funds were given to Front Range, a commercial passenger plane could not be prevented from setting up shop in Front Range. The threat to United was the low-cost carriers such as Southwest Airlines. Because costs were fixed, fewer passengers traveling through DIA meant less profits for the airlines. United simply did not want any airline activities removed from DIA!

August 1993

Plans for a train to connect downtown Denver to DIA were underway. A $450,000 feasibility study and federal environmental assessment were being conducted, with the results due November 30, 1993. Union Pacific had spent $350,000 preparing a design for the new track, which could be constructed in 13 to 16 months.

The major hurdle would be the financing, which was estimated between $70 million and $120 million, based upon hourly trips or 20-minute trips. The more frequent the trips, the higher the cost.

The feasibility study also considered the possibility of baggage check-in at each of the stops. This would require financial support and management assistance from the airlines.

September 1993

Denver officials disclosed plans for transfering airport facilities and personnel from Stapleton to DIA. The move would be stage-managed by Larry Sweat, a retired military officer who had coordinated troop movements for Operation Desert Shield. Bechtel Corporation would be responsible for directing the transport and setup of machinery, computer systems, furniture, and service equipment, all of which had to be accomplished overnight since the airport had to be operational again in the morning.

October 1993 DIA, which was already $1.1 billion over budget, was to be delayed again. The new opening date would be March 1994. The city blamed the airlines for the delays, citing the numerous scope changes required. Even the fire safety system hadn't been completed.

Financial estimates became troublesome. Airlines would have to charge a $15 per person tax, the largest in the nation. Fees and rent charged the airlines would triple from $74 million at Stapleton to $247 million at DIA.

January 1994 Front Range Airport and DIA were considering the idea of being designated as one system by the FAA. Front Range could legally be limited to cargo only. This would also prevent low-cost carriers from paying lower landing fees and rental space at Front Range.

February 1994 Southwest Airlines, being a low-cost no-frills carrier, said that it would not service DIA. Southwest wanted to keep its airport fees below $3 a passenger. Current projections indicated that DIA would have to charge between $15 and $20 per passenger in order to service its debt. This was based upon a March 9 opening day.

Continental announced that it would provide a limited number of low-frill service flights in and out of Denver. Furthermore, Continental said that because of the high landing fees, it would cancel 23 percent of its flights through Denver and relocate some of its maintenance facilities.

United Airlines expected its operating cost to be $100 million more per year at DIA than at Stapleton. With the low-cost carriers either pulling out or reducing service to Denver, United was under less pressure to lower airfares.

March 1994 The city of Denver announced the fourth delay in opening DIA, from March 9 to May 15. The cost of the delay, $100 million, would be paid mostly by United and Continental. As of March, only Concourse C, which housed the carriers other than United and Continental, was granted a temporary certificate of occupancy (TCO) by the city.

As the finger-pointing began, blame for this delay was given to the baggage handling system, which was experiencing late changes, restricted access flow, and a slowdown in installation and testing. A test by Continental Airlines indicated that only 39 percent of baggage was delivered to the correct location. Other problems also existed. As of December 31, 1993, there were 2,100 design changes. The city of Denver had taken out insurance for construction errors and omissions. The city's insurance claims cited failure to coordinate design of the ductwork with ceiling and structure, failure to properly design the storm draining systems for the terminal to prevent freezing, failure to coordinate mechanical and structural designs of the terminal, and failure to design an adequate subfloor support system.

Consultants began identifying potential estimating errors in DIA's operations. The runways at DIA were six times longer than the runways at Stapleton, but DIA had purchased only 25 percent more equipment. DIA's cost projections would be $280 million for debt service and $130 million for operating costs, for a total of $410 million per year. The total cost at Stapleton was $120 million per year.

April 1994 Denver International Airport began having personnel problems. According to DIA's personnel officer, Linda Rubin Royer, moving 17 miles away from its present site was creating serious problems. One of the biggest issues was the additional 20-minute drive that employees had to bear. To resolve this problem, she proposed a car/van pooling scheme and tried to get the city bus company to transport people to and from the new airport. There was also the problem of transfering employees to similar jobs elsewhere if they truly disliked working at DIA. The scarcity of applicants wanting to work at DIA was creating a problem as well.

May 1994 Standard and Poor's Corporation lowered the rating on DIA's outstanding debt to the noninvestment grade of BB, citing the problems with the baggage handling system and no immediate cure in sight. Denver was currently paying $33.3 million per month to service debt. Stapleton was generating $17 million per month and United Airlines had agreed to pay $8.8 million in cash for the next three months only. That left a current shortfall of $7.5 million each month that the city would have to fund. Beginning in August 1994, the city would be burdened with $16.3 million each month.

BAE Automated Systems personnel began to complain that they were pressured into doing the impossible. The only other system of this type in the world was in Frankfurt, Germany. That system required six years to install and two years to debug. BAE was asked to do it all in two years.

BAE underestimated the complexity of the routing problems. During trials, cars crashed into one another, luggage was dropped at the wrong location, cars that were needed to carry luggage were routed to empty waiting pens, and some cars traveled in the wrong direction. Sensors became coated with dirt, throwing the system out of alignment, and luggage was dumped prematurely because of faulty latches, jamming cars against the side of a tunnel. By the end of May, BAE was conducting a worldwide search for consultants who could determine what was going wrong and how long it would take to repair the system.

BAE conducted an end-of-month test with 600 bags. Outbound (terminal to plane), the sort accuracy was 94 percent and inbound the accuracy was 98 percent. The system had a zero downtime for both inbound and outbound testing. The specification requirements called for 99.5 percent accuracy.

BAE hired three technicians from Germany's Logplan, which helped solve similar problems with the automated system at Frankfurt, Germany. With no opening date set, DIA contemplated opening the east side of the airport for general aviation and air cargo flights. That would begin generating at least some revenue.

June 1994 The cost for DIA was now approaching $3.7 billion and the jokes about DIA appeared everywhere. One common joke as that when you fly to Denver, you will have to stop in Chicago to pick up your luggage. Other common jokes included the abbreviation, DIA. Exhibit B provides a listing of some 152 of the jokes.

The people who did not appear to be laughing at these jokes were the concessionaires, including about 50 food service operators, who had been forced to rehire, retrain, and reequip at considerable expense. Several small businesses were forced to call it quits because of the eight-month delay. Red ink was flowing despite the fact that the $45-a-square foot rent would not have to be paid until DIA officially opened. Several of the concessionaires had requested that the rent be cut by $10 a square foot for the first six months or so, after the airport opened. A merchant's association was formed at DIA to fight for financial compensation.

The Project's Work Breakdown Structure (WBS)

The city had managed the design and construction of the project by grouping design and construction activities into seven categories or "areas":

Area #0 Program management/preliminary design
Area #1 Site development
Area #2 Roadways and on-grade parking
Area #3 Airfield
Area #4 Terminal complex
Area #5 Utilites and specialty systems
Area #6 Other

Since the fall of 1992, the project budget had increased by $224 million (from $2,700 million to $2,924 million), principally as a result of scope changes.

- Structural modifications to the terminal buildings (primarily in the Landside Terminal and Concourse B) to accommodate the automated baggage system
- Changes in the interior configuration of Concourse B
- Increases in the scope of various airline tenant finished, equipment, and systems, particularly in Concourse B
- Grading, drainage, utilities, and access costs associated with the relocation of air cargo facilities to the south side of the airport
- Increases in the scope and costs of communication and control systems, particularly premises wiring
- Increases in the costs of runway, taxiway, and apron paving and change orders as a result of changing specifications for the runway lighting system
- Increased program management costs because of schedule delays

Yet even with all of these design changes, the airport was ready to open except for the baggage handling system.

July 1994

The Securities and Exchange Commission (SEC) disclosed that DIA was one of 30 municipal bond issuers that were under investigation for improper contributions to the political campaigns of Pena and his successor, Mayor Wellington Webb. Citing public records, Pena was said to have received $13,900 and Webb's campaign fund increased by $96,000. The SEC said that the contributions may have been in exchange for the right to underwrite DIA's muncipal bond offerings. Those under investigation included Merrill Lynch, Goldman Sachs & Co., and Lehman Brothers, Inc.

August 1994

Continental confirmed that as of November 1, 1994, it would reduce its flights out of Denver from 80 to 23. At one time, Continental had 200 flights out of Denver.

Denver announced that it expected to sell $200 million in new bonds. Approximately $150 million would be used to cover future interest payments on existing DIA debt and to replenish interest and other money paid due to the delayed opening.

Approximately $50 million would be used to fund the construction of an interim baggage handling system of the more conventional tug-and-conveyor type. The interim system would require 500–600 people rather than the 150–160 people needed for the computerized system. Early estimates said that the conveyor belt/tug-and-cart system would be at least as fast as the system at Stapleton and would be built using proven technology and off-the-shelf parts. However, modifications would have to be made to both the terminal and the concourses.

United Airlines asked for a 30-day delay in approving the interim system for fear that it would not be able to satisfy their requirements. The original lease agreement with DIA and United stipulated that on opening day there would be a fully operational automated baggage handling system in place. United had 284 flights a day out of Denver and had to be certain that the interim system would support a 25-minute turnaround time for passenger aircraft.

The city's District Attorney's Office said it was investigating accusations of falsified test data and shoddy workmanship at DIA. Reports had come in regarding fraudulent construction and contracting practices. No charges were filed at that time.

DIA began repairing cracks, holes, and fissures that had emerged in the runways, ramps, and taxiways. Officials said that the cracks were part of the normal settling problems and might require maintenance for years to come.

United Airlines agreed to invest $20 million and act as the project manager to the baggage handling system at Concourse B. DIA picked February 28, 1995 as the new opening date as long as either the primary or secondary baggage handling systems was operational.

United Benefits from Continental's Downsizing

United had been building up its Denver hub since 1991, increasing its total departures 9 percent in 1992, 22 percent in 1993, and 9 percent in the first six months of 1994. Stapleton is United's second largest connecting hub after Chicago O'Hare (ORD), ahead of San Francisco (SFO), Los Angeles (LAX), and Washington Dulles (IAD) International Airports, as shown in Exhibit VII.

In response to the downsizing by Continental, United is expected to absorb a significant portion of Continental's Denver traffic by means of increased load factors and increased service (i.e. capacity), particularly in larger markets where significant voids in service might be left by Continental. United served 24 of the 28 cities served by Continental from Stapleton in June,

Exhibit VII. Comparative United Airlines service at hub airports, June 1983 and June 1994

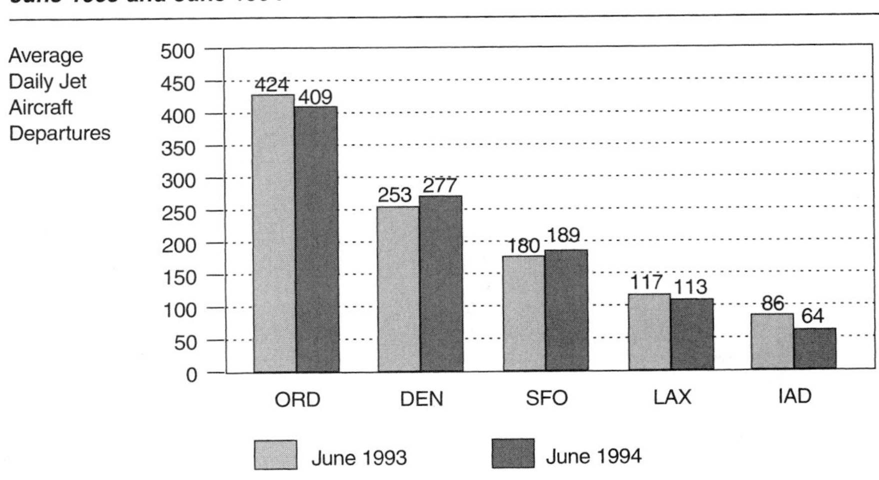

Note: Does not include activity by United Express.

Source: Official Airline Guides, Inc. (On-line Database), for periods shown.

1994, with about 79 percent more total available seats to those cities—23,937 seats provided by United compared with 13,400 seats provided by Continental. During 1993, United's average load factor from Denver was 63 percent, indicating that, with its existing service and available capacity, United had the ability to absorb many of the passengers abandoned by Continental. In addition, United had announced plans to increase service at Denver to 300 daily flights by the end of the calendar year.

As a result of its downsizing in Denver, Continental was forecasted to lose more than 3.9 million enplaned passengers from 1993 to 1995—a total decrease of 80 percent. However, this decrease was expected to be largely offset by the forecasted 2.2 million increase in enplaned passengers by United and 1.0 million by the other airlines, resulting in a total of 15,877,000 enplaned passengers at Denver in 1995. As discussed earlier, it was assumed that, in addition to a continuation of historical growth, United and the other airlines would pick up much of the traffic abandoned by Continental through a combination of added service, larger average aircraft size, and increased load factors.

From 1995 to 2000, the increase in total enplaned passengers is based on growth rates of 2.5 percent per year in originating passengers and 3.0 percent per year in connecting passengers. Between 1995 and 2000, United's emerging dominance at the airport (with almost twice the number of passengers of all other airlines combined) should result in somewhat higher fare levels in the Denver markets, and therefore may dampen traffic growth. As shown in Exhibit VIII,

Exhibit VIII. Enplaned passenger market shares at Denver Airports

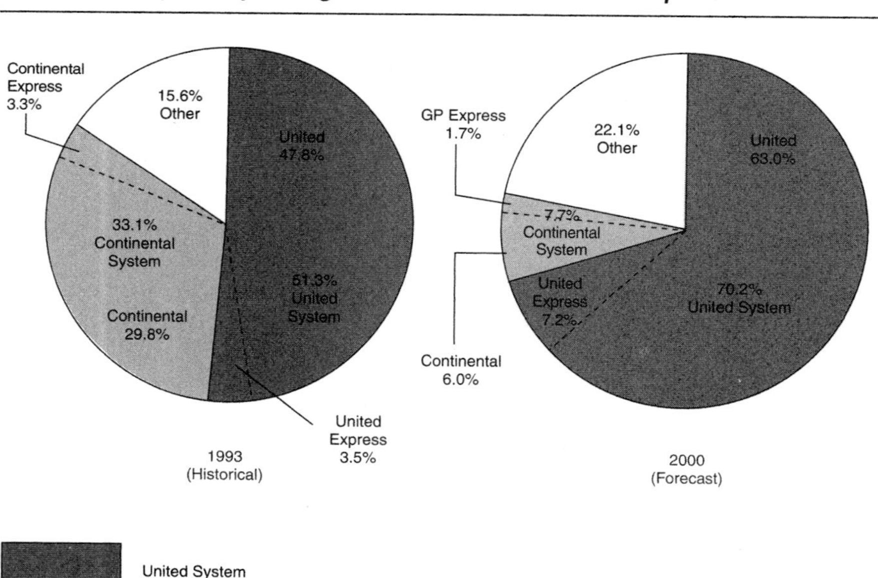

Source: 1993: Airport Management Records.

of the 18.2 million forecasted enplaned passengers in 2000, United and United Express together are forecasted to account for 70 percent of total passengers at the airport—up from about 51 percent in 1993—while Continental's share, including GP Express, is forecasted to be less than 8 percent—down from about 33 percent in 1993.

Total connecting passengers at Stapleton increased from about 6.1 million in 1990 to about 8.2 million in 1993—an average increase of about 10 percent per year. The number of connecting passengers was forecast to decrease in 1994 and 1995, as a result of the downsizing by Continental, and then return to steady growth of 3.0 percent per year through 2000, reflecting expected growth in passenger traffic nationally and a stable market share by United in Denver. Airline market share of connecting passengers in 1993 and 1995 are shown in Exhibit IX.

September 1994

Denver began discussions with cash-strapped MarkAir of Alaska to begin service at DIA. For an undercapitalized carrier, the prospects of tax breaks, favorable rents, and a $30 million guaranteed city loan were enticing.

Exhibit IX. Connecting passenger market shares at Denver Airports

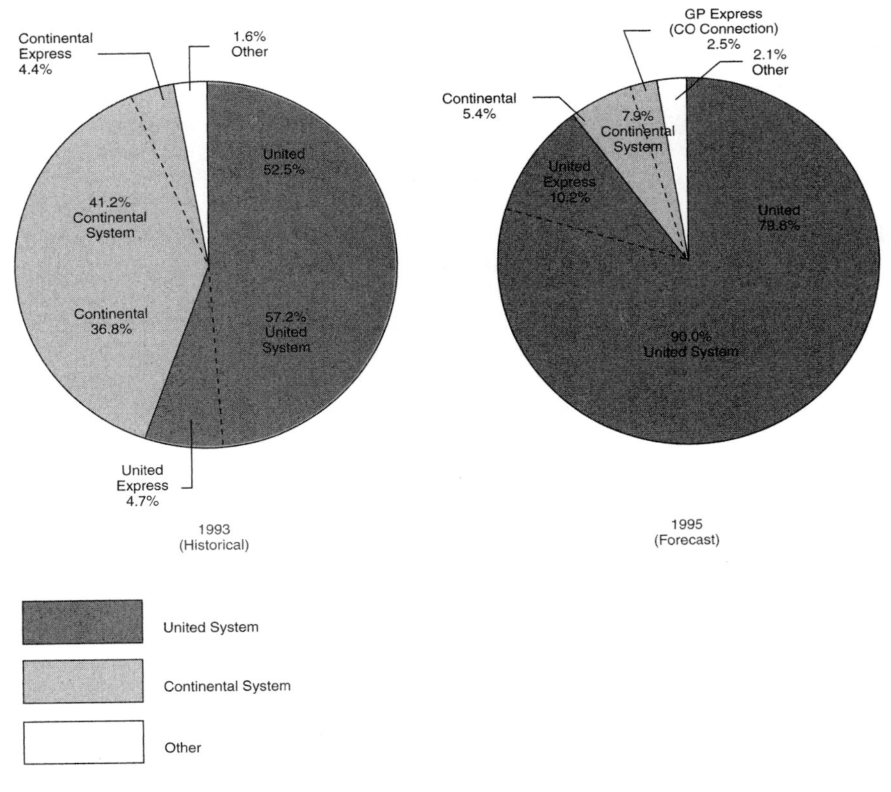

Source: 1993: Airport Management Records and U.S. Department of Transportation

DIA officials estimated an $18 per person charge on opening day. Plans to allow only cargo carriers and general aviation to begin operations at DIA were canceled.

Total construction cost for the main terminal exceeded $455 million (including the parking structure and the airport office building). .

General site expenses, commission	$38,667,967
Sitework, building excavations	15,064,817
Concrete	89,238,296
Masonry	5,501,608
Metals	40,889,411
Carpentry	3,727,408
Thermal, moisture protection	8,120,907
Doors and windows	13,829,336
Finishes	37,025,019
Specialties	2,312,691
Building equipment	227,720
Furnishings	3,283,852
Special construction	39,370,072
Conveying systems	23,741,336
Mechanical	60,836,566
Electrical	73,436,575
Total	$455,273,581

October 1994 A federal grand jury convened to investigate faulty workmanship and falsified records at DIA. The faulty workmanship had resulted in falling ceilings, buckling walls, and collapsing floors.

November 1994 The baggage handling system was working, but only in segments. Frustration still existed in not being able to get the whole system to work at the same time. The problem appeared to be with the software required to get computers to talk to computers. The fact that a mere software failure could hold up Denver's new airport for more than a year put in question the project's risk management program.

Jerry Waddles was the risk manager for Denver. He left that post to become risk manager for the State of Colorado. Eventually the city found an acting risk manager, Molly Austin Flaherty, to replace Mr. Waddles, but for the most part, DIA construction over the past several months had continued without a full-time risk manager.

The failure of the baggage handling system had propelled DIA into newspaper headlines around the country. The U.S. Securities and Exchange Commission had launched a probe into whether Denver officials had deliberately deceived bondholders about how equipment malfunctions would affect the December 19, 1993 opening. The allegations were made by Denver's KCNC-TV. Internal memos indicated that in the summer of 1993 city engineers believed it would take at least until March, 1994 to get the system working. However, Mayor Wellington Webb did not announce the delayed opening until October 1993. The SEC was investigating whether the last postponement misled investors holding $3 billion in airport bonds.

Under a new agreement, the city agreed to pay BAE an additional $35 million for modifications *if* the system was working for United Airlines by February 28, 1995. BAE would then have until August 1995 to complete the rest of the system for the other tenants. If the system was not operational by February 28, the city could withhold payment of the $35 million.

BAE lodged a $40 million claim against the city, alleging that the city caused the delay by changing the system's baseline configuration after the April 1, 1992 deadline. The city filed a $90 million counterclaim, blaming BAE for the delays.

The lawsuits were settled out of court when BAE agreed to pay $12,000 a day in liquidated damages dating from December 19, 1993 to February 28, 1995, or approximately $5 million. The city agreed to pay BAE $6.5 million to cover some invoices submitted by BAE for work already done to repair the system.

Under its DIA construction contract, BAE's risks were limited. BAE's liability for consequential damages resulting from its failure to complete the baggage handling system on time was capped at $5 million. BAE had no intention of being held liable for changes to the system. The system as it was at the time was not the system that BAE had been hired to install.

Additional insurance policies also existed. Builder's risk policies generally pay damages caused by defective parts or materials, but so far none of the parts used to construct the system had been defective. BAE was also covered for design errors or omissions. The unknown risk at that point was who would be responsible if the system worked for Concourse B (i.e., United) but then failed when it was expanded to cover all concourses.

A study was underway to determine the source of respiratory problems suffered by workers at the construction site. The biggest culprit appeared to be the use of concrete in a confined space.

The city and DIA were also protected from claims filed by vendors whose businesses were put on hold because of the delays under a hold-harmless agreement in the contracts. However, the city had offered to permit the concessionaires to charge higher fees and also to extend their leases for no charge to make up for lost income due to the delays.

December 1994

The designer of the baggage handling system was asked to reexamine the number of bags per minute that the BAE system was required to accommodate as per the specifications. The contract called for departing luggage to Concourse A to be delivered at a peak rate of 90 bags per minute. The designer estimated peak demand at 25 bags per minute. Luggage from Concourse A was contracted for at 223 bags per minute but again, the designer calculated peak demand at a lower rate of 44 bags per minute.

Airport Debt

By December 1994, DIA was more than $3.4 billion in debt, as shown below:

Series 1984 Bonds	$ 103,875,000
Series 1985 Bonds	175,930,000
Series 1990A Bonds	700,003,843
Series 1991A Bonds	500,003,523
Series 1991D Bonds	600,001,391
Series 1992A Bonds	253,180,000
Series 1992B Bonds	315,000,000
Series 1992C Bonds	392,160,000
Series 1992D–G Bonds	135,000,000
Series 1994A Bonds	257,000,000
	$3,432,153,757

Airport Revenue

Airports generally have two types of contracts with their tenants. The first type is the residual contract where the carriers guarantee that the airport will remain solvent. Under this contract, the carriers absorb the majority of the risk. The airport maintains the right to increase rents and landing fees to cover operating expenses and debt coverage. The second type of contract is the compensatory contract where the airport is at risk. DIA has a residual contract with its carriers.

Airports generate revenue from several sources. The most common breakdown includes landing fees and rent from the following entities: airline carriers, passenger facilities, rental car agencies, concessionary stores, food and beverage services, retail shops, and parking garages. Retail shops and other concessionary stores also pay a percent of sales.

Airline Costs per Enplaned Passenger

Revenues derived from the airlines are often expressed on a per enplaned passenger basis. The average airline cost per enplaned passenger at Stapleton in 1993 was $5.02. However, this amount excludes costs related to major investments in terminal facilities made by United Airlines in the mid-1980s and, therefore, understates the true historical airline cost per passenger.

Average airline costs per enplaned passenger at the airport in 1995 and 2000 are forecast to be as follows:

	Total Average Airline Costs per Enplaned Passenger	
Year	**Current Dollars**	**1990 Dollars**
1995	$18.15	$14.92
2000	17.20	11.62

The forecasted airline costs per enplaned passenger at the airport are considerably higher than costs at Stapleton today and the highest of any major airport in the United States. (The cost per enplaned passenger at Cleveland Hopkins is $7.50). The relatively high airline cost per passenger is attributable, in part, to (1) the unusually large amount of tenant finishes, equipment, and systems costs being financed as part of the project relative to other airport projects and (2) delayed costs incurred since the original opening date for purposes of the Plan of Financing (January 1, 1994).

The City estimates that, as a result of the increased capacity and efficiency of the airfield, operation of the airport will result in annual delay savings to the airlines of $50 million to $100 million per year (equivalent to about $3 to $6 per enplaned passenger), and that other advanced technology and systems incorporated into the design of the airport will result in further operational savings. In the final analysis, the cost effectiveness of operating at the airport is a judgment that must be made by the individual airlines in deciding to serve the Denver market.

It is assumed for the purposes of this analysis that the city and the airlines will resolve the current disputes regarding cost allocation procedures and responsibility for delay costs, and that the airlines will pay rates generally in accordance with the procedures of the use and lease agreements as followed by the city and as summarized in the accompanying exhibits.

February 28, 1995

The airline opened as planned on February 28, 1995. However, several problems became apparent. First, the baggage handling system did have "bad days." Passengers traveling to and from Denver felt more comfortable carrying bags than having them transfered by the computerized baggage handling system. Large queues began to form at the end of the escalators in the main terminal going down to the concourse trains. The trains were not running frequently enough, and the number of cars in each train did not appear to be sufficient to handle the necessary passenger traffic.

The author flew from Dallas–Ft. Worth to Denver in one hour and 45 minutes. It then took one hour and 40 minutes to catch the airport shuttles (which stop at all the hotels) and arrive at the appropriate hotel in downtown Denver. Passengers began to balk at the discomfort of the remote rental car facilities, the additional $3 tax per day for each rental car, and the fact that the nearest gas station was 15 miles away. How does one return a rental car with a full tank of gas?

Departing passengers estimated it would take two hours to drive to the airport from downtown Denver, unload luggage, park their automobile, check in, and take the train to the concourse.

Faults in the concourse construction were becoming apparent. Tiles that were supposed to be 5/8 inches thick were found to be 1/2 inch thick. Tiles began to crack. During rainy weather, rain began seeping in through the ceiling.

EXHIBIT A[2]
Municipal Bond Prospectus

$261,415,000
City and County Of Denver, Colorado
6.875% Special Facilities Airport Revenue Bonds
(United Airlines Project)
Series 1992A
Date: October 1, 1992
Due: October 1, 2032
Rating: Standard & Poor's BBB−
Moody's Baa2

Introduction

This official statement is provided to furnish information in connection with the sale by the City and County of Denver, Colorado (the "City") of 6.875% Special Facilities Airport Revenue Bonds (United Airlines Project) series 1992A in the aggregate principle amount of $261,415,000 (the "Bonds"). The bonds will be dated, mature, bear interest, and be subject to redemption prior to maturity as described herein.

The Bonds will be issued pursuant to an Ordinance of the City and County of Denver, Colorado (the "Ordinance").

The proceeds received by the City from the sale of the Bonds will be used to acquire, construct, equip, or improve (or a reimbursement of payments for the acquisition, construction, equipping, or improvement of) certain terminals, Concourse B, aircraft maintenance, ground equipment maintenance, flight kitchen, and air freight facilities (the "Facilities") at the new Denver International Airport (the "New Airport").

The City will cause such proceeds to be deposited, distributed, and applied in accordance with the terms of a Special Facilities and Ground Lease, dated as of October 1, 1992 (the "Lease") between United Airlines and the City. Under the Lease, United has agreed to make payments sufficient to pay the principal, premium, if any, and interest on the Bonds. Neither the Facilities nor the ground rental payments under the Lease are pledged as security for the payment of principal, premium, if any, and interest on the bonds.

Agreement between United and the City

On June 26, 1991, United and the City entered into an agreement followed by a second agreement on December 12, 1991, which, among other things, collectively provide for the use and lease by United of certain premises and facilities at the New Airport. In the United Agreement, United agrees among other things, to (1) support the construction of the New

2. Only excerpts from the prospectus are included here.

Airport, (2) relocate its present air carrier operations from Stapleton to the New Airport, (3) occupy and lease certain facilities at the New Airport, including no less than 45 gates on Concourse B within two years of the date of beneficial occupancy as described in the United Agreement, and (4) construct prior to the date of beneficial occupancy, a regional reservation center at a site at Stapleton.

In conjunction with the execution of the United Agreement, United also executes a 30-year use and lease agreement. United has agreed to lease, on a preferential use basis, Concourse B, which is expected to support 42 jet aircraft with up to 24 commuter aircraft parking positions at the date of beneficial occupancy, and, on an exclusive use basis, certain ticket counters and other areas in the terminal complex of the New Airport.

The Facilities

The proceeds of the bonds will be used to finance the acquisition, construction, and equipping of the Facilities, as provided under the Lease. The Facilities will be located on approximately 100 acres of improved land located within the New Airport, which United will lease from the City. The Facilities will include an aircraft maintenance facility capable of housing ten jet aircraft, a ground equipment support facility with 26 maintenance bays, an approximately 55,500-square-foot air freight facility, and an approximately 155,000-square-foot flight kitchen. Additionally, the proceeds of the Bonds will be used to furnish, equip, and install certain facilities to be used by United in Concourse B and in the terminal of the New Airport.

Redemption of Bonds

The Bonds will be subject to optional and mandatory redemption prior to maturity in the amounts, at the times, at the prices, and in the manner as provided in the Ordinance. If less than all of the Bonds are to be redeemed, the particular Bonds to be called for redemption will be selected by lot by the Paying Agent in any manner deemed fair and reasonable by the Paying Agent.

The bonds are subject to redemption prior to maturity by the City at the request of United, in whole or in part, by lot, on any date on or after October 1, 2002 from an account created pursuant to the Ordinance used to pay the principal, premium, if any, and interest on the Bonds (the "Bond Fund") and from monies otherwise available for such purpose. Such redemptions are to be made at the applicable redemption price shown below as a percentage of the principal amount thereof, plus interest accrued to the redemption date:

Redemption Period	Optional Redemption Price
October 1, 2002 through September 30, 2003	102%
October 1, 2003 through September 30, 2004	101%
October 1, 2004 and thereafter	100%

The Bonds are subject to optional redemption prior to maturity, in whole or in part by lot, on any date, upon the exercise by United of its option to prepay Facilities Rentals under the Lease at a redemption price equal to 100% of the principal amount thereof plus interest accrued to the redemption date, if one or more of the following events occurs with respect to one or more of the units of the Leased Property:

(a) the damage or destruction of all or substantially all of such unit or units of the Leased Property to such extent that, in the reasonable opinion of United, repair and restoration would not be economical and United elects not to restore or replace such unit or units of the Leased Property; or,

(b) the condemnation of any part, use, or control of so much of such unit or units of the Leased Property that such unit or units cannot be reasonably used by United for carrying on, at substantially the same level or scope, the business theretofore conducted by United on such unit or units.

In the event of a partial extraordinary redemption, the amount of the Bonds to be redeemed for any unit of the Leased Property with respect to which such prepayment is made shall be determined as set forth below (expressed as a percentage of the original principal amount of the Bonds) plus accrued interest on the Bonds to be redeemed to the redemption date of such Bonds provided that the amount of Bonds to be redeemed may be reduced by the aggregate principal amount (valued at par) of any Bonds purchased by or on behalf of United and delivered to the Paying Agent for cancelation:

Terminal Concourse B Facility	Aircraft Maintenance Facility	Ground Equipment Maintenance Facility	Flight Kitchen	Air Freight Facility
20%	50%	10%	15%	5%

The Bonds shall be subject to mandatory redemption in whole prior to maturity, on October 1, 2023, at a redemption price equal to 100% of the principal amount thereof, plus accrued interest to the redemption date if the term of the Lease is not extended to October 1, 2032 in accordance with the provisions of the Lease and subject to the conditions in the Ordinance.

Limitations

Pursuant to the United Use and Lease Agreement, if costs at the New Airport exceed $20 per revenue enplaned passenger, in 1990 dollars, for the preceding calendar year, calculated in accordance with such agreement, United can elect to terminate its Use and Lease Agreement. Such termination by United would not, however, be an event of default under the Lease.

If United causes an event of default under the Lease and the City exercises its remedies thereunder and accelerates Facilities Rentals, the City is not obligated to relet the Facilities. If the City relets the Facilities, it is not obligated to use any of the payments received to pay principal, premium, if any, or interest on the Bonds.

Application of the Bond Proceeds

It is estimated that the proceeds of the sale of the Bonds will be applied as follows:

Cost of Construction	$226,002,433
Interest on Bonds During Construction	22,319,740
Cost of Issuance Including Underwriters' Discount	1,980,075
Original Issue Discount	11,112,742
Principal Amount of the Bonds	261,415,000

Tax Covenant

Under the terms of the lease, United has agreed that it will not take or omit to take any action with respect to the Facilities or the proceeds of the bonds (including any investment earnings thereon), insurance, condemnation, or any other proceeds derived in connection with the Facilities, which would cause the interest on the Bonds to become included in the gross income of the Bondholder for federal income tax purposes.

Exhibit B
Jokes about the Abbreviation DIA

DENVER—The Denver International Airport, whose opening has been delayed indefinitely because of snafus, has borne the brunt of joke writers

Punsters in the aviation and travel community have done their share of work on one particular genre, coming up with new variations on the theme of DIA, the star-crossed airport's new and as-yet-unused city code.

Here's what's making the rounds on electronic bulletin boards; it originated in the May 15 issue of the *Boulder* (Colo.) *Camera* newspaper.

1. Dis Is Awful
2. Doing It Again
3. Dumbest International Airport
4. Dinosaur In Action
5. Debt In Arrival
6. Denver's Intense Adventure
7. Darn It All
8. Dollar Investment Astounding
9. Delay It Again
10. Denver International Antique
11. Date Is AWOL
12. Denver Intellects Awry
13. Dance Is Autumn
14. Dopes In Authority
15. Don't Ice Attendance
16. Drop In Asylum
17. Don't Immediately Assume
18. Don't Ignore Aspirin
19. Dittohead Idle Again
20. Doubtful If Atall
21. Denver In Action
22. Deces, l'Inaugural Arrivage (means "dead on arrival" in French)
23. Dummies In Action
24. Dexterity In Action
25. Display In Arrogance
26. Denver Incomplete Act
27. D'luggage Is A'coming
28. Defect In Automation
29. Dysfunctional Itinerary Apparatus
30. Dis Is Absurd
31. Delays In Abundance
32. Did It Arrive?
33. Denver's Infamous Air-or-port (sounds like "error")
34. Dopes In Action
35. Doubtful Intermittent Access
36. Don't Intend Atall
37. Damned Inconvenient Airport
38. Duped In Anticipation
39. Delay In Action
40. Delirious In Accounting
41. Date Indeterminate, Ah?
42. Denver's Indisposed Access
43. Detained Interphase Ahead
44. Denver's Interminably Aground
45. Deceit In Action
46. Delay Institute America
47. Denver's Intractable Airport
48. Delayed Indefinitely Again
49. Delayed Introduction Again
50. Disaster In Arrears
51. Denver International Amusementpark
52. Debacle In Action
53. Deadline (of) Incomprehensible Attainment
54. Duffel Improbable Arrival
55. Delay In America
56. Dying In Anticipation
57. Dazzling Inaccessible Absurdity
58. Damned Intractable Automation
59. Da Infamous Annoyance
60. Dare I Ask?
61. Done In Arrears
62. Done In Ancestral
63. Denver International Accident
64. Dumb Idea Anyway
65. Diversion In Accounting
66. Doesn't Include Airlines
67. Disparate Instruments in Action
68. Delay International Airport
69. Dumb Idea Askew
70. Delayed Indefinitely Airport
71. Delays In Arrival
72. Deja In Absentee
73. Done In Aminute
74. Done In August
75. Denver's Inordinate Airport
76. Denver's Imaginary Airport
77. Debentures In Arrears
78. Denver Isn't Airborne
79. Descend Into Abyss
80. Done In April 2000
81. Disaster In Aviation
82. Denver's Interminable Airport
83. Denver In Arrears
84. Dallying Is Aggravating
85. Don't In Angst
86. Distress Is Acute
87. Development Is Arrested
88. Darned Inevitable Atrocity
89. Debt In Airport
90. Devastation In Aviation
91. Debacle in Automation
92. Denver's Inconstructable Airport
93. Denver Is Awaitin'
94. DIsAster
95. Denver's Inoperable Airport
96. Delay, Impede, Await
97. Date Isn't Available
98. Delayed International Airport
99. Denver Irrational Airport
100. Denver Irate Association
101. Denver's Ignominious Atrocity
102. Daytrippers Invitational Airport
103. Delay Is Anticipated
104. Doofis, Interruptness, Accidentalis
105. Denver International Arrival
106. Denver's Interminable Apparition
107. Distance Is Astronomical
108. Doubtful It's Able
109. Dreadfully Ineffective Automation
110. Do It Again
111. Did it, Installed it, Ate it
112. Drowned In Apoplexy
113. Dodo International Airport (the dodo is an extinct, flightless bird)
114. Dead In the Air
115. Denouement In Ambiguity
116. Deserted, Inactive Airport
117. Definitely Incapable of Activation
118. Democracy In Action
119. Dysfunction Imitating Art
120. Design In Alabaster
121. Desperately In Arrears
122. Dazzling, If Anything
123. Delays In Aeternum
124. Delighted If Actualized
125. Destination: Imagine Arabia
126. Dumb Idea: Abandoned?
127. Deem It Apiary
128. Dollars In Action
129. Definitely Iffy Achievement
130. Dreadfully Incompetent Architects
131. Denver International Ain't
132. Delayed In Automation
133. Dragging Its Ass
134. Driving Is Advantageous
135. Dang It All
136. Druggies Installing Automation
137. Dumb Idea Approved
138. Didn't Invite Airplanes
139. Died In April
140. Deplane In Albuquerque
141. Departure Is Agonizing
142. Denver's Infuriating Abscess
143. Denver's Ill-fated Airport
144. Domestic International Aggravation
145. Duffels In Anchorage
146. Denver's Indeterminate Abomination
147. Damn It All
148. Darn Idiotic Airport
149. Delay Is Acceptable
150. Denver's Idle Airport
151. Does It Arrive?
152. Damned Inconvenient Anyway

Source: Reprinted from *Boulder* (Colorado) *Camera* newspaper, May 15, 1991.

Other Material Covenants

United has agreed to acquire, construct, and install the Facilities to completion pursuant to the terms of the Lease. If monies in the Construction Fund are insufficient to pay the cost of such acquisition, construction, and installation in full, then United shall pay the excess cost without reimbursement from the City, the Paying Agent, or any Bondholder.

United has agreed to indemnify the City and the Paying Agent for damages incurred in connection with the occurrence of certain events, including without limitation, the construction of the Facilities, occupancy by United of the land on which the Facilities are located, and violation by United of any of the terms of the Lease or other agreements related to the Leased Property.

During the Lease Term, United has agreed to maintain its corporate existence and its qualifications to do business in the state. United will not dissolve or otherwise dispose of its assets and will not consolidate with or merge into another corporation provided, however, that United may, without violating the Lease, consolidate or merge into another corporation.

Additional Bonds

At the request of United, the City may, at its option, issue additional bonds to finance the cost of special Facilities for United upon the terms and conditions in the Lease and the Ordinance.

The Guaranty

Under the Guaranty, United will unconditionally guarantee to the Paying Agent, for the benefit of the Bondholders, the full and prompt payment of the principal, premium, if any, and interest on the Bonds, when and as the same shall become due whether at the stated maturity, by redemption, acceleration, or otherwise. The obligations of United under the Guaranty are unsecured, but are stated to be absolute and unconditional, and the Guaranty will remain in effect until the entire principal, premium, if any, and interest on the Bonds has been paid in full or provision for the payment thereof has been made in accordance with the Ordinance.

Bibliography (in Chronological Order)

1. David A. Brown, "Denver Aims for Global Hub Status with New Airport Under Construction," *Aviation Week & Space Technology,* March 11, 1991, pp. 42–45
2. "Satellite Airport to Handle Corporate, General Aviation for Denver Area," *Aviation Week & Space Technology,* March 11, 1991, pp. 44–45.
3. "Denver to Seek Bids This Spring for Wide-Open Terminal Building," *Aviation Week & Space Technology,* March 11, 1991, p. 50.
4. "Denver City Council Supports Airport Despite Downgrade," *The Wall Street Journal,* March 20, 1991, p. A1D.
5. "Denver Airport Bonds' Rating Is Confirmed by Moody's Investors," *The Wall Street Journal,* March 22, 1991, p. C14.
6. "Bonds for Denver Airport Priced to Yield up to 9.185%," *New York Times,* April 10, 1991, p. D16.
7. Marj Charlier, "Denver Reports a Tentative Agreement with United over Hub at New Airport," *The Wall Street Journal,* May 3, 1991, p. B2.
8. Brad Smith, "New Airport Has Its Ups and Downs," *Los Angeles Times,* July 9, 1991, p. A5.
9. Christopher Wood, "Hotel Development at New Airport Not Likely Until After '93," *Denver Business Journal,* August 2, 1991, p. 8S.
10. Christopher Wood, "FAA: Link Air Cargo, Passengers," *Denver Business Journal,* November 1–7, 1991, p. 3.
11. Christopher Wood, "Airport May Move Cargo Operations, Offer Reserve Funds," *Denver Business Journal,* December 6–12, 1991, pp. 1, 34.
12. "UAL in Accord on Denver," *The New York Times,* December 7, 1991, p. 39L.
13. Thomas Fisher, "Projects Flights of Fantasy," *Progressive Architecture,* March 1992, p. 103.

14. Tom Locke, "Disconnected," *Denver Business Journal,* June 12–18, 1992, p. 19.
15. "Big Ain't Hardly the Word for It," *ENR,* September 7, 1992, pp. 28–29.
16. Christopher Wood, "Adams Seeks Action," *Denver Business Journal,* September 4–10, 1992, pp. 1, 13.
17. "Denver Airport Rises under Gossamer Roof," *The Wall Street Journal,* November 17, 1992, p. B1.
18. Mark B. Solomon, "Denver Airport Delay Angers Cargo Carriers," *Journal of Commerce,* March 17, 1993, p. 3B.
19. "Denver Airport Opening Delayed Until December," *Aviation Week & Space Technology,* May 10, 1993, p. 39.
20. Aldo Svaldi, "DIA Air Train Gathering Steam as Planners Shift Possible Route," *Denver Business Journal,* August 27–September 2, 1993, p. 74.
21. Dirk Johnson, "Opening of New Denver Airport is Delayed Again," *The New York Times,* October 26, 1993, p. A19.
22. "Denver's Mayor Webb Postpones Opening International Airport," *The Wall Street Journal,* October 26, 1993, p. A9.
23. "An Airport Comes to Denver," *Skiing,* December 1993, p. 66.
24. Ellis Booker, "Airport Prepares for Takeoff," *Computerworld,* January 10, 1994.
25. Aldo Svaldi, "Front Range, DIA Weigh Merging Airport Systems," *Denver Business Journal,* January 21–27, 1994, p. 3.
26. Don Phillips, "$3.1 Billion Airport at Denver Preparing for a Rough Takeoff," *The Washington Post,* February 13, 1994, p. A10.
27. "New Denver Airport Combines Several State-of-the-Art Systems," *Travel Weekly,* February 21, 1994, p. 20.
28. Steve Munford, "Options in Hard Surface Flooring," *Buildings,* March 1994, p. 58.
29. Mars Charles, "Denver's New Airport, Already Mixed in Controversy, Won't Open Next Week," *The Wall Street Journal,* March 2, 1994, pp. B1, B7.
30. "Denver Grounded for Third Time," *ENR,* March 7, 1994, p. 6.
31. Shannon Peters, "Denver's New Airport Creates HR Challenges," *Personnel Journal,* April 1994, p. 21.
32. Laura Del Rosso, "Denver Airport Delayed Indefinitely," *Travel Weekly,* May 5, 1994, p. 37.
33. "DIA Bond Rating Cut," *Aviation Week & Space Technology,* May 16, 1994, p. 33.
34. Robert Scheler, "Software Snafu Grounds Denver's High-Tech Airport," *PC Week,* May 16, 1994, p. 1.
35. John Dodge, "Architects Take a Page from Book on Denver Airport-Bag System," *PC Week,* May 16, 1994, p. 3.
36. Jean S. Bozman, "Denver Airport Hits Systems Layover," *Computerworld,* May 16, 1994, p. 30.
37. Richard Woodbury, "The Bag Stops Here," *Time,* May 16, 1994, p. 52.
38. "Consultants Review Denver Baggage Problems," *Aviation Week & Space Technology,* June 6, 1994, p. 38.
39. "Doesn't It Amaze? The Delay that Launched a Thousand Gags," *Travel Weekly,* June 6, 1994, p. 16.
40. Michael Romano, "This Delay Is Costing Business a Lot of Money," *Restaurant Business,* June 10, 1994, p. 26.
41. Scott Armstrong, "Denver Builds New Airport, Asks 'Will Planes Come?'," *The Christian Science Monitor,* June 21, 1994, p. 1.
42. Benjamin Weiser, "SEC Turns Investigation to Denver Airport Financing," *The Washington Post,* July 13, 1994, p. D1.
43. Bernie Knill, "Flying Blind at Denver International Airport," *Material Handling Engineering,* July 1994, p. 47.
44. Keith Dubay, "Denver Airport Seeks Compromise on Baggage Handling," *American Banker Washington Watch,* July 25, 1994, p. 10.
45. Dirk Johnson, "Denver May Open Airport in Spite of Glitches," *The New York Times,* July 27, 1994, p. A14.

46. Jeffrey Leib, "Investors Want a Plan," *The Denver Post,* August 2, 1994, p. A1.
47. Marj Charlier, "Denver Plans Backup Baggage System for Airport's Troubled Automated One," *The Wall Street Journal,* August 5, 1994, p. B2.
48. Louis Sahagun, "Denver Airport to Bypass Balky Baggage Mover," *Los Angeles Times,* August 5, 1994, p. A1.
49. Len Morgan, "Airports Have Growing Pains," *Flying,* August 1994, p. 104.
50. Adam Bryant, "Denver Goes Back to Basics for Baggage," *The New York Times,* August 6, 1994, pp. 5N, 6L.
51. "Prosecutors Scrutinize New Denver Airport," *The New York Times,* August 21, 1994, p. 36L.
52. Kevin Flynn, "Panic Drove New DIA Plan," *Rocky Mountain News,* August 7, 1994, p. 5A.
53. David Hughes, "Denver Airport Still Months from Opening," *Aviation Week & Space Technology,* August 8, 1994, p. 30.
54. "Airport May Open in Early '95," *Travel Weekly,* August 8, 1994, p. 57.
55. Michael Meyer, and Daniel Glick, "Still Late for Arrival," *Newsweek,* August 22, 1994, p. 38.
56. Andrew Bary, "A $3 Billion Joke," *Barron's,* August 22, 1994, p. MW10.
57. Jean Bozman, "Baggage System Woes Costing Denver Airport Millions," *Computerworld,* August 22, 1994, p. 28.
58. Edward Phillips, "Denver, United Agree on Baggage System Fixes," *Aviation Week & Space Technology,* August 29, 1994.
59. Glenn Rifkin, "What Really Happened at Denver's Airport," *Forbes,* August 29, 1994, p. 110.
60. Andrew Bary, "New Denver Airport Bond Issue Could Face Turbulence from Investors," *Barron's,* August 29, 1994, p. MW9.
61. Andrew Bary, "Denver Airport Bonds Take Off as Investors Line Up for Higher Yields," *Barron's,* August 29, 1994, p. MW9.
62. Susan Carey, "Alaska's Cash-Strapped MarkAir Is Wooed by Denver," *The Wall Street Journal,* September 1, 1994, p. B6.
63. Dana K. Henderson, "It's in the Bag(s)," *Air Transport World,* September 1994, p. 54.
64. Dirk Johnson, "Late Already, Denver Airport Faces More Delays," *The New York Times,* September 25, 1994, p. 26L.
65. Gordon Wright, "Denver Builds a Field of Dreams," *Building Design and Construction,* September 1994, p. 52.
66. Alan Jabez, "Airport of the Future Stays Grounded," *Sunday Times,* October 9, 1994, Features Section.
67. Jean Bozman, "United to Simplify Denver's Troubled Baggage Project," *Computerworld,* October 10, 1994, p. 76.
68. "Denver Aide Tells of Laxity in Airport Job," *The New York Times,* October 17, 1994, p. A12.
69. Brendan Murray, "In the Bags: Local Company to Rescue Befuddled Denver Airport," *Marietta Daily Journal,* October 21, 1994, p. C1.
70. Joanne Wojcik, "Airport in Holding Pattern, Project Is Insured, but Denver to Retain Brunt of Delay Costs," *Business Insurance,* November 7, 1994, p. 1.
71. James S. Russell, "Is This Any Way to Build an Airport?," *Architectural Record,* November 1994, p. 30.

Questions

1. Is the decision to build a new airport at Denver strategically a sound decision?
2. Perform an analysis for strengths, weaknesses, opportunities, and threats (SWOT) on the decision to build DIA.
3. Who are the stakeholders and what are their interests or objectives?
4. Did the airlines support the decision to build DIA?
5. Why was United Opposed to expansion at Front Range Airport?
6. Why was the new baggage handling system so important to United?
7. Is DIA a good strategic fit for Continental?

8. What appears to be the single greatest risk in the decision to build DIA?
9. United is a corporation in business to make money. How can United issue tax-free municipal bonds?
10. What impact do the rating agencies (i.e., Moody's and Standard & Poor's) have in the financing of the airport?
11. According to the prospectus, the DIA bonds were rated as BBB− by Standard & Poor's Corporation. Yet, at the same time, the City of Denver was given a rating of AA. How can this be?
12. On October 1, 1992, the United bonds were issued at an interest rate of 6.875 percent. Was this an appropriate coupon for the bonds?
13. There are numerous scenarios that can occur once the airport opens. The following questions are "what if" exercises and may not have a right or wrong answer. The questions are used to stimulate classroom discussion. The students must use the prospectus excerpts in the exhibit at the end of the case study. For each situation, what will be the possible outcome and what impact is there upon the bondholders?
14. Assume that DIA finally opens and with a debt of $3 billion. Is the revenue stream sufficient to pay interest each year *and* pay the principal at maturity?
15. What options are available to DIA if the coverage falls below 100 percent?
16. If the debt coverage were actually this good, why would the ratings on the bonds be BB?
17. One of the critical parameters that airlines use is the cost per enplaned passenger. Using Exhibit V, determine whether the cost per enplaned passenger can be lowered.
18. Is there additional revenue space available (i.e., unused capacity)?
19. What is the function of the project management team (PMT) and why were two companies involved?
20. When did the effectiveness of the project management team begin to be questioned?
21. Did it sound as though the statement of work/specifications provided by the city to the PMT was "vague" for the design phase?
22. During the design phase, contractors were submitting reestimates for work, 30 days after their original estimates, and the new estimates were up to $50 million larger than the prior estimate. Does this reflect upon the capabilities of the PMT?
23. Should the PMT be qualified to perform risk analyses?
24. Why were the architects coordinating the changes at the construction site?
25. Should the PMT have been replaced?
26. Do scope changes reflect upon the ineffectiveness of a project management team?
27. Why did United Airlines decide to act as the project manager for the baggage handling system on Concourse B?

Case 24

MIS Project Management at First National Bank*

During the last five years, First National Bank (FNB) has been one of the fastest-growing banks in the Midwest. The holding company of the bank has been actively involved in purchasing small banks thoughout the state of Ohio. This expansion and the resulting increase of operations had been attended by considerable growth in numbers of employees and in the complexity of the organizational structure. In five years the staff of the bank has increased by 35 percent, and total assets have grown by 70 percent. FNB management is eagerly looking forward to a change in the Ohio banking laws that will allow statewide branch banking.

Information Services Division (ISD) History

Data processing at FNB has grown at a much faster pace than the rest of the bank. The systems and programming staff grew from 12 in 1970 to over 75 during the first part of 1977. Because of several future projects, the staff is expected to increase by 50 percent during the next two years.

Prior to 1972, the Information Services Department reported to the executive vice president of the Consumer Banking and Operations Division. As a result, the first banking applications to be computerized were in the demand deposit, savings, and consumer credit banking areas. The computer was seen as a tool to speed up the processing of consumer transactions. Little effort was expended to meet the informational requirements of the rest of the bank. This caused a high-level conflict, since each major operating organization of the bank did not have equal access to systems and programming resources. The management of FNB became increasingly aware of the benefits that could accrue from a realignment of the bank's organization into one that would be better attuned to the total information requirements of the corporation.

In 1982 the Information Services Division (ISD) was created. ISD was removed from the Consumer Banking and Operations Division to become a separate division reporting directly to the president. An organizational chart depicting the Information Services Division is shown in Exhibit I.

---

*Reprinted from H. Kerzner, *Project Management: A Systems Approach to Planning, Scheduling, and Controlling,* 6th ed. New York: Wiley, 1998, pp. 359–367.

Exhibit I. Information Services Division organizational chart

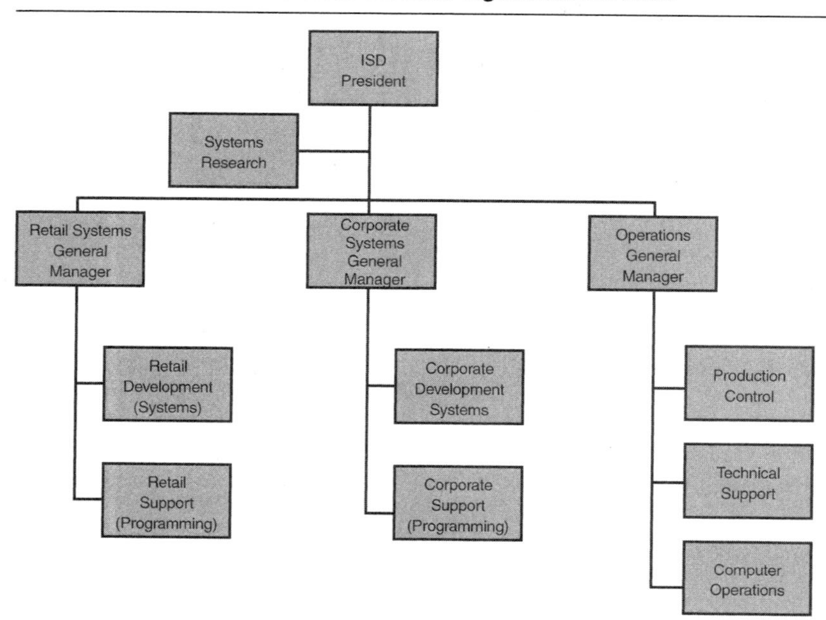

Priorities Committee

During 1982 the Priorities Committee was formed. It consists of the chief executive officer of each of the major operating organizations whose activities are directly affected by the need for new or revised information systems. The Priorities Committee was established to ensure that the resources of systems and programming personnel and computer hardware would be used only on those information systems that can best be cost justified. Divisions represented on the committee are included in Exhibit II.

The Priorities Committee meets monthly to reaffirm previously set priorities and rank new projects introduced since the last meeting. Bank policy states that the only way to obtain funds for an information development project is to submit a request to the Priorities Committee and have it approved and ranked in overall priority order for the bank. Placing potential projects in ranked sequence is done by the senior executives. The primary document used for Priorities Committee review is called the project proposal.

The Project Proposal Life Cycle

When a user department determines a need for the development or enhancement of an information system, it is required to prepare a draft containing a statement of the problem from its functional perspective. The problem statement is sent to the president of ISD, who authorizes Systems Research (see Exhibit I) to prepare an impact statement. This impact statement will include a general overview from ISD's perspective of:

- Project feasibility
- Project complexity
- Conformity with long-range ISD plans
- Estimated ISD resource commitment

Exhibit II. First National Bank organizational chart

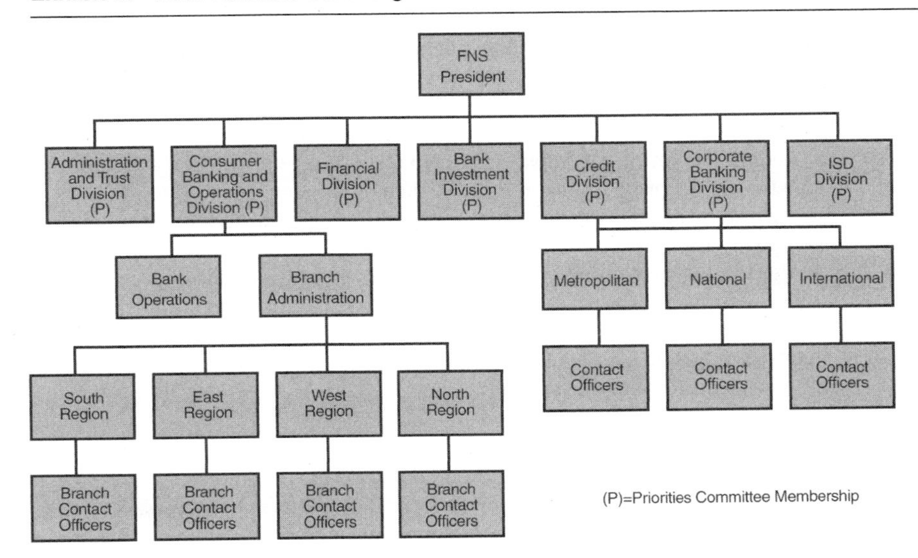

- Review of similar requests
- Unique characteristics/problems
- Broad estimate of total costs

The problem and impact statements are then presented to the members of the Priorities Committee for their review. The proposals are preliminary in nature, but they permit the broad concept (with a very approximate cost attached to it) to be reviewed by the executive group to see if there is serious interest in pursuing the idea. If the interest level of the committee is low, then the idea is rejected. However, if the Priorities Committee members feel the concept has merit, they authorize the Systems Research Group of ISD to prepare a full-scale project proposal that contains:

- A detailed statement of the problem
- Identification of alternative solutions
- Impact of request on:
  - User division
  - ISD
  - Other operating divisions
- Estimated costs of solutions
- Schedule of approximate task duration
- Cost-benefit analysis of solutions
- Long-range implications
- Recommended course of action

After the project proposal is prepared by systems research, the user sponsor must review the proposal and appear at the next Priorities Committee meeting to speak in favor of the approval and priority level of the proposed work. The project proposal is evaluated by the committee and either dropped, tabled for further review, or assigned a priority relative to ongoing projects and available resources.

The final output of a Priorities Committee meeting is an updated list of project proposals in priority order with an accompanying milestone schedule that indicates the approximate time span required to implement each of the proposed projects.

The net result of this process is that the priority-setting for systems development is done by a cross section of executive management; it does not revert by default to data processing management. Priority-setting, if done by data processing, can lead to misunderstanding and dissatisfaction by sponsors of the projects that did not get ranked high enough to be funded in the near future. The project proposal cycle at FNB is diagrammed in Exhibit III. Once a project has risen to the top of the ranked priority list, it is assigned to the appropriate systems group for systems definition, system design and development, and system implementation.

The time spent by systems research in producing impact statements and project proposals is considered to be overhead by ISD. No systems research time is directly charged to the development of information systems.

Project Life Cycle

As noted before, the systems and programming staff of ISD has increased in size rapidly and is expected to expand by another 50 percent over the next two years. As a rule, most new employees have previous data processing experience and training in various systems methodologies. ISD management recently implemented a project management system dedicated to providing a uniform step-by-step methodology for the development of management information systems. All project work is covered by tasks that make up the information project development life cycle at FNB. The subphases used by ISD in the project life cycle are:

1. Systems definition
   a. Project plan
   b. User requirements
   c. Systems definition
   d. Advisability study
2. Systems design and development
   a. Preliminary systems design
   b. Subsystems design
   c. Program design
   d. Programming and testing
3. System implementation
   a. System implementation
   b. System test
   c. Production control turnover
   d. User training
   e. System acceptance

Project Estimating

The project management system contains a list of all normal tasks and subtasks (over 400) to be performed during the life cycle of a development project. The project manager must examine all the tasks to determine if they apply to a given project. The manager must insert additional tasks if required and delete tasks that do not apply. The project manager next estimates the amount of time (in hours) to complete each task of each subphase of the project life cycle.

The estimating process of the project management system uses a "moving window" concept. ISD management feels that detailed cost estimating and time schedules are only mean-

Exhibit III. The project proposal cycle

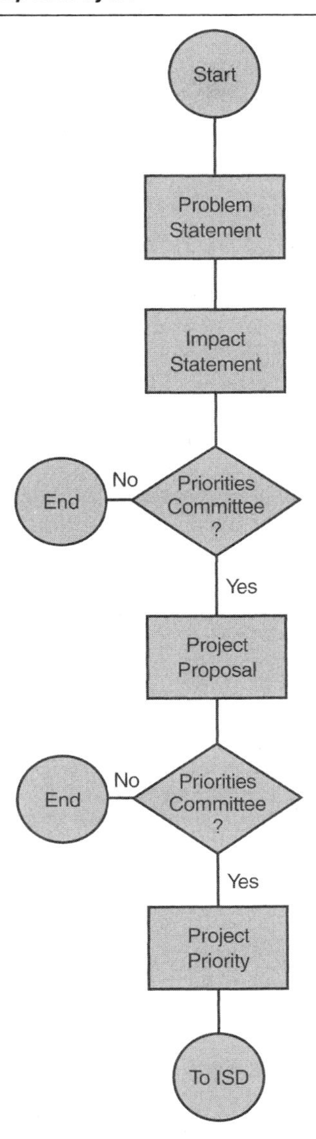

ingful for the next subphase of a project, where the visibility of the tasks to be performed is quite clear. Beyond that subphase, a more summary method of estimating is relied on. As the project progresses, new segments of the project gain visibility. Detailed estimates are made for the next major portion of the project, and summary estimates are done beyond that until the end of the project.

Estimates are performed at five intervals during the project life cycle. When the project is first initiated, the funding is based on the original estimates, which are derived from the list of normal tasks and subtasks. At this time, the subphases through the advisability study are estimated in detail, and summary estimates are prepared for the rest of the tasks in the project. Once the project has progressed through the advisability study, the preliminary systems design is estimated in detail, and the balance of the project is estimated in a more summary fashion. Estimates are conducted in this manner until the systems implementation plan is completed and the scope of the remaining subphases of the project is known. This multiple estimating process is used because it is almost impossible at the beginning of many projects to be certain of what the magnitude of effort will be later on in the project life cycle.

Funding of Projects

The project plan is the official document for securing funding from the sponsor in the user organization. The project plan must be completed and approved by the project manager before activity can begin on the user requirements subphase (1b). An initial stage in developing a project plan includes the drawing of a network that identifies each of the tasks to be done in the appropriate sequence for their execution. The project plan must include a milestone schedule, a cost estimate, and a budget request. It is submitted to the appropriate general manager of systems and programming for review so that an understanding can be reached of how the estimates were prepared and why the costs and schedules are as shown. At this time the general manager can get an idea of the quantity of systems and programming resources required by the project. The general manager next sets up a meeting with the project manager and the user sponsor to review the project plan and obtain funding from the user organization.

The initial project funding is based on an estimate that includes a number of assumptions concerning the scope of the project. Once certain key milestones in the project have been achieved, the visibility on the balance of the project becomes much clearer, and reestimates are performed. The reestimates may result in refunding if there has been a significant change in the project. The normal milestone refunding points are as follows:

1. After the advisability study (1d)
2. After the preliminary systems design (2a)
3. After the program design (2c)
4. After system implementation (3a)

The refunding process is similar to the initial funding with the exception that progress information is presented on the status of the work and reasons are given to explain deviations from project expenditure projections. A revised project plan is prepared for each milestone refunding meeting.

During the systems design and development stage, design freezes are issued by the project manager to users announcing that no additional changes will be accepted to the project beyond that point. The presence of these design freezes is outlined at the beginning of the project. Following the design freeze, no additional changes will be accepted unless the project is reestimated at a new level and approved by the user sponsor.

System Quality Reviews

The key element in ensuring user involvement in the new system is the conducting of quality reviews. In the normal system cycles at FNB, there are ten quality reviews, seven of which are participated in jointly by users and data processing personnel, and three of which are technical reviews by data processing (DP) personnel only. An important side benefit of this review process is that users of a new system are forced to become involved in and are permitted to make a contribution to the systems design.

Each of the quality review points coincides with the end of a subphase in the project life cycle. The review must be held at the completion of one subphase to obtain authorization to begin work on the tasks of the next subphase of the project.

All tasks and subtasks assigned to members of the project team should end in some "deliverable" for the project documentation. The first step in conducting a quality review is to assemble the documentation produced during the subphase for distribution to the Quality Review Board. The Quality Review Board consists of between two and eight people who are appointed by the project manager with the approval of the project sponsor and the general manager of systems and programming. The minutes of the quality review meeting are written either to express "concurrence" with the subsystem quality or to recommend changes to the system that must be completed before the next subphase can be started. By this process the system is fine-tuned to the requirements of the members of the review group at the end of each subphase in the system. The members of the Quality Review Board charge their time to the project budget.

Quality review points and review board makeup are as follows:

Review	Review Board
User requirements	User oriented
Systems definition	User oriented
Advisability study	User oriented
Preliminary systems design	User oriented
Subsystems design	Users and DP
Program design	DP
Programming and testing	DP
System implementation	User oriented
System test	User oriented
Production control turnover	DP

To summarize, the quality review evaluates the quality of project subphase results, including design adequacy and proof of accomplishment in meeting project objectives. The review board authorizes work to progress based on their detailed knowledge that all required tasks and subtasks of each subphase have been successfully completed and documented.

Project Team Staffing

Once a project has risen to the top of the priority list, the appropriate manager of systems development appoints a project manager from his or her staff of analysts. The project manager has a short time to review the project proposal created by systems research before developing a project plan. The project plan must be approved by the general manager of systems and programming and the user sponsor before the project can be funded and work started on the user requirements subphase.

The project manager is "free" to spend as much time as required in reviewing the project proposal and creating the project plan; however, this time is "charged" to the project at a rate of $26 per hour. The project manager must negotiate with a "supervisor," the manager of systems development, to obtain the required systems analysts for the project, starting with the user requirements subphase. The project manager must obtain programming resources from the manager of systems support. Schedule delays caused by a lack of systems or programming resources are to be communicated to the general manager by the project manager. All ISD personnel working on a project charge their time at a rate of $26 per hour. All computer time is billed at a rate of $64 per hour.

There are no user personnel on the project team; all team members are from ISD.

Corporate Database

John Hart had for several years seen the need to use the computer to support the corporate marketing effort of the bank. Despite the fact that the majority of the bank's profits were from corporate customers, most information systems effort was directed at speeding up transactions handling for small unprofitable accounts.

Mr. Hart had extensive experience in the Corporate Banking Division of the bank. He realized the need to consolidate information about corporate customers from many areas of the bank into one corporate database. From this information corporate banking services could be developed not only to better serve the corporate customers, but also to contribute heavily to the profit structure of the bank through repricing of services.

The absence of a corporate database meant that no one individual knew what total banking services a corporate customer was using, because corporate services were provided by many banking departments. It was also impossible to determine how profitable a corporate customer was to the bank. Contact officers did not have regularly scheduled calls. They serviced corporate customers almost on a hit-or-miss basis. Unfortunately, many customers were "sold" on a service because they walked in the door and requested it. Mr. Hart felt that there was a vast market of untapped corporate customers in Ohio who would purchase services from the bank if they were contacted and "sold" in a professional manner. A corporate database could be used to develop corporate profiles to help contact officers sell likely services to corporations.

Mr. Hart knew that data about corporate customers was being processed in many departments of the bank, but mainly in the following divisions:

- Corporate Banking
- Corporate Trust
- Consumer banking

He also realized that much of the information was processed in manual systems, some was processed by time-sharing at various vendors, and other information was computerized in many internal information systems.

The upper management of FNB must have agreed with Mr. Hart because in December of 1986 the Corporate Marketing Division was formed with John Hart as its executive vice president. Mr. Hart was due to retire within the year but was honored to be selected for the new position. He agreed to stay with the bank until "his" new system was "off the ground." He immediately composed a problem statement and sent it to the ISD. Systems Research compiled a preliminary impact statement. At the next Priorities Committee meeting, a project proposal was authorized to be done by Systems Research.

The project proposal was completed by Systems Research in record time. Most information was obtained from Mr. Hart. He had been thinking about the systems requirements for years and possessed vast experience in almost all areas of the bank. Other user divisions and departments were often "too busy" when approached for information. A common reply to a request for information was, "That project is John's baby; he knows what we need."

The project proposal as prepared by Systems Research recommended the following:

- Interfaces should be designed to extract information from existing computerized systems for the corporate database (CDB).
- Time-sharing systems should be brought in-house to be interfaced with the CDB.
- Information should be collected from manual systems to be integrated into the CDB on a temporary basis.
- Manual systems should be consolidated and computerized, potentially causing a reorganization of some departments.
- Information analysis and flow for all departments and divisions having contact with corporate customers should be coordinated by the Corporate Marketing Division.

- All corporate database analysis should be done by the Corporate Marketing Division staff, using either a user-controlled report writer or interactive inquiry.

The project proposal was presented at the next Priorities Committee meeting where it was approved and rated as the highest priority MIS development project in the bank. Mr. Hart became the user sponsor for the CDB project.

The project proposal was sent to the manager of corporate development, who appointed Jim Gunn as project manager from the staff of analysts in corporate development. Jim Gunn was the most experienced project manager available. His prior experience consisted of successful projects in the Financial Division of the bank.

Jim reviewed the project proposal and started to work on his project plan. He was aware that the corporate analyst group was presently understaffed but was assured by his manager, the manager of corporate development, that resources would be available for the user requirements subphase. He had many questions concerning the scope of the project and the interrelationship between the Corporate Marketing Division and the other users of corporate marketing data. But each meeting with Mr. Hart ended with the same comment: "This is a waste of time. I've already been over this with Systems Research. Let's get moving." Jim also was receiving pressure from the general manager to "hurry up" with the project plan. Jim therefore quickly prepared his project plan, which included a general milestone schedule for subphase completion, a general cost estimate, and a request for funding. The project plan was reviewed by the general manager and signed by Mr. Hart.

Jim Gunn anticipated the need to have four analysts assigned to the project and went to his manager to see who was available. He was told that two junior analysts were available now and another analyst should be free next week. No senior analysts were available. Jim notified the general manager that the CDB schedule would probably be delayed because of a lack of resources, but received no response.

Jim assigned tasks to the members of the team and explained the assignments and the schedule. Since the project was understaffed, Jim assigned a heavy load of tasks to himself.

During the next two weeks the majority of the meetings set up to document user requirements were canceled by the user departments. Jim notified Mr. Hart of the problem and was assured that steps would be taken to correct the problem. Future meetings with the users in the Consumer Banking and Corporate Banking Divisions became very hostile. Jim soon discovered that many individuals in these divisions did not see the need for the corporate database. They resented spending their time in meetings documenting the CDB requirements. They were afraid that the CDB project would lead to a shift of many of their responsibilities and functions to the Corporate Marketing Division.

Mr. Hart was also unhappy. The CDB team was spending more time than was budgeted in documenting user requirements. If this trend continued, a revised budget would have to be submitted to the Priorities Committee for approval. He was also growing tired of ordering individuals in the user departments to keep appointments with the CDB team. Mr. Hart could not understand the resistance to his project.

Jim Gunn kept trying to obtain analysts for his project but was told by his manager that none were available. Jim explained that the quality of work done by the junior analysts was not "up to par" because of lack of experience. Jim complained that he could not adequately supervise the work quality because he was forced to complete many of the analysis tasks himself. He also noted that the quality review of the user requirements subphase was scheduled for next month, making it extremely critical that experienced analysts be assigned to the project. No new personnel were assigned to the project. Jim thought about contacting the general manager again to explain his need for more experienced analysts, but did not. He was due for a semiyearly evaluation from his manager in two weeks.

Even though he knew the quality of the work was below standards, Jim was determined to get the project done on schedule with the resources available to him. He drove both himself and the team very hard during the next few weeks. The quality review of the user requirement subphase was held on schedule. Over 90 percent of the assigned tasks had to be redone before the Quality Review Board would sign-off on the review. Jim Gunn was removed as project manager.

Three senior analysts and a new project manager were assigned to the CDB project. The project received additional funding from the Priorities Committee. The user requirements subphase was completely redone despite vigorous protests from the Consumer Banking and Corporate Banking divisions.

Within the next three months the following events happened:

- The new project manager resigned to accept a position with another firm.
- John Hart took early retirement.
- The CDB project was tabled.

Synopsis

All projects at First National Bank (FNB) have project managers assigned and are handled through the Information Services Division (ISD). The organizational structure is not a matrix, although some people think that it is. The case describes one particular project, the development of a corporate database, and the resulting failure. The problem at hand is to investigate why the project failed.

Questions

1. What are the strengths of FNB?
2. What are the major weaknesses?
3. What is the major problem mentioned above? Defend your answer.
4. How many people did the project manager have to report to?
5. Did the PM remain within vertical structure of the organization?
6. Is there anything wrong if a PM is a previous co-worker of some team members before the team is formed?
7. Who made up the project team?
8. Was there any resistance to the project by comanagement?
9. Was there an unnecessary duplication of work?
10. Was there an increased resistance to change?
11. Was the communication process slow or fast?
12. Was there an increased amount of paperwork?
13. What are reasonable recommendations?

Case 25

Concrete Masonry Corporation*

Introduction

The Concrete Masonry Corporation (CMC), after being a leader in the industry for over 25 years, decided to get out of the prestressed concrete business. Although there had been a boom in residential construction in recent years, commercial work was on the decline. As a result, all the prestressed concrete manufacturers were going farther afield to big jobs. In order to survive, CMC was forced to bid on jobs previously thought to be out of their geographical area. Survival depended upon staying competitive.

In 1975, the average selling price of a cubic foot of concrete was $8.35, and in 1977, the average selling price had declined to $6.85. As CMC was producing at a rate of a million cubic feet a year, not much mathematics was needed to calculate they were receiving one-and-a-half million dollars per year less than they had received a short two years before for the same product.

Product management was used by CMC in a matrix organizational form. CMC's project manager had total responsibility from the design to the completion of the construction project. However, with the declining conditions of the market and the evolution that had drastically changed the character of the marketplace, CMC's previously successful approach was in question.

History—The Concrete Block Business

CMC started in the concrete block business in 1946. At the beginning, CMC became a leader in the marketplace for two reasons: (1) advanced technology of manufacturing and (2) an innovative delivery system. With modern equipment, specifically the flat pallet block machine, CMC was able to make different shapes of block without having to make major changes in the machinery. This change, along with the pioneering of the self-unloading boom truck, which

*Reprinted from H. Kerzner, *Project Management: A Systems Approach to Planning, Scheduling, and Controlling,* 6th ed. New York: Wiley, 1998, pp. 368–373.

permitted efficient, cost-saving delivery, contributed to the success of CMC's block business. Consequently, the block business success provided the capital needed for CMC to enter the prestressed concrete business.

History—The Prestressed Concrete Business

Prestressed concrete is made by casting concrete around steel cables that are stretched by hydraulic jacks. After the concrete hardens, the cables are releasd, thus compressing the concrete. Concrete is strongest when it is compressed. Steel is strongest when it is stretched, or in tension. In this way, CMC combined the two strongest qualities of the two materials. The effectiveness of the technique can be readily demonstrated by lifting a horizontal row of books by applying pressure at each end of the row at a point below the center of gravity.

Originally, the concrete block manufacturing business was a natural base from which to enter the prestressed concrete business because the very first prestressed concrete beams were made of a row of concrete block, prestressed by using high tension strength wires through the cores of the block. The wire was pulled at a high tension, and the ends of the beams were grouted. After the grout held the wires or cables in place, the tension was released on the cables, with resultant compression on the bottom portion of the beams. Thus the force on the bottom of the beam would tend to counteract the downward weight put on the top of the beam. By this process, these prestressed concrete beams could cover three to four times the spans possible with conventional reinforced concrete.

In 1951, after many trips to Washington, DC, and an excellent selling job by CMC's founder, T. L. Goudvis, CMC was able to land their first large-volume prestressed concrete project with the Corps of Engineers. The contract authorized the use of prestressed concrete beams, as described, with concrete block for the roofs of warehouses in the large Air Force depot complex being built in Shelby, Ohio. The buildings were a success, and CMC immediately received prestige and notoriety as a leader in the prestressed concrete business.

Wet-cast beams were developed next. For wet-cast beams, instead of concrete block, the cables were placed in long forms and pulled to the desired tension, after which concrete was poured in the forms to make beams. As a result of wet-cast beams, prestressed concrete was no longer dependent on concrete block.

At first, prestressed concrete was primarily for floors and roofs, but, in the early 1960s, precasters became involved in more complicated structures. CMC started designing and making not only beams, but columns and whatever other components it took to put together a whole structure. Parking garages became a natural application for prestressed concrete structures. Eventually an entire building could be precast out of prestressed concrete.

Project Management

Constructing the entire building, as in the case of a parking garage, meant that jobs were becoming more complex with respect to interdependence of detailed task accomplishment. Accordingly, in 1967, project management was established at CMC. The functional departments did the work, but the project managers saw to it that the assigned projects were completed on schedule and within budget and specifications. A matrix organization, as illustrated in Exhibit I, was adopted and used effectively by CMC. The concept of a matrix organization, as applied at CMC, entailed an organizational system designed as "web of relationships" rather than a line and staff relationship for work performance.

Each project manager was assigned a number of personnel with the required qualification from the functional departments for the duration of the project. Thus the project organization was composed of the project manager and functional personnel groups. The project manager had not only the responsibility and accountability for the successful completion of the contract,

Exhibit I. Matrix organization of Concrete Masonry Corporation

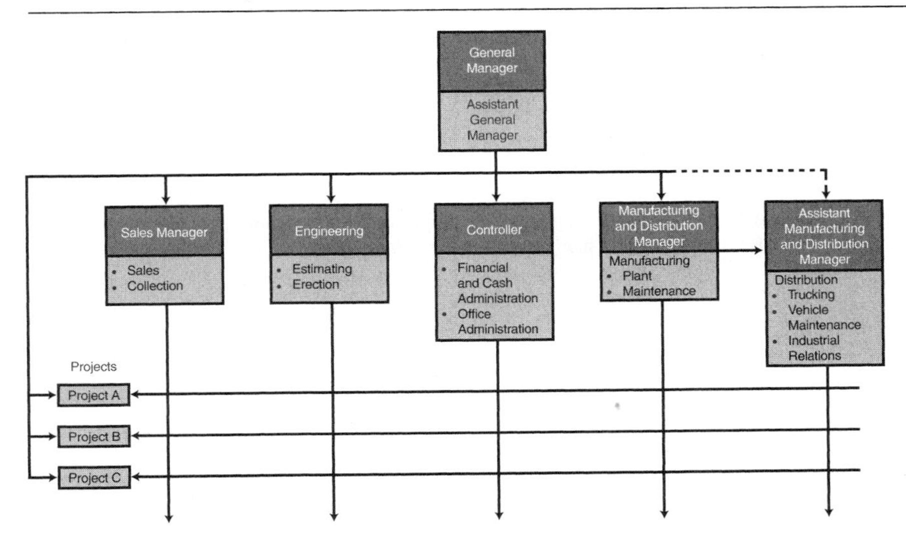

but also the delegated authority for work design, assignments of functional group personnel, and the determination of procedural relationships.

The most important functional area for the project manager was the engineering department, since prestressed concrete is a highly engineered product. A great deal of coordination and interaction was required between the project manager and the engineering department just to make certain that everything fit together and was structurally sound. A registered engineer did the design. The project manager's job was to see that the designing was done correctly and efficiently. Production schedules were made up by the project manager subject, of course, to minor modifications by the plant. The project manager was also required to do all the coordination with the customer, architect, general contractor, and the erection force. The project manager was also required to have interaction with the distribution manager to be certain that the product designed could be shipped by trucks. Finally, there had to be interaction between the project manager and the sales department to determine that the product the project manager was making was what the sales department had sold.

Estimating—Which Department?

At one time or another during CMC's history, the estimating function had been assigned to nearly every functional area of the organization, including sales, engineering, manufacturing, and administration. Determining which functional area estimating was to be under was a real problem for CMC. There was a short time when estimating was on its own, reporting directly to the general manager.

Assignment of this function to any one department carried with it some inherent problems, not peculiar to CMC, but simply related to human nature. For example, when the estimating was supervised in the sales department, estimated costs would tend to be low. In sales, the estimator knows the boss wants to be the low bidder on the job and therefore believes he or she is right to say, "It is not going to take us ten days to cast this thing; we could run three at a time."

When estimating was performed by production, the estimate would tend to be high. This was so because the estimator did not want the boss, the production manager, coming back and saying, "How come you estimated this thing at $5 a cubic foot and it's costing us $6? It's not the cost of production that's wrong, it's the estimate."

W. S. Lasch, general manager of CMC, had this comment about estimating in a project management situation:

> It is very difficult to get accountability for estimating a project. When many of your projects are new ballgames, a lot of your information has to come from . . . well, let's just say there is a lot of art to it as well as science. You never can say with 100 percent certainty that costs were high because you could have just as easily said the estimate was too low.
>
> So, as a compromise, most of the time we had our estimating done by engineering. While it solved some problems, it also created others. Engineers would tend to be more fair; they would call the shots as they saw them. However, one problem was that they still had to answer to sales as far as their workload was concerned. For example, an engineer is in the middle of estimating a parking garage, a task that might take several days. All of a sudden, the sales department wants him to stop and estimate another job. The sales department had to be the one to really make that decision because they are the ones that know what the priorities are on the bidding. So even though the estimator was working in engineering, he was really answering to the sales manager as far as his workload was concerned.

Estimating—Costing

Estimating was accomplished through continual monitoring and comparison of actual versus planned performance, as shown in Exhibit II.

The actual costing process was not a problem for CMC. In recent years, CMC had eliminated as much as possible the actual dollars and cents from the estimator's control. A great deal of the ódrudge work£ was done on the computer. The estimator, for example, would predict how much the prestressed concrete must span, and how many cubic feet of concrete was needed. Once that information was in hand, the estimator entered it in the computer. The computer would then come up with the cost. This became an effective method because the estimator would not be influenced by either sales or production personnel.

The Evolution of the Prestressed Concrete Marketplace

During the 20 or more years since prestressing achieved wide acceptance in the construction industry, an evolution has been taking place that has drastically changed the character of the marketplace and thus greatly modified the role of the prestresser.

Lasch had the following comments about these changes that occurred in the marketplace:

> In the early days, designers of buildings looked to prestressers for the expertise required to successfully incorporate the techniques and available prestressed products into their structures. A major thrust of our business in those days was to introduce design professionals, architects, and engineers to our fledgling industry and to assist them in making use of the many advantages that we could offer over other construction methods. These advantages included fire resistance, long spans, permanence, factory-controlled quality, speed of erection, aesthetic desirability, virtual elimination of maintenance costs, and, last but of prime importance, the fact that we were equipped to provide the expertise and coordination necessary to successfully integrate our product into the building. Many of our early jobs were bid from sketches. It was then up to our in-house experts, working closely with the owner's engineer and architect, to develop an appropriate, efficient structure that satisfied the aesthetic and functional requirements and hopefully maximized production and erection efficiency, thereby providing maximum financial return to

Exhibit II. Actual versus planned performance

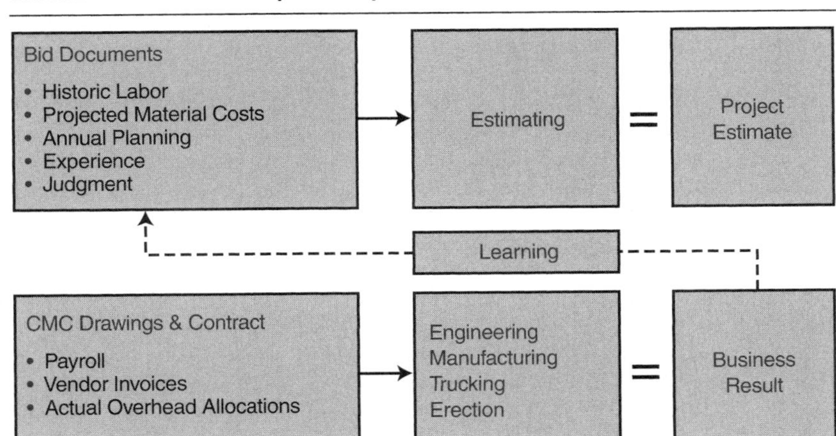

CMC. It should be noted that, although our contract was normally with the project's general contractor, most of our design coordination was through the owner's architect or engineer and, more often than not, it was our relationship with the owner and his design professional that determined our involvement in the project in the first place. It should be readily seen that, in such an environment, only organizations with a high degree of engineering background and a well-organized efficient team of professionals, could compete successfully. CMC was such an organization.

There are, however, few, if any, proprietary secrets in the prestressing industry, and it was inevitable that this would in later years be largely responsible for a dramatic change in the marketplace. The widespread acceptance of the product, which had been achieved through the success of companies like CMC, carried with it a proliferation of the technical knowledge and production techniques which design professionals had previously relied upon the producer to provide. In the later 1960s, some colleges and universities began to include prestressed concrete design as a part of their structural engineering programs. Organizations, such as the Portland Cement Association, offered seminars for architects and engineers to promote the prestressing concept. As a result, it is now common for architects and engineers to incorporate prestressed concrete products in bid drawings for their projects, detailing all connections, reinforcement, mix designs, and so on. This, obviously, makes it possible for any organization capable of reading drawings and filling forms to bid on the project. We have found ourselves bidding against companies with a few molds in an open field and, in several cases, a broker with no equipment or organization at all! The result of all this, of course, is a market price so low as to prohibit the involvement of professional prestressing firms with the depth of organization described earlier.

Obtaining a Prestressed Concrete Job

The author believes the following example demonstrates the change in market conditions and best illustrates one of the reasons CMC decided not to remain in the prestressed concrete business. A large insurance company in Columbus, Ohio, was planning a parking garage for 2,500 cars. CMC talked to the owner and owner's representative (a construction management firm) about using prestressed concrete in the design of their project rather than the poured-in-place concrete, steel, or whatever options they had. Just by doing this, CMC had to give away some knowledge. You just cannot walk in and say, "Hey, how about using prestressed concrete?" You have to tell them what is going to be saved and how, because the architect has to make the drawings. Once CMC felt there was an open door, and that the architect and owner would possibly incorporate their product, then sales would consult engineering to come up with a proposal. A proposal in the early stages was simply to identify what the costs were going to be, and to show the owner and architect photographs or sketches of previous jobs. As time went by, CMC had to go into more detail and provide more and more information, including detailed drawings of several proposed layouts. CMC illustrated connection details, reinforcing details, and even computer design of some of the pieces for the parking garage. Receiving all this engineering information, the owner and the construction management firm became convinced that using this product was the most inexpensive way for them to go. In fact, CMC demonstrated to the insurance company that they could save over $1 million over any other product. At this point, CMC had spent thousands of dollars to come up with the solution for the problem of designing the parking garage.

Months and years passed until the contract manager chose to seek bids from other precasters, who, up to this time, had little or no investment in the project. CMC had made available an abundance of free information that could be used by the competition. The competition only had to put the information together, make a material takeoff, calculate the cost, and put a price on it. Without the costly depth of organization required to support the extensive promotional program conducted by CMC, the competition could naturally bid the job lower.

Lasch felt that, as a result of present-day market conditions, there were only two ways that one survives in the prestressed concrete business:

> Face the fact that you are going to be subservient to a general contractor and that you are going to sell not your expertise but your function as a "job shop" manufacturer producing concrete products according to someone else's drawings and specifications. If you do that, then you no longer need, for example, an engineering department or a technically qualified sales organization. All you are going to do is look at drawings, have an estimator who can read the drawings, put a price tag on them, and give a bid. It is going to be a low bid because you have eliminated much of your overhead. We simply do not choose to be in business in this manner.
>
> The other way to be in the business is that you are not going to be subservient to a general contractor, or owner's architect, or engineer. What you are going to do is to deal with owners or users. That way a general contractor may end up as a subcontractor to the prestresser. We might go out and build a parking garage or other structure and assume the role of developer or builder or even owner/leaser. In that way, we would control the whole job. After all, in most cases the precast contract on a garage represents more than half the total cost. It could be argued with great justification that the conventional approach (i.e., precaster working for general contractor) could be compared to the tail wagging the dog.
>
> With complete control of design, aesthetics, and construction schedule, it would be possible to achieve maximum efficiency of design, plant usage, and field coordination which, when combined, would allow us to achieve that most important requirement—that of providing the eventual user with maximum value for a minimum investment. Unless this can be achieved, the venture would not be making a meaningful contribution to society, and there would be no justification for being in business.

Synopsis

Concrete Masonry Corporation's (CMC) difficulties do not arise from the fact that the organization employs a matrix structure, but rather from the failure of the corporation's top management to recognize, in due time, the changing nature (with respect to the learning curve of the competition and user of the product and services of CMC) of the prestressed concrete business.

At the point in time when prestressed concrete gained wide industry acceptance, and technical schools and societies began offering courses in the techniques for utilizing this process, CMC should have begun reorganizing its prestressed concrete business activities in two separate functional costing groups. Marketing and selling CMC's prestressed concrete business services and utilizing the company's experience, technical expertise, judgment, and job estimating abilities should satisfy the responsibilities of one of these groups, to perform the actual prestressed concrete engineering and implementation of the other.

With the responsibilities and functions separated as noted above, the company is able to determine more precisely how competitive they really are and which (if either) phase of the concrete business to divest themselves of.

Project management activities are best performed when complex tasks are of a limited life. Such is not the case in securing new or continuing business in the prestressed concrete business but rather is an effort or activity that should continue as long as CMC is in the business. This phase of the business should therefore be assigned to a functional group. However, it may be advantageous at times to form or utilize a project management structure in order to assist the functional group in satisfying a task's requirements when the size of the task is large and complex.

The engineering and implementation phase of the business should continue to be performed through the project management–matrix structure because of the limited life of such tasks and the need for concentrated attention to time, cost, and performance constraints inherent in these activities.

Questions

1. Did CMC have long-range planning?
2. What are the problems facing CMC?
3. Did CMC utilize the matrix effectively?
4. Where should project estimating be located?
5. Does the shifting of the estimating function violate the ground rules of the matrix?
6. What are the alternatives for CMC?
7. Will they be successful as a job shop?
8. Should companies like CMC utilize a matrix?
9. How does the company plan to recover R&D and bid and proposal costs?
10. Has CMC correctly evaluated the marketplace?
11. Do they respond to changes in the marketplace?
12. With what speed is monitoring done? (Exhibit II). How many projects must be estimated, bid, and sold before actuals catch up to and become historical data?

Case 26

Construction of a Gas Testing Laboratory in Iran*

With the increase in the availability of natural gas, the country of Iran had decided to embark on an extensive development program to test and evaluate the gas-utilizing appliances and accessories that might be required to satisfy future demands. The Iranian government desired to have all such items tested prior to use. The responsibility for this testing was delegated to the National Iranian Gas Company (NIGC).

Testing requires a facility. NIGC employeed the American Gas Association (AGA), a nonprofit organization. NIGC contracted AGA for engineering services, training, technical assistance, and special equipment fabrication work required for establishment of a testing lab. Testing for safety and performance would be in compliance with Iranian National Standards. Except for equipment installation and program start-up, all work would be accomplished in the United States. The final assembly would be at the NIGC city gate station in Rey, Iran.

The project is a technical assistance contract to provide program planning, building design (but not construction), instruments purchasing, special equipment fabrication, operations personnel training, equipment installation, and program start-up aid.

The contractor will furnish general design specifications and layouts including mechanical and electrical drawings. Architectural details, building construction, and site preparation are furnished by the customer.

The project consists of five phases with a total time frame of 21 months. The five phases are:

1. Program plan and building design
2. Equipment purchase
3. Equipment construction

---

*Reprinted from H. Kerzner, *Project Management: A Systems Approach to Planning, Scheduling, and Controlling,* 6th ed. New York: Wiley, 1998, pp. 374–379.

4. Training
5. Plant start-up

Work Breakdown Structure

The analysis of the cost associated with building the project begins with the separation of the program into its basic tasks. The tasks involved with the project are defined as engineering, procurement, and training. The costs associated with each of the basic tasks are broken out and allocated to specific cost centers.

Project 1–1–00 Program Plan and Building Design

Project #1 consists of engineering and program management in the following areas:

Task 1–1–1 Engineering Engineering time required to design the testing building.
Task 1–1–2 Program Management Management time allocated to project planning and building design.

Project 1–2–00 Equipment Purchase

Project #2 included time required to specify and purchase the equipment for the testing laboratory as follows:

Task 1–2–1 Program Management Time allocated for management of the purchasing function.
Task 1–2–2 Engineering Provides the basic specifications for the equipment to be purchased.
Task 1–2–3 Testing and Inspection Ensures that all equipment meets the established specifications.
Task 1–2–4 Shipping Includes packing and storage for foreign shipment of the equipment.
Task 1–2–5 Procurement Purchase of all equipment and establishing dates that it will be shipped.

Project 1–3–00 Equipment Construction

Project #3 is required to specify, purchase, and fabricate equipment that is not on the market.

Task 1–3–1 Program Management Overall management of the special equipment function.
Task 1–3–2 Engineering Developing the specifications for the special equipment.
Task 1–3–3 Procurement The purchase and evaluation of special equipment, and the establishment of dates for shipment of the equipment.
Task 1–3–4 shipping The packing and shipment of the special equipment for foreign shipment.
Task 1–3–5 Fabrication Building and testing the equipment at selected vendors.

Project 1–4–00 Training

Project #4 involves the preparation of material and equipment to train the testing laboratory personnel.

Task 1–4–1 Program Management Overall management of the training function.
Task 1–4–2 Engineering Developing the materials used to train the laboratory personnel.
Task 1–4–3 Training The actual time required to make the testing people proficient with their new equipment.

Project 1–5–00 Plant Startup

Project #5 provides the time for field personnel to put the laboratory in operation.
Task 1–5–1 Program Management Provides the overall coordination of the plant start-up function.
Task 1–5–2 Field Engineering The time involved to provide the laboratory with start-up personnel.

Other Costs Associated with the Project

Purchased parts—Testing equipment
Freight—Shipping and packing
Travel—Procurement
Other—Purchased goods, freight, subcontracts, materials

Exhibit I contains the program evaluation and review technique (PERT) chart for the program. Exhibit II shows the bar chart for the total program, together with monthly hours and initial salary structures.

Base Case Discussion

The parameters used in the strategic planning model are listed below:

1. Salary costs will increase 6 percent per year with the increase beginning January 1 of each year.
2. Raw material costs will increase 10 percent per year with the increase beginning January 1 of each year.
3. Demanning ratio is 10 percent of following month's labor and man-hour costs.
4. Termination liability on materials is 0 percent, and material commitments are based on 6 months or less.
5. Indirect cost for each project is 14 percent of total cost of labor and materials.
6. Corporate cost for each project is 1 percent of total cost of labor and materials.
7. A profit of 12 percent is used for each project in the base case.
8. No delays or additional increased costs are assumed in the base case.
9. A separate overhead rate is included for each task (see Exhibit I).

The output obtained from the model using the parameters established for the program are listed below:

1. The total cost of establishing a natural gas testing laboratory is $920,322 including a 12 percent profit, overhead rates, and corporate costs.
2. Cash flow for the first year, 1978, is $642,718 and $178,999 for the second year. Profit is not included in these figures.

Exhibit I. Critical path diagram

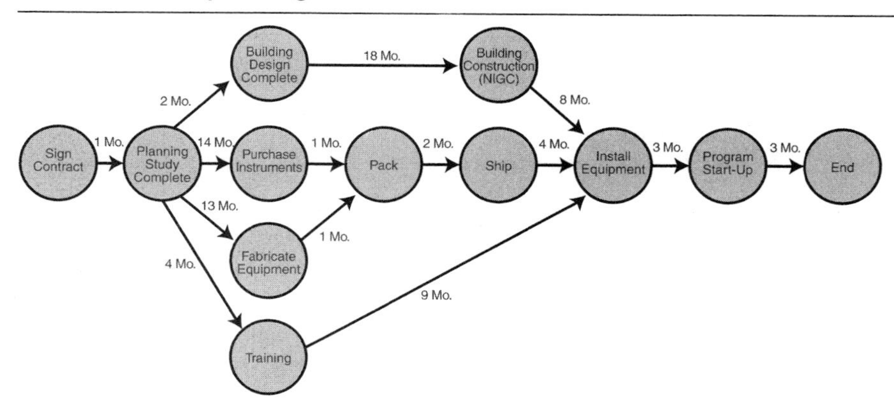

3. The cost for each project and total program percentage is shown below:

Project	*Cost*	*% of Program Cost*
Project 1	$ 17,649	2%
Project 2	$250,978	31%
Project 3	$496,005	60%
Project 4	$ 25,414	3%
Project 5	$ 31,671	4%

The two projects representing 91 percent of the total program are equipment purchasing and equipment construction. (Profit is not included.)

4. Total labor expenditure for the project is 23,830 hours, costing $583,410. This breaks down into 16,604 hours, costing $404,399 in 1978, and 7,226 hours, costing $179,011 in 1979.
5. Total material expenditure is $238,307 with the total commitment established in July of 1978.
6. A functional cost summary has the program divided into four major divisions, excluding material costs. The divisions and associated costs (including overhead) are as follows:

Division	*Cost*	*% of Program Cost*
Engineering	$ 96,732	19%
Program management	$ 26,349	5%
Finance	$ 14,654	3%
Operations	$368,953	73%

Problems

1. *Construction of Special Equipment:* There is special test eqiupment required for the laboratory. Since many of the tests are unique to gas appliance testing under the

American National Standards, the equipment cannot be purchased on the open market and must be constructed. The costs of construction are very difficult to estimate accurately, since many of these items are unique.

2. *Building Construction:* The erection of the laboratory building is the critical path in the Program. The construction is being supervised and contracted for in Iran by the overburdened engineering department of NIGC. Delay of building opening shifts the entire program to the right (except for the planning stage), resulting in increased costs and scheduling problems as follows:
   a. Materials costs
   b. Labor costs
   c. Scheduling problems
   d. Storage costs
3. *Transportation of Equipment:* The large amount of construction in Iran will result in long delays in unloading at Persian Gulf ports. On a similar program conducted previously, just before shipment of completed equipment and instruments, the contractor was made aware by the customer that there would be a delay of up to eighteen months. Alternate shipping plans were worked out by the customer (who was responsible for shipment from New York to the site in Tehran), involving shipment to Hamburg, Germany, and overland shipment by truck from Hamburg to Tehran. This shipping routing consumed about four months.
4. *Language Difficulties:* There was great uncertainty at the outset of the contract about accurate communication because of the language problem. All parties at the customer could speak English, but not well. This resulted in occasional misunderstandings.
5. *Building Subcontractor:* The building work was contracted out by NIGC to a local construction firm.

Results

The company's strategy for shipping the building materials was to fabricate components prior to shipping. The shipping schedule was of prime importance because a delay in shipping, for whatever cause, would immediately put the project behind. It was projected that any sizable delay would increase costs in the area of materials and labor.

Plans were initiated to commence with a screening of Iranian labor and for training of selected personnel in the event of a shipping delay. An analysis of costs was made based on an extension of the construction period over an additional twenty months; otherwise, the training would take place at the completion of the construction. Iran had a shortage of labor and might be hard-pressed to find an ample workforce for this project.

Alternative plans were set up in the event of a delay in the construction schedule, problems involving scheduling personnel for transportation to the project, and storage of materials and supplies.

Exhibit III is a comparative cost analysis of the two schedules.

Plans were made to construct energy supply generators to combat the possibility of a power shortage, since there is a utility shortage in Iran.

A comparison of the base cost is estimated on an 18-month construction period extended to the 38 months. The above analysis for the two periods does not reflect a profit factor. However, there is a 10 percent escalation factor for raw materials and a demanning ratio of 10 percent. The termination liability ratio is 0 percent, and salary increases are equal to 6 percent per year. Corporate costs and indirect costs are 1 percent and 14 percent, respectively.

Exhibit II. Program bar chart

PROJECTS/TASKS	1977 JAN.	FEB.	MAR.	APR.	MAY	JUNE	JULY	AUG.	SEPT.	OCT.	NOV.	DEC.	1978 JAN.	FEB.	MAR.	APR.	MAY	JUNE	JULY	AUG.	SEPT.
I. PROGRAM PLANS & BUILDING DESIGN																					
1. PROGRAM MANAGEMENT	102	50	48																		
2. ENGINEERING	20	174	283																		
II. EQUIPMENT PURCHASE																					
1. PROGRAM MANAGEMENT		52	40	40	30	30	30	30	30	30	20	20	20	20	20						
2. ENGINEERING		267	232	200	185	131	95	60	40												
3. TESTING & INSPECTION								154	154	266	266	266	154	112							
4. SHIPPING												154	154	308	308						
5. PROCUREMENT					154	154	154	77	77												
III. EQUIPMENT CONSTRUCTION																					
1. PROGRAM MANAGEMENT			20	20	30	30	30	30	30	30	20	20	20	20	20						
2. ENGINEERING			154	154	154	154	154	154	154	154	77										
3. PROCUREMENT					154	308	308	308	308	308	308	308	308	308							
4. SHIPPING													47	100	308						
5. FABRICATION				154	308	462	616	770	924	1078	1386	1848	1848	1848	154						
IV. TRAINING																					
1. PROGRAM MANAGEMENT											20	20	20	20							
2. ENGINEERING											60	60	60	60							
3. TRAINING											154	154	154	154							
V. PLANT START-UP																					
1. PROGRAM MANAGEMENT														20	5	5	5	5	5	20	
2. FIELD ENGINEERING														154	154	154	308	154	154	154	

II. OTHER COSTS:

PURCHASED GOODS:	$121,981.
FREIGHT:	1,988.
OTHER:	3,049.
OVERSEAS PACKING:	6,242.
	$133,260.

III. OTHER COSTS:

PURCHASED MAT'LS:	39,527.
SUBCONTRACTS:	28,082.
OVERSEAS PACKING:	3,520.
FREIGHT:	2,598.
	$ 73,727.

IV. OTHER COSTS:

SUPPLIES:	$ 980.

	RATE	OH	
PROGRAM MANAGEMENT	11.00	120%	
ENGINEERING	10.00	120%	
TESTING	8.00	117%	125%
PROCUREMENT	8.00	110%	
SHIPPING	5.70	100%	
FABRICATION	10.00	125%	
TRAINING	10.50	120%	12%
FIELD ENGINEERING	9.00	80%	

INDIRECT COSTS	14%
CORPORATE COSTS	1%
PROFIT	12%

RAW MATERIALS ESCALATION	10%
DEMANNING RATIO	10%
TERMINATION LIABILITY	0%
SALARY INCREASES	6%

NOTE: ONE MAN MONTH = 154 MAN HOURS

Exhibit III. Comparative cost analysis

	18 Months	38 Months	Unfavorable Variance
Engineering	$ 96,731	$105,382	$ 8,651
Program management	26,349	28,424	2,075
Finance	14,654	15,980	1,326
Operations	368,953	406,567	37,614
Materials	206,987	216,749	9,762
Indirect costs	99,916	108,234	8,318
Corporate costs	8,127	8,806	679
Total costs	$821,717	$890,142	$68,425

Synopsis

This is a factual case study (but disguised). The objective of the case study is to show the student the difficulties with managing international projects. There are no right or wrong answers to this case study.

Questions

1. What are the differences between managing domestic versus international projects?
2. What mistakes did the company make?
3. Could any of these problems have been anticipated?
4. If you had to manage another project similar to this one, what would you do differently? Would the PERT chart change?

Case 27

The Space Shuttle Challenger Disaster

On January 28, 1986, the Space Shuttle Challenger lifted off the launch pad at 11:38 A.M., beginning the flight of mission 51-L.[1] Approximately 74 seconds into the flight, the Challenger was engulfed in an explosive burn and all communication and telemetry ceased. Seven brave crewmembers lost their lives. On board the Challenger were Francis R. (Dick) Scobee (commander), Michael John Smith (pilot), Ellison S. Onizuka (mission specialist one), Judith Arlene Resnik (mission specialist two), Ronald Erwin McNair (mission specialist three), S. Christa McAuliffe (payload specialist one), and Gregory Bruce Jarvis (payload specialist two). A faulty seal, or O-ring, on one of the two solid rocket boosters caused the accident.

Following the accident, significant energy was expended trying to ascertain whether or not the accident had been predictable. Controversy arose from the desire to assign, or to avoid, blame. Some publications called it a management failure, specifically in risk management, while others called it a technical failure.

Whenever accidents had occurred in the past at the National Aeronautics and Space Administration (NASA), an internal investigation team had been formed. But in this case, perhaps because of the visibility, the White House took the initiative in appointing an independent commission. There did exist significant justification for the commission. NASA was in a state of disarray, especially in the management ranks. The agency had been without a permanent administrator for almost four months. The turnover rate at the upper echelons of management was significantly high, and there seemed to be a lack of direction from the top down.

---

1. The first digit indicates the fiscal year of the launch (i.e., "5" means 1985). The second number indicates the launch site (i.e., "1" is the Kennedy Space Center in Florida, "2" is Vandenberg Air Force Base in California). The letter represents the mission number (i.e., "C" would be the third mission scheduled). This designation system was implemented after Space Shuttle flights one through nine, which were designated STS-X. STS is the Space Transportation System and X would indicate the flight number.

Another reason for appointing a Presidential Commission was the visibility of this mission. This mission had been known as the Teacher in Space mission, and Christa McAuliffe, a Concord, New Hampshire, schoolteacher, had been selected from a list of over 10,000 applicants. The nation knew the names of all of the crewmembers on board Challenger. The mission had been highly publicized for months, stating that Christa McAuliffe would be teaching students from aboard the Challenger on day four of the mission.

The Presidential Commission consisted of the following members:

- *William P. Rogers,* Chairman: Former secretary of state under President Nixon and attorney general under President Eisenhower.
- *Neil A. Armstrong,* Vice Chairman: Former astronaut and spacecraft commander for Apollo 11.
- *David C. Acheson:* Former senior vice president and general counsel, Communications Satellite Corporation (1967–1974), and a partner in the law firm of Drinker Biddle & Reath.
- *Dr. Eugene E. Covert:* Professor and head, Department of Aeronautics and Astronautics at Massachusetts Institute of Technology.
- *Dr. Richard P. Feynman:* Physicist and professor of theoretical physics at California Institute of Technology; Nobel Prize winner in Physics, 1965.
- *Robert B. Hotz:* Editor-in-chief of *Aviation Week & Space Technology* magazine (1953–1980).
- *Major General Donald J. Kutyna,* USAF: Director of Space Systems and Command, Control, Communications.
- *Dr. Sally K. Ride:* Astronaut and mission specialist on STS-7, launched on June 18, 1983, making her the first American woman in space. She also flew on mission 41-G, launched October 5, 1984. She holds a Doctorate in Physics from Stanford University (1978) and was still an active astronaut.
- *Robert W. Rummel:* Vice president of Trans World Airlines and president of Robert W. Rummel Associates, Inc., of Mesa, Arizona.
- *Joseph F. Sutter:* Executive vice president of the Boeing Commercial Airplane Company.
- *Dr. Arthur B. C. Walker, Jr.:* Astronomer and professor of Applied Physics; formerly associate dean of the Graduate Division at Stanford University, and consultant to Aerospace Corporation, Rand Corporation, and the National Science Foundation.
- *Dr. Albert D. Wheelon:* Executive vice president, Hughes Aircraft Company.
- *Brigadier General Charles Yeager,* USAF (Retired): Former experimental test pilot. He was the first person to break the sound barrier and the first to fly at a speed of more than 1,600 miles an hour.
- *Dr. Alton G. Keel, Jr.,* Executive Director: Detailed to the Commission from his position in the Executive Office of the President, Office of Management and Budget, as associate director for National Security and International Affairs; formerly assistant secretary of the Air Force for Research, Development and Logistics, and Senate Staff.

The Commission interviewed more than 160 individuals, and more than 35 formal panel investigative sessions were held generating almost 12,000 pages of transcript. Almost 6,300 documents totaling more than 122,000 pages, along with hundreds of photographs, were examined and made a part of the Commission's permanent database and archives. These sessions and all the data gathered added to the 2,800 pages of hearing transcript generated by the Commission in both closed and open sessions. Unless otherwise stated, all of the quotations and memos in this case study come from the direct testimony cited in the *Report by the Presidential Commission (RPC).*

Background to the Space Transportation System

During the early 1960s, NASA's strategic plans for post-Apollo manned space exploration rested upon a three-legged stool. The first leg was a reusable space transportation system, the Space Shuttle, which could transport people and equipment to low earth orbits and then return to earth in preparation for the next mission. The second leg was a manned space station that would be resupplied by the Space Shuttle and serve as a launch platform for space research and planetary exploration. The third leg would be planetary exploration to Mars. But by the late 1960s, the United States was involved in the Vietnam War, which was becoming costly. In addition, confidence in the government was eroding because of civil unrest and assassinations. With limited funding due to budgetary cuts, and with the lunar landing missions coming to an end, prioritization of projects was necessary. With a Democratic Congress continuously attacking the cost of space exploration, and minimal support from President Nixon, the space program was left standing on one leg only, the Space Shuttle.

President Nixon made it clear that funding all the programs NASA envisioned would be impossible, and that funding for even one program on the order of the Apollo Program was likewise not possible. President Nixon seemed to favor the space station concept, but this required the development of a reusable Space Shuttle. Thus NASA's Space Shuttle Program became the near-term priority.

One of the reasons for the high priority given to the Space Shuttle program was a 1972 study completed by Dr. Oskar Morgenstern and Dr. Klaus Heiss of the Princeton-based Mathematica organization. The study showed that the Space Shuttle would be able to orbit payloads for as little as $100 per pound based on 60 launches per year with payloads of 65,000 pounds. This provided tremendous promise for military applications such as reconnaissance and weather satellites, as well as for scientific research.

Unfortunately, the pricing data were somewhat tainted. Much of the cost data were provided by companies who hoped to become NASA contractors and who therefore provided unrealistically low cost estimates in hopes of winning future bids. The actual cost per pound would prove to be more than twenty times the original estimate. Furthermore, the main engines never achieved the 109 percent of thrust that NASA desired, thus limiting the payloads to 47,000 pounds instead of the predicted 65,000 pounds. In addition, the European Space Agency began successfully developing the capability to place satellites into orbit and began competing with NASA for the commercial satellite business.

NASA Succumbs to Politics and Pressure

To retain Shuttle funding, NASA was forced to make a series of major concessions. First, facing a highly constrained budget, NASA sacrificed the research and development necessary to produce a truly reusable Shuttle, and instead accepted a design that was only partially reusable, eliminating one of the features that had made the Shuttle attractive in the first place. Solid rocket boosters (SRBs) were used instead of safer liquid fueled boosters because they required a much smaller research and development effort. Numerous other design changes were made to reduce the level of research and development required.

Second, to increase its political clout and to guarantee a steady customer base, NASA enlisted the support of the United States Air Force. The Air Force could provide the considerable political clout of the Department of Defense and it used many satellites, which required launching. However, Air Force support did not come without a price. The Shuttle payload bay was required to meet Air Force size and shape requirements, which placed key constraints on the ultimate design. Even more important was the Air Force requirement that the Shuttle be able to launch from Vandenburg Air Force Base in California. This constraint required a larger cross range than the Florida site, which in turn decreased the total allowable vehicle weight. The weight reduction required the elimination of the design's air breathing engines, resulting in a single-pass unpowered landing. This greatly limited the safety and landing versatility of the vehicle.[2]

As the year 1986 began, there was extreme pressure on NASA to "Fly out the Manifest." From its inception, the Space Shuttle Program had been plagued by exaggerated expectations, funding inconsistencies, and political pressure. The ultimate vehicle and mission design were shaped almost as much by politics as by physics. President Kennedy's declaration that the United States would land a man on the moon before the end of the decade (the 1960s) had provided NASA's Apollo Program with high visibility, a clear direction, and powerful political backing. The Space Shuttle Program was not as fortunate; it had neither a clear direction nor consistent political backing.

Cost containment became a critical issue for NASA. In order to minimize cost, NASA designed a Space Shuttle system that utilized both liquid and solid propellants. Liquid propellant engines are more easily controllable than solid propellant engines. Flow of liquid propellant from the storage tanks to the engine can be throttled and even shut down in case of an emergency. Unfortunately, an all-liquid-fuel design was prohibitive because a liquid fuel system is significantly more expensive to maintain than a solid fuel system.

Solid fuel systems are less costly to maintain. However, once a solid propellant system is ignited, it cannot be easily throttled or shut down. Solid propellant rocket motors burn until all of the propellant is consumed. This could have a significant impact on safety, especially during launch, at which time the solid rocket boosters are ignited and have maximum propellant loads. Also, solid rocket boosters can be designed for reusability, whereas liquid engines are generally used only once.

The final design that NASA selected was a compromise of both solid and liquid fuel engines. The Space Shuttle would be a three-element system composed of the Orbiter vehicle, an expendable external liquid fuel tank carrying liquid fuel for the Orbiter's engines, and two recoverable solid rocket boosters.[3] The Orbiter's engines were liquid fuel because of the necessity for throttle capability. The two solid rocket boosters would provide the added thrust necessary to launch the Space Shuttle into its orbiting altitude.

In 1972, NASA selected Rockwell as the prime contractor for building the Orbiter. Many industry leaders believed that other competitors who had actively participated in the Apollo Program had a competitive advantage. Rockwell, however, was awarded the contract. Rockwell's proposal did not include an escape system. NASA officials decided against the launch escape system since it would have added too much weight to the Shuttle at launch and was very expensive. There was also some concern on how effective an escape system would be if an accident occurred during launch while all of the engines were ignited. Thus the Space Shuttle Program became the first U.S. manned spacecraft without a launch escape system for the crew.

In 1973, NASA went out for competitive bidding for the solid rocket boosters. The competitors were Morton-Thiokol, Inc. (MTI) [henceforth called Thiokol], Aerojet General, Lockheed, and United Technologies. The contract was eventually awarded to Thiokol because of its low cost, $100 million lower than the nearest competitor. Some believed that other competitors, who ranked higher in technical design and safety, should have been given the contract. NASA believed that Thiokol-built solid rocket motors would provide the lowest cost per flight.

---

2. Kurt Hoover and Wallace T. Fowler (The University of Texas at Austin and The Texas Space Grant Consortium), "Studies in Ethics, Safety and Liability for Engineers," *www.tsgc.utexas. edu/archive/general/ethics/shuttle.html*, page 2.

3. The terms "solid rocket booster" (SRB) and "solid rocket motor" (SRM) will be used interchangeably.

The Solid Rocket Boosters

Thiokol's solid rocket boosters had a height of approximately 150 feet and a diameter of 12 feet. The empty weight of each booster was 192,000 pounds and the full weight was 1,300,000 pounds. Once ignited, each booster provided 2.65 million pounds of thrust, which is more than 70 percent of the thrust needed to lift off the launch pad.

Thiokol's design for the boosters was criticized by some of the competitors, and even by some NASA personnel. The boosters were to be manufactured in four segments and then shipped from Utah to the launch site, where the segments would be assembled into a single unit. The Thiokol design was largely based upon the segmented design of the Titan III solid rocket motor produced by United Technologies in the 1950s for Air Force satellite programs. Satellite programs were unmanned efforts.

The four solid rocket sections made up the case of the booster, which essentially encased the rocket fuel and directed the flow of the exhaust gases. This is shown in Exhibit I. The cylindrical

Exhibit I. Solid rocket booster (SRB)

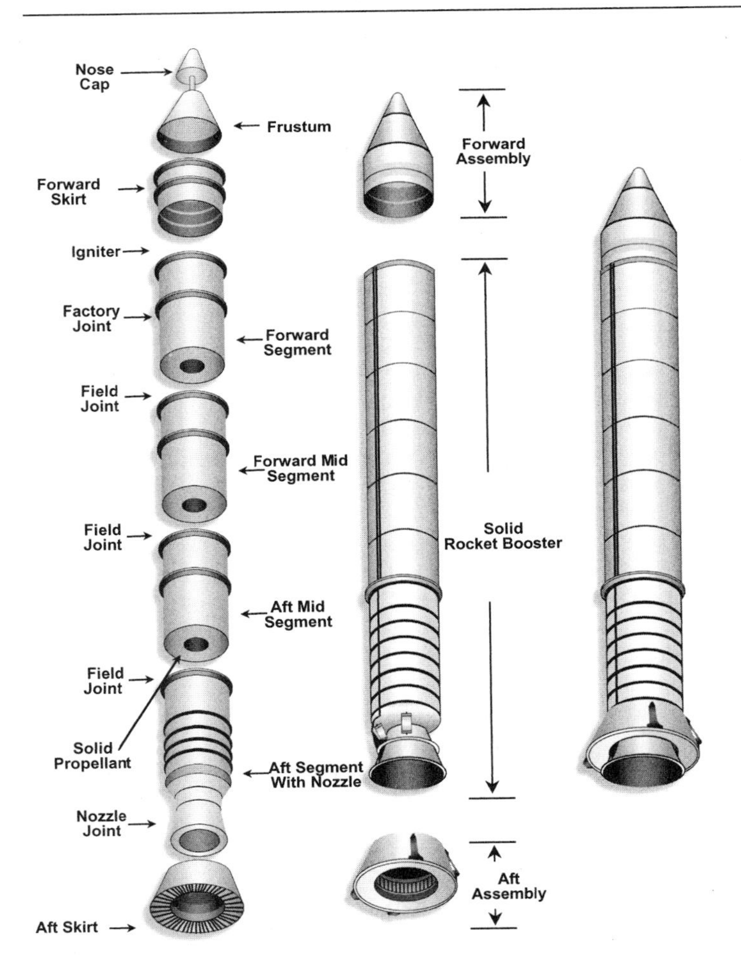

Exhibit II. Location of the O-rings

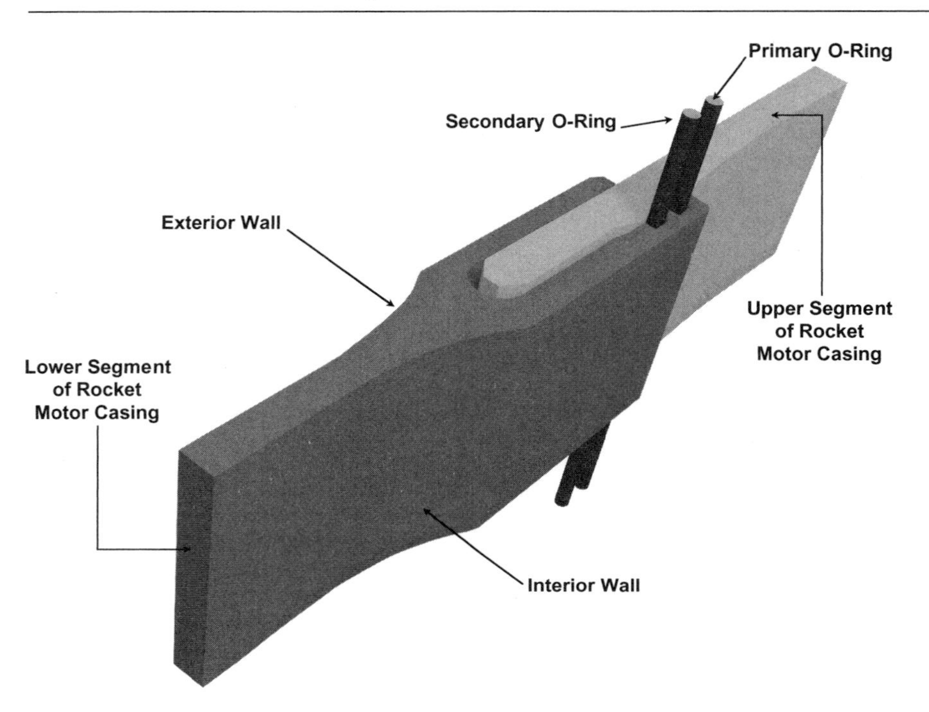

shell of the case is protected from the propellant by a layer of insulation. The mating sections of the field joint are called the tang and the clevis. One hundred and seventy-seven pins spaced around the circumference of each joint hold the tang and the clevis together. The joint is sealed in three ways. First, zinc chromate putty is placed in the gap between the mating segments and their insulation. This putty protects the second and third seals, which are rubber-like rings, called O-rings. The first O-ring is called the primary O-ring and is lodged in the gap between the tang and the clevis. The last seal is called the secondary O-ring, which is identical to the primary O-ring except it is positioned further downstream in the gap. Each O-ring is 0.280 inches in diameter. The placement of each O-ring can be seen in Exhibit II. Another component of the field joint is called the leak check port, which is shown in Exhibit III. The leak check port is designed to allow technicians to check the status of the two O-ring seals. Pressurized air is inserted through the leak check port into the gap between the two O-rings. If the O-rings maintain the pressure, and do not let the pressurized air past the seal, the technicians know the seal is operating properly.[4]

In the Titan III assembly process, the joints between the segmented sections contained one O-ring. Thiokol's design had two O-rings instead of one. The second O-ring was initially considered as redundant, but included to improve safety. The purpose of the O-rings was to seal the

4. University of Texas, "The Challenger Accident: Mechanical Causes of the Challenger Accident," *www.me.utexas.edu/~uer/challenger/chall2.html,* pages 1–2.

Exhibit III. Cross section showing the leak test port

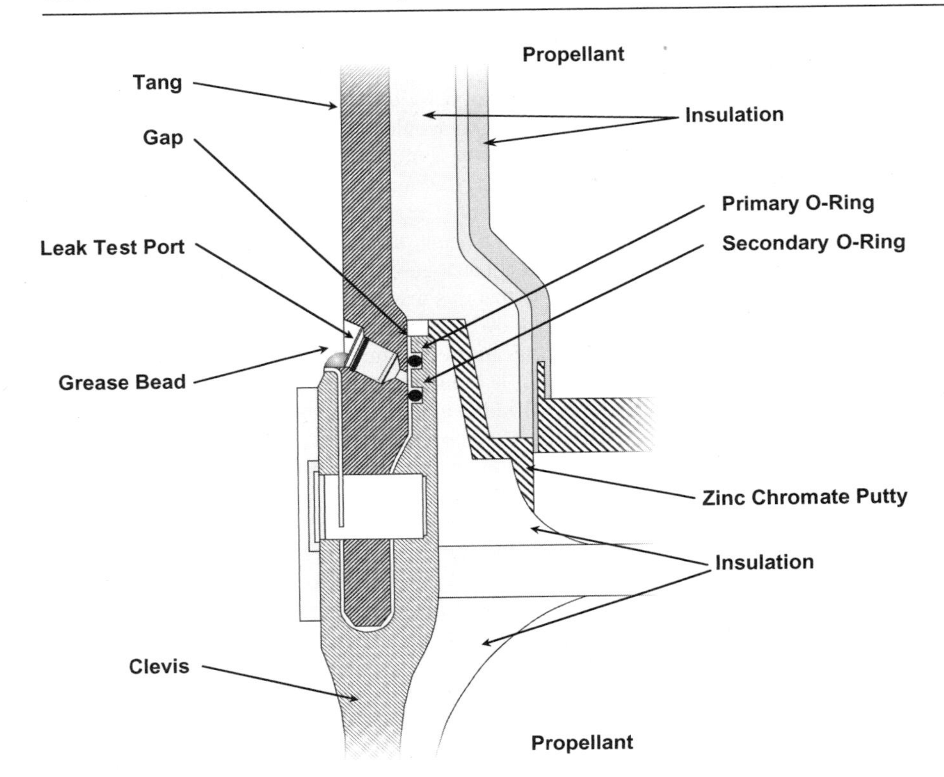

space in the joints such that the hot exhaust gases could not escape and damage the case of the boosters.

Both the Titan III and Shuttle O-rings were made of Viton rubber, which is an elastomeric material. For comparison, rubber is also an elastomer. The elastomeric material used is a fluoroelastomer, which is an elastomer that contains fluorine. This material was chosen because of its resistance to high temperatures and its compatibility with the surrounding materials. The Titan III O-rings were molded in one piece, whereas the Shuttle's SRB O-rings would be manufactured in five sections and then glued together. Routinely, repairs would be necessary for inclusions and voids in the rubber received from the material suppliers.

Blowholes

The primary purpose of the zinc chromate putty was to act as a thermal barrier that protected the O-rings from the hot exhaust. As mentioned before, the O-ring seals were tested using the leak check port to pressurize the gap between the seals. During the test, the secondary seal was pushed down into the same, seated position as it occupied during ignition pressurization. However, because the leak check port was between the two O-ring seals, the primary O-ring was pushed up and seated against the putty. The position of the O-rings during flight and their position during the leak check test is shown in Exhibit III.

During early flights, engineers worried that, because the putty above the primary seal could withstand high pressures, the presence of the putty would prevent the leak test from identifying problems with the primary seal. They contended that the putty would seal the gap during testing regardless of the condition of the primary seal. Since the proper operation of the primary seal was essential, engineers decided to increase the pressure used during the test to above the pressure that the putty could withstand. This would ensure that the primary O-ring was properly sealing the gap without the aid of the putty. Unfortunately, during this new procedure, the high-test pressures blew holes through the putty before the primary O-ring could seal the gap.

Since the putty was on the interior of the assembled solid rocket booster, technicians could not mend the blowholes in the putty. As a result, this procedure left small, tunneled holes in the putty. These holes would allow focused exhaust gases to contact a small segment of the primary O-ring during launch. Engineers realized that this was a problem, but decided to test the seals at the high pressure despite the formation of blowholes, rather than risking a launch with a faulty primary seal.

The purpose of the putty was to prevent the hot exhaust gases from reaching the O-rings. For the first nine successful Shuttle launches, NASA and Thiokol used asbestos-bearing putty manufactured by the Fuller-O'Brien Company of San Francisco. However, because of the notoriety of products containing asbestos, and the fear of potential lawsuits, Fuller-O'Brien stopped manufacturing the putty that had served the Shuttle so well. This created a problem for NASA and Thiokol.

The new putty selected came from Randolph Products of Carlstadt, New Jersey. Unfortunately, with the new putty, blowholes and O-ring erosion were becoming more common to a point where the Shuttle engineers became worried. Yet the new putty was still used on the boosters. Following the Challenger disaster, testing showed that, at low temperatures, the Randolph putty became much stiffer than the Fuller-O'Brien putty and lost much of its stickiness.[5]

O-Ring Erosion

If the hot exhaust gases penetrated the putty and contacted the primary O-ring, the extreme temperatures would break down the O-ring material. Because engineers were aware of the possibility of O-ring erosion, the joints were checked after each flight for evidence of erosion. The amount of O-ring erosion found on flights before the new high-pressure leak check procedure was around 12 percent. After the new high-pressure leak test procedure, the percentage of O-ring erosion was found to increase by 88 percent. High percentages of O-ring erosion in some cases allowed the exhaust gases to pass the primary O-ring and begin eroding the secondary O-ring. Some managers argued that some O-ring erosion was "acceptable" because the O-rings were found to seal the gap even if they were eroded by as much as one-third their original diameter.[6] The engineers believed that the design and operation of the joints were an acceptable risk because a safety margin could be identified quantitatively. This numerical boundary would become an important precedent for future risk assessment.

Joint Rotation

During ignition, the internal pressure from the burning fuel applies approximately 1000 pounds per square inch on the case wall, causing the walls to expand. Because the joints are generally stiffer than the case walls, each section tends to bulge out. The swelling of the solid rocket sections causes the tang and the clevis to become misaligned; this misalignment is called joint rotation. A diagram showing a field joint before and after joint rotation is seen in Exhibit IV. The problem with joint rotation is that it increases the gap size near the O-rings. This increase in

5. "The Challenger Accident," page 3.

6. "The Challenger Accident," page 4.

Exhibit IV. Field joint rotation

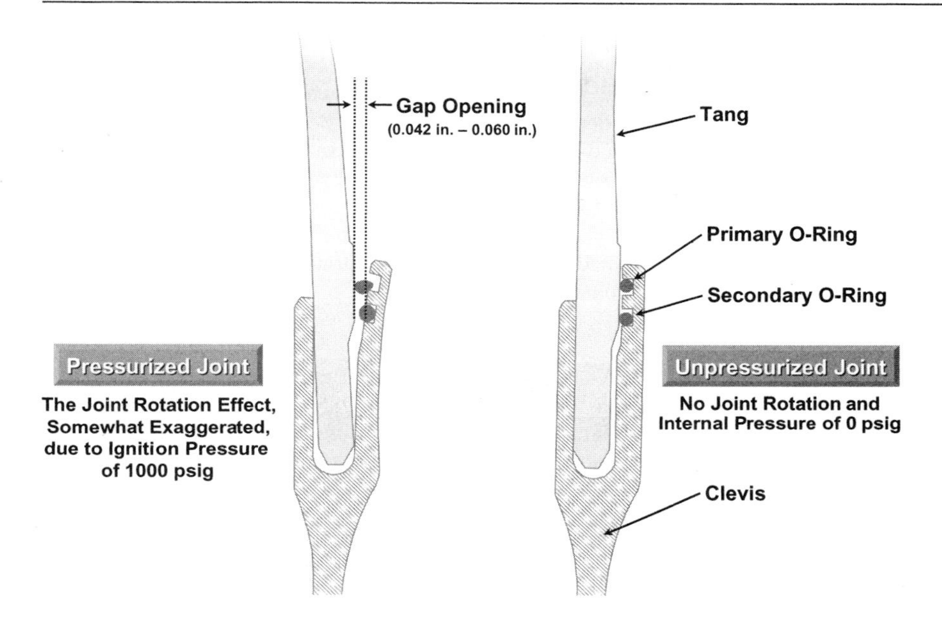

size is extremely fast, which makes it difficult for the O-rings to follow the increasing gap and keep the seal.[7]

Prior to ignition, the gap between the tang and the clevis is approximately 0.004 inches. At ignition, the gap will enlarge to between 0.042 and 0.060 inches, *but for a maximum of 0.60 second,* and then return to its original position.

O-Ring Resilience

The term "O-ring resilience" refers to the ability of the O-ring to return to its original shape after it has been deformed. This property is analogous to the ability of a rubber band to return to its original shape after it has been stretched. As with a rubber band, the resiliency of an O-ring is directly related to its temperature. As the temperature of the O-ring gets lower, the O-ring material becomes stiffer. Tests have shown that an O-ring at 75°F is five times more responsive in returning to its original shape than an O-ring at 30°F. This decrease in O-ring resiliency during a cold weather launch would make the O-ring much less likely to follow the increasing gap size during joint rotation. As a result of poor O-ring resiliency, the O-ring would not seal properly.[8]

The External Tank

The solid rockets are each joined forward and aft to the external liquid fuel tank. They are not connected to the Orbiter vehicle. The solid rocket motors are mounted first, and the external liquid fuel tank is put between them and connected. Then the Orbiter is mounted to the external

7. "The Challenger Accident," page 4.

8. "The Challenger Accident," pages 4–5.

tank at two places in the back and one place forward, and those connections carry all of the structural loads for the entire system at liftoff and through the ascent phase of flight. Also connected to the Orbiter, under the Orbiter's wing, are two large propellant lines 17 inches in diameter. The one on the port side carries liquid hydrogen from the hydrogen tank in the back part of the external tank. The line on the right side carries liquid oxygen from the oxygen tank at the forward end, inside the external tank.[9]

The external tank contains about 1.6 million pounds of propellant, or about 526,000 gallons. The Orbiter's three engines burn the liquid hydrogen and liquid oxygen at a ratio of 6:1 and at a rate equivalent to emptying out a family swimming pool every 10 seconds! Once ignited, the exhaust gases leave the Orbiter's three engines at approximately 6,000 miles per hour. After the fuel is consumed, the external tank separates from the Orbiter, falls to earth, and disintegrates in the atmosphere on reentry.

The Spare Parts Problem

In March 1985, NASA's administrator, James Beggs, announced that there would be one Shuttle flight per month for all of fiscal year 1985. In actuality, there were only six flights. Repairs became a problem. Continuous repairs were needed on the heat tiles required for reentry, the braking system, and the main engines' hydraulic pumps. Parts were routinely borrowed from other Shuttles. The cost of spare parts was excessively high, and NASA was looking for cost containment.

Risk Identification Procedures

The necessity for risk management was apparent right from the start. Prior to the launch of the first Shuttle in April of 1981, hazards were analyzed and subjected to a formalized hazard reduction process as described in NASA Hand Book, NHB5300.4. The process required that the credibility and probability of the hazards be determined. A Senior Safety Review Board was established for overseeing the risk assessment process. For the most part, the risks assessment process was qualitative. The conclusion reached was that no single hazard or combination of hazards should prevent the launch of the first Shuttle *as long as the aggregate risk remained acceptable.*

NASA used a rather simplistic Safety (Risk) Classification System. A quantitative method for risk assessment was not in place at NASA because gathering the data needed to generate statistical models would be expensive and labor-intensive. If the risk identification procedures were overly complex, NASA would have been buried in paperwork due to the number of components on the Space Shuttle. The risk classification system selected by NASA is shown below:

Level	Description
Criticality 1 (C1)	Loss of life and/or vehicle if the component fails
Criticality 2 (C2)	Loss of mission if the component fails
Criticality 3 (C3)	All others
Criticality 1R (C1R)	Redundant components exist. The failure of both could cause loss of life and/or vehicle
Criticality 2R (C2R)	Redundant components exist. The failure of both could cause loss of mission

From 1982 on, the O-ring seal was labeled Criticality 1. By 1985, there were 700 components identified as Criticality 1.

9. *RPC,* page 50.

Teleconferencing

The Space Shuttle Program involves a vast number of people at both NASA and the contractors. Because of the geographical separation between NASA and the contractors, it became impractical to have continuous meetings. Travel between Thiokol in Utah and the Cape in Florida took one day each way. Therefore, teleconferencing became the primary method of communication and a way of life. Interface meetings were still held, but the emphasis was on teleconferencing. All locations could be linked together in one teleconference and data could be faxed back and forth as needed.

The Paperwork Constraints

With the rather optimistic flight schedule provided to the news media, NASA was under scrutiny and pressure to deliver. For fiscal 1986, the mission manifest called for sixteen flights. The pressure to meet schedule was about to take its toll. Safety problems had to be resolved quickly.

As the number of flights scheduled began to increase, so did the requirements for additional paperwork. The majority of the paperwork had to be completed prior to NASA's Flight Readiness Review (FRR) meetings. Approximately one week, prior to every flight, flight operations and cargo managers were required to endorse the commitment of flight readiness to the NASA associate administrator for space flight at the FRR meeting. The responsible project/element managers would conduct pre-FRR meetings with their contractors, center managers, and the NASA Level II manager. The content of the FRR meetings included:

- Determine overall status, as well as establish the baseline in terms of significant changes since the last mission.
- Review significant problems resolved since the last review, and significant anomalies from the previous flight.
- Review all open items and constraints remaining to be resolved before the mission.
- Present all new waivers since the last flight.

NASA personnel were working excessive overtime, including weekends, to fulfill the paperwork requirements and prepare for the required meetings. As the number of space flights increased, so did the paperwork and overtime.

The paperwork constraints were affecting the contractors as well. Additional paperwork requirements existed for problem solving and investigations. On October 1, 1985, an Interoffice Memo was sent from Scott Stein, space booster project engineer at Thiokol, to Bob Lund, vice president for engineering at Thiokol, and to other selected managers concerning the O-Ring Investigation Task Force:

> We are currently being hog-tied by paperwork every time we try to accomplish anything. I understand that for production programs, the paperwork is necessary. However, for a priority, short schedule investigation, it makes accomplishment of our goals in a timely manner extremely difficult, if not impossible. We need the authority to bypass some of the paperwork jungle. As a representative example of problems and time that could easily be eliminated, consider assembly or disassembly of test hardware by manufacturing personnel. . . . I know the established paperwork procedures can be violated if someone with enough authority dictates it. We did that with the DR system when the FWC hardware "Tiger Team" was established. If changes are not made to allow us to accomplish work in a reasonable amount of time, then the O-ring investigation task force will never have the potency necessary to resolve problems in a timely manner.

Both NASA and the contractors were now feeling the pressure caused by the paperwork constraints.

Issuing Waivers

One quick way of reducing paperwork and meetings was to issue a waiver. Historically a waiver was a formalized process that allowed an exception to either a rule, a specification, a technical criterion, or a risk. Waivers were ways to reduce excessive paperwork requirements. Project managers and contract administrators had the authority to issue waivers, often with the intent of bypassing standard protocols in order to maintain a schedule. The use of waivers had been in place well before the manned space program even began. What is important here was *not* NASA's use of the waiver, but the *justification* for the waiver given the risks.

NASA had issued waivers on both Criticality 1 status designations and launch constraints. In 1982, the solid rocket boosters were designated C1 by the Marshall Space Flight Center because failure of the O-rings could have caused loss of crew and the Shuttle. This meant that the secondary O-rings were not considered redundant. The SRB project manager at Marshall, Larry Mulloy, issued a waiver just in time for the next Shuttle launch to take place as planned. Later, the O-rings designation went from C1 to C1R (i.e., a redundant process), thus partially avoiding the need for a waiver. The waiver was a necessity to keep the Shuttle flying according to the original manifest.

Having a risk identification of C1 was not regarded as a sufficient reason to cancel a launch. It simply meant that component failure could be disastrous. It implied that this might be a potential problem that needed attention. If the risks were acceptable, NASA could still launch. A more serious condition was the issuing of launch constraints. Launch constraints were official NASA designations for situations in which mission safety was a serious enough problem to justify a decision not to launch. But once again, a launch constraint did not imply that the launch should be delayed. It meant that this was an important problem and needed to be addressed.

Following the 1985 mission that showed O-ring erosion and exhaust gas blow-by, a launch constraint was imposed. Yet on each of the next five Shuttle missions, NASA's Mulloy issued a launch constraint waiver allowing the flights to take place on schedule without any changes to the O-rings.

Were the waivers a violation of serious safety rules just to keep the Shuttle flying? The answer is *no*! NASA had protocols such as policies, procedures, and rules for adherence to safety. Waivers were also protocols but for the purpose of deviating from other existing protocols. Larry Mulloy, his colleagues at NASA, and the contractors had no intentions of doing evil. Waivers were simply a way of saying that we believe that the risk is an *acceptable risk.*

The lifting of launch constraints and the issuance of waivers became the norm—standard operating procedure. Waivers became a way of life. If waivers were issued and the mission was completed successfully, then the same waivers would exist for the next flight and did not have to be brought up for discussion at the Flight Readiness Review meeting. The justification for the waivers seemed to be the similarity between flight launch conditions, temperature, and so on. Launching under similar conditions seemed to be important for the engineers at NASA and Thiokol because it meant that the forces acting on the O-rings were within their region of experience and could be correlated to existing data. The launch temperature effect on the O-rings was considered predictable, and therefore constituted an acceptable risk to both NASA and Thiokol, thus perhaps eliminating costly program delays that would have resulted from having to redesign the O-rings. The completion of each Shuttle mission added another data point to the region of experience, thus guaranteeing the same waivers on the next launch. Flying with acceptable risk became the norm in NASA's culture.

Launch Liftoff Sequence Profile: Possible Aborts

During the countdown to liftoff, the launch team closely monitors weather conditions, not only at the launch site, but also at touchdown sites should the mission need to be prematurely aborted.

Dr. Feynman: Would you explain why we are so sensitive to the weather?

Mr. Moore (NASA's deputy administrator for space flight): Yes, there are several reasons. I mentioned the return to the landing site. We need to have visibility if we get into a situation where we need to return to the landing site after launch, and the pilots and the commanders need to be able to see the runway and so forth. So, you need a ceiling limitation on it [i.e., weather].

We also need to maintain specifications on wind velocity so we don't exceed crosswinds. Landing on a runway and getting too high of a crosswind may cause us to deviate off of the runway and so forth, so we have a crosswind limit. During ascent, assuming a normal flight, a chief concern is damage to tiles due to rain. We have had experiences in seeing what the effects of a brief shower can do in terms of the tiles. The tiles are thermal insulation blocks, very thick. A lot of them are very thick on the bottom of the Orbiter. But if you have a raindrop and you are going at a very high velocity, it tends to erode the tiles, pock the tiles, and that causes us a grave concern regarding the thermal protection.

In addition to that, you are worried about the turnaround time of the Orbiters as well, because with the kind of tile damage that one could get in rain, you have an awful lot of work to do to go back and replace tiles back on the system. So, there are a number of concerns that weather enters into, and it is a major factor in our assessment of whether or not we are ready to launch.[10]

Approximately six to seven seconds prior to the liftoff, the Shuttle's main engines (liquid fuel) ignite. These engines consume one-half million gallons of liquid fuel. It takes nine hours prior to launch to fill the liquid fuel tanks. At ignition, the engines are throttled up to 104 percent of rated power. Redundancy checks on the engines' systems are then made. The launch site ground complex and the Orbiter's onboard computer complex check a large number of details and parameters about the main engines to make sure that everything is proper and that the main engines are performing as planned.

If a malfunction is detected, the system automatically goes into a shutdown sequence, and the mission is scrubbed. The primary concern at this point is to make the vehicle "safe." The crew remains on board and performs a number of functions to get the vehicle into a safe mode. These functions include making sure that all propellant and electrical systems are properly safed. Ground crews at the launch pad begin servicing the launch pad. Once the launch pad is in a safe condition, the hazard and safety teams begin draining the remaining liquid fuel out of the external tank.

If no malfunction is detected during this six-second period of liquid fuel burn, then a signal is sent to ignite the two solid rocket boosters, and liftoff occurs. For the next two minutes, with all engines ignited, the Shuttle goes through a Max Q, or high dynamic pressure phase, that exerts maximum pressure loads on the Orbiter vehicle. Based upon the launch profile, the main engines may be throttled down slightly during the Max Q phase to lower the loads.

After 128 seconds into the launch sequence, all of the solid fuel is expended and the solid rocket boosters (SRBs) staging occurs. The SRB parachutes are deployed. The SRBs then fall back to earth 162 miles from the launch site and are recovered for examination, cleaning, and

10. *RPC,* page 18.

reuse on future missions. The main liquid fuel engines are then throttled up to maximum power. After 523 seconds into the liftoff, the external liquid fuel tanks are essentially expended of fuel. The main engines are shut down. Ten to eighteen seconds later, the external tank is separated from the Orbiter and disintegrates on reentry into the atmosphere.

From a safety perspective, the most hazardous period is the first 128 seconds when the SRBs are ignited. Here's what Arnold Aldrich, manager of NASA's STS Program, Johnson Space Center, had to say:

Mr. Aldrich: Once the Shuttle System starts off the launch pad, there is no capability in the system to separate these [solid propellant] rockets until they reach burnout. They will burn for two minutes and eight or nine seconds, and the system must stay together. There is not a capability built into the vehicle that would allow these to separate. There is a capability available to the flight crew to separate at this interface the Orbiter from the tank, but that is thought to be unacceptable during the first stage when the booster rockets are on and thrusting. So, essentially the first two minutes and a little more of flight, the stack is intended and designed to stay together, and it must stay together to fly successfully.

Mr. Hotz: Mr. Aldrich, why is it unacceptable to separate the Orbiter at that stage?

Mr. Aldrich: It is unacceptable because of the separation dynamics and the rupture of the propellant lines. You cannot perform the kind of a clean separation required for safety in the proximity of these vehicles at the velocities and the thrust levels they are undergoing, [and] the atmosphere they are flying through. In that regime, it is the design characteristic of the total system.[11]

If an abort is deemed necessary during the first 128 seconds, the actual abort will not begin until *after* SRB staging has occurred, which is after 128 seconds into the launch sequence. Based upon the reason and timing of an abort, options include:

Type of Abort	Landing Site
Once-around abort	Edwards Air Force Base
Trans-Atlantic abort	DaKar
Trans-Atlantic abort	Casablanca
Return-to-landing-site (RTLS)	Kennedy Space Center

Arnold Aldrich commented on different abort profiles:

Chairman Rogers: During the two-minute period, is it possible to abort through the Orbiter?

Mr. Aldrich: You can abort for certain conditions. You can start an abort, but the vehicle won't do anything yet, and the intended aborts are built around failures in the main engine system, the liquid propellant systems and their controls. If you have a failure of a main engine, it is well detected by the crew and by the ground support, and you can call for a return-to-launch-site abort. That would be logged in the computer. The computer would be set up to execute it, but everything waits until the solids take you to altitude. At that time, the solids will separate in the sequence I described, and then the vehicle flies downrange some 400 miles, maybe 10 to 15 additional minutes, while all of the tank propellant is expelled through these engines.

11. *RPC,* page 51.

As a precursor to setting up the conditions for this return-to-launch-site abort to be successful towards the end of that burn downrange, using the propellants and the thrust of the main engines, the vehicle turns and actually points heads up back towards Florida. When the tank is essentially depleted, automatic signals are sent to close off the [liquid] propellant lines and to separate the Orbiter, and the Orbiter then does a similar approach to the one we are familiar with with orbit back to the Kennedy Space Center for approach and landing.

Dr. Walker: So, the propellant is expelled but not burned?

Mr. Aldrich: No, it is burned. You burn the system on two engines all the way down-range until it is gone, and then you turn around and come back because you don't have enough to burn to orbit. That is the return-to-launch-site abort, and it applies during the first 240 seconds of—no, 240 is not right. It is longer than that—the first four minutes, either before or after separation you can set that abort up, but it will occur after the solids separate, and if you have a main engine anomaly after the solids separate, at that time you can start the RTLS, and it will go through that same sequence and come back.

Dr. Ride: And you can also only do an RTLS if you have lost just one main engine. So if you lose all three main engines, RTLS isn't a viable abort mode.

Mr. Aldrich: Once you get through the four minutes, there's a period where you now don't have the energy conditions right to come back, and you have a forward abort, and Jesse mentioned the sites in Spain and on the coast of Africa. We have what is called a trans-Atlantic abort, and where you can use a very similar sequence to the one I just described. You still separate the solids, you still burn all the propellant out of the tanks, but you fly across and land across the ocean.

Mr. Hotz: Mr. Aldrich, could you recapitulate just a bit here? Is what you are telling us that for two minutes of flight, until the solids separate, there is no practical abort mode?

Mr. Aldrich: Yes, sir.

Mr. Hotz: Thank you.

Mr. Aldrich: A trans-Atlantic abort can cover a range of just a few seconds up to about a minute in the middle where the across-the-ocean sites are effective, and then you reach this abort once-around capability where you go all the way around and land in California or back to Kennedy by going around the earth. And finally, you have abort-to-orbit where you have enough propulsion to make orbit but not enough to achieve the exact orbital parameters that you desire. That is the way that the abort profiles are executed.

There are many, many nuances of crew procedure and different conditions and combinations of sequences of failures that make it much more complicated than I have described it.[12]

The O-Ring Problem

There were two kinds of joints on the Shuttle—field joints that were assembled at the launch site connecting together the SRB's cylindrical cases, and nozzle joints that connected the aft end of the case to the nozzle. During the pressure of ignition, the field joints could become bent such that the secondary O-ring could lose contact within an estimated 0.17 to 0.33 seconds after ignition. If the primary O-ring failed to seal properly before the gap within the joints opened up and the secondary seal failed, the results could be disastrous.

---

12. *RPC,* pages 51–52

When the solid propellant boosters are recovered after separation, they are disassembled and checked for damage. The O-rings could show evidence of coming into contact with heat. Hot gases from the ignition sequence could blow by the primary O-ring briefly before sealing. This "blow-by" phenomenon could last for only a few milliseconds before sealing and result in no heat damage to the O-ring. If the actual sealing process takes longer than expected, then charring and erosion of the O-rings can occur. This would be evidenced by gray or black soot and erosion to the O-rings. The terms used are impingement erosion and "by-pass" erosion, with the latter identified also as sooted "blow-by."

Roger Boisjoly of Thiokol describes blow-by erosion and joint rotation as follows:

> O-ring material gets removed from the cross section of the O-ring much, much faster than when you have bypass erosion or blow-by, as people have been terming it. We usually use the characteristic blow-by to define gas past it, and we use the other term [bypass erosion] to indicate that we are eroding at the same time. And so you can have blow-by without erosion, [and] you [can] have blow-by with erosion.[13]
>
> At the beginning of the transient cycle [initial ignition rotation, up to 0.17 seconds] . . . [the primary O-ring] is still being attacked by hot gas, and it is eroding at the same time it is trying to seal, and it is a race between, will it erode more than the time allowed to have it seal.[14]

On January 24, 1985, STS 51-C [Flight No. 15] was launched at 51°F, which was the lowest temperature of any launch up to that time. Analyses of the joints showed evidence of damage. Black soot appeared between the primary and secondary O-rings. The engineers concluded that the cold weather had caused the O-rings to harden and move more slowly. This allowed the hot gases to blow by and erode the O-rings. This scorching effect indicated that low temperature launches could be disastrous.

On July 31, 1985, Roger Boisjoly of Thiokol sent an interoffice memo to R. K. Lund, vice president for engineering at Thiokol:

> This letter is written to insure that management is fully aware of the seriousness of the current O-ring erosion problem in the SRM joints from an engineering standpoint.
>
> The mistakenly accepted position on the joint problem was to fly without fear of failure and to run a series of design evaluations which would ultimately lead to a solution or at least a significant reduction of the erosion problem. This position is now drastically changed as a result of the SRM 16A nozzle joint erosion which eroded a secondary O-ring with the primary O-ring never sealing.
>
> If the same scenario should occur in a field joint (and it could), then it is a jump ball as to the success or failure of the joint because the secondary O-ring cannot respond to the clevis opening rate and may not be capable of pressurization. The result would be a catastrophe of the highest order—loss of human life.
>
> An unofficial team (a memo defining the team and its purpose was never published) with [a] leader was formed on 19 July 1985 and was tasked with solving the problem for both the short and long term. This unofficial team is essentially nonexistent at this time. In my opinion, the team must be officially given the responsibility and the authority to execute the work that needs to be done on a non-interference basis (full time assignment until completed).
>
> It is my honest and very real fear that if we do not take immediate action to dedicate a team to solve the problem with the field joint having the number one priority, then we stand in jeopardy of losing a flight along with all the launch pad facilities.[15]

---

13. *RPC,* pages 784–785.

14. *RPC,* page 136.

15. *RPC,* pages 691–692.

On August 9, 1985, a letter was sent from Brian Russell, manager of the SRM Ignition System, to James Thomas at the Marshall Space Flight Center. The memo addressed the following:

Per your request, this letter contains the answers to the two questions you asked at the July Problem Review Board telecon.

1. *Question:* If the field joint secondary seal lifts off the metal mating surfaces during motor pressurization, how soon will it return to a position where contact is re-established?

   Answer: Bench test data indicates that the O-ring resiliency (its capability to follow the metal) is a function of temperature and rate of case expansion. MTI [Thiokol] measured the force of the O-ring against Instron plattens, which simulated the nominal squeeze on the O-ring and approximated the case expansion distance and rate.

   At 100°F, the O-ring maintained contact. At 75°F, the O-ring lost contact for 2.4 seconds. At 50°F, the O-ring did not re-establish contact in 10 minutes at which time the test was terminated.

   The conclusion is that secondary sealing capability in the SRM field joint cannot be guaranteed.

2. *Question:* If the primary O-ring does not seal, will the secondary seal seat in sufficient time to prevent joint leakage?

   Answer: MTI has no reason to suspect that the primary seal would ever fail after pressure equilibrium is reached; i.e., after the ignition transient. If the primary O-ring were to fail from 0 to 170 milliseconds, there is a very high probability that the secondary O-ring would hold pressure since the case has not expanded appreciably at this point. If the primary seal were to fail from 170 to 330 milliseconds, the probability of the secondary seal holding is reduced. From 330 to 600 milliseconds the chance of the secondary seal holding is small. This is a direct result of the O-ring's slow response compared to the metal case segments as the joint rotates.[16]

At NASA, the concern for a solution to the O-ring problem became not only a technical crisis, but also a budgetary crisis. In a July 23, 1985, memorandum from Richard Cook, program analyst, to Michael Mann, chief of the STS Resource Analysis Branch, the impact of the problem was noted:

Earlier this week you asked me to investigate reported problems with the charring of seals between SRB motor segments during flight operations. Discussions with program engineers show this to be a potentially major problem affecting both flight safety and program costs.

Presently three seals between SRB segments use double O-rings sealed with putty. In recent Shuttle flights, charring of these rings has occurred. The O-rings are designed so that if one fails, the other will hold against the pressure of firing. However, at least in the joint between the nozzle and the aft segment, not only has the first O-ring been destroyed, but the second has been partially eaten away.

Engineers have not yet determined the cause of the problem. Candidates include the use of a new type of putty (the putty formerly in use was removed from the market by EPA because it contained asbestos), failure of the second ring to slip into the groove which must engage it for it to work properly, or new, and as yet unidentified, assembly procedures at Thiokol. MSC is

16. *RPC,* pages 1568–1569.

> trying to identify the cause of the problem, including on-site investigation at Thiokol, and OSF hopes to have some results from their analysis within 30 days. There is little question, however, that flight safety has been and is still being compromised by potential failure of the seals, and it is acknowledged that failure during launch would certainly be catastrophic. There is also indication that staff personnel knew of this problem sometime in advance of management's becoming apprised of what was going on.
>
> The potential impact of the problem depends on the as yet undiscovered cause. If the cause is minor, there should be little or no impact on budget or flight rate. A worst case scenario, however, would lead to the suspension of Shuttle flights, redesign of the SRB, and scrapping of existing stockpiled hardware. The impact on the FY 1987-8 budget could be immense.
>
> It should be pointed out that Code M management [NASA's associate administrator for space flight] is viewing the situation with the utmost seriousness. From a budgetary standpoint, I would think that any NASA budget submitted this year for FY 1987 and beyond should certainly be based on a reliable judgment as to the cause of the SRB seal problem and a corresponding decision as to budgetary action needed to provide for its solution.[17]

On October 30, 1985, NASA launched Flight STS 61-A [Flight no. 22] at 75°F. This flight also showed signs of sooted blow-by, but the color was significantly blacker. Although there was some heat effect, there was no measurable erosion observed on the secondary O-ring. Since blow-by and erosion had now occurred at a higher launch temperature, the original premise that launches under cold temperatures were a problem was now being questioned. Exhibit V shows the temperature at launch of all the Shuttle flights up to this time and the O-ring damage, if any.

Management at both NASA and Thiokol wanted *concrete* evidence that launch temperature was directly correlated to blow-by and erosion. Other than simply a "gut feel," engineers were now stymied on how to show the direct correlation. NASA was not ready to cancel a launch simply due to an engineer's "gut feel."

William Lucas, director of the Marshall Space Center, made it clear that NASA's manifest for launches would be adhered to. Managers at NASA were pressured to resolve problems internally rather than to escalate them up the chain of command. Managers became afraid to inform anyone higher up that they had problems, even though they knew that one existed.

Richard Feynman, Nobel laureate and member of the Rogers Commission, concluded that a NASA official altered the safety criteria so that flights could be certified on time under pressure imposed by the leadership of William Lucas. Feynman commented:

> . . . They, therefore, fly in a relatively unsafe condition with a chance of failure of the order of one percent. Official management claims to believe that the probability of failure is a thousand times less.

Without concrete evidence of the temperature effect on the O-rings, the secondary O-ring was regarded as a redundant safety constraint and the criticality factor was changed from C1 to C1R. Potentially serious problems were treated as anomalies peculiar to a given flight. Under the guise of anomalies, NASA began issuing waivers to maintain the flight schedules. Pressure was placed upon contractors to issue closure reports. On December 24, 1985, L. O. Wear, NASA's SRM Program Office manager, sent a letter to Joe Kilminster, Thiokol's vice president for the Space Booster Program:

> During a recent review of the SRM Problem Review Board open problem list I found that we have 20 open problems, 11 opened during the past 6 months, 13 open over 6 months, 1 three years old, 2 two years old, and 1 closed during the past six months. As you can see our closure

---

17. *RPC,* pages 391–392.

Exhibit V. Erosion and blow-by history (temperature in ascending order from coldest to warmest)

Flight	Date	Temperature (°F)	Erosion Incidents	Blow-by Incidents	Comments
51-C	01/24/85	53	3	2	Most erosion any flight; blow-by; secondary O-rings heated up
41-B	02/03/84	57	1		Deep, extensive erosion
61-C	01/12/86	58	1		O-rings erosion
41-C	04/06/84	63	1		O-rings heated but no damage
1	04/12/81	66			Coolest launch without problems
6	04/04/83	67			
51-A	11/08/84	67			
51-D	04/12/85	67			
5	11/11/82	68			
3	03/22/82	69			
2	11/12/81	70	1		Extent of erosion unknown
9	11/28/83	70			
41-D	08/30/84	70	1		
51-G	06/17/85	70			
7	06/18/83	72			
8	08/30/83	73			
51-B	04/29/85	75			
61-A	10/20/85	75		2	No erosion but soot between O-rings
51-1	08/27/85	76			
61	11/26/85	76			
41-G	10/05/84	78			
51-J	10/03/85	79			
4	06/27/82	80			No data; casing lost at sea
51-F	07/29/85	81			

record is very poor. You are requested to initiate the required effort to assure more timely closures and the MTI personnel shall coordinate directly with the S&E personnel the contents of the closure reports.[18]

Pressure, Paperwork, and Waivers

To maintain the flight schedule, critical issues such as launch constraints had to be resolved or waived. This would require extensive documentation. During the Rogers Commission investigation, it seemed that there had been a total lack of coordination between NASA's Marshall Space Center and Thiokol prior to the Challenger disaster. Joe Kilminster, Thiokol's vice president for the Space Booster Program, testified:

Mr. Kilminster: Mr. Chairman, if I could, I would like to respond to that. In response to the concern that was expressed—and I had discussions with the team leader, the task force team leader, Mr. Don Ketner, and Mr. Russell and Mr. Ebeling. We held a meeting in my office and that was done in the October time period where we called the people who were in a support role to the task team, as well as the task force members themselves.

18. *RPC,* page 1554.

In that discussion, some of the task force members were looking to circumvent some of our established systems. In some cases, that was acceptable; in other cases, it was not. For example, some of the work that they had recommended to be done was involved with full-scale hardware, putting some of these joints together with various putty layup configurations; for instance, taking them apart and finding out what we could from that inspection process.

Dr. Sutter: Was that one of these things that was outside of the normal work, or was that accepted as a good idea or a bad idea?

Mr. Kilminster: A good idea, but outside the normal work, if you will.

Dr. Sutter: Why not do it?

Mr. Kilminster: Well, we were doing it. But the question was, can we circumvent the system, the paper system that requires, for instance, the handling constraints on those flight hardware items? And I said no, we can't do that. We have to maintain our handling system, for instance, so that we don't stand the possibility of injuring or damaging a piece of flight hardware.

I asked at that time if adding some more people, for instance, a safety engineer—that was one of the things we discussed in there. The consensus was no, we really didn't need a safety engineer. We had the manufacturing engineer in attendance who was in support of that role, and I persuaded him that, typical of the way we normally worked, that he should be calling on the resources from his own organization, that is, in Manufacturing, in order to get this work done and get it done in a timely fashion.

And I also suggested that if they ran across a problem in doing that, they should bubble that up in their management chain to get help in getting the resources to get that done. Now, after that session, it was my impression that there was improvement based on some of the concerns that had been expressed, and we did get quite a bit of work done. For your evaluation, I would like to talk a little bit about the sequence of events for this task force.

Chairman Rogers: Can I interrupt? Did you know at that time it was a launch constraint, a formal launch constraint?

Mr. Kilminster: Not an overall launch constraint as such. Similar to the words that have been said before, each Flight Readiness Review had to address any anomalies or concerns that were identified at previous launches and in that sense, each of those anomalies or concerns were established in my mind as launch constraints unless they were properly reviewed and agreed upon by all parties.

Chairman Rogers: You didn't know there was a difference between the launch constraint and just considering it an anomaly? You thought they were the same thing?

Mr. Kilminster: No, sir. I did not think they were the same thing.

Chairman Rogers: My question is: Did you know that this launch constraint was placed on the flights in July 1985?

Mr. Kilminster: Until we resolved the O-ring problem on that nozzle joint, yes. We had to resolve that in a fashion for the subsequent flight before we would be okay to fly again.

Chairman Rogers: So you did know there was a constraint on that?

Mr. Kilminster: On a one flight per one flight basis; yes, sir.

Chairman Rogers: What else would a constraint mean?

Mr. Kilminster: Well, I get the feeling that there's a perception here that a launch constraint means all launches, whereas we were addressing each launch through the Flight Readiness Review process as we went.

Chairman Rogers: No, I don't think—the testimony that we've had is that a launch constraint is put on because it is a very serious problem and the constraint means don't fly unless it's fixed or taken care of, but somebody has the authority to waive it for a particular flight. And in this case, Mr. Mulloy was authorized to waive it, which he did, for a number of flights before 51-L. Just prior to 51-L, the papers showed the launch constraint was closed out, which I guess means no longer existed. And that was done on January 23, 1986. Now, did you know that sequence of events?

Mr. Kilminster: Again, my understanding of closing out, as the term has been used here, was to close it out on the problem actions list, but not as an overall standard requirement. We had to address these at subsequent Flight Readiness Reviews to insure that we were all satisfied with the proceeding to launch.

Chairman Rogers: Did you understand the waiver process, that once a constraint was placed on this kind of a problem, that a flight could not occur unless there was a formal waiver?

Mr. Kilminster: Not in the sense of a formal waiver, no, sir.

Chairman Rogers: Did any of you? Didn't you get the documents saying that?

Mr. McDonald: I don't recall seeing any documents for a formal waiver.[19]

Mission 51-L

On January 25, 1986, questionable weather caused a delay of Mission 51-L to January 27. On January 26, the launch was reconfirmed for 9:37 A.M. on the 27th. However, on the morning of January 27, a malfunction with the hatch, combined with high crosswinds, caused another delay. All preliminary procedures had been completed and the crew had just boarded when the first problem appeared. A microsensor on the hatch indicated that the hatch was not shut securely. It turned out that the hatch was shut securely but the sensor had malfunctioned. Valuable time was lost in determining the problem.

After the hatch was finally closed, the external handle could not be removed. The threads on the connecting bolt were stripped and instead of cleanly disengaging when turned, simply spun around. Attempts to use a portable drill to remove the handle failed. Technicians on the scene asked Mission Control for permission to saw off the bolt. Fearing some form of structural stress to the hatch, engineers made numerous time-consuming calculations before giving the go-ahead to cut off the bolt. The entire process consumed almost two hours before the countdown resumed.

However, the misfortunes continued. During the attempts to verify the integrity of the hatch and remove the handle, the wind had been steadily rising. Chief Astronaut John Young flew a series of approaches in the Shuttle training aircraft and confirmed the worst fears of mission control. The crosswinds at the Cape were in excess of the level allowed for the abort contingency. The opportunity had been missed. The mission was then reset to launch the next day, January 28, at 9:38 A.M. Everyone was quite discouraged since extremely cold weather was forecast for Tuesday that could further postpone the launch.[20]

---

19. *RPC,* pages 1577–1578.

20. Hoover and Wallace, pages 3–4.

Weather conditions indicated that the temperature at launch could be as low as 26°F. This would be much colder and well below the temperature range that the O-rings were designed to operate in. The components of the solid rocket motors were qualified only to 40°F at the lower limit. Undoubtedly, when the sun came up and launch time approached, both the air temperature and vehicle would warm up, but there was still concern. Would the ambient temperature be high enough to meet the launch requirements? NASA's Launch Commit Criteria stated that no launch should occur at temperatures below 31°F. There were also worries over any permanent effects on the Shuttle due to the cold overnight temperatures. NASA became concerned and asked Thiokol for their recommendation on whether or not to launch. NASA admitted under testimony that if Thiokol had recommended not launching, then the launch would not have taken place.

At 5:45 P.M. eastern standard time, a teleconference was held between the Kennedy Space Center, Marshall Space Flight Center, and Thiokol. Bob Lund, vice president for engineering, summarized the concerns of the Thiokol engineers that in Thiokol's opinion, the launch should be delayed until noontime or even later such that a launch temperature of at least 53°F could be achieved. Thiokol's engineers were concerned that no data were available for launches at this temperature of 26°F. This was the first time in fourteen years that Thiokol had recommended not to launch.

The design validation tests originally done by Thiokol covered only a narrow temperature range. The temperature data did not include any temperatures below 53°F. The O-rings from Flight 51-C, which had been launched under cold conditions the previous year, showed very significant erosion. These were the only data available on the effects of cold, but all of the Thiokol engineers agreed that the cold weather would decrease the elasticity of the synthetic rubber O-rings, which in turn might cause them to seal slowly and allow hot gases to surge through the joint.[21]

Another teleconference was set up for 8:45 P.M. to invite more parties to be involved in the decision. Meanwhile, Thiokol was asked to fax all relevant and supporting charts to all parties involved in the 8:45 P.M. teleconference.

The following information was included in the pages that were faxed:

Blow-by History

SRM-15 Worst Blow-by
- 2 case joints (80°), (110°) *Arc*
- Much worse visually than SRM-22

SRM-22 Blow-by
- 2 case joints (30–40°)

SRM-13A, 15, 16A, 18, 23A, 24A
- Nozzle blow-by

Field Joint Primary Concerns—SRM-25
- A temperature lower than the current database results in changing primary O-ring sealing timing function
- SRM-15A—80° arc black grease between O-rings
  SRM-15B—110° arc black grease between O-rings
- Lower O-ring squeeze due to lower temp

21. Hoover and Wallace, page 4.

- Higher O-ring shore hardness
- Thicker grease viscosity
- Higher O-ring pressure activation time
- If actuation time increases, threshold of secondary seal pressurization capability is approached
- If threshold is reached then secondary seal may not be capable of being pressurized

Conclusions:

- Temperature of O-ring is not only parameter controlling blow-by
  - SRM-15 with blow-by had an O-ring temp at 53°F
  - SRM-22 with blow-by had an O-ring temp at 75°F
  - Four development motors with no blow-by were tested at O-ring temp of 47° to 52°F
  - Development motors had putty packing which resulted in better performance
- At about 50°F blow-by could be experienced in case joints
- Temp for SRM-25 on 1-28-86 launch will be: 29°F 9 A.M.
38°F 2 P.M.
- Have no data that would indicate SRM-25 is different than SRM-15 other than temp

Recommendations:

- O-ring temp must be ≥ 53°F at launch
  - Development motors at 47° to 52°F with putty packing had no blow-by
  - SRM-15 (the best simulation) worked at 53°
- Project ambient conditions (temp & wind) to determine launch time

From NASA's perspective, the launch window was from 9:30 A.M. to 12:30 P.M. on January 28th. This was based upon weather conditions and visibility, not only at the launch site but also at the landing sites should an abort be necessary. An additional consideration was the fact that the temperature might not reach 53°F prior to the launch window closing. Actually, the temperature at the Kennedy Space Center was not expected to reach 50°F until two days later. NASA was hoping that Thiokol would change their minds and recommend launch.

The Second Teleconference

At the second teleconference, Bob Lund once again asserted Thiokol's recommendation not to launch below 53°F. NASA's Mulloy then burst out over the teleconference network:

My God, Morton Thiokol! When do you want me to launch—next April?

NASA challenged Thiokol's interpretation of the data and argued that Thiokol was inappropriately attempting to establish a new Launch Commit Criterion just prior to launch. NASA asked Thiokol to reevaluate their conclusions. Crediting NASA's comments with some validity, Thiokol then requested a five-minute *off-line* caucus. In the room at Thiokol were 14 engineers, namely:

- Jerald Mason, senior vice president, Wasatch Operations
- Calvin Wiggins, vice president and general manager, Space Division
- Joe C. Kilminster, vice president, Space Booster Programs

- Robert K. Lund, vice president, Engineering
- Larry H. Sayer, director, Engineering and Design
- William Macbeth, manager, Case Projects, Space Booster Project
- Donald M. Ketner, supervisor, Gas Dynamics Section, and head, Seal Task Force
- Roger Boisjoly, member, Seal Task Force
- Arnold R. Thompson, supervisor, Rocket Motor Cases
- Jack R. Kapp, manager, Applied Mechanics Department
- Jerry Burn, associate engineer, Applied Mechanics
- Joel Maw, associate scientist, Heat Transfer Section
- Brian Russell, manager, Special Projects, SRM Project
- Robert Ebeling, manager, Ignition System and Final Assembly, SRB Project

There were no safety personnel in the room because nobody thought to invite them. The caucus lasted some 30 minutes. Thiokol (specifically Joe Kilminster) then returned to the teleconference stating that they were unable to sustain a valid argument that temperature affects O-ring blow-by and erosion. *Thiokol then reversed its position and was now recommending launch.*

NASA stated that the launch of the Challenger would not take place without Thiokol's approval. But when Thiokol reversed its position following the caucus and agreed to launch, NASA interpreted this as an acceptable risk. The launch would now take place.

Mr. McDonald (Thiokol): "The assessment of the data was that the data was not totally conclusive, that the temperature could affect everything relative to the seal. But there was data that indicated that there were things going in the wrong direction, and this was far from our experience base.

The conclusion being that Thiokol was directed to reassess all the data because the recommendation was not considered acceptable at that time of [waiting for] the 53 degrees [to occur]. NASA asked us for a reassessment and some more data to show that the temperature in itself can cause this to be a more serious concern than we had said it would be. At that time Thiokol in Utah said that they would like to go off-line and caucus for about five minutes and reassess what data they had there or any other additional data.

And that caucus lasted for, I think, a half hour before they were ready to go back on. When they came back on they said they had reassessed all the data and had come to the conclusions that the temperature influence, based on the data they had available to them, was inconclusive and therefore they recommended a launch.[22]

During the Rogers Commission testimony, NASA's Mulloy stated his thought process in requesting Thiokol to rethink their position:

General Kutyna: You said the temperature had little effect?

Mr. Mulloy: I didn't say that. I said I can't get a correlation between O-ring erosion, blow-by and O-ring, and temperature.

General Kutyna: 51-C was a pretty cool launch. That was January of last year.

Mr. Mulloy: It was cold before then but it was not that much colder than other launches.

---

22. *RPC,* page 300.

General Kutyna: So it didn't approximate this particular one?

Mr. Mulloy: Unfortunately, that is one you look at and say, aha, is it related to a temperature gradient and the cold. The temperature of the O-ring on 51-C, I believe, was 53 degrees. We have fired motors at 48 degrees.[23]

Mulloy asserted he had not pressured Thiokol into changing their position. Yet, the testimony of Thiokol's engineers stated they believed they were being pressured.

Roger Boisjoly, one of Thiokol's experts on O-rings, was present during the caucus and vehemently opposed the launch. During testimony, Boisjoly described his impressions of what occurred during the caucus:

> The caucus was started by Mr. Mason stating that a management decision was necessary. Those of us who were opposed to the launch continued to speak out, and I am specifically speaking of Mr. Thompson and myself because in my recollection, he and I were the only ones who vigorously continued to oppose the launch. And we were attempting to go back and rereview and try to make clear what we were trying to get across, and we couldn't understand why it was going to be reversed.
>
> So, we spoke out and tried to explain again the effects of low temperature. Arnie actually got up from his position which was down the table and walked up the table and put a quad pad down in front of the table, in front of the management folks, and tried to sketch out once again what his concern was with the joint, and when he realized he wasn't getting through, he just stopped.
>
> I tried one more time with the photos. I grabbed the photos and I went up and discussed the photos once again and tried to make the point that it was my opinion from actual observations that temperature was indeed a discriminator, and we should not ignore the physical evidence that we had observed.
>
> And again, I brought up the point that SRM-15 had a 110 degree arc of black grease while SRM-22 had a relatively different amount, which was less and wasn't quite as black. I also stopped when it was apparent that I could not get anybody to listen.

Dr. Walker: At this point did anyone else [i.e., engineers] speak up in favor of the launch?

Mr. Boisjoly: No, sir. No one said anything, in my recollection. Nobody said a word. It was then being discussed amongst the management folks. After Arnie and I had our last say, Mr. Mason said we have to make a management decision. He turned to Bob Lund and asked him to take off his engineering hat and put on his management hat. From this point on, management formulated the points to base their decision on. There was never one comment in favor, as I have said, of launching by any engineer or other nonmanagement person in the room before or after the caucus. I was not even asked to participate in giving any input to the final decision charts.

I went back on the net with the final charts or final chart, which was the rationale for launching, and that was presented by Mr. Kilminster. It was handwritten on a notepad, and he read from that notepad. I did not agree with some of the statements that were being made to support the decision. I was never asked nor polled, and it was clearly a management decision from that point.

I must emphasize, I had my say, and I never take any management right to take the input of an engineer and then make a decision based upon that input, and I truly believe that. I have worked at a lot of companies, and that has been done from time to time, and I truly believe that, and so there was no point in me doing anything any further [other] than [what] I had already attempted to do.

---

23. *RPC,* page 290.

I did not see the final version of the chart until the next day. I just heard it read. I left the room feeling badly defeated, but I felt I really did all I could to stop the launch. I felt personally that management was under a lot of pressure to launch, and they made a very tough decision, but I didn't agree with it.

One of my colleagues who was in the meeting summed it up best. This was a meeting where the determination was to launch, and it was up to us to prove beyond a shadow of a doubt that it was not safe to do so. This is in total reverse to what the position usually is in a preflight conversation or a Flight Readiness Review. It is usually exactly opposite that.

Dr. Walker: Do you know the source of the pressure on management that you alluded to?

Mr. Boisjoly: Well, the comments made over the net are what I felt. I can't speak for them, but I felt it. I felt the tone of the meeting exactly as I summed up, that we were being put in a position to prove that we should not launch rather than being put in the position and prove that we had enough data to launch.[24]

General Kutyna: What was the motivation driving those who were trying to overturn your opposition?

Mr. Boisjoly: They felt that we had not demonstrated, or I had not demonstrated, because I was the prime mover in SRM-15. Because of my personal observations and involvement in the Flight Readiness Reviews, they felt that I had not conclusively demonstrated that there was a tie-in between temperature and blow-by.

My main concern was if the timing function changed and that seal took longer to get there, then you might not have any seal left because it might be eroded before it seats. And then, if that timing function is such that it pushes you from the 170 millisecond region into the 330 second region, you might not have a secondary seal to pick up if the primary is gone. That was my major concern.

I can't quantify it. I just don't know how to quantify that. But I felt that the observations made were telling us that there was a message there telling us that temperature was a discriminator, and I couldn't get that point across. I basically had no direct input into the final recommendation to launch, and I was not polled.

I think Astronaut Crippin hit the tone of the meeting exactly right on the head when he said that the opposite was true of the way the meetings were normally conducted. We normally have to absolutely prove beyond a shadow of a doubt that we have the ability to fly, and it seemed like we were trying to prove, have proved that we had data to prove that we couldn't fly at this time, instead of the reverse. That was the tone of the meeting, in my opinion.[25]

Jerald Mason, senior vice president at Thiokol's Wasatch Division, directed the caucus at Thiokol. Mason continuously asserted that a management decision was needed and instructed Bob Lund, vice president for engineering, to take off his engineering hat and put on his management hat. During testimony, Mason commented on his interpretation of the data:

Dr. Ride [a member of the Commission]: You know, what we've seen in the charts so far is that the data was inconclusive and so you said go ahead.

---

24. *RPC,* pages 793–794.

25. *RPC,* page 676.

Mr. Mason: . . . I hope I didn't convey that. But the reason for the discussion was the fact that we didn't have enough data to quantify the effect of the cold, and that was the heart of our discussion . . . We have had blow-by on earlier flights. We had not had any reason to believe that we couldn't experience it again at any temperature. . . .[26]

At the end of the second teleconference, NASA's Hardy at Marshall Space Flight Center requested that Thiokol put their recommendation to launch in writing and fax it to both Marshall Space Flight Center and Kennedy Space Center. The memo (shown below) was signed by Joe Kilminster, Vice President for Thiokol's Space Booster Program, and faxed at 11:45 P.M. the night before the launch.

- Calculations show that SRM-25 O-rings will be 20° colder than SRM-15 O-rings
- Temperature data not conclusive on predicting primary O-ring blow-by
- Engineering assessment is that:
  - Colder O-rings will have increased effective durometer ("harder")
  - "Harder" O-rings will take longer to "seat"
    - More gas may pass primary O-ring before the primary seal seats (relative to SRM-15)
      - Demonstrated sealing threshold is 3 times greater than 0.038" erosion experienced on SRM-15
  - **If the primary seal does not seat, the secondary seal will seat**
    - **Pressure will get to secondary seal before the metal parts rotate**
      - **O-ring pressure leak check places secondary seal in outboard position which minimizes sealing time**
- **MTI recommends STS-51L launch proceed on 28 January 1986**
  - **SRM-25 will not be significantly different from SRM-15**[27]

The Ice Problem

At 1:30 A.M. on the day of the launch, NASA's Gene Thomas, launch director, ordered a complete inspection of the launch site due to cold weather and severe ice conditions. The prelaunch inspection of the Challenger and the launch pad by the ice-team was unusual to say the least. The ice-team's responsibility was to remove any frost or ice on the vehicle or launch structure. What they found during their inspection looked like something out of a science fiction movie. The freeze protection plan implemented by Kennedy personnel had gone very wrong. Hundreds of icicles, some up to 16 inches long, clung to the launch structure. The handrails and walkways near the Shuttle entrance were covered in ice, making them extremely dangerous if the crew had to make an emergency evacuation. One solid sheet of ice stretched from the 195 foot level to the 235 foot level on the gantry. However, NASA continued to cling to its calculations that there would be no damage due to flying ice shaken loose during the launch.[28] A decision was then made to delay the launch from 9:38 A.M. to 11:30 A.M. so that the ice on the launch pad could melt. The delay was still within the launch window of 9:30 A.M.–12:30 P.M.

At 8:30 A.M., a second ice inspection was made. Ice was still significantly present at the launch site. Robert Glaysher, vice president for orbital operations at Rockwell, stated that

---

26. *RPC,* page 764.

27. *RPC,* page 764.

28. Hoover and Wallace, page 5.

the launch was unsafe. Rockwell's concern was that falling ice could damage the heat tiles on the Orbiter. This could have a serious impact during reentry.

At 10:30 A.M., a third ice inspection was made. Though some of the ice was beginning to melt, there was still significant ice on the launch pad. The temperature of the left solid rocket booster was measured at 33°F and the right booster was measured at 19°F. Even though the right booster was 34 degrees colder than Thiokol's original recommendation for a launch temperature (i.e., 53°F), no one seemed alarmed. Rockwell also agreed to launch even though their earlier statement had been that the launch was unsafe.

Arnold Aldrich, manager of the STS Program at the Johnson Space Center, testified on the concern over the ice problem:

Mr. Aldrich: Kennedy facility people at that meeting, everyone in that meeting, voted strongly to proceed and said they had no concern, except for Rockwell. The comment to me from Rockwell, which was not written specifically to the exact words, and either recorded or logged, was that they had some concern about the possibility of ice damage to the Orbiter. Although it was a minor concern, they felt that we had no experience base launching in this exact configuration before, and therefore they thought we had some additional risk of Orbiter damage from ice than we had on previous meetings, or from previous missions.

Chairman Rogers: Did they sign off on it or not?

Mr. Aldrich: We don't have a sign-off at that point. It was not—it was not maybe 20 minutes, but it was close to that. It was within the last hour of launch.

Chairman Rogers: But they still objected?

Mr. Aldrich: They issued what I would call a concern, a less than 100 percent concurrence in the launch. They did not say we do not want to launch, and the rest of the team overruled them. They issued a more conservative concern. They did not say don't launch.

General Kutyna: I can't recall a launch that I have had where there was 100 percent certainty that everything was perfect, and everyone around the table would agree to that. It is the job of the launch director to listen to everyone, and it's our job around the table to listen and say there is this element of risk, and you characterize this as 90 percent, or 95, and then you get a consensus that that risk is an acceptable risk, and then you launch.

So I think this gentleman is characterizing the degree of risk, and he's honest, and he had to say something.

Dr. Ride: But one point is that their concern is a specific concern, and they weren't concerned about the overall temperature or damage to the solid rockets or damage to the external tank. They were worried about pieces of ice coming off and denting the tile.[29]

Following the accident, the Rogers Commission identified three major concerns about the ice-on-the-pad issue:

1. An analysis of all of the testimony and interviews established that Rockwell's recommendation on launch was ambiguous. The Commission found it difficult, as did Mr. Aldrich, to conclude that there was a no-launch recommendation. Moreover, all parties were asked specifically to contact Aldrich or Moore about launch objections due to

29. *RPC,* pages 237–238.

weather. Rockwell made no phone calls or further objections to Aldrich or other NASA officials after the 9:00 A.M. Mission Management Team meeting and subsequent to the resumption of the countdown.

2. The Commission was also concerned about the NASA response to the Rockwell position at the 9:00 A.M. meeting. While it was understood that decisions have to be made in launching a Shuttle, the Commission was not convinced Levels I and II [of NASA's management] appropriately considered Rockwell's concern about the ice. However ambiguous Rockwell's position was, it was clear that they did tell NASA that the ice was an unknown condition. Given the extent of the ice on the pad, the admitted unknown effect of the Solid Rocket Motor and Space Shuttle Main Engines ignition on the ice, as well as the fact that debris striking the Orbiter was a potential flight safety hazard, the Commission found the decision to launch questionable under those circumstances. In this situation, NASA appeared to be requiring a contractor to prove that it was not safe to launch, rather than proving it was safe. Nevertheless, the Commission had determined that the ice was not a cause of the 51-L accident and does not conclude that NASA's decision to launch specifically overrode a no-launch recommendation by an element contractor.
3. The Commission concluded that the freeze protection plan for launch pad 39B was inadequate. The Commission believed that the severe cold and presence of so much ice on the fixed service structure made it inadvisable to launch on the morning of January 28, and that margins of safety were whittled down too far.

It became obvious that NASA's management knew of the ice problem, but did they know of Thiokol's original recommendation not to launch and then their reversal? Larry Mulloy, the SRB Project manager for NASA, and Stanley Reinartz, NASA's manager of the Shuttle Office, both admitted that they told Arnold Aldrich, manager of the STS program, Johnson Space Center, about their concern for the ice problem but there was no discussion about the teleconferences with Thiokol over the O-rings. It appeared that Mulloy and Reinartz considered the ice as a potential problem whereas the O-rings constituted an acceptable risk. Therefore, only potential problems went up the chain of command, not the components of the "aggregate acceptable launch risk." It became common practice in Flight Readiness Review documentation to use the term "acceptable risk." This became the norm at NASA and resulted in insulating senior management from certain potential problems. It was the culture that had developed at NASA that created the flawed decision-making process rather than an intent by individuals to withhold information and jeopardize safety.

The Accident

Just after liftoff at 0.678 seconds into the flight, photographic data showed a strong puff of gray smoke spurting from the vicinity of the aft field joint on the right solid rocket booster. The two pad 39B cameras that would have recorded the precise location of the puff were inoperative. Computer graphic analysis of film from other cameras indicated the initial smoke came from the 270- to 310-degree sector of the circumference of the aft field joint of the right solid rocket booster. This area of the solid booster faced the external tank. The vaporized material streaming from the joint indicated there was incomplete sealing action within the joint.

Eight more distinctive puffs of increasingly blacker smoke were recorded between 0.836 and 2.500 seconds. The smoke appeared to puff upwards from the joint. While each smoke puff was being left behind by the upward flight of the Shuttle, the next fresh puff could be seen near the level of the joint. The multiple smoke puffs in this sequence occurred about four times per second, approximating the frequency of the structural load dynamics and resultant joint flexing. Computer graphics applied to NASA photos from a variety of cameras in this sequence again placed the smoke puffs' origin in the same 270- to 310-degree sector of the circumference as the original smoke spurt.

As the Shuttle Challenger increased its upward velocity, it flew past the emerging and expanding smoke puffs. The last smoke was seen above the field joint at 2.733 seconds.

The black color and dense composition of the smoke puffs suggested that the grease, joint insulation, and rubber O-rings in the joint seal were being burned and eroded by the hot propellant gases.

At approximately 37 seconds, Challenger encountered the first of several high altitude wind shear conditions that lasted about 64 seconds. The wind shear created forces of relatively large fluctuations on the vehicle itself. These were immediately sensed and countered by the guidance, navigation, and control systems.

The steering system (thrust vector control) of the solid rocket booster responded to all commands and wind shear effects. The wind shear caused the steering system to be more active than on any previous flight.

Both the Challenger's main engines and the solid rockets operated at reduced thrust approaching and passing through the area of maximum dynamic pressure of 720 pounds per square foot. Main engines had been throttled up to 104 percent thrust and the solid rocket boosters were increasing their thrust when the first flickering flame appeared on the right solid rocket booster in the area of the aft field joint. This first very small flame was detected on image-enhanced film at 58.788 seconds into the flight. It appeared to originate at about 305 degrees around the booster circumference at or near the aft field joint.

One film frame later from the same camera, the flame was visible without image enhancement. It grew into a continuous, well-defined plume at 59.262 seconds. At approximately the same time (60 seconds), telemetry showed a pressure differential between the chamber pressures in the right and left boosters. The right booster chamber pressure was lower, confirming the growing leak in the area of the field joint.

As the flame plume increased in size, it was deflected rearward by the aerodynamic slipstream and circumferentially by the protruding structure of the upper ring attaching the booster to the external tank. These deflections directed the flame plume onto the surface of the external tank. This sequence of flame spreading is confirmed by analysis of the recovered wreckage. The growing flame also impinged on the strut attaching the solid rocket booster to the external tank.

The first visual indication that swirling flame from the right solid rocket booster breached the external tank was at 64.660 seconds when there was an abrupt change in the shape and color of the plume. This indicated that it was mixing with leaking hydrogen from the external tank. Telemetered changes in the hydrogen tank pressurization confirmed the leak. Within 45 milliseconds of the breach of the external tank, a bright, sustained glow developed on the black tiled underside of the Challenger between it and the external tank.

Beginning around 72 seconds, a series of events occurred extremely rapidly that terminated the flight. Telemetered data indicated a wide variety of flight system actions that supported the visual evidence of the photos as the Shuttle struggled futilely against the forces that were destroying it.

At about 72.20 seconds, the lower strut linking the solid rocket booster and the external tank was severed or pulled away from the weakened hydrogen tank, permitting the right solid rocket booster to rotate around the upper attachment strut. This rotation was indicated by divergent yaw and pitch rates between the left and right solid rocket boosters.

At 73.124 seconds, a circumferential white vapor pattern was observed blooming from the side of the external tank bottom dome. This was the beginning of the structural failure of the hydrogen tank that culminated in the entire aft dome dropping away. This released massive amounts of liquid hydrogen from the tank and created a sudden forward thrust of about 2.8 million pounds, pushing the hydrogen tank upward into the intertank structure. About the same time, the rotating right solid rocket booster impacted the intertank structure and the lower part of the liquid oxygen tank. These structures failed at 73.137 seconds as evidenced by the white vapors appearing in the intertank region.

Within milliseconds there was massive, almost explosive, burning of the hydrogen streaming from the failed tank bottom and the liquid oxygen breach in the area of the intertank.

At this point in its trajectory, while traveling at a Mach number of 1.92 at an altitude of 46,000 feet, the Challenger was totally enveloped in the explosive burn. The Challenger's reaction control system ruptured, and a hypergolic burn of its propellants occurred, producing the oxygen-hydrogen flames. The reddish brown colors of the hypergolic fuel burn were visible on the edge of the main fireball. The Orbiter, under severe aerodynamic loads, broke into several large sections, which emerged from the fireball. Separate sections that can be identified on film include the main engine/tail section with the engines still burning, one wing of the Orbiter, and the forward fuselage trailing a mass of umbilical lines pulled loose from the payload bay.

The consensus of the Commission and participating investigative agencies was that the loss of the Space Shuttle Challenger was caused by a failure in the joint between the two lower segments of the right solid rocket motor. The specific failure was the destruction of the seals that were intended to prevent hot gases from leaking through the joint during the propellant burn of the rocket motor. The evidence assembled by the Commission indicates that no other element of the Space Shuttle system contributed to this failure.

In arriving at this conclusion, the Commission reviewed in detail all available data, reports, and records; directed and supervised numerous tests, analyses, and experiments by NASA, civilian contractors, and various government agencies; and then developed specific failure scenarios and the range of most probably causative factors.

The failure was due to a faulty design unacceptably sensitive to a number of factors. These factors were the effects of temperature, physical dimensions, the character of materials, the effects of reusability, processing, and the reaction of the joint to dynamic loading.

NASA and the Media

Following the tragedy, many believed that NASA's decision to launch had been an attempt to minimize further ridicule by the media. Successful Shuttle flights were no longer news because they were almost ordinary. However, launch aborts and delayed landings were more newsworthy because they were less common. The Columbia launch, which had immediately preceded the Challenger mission, had been delayed seven times. The Challenger launch had gone through four delays already. News anchor personnel were criticizing NASA. Some believed that NASA felt it had to do something quickly to dispel its poor public image.

The Challenger mission had had more media coverage and political ramifications than other recent missions. This would be the launch of the Teacher in Space Project. The original launch date of the Challenger had been scheduled just before President Reagan's State of the Union message, that was to be delivered the evening of January 28. Some believed that the president had intended to publicly praise NASA for the Teacher in Space Project and possibly even talk to Ms. McAuliffe live during his address. This would certainly have enhanced NASA's image. Following the tragedy, there were questions as to whether or not the White House had pressured NASA into launching the Shuttle because of President Reagan's (and NASA's) love of favorable publicity. The commission, however, found no evidence of White House intervention in the decision to launch.

Findings of the Commission

Determining the cause of an engineering disaster can take years of investigation. The Challenger disaster arose from many factors, including launch conditions, mechanical failure, faulty communication, and poor decision-making. In the end, the last minute decision to launch combined all possible factors into a lethal action.

The Commission concluded that the accident was rooted in history. The space Shuttle's solid rocket booster problem began with the faulty design of its joint and increased as both NASA and contractor management first failed to recognize that they had a problem, then failed to fix it, and finally treated it as an acceptable flight risk.

Morton Thiokol, Inc., the contractor, did not accept the implication of tests early in the program that the design had a serious and unanticipated flaw. NASA did not accept the judgment of its engineers that the design was unacceptable, and as the joint problems grew in number and severity, NASA minimized them in management briefings and reports. Thiokol's stated position was that "the condition is not desirable but is acceptable."

Neither Thiokol nor NASA expected the rubber O-rings sealing the joints to be touched by hot gases of motor ignition, much less to be partially burned. However, as tests and then flights confirmed damage to the sealing rings, the reaction by both NASA and Thiokol was to increase the amount of damage considered "acceptable." At no time did management either recommend a redesign of the joint or call for the Shuttle's grounding until the problem was solved.

The genesis of the Challenger accident—the failure of the joint of the right solid rocket motor—lay in decisions made in the design of the joint and in the failure by both Thiokol and NASA's Solid Rocket Booster project office to understand and respond to facts obtained during testing.

The Commission concluded that neither Thiokol nor NASA had responded adequately to internal warnings about the faulty seal design. Furthermore, Thiokol and NASA did not make a timely attempt to develop and verify a new seal after the initial design was shown to be deficient. Neither organization developed a solution to the unexpected occurrences of O-ring erosion and blow-by, even though this problem was experienced frequently during the Shuttle flight history. Instead, Thiokol and NASA management came to accept erosion and blow-by as unavoidable and an acceptable flight risk. Specifically, the Commission found that:

1. The joint test and certification program was inadequate. There was no requirement to configure the qualifications test motor as it would be in flight, and the motors were static tested in a horizontal position, not in the vertical flight position.
2. Prior to the accident, neither NASA nor Thiokol fully understood the mechanism by which the joint sealing action took place.
3. NASA and Thiokol accepted escalating risk apparently because they "got away with it last time." As Commissioner Feynman observed, the decision-making was:

> A kind of Russian roulette. . . . [The Shuttle] flies [with O-ring erosion] and nothing happens. Then it is suggested, therefore, that the risk is no longer so high for the next flights. We can lower our standards a little bit because we got away with it last time. . . . You got away with it, but it shouldn't be done over and over again like that.

4. NASA's system for tracking anomalies for Flight Readiness Reviews failed in that, despite a history of persistent O-ring erosion and blow-by, flight was still permitted. It failed again in the strange sequence of six consecutive launch constraint waivers prior to 51-L, permitting it to fly without any record of a waiver, or even of an explicit constraint. Tracking and continuing only anomalies that are "outside the data base" of prior flight allowed major problems to be removed from, and lost by, the reporting system.

5. The O-ring erosion history presented to Level I at NASA Headquarters in August 1985 was sufficiently detailed to require corrective action prior to the next flight.
6. A careful analysis of the flight history of O-ring performance would have revealed the correlation of O-ring damage and low temperature. Neither NASA nor Thiokol carried out such an analysis; consequently, they were unprepared to properly evaluate the risks of launching the 51-L mission in conditions more extreme than they had encountered before.

The Commission also identified a concern for the "silent" safety program. The Commission was surprised to realize after many hours of testimony that NASA's safety staff was never mentioned. No witness related the approval or disapproval of the reliability engineers, and none expressed the satisfaction or dissatisfaction of the quality assurance staff. No one thought to invite a safety representative or a reliability and quality assurance engineer to the January 27, 1986, teleconference between Marshall and Thiokol. Similarly, there was no safety representative on the Mission Management Team that made key decisions during the countdown on January 28, 1986.

The unrelenting pressure to meet the demands of an accelerating flight schedule might have been adequately handled by NASA if it had insisted upon the exactingly thorough procedures that had been its hallmark during the Apollo program. An extensive and redundant safety program comprising interdependent safety, reliability, and quality assurance functions had existed during the lunar program to discover any potential safety problems. Between that period and 1986, however, the safety program had become ineffective. This loss of effectiveness seriously degraded the checks and balances essential for maintaining flight safety.

On April 3, 1986, Arnold Aldrich, the Space Shuttle program manager, appeared before the Commission at a public hearing in Washington, D.C. He described five different communication or organization failures that affected the launch decision on January 28, 1986. Four of those failures related directly to faults within the safety program. These faults included a lack of problem reporting requirements, inadequate trend analysis, misrepresentation of criticality, and lack of involvement in critical discussions. A robust safety organization that was properly staffed and supported might well have avoided these faults, and thus eliminated the communication failures.

NASA had a safety program to ensure that the communication failures to which Mr. Aldrich referred did not occur. In the case of mission 51-L, however, that program fell short.

The Commission concluded that there were severe pressures placed on the launch decision-making system to maintain a flight schedule. These pressures caused rational men to make irrational decisions.

With the 1982 completion of the orbital flight test series, NASA began a planned acceleration of the Space Shuttle launch schedule. One early plan contemplated an eventual rate of a mission a week, but realism forced several downward revisions. In 1985, NASA published a projection calling for an annual rate of 24 flights by 1990. Long before the Challenger accident, however, it was becoming obvious that even the modified goal of two flights a month was overambitious.

In establishing the schedule, NASA had not provided adequate resources. As a result, the capabilities of the launch decision-making system were strained by the modest nine-mission rate of 1985, and the evidence suggested that NASA would not have been able to accomplish the 15 flights scheduled for 1986. These were the major conclusions of a Commission examination of the pressures and problems attendant upon the accelerated launch schedule:

1. The capabilities of the launch decision-making system were stretched to the limit to support the flight rate in winter 1985/1986. Projections into the spring and summer of 1986 showed a clear trend; the system, as it existed, would have been unable to deliver crew training software

for scheduled flights by the designated dates. The result would have been an unacceptable compression of the time available for the crews to accomplish their required training.

2. Spare parts were in critically short supply. The Shuttle program made a conscious decision to postpone spare parts procurements in favor of budget items of perceived higher priority. Lack of spare parts would likely have limited flight operations in 1986.
3. Stated manifesting policies were not enforced. Numerous late manifest changes (after the cargo integration review) had been made to both major payloads and minor payloads throughout the Shuttle program
   - Late changes to major payloads or program requirements required extensive resources (money, manpower, facilities) to implement.
   - If many late changes to "minor" payloads occurred, resources were quickly absorbed.
   - Payload specialists frequently were added to a flight well after announced deadlines.
   - Late changes to a mission adversely affected the training and development of procedures for subsequent missions.
4. The scheduled flight rate did not accurately reflect the capabilities and resources.
   - The flight rate was not reduced to accommodate periods of adjustment in the capacity of the workforce. There was no margin for error in the system to accommodate unforeseen hardware problems.
   - Resources were primarily directed toward supporting the flights and thus not enough were available to improve and expand facilities needed to support a higher flight rate.
5. Training simulators may have been the limiting factor on the flight rate: the two simulators available at that time could not train crews for more than twelve to fifteen flights per year.
6. When flights came in rapid succession, the requirements then current did not ensure that critical anomalies occurring during one flight would be identified and addressed appropriately before the next flight.

Chain-of-Command Communication Failure

The Commission also identified a communication failure within the reporting structure at both NASA and Thiokol. Part of the problem with the chain of command structure was the idea of the proper reporting channel. Engineers report only to their immediate managers, while those managers report only to their direct supervisors. Engineers and managers believed in the chain of command structure; they felt reluctant to go above their superiors with their concerns. Boisjoly at Thiokol and Powers at Marshall felt that they had done all that they could as far as voicing their concerns. Anything more could have cost them their jobs. When questioned at the Rogers Commission hearing about why he did not voice his concerns to others, Powers replied, "That would not be my reporting channel." The chain of command structure dictated the only path that information could travel at both NASA and Thiokol. If information was modified or silenced at the bottom of the chain, there was not an alternate path for it to take to reach high-level officials at NASA. The Rogers Commission concluded that there was a breakdown in communication between Thiokol engineers and top NASA officials and faulted the management structure for not allowing important information about the SRBs to flow to the people who needed to know it. The Commission reported that the "fundamental problem was poor technical decision-making over a period of several years by top NASA and contractor personnel."

Bad news does not travel well in organizations like NASA and Thiokol. When the early signs of problems with the SRBs appeared, Thiokol managers did not believe that the problems were serious. Thiokol did not want to accept the fact that there could be a problem with their boosters. When Marshall received news of the problems, they considered it Thiokol's problem and did not pass the bad news upward to NASA headquarters. At Thiokol, Boisjoly described his managers as shutting out the bad news. He claims that he argued about the importance of the O-ring seal problems until he was convinced that "no one wanted to hear what he had to say." When Lund finally decided to recommend delay of the launch to Marshall, managers at Marshall rejected the bad news and refused to accept the recommendation not to launch. As with any information going up the chain of command at these two organizations, bad news was often modified so that it had less impact, perhaps skewing its importance.[30]

On January 31, 1986, President Ronald Reagan stated:

> The future is not free: the story of all human progress is one of a struggle against all odds. We learned again that this America, which Abraham Lincoln called the last, best hope of man on Earth, was built on heroism and noble sacrifice. It was built by men and women like our seven star voyagers, who answered a call beyond duty, who gave more than was expected or required and who gave it with little thought of worldly reward.

Epilogue

Following the tragic accident, virtually every senior manager that was involved in the Space Shuttle Challenger decision-making processes, at both NASA and Thiokol, accepted early retirement. Whether this was the result of media pressure, peer pressure, fatigue, or stress we can only postulate. The only true failures are the ones from which nothing is learned. Lessons on how to improve the risk management process were learned, unfortunately at the expense of human life.

On January 27, 1967, Astronauts Gus Grissom, Edward White, and Roger Chaffee were killed on board a test on Apollo-Saturn 204. James Webb, NASA's Administrator at that time, was allowed by President Johnson to conduct an internal investigation of the cause. The investigation was primarily a technical investigation. NASA was fairly open with the media during the investigation. As a result of the openness, the credibility of the agency was maintained.

With the Challenger accident, confusion arose as to whether it had been a technical failure or a management failure. There was no question in anyone's mind that the decision-making process was flawed. NASA and Thiokol acted independently in their response to criticism. Critical information was withheld, at least temporarily, and this undermined people's confidence in NASA. The media, as might have been expected, began vengeful attacks on NASA and Thiokol.

Following the Apollo-Saturn 204 fire, there were few changes made in management positions at NASA. Those changes that did occur were the result of a necessity for improvement and where change was definitely warranted. Following the Challenger accident, almost every top management position at NASA underwent a change of personnel.

How an organization fares after an accident is often measured by how well it interfaces with the media. Situations such as the Tylenol tragedy (subject of another case study in this volume) and the Apollo-Saturn 204 fire bore this out.

Following the accident, and after critical data were released, papers were published showing that the O-ring data correlation was indeed possible. In one such paper, Lighthall[31] showed

---

30. "The Challenger Accident: Administrative Causes of the Challenger Accident," www.me.utexas.edu/~uer/challenger/chall3.html pages 8–9.

31. Frederick F. Lighthall, "Launching The Space Shuttle Challenger: Disciplinary Deficiencies in the Analysis of Engineering Data," *IEEE Transactions on Engineering Management,* Vol. 38, No. 1, February 1991, pp. 63–74.

that not only was a correlation possible, but the real problem may be a professional weakness shared by many people, but especially engineers, who have been required to analyze technical data. Lighthall's argument was that engineering curriculums might not provide engineers with strong enough statistical education, especially in covariance analysis. The Rogers Commission also identified this conclusion when they found that there were no engineers at NASA trained in statistical sciences.

Almost all scientific achievements require the taking of risks. The hard part is deciding which risk is worth taking and which is not. Every person who has ever flown in space, whether military or civilian, was a volunteer. They were all risk-takers who understood that safety in space can never be guaranteed with 100 percent accuracy.

Discussion Questions

Below are a series of questions categorized according to the principles of risk management. There may not be any single right or wrong answer to these questions.

Risk Management Plan

1. Does it appear, from the data provided in the case, that a risk management plan was in existence?
2. If such a plan did exist, then why wasn't it followed, or was it followed?
3. Is there a difference between a risk management plan, a quality assurance plan, and a safety plan, or are they the same?
4. Would there have been a better way to handle risk management planning at NASA assuming sixteen flights per year, twenty-five flights per year, or as originally planned, sixty flights per year? Why is the number of flights per year critical in designing a formalized risk management plan?

Risk Identification

5. What is the difference between a risk and an anomaly? Who determines the difference?
6. Does there appear to have been a structured process in place for risk identification at either NASA or Thiokol?
7. How should problems with risk identification be resolved if there exist differences of opinion between the customer and the contractors?
8. Should senior management or sponsors be informed about all risks identified or just the overall "aggregate" risk?
9. How should one identify or classify the risks associated with using solid rocket boosters on manned spacecraft rather than the conventional liquid fuel boosters?
10. How should one identify or classify trade-off risks such as trading off safety for political acceptability?
11. How should one identify or classify the risks associated with pressure resulting from making promises that may be hard to keep?
12. Suppose that a risk identification plan had been established at the beginning of the space program when the Shuttle was still considered an experimental design. If the Shuttle is now considered as an operational vehicle rather than as an experimental design, could that affect the way that risks were identified to the point where the risk identification plan would need to be changed?

Risk Quantification

13. Given the complexity of the Space Shuttle Program, is it feasible and/or practical to develop a methodology for quantifying risks, or should each situation be addressed individually? Can we have both a quantitative and qualitative risk evaluation system in place at the same time?
14. How does one quantify the dangers associated with the ice problem?
15. How should risk quantification problems be resolved if there exist differences of opinion between the customer and the contractors?
16. If a critical risk is discovered, what is the proper way for the project manager to present to senior management the impact of the risk? How do you as a project manager make sure that senior management understand the ramifications?
17. How were the identified risks quantified at NASA? Is the quantification system truly quantitative or is it a qualitative system?
18. Were probabilities assigned to any of the risks? Why or why not?

Risk Response (Risk Handling)

19. How does an organization decide what is or is not an acceptable risk?
20. Who should have final say in deciding upon the appropriate response mechanism for a risk?
21. What methods of risk response were used at NASA?
22. Did it appear that the risk response method selected was dependent on the risk or on other factors?
23. How should an organization decide whether or not to accept a risk and launch if the risks cannot be quantified?
24. What should be the determining factors in deciding which risks are brought upstairs to the executive levels for review before selecting the appropriate risk response mechanism?
25. Why weren't the astronauts involved in the launch decision (i.e., the acceptance of the risk)? Should they have been involved?
26. What risk response mechanism did NASA administrators use when they issued waivers for the Launch Commit Criteria?
27. Are waivers a type of risk response mechanism?
28. Did the need to maintain a flight schedule compromise the risk response mechanism that would otherwise have been taken?
29. What risk response mechanism were managers at Thiokol and NASA using when they ignored the recommendations of their engineers?
30. Did the engineers at Thiokol and NASA do all they could to convince their own management that the wrong risk response mechanism was about to be taken?
31. When NASA pressed its contractors to recommend a launch, did NASA's risk response mechanism violate their responsibility to ensure crew safety?
32. When NASA discounted the effects of the weather, did NASA's risk response mechanism violate their responsibility to ensure crew safety?

Risk Control

33. How much documentation should be necessary for the tracking of a risk management plan? Can this documentation become overexcessive and create decision-making problems?

34. Risk management includes the documentation of lessons-learned. In the case study, was there an audit trail of lessons learned or was that audit trail simply protection memos?
35. How might Thiokol engineers have convinced both their own management and NASA to postpone the launch?
36. Should someone have stopped the Challenger launch and, if so, how could this have been accomplished without risking one's job and career?
37. How might an engineer deal with pressure from above to follow a course of action that the engineer knows to be wrong?
38. How could the chains of communication and responsibility for the Shuttle Program have been made to function better?
39. Because of the ice problem, Rockwell could not guarantee the Shuttle's safety, but did nothing to veto the launch. Is there a better way for situations as this to be handled in the future?
40. What level of risk should have been acceptable for launch?
41. How should we handle situations where people in authority believe that the potential rewards justify what they believe to be relatively minor risks?
42. If you were on a jury attempting to place liability, whom would you say was responsible for the Challenger disaster?

Case 28

Philip Condit and the Boeing 777: From Design and Development to Producton and Sales*

Following his promotion to Boeing CEO in 1988, Frank Shrontz looked for ways to stretch and upgrade the Boeing 767—an eight-year-old wide-body twin jet—in order to meet Airbus competition. Airbus had just launched two new 300-seat wide-body models, the two-engine A330 and the four-engine A340. Boeing had no 300-seat jetliner in service, nor did the company plan to develop such a jet.

To find out whether Boeing's customers were interested in a double-decker 767, Philip Condit, Boeing Executive Vice President and future CEO (1996) met with United Airlines Vice President Jim Guyette. Guyette rejected the idea outright, claiming that an upgraded 767 was no match to Airbus' new model transports. Instead, Guyette urged Boeing to develop a brand new commercial jet, the most advanced airplane of its generation.[1] Shrontz had heard similar suggestions from other airline carriers. He reconsidered Boeing's options, and decided to abandon the 767 idea in favor of a new aircraft program. In December 1989, accordingly, he announced the 777 project and put Philip Condit in charge of its management. Boeing had launched the 777 in 1990, delivered the first jet in 1995, and by February 2001, 325 B-777s were flying in the services of the major international and U.S. airlines.[2]

Condit faced a significant challenge in managing the 777 project. He wanted to create an airplane that was preferred by the airlines at a price that was truly competitive. He sought to attract airline customers as well as cut production costs, and he did so by introducing several innovations—both technological and managerial—in aircraft design, manufacturing, and assembly. He looked for ways to revitalize Boeing's outmoded engineering production system, and update Boeing's manufacturing strategies. And to achieve these goals, Condit made continual efforts to spread the 777 program-innovations company wide.

---

* This case was presented by Isaac Cohen, San Jose State University, at the 2000 North American Case Research Association (NACRA) workshop. Copyright (c) 2000 by Isaac Cohen and NACRA. Reprinted with permission. All rights reserved.

Looking back at the 777 program, this case focuses on Condit's efforts. Was the 777 project successful and was it cost effective? Would the development of the 777 allow Boeing to diffuse the innovations in airplane design and production beyond the 777 program? Would the development of the 777's permit Boeing to revamp and modernize its aircraft manufacturing system? Would the making and selling of the 777 enhance Boeing competitive position relative to Airbus, its only remaining rival?

The Aircraft Industry

Commercial aircraft manufacturing was an industry of enormous risks where failure was the norm, not the exception. The number of large commercial jet makers had been reduced from four in the early 1980s—Boeing, McDonnell Douglas, Airbus, and Lockheed—to two in late 1990s, turning the industry into a duopoly, and pitting the two survivors—Boeing and Airbus—one against the other. One reason why aircraft manufacturers so often failed was the huge cost of product development.

Developing a new jetliner required an up front investment of up to $15 billion (2001 dollars), a lead time of five to six years from launch to first delivery, and the ability to sustain a negative cash flow throughout the development phase. Typically, to break even on an entirely new jetliner, aircraft manufacturers needed to sell a minimum of 300 to 400 planes and at least 50 planes per year.[3] Only a few commercial airplane programs had ever made money.

The price of an aircraft reflected its high development costs. New model prices were based on the average cost of producing 300 to 400 planes, not a single plane. Aircraft pricing embodied the principle of learning by doing, the so called "learning curve"[4]: workers steadily improved their skills during the assembly process, and as a result, labor cost fell as the number of planes produced rose.

The high and increasing cost of product development prompted aircraft manufacturers to utilize subcontracting as a risk-sharing strategy. For the 747, the 767, and the 777, the Boeing Company required subcontractors to share a substantial part of the airplane's development costs. Airbus did the same with its own latest models. Risk sharing subcontractors performed detailed design work and assembled major subsections of the new plane while airframe integrators (i.e. aircraft manufacturers) designed the aircraft, integrated its systems and equipment, assembled the entire plane, marketed it, and provided customer support for twenty to thirty years. Both the airframe integrators and their subcontractors were supplied by thousands of domestic and foreign aircraft components manufacturers.[5]

Neither Boeing, nor Airbus, nor any other post-war commercial aircraft manufacturer produced jet engines. A risky and costly venture, engine building had become a highly specialized business. Aircraft manufacturers worked closely with engine makers—General Electric, Pratt and Whitney, and Rolls Royce—to set engine performance standards. In most cases, new airplanes were offered with a choice of engines. Over time, the technology of engine building had become so complex and demanding that it took longer to develop an engine than an aircraft. During the life of a jetliner, the price of the engines and their replacement parts was equal to the entire price of the airplane.[6]

A new model aircraft was normally designed around an engine, not the other way around. As engine performance improved, airframes were redesigned to exploit the engine's new capabilities. The most practical way to do so was to stretch the fuselage and add more seats in the cabin. Aircraft manufacturers deliberately designed flexibility into the airplane so that future engine improvements could facilitate later stretching. Hence the importance of the "family concept" in aircraft design, and hence the reason why aircraft manufacturers introduced families of planes made up of derivative jetliners built around a basic model, not single, standardized models.[7]

The commercial aircraft industry, finally, gained from technological innovations in two other industries. More than any other manufacturing industry, aircraft construction benefited from advances in material applications and electronics. The development of metallic and non-metallic composite materials played a key role in improving airframe and engine performance. On the one hand, composite materials that combined light weight and great strength were utilized by aircraft manufacturers; on the other, heat-resisting alloys that could tolerate temperatures of up to 3,000 degrees were used by engine makers. Similarly, advances in electronics revolutionized avionics. The increasing use of semiconductors by aircraft manufacturers facilitated the miniaturization of cockpit instruments, and more important, it enhanced the use of computers for aircraft communication, navigation, instrumentation, and testing.[8] The use of computers contributed, in addition, to the design, manufacture, and assembly of new model aircraft.

The Boeing Company

The history of the Boeing company may be divided into two distinct periods: the piston era and the jet age. Throughout the piston era, Boeing was essentially a military contractor producing fighter aircraft in the 1920s and 1930s, and bombers during World War II. During the jet age, beginning in the 1950s, Boeing had become the world's largest manufacturer of commercial aircraft, deriving most of its revenues from selling jetliners.

Boeing's first jet was the 707. The introduction of the 707 in 1958 represented a major breakthrough in the history of commercial aviation; it allowed Boeing to gain a critical technological lead over the Douglas Aircraft Company, its closer competitor. To benefit from government assistance in developing the 707, Boeing produced the first jet in two versions: a military tanker for the Air Force (k-135) and a commercial aircraft for the airlines (707-120). The company, however, did not recoup its own investment until 1964, six years after it delivered the first 707, and twelve years after it had launched the program. In the end, the 707 was quite profitable, selling 25 percent above its average cost.[9] Boeing retained the essential design of the 707 for all its subsequent narrow-body single-aisle models (the 727, 737, and 757), introducing incremental design improvements, one at a time.[10] One reason why Boeing used shared design for future models was the constant pressure experienced by the company to move down the learning curve and reduce overall development costs.

Boeing introduced the 747 in 1970. The development of the 747 represented another breakthrough; the 747 wide body design was one of a kind; it had no real competition anywhere in the industry. Boeing bet the entire company on the success of the 747, spending on the project almost as much as the company's total net worth in 1965, the year the project started.[11] In the short-run, the outcome was disastrous. As Boeing began delivering its 747s, the company was struggling to avoid bankruptcy. Cutbacks in orders as a result of a deep recession, coupled with production inefficiencies and escalating costs, created a severe cash shortage that pushed the company to the brink. As sales dropped, the 747's break-even point moved further and further into the future.

Yet, in the long run, the 747 program was a triumph. The Jumbo Jet had become Boeing's most profitable aircraft and the industry's most efficient jetliner. The plane helped Boeing solidify its position as the industry leader for years to come, leaving McDonnell Douglas far behind, and forcing the Lockheed Corporation to exit the market. The new plane, furthermore, contributed to Boeing's manufacturing strategy in two ways. First, as Boeing increased its reliance on outsourcing, six major subcontractors fabricated 70 percent of the value of the 747 airplane,[12] thereby helping Boeing reduce the project's risks. Second, for the first time, Boeing applied the family concept in aircraft design to a wide-body jet, building the 747 with wings large enough to support a stretched fuselage with bigger engines, and offering a variety of other modifications in the 747's basic design. The 747-400 (1989) is a case in point. In 1997, Boeing sold the stretched and upgraded 747-400 in three versions, a standard jet, a freighter, and a "combi" (a jetliner whose main cabin was divided between passenger and cargo compartments).[13]

Boeing developed other successful models. In 1969, Boeing introduced the 737, the company's narrow-body flagship, and in 1982 Boeing put into service two additional jetliners, the 757 (narrow-body) and the 767 (wide-body). By the early 1990s, the 737, 757, and 767 were all selling profitably. Following the introduction of the 777 in 1995, Boeing's families of planes included the 737 for short-range travel, the 757 and 767 for medium-range travel, and the 747 and 777 for medium- to long-range travel (Exhibit I).

In addition to building jetliners, Boeing also expanded its defense, space, and information businesses. In 1997, the Boeing Company took a strategic gamble, buying the McDonnell Douglas Company in a $14 billion stock deal. As a result of the merger, Boeing had become the world's largest manufacturer of military aircraft, NASA'S largest supplier, and the Pentagon's second largest contractor (after Lockheed). Nevertheless, despite the growth in its defense and space businesses, Boeing still derived most of its revenues from selling jetliners. Commercial aircraft revenues accounted for 59 percent of Boeing's $49 billion sales in 1997 and 63 percent of Boeing's $56 billion sales in 1998.[14]

Following its merger with McDonnell, Boeing had one remaining rival: Airbus Industrie.[15] In 1997, Airbus booked 45 percent of the worldwide orders for commercial jetliners[16] and delivered close to 1/3 of the worldwide industry output. In 2000, Airbus shipped nearly 2/5 of the worldwide industry output (Exhibit II).

Airbus' success was based on a strategy that combined cost leadership with technological leadership. First, Airbus distinguished itself from Boeing by incorporating the most advanced technologies into its planes. Second, Airbus managed to cut costs by utilizing a flexible, lean production manufacturing system that stood in a stark contrast to Boeing's mass production system.[17]

As Airbus prospered, the Boeing company was struggling with rising costs, declining productivity, delays in deliveries, and production inefficiencies. Boeing Commercial Aircraft Group lost $1.8 billion in 1997 and barely generated any profits in 1998.[18] All through the 1990s, the Boeing Company looked for ways to revitalize its outdated production manufactur-

Exhibit I. Total number of commercial jetliners delivered by the Boeing Company, 1958–2/2001[a]

Model	No. Delivered	First Delivery
B-707	1,010 (retired)	1958
B-727	1,831 (retired)	1963
B-737	3,901	1967
B-747	1,264	1970
B-757	953	1982
B-767	825	1982
B-777	325	1995
B-717	49	2000
Total:	**10,158**	

[a]McDonnell Douglas commercial jetliners (the MD-11, MD-80, and MD-90) are excluded.

Sources: Boeing Commercial Airplane Group, *Announced Orders and Deliveries as of 12/31/97; The Boeing Company 1998 Annual Report,* p. 35.

"Commercial Airplanes: Order and Delivery Summary," *http://www.Boeing com/commercial/orders/index.html.* Retrieved from Web 3/20/2001.

Exhibit II. Market share of shipments of commercial aircraft, Boeing, McDonnell Douglas (MD), Airbus, 1992–2000

	1992	1993	1994	1995	1996	1997	1998	1999	2000
Boeing	61%	61%	63%	54%	55%	67%	71%	68%	61%
MD	17	14	9	13	13				
Airbus	22	25	28	33	32	33	29	32	39

Source: Aerospace Facts and Figures, 1997–98, p. 34; *Wall Street Journal,* December 3, 1998, and January 12, 1999; *The Boeing Company 1997 Annual Report,* p. 19; data supplied by Mark Luginbill, Airbus Communication Director, November 16, 1998, February 1, 2000, and March 20, 2001.

ing system on the one hand, and to introduce leading edge technologies into its jetliners on the other. The development and production of the 777, first conceived of in 1989, was an early step undertaken by Boeing managers to address both problems.

The 777 Program

The 777 program was Boeing's single largest project since the completion of the 747. The total development cost of the 777 was estimated at $6.3 billion and the total number of employees assigned to the project peaked at nearly 10,000. The 777's twin-engines were the largest and most powerful ever built (the diameter of the 777's engine equaled the 737's fuselage), the 777's construction required 132,000 uniquely engineered parts (compared to 70,000 for the 767), the 777's seat capacity was identical to that of the first 747 that had gone into service in 1970, and its manufacturer empty weight was 57 percent greater than the 767's. Building the 777 alongside the 747 and 767 at its Everett plant near Seattle, Washington, Boeing enlarged the plant to cover an area of seventy-six football fields.[19]

Boeing's financial position in 1990 was unusually strong. With a 21 percent rate of return on stockholder equity, a long term debt of just 15 percent of capitalization, and a cash surplus of $3.6 billion, Boeing could gamble comfortably.[20] There was no need to bet the company on the new project as had been the case with the 747, or to borrow heavily, as had been the case with the 767. Still, the decision to develop the 777 was definitely risky; a failure of the new jet might have triggered an irreversible decline of the Boeing Company and threatened its future survival.

The decision to develop the 777 was based on market assessment—the estimated future needs of the airlines. During the fourteen year period, 1991–2005, Boeing market analysts forecasted an +100 percent increase in the number of passenger miles traveled worldwide, and a need for about 9,000 new commercial jets. Of the total value of the jetliners needed in 1991–2005, Boeing analysts forecasted a $260 billion market for wide body jets smaller than the 747. An increasing number of these wide body jets were expected to be larger than the 767.[21]

A Consumer Driven Product

To manage the risk of developing a new jetliner, aircraft manufacturers had first sought to obtain a minimum number of firm orders from interested carriers, and only then commit to the project. Boeing CEO Frank Shrontz had expected to obtain 100 initial orders of the 777 before asking the Boeing board to launch the project, but as a result of Boeing's financial strength on the one hand, and the increasing competitiveness of Airbus on the other, Schrontz decided to seek the board's approval earlier. He did so after securing only one customer: United Airlines.

On October 12, 1990, United had placed an order for thirty-four 777s and an option for an additional thirty-four aircraft, and two weeks later, Boeing's board of directors approved the project.[22] Negotiating the sale, Boeing and United drafted a handwritten agreement (signed by Philip Condit and Richard Albrecht, Boeing's executive vice presidents, and Jim Guyette, United's Executive Vice President) that granted United a larger role in designing the 777 than the role played by any airline before. The two companies pledged to cooperate closely in developing an aircraft with the "best dispatch reliability in the industry" and the "greatest customer appeal in the industry." "We will endeavor to do it right the first time with the highest degree of professionalism" and with "candor, honesty, and respect" [the agreement read]. Asked to comment on the agreement, Philip Condit, said: "We are going to listen to our customers and understand what they want. Everybody on the program has that attitude."[23] Gordon McKinzie, United's 777 program director agreed: "In the past we'd get brochures on a new airplane and its options. . . wait four years for delivery, and hope we'd get what we ordered. This time Boeing really listened to us."[24]

Condit invited other airline carriers to participate in the design and development phase of the 777. Altogether, eight carriers from around the world (United, Delta, American, British Airways, Qantas, Japan Airlines, All Nippon Airways, and Japan Air System) sent full-time representatives to Seattle; British Airways alone assigned 75 people at one time. To facilitate interaction between its design engineers and representatives of the eight carriers, Boeing introduced an initiative called "Working Together." "If we have a problem," a British Airways production manager explained, "we go to the source—design engineers on the IPT [Integrated Product Teams], not service engineer(s). One of the frustrations on the 747 was that we rarely got to talk to the engineers who were doing the work."[25]

"We have definitely influenced the design of the aircraft," a United 777 manager said, mentioning changes in the design of the wing panels that made it easier for airline mechanics to access the slats (slats, like flaps, increased lift on takeoffs and landings), and new features in the cabin that made the plane more attractive to passengers.[26] Of the 1,500 design features examined by representatives of the airlines, Boeing engineers modified 300. Among changes made by Boeing was a redesigned overhead bin that left more stand-up headroom for passengers (allowing a six-foot-three tall passenger to walk from aisle to aisle), "flattened" side walls which provided the occupant of the window seat with more room, overhead bin doors which opened down and made it possible for shorter passengers to lift baggage into the overhead compartment, a redesigned reading lamp that enabled flight attendants to replace light bulbs, a task formerly performed by mechanics, and a computerized flight deck management system that adjusted cabin temperature, controlled the volume of the public address system, and monitored food and drink inventories.[27]

More important were changes in the interior configuration (layout plan) of the aircraft. To be able to reconfigure the plane quickly for different markets of varying travel ranges and passenger loads, Boeing's customers sought a flexible plan of the interior. On a standard commercial jet, kitchen galleys, closets, lavatories, and bars were all removable in the past, but were limited to fixed positions where the interior floor structure was reinforced to accommodate the "wet" load. On the 777, by contrast, such components as galleys and lavatories could be positioned anywhere within several "flexible zones" designed into the cabin by the joint efforts of Boeing engineers and representatives of the eight airlines. Similarly, the flexible design of the 777's seat tracks made it possible for carriers to increase the number of seat combinations as well as reconfigure the seating arrangement quickly. Flexible configuration resulted, in turn, in significant cost savings; airlines no longer needed to take the aircraft out of service for an extended period of time in order to reconfigure the interior.[28]

The airline carriers also influenced the way in which Boeing designed the 777 cockpit. During the program definition phase, representatives of United Airlines, British Airways, and

Exhibit III. The 777: Selected design features proposed by Boeing airline customers and adapted by the Boeing Company

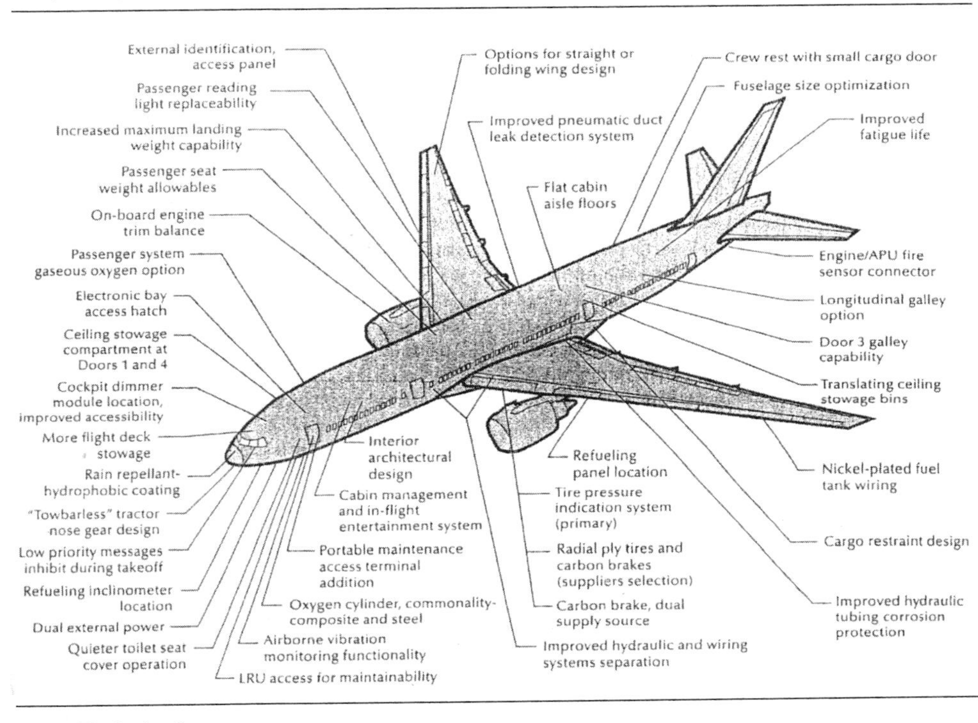

Source: The Boeing Company.

Qantas—three of Boeing's clients whose fleets included a large number of 747-400s—asked Boeing engineers to model the 777 cockpit on the 747-400's. In response to these requests, Boeing introduced a shared 747/777 cockpit design that enabled its airline customers to use a single pool of pilots for both aircraft types at a significant cost savings.[29]

Additionally, the airline carriers urged Boeing to increase its use of avionics for in-flight entertainment. The 777, as a consequence, was equipped with a fully computerized cabin. Facing each seat on the 777, and placed on the back of the seat in front, was a combined computer and video monitor that featured movies, video programs, and interactive computer games. Passengers were also provided with a digital sound system comparable to the most advanced home stereo available, and a telephone. About 40 percent of the 777's total computer capacity was reserved for passengers in the cabin.[30]

The 777 was Boeing's first fly by wire (FBW) aircraft, an aircraft controlled by a pilot transmitting commands to the moveable surfaces (rudder, flaps, etc.) electrically, not mechanically. Boeing installed a state of the art FBW system on the 777 partly to satisfy its airline customers, and partly to challenge Airbus' leadership in flight control technology, a position Airbus had held since it introduced the world's first FBW aircraft, the A-320, in 1988.

Lastly, Boeing customers were invited to contribute to the design of the 777's engine. Both United Airlines and All Nippon Airlines assigned service engineers to work with representatives of Pratt and Whitney (P&W) on problems associated with engine maintenance. P&W held three specially scheduled "airline conferences." At each conference, some forty airline representatives clustered around a full scale mock-up of the 777 engine and showed Pratt and Whitney engineers gaps in the design, hard-to-reach points, visible but inaccessible parts, and accessible but invisible components. At the initial conference, Pratt and Whitney picked up 150 airline suggestions, at the second, fifty, and at the third, ten more suggestions.[31]

A Globally Manufactured Product

Twelve international companies located in ten countries, and eighteen more U.S. companies located in twelve states, were contracted by Boeing to help manufacture the 777. Together, they supplied structural components as well as systems and equipment. Among the foreign suppliers were companies based in Japan, Britain, Australia, Italy, Korea, Brazil, Singapore, and Ireland; among the major U.S. subcontractors were the Grumman Corporation, Rockwell (later merged with Boeing), Honeywell, United Technologies, Bendix, and the Sunstrand Corporation (Exhibits IV and V). Of all foreign participants, the Japanese played the largest role. A consortium made up of Fuji Heavy Industries, Kawasaki Heavy Industries, and Mitsubishi Heavy Industries had worked with Boeing on its wide-body models since the early days of the 747. Together, the three Japanese subcontractors produced 20 percent of the value of the 777's airframe (up from 15 percent of the 767's). A group of 250 Japanese engineers had spent a year in Seattle working on the 777 alongside Boeing engineers before most of its members went back home to begin production. The fuselage was built in sections in Japan and then shipped to Boeing's huge plant at Everett, Washington for assembly.[32]

Boeing used global subcontracting as a marketing tool as well. Sharing design work and production with overseas firms, Boeing required overseas carriers to buy the new aircraft. Again, Japan is a case in point. In return for the contract signed with the Mitsubishi, Fuji, and Kawasaki consortium—which was heavily subsidized by the Japanese government—Boeing sold forty-six 777 jetliners to three Japanese air carriers: All Nippon Airways, Japan Airlines, and Japan Air System.[33]

A Family of Planes

From the outset, the design of the 777 was flexible enough to accommodate derivative jetliners. Because all derivatives of a given model shared maintenance, training, and operating procedures, as well as replacement parts and components, and because such derivatives enabled carriers to serve different markets at lower costs, Boeing's clients were seeking a family of planes built around a basic model, not a single 777. Condit and his management team, accordingly, urged Boeing's engineers to incorporate the maximum flexibility into the design of the 777.

The 777's design flexibility helped Boeing manage the project's risks. Offering a family of planes based on a single design to accommodate future changes in customers' preferences, Boeing spread the 777 project's risks among a number of models all belonging to the same family.

The key to the 777's design efficiency was the wing. The 777 wings, exceptionally long and thin, were strong enough to support vastly enlarged models. The first model to go into service, the 777-200, had a 209 foot-long fuselage, was designed to carry 305 passengers in three class configurations, and had a travel range of 5,900 miles in its original version (1995), and up to 8,900 miles in its extended version (1997). The second model to be introduced (1998), the 777-300, had a stretched fuselage of 242 ft (ten feet longer than the 747), was configured for 379 passengers (three-class), and flew to destinations of up to 6,800 miles away. In all-tourist class configuration, the stretched 777-300 could carry as many as 550 passengers.[34]

Digital Design

The 777 was the first Boeing jetliner designed entirely by computers. Historically, Boeing had designed new planes in two ways: paper drawings and full-size models called mock-ups. Paper

Exhibit IV. 777 supplier contracts

U.S. Suppliers of Structural Components		
Astech/MCI	Santa Ana, CA	Primary exhaust cowl assembly (plug and nozzle)
Grumman Aerospace	Bethpage, NY	Spoilers, inboard flaps
Kaman	Bloomfield, CT	Fixed training edge
Rockwell	Tulsa, OK	Floor beams, wing leading edge slats
International Suppliers of Structural Components		
AeroSpace Technologies of Australia	Australia	Rudder
Alenia	Italy	Wing outboard flaps, radome
Embrace-Empresa Brasiera de Aeronautica	Brazil	Dorsal fin, wingtip assembly
Hawker de Havilland	Australia	Elevators
Korean Air	Korea	Flap support fairings, wingtip assembly
Menasco Aerospace/ Messier-Bugatti	Canada/France	Main and nose landing gears
Mitsubishi Heavy Industries, Kawasaki Heavy Industries, and Fuji Heavy Industries[a]	Japan	Fuselage panels and doors, wing center section wing-to-body fairing, and wing in-spar ribs
Short Brothers	Ireland	Nose landing gear doors
Singapore Aerospace Manufacturing	Singapore	Nose landing gear doors
U.S. Suppliers of Systems and Equipment		
AlliedSignal Aerospace Company, AiResearch Divisions	Torrance, CA	Cabin pressure control system, air supply control system, integrated system controller, ram air turbine
Bendix Wheels and	South Bend, IN	Wheel and brakes
Garrett Divisions	Phoenix/Tempe, AZ	Auxillary power unit (APU), air-driven unit
BFGoodrich	Troy, OH	Wheel and brakes
Dowly Aerospace	Los Angeles, CA	Thrust reverser actuator system
Eldec	Lynnwood, WA	Power supply electronics
E-Systems, Montek Division	Salt Lake City, UT	Stabilizer trim control module, secondary hydraulic brake, optional folding wingtip system
Honeywell	Phoenix, AZ Coon Rapid, MN	Airplane information management system (AIMS), air data/inertial reference system (ADIRS)
Rockwell, Collins Division	Cedar Rapids, IA	Autopilot flight director system, electronic library system (ELS) displays
Sundstrand Corporation	Rockford, IL	Primary and backup electrical power systems
Teijin Seiki America	Redmond, WA	Power control units, actuator control electronics
United Technologies, Hamilton Standard Division	Windsor Lock, CT	Cabin air-conditioning and temperature control systems, ice protection system
International Suppliers of Systems and Equipment		
General Electric Company (GEC) Avionics	United Kingdom	Primary flight computers
Smiths Industries	United Kingdom	Integrated electrical management system (ELMS), throttle control system actuator, fuel quantity indicating system (FQIS)

[a]Program partners
Source: James Woolsey, "777, Boeing's New Large Twinjet," *Air Transport World,* April 1994, p. 24.

Exhibit V. The builders of the Boeing 777

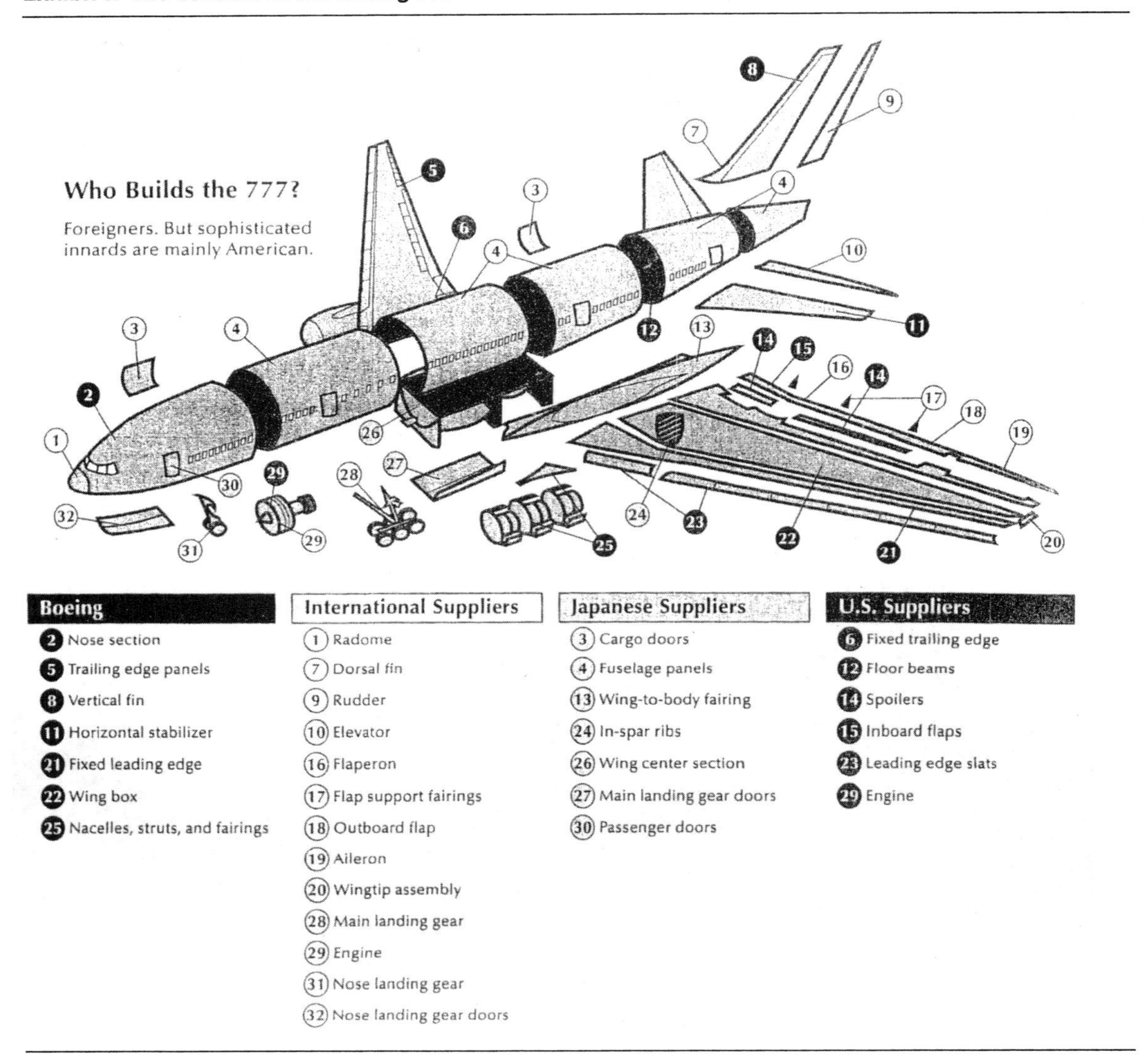

Source: Jeremy Main, "Corporate Performance: Betting on the 21st Century Jet," *Fortune,* April 20, 1992, p. 104.

drawings were two dimensional and therefore insufficient to account for the complex construction of the three dimensional airplane. Full-scale mock-ups served as a backup to drawings.

Boeing engineers used three classes of mock-ups. Made up of plywood or foam, class 1 mock-ups were used to construct the plane's large components in three dimensions, refine the design of these components by carving into the wood or foam, and feed the results back into the drawings. Made partly of metal, class 2 mock-ups addressed more complex problems such as the wiring and tubing of the airframe, and the design of the machine tools necessary to cut and shape the large components. Class 3 mock-ups gave the engineers one final opportunity to refine the model and thereby reduce the need to keep on changing the design during the actual assembly process or after delivery.[35]

Despite the engineers' efforts, many parts and components did not fit together on the final assembly line but rather "interfered" with each other, that is, overlapped in space. The problem was both pervasive and costly, Boeing engineers needed to rework and realign all overlapping parts in order to join them together.

A partial solution to the problem was provided by the computer. In the last quarter of the twentieth century, computer aided design was used successfully in car manufacture, building construction, machine production, and several other industries; its application to commercial aircraft manufacturing came later, both in the United States and in Europe. Speaking of the 777, Dick Johnson, Boeing chief engineer for digital design, noted the "tremendous advantage" of computer application:

> With mock-ups, the . . . engineer had three opportunities at three levels of detail to check his parts, and nothing in between. With Catia [Computer aided three dimensional, interactive application} he can do it day in and day out over the whole development of the airplane.[36]

Catia was a sophisticated computer program that Boeing bought from Dassault Aviation, a French fighter planes builder. IBM enhanced the program to improve image manipulation, supplied Boeing with eight of its largest mainframe computers, and connected the mainframes to 2,200 computer terminals that Boeing distributed among its 777 design teams. The software program showed on a screen exactly how parts and components fit together before the actual manufacturing process took place.[37]

A digital design system, Catia had five distinctive advantages. First, it provided the engineers with 100 percent visualization, allowing them to rotate, zoom, and "interrogate" parts geometrically in order to spotlight interferences. Second, Catia assigned a numerical value to each drawing on the screen and thereby helped engineers locate related drawings of parts and components, merge them together, and check for incompatibilities. Third, to help Boeing's customers service the 777, the digital design system created a computer simulated human—a Catia figure playing the role of the service mechanic—who climbed into the three dimensional images and showed the engineers whether parts were serviceable and entry accessible. Fourth, the use of Catia by all 777 design teams in the U.S., Japan, Europe, and elsewhere facilitated instantaneous communication between Boeing and its subcontractors and ensured the frequent updating of the design. And fifth, Catia provided the 777 assembly line workers with graphics that enhanced the narrative work instructions they received, showing explicitly on a screen how a given task should be performed.[38]

Design-Build Teams (DBT)

Teaming was another feature of the 777 program. About thirty integrated-level teams at the top and more than 230 design-build teams at the bottom worked together on the 777.[39] All team members were connected by Catia. The integrated-level teams were organized around large sections of the aircraft; the DBTs around small parts and components. In both cases, teams were cross-functional, as Philip Condit observed:

> If you go back . . . to earlier planes that Boeing built, the factory was on the bottom floor, and Engineering was on the upper floor. Both Manufacturing and Engineering went back and forth. When there was a problem in the factory, the engineer went down and looked at it. . . .

> With 10,000 people [working on the 777], that turns out to be really hard. So you start devising other tools to allow you to achieve that—the design-build team. You break the airplane down and bring Manufacturing, Tooling, Planning, Engineering, Finance, and Materials all together [in small teams].[40]

Under the design-build approach, many of the design decisions were driven by manufacturing concerns. As manufacturing specialists worked alongside engineers, engineers were less likely to design parts that were difficult to produce and needed to be re-designed. Similarly, under the design-build approach, customers' expectations as well as safety and weight considerations were all incorporated into the design of the aircraft; engineers no longer needed to "chain saw"[41] structural components and systems in order to replace parts that did not meet customers expectations, were unsafe, or were too heavy.

The design of the 777's wing provides an example. The wing was divided into two integration-level teams, the "leading edge" (the forward part of the wing) and the "trailing edge" (the back of the wing) team. Next, the trailing edge team was further divided into ten design-build teams, each named after a piece of the wing's trailing edge (Exhibit VI). Membership in these DBTs extended to two groups of outsiders: representatives of the customer airlines and engineers employed by the foreign subcontractors. Made up of up to twenty members, each DBT decided its own mix of insiders and outsiders, and each was led by a team leader. Each DBT included representatives from six functional disciplines: engineering, manufacturing, materials, customer support, finance, and quality assurance. The DBTs met twice a week for two hours to hear reports from team members, discuss immediate goals and plans, divide responsibilities, set time lines, and take specific notes of all decisions taken.[42] Described by a Boeing official as "little companies," the DBTs enjoyed a high degree of autonomy from management supervision; team members designed their own tools, developed their own manufacturing plans, and wrote their own contracts with the program management, specifying deliverables, resources, and schedules. John Monroe, a Boeing 777 senior project manager remarked:

> The team is totally responsible. We give them a lump of money to go and do th[eir] job. They decide whether to hire a lot of inexpensive people or to trade numbers for resources. It's unprecedented. We have some $100 million plus activities led by non-managers.[43]

Exhibit VI. The ten DBTs ("little companies") responsible for the wing's trailing edge

- Flap Supports Team
- Inboard Flap Team
- Outboard Flap Team
- Flaperon[a] Team
- Aileron[a] Team
- Inboard Fixed Wing and Gear Support Team
- Main Landing Gear Doors Team
- Spoilers[b] Team
- Fairings[c] Team

[a]The Flaperon and Aileron were moveable hinged sections of the trailing edge that helped the plane roll in flight. The Flaperon was used at high speed, the Aileron at low speed.
[b]The spoilers were the flat surfaces that lay on top of the trailing edge and extended during landing to slow down the plane.
[c]The fairing were the smooth parts attached to the outline of the wing's trailing edge. They helped reduce drag.
Source: Karl Sabbagh, *21st Century Jet: The Making and Marketing of the Boeing 777* (New York: Scribner, 1996), p. 73.

Employees' Empowerment and Culture

An additional aspect of the 777 program was the empowering of assembly line workers. Boeing managers encouraged factory workers at all levels to speak up, offer suggestions, and participate in decision making. Boeing managers also paid attention to a variety of "human relations" problems faced by workers, problems ranging from childcare and parking to occupational hazards and safety concerns.[44]

All employees entering the 777 program—managers, engineers, assembly line workers, and others—were expected to attend a special orientation session devoted to the themes of team work and quality control. Once a quarter, the entire "777 team" of up to 10,000 employees met offsite to hear briefings on the aircraft status. Dressed casually, the employees were urged to raise questions, voice complaints, and propose improvements. Under the 777 program, managers met frequently to discuss ways to promote communication with workers. Managers, for example, "fire fought" problems by bringing workers together and empowering them to offer solutions. In a typical "firefight" session, Boeing 777 project managers learned from assembly line workers how to improve the process of wiring and tubing the airframe's interior: "staffing" fuselage sections with wires, ducts, tubs, and insulation materials before joining the sections together was easier than installing the interior parts all at once in a preassembled fuselage.[45]

Under the 777 program, in addition, Boeing assembly line workers also were empowered to appeal management decisions. In a case involving middle managers, a group of Boeing machinists sought to replace a non-retractable jig (a large device used to hold parts) with a retractable one in order to ease and simplify their jobs. Otherwise they had to carry heavy equipment loads up and down stairs. Again and again, their supervisors refused to implement the change. When the machinists eventually approached a factory manager, he inspected the jig personally, and immediately ordered the change.[46]

Under the 777 program, work on the shop floor was ruled by the "Bar Chart." A large display panel placed at different work areas, the Bar Chart listed the name of each worker, his or her daily job description, and the time available to complete specific tasks. Boeing had utilized the Bar Chart system as a "management visibility system" in the past, but only under the 777 program was the system fully computerized. The chart whether assembly line workers were meeting or missing their production goals. Boeing industrial engineers estimated the time it took to complete a given task and fed the information back to the system's computer. Workers ran a scanner across their ID badges and supplied the computer with the data necessary to log their job progress. Each employee "sold" his/her completed job to an inspector, and no job was declared acceptable unless "bought" by an inspector.[47]

Leadership and Management Style

The team in charge of the 777 program was led by a group of five vice presidents, headed by Philip Condit, a gifted engineer who was described by one Wall Street analyst as "a cross between a grizzly bear and a teddy bear. Good people skills, but furious in the marketplace."[48] Each of the five vice presidents rose through the ranks, and each had a twenty-five to thirty years experience with Boeing. All were men.[49]

During the 777 design phase, the five VPs met regularly every Tuesday morning in a small conference room at Boeing's headquarters in Seattle in what was called the "Muffin Meeting." There were no agendas drafted, no minutes drawn, no overhead projectors used, and no votes taken. The homemade muffins served during the meeting symbolized the informal tone of the forum. Few people outside the circle of five had ever attended these weekly sessions. Acting as an informal chair, Condit led a freewheeling discussion of the 777 project, asking each VP to say anything he had on his mind.[50]

The weekly session reflected Boeing's sweeping new approach to management. Traditionally, Boeing had been a highly structured company governed by engineers. Its culture was secretive, formal, and stiff. Managers seldom interacted, sharing was rare, divisions kept to themselves, and engineers competed with each other. Under the 777 program, Boeing made serious efforts to abandon its secretive management style. Condit firmly believed that open communication among top executives, middle managers, and assembly line workers was indispensable for improving morale and raising productivity. He urged employees to talk to each other and share information, and he used a variety of management tools to do so: information sheets, orientation sessions, question and answer sessions, leadership meetings, regular workers as well as middle managers, Condit introduced a three-way performance review procedure whereby managers were evaluated by their supervisors, their peers, and their subordinates.[51] Most important, Condit made teamwork the hallmark of the 777 project. In an address entitled "Working Together: The 777 Story" and delivered in December 1992 to members of the Royal Aeronautics Society in London,[52] Condit summed up his team approach:

> [T]eam building is . . . very difficult to do well but when it works the results are dramatic. Teaming fosters the excitement of a shared endeavor and creates an atmosphere that stimulates creativity and problem solving. But building team[s] . . . is hard work. It doesn't come naturally. Most of us are taught from an early age to compete and excel as individuals. Performance in school and performance on the job are usually measured by individual achievement. Sharing your ideas with others, or helping others to enhance their performance, is often viewed as contrary to one's self interest.
>
> This individualistic mentality has its place, but . . . it is no longer the most useful attitude for a workplace to possess in today's world. To create a high performance organization, you need employees who can work together in a way that promotes continual learning and the free flow of ideas and information.

The Results of the 777 Project

The 777 entered revenue service in June 1995. Since many of the features incorporated into the 777's design reflected suggestions made by the airline carriers, pilots, mechanics, and flight attendants were quite enthusiastic about the new jet. Three achievements of the program, in airplane interior, aircraft design, and aircraft manufacturing, stood out.

Configuration Flexibility

The 777 offered carriers enhanced configuration flexibility. A typical configuration change took only seventy-two hours on the 777 compared to three weeks in competing aircraft. In 1992, the Industrial Design Society of America granted Boeing its Excellence Award for building the 777 passenger cabin, honoring an airplane interior for the first time.[53]

Digital Design

The original goal of the program was to reduce "change, error, and rework" by 50 percent, but engineers building the first three 777's managed to reduce such modification by 60 percent to 90 percent. Catia helped engineers identify more than 10,000 interferences that would have otherwise remained undetected until assembly, or until after delivery. The first 777 was only 0.023 inch short of perfect alignment, compared to as much as 0.5 inch on previous programs.[54] Assembly line workers confirmed the beneficial effects of the digital design system. "The parts snap together like Lego blocks," said one mechanics.[55] Reducing the need for reengineering, replanning, retooling, and retrofitting, Boeing's innovative efforts were recognized yet again. In 1993, the Smithsonian Institution honored the Boeing 777 division with its Annual Computerworld Award for the manufacturing category.[56]

Empowerment

Boeing 777 assembly line workers expressed a high level of job satisfaction under the new program. "It's a whole new world," a fourteen-year Boeing veteran mechanic said, "I even like going to work. It's bubbly. It's clean. Everyone has confidence."[57] "We never used to speak up," said another employee, "didn't dare. Now factory workers are treated better and are encouraged to offer ideas."[58] Although the Bar Chart system required Boeing 777 mechanics to work harder and faster as they moved down the learning curve, their principal union organization, the International Association of Machinists, was pleased with Boeing's new approach to labor-management relations. A union spokesman reported that under the 777 program, managers were more likely to treat problems as opportunities from which to learn rather than mistakes for which to blame. Under the 777 program, the union representative added, managers were more respectful of workers' rights under the collective bargaining agreement.[59]

Unresolved Problems and Lessons Learned

Notwithstanding Boeing's success with the 777 project, the cost of the program was very high. Boeing did not publish figures pertaining to the total cost of Catia. But a company official reported that under the 777 program, the 3D digital design process required 60 percent more engineering resources than the older, 2D drawing-based design process. One reason for the high cost of using digital design was slow computing tools: Catia's response time often lasted minutes. Another was the need to update the design software repeatedly. Boeing revised Catia's design software four times between 1990 and 1996, making the system easier to learn and use. Still, Catia continued to experience frequent software problems. Moreover, several of Boeing's outside suppliers were unable to utilize Catia's digital data in their manufacturing process.[60]

Boeing faced training problems as well. One challenging problem, according to Ron Ostrowski, Director of 777 engineering, was "to convert people's thinking from 2D to 3D. It took more time than we thought it would. I came from a paper world and now I am managing a digital program."[61] Converting people's thinking required what another manager called an "unending communication" coupled with training and retraining. Under the 777 program, Ostrowski recalled, "engineers had to learn to interact. Some couldn't, and they left. The young ones caught on" and stayed.[62]

Learning to work together was a challenge to managers too. Some managers were reluctant to embrace Condit's open management style, fearing a decline in their authority. Others were reluctant to share their mistakes with their superiors, fearing reprisals. Some other managers, realizing that the new approach would end many managerial jobs, resisted change when they could, and did not pursue it whole-heartedly when they could not. Even top executives were sometimes uncomfortable with Boeing's open management style, believing that sharing information with employees was likely to help Boeing's competitors obtain confidential 777 data.[63]

Teamwork was another problem area. Working under pressure, some team members did not function well within teams and had to be moved. Others took advantage of their new-born freedom to offer suggestions, but were disillusioned and frustrated when management either ignored these suggestions, or did not act upon them. Managers experienced different team-related problems. In several cases, managers kept on meeting with their team members repeatedly until they arrived at a solution desired by their bosses. They were unwilling to challenge senior executives, nor did they trust Boeing's new approach to teaming. In other cases, managers distrusted the new digital technology. One engineering manager instructed his team members to draft paper drawings alongside Catia's digital designs. When Catia experienced a problem, he followed the drawing, ignoring the computerized design, and causing unnecessary and costly delays in his team's part of the project.[64]

Extending the 777 Revolution

Boeing's learning pains played a key role in the company's decision not to implement the 777 program company wide. Boeing officials recognized the importance of team work and Catia in reducing change, error, and rework, but they also realized that teaming required frequent training, continuous reinforcement, and ongoing monitoring, and that the use of Catia was still too expensive, though its cost was going down (in 1997, Catia's "penalty" was down to 10 percent). Three of Boeing's derivative programs, the 737 Next Generation, the 757-300, and the 767-400, had the option of implementing the 777's program innovations, and only one, the 737, did so, adopting a modified version of the 777's cross-functional teams.[65]

Yet the 777's culture was spreading in other ways. Senior executives took broader roles as the 777 entered service, and their impact was felt through the company. Larry Olson, director of information systems for the 747/767/777 division, was a former 777 manager who believed that Boeing 777 employees "won't tolerate going back to the old ways." He expected to fill new positions on Boeing's next program—the 747X—with former 777 employees in their 40s.[66] Philip Condit, Boeing CEO, implemented several of his own 777's innovations, intensifying the use of meeting among Boeing's managers, and promoting the free flow of ideas throughout the company. Under Condit's leadership, all mid-level managers assigned to Boeing Commercial Airplane Group, about sixty people, met once a week to discuss costs, revenues, and production schedules, product by product. By the end of the meeting—which sometimes ran into the evening—each manager had to draft a detailed plan of action dealing with problems in his/her department.[67] Under Condit's leadership, more important, Boeing developed a new "vision" that grew out of the 777 project. Articulating the company's vision for the next two decades (1996–2016), Condit singled out "Customer satisfaction," "Team leadership," and "A participatory workplace," as Boeing's core corporate values.[68]

Conclusion: Boeing, Airbus, and the 777

Looking back at the 777 program twelve years after the launch and seven years after first delivery, it is now (2002) clear that Boeing produced the most successful commercial jetliner of its kind. Airbus launched the A330 and A340 in 1987, and McDonnell Douglas launched a new 300-seat wide body jet in the mid 1980s, the three-engine MD11. Coming late to market, the Boeing 777 soon outsold both models. The 777 had entered service in 1995, and within a year Boeing delivered more than twice as many 777s as the number of MD11s delivered by McDonnell Douglas. In 1997, 1998, 1999, and 2001, Boeing delivered a larger number of 777s than the combined number of A330s and A340s delivered by Airbus (Exhibit VII). A survey of nearly 6,000 European airline passengers who had flown both the 777 and the A330/A340 found that the 777 was preferred by more than three out of four passengers.[69] In the end, a key element in the 777's triumph was its popularity with the traveling public.

Exhibit VII. Total number of MD11, A330, A340, and 777 airplanes delivered during 1995–2001

	1995	1996	1997	1998	1999	2000	2001
McDonnell Douglas/ Boeing MD11	18	15	12	12	8	4	2
Airbus A330	30	10	14	23	44	43	35
Airbus A340	19	28	33	24	20	19	20
Boeing 777	13	32	59	74	83	55	61

Source: For Airbus, Mark Luginbill Airbus Communication Director, February 1, 2000, and March 11, 2002. For Boeing, *The Boeing Company Annual Report,* 1997, p. 35, 1998, p. 35; "Commerical Airplanes: Order and Delivery, Summary," *http//www.boeing.com/commercial/order/index.html.* Retreived from Web, February 2, 2000 and March 9, 2002.

Notes

1. Rodgers, Eugene. *Flying High: The Story of Boeing* (New York: Atlantic Monthly Press, 1996), 415–416; Michael Dornheim, "777 Twinjet Will Grow to Replace 747-200," *Aviation Week and Space Technology* (3 June 1991): 43.
2. "Commercial Airplanes: Order and Delivery, Summary," *http/www.boeing.com/commercial/orders/index.html.* Retrieved from Web, 2 February 2000.
3. Donlon,). P. "Boeing's Big Bet" (an interview with CEO Frank Shrontz), *Chief Executive* (November/December 1994): 42; Dertouzos, Michael, Richard Lester, and Robert Solow, *Made in America: Regaining the Productive Edge* (New York: Harper Perennial, 1990), 203.
4. John Newhouse, *The Sporty Game* (New York: Alfred Knopf, 1982), 21, but see also 10–20.
5. Mowery, David C. and Nathan Rosenberg. "The Commercial Aircraft Industry," in Richard R. Nelson, ed., *Government and Technological Progress: A Cross Industry Analysis* (New York: Pergamon Press, 1982), 116; Dertouzos et al, *Made in America,* 200.
6. Dertouzos et al., *Made in America,* 200.
7. Newhouse, *Sporty Game,* 188. Mowery and Rosenberg, "The Commercial Aircraft Industry," 124–125.
8. Mowery and Rosenberg, "The Commercial Aircraft Industry," 102–103, 126–128.
9. Rae, John B. *Climb to Greatness: The American Aircraft Industry, 1920–1960* (Cambridge, Mass.: MIT Press, 1968), 206–207; Rodgers, *Flying High,* 197–198.
10. Spadaro, Frank. "A Transatlantic Perspective," *Design Quarterly* (Winter 1992): 23.
11. Rodgers, *Flying High,* 279; Newhouse, *Sporty Game,* Ch. 7.
12. Hochmuth, M. S. "Aerospace," in Raymond Vernon, ed., *Big Business and the State* (Cambridge: Harvard University Press, 1974), 149.
13. Boeing Commercial Airplane Group, *Announced Orders and Deliveries as of 12/31/97,* Section A 1.
14. *The Boeing Company 1998 Annual Report,* 76.
15. Formed in 1970 by several European aerospacc firms, the Airbus Consortium had received generous assistance from the French, British, German, and Spanish governments for a period of over two decades. In 1992, Airbus had signed an agreement with Boeing that limited the amount of government funds each aircraft manufacturer could receive, and in 1995, at long last, Airbus had become profitable. "Airbus 25 Years Old," *Le Figaro,* October 1997 (reprinted in English by Airbus Industrie); Rodgers, *Flying High,* Ch. 12; *Business Week* (30 December 1996): 40.
16. Charles Goldsmith, "Re-engineering, After Trailing Boeing for Years, Airbus Aims for 50% of the Market," *Wall Street Journal,* 16 March 1998.
17. "Hubris at Airbus, Boeing Rebuild," *Economist,* 28 November 1998.
18. *The Boeing Company 1997 Annual Report,* 19; *The Boeing Company 1998 Annual Report,* 51.
19. Donlon, "Boeing's Big Bet," 40; John Mintz, "Betting It All on 777" *Washington Post,* 26 March 1995; James Woolsey, "777: A Program of New Concepts," *Air Transport World,* (April 1991): 62; Jeremy Main, "Corporate Performance: Betting on the 21st Century Jet," *Fortune,* 20 April 1992, 104; James Woolsey, "Crossing New Transport Frontiers," *Air Transport World* (March 1991): 21; James Woolsey, "777: Boeing's New Large Twinjet," *Air Transport World* (April 1994): 23; Michael Dornheim, "Computerized Design System Allows Boeing to Skip Building 777 Mockup," *Aviation Week and Space Technology* (3 June 1991): 51; Richard O'Lone, "Final Assembly of 777 Nears," *Aviation Week and Space Technology* (2 October 1992): 48.
20. Rodgers, *Flying High,* 42.

21. *Air Transport World* (March 1991): 20; *Fortune,* 20 April 1992, 102–103.
22. Rodgers, *Flying High,* 416, 420–24.
23. Richard O'Lone and James McKenna, "Quality Assurance Role was Factor in United's 777 Launch Order," *Aviation Week and Space Technology* (29 October 1990): 28–29; *Air Transport World* (March 1991): 20.
24. Quoted in the *Washington Post,* 25 March 1995.
25. Quoted in Bill Swectman, "As Smooth as Silk: 777 Customers Applaud the Aircraft's First 12 Months in Service," *Air Transport World* (August 1996): 71, but see also *Air Transport World* (April 1994): 24, 27.
26. Quoted in *Fortune,* 20 April 1992, 112.
27. Rodgers, *Flying High,* 426; *Design Quarterly* (Winter 1992): 22; Polly Lane, "Boeing Used 777 to Make Production Changes," *Seattle Times,* 7 May 1995.
28. *Design Quarterly* (Winter 1992): 22; The Boeing Company, *Backgrounder: Pace Setting Design Value-Added Features Boost Boeing 777 Family,* 15 May 1998.
29. Boeing, *Backgrounder,* 15 May 1998; Sabbagh, *21st Century Jet,* p. 49.
30. Karl Sabbagh, *21st Century Jet: The Making and Marketing of the Boeing 777* (New York: Scribner, 1996), 264, 266.
31. Sabbagh, *21st Century Jet,* 131–132
32. *Air Transport World* (April 1994): 23; *Fortune,* 20 April 1992, 116.
33. *Washington Post,* 26 March 1995; Boeing Commercial Airplane Group, 777 Announced Order and Delivery Summary...As of 9/30/99.
34. Rodgers, *Flying High,* 420–426; *Air Transport World* (April 1994): 27, 31; "Leading Families of Passenger Jet Airplanes," Boeing Commercial Airplane Group, 1998.
35. Sabbagh, *21st Century Jet,* 58.
36. Quoted in Sabbagh, *21st Century Jet,* 63.
37. *Aviation Week and Space Technology* (3 June 1991): 50, 12 October 1992, p. 49; Sabbagh *21st Century Jet,* p. 62.
38. George Taninecz, "Blue Sky Meets Blue Sky," *Industry Week* (18 December 1995); 49–52; Paul Proctor, "Boeing Rolls Out 777 to Tentative Market," *Aviation Week and Space Technology* (12 October 1992): 49.
39. *Aviation Week and Space Technology* (11 April 1994): 37; *Aviation Week and Space Technology* (3 June 1991): 35.
40. Quoted in Sabbagh, *21st Century Jet,* 68–69.
41. This was the phrase used by Boeing project managers working on the 777. See Sabbagh, *21st Century Jet,* Ch. 4.
42. *Fortune,* 20 April 1992,116; Sabbagh,*2lst Century Jet,* 69–73; Wolf L. Glende, "The Boeing 777: A Look Back," The Boeing Company, 1997, 4.
43. Quoted in *Air Transport World* (August 1996): 78.
44. Richard O'Lone, "777 Revolutionizes Boeing Aircraft Development Process" *Aviation Week and Space Technology* (3 June 1992): 34.
45. O. Casey Corr. "Boeing's Future on the Line: Company's Betting its Fortunes Not Just on a New Jet, But on a New Way of Making Jets," *Seattle Times,* 29 August 1993; Polly Lane, "Boeing Used 777 to Make Production Changes, Meet Desires of Its Customers," *Seattle Times,* 7 May 1995; *Aviation Week and Space Technology* (3 June 1991): 34.
46. *Seattle Times,* 29 August 1993.
47. *Seattle Times,* 7 May 1995, and 29 August 1993.
48. Quoted in Rodgers, *Flying High,* 419–420.
49. Sabbagh, *21st Century Jet,* 33.

50. Sabbagh, *21st Century Jet,* 99.
51. Dori Jones Young, "When the Going Gets Tough, Boeing Gets Touchy-Feely, *Business Week* (17 January 1994): 65–67; *Fortune,* 20 April 1992, 117.
52. Reprinted by The Boeing Company, Executive Communications, 1992.
53. Boeing, *Backgrounder,* 15 May 1998.
54. *Industry Week* (18 December 1995): 50–51; *Air Transport World* (April 1994).
55. *Aviation Week and Space Technology* (11 April 1994): 37.
56. Boeing, *Backgrounder,* "Computing & Design/Build Process Help Develop the 777." Undated.
57. *Seattle Times,* 29 August 1993.
58. *Seattle Times,* 7 May 1995.
59. *Seattle Times,* 29 August 1993.
60. Glende, "The Boeing 777: A Look Back," 1997, 10; *Air Transport World* (August 1996): 78.
61. *Air Transport World* (April 1994): 23.
62. *Washington Post,* 26 March 1995.
63. *Seattle Times,* 7 May 1995; Rodgers, *Flying High,* 441.
64. *Seattle Times,* 7 May 1995; Rodgers, *Flying High,* 441–442.
65. Glende, "The Boeing 777: A Look Back," 1997, 10.
66. *Air Transport World,* August 1996, 78.
67. "A New Kind of Boeing," *Economist,* 22 January 2000, 63.
68. "Vision 2016," The Boeing Company 1997.
69. "Study: Passengers Voice Overwhelming Preference for Boeing 777, *http/www.boeing.com/news/releases/1999.* Retrieved from Web 11/23/99.

Index

More great resources from Harold Kerzner, PhD.

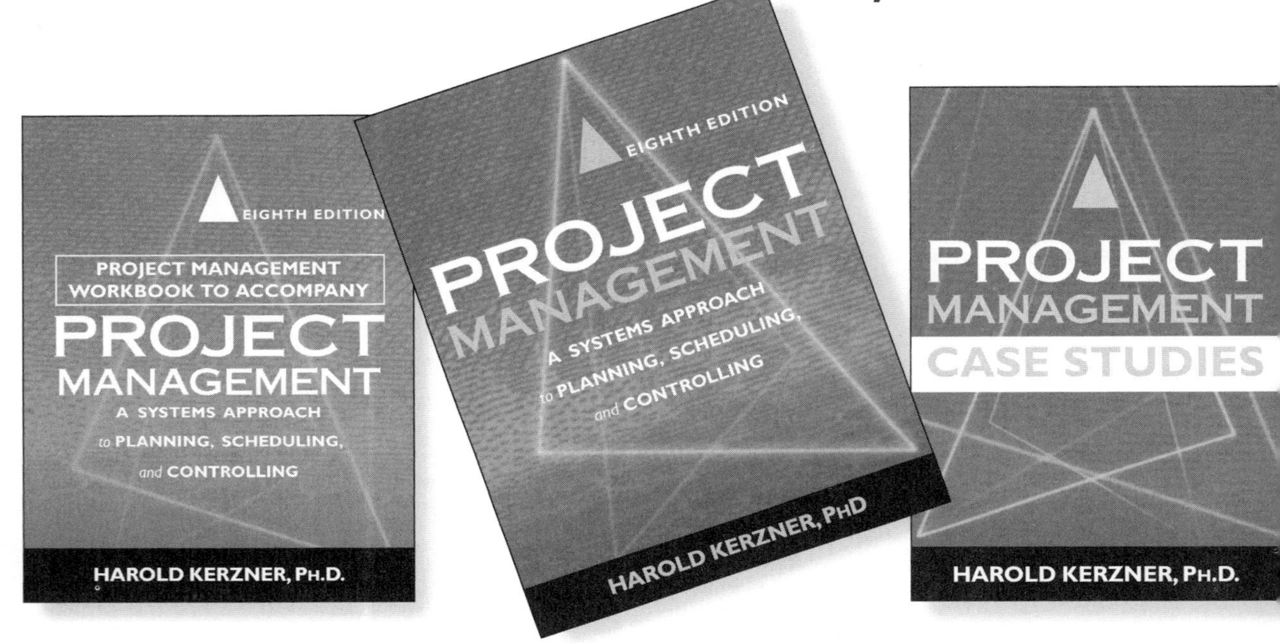

Project Management Workbook, To Accompany Project Management: A Systems Approach to Planning, Scheduling, and Controlling, Eighth Edition

0-471-22579-7 • Paper • 400 pages

An excellent preparation tool for the Project Management Professional certification exam

Project Management: A Systems Approach to Planning, Scheduling, and Controlling, Eigth Edition

0-471-22577-0 • Cloth • 912 pages

A quintessential model for project management–the streamlined edition of the landmark reference

Project Management Case Studies

0-471-22578-9 • Paper • 448 pages

A versatile, up-to-date selection of case studies from a leading innovator in project management

Bringing you the best in Project Management.

www.allpm.com

wiley.com